AF892780

OPTICAL AND FIBER COMMUNICATIONS REPORTS

Editorial Board: A. Bjarklev
H.J. Caulfield
A.K. Majumdar
G. Marowsky
M. Nakazawa
M.W. Sigrist
C.G. Someda
H.-G. Weber

For further volumes:
http://www.springer.com/series/4810

OPTICAL AND FIBER COMMUNICATIONS REPORTS

The Optical and Fiber Communications Reports (OFCR) book series provides a survey of selected topics at the forefront of research. Each book is a topical collection of contributions from leading research scientists that gives an up-to-date and broad-spectrum overview of various subjects. The main topics in this expanding field will cover for example:

- specialty fibers (periodic fibers, holey fibers, erbium-doped fibers)
- broadband lasers
- optical switching (MEMS or others)
- polarization and chromatic mode dispersion and compensation
- long-haul transmission
- optical networks (LAN, MAN, WAN)
- protection and restoration
- further topics of contemporary interest.

Including both general information and a highly technical presentation of the results, this series satisfies the needs of experts as well as graduates and researchers starting in the field. Books in this series establish themselves as comprehensive guides and reference texts following the impressive evolution of this area of science and technology.

The editors encourage prospective authors to correspond with them in advance of submitting a manuscript. Submission of manuscripts should be made to one of the editors. See also http://springeronline.com/series/4810.

Editorial Board

Anders Bjarklev
COM, Technical University of Denmark
DTU Building 345V
2800 Ksg. Lyngby, Denmark
Email: ab@com.dtu.dk

H. John Caulfield
Fisk University
Department of Physics
1000 17th Avenue North
Nashville, TN 37208
USA
Email: hjc@fisk.edu

Arun K. Majumdar
LCResearch, Inc.
30402 Rainbow View Drive
Agoura Hills, CA 91301
Email: a.majumdar@IEEE.org

Gerd Marowsky
Laser-Laboratorium Göttingen e.V.
Hans-Adolf-Krebs-Weg 1
37077 Göttingen
Germany
Email: gmarows@gwdg.de

Masataka Nakazawa
Research Institute of Electrical Communication
Tohoku University
Katahira 2-1-1, Aoba-ku
980-8577 Sendai-shi, Miyagiken
Japan
Email: nakazawa@riec.tohoku.ac.jp

Markus W. Sigrist
ETH Zürich
Institut für Quantenelektronik
Lab. Laserspektroskopie – HPF D19
ETH Hönggerberg
8093 Zürich
Switzerland
Email: sigrist@iqu.phys.ethz.ch

Carlo G. Someda
DEI-Università di Padova
Via Gradenigo 6/A
35131 Padova, Italy
Email: someda@dei.unipd.it

Hans-Georg Weber
Heinrich-Hertz Institut (HHI)
Einsteinufer 37
10587 Berlin, Germany
Email: hgweber@hhi.de

Shiva Kumar
Editor

Impact of Nonlinearities on Fiber Optic Communications

Springer

Editor
Shiva Kumar
Department of Electrical
& Computer Engineering
McMaster University
Main Street West 1280
L8S 4K1 Hamilton Ontario
Canada
kumars@mail.ece.mcmaster.ca

ISBN 978-1-4419-8138-7 e-ISBN 978-1-4419-8139-4
DOI 10.1007/978-1-4419-8139-4
Springer New York Dordrecht Heidelberg London

Library of Congress Control Number: 2011922498

© Springer Science+Business Media, LLC 2011
All rights reserved. This work may not be translated or copied in whole or in part without the written permission of the publisher (Springer Science+Business Media, LLC, 233 Spring Street, New York, NY 10013, USA), except for brief excerpts in connection with reviews or scholarly analysis. Use in connection with any form of information storage and retrieval, electronic adaptation, computer software, or by similar or dissimilar methodology now known or hereafter developed is forbidden.
The use in this publication of trade names, trademarks, service marks, and similar terms, even if they are not identified as such, is not to be taken as an expression of opinion as to whether or not they are subject to proprietary rights.

Printed on acid-free paper

Springer is part of Springer Science+Business Media (www.springer.com)

Preface

Nonlinear effects occur in optical communication systems at the transmitter, fiber channel, and receiver. First, at the transmitter, when a Mach–Zehnder modulator is used to modulate the optical carrier by electrical data, its transfer function is not linear. Second, the nonlinear effects in fibers such as the Kerr effect and the Raman effect lead to interaction among signals propagating down the fiber. Finally, in direct-detection systems, the nonlinearity occurs in the photodetector, which is a square-law device. However, with coherent detection, the linear translation of information in optical domain into electrical domain can be achieved. This book covers the various types of nonlinear effects that occur in fiberoptic communication systems. The performance degradations caused by the nonlinear effects and how to mitigate them are also discussed in various chapters.

The first chapter, by X. Liu and M. Nazarathy, introduces the recent developments in self-coherent, differentially coherent, and coherent fiberoptic transmission systems. The benefits of advanced detection schemes and the impact of fiber nonlinearity are also discussed. The second chapter, by Qi Yang, A.A. Amin and W. Shieh, reviews the basic principles of orthogonal frequency division multiplexing (OFDM). The authors discuss the recent experimental demonstrations of coherent optical OFDM systems with bit rates ranging from 100 Gb s^{-1} to 1 Tb s^{-1} and with off-line as well as real-time signal processing. These two chapters provide the basis for nonlinear impairment issues discussed in later chapters. Chapter 3, by M. Nazarathy and R. Weidenfeld, addresses the impact of fiber nonlinear effects on coherent OFDM systems and discusses electrical equalizing techniques to mitigate these nonlinear impairments. The authors analyze the impact of nonlinear effects using the Volterra approach and later, based on the analytical tools, they develop effective nonlinear compensators for OFDM systems.

Coherent technologies have enabled novel spectrally efficient and power-efficient modulation formats. The spectrally efficient formats allow upgrading to higher channel data rates using the existing lower speed transmission equipments. Chapter 4, by M. Seimetz, reviews the basics of modulation schemes, and optical implementation of novel modulation schemes and their detection techniques are discussed. The author provides the details of long-haul optical transmission experiments with RZ-QPSK, RZ-8PSK, and RZ-16QAM signals.

Single-mode fiber (SMF) is actually bimodal due to the x- and y-polarization components, and an optical carrier propagating in SMF has four degrees of freedom. They are in-phase (I) and quadrature (Q) components of the x- and y-polarizations. Chapter 5, by M. Karlsson and E. Agrell, discusses the modulation formats in the four-dimensional space. The authors explain the relation between dense sphere packing and power-efficient constellations. Fundamental sensitivity limits for the four-dimensional channel and influence of fiber nonlinearities are also presented in Chap. 5.

The novel modulation/multiplexing schemes have enabled high spectral efficiencies. However, as the spectral efficiency increases, typically the system reach reduces mainly because of nonlinear effects. Chaps. 6–9 focus on the various aspects of fiber nonlinearities and performance degradation caused by them. Chapter 6, by A. Mecozzi, discusses the intrachannel nonlinearities in pseudolinear systems. The full details of the first-order perturbation theory for the calculations of intrachannel nonlinear impairments in coherent and direct-detection systems are provided in this chapter. Although the main results obtained using a perturbation theory for direct-detection systems were published earlier by the author and his collaborators, the details of the theory and its derivations were never published before in the open literature.

Fiber nonlinearity translates the amplitude fluctuations caused by amplifier noise into phase fluctuations, which leads to nonlinear phase noise. Although the digital back-propagation can undo the deterministic and bit-pattern-dependent nonlinear effects, nonlinear phase noise cannot be compensated and it sets a fundamental limit on the achievable capacity. Chapters 7 and 8 focus on the impairments due to nonlinear phase noise. Chapter 7, by S. Kumar and X. Zhu, deals with nonlinear phase noise caused by self-phase modulation in single carrier and OFDM systems. Chapter 8, by K.-P. Ho, discusses the nonlinear phase noise due to cross-phase modulation (XPM) in quadriphase-shift keying (QPSK) and differential QPSK (DQPSK) systems. The author explains the impact of penalty caused by the XPM-induced nonlinear phase noise from the adjacent on-off keying (OOK) channel for DQPSK signals.

Polarization division multiplexing (PDM), in which two sets of data are encoded onto x- and y-polarization components separately, could double the capacity of a fiberoptic transmission system in the absence of fiber nonlinearity. However, the nonlinear interaction between x- and y- polarization components leads to signal distortions and impairments. Chapter 9, by C. Xie, deals with nonlinear polarization scattering in PDM systems. Although the digital signal processing (DSP) can equalize the distortions due to polarization mode dispersion (PMD) and polarization-dependent loss (PDL), it is hard to compensate nonlinear polarization scattering as the state of polarization (SOP) changes caused by nonlinear effects are typically in the scale of a symbol period. The author also discusses the techniques to mitigate the nonlinear polarization scattering.

To assess the quality of the received signal, the Monte-Carlo simulation of the fiberoptic transmission system needs to be carried out. This simulation takes too much time because of fiber nonlinearities especially when the bit error rate (BER) is low. Chapter 10, by A. Bononi and L.A. Rusch, deals with the multicanonical Monte-Carlo (MMC), which is a simulation-acceleration technique for the estimation of the statistical distribution of a desired system output variable. The authors present several examples from optical communication, where MMC techniques have provided accurate performance predictions.

In a fiberoptic transmission system, the noise accumulation can be suppressed by introducing optical regenerators at certain locations on the transmission line. Typically, optical regenerators suppress the amplitude noise rather than the phase noise and therefore, they cannot be used directly for phase-modulated systems. Chapter 11, by M. Matsumoto, reviews the all-optical regeneration schemes for phase-encoded signals. The author discusses various regeneration schemes for the suppression of linear and nonlinear phase noise in systems based on (D)BPSK and (D)QPSK.

Chapter 12, by I.B. Djordjevic, reviews the basics of forward error correction (FEC), coded modulation, and turbo equalization for high speed optical communication system. The details of low-density parity-check (LDPC)-coded turbo equalizer to compensate for dispersion, PMD, and fiber nonlinearities are provided in this chapter. The author also addresses the limits on channel capacity of fiberoptic systems with coded modulation schemes.

The understanding of the ultimate limits on the capacity of fiberoptic communication system is of fundamental importance. The last chapter, by A. Ellis and J. Zhao, explores the system design trade-offs to maximize the channel capacity of the nonlinear fiberoptic channel. The authors discuss various techniques that promise to allow the capacity limits to be extended.

I thank the authors for all the trouble they have taken to make their work accessible to a wide readership.

Hamilton, Canada *Shiva Kumar*
February 2011

Contents

1 Coherent, Self-Coherent, and Differential Detection Systems 1
Xiang Liu and Moshe Nazarathy

2 Optical OFDM Basics 43
Qi Yang, Abdullah Al Amin, and William Shieh

3 Nonlinear Impairments in Coherent Optical OFDM Systems and Their Mitigation 87
Moshe Nazarathy and Rakefet Weidenfeld

4 Systems with Higher-Order Modulation 177
Matthias Seimetz

5 Power-Efficient Modulation Schemes 219
Magnus Karlsson and Erik Agrell

6 A Unified Theory of Intrachannel Nonlinearity in Pseudolinear Transmission 253
Antonio Mecozzi

7 Analysis of Nonlinear Phase Noise in Single-Carrier and OFDM Systems 293
Shiva Kumar and Xianming Zhu

8 Cross-Phase Modulation-Induced Nonlinear Phase Noise for Quadriphase-Shift-Keying Signals 325
Keang-Po Ho

9 Nonlinear Polarization Scattering in Polarization-Division-Multiplexed Coherent Communication Systems 343
Chongjin Xie

10 Multicanonical Monte Carlo for Simulation of Optical Links 373
Alberto Bononi and Leslie A. Rusch

11 Optical Regenerators for Novel Modulation Schemes 415
Masayuki Matsumoto

12 Codes on Graphs, Coded Modulation and Compensation of Nonlinear Impairments by Turbo Equalization 451
Ivan B. Djordjevic

13 Channel Capacity of Non-Linear Transmission Systems 507
Andrew D. Ellis and Jian Zhao

Index 539

Contributors

Erik Agrell Communication Systems Group, Department of Signals and Systems, Chalmers University of Technology, SE-412 96 Göteborg, Sweden, agrell@chalmers.se

Abdullah Al Amin Center for Ultra-broadband Information Networks, Department of Electrical and Electronic Engineering, University of Melbourne, Melbourne, VIC 3010, Australia, aalamin@unimelb.edu.au

Alberto Bononi Dipartimento di Ingegneria dell'Informazione, Università di Parma, 43100 Parma, Italy, alberto.bononi@unipr.it

Ivan B. Djordjevic Department of Electrical and Computer Engineering, University of Arizona, Tucson, AZ 85721, USA, ivan@ece.arizona.edu

Andrew D. Ellis Tyndall National Institute and Department of Physics, University College Cork, Cork, Ireland, andrew.ellis@tyndall.ie

Keang-Po Ho SiBEAM, Sunnyvale, CA 94085, USA, kpho@ieee.org

Magnus Karlsson Photonics Laboratory, Department of Microtechnology and Nanoscience, Chalmers University of Technology, SE-412 96 Göteborg, Sweden, magnus.karlsson@chalmers.se

Shiva Kumar Electrical and Computer Engineering, McMaster University, ITBA 322, 1280 Main St. West, Hamilton, ON-L8S 4K1, Canada, kumars@mail.ece.mcmaster.ca

Xiang Liu Bell Laboratories, Alcatel-Lucent, Holmdel, NJ 07733, USA, Xiang.Liu@alcatel-lucent.com

Masayuki Matsumoto Graduate School of Engineering, Osaka University, Osaka 565-0871, Japan, matumoto@comm.eng.osaka-u.ac.jp

Antonio Mecozzi University of L'Aquila, 67100 L'Aquila, Italy, antonio.mecozzi@univaq.it

Moshe Nazarathy Electrical Engineering Department, Technion, Israel Institute of Technology, Israel, nazarat@ee.technion.ac.il

Leslie A. Rusch Electrical and Computer Engineering Department, Université Laval, Québec City, QC, Canada G1V 0A6, rusch@gel.ulaval.ca

Matthias Seimetz Beuth Hochschule für Technik Berlin, FB VII: Elektrotechnik und Feinwerktechnik, Luxemburger Str. 10, 13353 Berlin, Germany, matthias.seimetz@beuth-hochschule.de

William Shieh Center for Ultra-broadband Information Networks, Department of Electrical and Electronic Engineering, University of Melbourne, Melbourne, VIC 3010, Australia, shiehw@unimelb.edu.au

Rakefet Weidenfeld Electrical Engineering Department, Technion, Israel Institute of Technology, Israel, wrakefet@gmail.com

Chongjin Xie Transmission Systems and Networking Research, Bell Laboratories, Alcatel-Lucent, 791 Holmdel-Keyport Road, Holmdel, NJ 07733, USA, chongjin.xie@alcatel-lucent.com

Qi Yang State Key Lab. of Opt. Commu. Tech. and Networks, Wuhan Research Institute of Post & Telecomnunication, Wuhan, China, qyang@wri.com.cn

Jian Zhao Tyndall National Institute and Department of Physics, University College Cork, Cork, Ireland, jian.zhao@tyndall.ie

Xianming Zhu Science and Technology, Corning Incorporated, SP-TD-01-1, Science Center Drive, Corning, NY 14831, USA, zhux@corning.com

Chapter 1
Coherent, Self-Coherent, and Differential Detection Systems

Xiang Liu and Moshe Nazarathy

1.1 Introduction

In order to meet the ever-increasing demand in telecommunication capacity, fiberoptic communication systems have been evolving dramatically over the past decade [1, 2]. The fiberoptic communication traffic growth has been at a rate of about 2 dB per year, representing a traffic increase of a factor of 100 in 10 years [1,2]. The capacity increase in fiberoptic communication systems has been achieved mainly by deploying more fiber links, populating more wavelength channels per fiber link through dense wavelength-division-multiplexing (DWDM), and increasing the data rate per wavelength channel. In addition to increased capacity, the cost per bit in terms of both capital and operational expenditure has been decreased to sustain the traffic growth. Increasing the data rate per wavelength channel is regarded as an effective way to provide both increased capacity and lowered cost per bit. Indeed, in most fiberoptic transmission systems, the channel data rate has been upgraded from 2.5 Gb s^{-1} to 10 Gb s^{-1}, and 40 Gb s^{-1} is under active deployment. The 100-Gb s^{-1} channel data rate is accepted as the next-generation standard for optical transport and Ethernet (see, e.g., IEEE P802.3ba 40 Gb s^{-1} and 100 Gb s^{-1} Ethernet Task Force, http://www.ieee802.org/3/ba/).

Several recent technological advances constitute the enablers of increased data rate per wavelength. Among these, advanced detection schemes such as differential detection [3–5], self-coherent detection (SCD) [5], and digital coherent detection (DCD) [6–10], provide major breakthroughs. These advanced detection schemes, together with advanced optical modulation formats, increase system tolerance to optical noise and/or transmission impairments such as chromatic dispersion (CD), polarization-mode dispersion (PMD), and fiber nonlinearity, which are limiting factors for high-speed optical transmission. Moreover, advanced detection schemes

X. Liu (✉)
Bell Laboratories, Alcatel-Lucent, Holmdel, NJ 07733, USA
e-mail: Xiang.Liu@alcatel-lucent.com

M. Nazarathy
Electrical Engineering Department, Technion, Israel Institute of Technology, Israel
e-mail: nazarat@ee.technion.ac.il

S. Kumar (ed.), *Impact of Nonlinearities on Fiber Optic Communications*, Optical and Fiber Communications Reports 7, DOI 10.1007/978-1-4419-8139-4_1,
© Springer Science+Business Media, LLC 2011

enable high spectral-efficiency (SE) optical modulation formats supporting higher data rates in systems originally designed for lower data rates.

In this chapter, we review recent progress in coherent, self-coherent, and differential detection-based fiberoptic communication systems. Particular emphasis is placed on the system benefits of the advanced detection schemes and the impact of fiber nonlinearity. This chapter is organized as follows. In Sect. 1.2, we review recent research demonstrations of advanced detection schemes for high-speed high-SE optical transmission. Highlights include long-haul transmission with channel data rates of 400 Gb s^{-1} and 1 Tb s^{-1}, system SE reaching 8 b s^{-1} Hz^{-1}, and per-fiber transmission capacities of up to 69 Tb s^{-1}. Section 1.3 describes recent progress in differential-detection and SCD-based optical communication systems, addressing fiber nonlinear interactions in data-rate-mixed DWDM transmission, combining 10-Gb s^{-1}, 40-Gb s^{-1}, and 100-Gb s^{-1} channels. Section 1.4 presents recent progresses in DCD-based systems. State-of-the-art research demonstrations of 400-Gb s^{-1}, 1-Tb s^{-1} transmission, and high-SE transmission are reviewed. Section 1.5 concludes this chapter discussing future evolution of fiberoptic transmission systems.

1.2 Recent Advances in Fiberoptic Communication Systems

The last few years have witnessed many record-breaking high-speed and high-SE optical transmission demonstrations, enabled by advanced detection schemes. Table 1.1 summarizes highlights of the state-of-the-art high-speed high-SE transmission, sorted roughly in order of the channel data rate and SE. The achieved SE-distance product (SEDP) is also listed. SEDP is a key system performance indicator in that it is directly related to the transmission capacity-distance product for a given optical bandwidth allocation.

1.2.1 40-Gb s^{-1} Transmission

With direct differential detection (DDD), differential binary phase-shift keying (DBPSK) was first demonstrated at 43 Gb s^{-1} per wavelength, with long-haul transmission capability [11]. DWDM transmission of sixty-four 43-Gb s^{-1} DBPSK channels on a 100-GHz grid over 4,000 km (forty 100-km spans) of nonzero dispersion-shifted fiber (NZDSF) with distributed Raman amplification (DRA) was demonstrated. The achieved net system SE and SEDP were 0.4 b s^{-1} Hz^{-1} and 1,600 b km s^{-1} Hz^{-1}, respectively. Although these values are modest compared to more recent research demonstrations, this DD-based DBPSK demonstration is often regarded as the first major step toward to use of advanced modulation formats and detection schemes in optical fiber transmission [3–5]. Prior to this demonstration, the modulation and detection scheme used in fiberoptic transmission had overwhelmingly been intensity modulation direct detection (IM-DD) based, using on-off-keying (OOK).

Table 1.1 Summary of recent high-speed optical transmission demonstrations

Channel data rate (Gb s^{-1})	SE (b s^{-1} Hz^{-1})	Modulation format/detection scheme	Reach (km)	Fiber type/amplification	SEDP (km-b s^{-1} Hz^{-1})
40-Gb s^{-1} class					
43 [11]	0.4	DBPSK/DDD	4,000	NZDSF/DRA	1,600
43 [12]	0.8	DBPSK and DQPSK/DDD	1,280	SSMF/EDFA	1,024
40 [13]	0.8	PDM-QPSK/DCD	3,200	SSMF/EDFA	2,560
40 [14]	~2	16-QAM/SCD	160	SSMF/EDFA	320
100-Gb s^{-1} class					
112 [15]	2	PDM-QPSK/DCD	7,040	LCF/DRA	14,080
114 [16]	4	PDM-8-QAM/DCD	580	ULLF/EDFA	2,320
112 [17]	6.2	PDM-16-QAM/DCD	630	SSMF/DRA	3,906
171 [18]	6.4	PDM-16-QAM/DCD	240	PSCF/DRA	1,536
107 [19]	8	PDM-36-QAM/DCD	320	SSMF/DRA	2,560
200-Gb s^{-1} and beyond					
224 [20]	4	PDM-16-QAM/DCD	1,200	ULAF/DRA	4,800
448 [21]	5	RGI-CO-OFDM-16-QAM/B-DCD	2,000	ULAF/DRA	10,000
1,000 [22]	3.3[a]	CO-OFDM-QPSK/B-DCD	600	SSMF/EDFA	1,980
1,200 [23]	3.7[a]	NGI-CO-OFDM-QPSK/B-DCD	7,200	ULAF/DRA	27,000

DDD Direct differential detection; *SCD* Self-coherent detection; *DCD* Digital coherent detection; *B-DCD* Banded digital coherent detection; *DBPSK* Differential binary phase-shift keying; *DQPSK* Differential quadrature phase-shift keying; *PDM* Polarization-division multiplexed; *CO-OFDM* Coherent optical orthogonal frequency-division multiplexing; *RGI* Reduced-guard-interval; *NGI* No-guard-interval; *EDFA* Erbium-doped fiber amplifier; *DRA* Distributed Raman amplification; *NZDSF* Non-zero-dispersion-shifted fiber; *SSMF* Standard single-model fiber; *LCF* Large-core fiber; *PSCF* Pure silica core fiber; *ULLF* Ultra-low-loss fiber; *ULAF* Ultra-large-area fiber

[a]In these two Tb s^{-1} superchannel demonstrations, the quoted SE values do not include the spectral gap between the channels, so the actual system SE in DWDM configuration will be lower

At 43-Gb s^{-1} per-channel data rate, 0.8-b s^{-1} Hz^{-1} SE was demonstrated by co-propagating DBPSK and differential quadrature phase-shift keying (DQPSK) channels in a single DWDM system with 50-GHz channel spacing [12]. Transmission over a 1,280-km standard single-mode fiber (SSMF) link including four reconfigurable optical add/drop multiplexer (ROADM) passes was achieved. The optical amplification solely consisted of cost-effective Erbium-doped fiber amplifiers (EDFAs) in the C-band. The achieved SEDP was 1,024 km-b s^{-1} Hz^{-1}.

With DCD, polarization-division-multiplexed quadrature phase-shift keying (PDM-QPSK) was used to transmit forty 40-Gb s^{-1} channels on a 50-GHz grid over 3,200 km of CD-uncompensated SSMF, achieving an SE of 0.8 b s^{-1} Hz^{-1} SE and an SEDP of 2,560 km-b s^{-1} Hz^{-1} [13]. High PMD tolerance of 33-ps mean differential group delay (DGD) at an outage probability of 10^{-5} was also demonstrated.

With SCD, quadrature amplitude modulation (QAM) with 16 constellation points (16-QAM) was used to transmit a 40-Gb s^{-1} channel over 160 km of SSMF without optical CD compensation [14]. The expected achievable SE and SEDP are about 2 b s^{-1} Hz^{-1} and 320 km-b s^{-1} Hz^{-1}, respectively.

1.2.2 100-Gb s^{-1} Transmission

For 100-Gb s^{-1} per-channel transmission, DCD is the primary detection scheme of choice, due to its capability to digitally compensate for CD and PMD. Moreover, DCD enables straightforward PDM implementation, providing a highly sought factor-of-two in bit rate. At 2-b s^{-1} Hz^{-1} SE, seventy-two 112-Gb s^{-1} PDM-QPSK channels were transmitted on a 50-GHz grid over a 7,040-km fiber link consisting of large-core fiber (LCF) spans with 120-μm^2 effective area, achieving an impressive SEDP of 14,080 km-b s^{-1} Hz^{-1} [15].

At 4-b s^{-1} Hz^{-1} SE, 320 114-Gb s^{-1} PDM-8QAM channels on a 25-GHz channel grid were transmitted over 580 km of ultra-low-loss fiber (ULLF) with an average loss coefficient of 0.176 dB km^{-1}, achieving an SEDP of 2,320 km-b s^{-1} Hz^{-1} [16].

At 6.2-b s^{-1} Hz^{-1} SE, ten 112-Gb s^{-1} PDM-16QAM channels on a 16.7-GHz grid were transmitted over 630 km of SSMF, achieving an SEDP of 3,906 km-b s^{-1} Hz^{-1} [17].

Remarkably, a record single-fiber capacity of 69.1 Tb s^{-1} was recently demonstrated by transmitting 432 171-Gb s^{-1} PDM-16-QAM channels on a 25-GHz grid in the C- and extended L-band [18]. The achieved SE and transmission distance were 6.4 b s^{-1} Hz^{-1} and 240 km, respectively, resulting in an SEDP of 1,536 km-b s^{-1} Hz^{-1}.

The highest SE demonstrated so far for long-haul transmission is 8 b s^{-1} Hz^{-1}, achieved by using 107-Gb s^{-1} PDM-36QAM channels on a 12.5-GHz grid [19]. DWDM transmission of 640 107-Gb s^{-1} PDM-36QAM channels over 320 km of ultra-large-area fiber (ULAF) with 127-μm^2 effective area and 0.179-dB km^{-1} loss 64-Tb s^{-1} (640 × 107-Gb s^{-1}) was demonstrated, achieving an SEDP of 2,560 km-b s^{-1} Hz^{-1}.

In the demonstrations surveyed above, different fiber types, span lengths, optical amplification schemes, and/or forward-error correction (FEC) thresholds were used; hence, the comparison of the attained SEDP values merely provides a rough indication of comparative performance. The general trend is that the achievable transmission distance and SEDP decrease as the SE increases. This is understandable as tolerance to both noise and fiber nonlinearity is generally lowered when the number of signal constellation points is increased in order to achieve higher SE.

1.2.3 200-Gb s^{-1} Transmission and Beyond

As 100-Gb s^{-1} technology has been maturing, research effort has recently been diverted to transmission beyond 100-Gb s^{-1}. At 224-Gb s^{-1} per-channel data rate, DWDM transmission of ten 224-Gb s^{-1} PDM-16-QAM channels on a 50-GHz grid over 1,200 km of ULAF was demonstrated, achieving a net SE of 4 b s^{-1} Hz^{-1} and an SEDP of 4,800 km-b s^{-1} Hz^{-1} [20]. Notably, these 224-Gb s^{-1} channels also traversed three wavelength-selective switches (WSSs), indicating the potential to transport such channels over transparent mesh optical networks.

At 448-Gb s^{-1} per-channel data rate, a novel reduced-guard-interval (RGI) coherent optical orthogonal frequency-division multiplexing (CO-OFDM) format with 16-QAM subcarrier modulation was recently introduced [21]. At 448-Gb s^{-1}, an RGI-CO-OFDM-16QAM channel was transmitted over 2,000 km of ULAF and five 80-GHz-grid WSSs, potentially allowing for an SE of 5 b s^{-1} Hz^{-1} and an SEDP of 10,000 km-b s^{-1} Hz^{-1} [21]. The optical bandwidth of the 448-Gb s^{-1} channel (60 GHz) was wider than the bandwidth of the analog-to-digital converters (ADCs) used in the DCD, therefore banded digital coherent detection (B-DCD) was introduced, based on two optical frontends with two optical local oscillators (OLOs) separated by 30 GHz.

At 1-Tb s^{-1} per-channel data rate, orthogonal-band-multiplexing (OBM) of multiple CO-OFDM bands with QPSK subcarrier modulation was used to realize 600-km transmission in SSMF, achieving an intrachannel SE of 3.3 b s^{-1} Hz^{-1} and an SEDP of 1,980 km-b s^{-1} Hz^{-1} [22]. In a multiband (multicarrier) channel, the intrachannel SE is defined as the ratio of the net bit rate per band (subcarrier) to the band (subcarrier) spacing [22, 23]. The intrachannel SE constitutes an upper bound on the SE achievable in WDM operation. The OBM is a technique wherein multiple OFDM bands are coherently locked onto a common grid to form an extended OFDM spectrum.

At 1.2-Tb s^{-1} data rate per channel, a multicarrier non-guard-interval (NGI) CO-OFDM scheme was reported for 7,200-km transmission over ULAF, achieving an intrachannel SE of 3.7 b s^{-1} Hz^{-1} and a record SEDP of 27,000 km-b s^{-1} Hz^{-1} [23]. This 1.2-Tb s^{-1} NGI-CO-OFDM channel consisted of twenty-four

12.5-Gbaud PDM-QPSK carriers spaced at 12.5 GHz, occupying an optical bandwidth of 312.5 GHz. The receiver comprised 50-Gsamp s^{-1} ADC-based B-DCD with twelve different OLO frequencies.

Note that OBM [24, 25], multicarrier modulation [25, 26], and B-DCD provide attractive solutions to alleviate the bandwidth limitation imposed by optical modulator, ADC, and digital signal processor (DSP) in detecting 400-Gb s^{-1} and 1-Tb s^{-1} channels, as shown in the above demonstrations. In a sense, these high-speed channels can be regarded as OFDM-based *superchannels*, wherein multiple modulated carriers or bands are optically multiplexed retaining the OFDM condition [24–26] to achieve maximum SE without coherent crosstalk in both the generation and detection stages. We note that each individual OFDM subchannel forming the superchannel aggregate may be of the single-carrier type, or of the OFDM type [24–26].

1.2.4 From Research Demonstration to Commercial Reality

Forty-Gb s^{-1} transceivers based on DDD and DCD have been commercially realized and deployed in real-world optical transport systems. Due to its relatively simple design, DDD-based DBPSK and DQPSK systems have been widely deployed. For 40-Gb s^{-1} DCD-based receivers, the ADC and DSP modules were integrated in a single application-specific integrated circuit (ASIC) based on 90-nm CMOS technology [27]. The ADC-DSP engine uses 20 million gates, and is capable of executing 12 trillion integer operations per second to implement linear of transmission impairments such as CD and PMD and even some nonlinear compensation. The ASIC has a size of approximately 12 mm × 16 mm, and dissipates a total power of 21 W [27].

In all the 100-Gb s^{-1} research demonstrations listed in Table 1.1, offline DSP was used due to the lack of high-speed DSP with sufficient processing power to receive these high data rate signals. The real-time detection of a 100-Gb s^{-1} 2-carrier PDM-QPSK signal with 20-GHz carrier spacing was recently reported [27] with two independent DCD-based receivers. Nevertheless, to save cost, power, and size, it is desirable to use a single DCD receiver per 100-Gb s^{-1} channel. This would require the use of ADC with sampling speed in the neighborhood of 56 G Samples s^{-1} and a DSP capable of executing multitrillion operations per second. New ADC and DSP techniques have recently made it feasible to realize single-chip 100-Gb s^{-1} DCD-based receivers in 65-nm CMOS, meeting the performance and power requirements of commercial fiberoptic transport systems [28]. More recently, two field trials have been reported regarding single-carrier 100-Gb s^{-1} transmission with real-time DCD. In the first field trial, a 126.5-Gb s^{-1} single-carrier PDM-QPSK channel was transmitted over 1,800 km of SSMF in AT& T's installed network with a field-programmable gate array (FPGA)-based DSP [29]. The mean bit-error ratio (BER) measured after transmission was $\sim 4.5 \times 10^{-3}$, which could yield error-free (BER $< 10^{-12}$) performance once a 20%-overhead FEC is used [29]. In the second field trial, a 112-Gb s^{-1} single-carrier real-time PDM-QPSK transceiver was demonstrated with FPGA-based DSP, and the link was used to carry native IP packet traffic over 1,520 km of SSMF in Verizon's installed network [30].

Proceeding beyond 100-Gb s^{-1} per-channel data rate, higher level modulation formats such as 16-QAM and/or optical multiplexing may be needed. The use of OFDM-based *superchannels* to achieve highest possible SEs without coherent crosstalk may be a promising approach. The use of banded detection to relax ADC/DSP complexity per chip may be required. More advanced ADC and DSP based on 40-nm CMOS or beyond would also be key enablers for beyond-100-Gb s^{-1} applications.

1.3 Self-Coherent and Differential Detection-Based Systems

Differentially coherent and self-coherent optical transmission based on differential phase-shift keying (DPSK) and DDD have recently emerged as attractive vehicles for supporting high-speed optical transmission. A large portion of current 40-Gb s^{-1} optical transceivers is based on DDD DPSK, such as DBPSK and DQPSK. In this section, we first review recent progress on mixing 40-Gb s^{-1} DBPSK and DQPSK channels with 10-Gb s^{-1} OOK channels in the same DWDM system for capacity upgrades. We then describe SCD and the benefits it brings relative to plain differential detection. The limitations of SCD are also discussed.

1.3.1 Upgrading 10-Gb s^{-1}-Based DWDM System to 40-Gb s^{-1} DBPSK and DQPSK

Most current DWDM optical transport systems are populated with 10-Gb s^{-1} OOK channels on a 50-GHz channel grid. A capacity upgrade of these systems calls for 40-Gb s^{-1} or 100-Gb s^{-1} wavelength channels to be carried over the same system [31, 32], as illustrated in Fig. 1.1. To achieve this, several technical challenges are to be addressed. First, the optical spectral extent of the 40-Gb s^{-1} or 100-Gb s^{-1} channel needs to be similar to that of the 10-Gb s^{-1} channel to fit onto

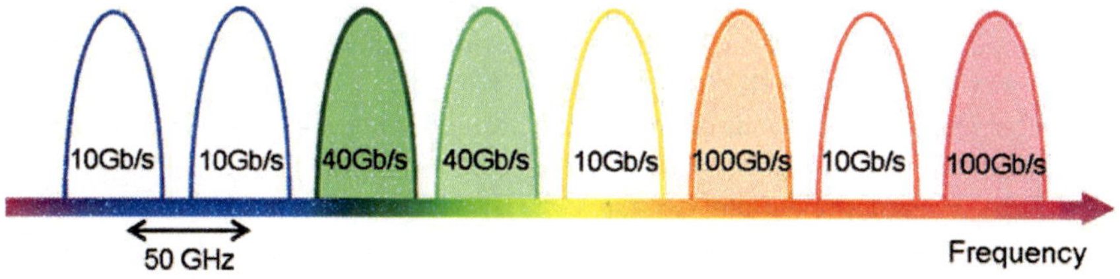

Fig. 1.1 Illustration of a channel plan with 10-Gb s^{-1}, 40-Gb s^{-1}, and 100-Gb s^{-1} wavelength channels coexisting in a 50-GHz spaced DWDM system for in-service capacity upgrade

the same channel grid. Second, it is desired that the transmission distance of the 40-Gb s^{-1} and 100-Gb s^{-1} channels be comparable to that of current 10-Gb s^{-1} OOK channels. Third, the 40-Gb s^{-1} and 100-Gb s^{-1} channels should have similar tolerance to CD and PMD as the 10-Gb s^{-1} OOK channel. Finally, the nonlinear crosstalk among adjacent channels with different data rates should not be excessive. To address these technical challenges, advanced modulation formats and detection schemes are required.

1.3.1.1 SE Consideration

To allow 40-Gb s^{-1} and 100-Gb s^{-1} channels to be added in a 50-GHz DWDM system carrying 10-Gb s^{-1} OOK channels, the optical spectral bandwidth of each of the higher speed channels should be similar to that of the 10-Gb s^{-1} channel, especially when multiple ROADM nodes are used. To achieve this, spectrally efficient optical modulation formats [2–5,33,34] have to be used. These formats include optical duobinary or phase-shaped binary transmission [35], DBPSK with partial-delay demodulation (P-DPSK) [36,37], DQPSK [38,39], and PDM-DQPSK [40].

Transmission with mixed 10-Gb s^{-1} and 40-Gb s^{-1} channels on a 50-GHz grid has been demonstrated over a nationwide optical transport network [31], in which the 10-Gb s^{-1} channels are in the OOK format and the 40-Gb s^{-1} channels are in the non-return-to-zero (NRZ) P-DBPSK format. This network incorporates an ROADM node architecture that uses 50-GHz-spaced asymmetric-bandwidth interleavers to allocate a wide-bandwidth path for 40-Gb s^{-1} P-DBPSK channels and a narrow bandwidth for 10-Gb s^{-1} OOK channels, without sacrificing the performance of the 10-Gb s^{-1} channels. The 10-Gb s^{-1} OOK signal passes through more than ten intermediate ROADM nodes with less than 1 dB penalty due to optical filtering, and the 40-Gb s^{-1} DBPSK channels can pass through more than four intermediate ROADM nodes with small filtering penalty (~1 dB). To further increase the capacity of such a deployed network, hybrid transmission of 40-Gb s P-DBPSK and return-to-zero (RZ) DQPSK channels with an SE of 0.8 b s^{-1} Hz^{-1} was demonstrated [41]. Twenty-five DWDM channels carrying an overall capacity of 1 Tb s^{-1} were transmitted over 16 × 80-km SSMF spans with EDFA-only amplification and four passes through bandwidth-managed ROADM nodes. The nonlinear crosstalk among the WDM channels was found to be small (<2 dB).

P-DPSK and DQPSK channels at 40-Gb s^{-1} have also been carried on a standard 50-GHz grid with a symmetric ROADM node architecture. A systematic study of their performance under tight optical filtering, coherent crosstalk, and PMD has been conducted [42]. Both formats have strengths and weaknesses; hence, the choice between them depends on the system requirements, e.g., nonlinear and PMD tolerance [43].

Optical transmission of ten 107-Gb s^{-1} NRZ-DQPSK channels on a 100-GHz channel grid was recently demonstrated [44]. To further increase SE, 107-Gb s^{-1} PDM-DQPSK channels were transmitted with 43-Gb s^{-1} RZ-DQPSK channels on the same 50-GHz-grid DWDM system, achieving a net system SE of 1.4 b s^{-1} Hz^{-1}

[45]. A reach of 1,280 km of SSMF including 4 ROADM passes was also achieved. In this experiment, polarization demultiplexing was performed by means of a polarization beam splitter (PBS) following a manually adjusted polarization controller (PC). For practical implementation, an automatic polarization demultiplexing scheme was recently demonstrated [46].

1.3.1.2 Transmission Distance Consideration

The transmission distances of 40-Gb s^{-1} channels should preferably be comparable to those of current 10-Gb s^{-1} OOK channels, to ensure smooth system upgrades and high cost effectiveness. In long-haul fiberoptic transmission, a key limiting factor on signal quality is the optical amplifiers noise. With a given modulation format and detection scheme, scaling the data rate from 10 Gb s^{-1} to 40 Gb s^{-1} would require 6 dB higher optical signal-to-noise ratio (OSNR) for a given BER in a back-to-back configuration. By introducing power-efficient modulation formats and detection schemes such as direct-detection DBPSK, the OSNR requirement can be relaxed by about 3 dB [2–5]. Another effective method to relax the OSNR requirement is to use advanced FEC with higher coding gain [47–49]. Further upgrading capacity by using channels at 100 Gb s^{-1} and beyond requires even higher OSNR performance. We shall elaborate on long-haul 100-Gb s^{-1} and beyond transmission in Sect. 1.4 on DCD-based systems.

1.3.1.3 CD and PMD Consideration

Forty-Gb s^{-1} and 100-Gb s^{-1} channels ought to accommodate similar amounts of CD and PMD as 10-Gb s^{-1} OOK channels. CD and PMD are two linear transmission impairments particularly impacting high-speed optical signals with wide spectral bandwidth. In 10-Gb s^{-1}-based long-haul DWDM systems, fiber CD is usually compensated for by using inline dispersion compensation modules (DCM), typically comprising dispersion-compensating fibers (DCFs). For DDD-based 40-Gb s^{-1} and 100-Gb s^{-1} channels, tunable optical dispersion compensators (TDC) are usually used on a per-channel basis, to bring the net CD experienced by a signal to within the receiver's dispersion tolerance. To tolerate more PMD, optical PMD compensators (PMDCs) may also be used on a per-channel basis [46]. Advanced signal processing at the transmitter can pre-compensate the CD experienced by a signal during fiber transmission [50–53]. For DCD-based channels, CD and PMD can be compensated digitally, so optical CD and PMD compensations are not required. This will be elaborated in the following section.

1.3.1.4 Nonlinear Tolerance Consideration

It is important to assess the nonlinear tolerance (NLT) of 40-Gb s^{-1} and 100-Gb s^{-1} signals in the presence of 10-Gb s^{-1} OOK channels, especially for inline-dispersion

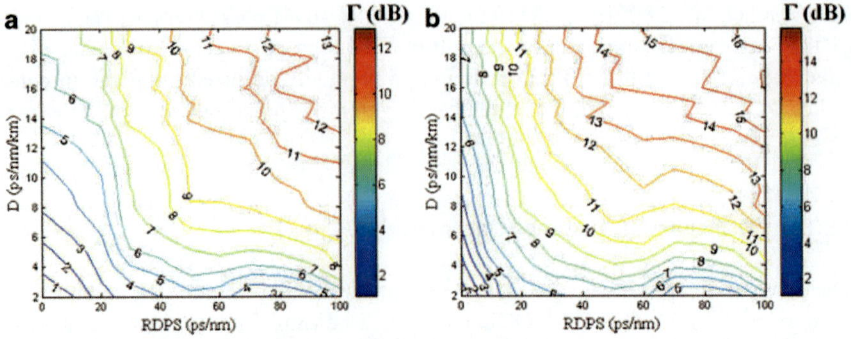

Fig. 1.2 The suppression factor (in dB) for the XPM induced by a 10-Gb s^{-1} OOK channel into a 40-Gb s^{-1} DQPSK channel, as a function of RDPS and D for $N = 10$ and $L_{\mathrm{eff}} = 20$ km (**a**), and for $N = 28$ and $L_{\mathrm{eff}} = 25$ km (**b**) [57]. The channel spacing is 50 GHz

compensated transmission. It was found that the interchannel cross-phase modulation (XPM) from 10-Gb s^{-1} OOK channels to 40-Gb s^{-1} DBPSK and DQPSK channels is a major nonlinear impairment [54–57]. It was also found that intrachannel nonlinear effects induced nonlinear penalties are usually not as severe as those caused by the interchannel XPM from neighboring 10-Gb s^{-1} OOK channels, particularly when inline optical DCMs are used [54–57]. Comprehensive surveys on intrachannel nonlinear effects in homogeneous DWDM systems can be found in Chaps. 6–8.

A systematic experimental investigation [57] showed that the XPM penalty strongly depends on system configurations such as the dispersion map and channel plan. Recently, an analytical model [58] was reported to provide an efficient tool to approximately assess the XPM impact under various system conditions. It was found that the interchannel XPM effect can be suppressed by proper dispersion management [58]. Figure 1.2 shows the XPM suppression factor (Γ) as a function of the residual dispersion per span (RDPS) and fiber dispersion coefficient (D) for two different sets of link conditions. The XPM suppression factor tends to increase with the increase of RDPS or fiber dispersion. This was explained by the dispersion-induced temporal walk-off among the channels, which helps to reduce the XPM penalty. In Fig. 1.2a, the total number of amplified fiber spans (N) is 10 and the effective fiber length (L_{eff}) of each span is 20 km, similar to the conditions used in [55]. In Fig. 1.2b, $N = 28$, and $L_{\mathrm{eff}} = 25$ km, similar to those used in [54]. The XPM suppression factor [54] is ~3 dB lower in system (a) relative to system (b) with system (a) based on NZDSF having $D = 4.25$ ps nm^{-1} km^{-1} and RDPS $\approx$ 50 ps nm^{-1} and with system (b) SSMF spans having $D = 17$ ps nm^{-1} km^{-1} and a mean RDPS of 12.5 ps nm^{-1}. Since the nonlinear coefficient of SSMF is ~1.6 dB lower than that of the NZDSF used, the XPM effect in system (b) is expected to be ~4.6 dB lower than that in system (a). These numerical results are in good agreement with the experimental results, explaining why negligible XPM penalty was measured in system (b) [54], whereas the XPM penalty was quite severe in system (a) [55].

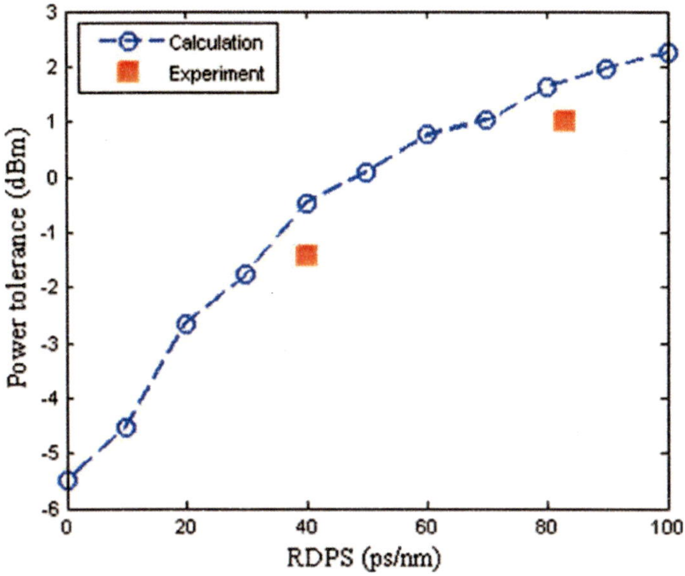

Fig. 1.3 The power tolerance of a 40-Gb s^{-1} DQPSK signal vs. RDPS in a link with 12×80-km SSMF spans. The solid squares are from experiment [29]. $D = 17\,\text{ps nm}^{-1}\,\text{km}^{-1}$ and $\gamma = 1.22\,\text{W}^{-1}\,\text{km}^{-1}$

Figure 1.3 shows the power tolerance as a function of RDPS in a transmission link consisting of 12 optically amplified 80-km SSMF spans, similar to that reported in [57]. Here, the power tolerance refers to the power of each of the neighboring 10-Gb s^{-1} OOK channels, for which the XPM penalty onto the 40-Gb s^{-1} DQPSK channel, induced by these 10-Gb s^{-1} channels is 1 dB at BER $= 10^{-3}$. Again, the simulated results are in reasonable agreement with the experimental ones. The calculated power tolerances are slightly higher than the experimental ones, since the intrachannel nonlinear effects were not considered in the analytical model [58]. Clearly, the interchannel XPM penalty can be reduced by suitably increasing the RDPS.

Figure 1.4 shows the XPM suppression factor Γ as a function of the number of neighboring OOK channels, on a 50-GHz grid using different channel plans. For the baseline case where no guard-band is used, the nearest neighbors that are 50 GHz away contribute the bulk of the XPM effect, whereas the presence of other 10-Gb s^{-1} OOK channels only slightly reduces Γ. This suggests that the XPM penalty may be effectively mitigated by introducing a guard band between the DQPSK channel and its nearest OOK neighbors. With a 100-GHz guard-band, Γ increases by $\sim$3.5 dB, which is in good agreement with the experimental observation [57]. With a 150-GHz guard-band, Γ further increases by $\sim$2 dB; however, the insertion of guard-bands reduces the overall system capacity. The capacity reduction can be minimized by grouping the 40-Gb s^{-1} channels together into a subband

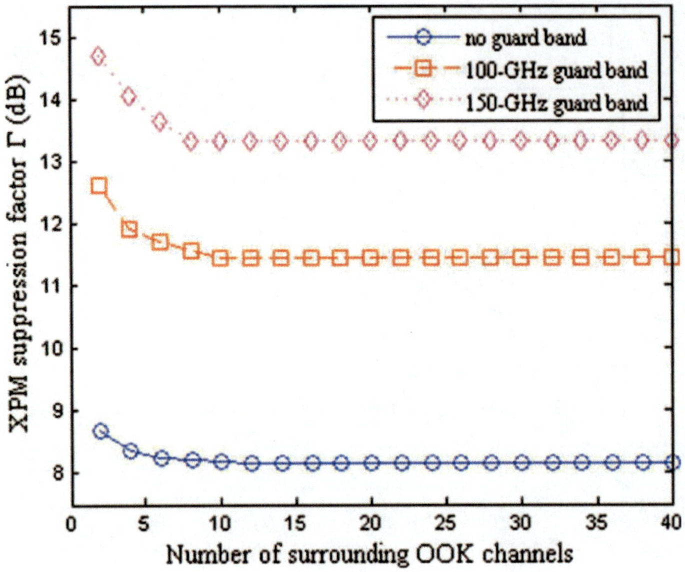

Fig. 1.4 The XPM suppression factor vs. the number of neighboring OOK channels in a link with 12 × 80-km SSMF spans. $D = 17\,\mathrm{ps\,km^{-1}\,nm^{-1}}$, $\gamma = 1.22\,\mathrm{W^{-1}\,km^{-1}}$, and RDPS = $40\,\mathrm{ps\,nm^{-1}}$

and allocating a pair of guard-bands at the edges of the subband. Thus, there is a trade-off between performance and system capacity and/or flexibility in mixed-data-rate transmission involving 10-Gb s^{-1} OOK and 40-Gb s^{-1} DPSK channels.

Figure 1.5 shows the power tolerance as a function of RDPS in a transmission link consisting of twelve optically amplified 80-km NZDSF spans with $D = 4\,\mathrm{ps\,km^{-1}\,nm^{-1}}$ and $\gamma = 1.79\,\mathrm{W^{-1}\,km^{-1}}$. The power tolerance with the NZDSF spans is over 6 dB lower than in the case of SSMF spans. This can be attributed to the smaller dispersion coefficient and higher nonlinear coefficient of the NZDSF. Figure 1.6 shows the XPM suppression factor Γ as a function of the number of neighboring OOK channels on a 50-GHz grid. Again, the XPM penalty can be reduced by introducing a guard-band between the DQPSK channel and its nearest OOK neighbors. With a guard band of 100(150) GHz, Γ increases by ~2(3) dB.

Compared to 40-Gb s^{-1} DQPSK, 40-Gb s^{-1} DBPSK has its symbol period halved and the minimum phase difference between symbols doubled. It is thus expected that the power tolerance of a 40-Gb s^{-1} DBPSK signal (namely the power of each of the neighboring 10-Gb s^{-1} OOK channels) would be ~6 dB higher than that of a 40-Gb s^{-1} DQPSK signal. Figures 1.7 and 1.8 plot power tolerance as a function of RDPS in SSMF-based and NZDSF-based links, respectively. Compared to 40-Gb s^{-1} DQPSK, the power tolerance of 40-Gb s^{-1} DBPSK is about 6 dB higher in the SSMF link, and about 4 dB higher in the NZDSF link, at a typical RDPS of 40 ps nm^{-1}.

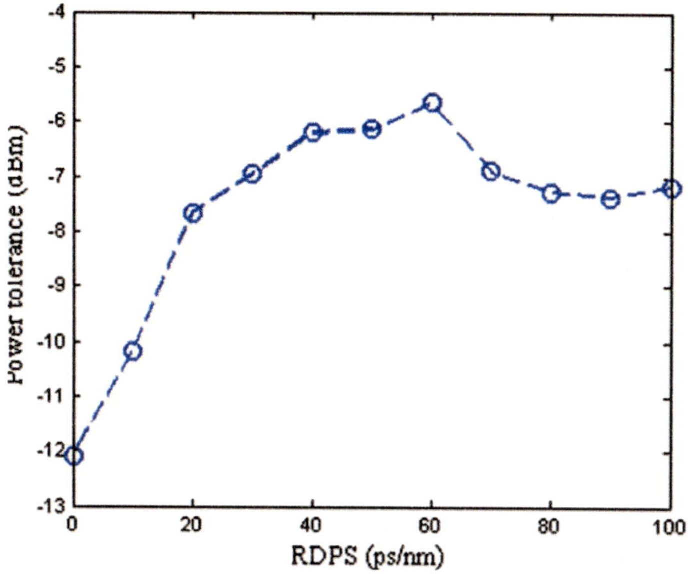

Fig. 1.5 The power tolerance of a 40-Gb s^{-1} DQPSK signal vs. RDPS in a link with 12×80-km NZDSF spans. $D = 4\,\mathrm{ps\,nm^{-1}\,km^{-1}}$ and $\gamma = 1.79\,\mathrm{W^{-1}\,km^{-1}}$

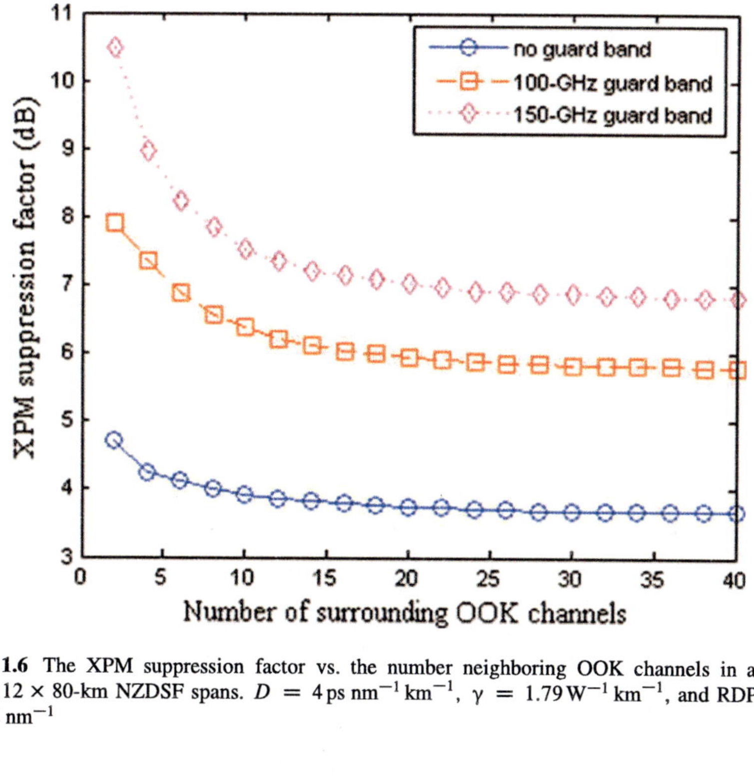

Fig. 1.6 The XPM suppression factor vs. the number neighboring OOK channels in a link with 12×80-km NZDSF spans. $D = 4\,\mathrm{ps\,nm^{-1}\,km^{-1}}$, $\gamma = 1.79\,\mathrm{W^{-1}\,km^{-1}}$, and RDPS = $40\,\mathrm{ps\,nm^{-1}}$

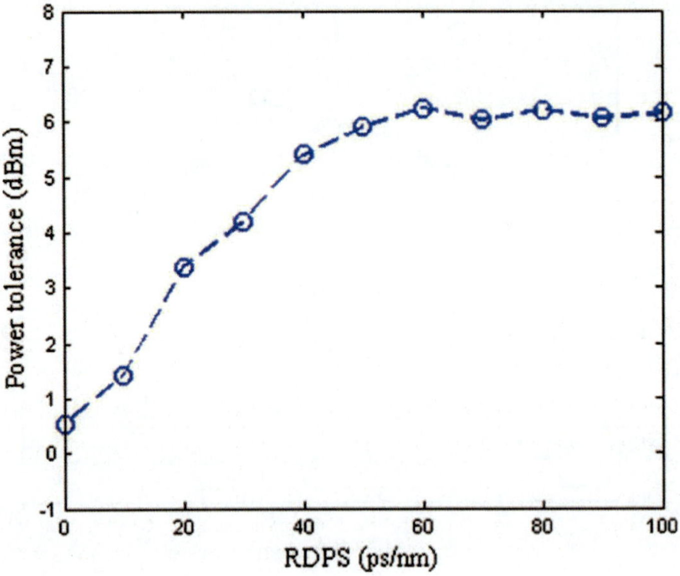

Fig. 1.7 The power tolerance of a 40-Gb s^{-1} DBPSK signal vs. RDPS in a link with 12 × 80-km SSMF spans. $D = 17\,\text{ps nm}^{-1}\,\text{km}^{-1}$ and $\gamma = 1.22\,\text{W}^{-1}\,\text{km}^{-1}$

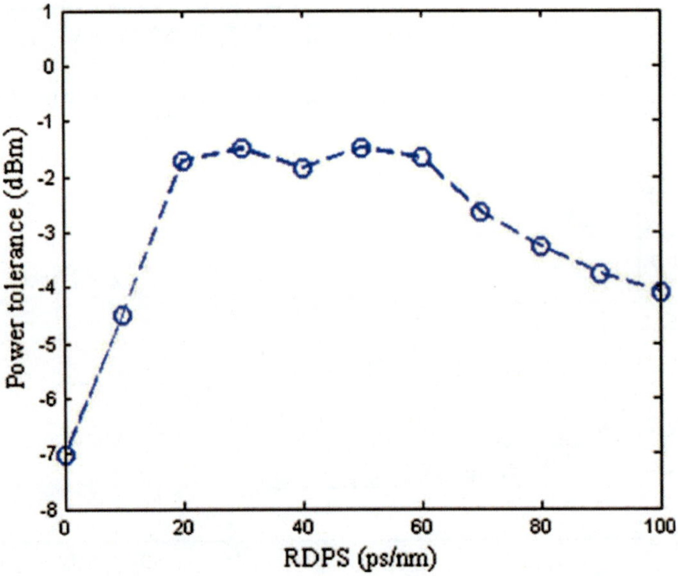

Fig. 1.8 The power tolerance of a 40-Gb s^{-1} DBPSK signal vs. RDPS in a link with 12 × 80-km NZDSF spans. $D = 4\,\text{ps nm}^{-1}\,\text{km}^{-1}$, $\gamma = 1.79\,\text{W}^{-1}\,\text{km}^{-1}$, and RDPS = $40\,\text{ps nm}^{-1}$

It has been shown that the XPM effect from neighboring 10-Gb s^{-1} OOK channels can also cause severe penalties for a 40-Gb s^{-1} PDM-QPSK signal with DCD [58]. For inline-dispersion compensated transmission, even the XPM effect from neighboring 40-Gb s^{-1} PDM-QPSK channels was found to generate a severe impairment [59]. There are three potential reasons for the degraded NLT. First, the symbol period of 40-Gb s^{-1} PDM-QPSK is twice as large as that of 40-Gb s^{-1} DQPSK, which reduces the temporal walk-off relative to a 10-Gb s^{-1} OOK channel during transmission and thus yields a higher XPM impact. Second, DCD relies on multiple adjacent symbols to perform phase estimation, and may be more susceptible to XPM-induced phase wandering than DDD, for which only two adjacent symbols are processed together. Indeed, the NLT of DCD-based PDM-QPSK was found to be improved upon reducing the number of symbols used for phase estimation [59]. Finally, PDM-QPSK is also susceptible to interchannel XPM-induced nonlinear polarization scattering. The XPM impairment is expected to be less severe for 40-Gb s^{-1} PDM-BPSK [60] and 100 Gb s^{-1} or beyond PDM-QPSK signals [61] due to shortened symbol period. A comprehensive survey on the effect of nonlinear polarization scattering in PDM systems is given in Chap. 9.

1.3.1.5 Overall Comparison

Based on the technical considerations surveyed above and the multiple references cited in this section, Table 1.2 attempts to provide a rough comparison among various 40-Gb s^{-1} and 100-Gb s^{-1} signal formats. DCD-based formats, to be discussed in the following section, are also included for completeness. More details on CO-OFDM and its NLT can be found in Chaps. 2 and 3.

Practical considerations such as complexity and commercial availability are also relevant when designing a DWDM system, but those tend to evolve with advances in relevant technologies. From Table 1.2, it is reasonable to conclude that seamless capacity upgrades, populating 40-Gb s^{-1} channels in a DWDM system originally designed for 10-Gb s^{-1} OOK, have become feasible through the use of DDD-based 40-Gb s^{-1} P-DPSK and RZ-DQPSK formats. Evidently, DCD-based formats would be the choice for 100-Gb s^{-1} and beyond. Future key tasks seem to include cost-effective implementations of DCD-based formats at 100 Gb s^{-1} and beyond, along with optimum system designs to best incorporate these high data-rate channels, with the consideration of nonlinear transmission performance.

1.3.2 Self-Coherent Detection

DDD has slight worse receiver sensitivity as compared to homodyne coherent detection but its complexity is lower as a laser is not required in the receiver, the OLO laser phase noise impairments are drastically reduced and polarization diversity or polarization control is not necessary. To generate coherent gain without the

Table 1.2 A comparison among different 40-Gb s^{-1} and 100-Gb s^{-1} signal formats for 50-GHz spaced DWDM transmission based on the references cited in this section

	10-Gb s^{-1}	40-Gb s^{-1}			100-Gb s^{-1}		
Modulation Formats	OOK	P-DPSK	RZ-DQPSK	PDM-QPSK	PDM-BPSK	PDM-QPSK	CO-OFDM
Detection	DD	DDD	DDD	DCD	DCD	DCD	DCD
Relative Sensitivity[a]	0 dB	−3 dB	−3 dB	−2 dB	−2 dB	−6 dB	−6 dB
Filtering Tolerance	High	Medium	High	High	High	High	High
CD Tolerance	No need for TDC	Need TDC	Need TDC	EDC	EDC	EDC	EDC
PMD Tolerance[b]	15 ps	~3.5 ps	~7 ps	>25 ps	>25 ps	>25 ps	>25 ps
Nonlinear Tolerance[c]	High	High	Medium	Low	Medium	Medium	Low-Medium
Relative Complexity	Low	Medium	High	High	High	High	High
Availability	Yes	Yes	Yes	Yes	Yes	Yes[d]	No

[a] In terms of required OSNR (0.1-nm noise bandwidth) at BER $= 10^{-3}$ assuming typical implementation penalties
[b] In terms of the mean differential group delay (DGD) allowed for a 1-dB OSNR penalty at an outage probability of 10^{-5} (assuming no optical PMDC is used)
[c] Assuming 10-Gb s^{-1} OOK channels present in the same DWDM system
[d] See, e.g., "Analyst: AlcaLu's 100G Game-Changer," http://www.lightreading.com/document.asp?doc_id=192989
TDC Tunable optical dispersion compensator; *EDC* Electronic dispersion compensator

actual presence of a physical OLO, SCD was recently proposed, based either on optical signal processing [62–67] or on digital signal processing (DSP) [68, 69]. In this subsection, we review recent progress in SCD. Following a brief description of the principle of digital self-coherent detection (DSCD), we review DSP-based techniques such as data-aided multi-symbol phase estimation (MSPE) for receiver sensitivity enhancement [70–72], a unified detection scheme for multilevel DPSK signals, and some more advanced signal processing techniques used in SCD. The limitations of SCD as compared to DCD are also discussed.

1.3.2.1 Principle of Digital Self-Coherent Detection

A schematic DSCD architecture is shown in Fig. 1.9 [69]. The optical complexity of the DSCD is similar to that of conventional direct-detection DQPSK. The received signal, denoted as $r\,(t) = |r\,(t)|\exp[j \times \phi\,(t)]$, is first split into two branches, which are connected to a pair of optical delay interferometers (ODIs) with *orthogonal* phase offsets θ and $\theta - \pi/2$, where θ is an arbitrary phase value. The delay in each of the ODI, τ, is set to be approximately T/*sps*, where T is the signal symbol period and *sps* is the number of samples per symbol of the ADCs used to convert the two detected analog signal waveforms, referred to as the I and Q components, to digitized waveforms $u_{\mathrm{I}}(t)$ and $u_{\mathrm{Q}}(t)$. Forming a complex waveform out of the I and Q components, we have

$$u(t) = u_{\mathrm{I}}(t) + j \cdot u_{\mathrm{Q}}(t) = \mathrm{e}^{j \cdot \theta} r(t) \cdot r^*(t - \tau) = |r(t)| \cdot |r(t - \tau)|\, \mathrm{e}^{j[\phi(t) - \phi(t-\tau) + \theta]}. \tag{1.1}$$

In the special case when $sps = 1$, the delay in the orthogonal ODI pair equals the symbol period, and the I and Q decision variables for m-ary DPSK detection can be directly obtained by setting $\theta = \pi/m$, as discussed further below. Any demodulator

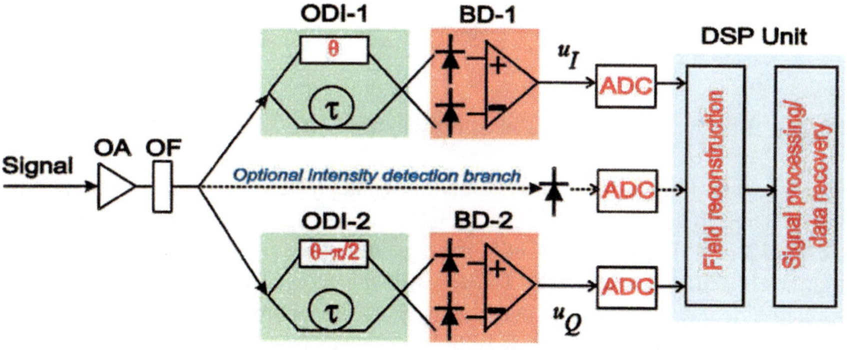

Fig. 1.9 Schematic DSCD architecture based on orthogonal differential direct-detection followed by ADC and DSP [69]. *OA* Optical pre-amplifier; *OF* Optical filter; *ODI* Optical delay interferometer; *BD* Balanced detector; *ADC* Analog-to-digital converter

phase error $\phi_e = \theta - \pi/m$ can be compensated by applying the following simple electronic demodulator error compensation (EDEC) process [69]

$$u(t) \to \mathrm{e}^{-j\cdot\varphi_e} u(t). \tag{1.2}$$

The optical phase difference between adjacent sampling locations is obtained from

$$q(t) = u(t)\mathrm{e}^{-j\cdot\theta} \Big/ \left|u(t)\mathrm{e}^{-j\cdot\theta}\right| = \mathrm{e}^{j\cdot[\varphi(t)-\varphi(t-\tau)]} = \mathrm{e}^{j\Delta\varphi(t)}, \tag{1.3}$$

where $\Delta\varphi(t) = \varphi(t) - \varphi(t-\tau)$.

With the differential phase information being available, a digital representation of the received signal field can be obtained by

$$\begin{aligned} r(t_0 + n\cdot\tau) &= |r(t_0 + n\cdot\tau)| \prod_{m=1}^{n} q(t_0 + m\cdot\tau) \\ &= |r(t_0 + n\cdot\tau)| \mathrm{e}^{j\cdot\phi(t_0)} \prod_{m=1}^{n} \mathrm{e}^{j\cdot\Delta\phi(t_0+m\cdot\tau)}, \end{aligned} \tag{1.4}$$

where t_0 is an arbitrary reference time, $\phi(t_0)$ is a reference phase which may be set to 0, and the amplitude $|r(t_0 + n\cdot\tau)|$ of the received signal can be obtained from an additional intensity detection branch, or approximating the amplitude samples from the ODIs complex output (1) as below

$$|r(t_0 + n\cdot\tau)| \approx |u(t_0 + n\cdot\tau)\cdot u(t_0 + n\cdot\tau + \tau)|^{1/4} \tag{1.5}$$

We note, however, that performance is degraded at sampling locations where the signal amplitude is close to zero, particularly when the sampling amplitude resolution is limited [69]. Also, note that DSCD can be designed to be polarization independent to readily receive a single-polarization signal in an arbitrary polarization state, while DCD usually requires polarization diversity.

1.3.2.2 Receiver Sensitivity Enhancement via Data-Aided MSPE

There is a well-known differential-detection penalty in receiver sensitivity for DPSK as compared to coherent PSK. This penalty can be substantially reduced by using a data-aided MSPE algorithm, utilizing the previously recovered data symbols to recursively extract a new phase reference, which is more accurate than that provided by the immediate past symbol alone. *Analog* implementations of this concept have been proposed for optical DQPSK [70], DQPSK/ASK [71], and m-ary DPSK [72]. Optical processing realizations have been introduced in [62–67]. The MSPE concept was recently extended to the *digital* domain [69,72]. An improved complex decision variable for m-ary DPSK can be written as [69]

$$x(n) = u(n) + \sum_{p=1}^{N} \left\{ w^p \mathrm{e}^{-jp\pi/m} u(n) \prod_{q=1}^{p} \left[u(n-q) \cdot \mathrm{e}^{-j\Delta\varphi(n-q)} \right] \right\}, \tag{1.6}$$

where $u(n)$ is the directly detected complex decision variable for the nth symbol, m is the number of phase states of the m-ary DPSK signal, N is the number of past decisions used in the MSPE process, w is a forgetting factor, and $\Delta\phi\,(n-q) = \phi\,(n-q) - \phi(n-q-1)$ is the optical phase difference between the $(n-q)$th and the $(n-q-1)$th symbols, which can be estimated based on the past decisions. An insightful analysis appears in [66]. The benefits of the MSPE and EDEC were recently confirmed in a 40-Gb s^{-1} DQPSK experiment with offline DSP [73].

1.3.2.3 Unified Detection of m-ary DPSK

The DSCD can be used to receive high SE m-ary DPSK signals [72]. An m-ary DPSK signal has $\log_2(\mathrm{m})$ binary data tributaries that are usually obtained from m/2 decision variables associated with m/4 ODI pairs having the following orthogonal phase offsets, $\left(\frac{\pi}{m}, \frac{\pi}{m} - \frac{\pi}{2}\right), \left(\frac{3\pi}{m}, \frac{3\pi}{m} - \frac{\pi}{2}\right), \ldots, \left(\frac{(m/2-1)\pi}{m}, -\frac{\pi}{m}\right)$. With DSP, the last (m/2–2) decision variables can be derived by linear combinations of the first two decision variables, u_I and u_Q. This dramatically reduces the optical complexity associated with the detection of m-ary DPSK, by using just two rather than m/2 ODIs. The decision variables associated with phase offset $\pi\mathrm{p}/\mathrm{m}$ $(\mathrm{p} = 3, 5, \ldots, \mathrm{m}/2 - 1)$ are expressed as

$$\nu(\pi p/m) = \cos\left(\frac{p-1}{m}\pi\right) u_\mathrm{I} - \sin\left(\frac{p-1}{m}\pi\right) u_\mathrm{Q}. \tag{1.7}$$

Similarly, we may express their orthogonal counterparts as

$$\nu\,(\pi p/m - \pi/2) = \sin\left(\frac{p-1}{m}\pi\right) u_\mathrm{I} + \cos\left(\frac{p-1}{m}\pi\right) u_\mathrm{Q}. \tag{1.8}$$

The data tributaries of an m-ary DPSK signal can then be retrieved by [72].

$$\begin{aligned}
c_1 = c_\mathrm{I} &= \left[u\left(\frac{\pi}{m}\right) > 0\right], \qquad c_2 = c_\mathrm{Q} = \left[u\left(\frac{\pi}{m} - \frac{\pi}{2}\right) > 0\right], \\
c_3 &= \left[u\left(\frac{\pi}{m} + \frac{\pi}{4}\right) > 0\right] \oplus \left[u\left(\frac{\pi}{m} - \frac{\pi}{4}\right) > 0\right], \ldots \\
c_{\log_2(m)} &= \left[u\left(\frac{3}{m}\pi\right) > 0\right] \oplus \left[u\left(\frac{7}{m}\pi\right) > 0\right] \ldots \oplus \left[u\left(\frac{m/2-1}{m}\pi\right) > 0\right] \\
&\oplus \left[u\left(\frac{3}{m}\pi - \frac{\pi}{2}\right) > 0\right] \oplus \left[u\left(\frac{7}{m} - \frac{\pi}{2}\right) > 0\right] \ldots \\
&\oplus \left[u\left(\frac{m/2-1}{m}\pi - \frac{\pi}{2}\right) > 0\right].
\end{aligned} \tag{1.9}$$

When the data-aided MSPE is applied, u_I and u_Q are to be replaced by their corresponding improved decision variables. In effect, the complex decision variable $u(n)$ or $x(n)$ contains complete information on the differential phase between adjacent symbols, providing sufficient statistics, allowing to derive all the required decision variables. The above formalism provides the basis of a simple yet universal DSCD receiver platform for m-ary DPSK using just one pair of orthogonal optical demodulators as shown in Fig. 1.9.

1.3.2.4 More Advanced DSCD Signal Processing

Recently, there have been several advanced DSP functions reported for DSCD systems to improve the system tolerance to transmission impairments and/or detection versatility. Pre-phase integration (PPI) is a newly introduced technique countering the effect of differential detection so that the signal phase information rather than the differential phase information is obtained upon differential detection [14, 74]. This technique facilitates the recovery of the signal phase information of QAM formats such as 8-QAM and 16-QAM, thereby increasing the DSCD versatility. In recent experiments [74], Kikuchi and Sasaki verified the PPI process for 30-Gb s^{-1} 8-QAM and 35.8-Gb s^{-1} 12-QAM transmission based on transmitter-side off-line DSP. In addition, CD pre-compensation was also implemented with a 53-stage digital FIR filter, mitigating up to 6,700 ps nm^{-1} worth of dispersion [74]. More recently, 40-Gb s^{-1} 16-QAM transmission over 160 km of SSMF has also been demonstrated with DSCD [14].

Due to differential detection, the noise-induced variance of the recovered single symbols along the angular direction in the signal constellation is larger than that along the radial direction. This nonisotropic noise distribution indicates that the commonly used Euclidean decision metric is no longer optimal for SCD. A computationally efficient non-Euclidean decision scheme was recently proposed, wherein the decision is based on a non-Euclidean distance metric, biased toward displacement along the radial direction [14, 75]. This technique was applied to DSCD of a 16-QAM signal, attaining an improvement of 2.2 dB in receiver sensitivity, relative to the Euclidean decision [14].

In fiberoptic transmission, phase-modulated signals are degraded by the Gordon–Mollenauer nonlinear phase noise [76] resulting from the interaction between the self-phase modulation (SPM) and amplified spontaneous emission (ASE) noise. It was found that Gordon–Mollenauer nonlinear phase noise can be substantially compensated by a lumped postcompensation process [77–79]. This can be achieved by replacing the directly measured complex decision variable, *u(n)*, with a compensated complex variable *v(n)* [65]

$$v(n) = u(n) \cdot \exp\left\{-j \cdot \frac{1}{2} c_{\mathrm{NL}} \cdot [P(n) - P(n-1)]\right\}, \tag{1.10}$$

where c_{NL} is a coefficient proportional to the average nonlinear phase shift experienced by the signal over the fiber transmission, P(n) is the normalized power of the nth symbol, and the factor of $1/2$ is for the 50% undercompensation that was found to be optimum in the lumped single-step postcompensation scheme [77]. Post nonlinear phase noise compensation was recently demonstrated in DSCD [80]. There are also alternative self-coherent approaches, making use of delay interferometers with delays, which are integer multiples of a fixed delay, T, but processing and decoding the photo-detected outputs digitally rather than in an analog manner [81, 82].

Although DSCD offers many attractive capabilities akin to those offered by DCD, there are some limitations of DSCD. Particularly, the DSP complexity needed for polarization demultiplexing and PMD compensation in DSCD is much higher than that in DCD due to the lack of the information on the phase difference between two reconstructed signal polarization components in DSCD [83]. In addition, the post-CD compensation capability of DSCD is limited as DSCD requires higher ADC resolution to mitigate the issue associated with the field reconstruction at "zero" intensity locations [69, 83]. Overall, it seems that DSCD is better suited for low-complexity single-polarization-based fiberoptical transmission systems, where long-range transmission effects such as CD and PMD are either pre-compensated or are sufficiently small. Remarkably, it is possible to port the mathematical techniques of MSPE, as applied to self-coherent direct detection in this section, for attaining improved carrier phase and frequency estimation performance for coherent (OLO-based) detection [84–86].

1.4 DCD-Based Systems

Digital coherent detection [6–10] has recently attracted extensive attention due to its capability to detect high SE signals with high receiver sensitivity and to digitally compensate transmission impairments such as CD and PMD. In DCD, polarization-diversity is usually required to align the signal's random received polarization state to that of the OLO; this makes DCD naturally suited for receiving PDM signals, while doubling SE as compared to their single-polarization counterparts, without requiring higher OSNR for a given signal data rate. Moreover, DCD can be used for both single-carrier and multi-carrier modulation formats. More details on single-carrier-based coherent transmission are provided in Chap. 4. CO-OFDM is a promising multi-carrier format that has attracted much attention recently, including the possibility of compensating for its nonlinear impairment. Reviews on CO-OFDM and its NLT are presented in Chaps. 2 and 3. In this section, a brief description of DCD is given, followed by a more extensive survey of recent DCD-based coherent transmission results at per-channel data rates of 100-Gb s^{-1} and beyond.

1.4.1 *Digital Coherent Detection*

Figure 1.10 shows a schematic of a typical polarization-diversity DCD receiver, consisting of an OLO, a polarization-diversity 2×8 optical hybrid, four balanced

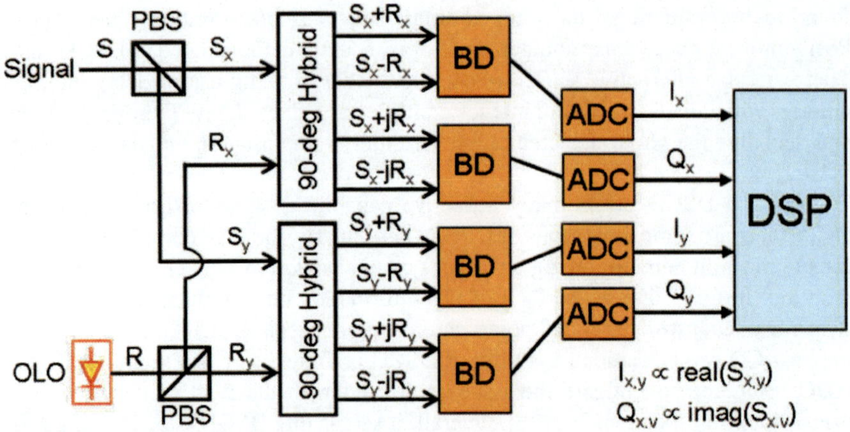

Fig. 1.10 Schematic of a typical polarization-diversity DCD receiver. *OLO* Optical local oscillator; *PBS* Polarization-beam splitter; *BD* Balanced detector; *ADC* Analog-to-digital converter; *DSP* Digital signal processor

detectors (BDs), four ADCs, and a DSP unit. The polarization-diversity optical hybrid mixes the incoming signal S with the reference source R generated by the OLO to obtain four pairs of mixed signals, $(S_x \pm R_x)$, $(S_x \pm jR_x)$, $(S_y \pm R_y)$, and $(S_y \pm jR_y)$. The power waveforms of each pair of the output mixed signals are photo-detected and differentially detected by a BD followed by an ADC. The resulting four digital signals $\mathrm{I}_{x,y}$ and $\mathrm{Q}_{x,y}$ are linearly related to the in-phase (I) and the quadrature (Q) components of each of the two orthogonal polarization components of the input signal, which is polarization-resolved by the PBS. These four digital signals are provided to a DSP unit for further processing to mitigate impairments and detect the amplitude and phase of the unknown incoming signal S.

PDM is an effective means to double the SE of a given modulation format without requiring additional OSNR for a same data rate. With the use of polarization-diversity digital coherent receiver, PDM is naturally supported. Indeed, most recent demonstrations with DCD [15–23] were using PDM. Polarization demultiplexing was performed in the digital domain by using adaptive algorithms such as the constant modulus algorithm (CMA) [5, 87], which effectively derotate the polarization transformation (Jones matrix) of the fiber link. In addition, CMA-based equalization is capable of compensating for PMD, making DCD attractive for high-speed optical transmission, where large system tolerance to PMD is desired.

Figure 1.11 shows the constellation diagrams of popular modulation formats commonly used with DCD, quadrature phase-shift keying (QPSK) [8, 18–23] or 4-point QAM, 16-QAM, 32-QAM, and 64-QAM, respectively carrying 2, 4, 5, and 6 bits per symbol per polarization. Recently, the generation and detection of PDM-32-QAM [88] and PDM-64-QAM [89] have been demonstrated at about 100 Gb s^{-1}.

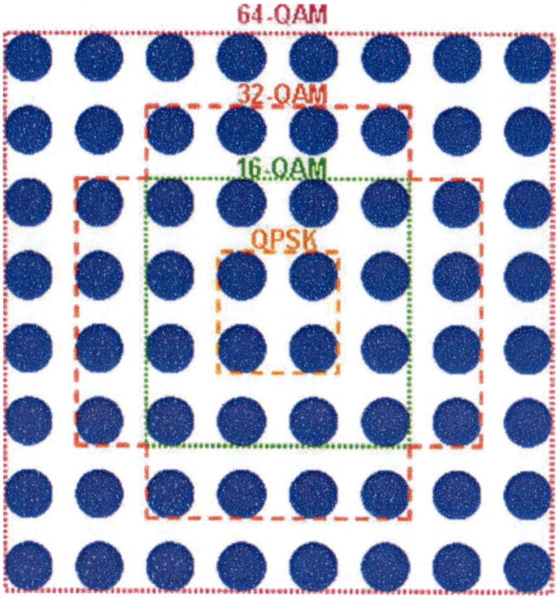

Fig. 1.11 Constellation diagrams of QPSK or 4-QAM, 16-QAM, 32-QAM, and 64-QAM, respectively carrying 2, 4, 5, and 6 bits per symbol per polarization

In optically amplified transmission, signal quality has a strong dependence on OSNR, which is commonly defined as the ratio between the signal power and the optical noise power in both orthogonal polarization states within a fixed bandwidth of 0.1 nm (or 12.5 GHz at a signal wavelength of about 1,550 nm). The OSNR required to achieve a given BER in an optical channel depends on its data rate, modulation format, and detection scheme. For a fixed data rate, the required OSNR at low BER values can be estimated from the minimum Euclidean distance between two closest symbols in the signal constellation diagram (with a normalized average signal power).

Using coherent homodyne detection binary phase-shift keying (BPSK) as the reference, the OSNR penalty (or additionally required OSNR in dB for a given BER) can be estimated. Figure 1.12 shows the OSNR penalties at low BER of the DCD- and DDD-based formats. PDM is assumed for DCD-based formats (as it essentially comes for free), but not for DDD-based formats. There are two important observations from Fig. 1.12. First, DCD-based formats offer substantially better OSNR performance than DDD-based formats, especially in the high-SE region. This is primarily because coherent detection offers higher receiver sensitivity or lower OSNR requirements relative to direct detection, and PDM allows coherent-detection formats to double the number of bits per symbol. The second observation is that the OSNR penalty quickly increases with the increase of the number of bits per symbol for both detection schemes. To achieve 5 bits/symbol with direct-detection D8PSK

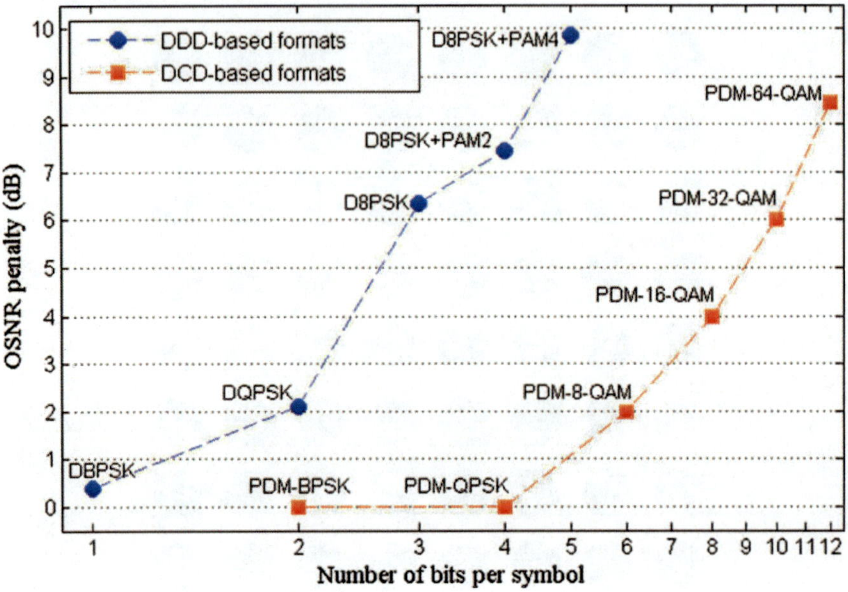

Fig. 1.12 OSNR penalties of DCD- and DDD-based formats with respect to homodyne-detection BPSK. *PAM* Pulse-amplitude modulation

in combination with 4-level pulse-amplitude modulation (PAM4), an OSNR penalty of almost 10 dB is incurred. To achieve 12 bits/symbol with PDM-64QAM, the OSNR penalty is about 8.5 dB. This means that a trade-off has to be made between the OSNR performance and the targeted SE. Moreover, modulation formats with larger number of phase and amplitude states are more susceptible to implementation imperfections such as intersymbol interference (ISI) due to transmitter and receiver bandwidth limitation and phase errors, stemming from laser phase noise and I/Q mismatch. In a recent 112.8-Gb s^{-1} PDM-64-QAM demonstration, the required OSNR at BER $= 10^{-3}$ was found to be $\sim$27 dB [89], which is $\sim$10.5 dB higher than that demonstrated for 112-Gb s^{-1} PDM-QPSK [8]. This indicates an additional implementation penalty of $\sim$2 dB, on top of the already large intrinsic OSNR penalty (8.5 dB), upon transitioning from PDM-QPSK to PDM-64-QAM. Moreover, the NLT of these higher-level formats is reduced due to the reduction in symbol spacing, further limiting their overall transmission performance.

For future high-speed optical transmission systems, the net channel data rates are expected to scale from 100 Gb s^{-1} to 200 Gb s^{-1}, 400 Gb s^{-1}, and even 1 Tb s^{-1}. It is known that PDM-QPSK-based 100-Gb s^{-1} channels can just fit onto a 50-GHz WDM grid with ROADM support. To fit 200-Gb s^{-1}, 400-Gb s^{-1}, and 1-Tb s^{-1} channels on a 50-GHz grid, PDM-16-QAM, PDM-256-QAM, and PDM-1048576(2^{20})-QAM would be needed, respectively. From the above discussion, it seems unlikely that future high-data-rate channels would be realized by scaling up the constellation size alone. OFDM-based *superchannels* and bandwidth-flexible ROADMs may be promising building blocks for future high-speed fiberoptic systems. Recent research demonstrations of 440-Gb s^{-1} and 1-Tb s^{-1} superchannels will be discussed in the following subsection.

1.4.2 State-of-the-Art DCD Demonstrations

1.4.2.1 100-Gb s^{-1} DCD-Based Field Trials

As briefly mentioned in Sect. 1.2, two field trials have recently been reported on single-carrier 100-Gb s^{-1} transmission with real-time DCD. In the first field trial, a 126.5-Gb s^{-1} single-carrier PDM-QPSK channel, assuming 20% overhead for FEC, was transmitted over 1,800 km of SSMF in AT&T's installed network with FPGA-based DSP [29]. In the second field trial, a 112-Gb s^{-1} single-carrier real-time PDM-QPSK transceiver, using FPGA-based DSP, carried native IP packet traffic over 1,520 km of SSMF in Verizon's installed network [30]. Figure 1.13 shows the configuration of the Verizon demonstration [30]. This trial shows the feasibility of interoperability between multi-suppliers' equipment for 100-Gb s^{-1} Ethernet (100GE) transport. This was also the first trial of end-to-end native IP data transport using 100G single-carrier coherent detection on field deployed fiber over a long haul distance. Key elements used in this trial over a 1,520-km deployed fiber link included a 112-Gb s^{-1} DP-QPSK transponder with real-time DSP, 100GE router cards, and 100GBASE-LR4 CFP interfaces. This successful field demonstration, which fully emulated a practical near-term deployment scenario, indicates that all key components needed for the deployment of high-performance DCD-based 100GE transport are on the verge of availability [30]. More recently, single-carrier 100-Gb s^{-1} transceivers using DCD-based PDM-QPSK have become commercially available (see, e.g., "Analyst: AlcaLu's 100G Game-Changer," http://www.lightreading.com/document.asp?doc_id=192989).

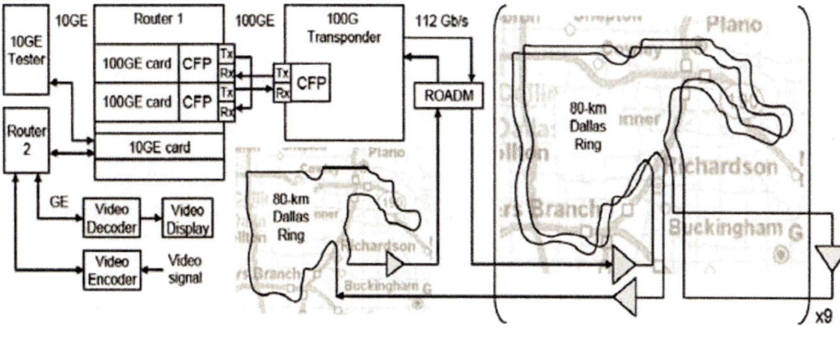

Fig. 1.13 Trial configuration of the end-to-end 100GE transport with a single-carrier PDM-QPSK transceiver using FPGA-based real-time DCD (After [30]. © 2010 IEEE/OSA)

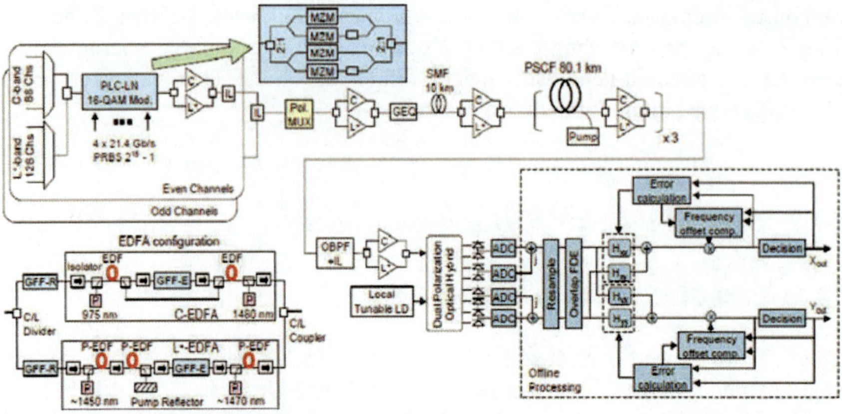

Fig. 1.14 Experiment setup used for demonstrating a record single-fiber transmission capacity of 68.1 Tb s^{-1} by using 432 171-Gb s^{-1} PDM-16-QAM channels (After [18]. © 2010 IEEE/OSA)

1.4.2.2 High-Capacity Transmission

In a recent *hero* experiment, a record single-fiber transmission capacity of 69.1 Tb s^{-1} was demonstrated by transmitting 432 171-Gb s^{-1} PDM-16-QAM channels on a 25-GHz grid in the C- and extended L-band [18]. Figure 1.14 shows the schematic of the experimental setup. Key enablers of this demonstration included a planar lightwave circuit (PLC)-based $LiNbO_3$ (LN) 16-QAM modulator, low-loss and low-nonlinear PSCF, and hybrid use of Raman/EDFA amplifiers to realize low-noise amplification over a wide optical bandwidth of 10.8 THz. Figure 1.15 shows the measured Q-factor performance after 240-km transmission. It was confirmed that the Q-factors of all 432 channels were better than 9.0 dB, which exceeds the Q-limit of 8.5 dB (dashed line) yielding BER below 1×10^{-12} with the use of today's commercial 10-Gb s^{-1} FEC techniques with 7% overhead [18]. This demonstration shows the potential of DCD and advanced fiber and amplification technologies in increasing the capacity of future fiberoptic communication systems.

1.4.2.3 High SE Transmission

The highest net system SE demonstrated so far for long-haul DWDM transmission is 8 b s^{-1} Hz^{-1}, achieved by using 107-Gb s^{-1} PDM-36-QAM channels on a 12.5-GHz grid [19]. DWDM transmission of 640 107-Gb s^{-1} PDM-36-QAM channels over 320 km of ULAF, having an effective core area of 127 μm^2 and a loss coefficient of 0.179 dB km^{-1}. An impressive total capacity of 64 Tb s^{-1} was demonstrated. Figure 1.16 shows the experimental setup and signal constellations and spectra. Low-noise hybrid Raman/EDFA amplification was used. It was

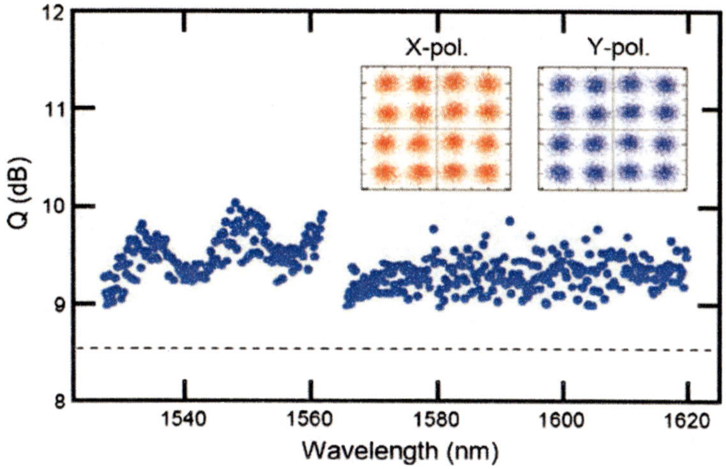

Fig. 1.15 Measured Q-factors after the 432-channel 240-km transmission. Inset: received constellation diagrams for the 1527.99-nm channel (After [18]. © 2010 IEEE/OSA)

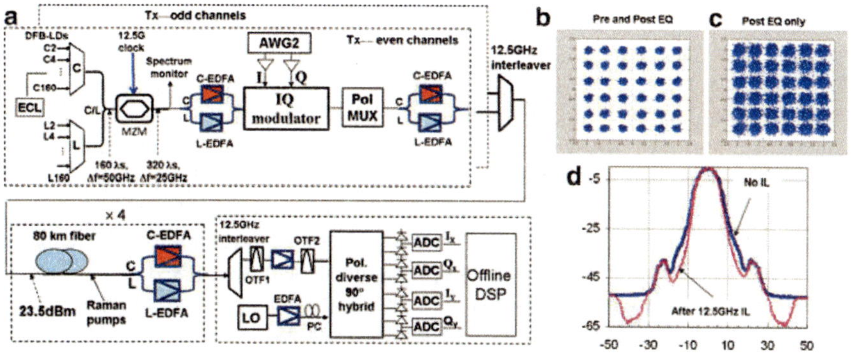

Fig. 1.16 (**a**) Experimental setup, (**b**) received constellation using both pre- and postequalization, (**c**) received constellation using purely postequalization, and (**d**) optical spectra of the generated 36-QAM signal. *AWG* Arbitrary waveform generator; *PC* Polarization controller; *OTF* Optical tunable filter; *IL* Wavelength interleaver (After [19]. © 2010 IEEE/OSA)

found that in addition to postequalization (post-EQ) at the receiver, pre-equalization (pre-EQ) at the transmitter also plays an important role in improving the quality of this high-level format. Figure 1.17 shows the measured BERs of all 640 channels, which are below the enhanced FEC threshold of 2×10^{-3}. This demonstration shows the possibility of realizing 8-b s^{-1} Hz^{-1} SE with advanced signal processing and improved fiber and amplification technologies.

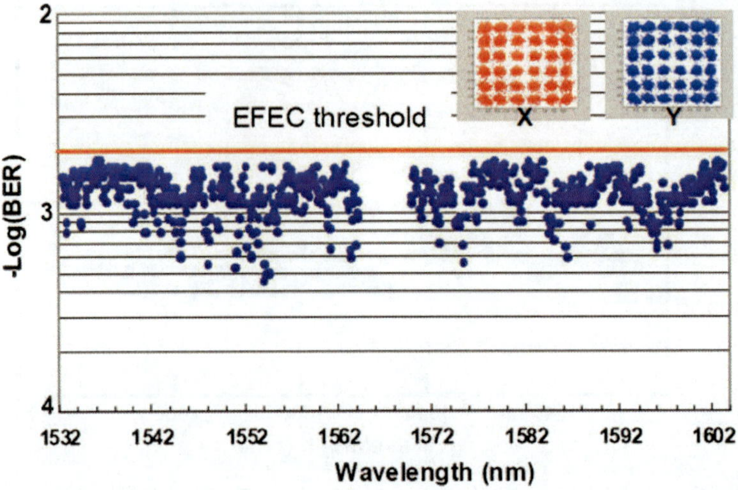

Fig. 1.17 Measured BER performance after the 320-km transmission. Inset: received constellation diagrams for the 1602-nm channel (After [19]. © 2010 IEEE/OSA)

1.4.2.4 448-Gb s^{-1} RGI-CO-OFDM Transmission

OFDM is a widely used modulation/multiplexing technology in wireless and data communications [90] that was recently introduced to optical fiber communications [91–93]. Enabled by DCD, coherent optical OFDM (CO-OFDM) [92–96] brings similar benefits as single-carrier-based coherent systems while additionally offering transmitter adaptation capability [97], efficient channel estimation and compensation [98], and unique nonlinear compensation capabilities [99–106]. A novel RGI-CO-OFDM format was recently introduced to take advantage of both DCD-enabled receive-side CD compensation and CO-OFDM-based transmitter signal processing [21]. The use of DCD-enabled receive-side CD compensation eliminates the need for a large guard interval (GI) or a cyclic prefix between adjacent symbols, as required in conventional CO-OFDM to accommodate large CD-induced ISI, thereby increasing SE and OSNR performance. The use of CO-OFDM-based transmitter signal processing facilitates the generation of high-speed high-level modulation formats. For example, the sampling speed of the digital-to-analog converters (DACs) required is usually smaller than that required for single-carrier transmission [96]. Also, the use of a small GI helps mitigate the ISI due to transmitter bandwidth limitations. A 448-Gb s^{-1} RGI-CO-OFDM signal with 16-QAM subcarrier modulation was transmitted over 2,000 km of ULAF and five 80-GHz-grid WSSs, potentially allowing for an SE of 5 b s^{-1} Hz^{-1} and an SEDP of 10,000 km-b s^{-1} Hz^{-1} [21].

Figure 1.18 shows the schematic of the experimental setup. Enabling technologies include efficient and fiber-nonlinearity tolerant CO-OFDM processing [107, 108], frequency-domain CD compensation [109], digital nonlinear compensation

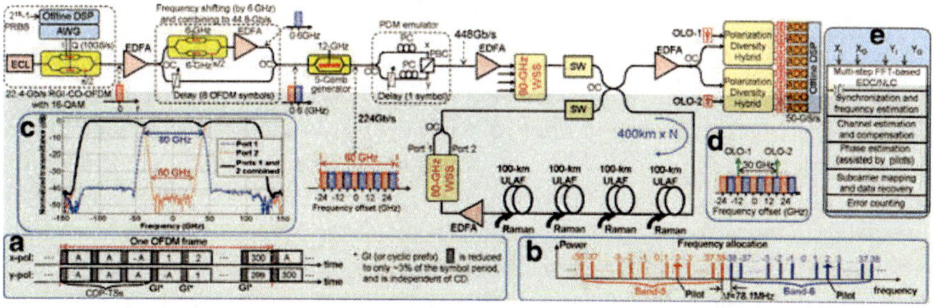

Fig. 1.18 Schematic of the experimental setup. Insets: (**a**) OFDM frame arrangement; (**b**) Frequency allocation of the OFDM subcarriers; (**c**) Passbands of the loop WSS configured for 80-GHz channel spacing; (**d**) Configuration of the banded digital coherent detection with 2 OLOs; (**e**) Block diagram of the receiver DSP. *OC* Optical coupler; *PC* Polarization controller; *SW* Optical switch (After [21]. © 2010 IEEE/OSA)

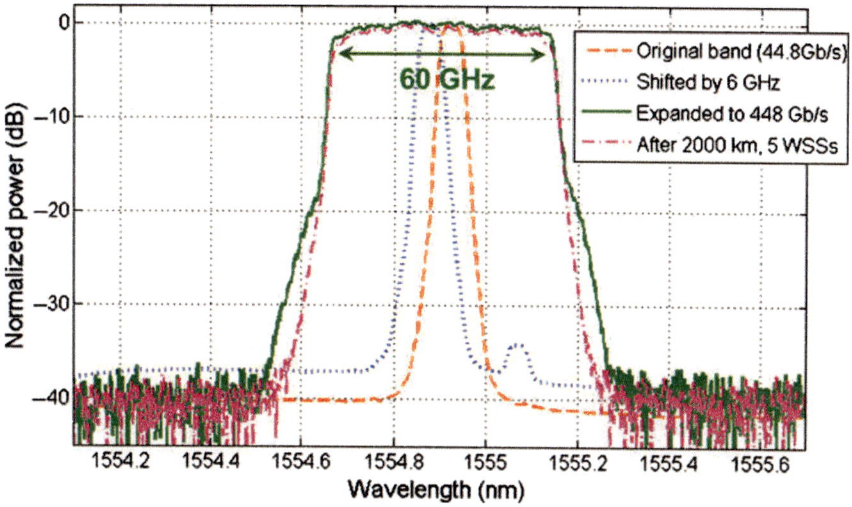

Fig. 1.19 Measured optical signal spectra at various stages (After [21]. © 2010 IEEE/OSA)

(NLC) [110–112], OBM [24], multicarrier modulation [26, 113, 114], and banded DCD. In addition, low-loss and low-nonlinearity ULAF fiber with low-noise DRA was used. Notably, the total overhead used in the RGI-CO-OFDM (excluding the FEC overhead) was only ~7% and was independent of CD.

The 448-Gb s^{-1} RGI-CO-OFDM signal consists of 10 44.8-Gb s^{-1} bands through OBM. Figure 1.19 shows the optical spectra of the 448-Gb s^{-1} signal, which exhibited a square-like profile with a 3-dB bandwidth of ~60 GHz. After passing five 80-GHz WSSs, the signal spectrum remained virtually unchanged, indicating the feasibility of transmission over an 80-GHz channel grid.

At the receiver, four 50-GS s^{-1} ADCs embedded in a real-time sampling oscilloscope with 16-GHz RF bandwidth were used. Due to the ADC bandwidth limitation, a banded DCD approach with two OLOs was used to recover the entire 448-Gb s^{-1} signal, as shown in inset (d) of Fig. 1.18. In the experiment, the lower (long-wavelength) and upper halves of the signal were sequentially detected with one optical frontend by switching one OLO between -15 GHz and $+15$ GHz relative to the signal center frequency. Figure 1.20 shows the RF spectra of the recovered two halves of the signal. Exemplary recovered SC constellations are shown as insets.

Figure 1.21a shows the measured BER as a function of OSNR. At BER $= 1 \times 10^{-3}$, the required OSNR for the 448-Gb s^{-1} signal is 28.2 dB, which is 10.8 dB

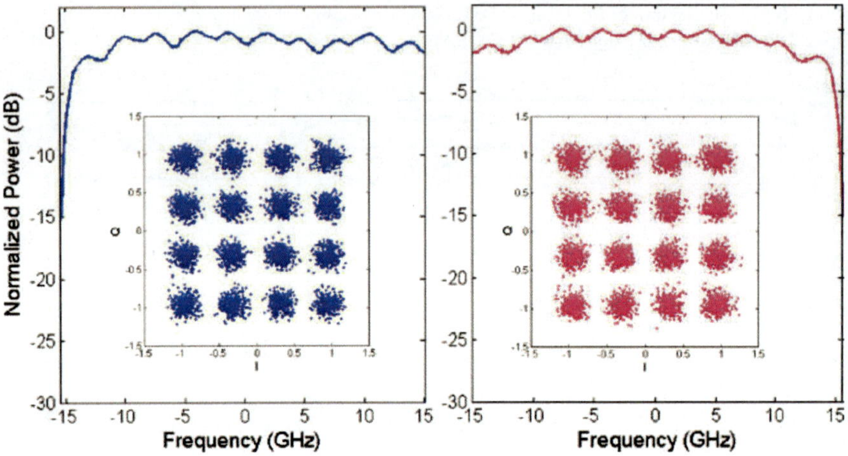

Fig. 1.20 RF spectra of the *lower* (*left*) and *upper* (*right*) halves of the 448-Gb s^{-1} signal. Insets: recovered constellations (After [21]. © 2010 IEEE/OSA)

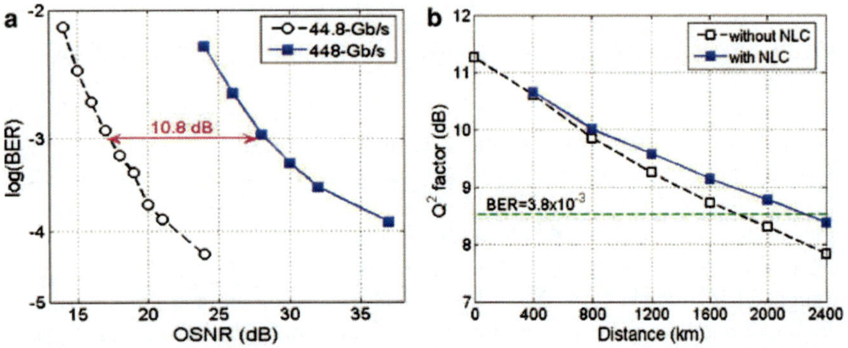

Fig. 1.21 (**a**) Measured BER performance of the multi-band 448-Gb s^{-1} RGI-CO-OFDM signal as compared to the original single-band 44.8-Gb s^{-1} signal; (**b**) Measured Q^2 factor as a function of transmission distance (After [21]. © 2010 IEEE/OSA)

higher than that for the original single-band 44.8-Gb s^{-1} signal, showing a small excess penalty of ~0.8 dB due to band multiplexing and simultaneous detection of five bands per sampling. At BER = 3.8×10^{-3}, the threshold of an advanced 7% FEC, the required OSNR is ~25 dB, within 3.5 dB from the theoretical limit. For 2,000-km transmission, the optimal signal launch power was found to be about 1.5 dBm, at which level the OSNR after transmission was 28.5 dB. Figure 1.21b shows the Q^2 factor as a function of transmission distance. With fiber nonlinearity compensation (NLC), the mean BER of the 448-Gb s^{-1} signal is below 3×10^{-3} after 2,000-km transmission and 5 WSS passes. The total transmission penalty is ~3 dB. The reach improvement due to NLC is ~25%. The performance of the ten bands performed similarly, indicating high signal tolerance to cascaded WSS filtering. This demonstration represents the longest transmission distance for >200-Gb s^{-1} transmission within an optical bandwidth allowing for SEs higher than 4 b s^{-1} Hz^{-1} and the lowest overhead (7.3%) for >100-Gb s^{-1} CO-OFDM transmission with $\geq$40, 000-ps nm^{-1} accumulated CD. This study also shows the feasibility of realizing spectrally efficient and optically transparent 400GE transport by using RGI-CO-OFDM.

1.4.2.5 1-Tb s^{-1} NGI-CO-OFDM Transmission

Terabit Ethernet (1TbE) was recently mentioned as a possible future Ethernet standard [115], and much research effort has been devoted to 1-Tb s^{-1} transmission [22, 23, 116, 117]. Limited by the transmitter and receiver bandwidths, both optical and electronic, the Tb/s channels demonstrated so far consist of multiple modulated carriers per channel to facilitate parallel modulation and detection. To attain high SE, the modulated carriers of such a multi-carrier signal are preferably arrayed under the orthogonal frequency-division multiplexing (OFDM) condition [22–26, 113]. Such type of multicarrier optical OFDM signal does not require a time-domain cyclic GI, as ISI is mitigated through equalization at the receiver, and is referred to as NGI-CO-OFDM [23, 26].

Figure 1.22 shows the schematic of a multicarrier NGI-CO-OFDM transmitter with multiple frequency-locked carriers, each modulated with PDM-QPSK. The multiple carriers can be generated by using a single laser followed by a multicarrier generator, which can be based on cascaded modulators [118] or recirculating frequency-shifting [23] or a $LiNbO_3$ ring resonator [119]. Alternatively, the laser and multicarrier generator may be replaced by a mode-locked-laser (MLL). The frequency-locked carriers are then separated by a wavelength demultiplexer (DMUX), before being individually modulated by an I/Q modulator array consisting of multiple I/Q modulators and polarization-beam combiners (PBCs). To achieve the *orthogonality* among the modulated carriers, all the carriers, in addition to being spaced at the modulation symbol rate, need to be synchronously modulated or symbol aligned [113]. The modulated carriers are then combined to form a special *superchannel*. Here, superchannel refers to a channel originating from a single laser source and consisting of multiple frequency-locked and synchronously modulated

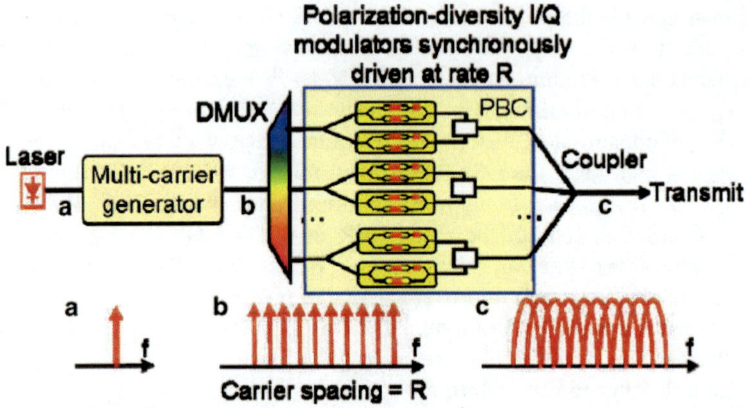

Fig. 1.22 Schematic of a multicarrier NGI-CO-OFDM transmitter with frequency-locked carriers. Optical spectra at locations **(a)**–**(c)** are illustrated. *DMUX* Wavelength demultiplexer; *PBC* Polarization beam combiner

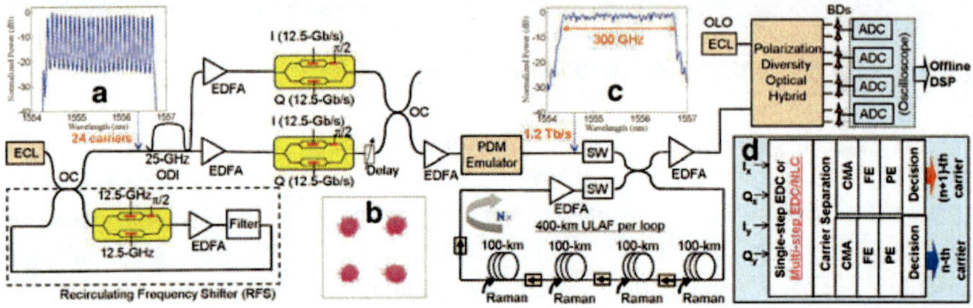

Fig. 1.23 Experimental setup for the 1.2-Tb s^{-1} NGI-CO-OFDM superchannel transmission [23]. Insets: (**a**) Optical spectrum of 24 frequency-locked 12.5-GHz-spaced carriers; (**b**) Sample back-to-back constellation of PDM-QPSK carrier modulation; (**c**) Optical spectrum of the 1.2-Tb s^{-1} superchannel; and (**d**) Block diagram of the receiver DSP. *OC* Optical coupler; *SW* Optical switch; *NLC* Nonlinearity compensation

carriers. Multi-carrier NGI-CO-OFDM is a special type of superchannel, offering the highest possible SE without coherent crosstalk among the carriers. Photonic integration of all or most of the optical elements in this type of multi-carrier transmitter is essential to enable cost-effective implementation.

A 1.2-Tb s^{-1} multi-carrier NGI-CO-OFDM signal was recently generated and transmitted over 7,200 km in ULAF, achieving an intra-channel SE of 3.7 b s^{-1} Hz^{-1} and a record SEDP of 27,000 b km s^{-1} Hz^{-1} [23]. Figure 1.23 shows the schematic of the experimental setup. This 1.2-Tb s^{-1} NGI-CO-OFDM channel consisted of twenty-four 12.5-Gbaud PDM-QPSK carriers spaced at 12.5 GHz, occupying an optical bandwidth of 312.5 GHz. Two modulated carriers were simultaneously received by a 50-Gsamples s^{-1} ADC based B-DCD, so 12 different OLO frequency settings were used to recover the entire 1.2-Tb s^{-1} superchannel.

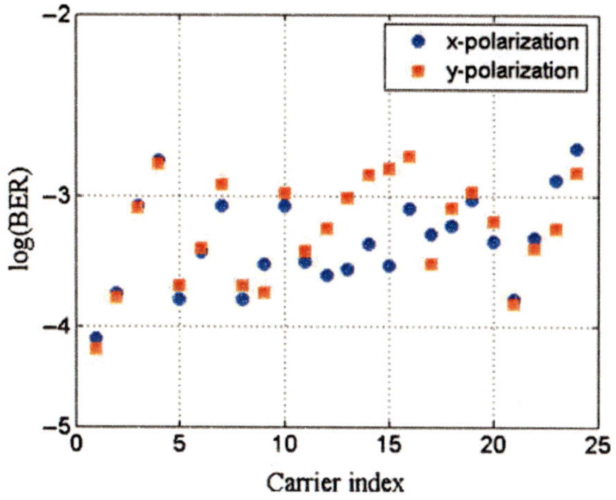

Fig. 1.24 Measured BER performance of a 1.2-Tb s^{-1} 24-carrier NGI-CO-OFDM superchannel after 7,200 km transmission in ULAF [23]

The required OSNR at BER $= 1 \times 10^{-3}$ was ~26 dB, 11 dB higher than that of a single-carrier 100-Gb s^{-1} PDM-QPSK signal, showing a small excess penalty of ~0.2 dB due to OFDM-based carrier multiplexing and B-DCD. Figure 1.24 shows the measured BER performances of all the 24 carriers of the 1.2-Tb s^{-1} superchannel after transmission over 7,200 km of ULAF. The mean BER was 6.8×10^{-4}, well below the threshold of enhanced FEC. More recently, simultaneous recovery of three modulated carriers was demonstrated with similar performance, leading to a low oversampling factor of 1.33 [120].

It is worth evaluating the NLT or power tolerance of the Tb s^{-1} superchannel. One way to evaluate the NLT is in terms of the nonlinear phase shift experienced by the signal at the optimal performance, given by $\Phi_{NL} = \gamma L_{eff} P_o N$, where γ is the fiber nonlinear coefficient, L_{eff} is the effective fiber span length, P_o is the optimum signal launch power, and N is the number of spans transmitted. Figure 1.25 shows the signal Q-factor (derived from the measured BER of a center carrier) after 7,200-km transmission as a function of the signal launch power (P_{in}) [121]. It was found that $P_o = 7.5$ dBm and $L_{eff} = 34.7$ km, so $\Phi_{NL} = 11.4$ rad, which is ~11.4 times larger than that for BPSK in the absence of dispersion [76]. This large NLT can be attributed to the large dispersive effect experienced by the superchannel [121], which is beneficial for mitigating the nonlinearity. Figure 1.25 also shows the signal Q-factor with an optimized 72-step NLC [121]. The optimal Q-factor is improved by ~0.7 dB, indicating small NLC benefit when the NLT is already improved by large dispersion. The high power tolerance of the Tb/s superchannel in dispersion-uncompensated long-haul transmission indicates the viability of future Tb/s/channel transmission in suitably designed optical links.

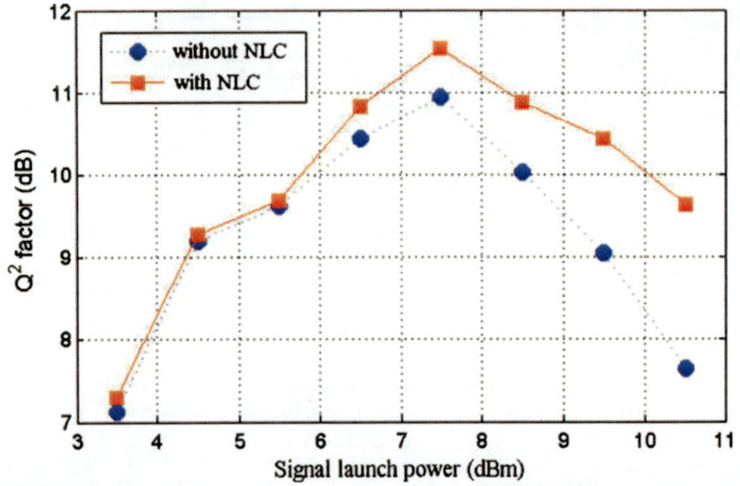

Fig. 1.25 Measured signal Q-factor after 7,200-km transmission vs. signal launch power without and with NLC [121]

1.5 Concluding Remarks

With the steady increase of fiberoptic transmission capacity in the foreseeable future, it is natural to pose the question whether there is a fundamental limit on the ultimate capacity. The search for fundamental bounds on transmission of information over various media has been an active area of research ever since Shannon published his pioneering paper in 1948 [122]. The answer to the above question is definitely yes, based on Shannon's theory and on more recent works accounting for the effect of fiber nonlinearity over the optical channel [123, 124]. In fact, according to R.-J. Essiambre et al. [104], recent fiberoptic transmission demonstrations are not too far away from the *Shannon limit of single-mode fiberoptic transmission*. A comprehensive survey on the nonlinear Shannon limit can be found in Chap. 13.

Some promising techniques assisting in further approaching the Shannon limit of single-mode fiberoptic transmission include advanced maximum likelihood sequence estimation (MLSE) techniques [125] and maximum likelihood carrier phase estimation [126, 127], and more advanced coding with higher coding gain and NLC [123, 124]. Detailed studies on these and related subjects may be found in Chap. 12, entitled "Coding/nonlinear impairments reduction by coding" by I. Djordjevic and Chap. 3, by M. Nazarathy and R. Weidenfeld. Recent advances in high-speed electronics, including ADC, DAC, and DSP, have dramatically advanced the field of fiberoptic communication. It is expected that riding on Moore's law, future advances in electronics will continue to enable the capacity growth of optical communication. It may also turn helpful to relax the nonlinear Shannon limit by using new fibers with lower loss and/or lower nonlinear coefficient, introducing better optical amplification schemes with lower ASE noise, and potentially utilizing the spatial degrees of freedom of new types of few-mode or multimode fiber by means of MIMO techniques [123, 124, 126, 127]. With the increase in capacity, the cost per bit needs to be reduced as well to sustain the capacity growth. Advances in areas such as photonic integrated circuits would also be essential.

While the strategies to meet the challenge imposed by *The coming capacity crunch* [1] may still be uncertain, what is certain is that *Research in this area is essential, challenging, and likely to be interesting* [2].

Acknowledgments X. Liu is deeply grateful to Dr. S. Chandrasekhar for close collaborations in recent years, generating many of the results reviewed in this chapter. He is also grateful to numerous current and past colleagues in Bell Laboratories, Alcatel-Lucent, for fruitful collaborations and valuable discussions. Among them are F. Buchali, C.R. Doerr, R. Essiambre, D.A. Fishman, D.M. Gill, A.H. Gnauck, I. Kang, Y.-H. Kao, N. Kaneda, S.K. Korotky, G. Kramer, A. Leven, C.J. McKinstrie, L.F. Mollenauer, A.J. van Wijngaarden, X. Wei, P.J. Winzer, C. Xie, and C. Xu. He also wishes to thank A.R. Chraplyvy, C.R. Giles, J.-P. Hamaide, and R.W. Tkach for their support.

M. Nazarathy would like to acknowledge: his former and current graduate students and his peers in the Technion EE Department, and in particular Prof. M. Orenstein; express deep gratitude to Profs. B. Fischer and G. Eisenstein who "enticed" Moshe to return to the academia, after having spent many years in the industry; national collaborators Prof. D. Sadot and Dr. D. Marom; US collaborators and in particular his co-author Xiang Liu, Prof. A.E. Willner and his past students Y.K. Lizé, and L. Christen and; EU collaborators: Prof. E. Forestieri and his group, and Prof. J. Prat and his group; his own family for their love and their infinite tolerance of imbalanced priorities.

Glossary

ADC	Analog-to-digital converter
ASIC	Application-specific integrated circuit
ASE	Amplified spontaneous emission
BER	Bit error ratio
B-DCD	Banded digital coherent detection
CD	Chromatic dispersion
CMA	Constant modulus algorithm
CO-OFDM	Coherent optical orthogonal frequency-division multiplexing
CP	Cyclic prefix
DAC	Digital-to-analog converter
DBPSK	Differential binary phase-shift keying
DCD	Digital coherent detection
DDD	Direct differential detection
DPSK	Differential phase-shift keying
DQPSK	Differential quadrature phase-shift keying
DRA	Distributed Raman amplifier
DSCD	Digital self-coherent detection
DSP	Digital signal processor
DWDM	Dense wavelength-division multiplexing
EDC	Electronic dispersion compensation

EDFA	Erbium-doped fiber amplifier
FEC	Forward error correction
FPGA	Field programmable gate array
FWM	Four-wave mixing
GI	Guard interval
ISI	Inter-symbol interference
J-SPMC	Joint self phase modulation compensation
MSPE	Multi-symbol phase estimation
MLSE	Maximum Likelihood Sequence Estimation
MZM	Mach-Zehnder modulator
NGI	No-guard-interval
NLC	Non-linear compensation
NRZ	Non-return-to-zero
OBM	Orthogonal band multiplexing
OFDM	Orthogonal frequency-division multiplexing
OLO	Optical local oscillator
OOK	On-off-keying
OSNR	Optical signal-to-noise ratio
PAM	Pulse amplitude modulation
PDM	Polarization-division multiplexing
P-DPSK	Partial DPSK
PMD	Polarization-mode dispersion
PSCF	Pure silica core fiber
PSK	Phase-shift keying
QAM	Quadrature amplitude modulation
RGI	Reduced-guard-interval
ROADM	Reconfigurable optical add/drop multiplexer
RZ	Return-to-zero
SCD	Self-coherent detection
SE	Spectral efficiency
SEDP	Spectral efficiency distance product
SPM	Self phase modulation
SPMC	Self phase modulation compensation
SSMF	Standard single-mode fiber
ULAF	Ultra-large-area fiber
WDM	Wavelength-division multiplexing
WSS	Wavelength-selective switch
XPM	Cross phase modulation

References

1. A.R. Chraplyvy, The Coming Capacity Crunch, ECOC Plenary Talk (2009)
2. R.W. Tkach, Bell Labs Tech. J. **14**, 3–10 (2010)

3. C. Xu, X. Liu, X. Wei, IEEE J. Select Topics Quant. Electron. **10**, 281–293 (2004)
4. A.H. Gnauck, P.J. Winzer, J. Lightwave Technol. **23**, 115–130 (2005)
5. X. Liu, S. Chandrasekhar, A. Leven, *Self-coherent optical transport systems*, chapter 4, ed. by I.P. Kaminov, T. Li, A.E. Willner. Optical Fiber Telecommunications V.B: Systems and Networks (Academic, San Diego 2008)
6. M.G. Taylor, IEEE Photon. Technol. Lett. **16**(2), 674–676 (2004)
7. Y. Han, G. Li, Opt. Express **13**(19), 7527–7534 (2005)
8. C.R.S. Fludger, T. Duthel, D. van den Borne, C. Schulien, E.D. Schmidt, T. Wuth, E. de Man, G.D. Khoe, H. de Waardt, 10 × 111 Gbit/s, 50 GHz spaced, POLMUX-RZ-DQPSK transmission over 2375 km employing coherent equalization. OFC'07, post-deadline paper PDP22, 2007
9. K. Kikuchi, *Coherent Optical Communication Systems*, chapter 3, ed. by I.P. Kaminov, T. Li, A.E. Willner. Optical Fiber Telecommunications V.B: Systems and Networks (Academic, San Diego, 2008)
10. E.M. Ip, A.P.T. Lau, D.J.F. Barros, J.M. Kahn, Opt. Express **16**, 753–791 (2008)
11. A.H. Gnauck, G. Raybon, S. Chandrasekhar, J. Leuthold, C. Doerr, L. Stulz, A. Agarwal, S. Banerjee, D. Grosz, S. Hunsche, A. Kung, A. Marhelyuk, D. Maywar, M. Movassaghi, X. Liu, C. Xu, X. Wei, D.M. Gill, 2.5 Tb/s (64 × 42.7 Gb/s) transmission over 40 × 100 km NZDSF using RZ-DPSK format and all-Raman-amplified spans. OFC'02, post-deadline paper FC2, 2002
12. S. Chandrasekhar, X. Liu, D. Kilper, C.R. Doerr, A.H. Gnauck, E.C. Burrows, L.L. Buhl, 0.8-bit/s/Hz terabit transmission at 42.7-Gb/s using hybrid RZ-DQPSK and NRZ-DBPSK formats over 16 × 80 km SSMF spans and 4 bandwidth-managed ROADMs. OFC'07, post-deadline paper PDP28, 2007
13. C. Laperle, B. Villeneuve, Z. Zhang, D. McGhan, H. Sun, M. O'Sullivan, Wavelength division multiplexing (WDM) and polarization mode dispersion (PMD) performance of a coherent 40Gbit/s dual-polarization quadrature phase shift keying (DP-QPSK) transceiver. OFC'07, post-deadline paper PDP16, 2007
14. N. Kikuchi, S. Sasaki, J. Lightwave Technol. **28**, 123–130 (2010)
15. G. Charlet, M. Salsi, P. Tran, M. Bertolini, H. Mardoyan, J. Renaudier, O. Bertran-Pardo, S. Bigo, 72 × 100Gb/s Transmission over transoceanic distance, using large effective area fiber, hybrid Raman-Erbium amplification and coherent detection. OFC'09, post-deadline paper PDPB6, 2009
16. X. Zhou, J. Yu, M.F. Huang, Y. Shao, T. Wang, P. Magill, M. Cvijetic, L. Nelson, M. Birk, G. Zhang, S.Y. Ten, H.B. Matthew, S.K. Mishra, 32Tb/s (320 × 114Gb/s) PDM-RZ-8QAM transmission over 580km of SMF-28 ultra-low-loss fiber. OFC'09, post-deadline paper PDPB4, 2009
17. A.H. Gnauck, P.J. Winzer, C.R. Doerr, L.L. Buhl, 10 × 112-Gb/s PDM 16-QAM transmission over 630 km of fiber with 6.2-b/s/Hz spectral efficiency. OFC'09, post-deadline paper PDPB8, 2009
18. A. Sano, H. Masuda, T. Kobayashi, M. Fujiwara, K. Horikoshi, E. Yoshida, Y. Miyamoto, M. Matsui, M. Mizoguchi, H. Yamazaki, Y. Sakamaki, 69.1-Tb/s (432 × 171-Gb/s) C- and extended L-band transmission over 240 km using PDM-16-QAM modulation and digital coherent detection. OFC'10 postdeadline paper PDPB7, 2010
19. X. Zhou, J. Yu, M.F. Huang, Y. Shao, T. Wang, L. Nelson, P. Magill, M. Birk, P.I. Borel, D.W. Peckham, R. Lingle, 64-Tb/s (640×107-Gb/s) PDM-36QAM transmission over 320km using both pre- and post-transmission digital equalization. OFC'10, post-deadline paper PDPB9, 2010
20. A.H. Gnauck, P.J. Winzer, S. Chandrasekhar, X. Liu, B. Zhu, D.W. Peckham, 10 × 224-Gb/s WDM transmission of 28-Gbaud PDM 16-QAM on a 50-GHz grid over 1,200 km of fiber. OFC'10, post-deadline paper PDPB8, 2010
21. X. Liu, S. Chandrasekhar, B. Zhu, P.J. Winzer, A.H. Gnauck, D.W. Peckham, Transmission of a 448-Gb/s reduced-guard-interval CO-OFDM signal with a 60-GHz optical bandwidth over 2000 km of ULAF and five 80–GHz–Grid ROADMs. OFC'10, post-deadline paper PDPC2, 2010

22. Y. Ma, Q. Yang, Y. Tang, S. Chen, W. Shieh, 1-Tb/s per channel coherent optical OFDM transmission with subwavelength bandwidth access. OFC'09, post-deadline paper PDPC1, 2009
23. S. Chandrasekhar, X. Liu, B. Zhu, D.W. Peckham, Transmission of a 1.2-Tb/s 24-carrier no-guard-interval coherent OFDM superchannel over 7200-km of ultra-large-area fiber. ECOC'09, post-deadline paper PD2.6, 2009
24. W. Shieh, Q. Yang, Y. Ma, Opt. Express **16**, 6378–6386 (2008)
25. M. Nazarathy, D.M. Marom, W. Shieh, Optical comb and filter bank (De)Mux enabling 1 Tb/s orthogonal sub-band multiplexed CO-OFDM free of ADC/DAC limits,. European conference on optical communications, Paper P3.12, ECOC'09, Vienna, September 2009
26. A. Sano, E. Yamada, H. Masuda, E. Yamazaki, T. Kobayashi, E. Yoshida, Y. Miyamoto, R. Kudo, K. Ishihara, Y. Takatori, J. Lightwave Technol. **27**, 3705–3713 (2009)
27. K. Roberts, M. O'Sullivan, K.T. Wu, H. Sun, A. Awadalla, D. Krause, C. Laperle, J. Lightwave Technol. **27**, 3546–3559 (2009)
28. I. Dedic, 56Gs/s ADC: Enabling 100GbE. OFC'10, invited paper OThT6, 2010
29. M. Birk, P. Gerard, R. Curto, L. Nelson, X. Zhou, P. Magill, T.J. Schmidt, C. Malouin, B. Zhang, E. Ibragimov, S. Khatana, M. Glavanovic, R. Lofland, R. Marcoccia, G. Nicholl, M. Nowell, F. Forghieri, Field trial of a real-time, single wavelength, coherent 100 Gbit/s PM-QPSK channel upgrade of an installed 1800km link. OFC'10, post-deadline paper PDPD1, 2010
30. T.J. Xia, G. Wellbrock, B. Basch, S. Kotrla, W. Lee, T. Tajima, K. Fukuchi, M. Cvijetic, J. Sugg, Y. Ma, B. Turner, C. Cole, C. Urricariet, End-to-end native IP data 100G single carrier real time DSP coherent detection transport over 1520–km field deployed fiber. OFC'10, post-deadline paper PDPD4, 2010
31. D.A. Fishman, W.A. Thompson, L. Vallone, Bell Labs Tech. J. 11, 27–53 (2006)
32. X. Liu, S. Chandrasekhar, High spectral-efficiency mixed 10G/40G/100G transmission. AOE'08, paper SuA2, 2008
33. K.P. Ho, *Phase-Modulated Optical Communication Systems* (Springer, New York, 2005)
34. P.J. Winzer, R.J. Essiambre, *Advanced Optical Modulation Formats*, chapter 2, ed. by I.P. Kaminov, T. Li, A.E. Willner. Optical Fiber Telecommunications V.B: Systems and Networks (Academic, San Diego, 2008)
35. A.J. Price, N. Le Mercier, Electron. Lett. **31**, 58–59 (1995)
36. X. Liu, A.H. Gnauck, X. Wei, Y.C. Hsieh, C. Ai, V. Chien, IEEE Photon. Technol. Lett. **17**, 2610–2612 (2005)
37. B. Mikkelsen, C. Rasmussen, P. Mamyshev, F. Liu, Electron. Lett. **42**, 1363–1364 (2006)
38. C. Wree, N. Hecker-Denschlag, E. Gottwald, P. Krummrich, J. Leibrich, E.D. Schmidt, B. Lankl, W. Rosenkranz, IEEE Photon. Technol. Lett. **15**, 1303–1305 (2003)
39. P.S. Cho, G. Harston, C. Kerr, A. Greenblatt, A. Kaplan, Y. Achiam, G. Yurista, M. Margalit, Y. Gross, J. Khurgin, IEEE Photon. Tech. Lett. **16**, 656–658 (2004)
40. D. van den Borne, S.L. Jansen, E. Gottwald, P.M. Krummrich, G.D. Khoe, H. de Waardt, J. Lightwave Technol. **25**, 222–232 (2007)
41. S. Chandrasekhar, X. Liu, D. Kilper, C.R. Doerr, A.H. Gnauck, E.C. Burrows, L.L. Buhl, J. Lightwave Technol. **26**, 85–90 (2008)
42. S. Chandrasekhar, X. Liu, Bell Labs Tech. J. **14**, 11–25 (2010)
43. C. Xie, D. Werner, H. Haunstein, R.M. Jopson, S. Chandrasekhar, X. Liu, y. Shi, S. Gronbach, T. Link, K. Czotscher, Bell Labs Tech. J. **14**, 115–129 (2010)
44. P.J. Winzer, G. Raybon, S. Chandrasekhar, C.R. Doerr, T. Kawanishi, T. Sakamoto, K. Higuma, 10×107-Gb/s NRZ-DQPSK transmission over 12×100 km including 6 routing nodes. OFC'07, post-deadline paper PDP24, 2007
45. S. Chandrasekhar, X. Liu, E.C. Burrows, L.L. Buhl, Hybrid 107-Gb/s polarization-multiplexed DQPSK and 42.7-Gb/s DQPSK transmission at 1.4 bits/s/Hz spectral efficiency over 1280 km of SSMF and 4 bandwidth-managed ROADMs. ECOC'07, post-deadline paper PD 1.9, 2007
46. X. Liu, S. Chandrasekhar, Direct Detection of 107-Gb/s polarization-multiplexed DQPSK with electronic polarization demultiplexing. OFC'08, paper OTuG4, 2008

47. G. Kramer, A. Ashikhmin, A.J. van Wijngaarden, X. Wei, J. Lightwave Technol. **21**, 2438–2445 (2003)
48. T. Mizuochi, J. Select Topics Quant. Electron. **12**, 544–554 (2006)
49. H. Sun, K. Wu, K. Roberts, Opt. Express **16**, 873–879 (2008)
50. D. McGhan, C. Laperle, A. Savchenko, C. Li, G. Mak, M. O'Sullivan, 5120 km RZ-DPSK transmission over G652 fiber at 10 Gb/s with no optical dispersion compensation. OFC'05, postdeadline paper PDP 27, 2005
51. M.M. El Said, J. Sitch, M.I. Elmasry, J. Lightwave Technol. **23**, 388–400 (2005)
52. R.I. Killey, P.M. Watts, M. Glick, P. Bayvel, Electronic precompensation techniques to combat dispersion and nonlinearities in optical transmission. ECOC'05, paper Tu4.2.1, 2005
53. X. Liu, D.A. Fishman, A fast and reliable algorithm for electronic preequalization of SPM and chromatic dispersion. OFC' 06, paper OThD4, 2006
54. A.H. Gnauck, P.J. Winzer, S. Chandrasekhar, IEEE Photon. Tech. Lett. **17**, 2203–2205 (2005)
55. G. Charlet, H. Mardoyan, P. Tran, M. Lefrancois, S. Bigo, Nonlinear interactions between 10Gb/s NRZ channels and 40Gb/s channels with RZ-DQPSK or PSBT format, over low-dispersion fiber. ECOC'06, paper Mo3.2.6, 2006
56. M. LeFrancois, F. Houndonoughbo, T. Fauconnier, G. Charlet, S. Bigo, Cross comparison of the nonlinear impairments caused by 10Gbit/s neighboring channels on a 40Gbit/s channel modulated with various formats, and over various fiber types. OFC'07, paper JThA44, 2007
57. S. Chandrasekhar, X. Liu, IEEE Photon. Tech. Lett. **19**, 1801–1803 (2007)
58. X. Liu, S. Chandrasekhar, Suppression of XPM penalty on 40-Gb/s DQPSK resulting from 10-Gb/s OOK channels by dispersion management. OFC'08, paper OMQ6, 2008
59. D. van den Borne, C. Fludger, T. Duthel, C. Schulien, T. Wuth, E.D. Schmidt, E. Gottwald, G.D. Khoe, H. de Waardt, Carrier phase estimation for coherent equalization of 43-Gb/s POLMUX-NRZ-DQPSK transmission with 10.7-Gb/s NRZ neighbours. ECOC'07, paper 7.2.3, 2007
60. G. Charlet, M. Salsi, H. Mardoyan, P. Tran, J. Renaudier, S. Bigo, M. Astruc, P. Sillard, L. Provost, F. Cerou, Transmission of 81 channels at 40Gbit/s over a transpacific-distance erbium-only link, using PDM-BPSK modulation, coherent detection, and a new large effective area fibre. ECOC'08, paper Th.3.E.3, 2008
61. G. Charlet, The impact and mitigation of nonlinear effects in coherent optical transmission. OFC'09, paper NThB4, 2009
62. M. Nazarathy, X. Liu, L. Christen, Y. Lize, A. Willner, IEEE Photon. Technol. Lett. **19**, 828–839 (2007)
63. M. Nazarathy, Y. Yadin, Approaching coherent homodyne performance with direct detection low-complexity advanced modulation formats. Coherent Optical Technologies and Applications (COTA), Whisler, Canada, 28–30 June 2006
64. M. Nazarathy, X. Liu, Y. Yadin, M. Orenstein, Multi-chip detection of optical differential phase-shift keying and complexity reduction by interferometric decision feedback. European conference of optical communication ECOC'06, Cannes, France, Paper We3.P.79, 24–28 September 2006
65. M. Nazarathy, Y. Yadin, M. Orenstein, Y. Lize, L. Christen, A. Willner, Enhanced self-coherent optical decision-feedback-aided detection of multi-symbol m-DPSK/PolSK in particular 8-DPSK/BPolSK at 40 Gbps. OFC'07, Paper JWA43, 2007
66. M. Nazarathy, X. Liu, L. Christen, Y. Lize, A. Wilner, J. Lightwave Technol. **26**, 1921–1934 (2008)
67. A. Atzmon, M. Nazarathy, Self-coherent differential transmission with decision feedback – phase noise impairments. Coherent Optical Technologies and Applications (COTA), Boston, 2008
68. N. Kikuchi, K. Mandai, S. Sasaki, K. Sekine, Proposal and first experimental demonstration of digital incoherent optical field detector for chromatic dispersion compensation, in *Proceedings of European Conference on Optical Communications*, Post-deadline Paper Th4.4.4, 2006
69. X. Liu, S. Chandrasekhar, A. Leven, Opt. Express **16**, 792–803 (2008)

70. D. van den Borne, S. Jansen, G. Khoe, H. de Wardt, S. Calabro, E. Gottwald, Differential quadrature phase shift keying with close to homodyne performance based on multi-symbol phase estimation, IEE seminar on optical fiber comm. and electronic signal processing, ref. No. 2005–11310, 2005
71. X. Liu, Receiver sensitivity improvement in optical DQPSK and DQPSK/ASK through data-aided multi-symbol phase estimation, in *Proceedings of European Conference on Optical Communications 2006*, Paper We2.5.6, 2006
72. X. Liu, Opt. Express **15**, 2927–2939 (2007)
73. X. Liu, S. Chandrasekhar, A.H. Gnauck, C.R. Doerr, I. Kang, D. Kilper, L.L. Buhl, J. Centanni, DSP-enabled compensation of demodulator phase error and sensitivity improvement in direct-detection 40-Gb/s DQPSK, in *Proceedings of European Conference on Optical Communications 2006*, post-deadline paper Th4.4.5, 2006
74. N. Kikuchi, S. Sasaki, Optical dispersion-compensation free incoherent multilevel signal transmission over standard single-mode fiber with digital pre-distortion and phase pre-integration techniques. ECOC'08, paper Tu.1.E.2, 2008
75. N. Kikuchi, S. Sasaki, Sensitivity improvement of incoherent multilevel (30-Gbit/s 8QAM and 40-Gbit/s 16QAM) signaling with non-Euclidean metric and MSPE (multi symbol phase estimation). OFC'09, paper OWG1, 2009
76. J.P. Gordon, L.F. Mollenauer, Opt. Lett. **15**, 1351–1353 (1990)
77. X. Liu, X. Wei, R.E. Slusher, C.J. McKinstrie, Opt. Lett. **27**, 1616–1618 (2002)
78. K.P. Ho, J.M. Kahn, J. Lightwave Technol **22**, 779–783 (2004)
79. G. Charlet, N. Maaref, J. Renaudier, H. Mardoyan, P. Tran, S. Bigo, Transmission of 40Gb/s QPSK with coherent detection over ultra long haul distance improved by nonlinearity mitigation, in *Proceedings of European Conference on Optical Communications 2006*, Post-deadline Paper Th4.3.4, 2006
80. N. Kikuchi, K. Mandai, S. Sasaki, Compensation of non-linear phase-shift in incoherent multilevel receiver with digital signal processing, in *Proceedings of European Conference on Optical Communications 2007*, Paper 9.4.1, 2007
81. Y.K. Lizé, L. Christen, M. Nazarathy, S. Nuccio, X. Wu, A.E. Willner, R. Kashyap, Opt. Express **15**, 6831–6839 (2007)
82. Y.K. Lizé, L. Christen, M. Nazarathy, Y. Atzmon, S. Nuccio, P. Saghari, R. Gomma, J.-Y. Yang, R. Kashyap, A. Willner, L. Paraschis, Photon. Technol. Lett. **19**, 1874–1876 (2007)
83. X. Liu, Digital self-coherent detection and mitigation of transmission impairments, 2008 OSA summer topic meeting on coherent optical technologies and applications (COTA'08), paper CWB2, 2008
84. S. Zhang, P.Y. Kam, J. Chen, C. Yu, Opt. Express **17**, 704–715 (2009)
85. C. Yu, S. Zhang, P.Y. Kam, J. Chen, Opt. Express **18**, 12088–12103 (2010)
86. M. Nazarathy, A. Gorshtein, D. Sadot, Doubly-differential coherent 100 G transmission: multi-symbol decision-directed carrier phase estimation with intradyne frequency offset cancellation, Signal processing techniques in communication, signal processing in photonic communications (SPPCom), Advanced photonics OSA conference, Karlsruhe, Germany, 21–24 June, 2010
87. S.J. Savory, Opt. Express **16**, 804–817 (2008)
88. Y. Mori, C. Zhang, M. Usui, K. Igarashi, K. Katoh, K. Kikuchi, 200-km transmission of 100-Gbit/s 32-QAM dual-polarization signals using a digital coherent receiver. ECOC'09, paper 8.4.6, 2009
89. J. Yu, X. Zhou, S. Gupta, Y.K. Huang, M.F. Huang, IEEE Photon. Technol. Lett. **22**, 115–117 (2010)
90. See, for example, IEEE standards 802.11a, 802.11g, and 802.16
91. A.J. Lowery, L. Du, J. Armstrong, Orthogonal frequency division multiplexing for adaptive dispersion compensation in long haul WDM systems. OFC'06, post-deadline paper PDP39, 2006
92. W. Shieh, C. Athaudage, Electron. Lett. **42**, 587–589 (2006)

93. I.B. Djordjevic, B. Vasic, Opt. Express **14**, 3767–3775 (2006)
94. S.L. Jansen, I. Morita, T.C. Schenk, H. Tanaka, J. Opt. Netw. **7**, 173–182 (2008)
95. W. Shieh, X. Yi, Y. Ma, Q. Yang, J. Opt. Netw. **7**, 234–255 (2008)
96. W. Shieh, H. Bao, Y. Tang, Opt. Express **16**, 841–859 (2008)
97. A. Bocoi1, M. Schuster, F. Rambach, D.A. Schupke, C.A. Bunge, B. Spinnler, Cost comparison of networks using traditional 10 and 40 Gb/s transponders versus OFDM transponders. OFC'08, paper OThB4, 2008
98. B. Spinnler, F.N. Hauske, M. Kuschnerov, Adaptive equalizer complexity in coherent optical receivers. ECOC'08, paper We.2.E.4, 2008
99. E.M. Ip, J.M. Khan, J. Lightwave Technol. **28**(4), 502–519 (2010)
100. X. Liu, F. Buchali, R.W. Tkach, S. Chandrasekhar, Bell Labs Tech. J. **14**, 47–59 (2010)
101. M. Nazarathy, J. Khurgin, R. Weidenfeld, Y. Meiman, P. Cho, R. Noe, I. Shpantzer, The FWM impairment in coherent OFDM compounds on a phased-array basis over dispersive multi-span links, Coherent optical technologies and applications (COTA), Boston, 2008
102. M. Nazarathy, J. Khurgin, R. Weidenfeld, Y. Meiman, P.S. Pak, R. Noe, I. Shpantzer, V. Karagodsky, Opt. Express **16**(6), 4228–4236 (2008)
103. R. Weidenfeld, M. Nazarathy, R. Noe, I. Shpantzer, Volterra nonlinear compensation of 112 Gb/s ultra-long-haul coherent optical OFDM based on frequency-shaped decision feedback, European conference on optical communications, Paper 2.3.3, ECOC'09, Vienna, September 2009
104. R. Weidenfeld, M. Nazarathy, R. Noe, I. Shpantzer, Volterra nonlinear compensation of 100G coherent OFDM with baud-rate ADC, tolerable complexity and low intra-channel FWM/XPM error propagation. Paper OTuE3, OFC'10, San Diego, March 2010
105. D. Liang, B. Schmidt, A. Lowery, Efficient digital backpropagation for PDM-CO-OFDM optical transmission systems, Optical fiber communications (OFC 2010), San Diego, CA. Paper OTuE2, 23 March 2010
106. M. Nazarathy, Nonlinear impairments in coherent optical OFDM systems and their mitigation, Invited paper, Signal processing in photonic communications (SPPCom), Advanced photonics OSA conference, Karlsruhe, Germany, 21–24 June, 2010
107. X. Liu, F. Buchali, Opt. Express **16**, 21944–21957 (2008)
108. X. Liu, F. Buchali, R.W. Tkach, J. Lightwave Technol. **27**, 3632–3640 (2009)
109. K. Ishihara et al., Electron. Lett. **44**, 1480–1481 (2008)
110. A.J. Lowery, Opt. Express **15**, 12965 (2007)
111. S. Oda, T. Tanimura, T. Hoshida, C. Ohshima, H. Nakashima, Z. Tao, J.C. Rasmussen, 112Gb/s DP-QPSK transmission using a novel nonlinear compensator in digital coherent receiver. OFC'09, paper OThR6, 2009
112. D.S. Millar, S. Makovejs, V. Mikhailov, R.I. Killey, P. Bayvel, S.J. Savory, Experimental comparison of nonlinear compensation in long-haul PDM-QPSK transmission at 42.7 and 85.4 Gb/s. ECOC'09, paper 9.4.4, 2009
113. S. Chandrasekhar, X. Liu, Opt. Express **17**, 12350–12361 (2009)
114. A. Ellis, F.C.G. Gunning, IEEE Photon. Technol. Lett. **17**, 504–506 (2005)
115. R.M. Metcalfe, Toward terabit Ethernet. OFC'08, plenary talk 2, 2008
116. A.D. Ellis, F.C.G. Gunning, B. Cuenot, T.C. Healy, E. Pincemin, Towards 1TbE using coherent WDM, in *Proceedings of OECC/ACOFT 2008*, Paper WeA-1, Sydney, Australia, 2008
117. R. Dischler, F. Buchali, Transmission of 1.2 Tb/s continuous waveband PDM-OFDM-FDM signal with spectral efficiency of 3.3 but/s/Hz over 400 km of SSMF. OFC'09, post-deadline paper PDPC2, 2009
118. T. Healy, F.C. Garcia Gunning, A.D. Ellis, J. D, Bull, Opt. Express **15**, 2981–2986 (2007)
119. A. Kaplan, A. Greenblatt, G. Harston, P.S. Cho, Y. Achiam, I. Shpantzer, Fully tunable LiNbO3 ring resonator cavity for frequency comb generator (FCG). ECIO'07, 2007
120. X. Liu, S. Chandrasekhar, B. Zhu, D.W. Peckham, Efficient digital coherent detection of a 1.2-Tb/s 24-carrier no-guard-interval CO-OFDM signal by simultaneously detecting multiple carriers per sampling. OFC'10, paper OWO2, 2010
121. X. Liu, S. Chandrasekhar, Impact of fiber nonlinearity on Tb/s PDM-OFDM transmission, 2010 IEEE photonics society summer topicals, invited paper TuA3, 2010

122. C.E. Shannon, Bell Syst. Tech. J. **27**, 379–423 623–656 (1948)
123. R.J. Essiambre, G. Kramer, P.J. Winzer, G.J. Foschini, B. Goebel, J. Lightwave Technol. **28**, 662–701, (2010) and references therein
124. A.D. Ellis, J. Zhao, D. Cotter, J. Lightwave Technol. **28**, 424–433, (2010) and references therein
125. D. Gorshtein G. Sadot O. Katz Levy, Coherent CD equalization for 111Gbps DP-QPSK with one sample per symbol based on anti-aliasing filtering and MLSE. OFC/NFOEC'10, paper OThT2, 2010
126. A. Agmon, M. Nazarathy, Opt. Express **15**, 13123–13128 (2007)
127. M. Nazarathy, A. Agmon, J. Lightwave Technol. **26**, 2037–2045 (2008)

Chapter 2
Optical OFDM Basics

Qi Yang, Abdullah Al Amin, and William Shieh

2.1 Introduction

We have witnessed a dramatic increase of interest in orthogonal frequency-division multiplexing (OFDM) from optical communication community in recent years. The number of publications on optical OFDM has grown dramatically since it was proposed as an attractive modulation format for long-haul transmission either in coherent detection [1] or in direct detection [2, 3]. Over the last few years, net transmission data rates grew at a factor of 10 per year at the experimental level. To date, experimental demonstration of up to 1 Tb s^{-1} transmission in a single channel [4, 5] and 10.8 Tb s^{-1} transmission based on optical FFT have been accomplished [6], whereas the demonstration of real-time optical OFDM with digital signal processing (DSP) has surpassed 10 Gb s^{-1} [7]. These progresses may eventually lead to realization of commercial transmission products based on optical OFDM in the future, with the potential benefits of high spectral efficiency and flexible network design.

This chapter intends to give a brief introduction on optical OFDM, from its fundamental mathematical concepts to the up-to-date experimental results. This is organized into seven sections, including this introduction as Sect. 2.1. Section 2.2 reviews the historical developments of OFDM and its application in

W. Shieh (✉)
Center for Ultra-broadband Information Networks, Department of Electrical and Electronic Engineering, University of Melbourne, Melbourne, VIC 3010, Australia
e-mail: shiehw@unimelb.edu.au

Q. Yang
State Key Lab. of Opt. Commu. Tech. and Networks, Wuhan Research Institute of Post & Telecommunication, Wuhan, China
e-mail: qyang@wri.com.cn

A. Al Amin
Center for Ultra-broadband Information Networks, Department of Electrical and Electronic Engineering, University of Melbourne, Melbourne, VIC 3010, Australia
e-mail: aalamin@unimelb.edu.au

S. Kumar (ed.), *Impact of Nonlinearities on Fiber Optic Communications*, Optical and Fiber Communications Reports 7, DOI 10.1007/978-1-4419-8139-4_2,
© Springer Science+Business Media, LLC 2011

optical transmission. Section 2.3 describes the fundamentals and different flavors of optical OFDM. As this book focuses on optical nonlinearity, which is a major concern for long-haul transmission, the coherent optical OFDM (CO-OFDM) is mainly considered in this chapter. Section 2.4 gives an introduction on CO-OFDM. The procedures of the DSP are also discussed in detail in this section. Some promising research directions for CO-OFDM are presented in Sect. 2.5. Section 2.6 gives the summary of the chapter.

2.2 Historical Perspective of OFDM

OFDM plays a significant role in the modem telecommunications for both wireless and wired communications. The history of frequency-division multiplexing (FDM) began in 1870s when the telegraph was used to carry information through multiple channels [8]. The fundamental principle of orthogonal FDM was proposed by Chang [9] as a way to overlap multiple channel spectra within limited bandwidth without interference, taking consideration of the effects of both filter and channel characteristics. Since then, many researchers have investigated and refined the technique over the years and it has been successfully adopted in many standards. Table 2.1 shows some of the key milestones of the OFDM technique in radiofrequency (RF) domain.

Although OFDM has been studied in RF domain for over four decades, the research on OFDM in optical communication began only in the late 1990s [13]. The fundamental advantages of OFDM in an optical channel were first disclosed in [14]. In the late 2000s, long-haul transmission by optical OFDM has been investigated by a few groups. Two major research directions appeared, direct-detection optical OFDM (DDO-OFDM) [2,3] looking into a simple realization based on low-cost optical components and CO-OFDM [1] aiming to achieve high spectral efficiency and receiver sensitivity. Since then, the interest in optical OFDM has increased dramatically. In 2007, the world's first CO-OFDM experiment with line rate of 8 Gb s^{-1} was reported [15]. In the last few years, the transmission capacity continued to grow

Table 2.1 Historical development of RF OFDM

1966	R. Chang, foundation work on OFDM [9]
1971	S.B. Weinstein and P.M. Ebert, DFT implementation of OFDM [10]
1980	R. Peled and A. Ruiz, Introduction of cyclic prefix [11]
1985	L. Cimini, OFDM for mobile communications [12]
1995	DSL formally adopted discrete multi-tone (DMT), a variation of OFDM
1995 (1997)	ETSI digital audio (video) broadcasting standard, DAB(DVB)
1999 (2002)	Wireless LAN standard, 802.11 a (g), Wi-Fi
2004	Wireless MAN standard, 802.16, WiMax
2009	Long time evolution (LTE), 4 G mobile standard

Table 2.2 Progress of optical OFDM

1996	Pan and Green, OFDM for CATV [13]
2001	You and Kahn, OFDM in direct modulation (DD) systems [16]
	Dixon et al., OFDM over multimode fiber [14]
2005	Jolley et al., experiment of 10 Gb s^{-1} optical OFDM over multimode fiber (MMF) [17]
	Lowery and Armstrong, power-efficient optical OFDM in DD systems [18]
2006	Lowery and Armstrong [2], and Djordjevic and Vasic [3], long-haul direct-detection optical OFDM (DDO-OFDM)
	Shieh and Athaudage, long-haul coherent optical OFDM (CO-OFDM) [15]
2007	Shieh et al. [15], 8 Gb s^{-1} CO-OFDM transmission over 1,000 km
2008	Yang et al. [19], Jansen et al. [20], Yamada et al. [21], >100 Gb s^{-1} per single channel CO-OFDM transmission over 1,000 km
2009	Ma et al. [4], Dischler et al. [5], Chandrasekhar et al. [22], >1 Tb s^{-1} CO-OFDM long-haul transmission

about ten times per year. In 2009, up to 1 Tb s^{-1} optical OFDM was successfully demonstrated [4, 5]. Table 2.2 shows the development of optical OFDM in the last two decades.

Besides offline DSP, from 2009 onward, a few research groups started to investigate real-time optical OFDM transmission. The first real-time optical OFDM demonstration took place in 2009 [23], 3 years later than real-time single-carrier coherent optical reception [24, 25]. The pace of real-time OFDM development is fast, with the net rate crossing 10 Gb s^{-1} within 1 year [7]. Moreover, by using orthogonal-band-multiplexing (OBM), which is a key advantage for OFDM, up to 56 Gb s^{-1} [26] and 110-Gb s^{-1} [27] over 600-km standard signal mode fiber (SSMF) was successfully demonstrated. Most recently, 41.25 Gb s^{-1} per single-band was reported in [28]. As evidenced by the commercialization of single-carrier coherent optical receivers, it is foreseeable that real-time optical OFDM transmission with much higher net rate will materialize in the near future based on state-of-the-art ASIC design.

2.3 OFDM Fundamentals

Before moving onto the description of optical OFDM transmission, we will review some fundamental concepts and basic mathematic expressions of OFDM. It is well known that OFDM is a special class of multi-carrier modulation (MCM), a generic implementation of which is depicted in Fig. 2.1. The structure of a complex multiplier (IQ modulator/demodulator), which is commonly used in MCM systems, is also shown at the bottom of the Fig. 2.1. The key distinction of OFDM from general multicarrier transmission is the use of orthogonality between the individual subcarriers.

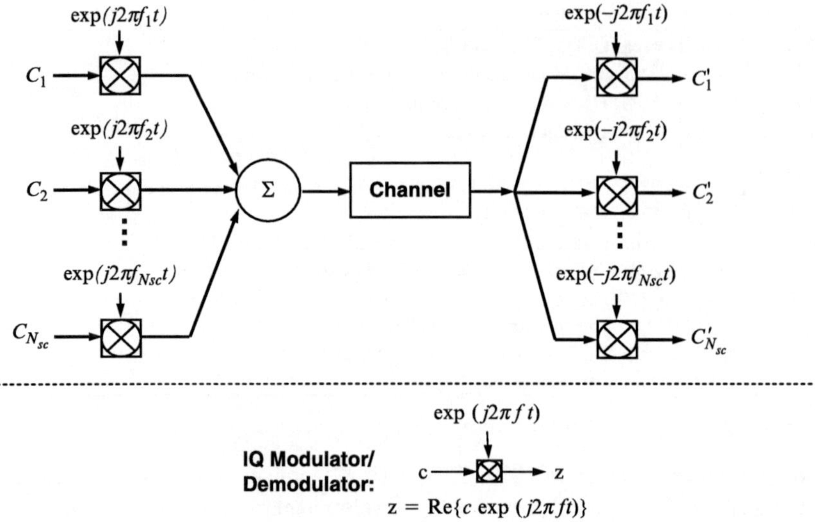

Fig. 2.1 Conceptual diagram for a multi-carrier modulation (MCM) system

2.3.1 Orthogonality Between OFDM Subcarriers and Subbands

The MCM transmitted signal $s(t)$ is represented as

$$s(t) = \sum_{i=-\infty}^{+\infty} \sum_{k=1}^{N_{\rm sc}} c_{ki} s_k \left(t - i T_{\rm s}\right) \tag{2.1}$$

$$s_k\left(t\right) = \Pi(t) e^{j2\pi f_k t} \tag{2.2}$$

$$\Pi\left(t\right) = \begin{cases} 1, & \left(0 < t \le T_{\rm s}\right) \\ 0, & \left(t \le 0, t > T_{\rm s}\right) \end{cases}, \tag{2.3}$$

where c_{ki} is the ith information symbol at the kth subcarrier, s_k is the waveform for the kth subcarrier, $N_{\rm sc}$ is the number of subcarriers, f_k is the frequency of the subcarrier, and $T_{\rm s}$ is the symbol period, $\Pi\left(t\right)$ is the pulse shaping function. The optimum detector for each subcarrier could use a filter that matches the subcarrier waveform, or a correlator matched with the subcarrier as shown in Fig. 2.1. Therefore, the detected information symbol c'_{ik} at the output of the correlator is given by

$$c'_{ki} = \frac{1}{T_{\rm s}} \int_0^{T_{\rm s}} r\left(t - i T_{\rm s}\right) s^*{}_k {\rm d}t = \frac{1}{T_{\rm s}} \int_0^{T_{\rm s}} r\left(t - i T_{\rm s}\right) {\rm e}^{-j2\pi f_k t} {\rm d}t, \tag{2.4}$$

where $r\left(t\right)$ is the received time-domain signal. The classical MCM uses nonoverlapped band-limited signals, and can be implemented with a bank of large number of oscillators and filters at both transmit and receive ends [29, 30]. The major disadvantage of MCM is that it requires excessive bandwidth. This is because in order to design the filters and oscillators cost-effectively, the channel spacing has to be multiple of the symbol rate, greatly reducing the spectral efficiency. A novel approach called OFDM was investigated by employing overlapped yet orthogonal signal set [9]. This orthogonality originates from straightforward correlation between any two subcarriers, given by

$$\begin{aligned}\delta_{kl} &= \frac{1}{T_s}\int_0^{T_s} s_k s^*{}_l \, \mathrm{d}t = \frac{1}{T_s}\int_0^{T_s} \exp\left(j2\pi\left(f_k - f_l\right)t\right)\mathrm{d}t \\ &= \exp\left(j\pi\left(f_k - f_l\right)T_s\right)\frac{\sin\left(\pi\left(f_k - f_l\right)T_s\right)}{\pi\left(f_k - f_l\right)T_s}. \end{aligned} \quad (2.5)$$

It can be seen that if the following condition

$$f_k - f_l = m\frac{1}{T_s} \quad (2.6)$$

is satisfied, then the two subcarriers are orthogonal to each other. This signifies that these orthogonal subcarrier sets, with their frequencies spaced at multiple of inverse of the symbol rate can be recovered with the matched filters in (2.5) without intercarrier interference (ICI), in spite of strong signal spectral overlapping. Moreover, the concept of this orthogonality can be extended to combine multiple OFDM bands into a signal with much larger spectral width. Such approach was first introduced in [19, 31] to flexibly expand the capacity of a single wavelength. This method of subdividing OFDM spectrum into multiple orthogonal bands is so-called "orthogonal-band-multiplexed OFDM" (OBM-OFDM).

Figure 2.2 shows the concept of orthogonal band multiplexing, where the entire spectrum is composed by N OFDM subbands. In order to maintain the orthogonality, the frequency spacing between two OFDM bands has to be a constant multiple of the subcarrier frequency spacing. The orthogonal condition between the different bands is given by $\Delta f_G = m\Delta f$, where m is an integer. This guarantees that each OFDM band is an orthogonal extension of another, and is a powerful method to increase channel capacity by adding OFDM subbands to the spectrum.

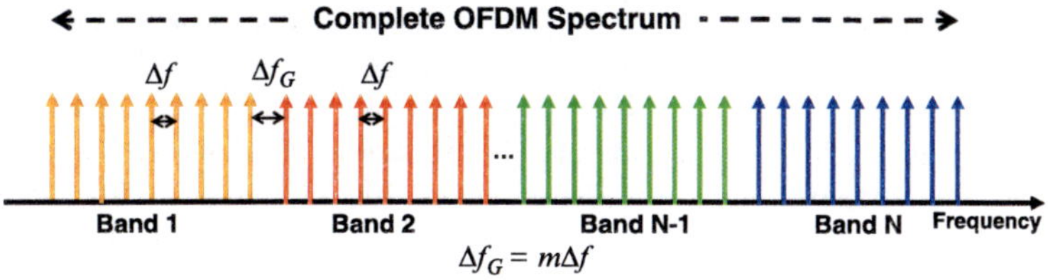

Fig. 2.2 Principle of orthogonal-band-multiplexed OFDM

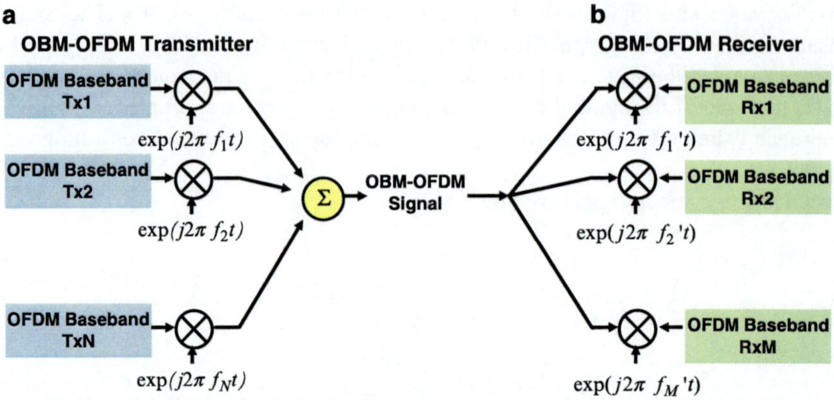

Fig. 2.3 Schematic of OBM-OFDM implementation in mixed-signal circuits for (**a**) the transmitter, and (**b**) the receiver

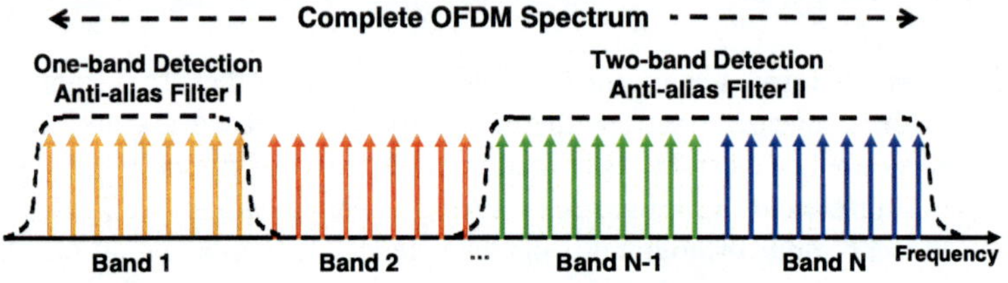

Fig. 2.4 Illustrations of one-band detection and two-band detection

A schematic of the transmitter and receiver configuration for OBM-OFDM is shown in the Fig. 2.3. The method has been first proposed in [32], where it is called cross-channel OFDM (XC-OFDM). The unique advantage of this method is that the data rate can be simply extended or modified to specification in a bandwidth-efficient manner.

Upon reception, the spectrum can be divided into multiple subbands. The band-partitioning at the receiver is not necessary to be the same as the transmitter. Figure 2.4 shows an example of single-band detection and multiband detection. In the former case, the receiver local oscillator laser is tuned to the center of each band, and an anti-aliasing filter (Filter I) selects a single OFDM band to be detected separately. In the latter case, the received laser tuned to the center of the guard band, and an anti-aliasing filter (Filter II) separates two OFDM bands, which are converted into digital symbols and separated by further digital down-converters to be detected simultaneously. In either case, the inter-band interference (IBI) is avoided because of the orthogonality between the neighboring bands, despite the "leakage" of the subcarriers from neighboring bands. Thus, CO-OFDM can achieve high net rate by employing OBM without requiring DAC/ADC operating at extremely high sampling rates.

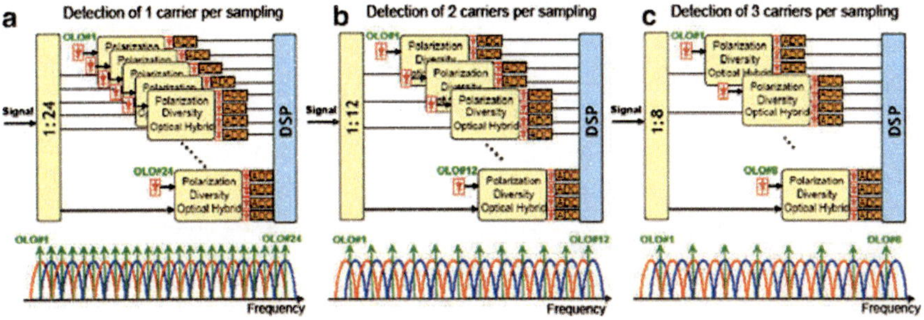

Fig. 2.5 Illustrations of three different methods used in [33] to detect a 1.2-Tb s^{-1} 24-carrier NGI-CO-OFDM signal having 12.5-Gbaud PDM-QPSK carriers with 50-GS s^{-1} ADC, (**a**) detecting 1 carrier per sampling with an oversampling factor of 4, (**b**) detecting 2 carriers per sampling with an oversampling factor of 2, and (**c**) detecting 3 carriers per sampling with an oversampling factor of 1.33. *OLO* Optical local oscillator

An additional advantage of the multi-band detection is its capability to save the number of required optical components at the receiver. One experimental demonstration of this has been shown in [33], where 24 orthogonal bands of OFDM are transmitted to generate a total of 1.2 Tb s^{-1} data rate. In the receiver, three schemes are used: (1) detecting 1 band per ADC with an oversampling factor of 4, (2) detecting 2 bands per ADC with an oversampling factor of 2, and (3) detecting 3 bands per ADC with an oversampling factor of 1.33. All three schemes can recover the received signal completely. Assuming the ADC bandwidth is sufficiently wide, the more the number of bands are detected simultaneously, the less the number of the optical receivers are required (Fig. 2.5).

As mentioned earlier, the orthogonality condition is satisfied when the guard band Δf_G is multiple of subcarrier spacing Δf. A generalized study of the influence of guard band to the system performance is shown in [34]. The validity of the orthogonality condition that minimizes the IBI was verified through experiment. Due to the IBI, the subcarriers at the edges of each band bear the largest inter-band penalty. Figure 2.6a, b show the received SNR of the "edge subcarriers" (the first and the last subcarrier of the band) as a function of the guard band normalized to the subcarrier spacing, at back-to-back and 1,000-km transmission, respectively. For simplicity, only one polarization is presented. The SNR oscillates as the guard spacing increases with a step size of half of the subcarrier spacing. It is shown in theory that ICI interference due to frequency spacing is a sinc function [35]. The SNR oscillation eventually stabilizes to a constant value, where effect of neighboring band can be considered negligible. By comparing with the stabilized SNR, the system penalty as a function of the guard band can be investigated. At 1,000 km transmission, when the guard band equals to a multiple of the subcarrier spacing, the SNR stabilizes at around a 10.5 dB, and the penalty almost decreases to zero, validating the assumption that guard band can be minimized for higher spectral efficiency using the orthogonal band multiplexing condition.

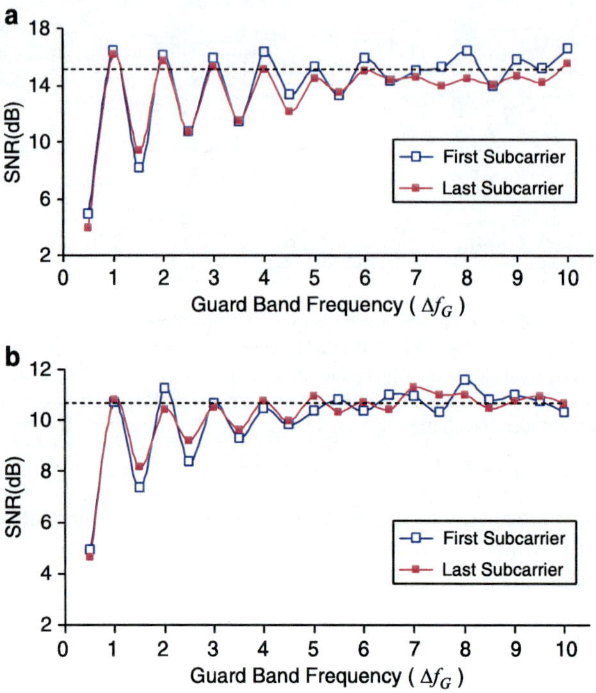

Fig. 2.6 SNR sensitivity performance of two edge subcarriers at (**a**) back-to-back transmission and (**b**) 1,000-km transmission. The guard band frequency is normalized to the subcarrier spacing [34]

2.3.2 Discrete Fourier Transform Implementation of OFDM

We rewrite the expression of (2.1)–(2.3)for one OFDM symbol as:

$$\tilde{s}(t) = \sum_{i=0}^{N-1} A_i \exp\left(j2\pi \frac{i}{T} t\right), \quad 0 \leq t \leq T, \tag{2.7}$$

which is the complex form of the OFDM baseband signal.

If we sample the complex signal with a sample rate of N/T, and add a normalization factor 1/N, then

$$S_n = \frac{1}{N} \sum_{i=0}^{N-1} A_i \exp\left(j2\pi \frac{i}{N} n\right), \quad n = 0, 1, \ldots, N-1 \tag{2.8}$$

where S_n is the nth time-domain sample. This is exactly the expression of inverse discrete Fourier transform (IDFT). It means that the OFDM baseband signal can be implemented by IDFT. The pre-coded signals are in the frequency domain, and output of the IDFT is in the time domain. Similarly, at the receiver side, the data is recovered by discrete Fourier transform (DFT), which is given by:

$$A_i = \sum_{i=0}^{N-1} R_n \exp\left(-j2\pi \frac{i}{N} n\right), \quad n = 0, 1, \ldots, N-1, \tag{2.9}$$

where R_n is the received sampled signal, and A_i is received information symbol for the ith subarrier. There are two fundamental advantages of DFT/IDFT implementation of OFDM. First, they can be implemented by (inverse) fast Fourier transform (I)FFT algorithm, where the number of complex multiplications is reduced from N^2 to $\frac{N}{2}\log_2(N)$, slightly higher than linear scaling with the number of subcarriers, N [36]. Second, a large number of orthogonal subcarriers can be modulated and demodulated without resorting to very complex array of RF oscillators and filters. This leads to a relatively simple architecture for OFDM implementation when large number of subcarriers is required.

2.3.3 *Cyclic Prefix for OFDM*

In addition to modulation and demodulation of many orthogonal subcarriers via (I)FFT, one has to mitigate dispersive channel effects such as chromatic and polarization mode dispersions for good performance. In this respect, one of the enabling techniques for OFDM is the insertion of cyclic prefix [37,38]. Let us first consider two consecutive OFDM symbols that undergo a dispersive channel with a delay spread of t_d. For simplicity, each OFDM symbol includes only two subcarriers with the fast delay and slow delay spread at t_d, represented by "fast subcarrier" and "slow subcarrier," respectively. Figure 2.7a shows that inside each OFDM symbol, the two subcarriers, "fast subcarrier" and "slow subcarrier" are aligned upon the transmission. Figure 2.7b shows the same OFDM signals upon the reception, where the "slow subcarrier" is delayed by t_d against the "fast subcarrier." We select a DFT window containing a complete OFDM symbol for the "fast subcarrier." It is apparent that due to the channel dispersion, the "slow subcarrier" has crossed the symbol boundary leading to the interference between neighboring OFDM symbols, formally, the so-called inter-symbol-interference (ISI). Furthermore, because the OFDM waveform in the DFT window for "slow subcarrier" is incomplete, the critical orthogonality condition for the subcarriers is lost, resulting in an inter-carrier-interference (ICI) penalty.

Cyclic prefix was proposed to resolve the channel dispersion-induced ISI and ICI [37]. Figure 2.7c shows insertion of a cyclic prefix by cyclic extension of the OFDM waveform into the guard interval Δ_G. As shown in Fig. 2.7c, the waveform in the guard interval is essentially an identical copy of that in the DFT window, with time-shifted by "t_s" forward. Figure 2.7d shows the OFDM signal with the guard interval upon reception. Let us assume that the signal has traversed the same dispersive channel, and the same DFT window is selected containing a complete

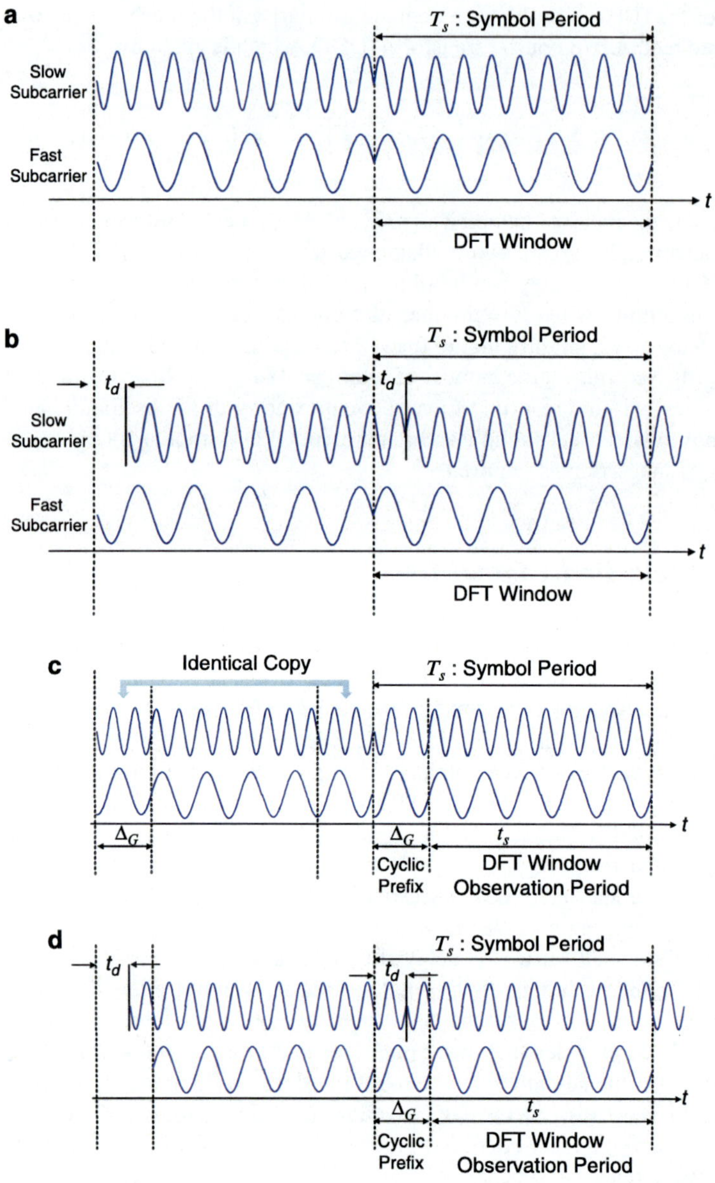

Fig. 2.7 OFDM signals (**a**) without cyclic prefix at the transmitter, (**b**) without cyclic prefix at the receiver, (**c**) with cyclic prefix at the transmitter, and (**d**) with cyclic prefix at the receiver

OFDM symbol for the "fast subcarrier" waveform. It can be seen from Fig. 2.7d, a complete OFDM symbol for "slow subcarrier" is also maintained in the DFT window, because a proportion of the cyclic prefix has moved into the DFT window to replace the identical part that has shifted out. As such, the OFDM symbol for "slow

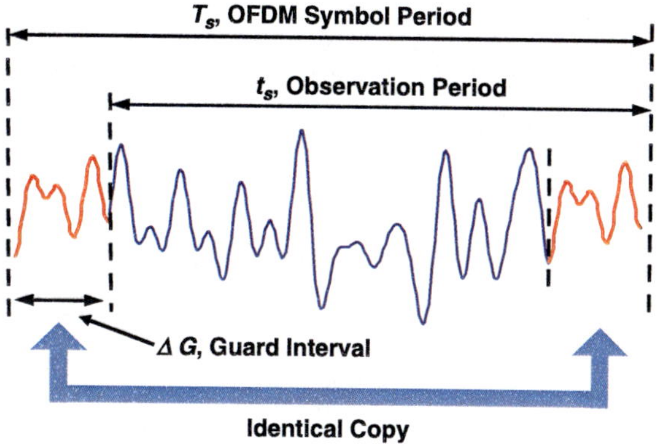

Fig. 2.8 Time-domain OFDM signal for one complete OFDM symbol

subcarrier" is an "almost" identical copy of the transmitted waveform with an additional phase shift. This phase shift is dealt with through channel estimation and will be subsequently removed for symbol decision. The important condition for ISI-free OFDM transmission is given by:

$$t_d < \Delta_G. \tag{2.10}$$

It can be seen that after insertion of the guard interval greater than the delay spread, two critical procedures must be carried out to recover the OFDM information symbol properly, namely, (1) selection of an appropriate DFT window, called DFT window synchronization, and (2) estimation of the phase shift for each subcarrier, called channel estimation or subcarrier recovery. Both signal processing procedures are actively pursued research topics, and their references can be found in both books and journal papers [37, 38].

The corresponding time-domain OFDM symbol is illustrated in Fig. 2.8, which shows one complete OFDM symbol composed of observation period and cyclic prefix. The waveform within the observation period will be used to recover the frequency-domain information symbols.

2.3.4 *Spectral Efficiency for Optical OFDM*

In DDO-OFDM systems, the electrical field of optical signal is usually not a linear replica of the baseband signal, and it requires a frequency guard band between the main optical carrier and OFDM spectrum, reducing the spectral efficiency. The net optical spectral efficiency is dependent on the implementation details. We will turn our attention to the optical spectral efficiency for CO-OFDM systems. In OFDM systems, N_{sc} subcarriers are transmitted in every OFDM symbol period of T_s. Thus, the total symbol rate R for OFDM systems is given by

$$R = N_{sc}/T_s. \tag{2.11}$$

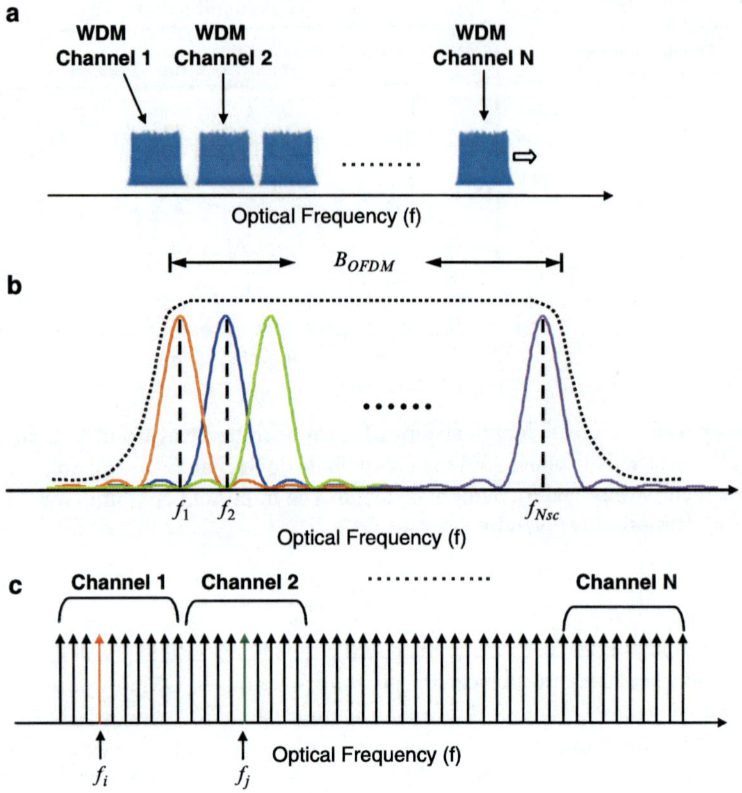

Fig. 2.9 Optical spectra for (**a**) N wavelength-division-multiplexed CO-OFDM channels, (**b**) zoomed-in OFDM signal for one wavelength, and (**c**) cross-channel OFDM (XC-OFDM) without guard band

Figure 2.9a shows the spectrum of wavelength-division-multiplexed (WDM) CO-OFDM channels, and Fig. 2.9b shows the zoomed-in optical spectrum for each wavelength channel. We use the frequency of the first null of the outermost subcarrier to denote the boundary of each wavelength channel. The OFDM bandwidth, B_{OFDM}, is thus given by

$$B_{\mathrm{OFDM}} = \frac{2}{T_{\mathrm{s}}} + \frac{N_{\mathrm{sc}} - 1}{t_{\mathrm{s}}}, \tag{2.12}$$

where t_{s} is the observation period (see Fig. 2.8). Assuming a large number of subcarriers used, the bandwidth efficiency of OFDM η is found to be

$$\eta = 2\frac{R}{B_{\mathrm{OFDM}}} = 2\alpha, \quad \alpha = \frac{t_{\mathrm{s}}}{T_{\mathrm{s}}}. \tag{2.13}$$

The factor of 2 accounts for two polarizations in the fiber. Using a typical value of 8/9, we obtain the optical spectral efficiency factor η of 1.8 Baud/Hz. The optical spectral efficiency gives 3.6 b s^{-1} Hz^{-1} if QPSK modulation is used for each subcarrier. The spectral efficiency can be further improved by using higher-order QAM modulation [39, 40]. To practically implement CO-OFDM systems, the optical spectral efficiency will be reduced by needing a sufficient guard band between WDM channels taking account of laser frequency drift about 2 GHz. This guard band can be avoided by using orthogonality across the WDM channels, which has been discussed in Sect. 2.3.1.

2.3.5 *Peak-to-Average Power Ratio for OFDM*

High peak-to-average-power ratio (PAPR) has been cited as one of the drawbacks of OFDM modulation format. In the RF systems, the major problem resides in the power amplifiers at the transmitter end, where the amplifier gain will saturate at high input power. One of the ways to avoid the relatively "peaky" OFDM signal is to operate the power amplifier at the so-called heavy "back-off" regime, where the signal power is much lower than the amplifier saturation power. Unfortunately, this requires an excess large saturation power for the power amplifier, which inevitably leads to low power efficiency. In the optical systems, interestingly enough, the optical power amplifier (predominately an Erbium-doped-amplifier today) is ideally linear regardless of its input signal power due to its slow response time in the order of millisecond. Nevertheless, the PAPR still poses a challenge for optical fiber communications due to the nonlinearity in the optical fiber [41–43].

The origin of high PAPR of an OFDM signal can be easily understood from its multicarrier nature. Because cyclic prefix is an advanced time-shifted copy of a part of the OFDM signal in the observation period (see Fig. 2.8), we focus on the waveform inside the observation period. The transmitted time-domain waveform for one OFDM symbol can be written as

$$s(t) = \sum_{k=1}^{N_{\mathrm{sc}}} c_k e^{j2\pi f_k t}, \quad f_k = \frac{k-1}{T_{\mathrm{s}}}. \tag{2.14}$$

The PAPR of the OFDM signal is defined as

$$\mathrm{PAPR} = \frac{\max\left\{|s(t)|^2\right\}}{E\left\{|s(t)|^2\right\}}, \quad t \in [0, T_{\mathrm{s}}]. \tag{2.15}$$

For the simplicity, we assume that an M-PSK encoding is used, where $|c_k| = 1$. The theoretical maximum of PAPR is $10\log_{10}(N_{\mathrm{sc}})$ in dB, by setting $c_k = 1$ and $t = 0$ in (2.14). For OFDM systems with 256 subcarriers, the theoretical maxim PAPR is

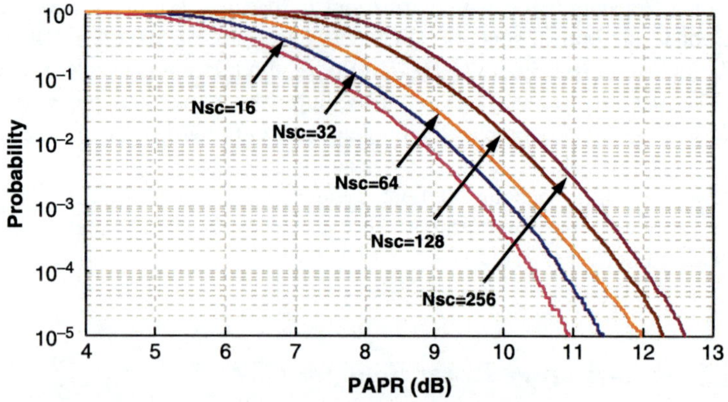

Fig. 2.10 Complementary cumulative distribution function (CCDF), P_c for the PAPR of OFDM signals with varying number of subcarriers. The oversampling factor is fixed at 2

24 dB, which obviously is excessively high. Fortunately, such a high PAPR is a rare event such that we do not need to worry about it. A better way to characterize the PAPR is to use complementary cumulative distribution function (CCDF) of PAPR, P_c, which is expressed as

$$P_c = \Pr\{\text{PAPR} > \zeta_P\}, \tag{2.16}$$

namely, P_c is the probability that PAPR exceeds a particular value of ζ_P.

Figure 2.10 shows CCDF with varying number of subcarriers. We have assumed QPSK encoding for each subcarrier. It can be seen that despite the theoretical maximum of PAPR is 24 dB for the 256-subcarrier OFDM systems, for the most interested probability regime, such as a CCDF of 10^{-3}, the PAPR is around 11.3 dB, which is much less than the maximum value of 24 dB. A PAPR of 11.3 dB is still very high as it implies that the peak value is about one order of magnitude stronger than the average, and some form of PAPR reduction should be used. It is also interesting to note that the PAPR of an OFDM signal increases slightly as the number of subcarriers increases. For instance, the PAPR increases by about 1.6 dB when the subcarrier number increases from 32 to 256.

The sampled waveform is used for PAPR evaluation, and subsequently the sampled points may not include the true maximum value of the OFDM signal. Therefore, it is essential to oversample the OFDM signal to obtain accurate PAPR. Assume that over-sampling factor is h, namely, number of the sampling points increases from N_{sc} to hN_{sc} with each sampling point given by

$$t_l = \frac{(l-1)\,T_s}{hN_{sc}}, \quad l = 1, 2, \ldots .hN_{sc}. \tag{2.17}$$

Substituting $f_k = \frac{k-1}{T_s}$ and (2.17) into (2.14), the lth sample of $s(t)$ becomes

$$s_l = s(t_l) = \sum_{k=1}^{N_{sc}} c_k e^{j2\pi \frac{(k-1)(l-1)}{hN_{sc}}}, \quad l = 1, 2, \ldots .hN_{sc}. \tag{2.18}$$

Expanding the number of subcarriers c_k from N_{sc} to hN_{sc} by appending zeros to the original set, the new subcarrier symbol c'_k after the zero padding is formally given by

$$\begin{aligned} c'_k &= c_k, \quad k = 1, 2, \ldots, N_{sc} \\ c'_k &= 0, \quad k = N_{sc}+1, N_{sc}+2, \ldots, hN_{sc}. \end{aligned} \tag{2.19}$$

Using the zero-padded new subcarrier set c'_k, (2.18) is rewritten as

$$s_l = \sum_{k=1}^{hN_{sc}} c'_k e^{j2\pi \frac{(k-1)(l-1)}{hN_{sc}}} = F^{-1}\left(c'_k\right), \quad l = 1, 2, \ldots .hN_{sc}. \tag{2.20}$$

From (2.20), it follows that the h times oversampling can be achieved by IFFT of a new subcarrier set that zero-pads the original subcarrier set to h times of the original size.

Figure 2.11 shows the CCDF of PAPR varying oversampling factors from 1 to 8. It can be seen that the difference between the Nyquist sampling ($h = 1$) and eight times oversampling is about 0.4 dB at the probability of 10^{-3}. However, most of the difference takes place below the oversampling factor of 4 and beyond this, PAPR changes very little. Therefore to use an oversampling factor of 4 for the purpose of PAPR, investigation seems to be sufficient.

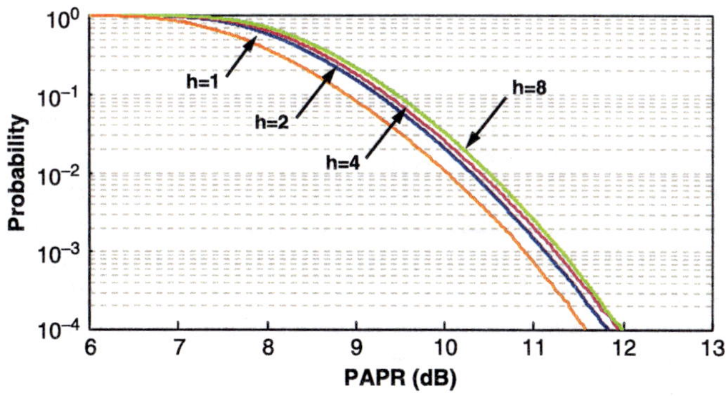

Fig. 2.11 Complementary cumulative distribution function (CCDF) for the PAPR of an OFDM signal with varying oversampling factors. The subcarrier number is fixed at 256

It is obvious that the PAPR of an OFDM signal is excessively high for either RF or optical systems. Consequently, PAPR reduction has been an intensely pursued field. Theoretically, for QPSK encoding, a PAPR smaller than 6 dB can be obtained with only a 4% redundancy [38]. Unfortunately, such code has not been identified so far. The PAPR reduction algorithms proposed so far allow for trade-off among three figure-of-merits of the OFDM signal: (1) PAPR, (2) bandwidth-efficiency, and (3) computational complexity. The most popular PAPR reduction approaches can be classified into two categories:

1. PAPR reduction with signal distortion. This is simply done by hard-clipping the OFDM signal [44–46]. The consequence of clipping is increased BER and out-of-band distortion. The out-of-band distortion can be mitigated through repeated filtering [46].
2. PAPR reduction without signal distortion. The idea behind this approach is to map the original waveform to a new set of waveforms that have a PAPR lower than the desirable value, most of the time, with some bandwidth reduction. Distortionless PAPR reduction algorithms include selective mapping (SLM) [47,48], optimization approaches such as partial transmit sequence (PTS) [49, 50], and modified signal constellation or active constellation extension (ACE) [51, 52].

2.3.6 Flavors of Optical OFDM

One of the major strengths of OFDM modulation format is its rich variation and ease of adaption to a wide range of applications. In wireless systems, OFDM has been incorporated in wireless LAN (IEEE 802. 11a/g, or better known as WiFi), wireless WAN (IEEE 802.16e, or better known as WiMax), and digital radio/video systems (DAB/DVB) adopted in most parts of the world. In RF cable systems, OFDM has been incorporated in ADSL and VDSL broadband access through telephone copper wiring or power line. This rich variation has something to do with the intrinsic advantages of OFDM modulation including dispersion robustness, ease of dynamic channel estimation and mitigation, high spectral efficiency and capability of dynamic bit and power loading. Recent progress in optical OFDM is of no exception. We have witnessed many novel proposals and demonstrations of optical OFDM systems from different areas of the applications that aim to benefit from the aforementioned OFDM advantages. Despite the fact that OFDM has been extensively studied in the RF domain, it is rather surprising that the first report on optical OFDM in the open literature only appeared in 1998 by Pan et al. [13], where they presented in-depth performance analysis of hybrid AM/OFDM subcarrier-multiplexed (SCM) fiberoptic systems. The lack of interest in optical OFDM in the past is largely due to the fact the silicon signal processing power had not reached the point, where sophisticated OFDM signal processing can be performed in a CMOS integrated circuitk (IC).

Optical OFDM are mainly classified into two main categories: coherent detection and direct detection according to their underlying techniques and applications. While direct detection has been the mainstay for optical communications over the last two decades, the recent progress in forward-looking research has unmistakably pointed to the trend that the future of optical communications is the coherent detection.

DDO-OFDM has much more variants than the coherent counterpart. This mainly stems from the broader range of applications for direct-detection OFDM due to its lower cost. For instance, the first report of the DDO-OFDM [13] takes advantage of that the OFDM signal is more immune to the impulse clipping noise in the CATV network. Other example is the single-side-band (SSB)-OFDM, which has been recently proposed by Lowery et al. and Djordjevic et al. for long-haul transmission [2, 3]. Tang et al. have proposed an adaptively modulated optical OFDM (AMOOFDM) that uses bit and power loading showing promising results for both multimode fiber and short-reach SMF fiber link [53, 54]. The common feature for DDO-OFDM is of course using the direct detection at the receiver, but we classify the DDO-OFDM into two categories according to how optical OFDM signal is being generated: (1) linearly mapped DDO-OFDM (LM-DDO-OFDM), where the optical OFDM spectrum is a replica of baseband OFDM, and (2) nonlinearly mapped DDO-OFDM (NLM-DDO-OFDM), where the optical OFDM spectrum does not display a replica of baseband OFDM [55].

CO-OFDM represents the ultimate performance in receiver sensitivity, spectral efficiency, and robustness against polarization dispersion, but yet requires the highest complexity in transceiver design. In the open literature, CO-OFDM was first proposed by Shieh and Authaudage [1], and the concept of the coherent optical MIMO-OFDM was formalized by Shieh et al. in [56]. The early CO-OFDM experiments were carried out by Shieh et al. for a 1,000 km SSMF transmission at 8 Gb s^{-1} [15], and by Jansen et al. for 4,160 km SSMF transmission at 20 Gb s^{-1} [57]. Another interesting and important development is the proposal and demonstration of thc no-guard interval CO-OFDM by Yamada et al. in [58], where optical OFDM is constructed using optical subcarriers without a need for the cyclic prefix. Nevertheless, the fundamental principle of CO-OFDM remain the same, which is to achieve high spectral efficiency by overlapping subcarrier spectrum yet avoiding the interference by using coherent detection and signal set orthogonality. As this book is primarily focused on fiber nonlinearity, coherent scheme will be mainly discussed in the following sections.

2.4 Coherent Optical OFDM Systems

Coherent optical communication was once intensively studied in late 1980s and early 1990s due to its high sensitivity [59–61]. However, with the invention of Erbium-doped fiber amplifiers (EDFAs), coherent optical communication has literally abandoned since the early of 1990s. Preamplified receivers using EDFA can achieve sensitivity within a few decibels of coherent receivers, thus making coherent detection less attractive, considering its enormous complexity. In the early twenty-first century, the impressive record-performance experimental demonstration using a differential-phase-shift-keying (DPSK) system [62], in spite of an incoherent form of modulation by itself, reignited the interest in coherent communications. The second wave of research on coherent communications is highlighted by the remarkable theoretical and experimental demonstrations from various groups around the world [56, 63, 64]. It is rather instructive to point out that the circumstances and the underlying technologies for the current drive for coherent communications are entirely different from those of a decade ago, thanks to the rapid technological advancement within the past decade in various fields. First, current coherent detection systems are heavily entrenched in silicon-based DSP for high-speed signal phase estimation and channel equalization. Second, multicarrier technology, which has emerged and thrived in the RF domain during the past decade, has gradually encroached into the optical domain [65, 66]. Third, in contrast to the optical system that was dominated by a low-speed, point-to-point, and single-channel system a decade ago, modern optical communication systems have advanced to massive wave-division-multiplexed (WDM) and reconfigurable optical networks with a transmission speed approaching 100 Gb s^{-1}. In a nutshell, the primary aim of coherent communications has shifted toward supporting these high-speed dynamic networks by simplifying the network installation, monitoring and maintenance.

When the modulation technique of OFDM combines with coherent detection, the benefits brought by these two powerful techniques are multifold [67]: (1) High spectral efficiency; (2) Robust to chromatic dispersion and polarization-mode dispersion; (3) High receiver sensitivity; (4) Dispersion Compensation Modules (DCM)-free operation; (5) Less DSP complexity; (6) Less oversampling factor; (7) More flexibility in spectral shaping and matched filtering.

2.4.1 Principle for CO-OFDM

Figure 2.12 shows the conceptual diagram of a typical coherent optical system setup. It contains five basic functional blocks: RF OFDM signal transmitter, RF to optical (RTO) up-converter, Fiber links, the optical to RF (OTR) down-converter, and the RF OFDM receiver. Such setup can be also used for single-carrier scheme, in which the DSP part in the transmitter and receiver needs to be modified, while all the hardware setup remains the same.

We will trace the signal flow end-to-end and illustrate each signal processing block. In the RF OFDM transmitter, the payload data is first split into multiple parallel branches. This is so-called "serial-to-parallel" conversion. The number of the multiple branches equals to the number of loaded subcarrier, including the pilot subcarriers. Then the converted signal is mapped onto various modulation formats, such as phase-shift keying (PSK), quadrature amplitude modulation (QAM), etc. The IDFT will convert the mapped signal from frequency domain into time domain. Two-dimensional complex signal is used to carry the information. The cyclic prefix is inserted to avoid channel dispersion. Digital-to-signal converters (DACs) are used to convert the time-domain digital signal to analog signal. A pair of electrical low-pass filters is used to remove the alias sideband signal. Figure 2.13 shows the effect of the anti-aliasing filter at the transmitter side.

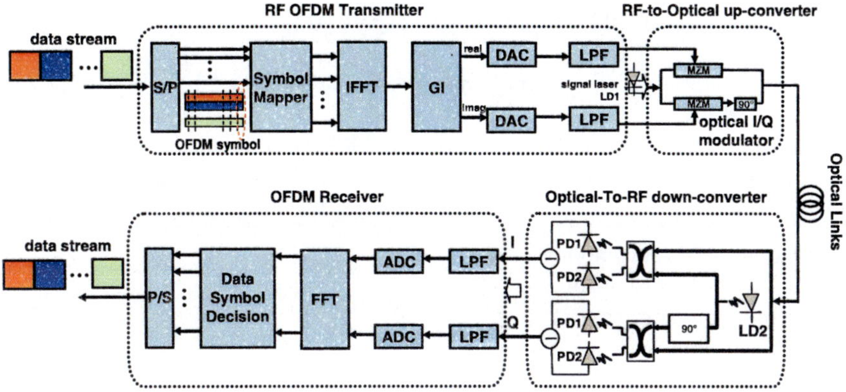

Fig. 2.12 Conceptual diagram of a coherent optical OFDM system

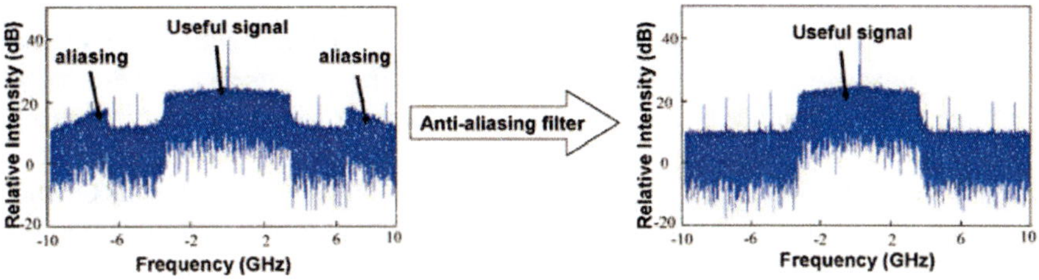

Fig. 2.13 Effect of the anti-aliasing filter

At the RTO up-converter, the baseband OFDM $S_B(t)$ signal is upshifted onto optical domain using an optical I/Q modulator, which is comprised by two Mach–Zehnder modulators (MZMs) with a 90° optical phase shifter. The up-converted OFDM signal in optical domain is given by

$$E(t) = \exp(j\omega_{LD1}t + \phi_{LD1})S_B(t), \tag{2.21}$$

where ω_{LD1} and ϕ_{LD1} are the frequency and phase of the transmitter laser, respectively. The optical signal $E(t)$ is launched into the optical fiber link, with an impulse response of $h(t)$. The received optical signal $E'(t)$ becomes

$$E'(t) = \exp(j\omega_{LD1}t + \phi_{LD1})S_B(t) \otimes h(t), \tag{2.22}$$

where $\otimes$ stands for the convolution operation.

When the optical signal is fed into the OTR converter, the optical signal $E'(t)$ is then mixed with a local laser at a frequency of ω_{LD2} and a phase of ϕ_{LD2}. Assume the frequency and phase difference between transmit and receiver lasers are

$$\Delta\omega = \omega_{LD1} - \omega_{LD2}, \quad \Delta\phi = \phi_{LD1} - \phi_{LD2} \tag{2.23}$$

Then the received RF OFDM signal can be expressed as

$$r(t) = \exp(j\Delta\omega t + \Delta\phi)S_{\mathrm{B}}(t) \otimes h(t) \tag{2.24}$$

In the RF OFDM receiver, the down-converted RF signal is first sampled by high speed analog-to-digital converter (ADC). The typical OFDM signal processing comprises five steps:

1. Window synchronization.
2. Frequency synchronization.
3. Discrete Fourier transform.
4. Channel estimation.
5. Phase noise estimation.

We here briefly describe the five DSP procedures [68]. Window synchronization aims to locate the beginning and end of an OFDM symbol correctly. One of the most popular methods was proposed by Schmidl and Cox [69] based on cross-correlation of detected symbols with a known pattern. A certain amount of frequency offset can be synchronized by a similar method, namely, the frequency offset can be estimated from the phase difference between two identical patterns with a known time offset. After window synchronization, OFDM signal is partitioned into blocks each containing a complete OFDM symbol. DFT is used to convert each block of OFDM signal from time domain to frequency domain. Then the channel and phase noise estimation are performed in the frequency domain using training symbols and pilot subcarriers, respectively. The details of these procedures are given in the following section. Note that the same procedures will also be followed for the real-time implementation.

2.4.2 OFDM Digital Signal Processing

2.4.2.1 Window Synchronization

The DSP begins with window synchronization in the OFDM reception. Its accuracy will influence the overall performance. Improper position of the DFT window on the OFDM signal will cause the inter-symbol interference (ISI) and ICI. In the worse case, the mis-synchronized symbol cannot be detected completely. The most commonly used method is Schmidl-Cox approach [69]. In this method, a preamble consisting of two identical patterns is inserted in the beginning of the multiple OFDM symbols, namely, an OFDM frame. Figure 2.14 shows the OFDM frame structure.

The Schmidl synchronization signal can be expressed as

$$s_m = s_{m+N\mathrm{sc}/2}, \quad m = 1, 2, \ldots, N\mathrm{sc}/2. \tag{2.25}$$

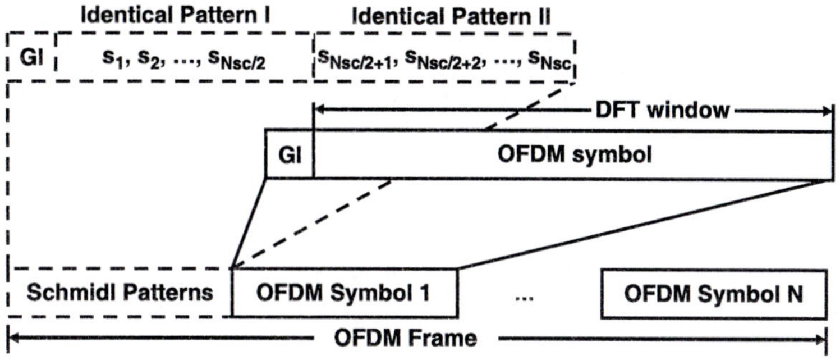

Fig. 2.14 OFDM frame structure showing Schmidl pattern for window synchronization

Considering the channel effect, from (2.24), the received samples will have the form as

$$r_m = \mathrm{e}^{j\Delta\omega t + \Delta\phi} s_m + n_m, \tag{2.26}$$

where $s_m = S_m(t) \otimes h(t)$. n_m stands for the random noise.

The delineation of OFDM symbol can be identified by studying the following correlation function defined as

$$R_d = \sum_{m=1}^{N_{sc}/2} r^*_{m+d} r_{m+d+N_{sc}/2}. \tag{2.27}$$

The principle is based on the fact that the second half of r_m is identical to the first half except for a phase shift. Assuming the frequency offset ω_{off} is small to start with, we anticipate that when $d = 0$, the correlation function R_d reaches its maximum value.

2.4.2.2 Frequency Offset Synchronization

In wireless communications, numerous approaches to estimate the frequency offset between transmitter and receiver have been proposed. In CO-OFDM systems, we use the correlation from the window synchronization to obtain the frequency offset. The phase difference from the sample s_m to $s_{m+N\text{sc}/2}$ is $\pi f_{\text{offset}} N_{\text{sc}}/S_{\text{sampling}}$, where S_{sampling} is the ADC sampling rate. The formula in Equation (2.27) can be re-written as

$$R_d = \sum_{m=1}^{N\text{sc}/2} |r_{m+d}|^2 \, \mathrm{e}^{\pi f_{\text{offset}} N_{\text{sc}}/S_{\text{sampling}}}. \tag{2.28}$$

Consequently, from the phase information of the correlation, the frequency offset can be derived as

$$f_{\text{offset}} = \frac{S_{\text{sampling}}}{\pi N_{\text{sc}}} \angle R_d, \tag{2.29}$$

where $\angle R_d$ stands for the angle of the correlation function of R_d. Because the phase information $\angle R_d$ ranges only from 0 to 2π, large frequency offset cannot be identified uniquely. Thus, this approach only supports the frequency offset range from $-f_{\mathrm{sub}}$ to f_{sub} where f_{sub} is the subcarrier spacing. To further increase the frequency offset compensation range, the synchronization symbol is further divided into $2^k (k > 1)$ segments [70]. The tolerable frequency offset can be enhanced to a few subcarrier spacing. Again, beside the Schmidl approach, there are other various approaches to perform the frequency offset estimation, such as the pilot-tone approach [71].

2.4.2.3 Channel Estimation

Assuming successful completion of window synchronization and frequency offset compensation, the RF OFDM signal after DFT operation is given by

$$r_{ki} = \mathrm{e}^{j\phi_i} h_{ki} s_{ki} + n_{ki}, \tag{2.30}$$

where s_{ki} (r_{ki}) is the transmitted (received) information symbol, ϕ_i is the OFDM common phase error (CPE), h_{ki} is the frequency domain channel transfer function, and n_{ki} is the noise. The common phase error is caused by the finite linewidth of the transmitter and receiver laser.

An OFDM frame usually contains a large number of OFDM symbols. Within each frame, the optical channel can be assumed to be invariant. There are various methods of channel estimation, such as time-domain pilot-assisted and the frequency-domain assisted approaches [3, 72]. Here, we are using the frequency domain pilot-symbol assisted approach. Figure 2.15 shows an OFDM frame in a time-frequency two-dimensional structure.

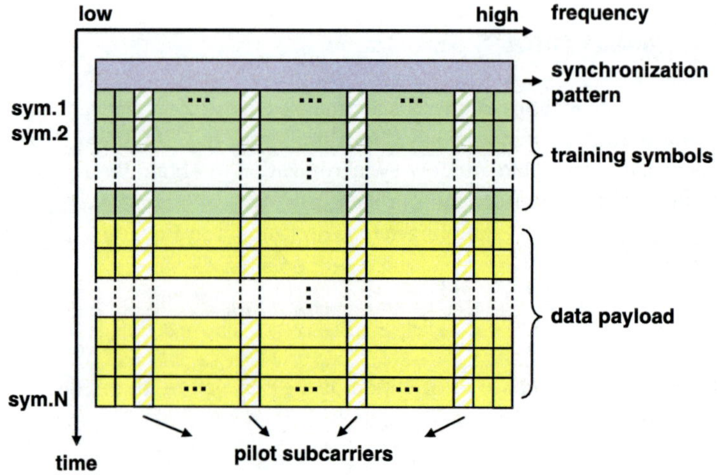

Fig. 2.15 Data structure of an OFDM frame

The first few symbols are the pilot-symbols or training symbols for which transmitted pattern is already known at the receiver side. The channel transfer function can be estimated as

$$h_{ki} = \mathrm{e}^{-j\phi_i} r_{ki}/s_{ki}. \tag{2.31}$$

Due to the presence of the random noise, the accuracy of the channel transfer function h is limited. To increase the accuracy of channel estimation, multiple training symbols are used. By performing averaging over multiple training symbols, the influence of the random noise can be much reduced. However, training symbols also leads to increase of overhead or decrease of the spectral efficiency. In order to obtain accurate channel information while still using little overhead, interpolation or frequency domain averaging algorithm [73] over one training symbol can be used.

2.4.2.4 Phase Estimation

As we mentioned above, the phase noise is due to the linewidth of the transmitter and receiver lasers. For CO-OFDM, we assume that Np subcarriers are used as pilot subcarrier to estimate the phase noise. The maximum likelihood CPE is given as [68]

$$\phi_i = \arg\left(\sum_{k=1}^{N_\mathrm{p}} r'_{ki} h_k^* s_{ki}^* / \delta_k^2\right), \tag{2.32}$$

where δ_k is the standard deviation of the constellation spread for the kth subcarrier. After the phase noise estimation and compensation, the constellation for every subcarrier can be constructed and symbol decision is made to recover the transmitted data.

2.4.3 Polarization-Diversity Multiplexed OFDM

In Sect. 2.4.2, the OFDM signal is presented in a scalar model. However, it is well known that SSMF supports two modes in polarization domain. To describe the multiple input multiple output (MIMO) model for CO-OFDM mathematically, Jones vector is introduced and the channel model is thus given by [56]

$$s(t) = \sum_{i=-\infty}^{+\infty} \sum_{k=1}^{N_{\mathrm{sc}}} c_{ki} \Pi(t - iT_{\mathrm{s}}) \exp(j2\pi f_k (t - iT_{\mathrm{s}})) \tag{2.33}$$

$$s(t) = \begin{bmatrix} s_x \\ s_y \end{bmatrix}, c_{ik} = \begin{bmatrix} c_x^{ik} \\ c_y^{ik} \end{bmatrix}$$

$$f_k = \frac{k-1}{t_{\mathrm{s}}}$$

$$s_k(t) = \Pi(t) \exp(j2\pi f_k t) \tag{2.34}$$

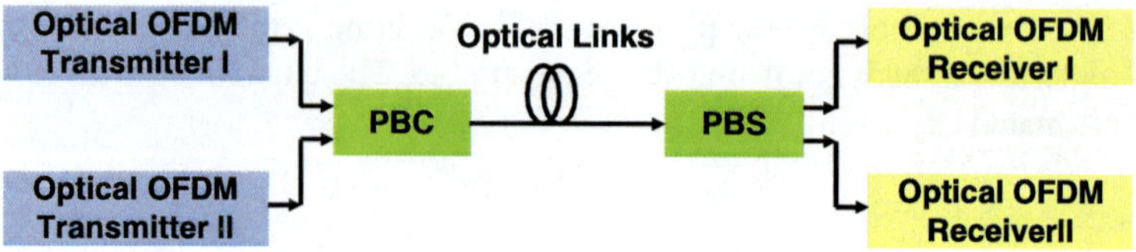

Fig. 2.16 PDM-OFDM conceptual diagram

$$\Pi(t) = \begin{cases} 1, & (0 < t \leq T_s) \\ 0, & (t \leq 0, t > T_s) \end{cases}, \tag{2.35}$$

where s_x and s_y are the two polarization components for s(t) in the time domain; c_{ik} is the transmitted OFDM information symbol in the form of Jones vector for the kth subcarrier in the ith OFDM symbol; c_x^{ik} and c_y^{ik} are the two polarization components for c_{ik}; f_k is the frequency for the kth subcarrier; Nsc is the number of OFDM subcarriers; and T_s and t_s are the OFDM symbol period and observation period, respectively [56]. In [56] four CO-MIMO-OFDM configurations are described: (1) (1×1) single-input signle-output, SISO-OFDM; (2) (1×2) single-input multiple-output SIMO-OFDM; (3) (2 × 1) multiple-input single-output MISO-OFDM; (4) (2 × 2) multiple-input multiple-output MIMO-OFDM. Among those configurations, SISO-OFDM and MIMO-OFDM are the preferred schemes. MIMO-OFDM is also called polarization diversity multiplexed (PDM) OFDM. Figure 2.16 shows the PDM-OFDM conceptual diagram.

In such scheme, the OFDM signal is transmitted via both polarizations, doubling the channel capacity compared to the SISO scheme. At the receiver, no hardware polarization tracking is needed as the channel estimation can help the OFDM receiver to recover the transmitted OFDM signals on two polarizations.

Some milestone experimental demonstrations for CO-OFDM are given in Table 2.2. Among these proof-of-concept demonstrations, two milestones are especially attention-grabbing – OFDM transmission at 100-Gb s^{-1} and 1-Tb s^{-1}. This is because 100 Gb s^{-1} Ethernet has recently been ratified as an IEEE standard and increasingly becoming a commercial reality, whereas 1-Tb s^{-1} Ethernet standard is anticipated to be available in the time frame as early as 2012–2013 [74]. In 2008, [19–21] demonstrated more than 100 Gb s^{-1} over 1,000 km SSMF transmission. In 2009, [4, 5] showed more than 1 Tb s^{-1} CO-OFDM transmission.

2.4.4 Real-Time Coherent Optical OFDM

The real-time optical OFDM has progressed rapidly in OFDM transmitter [75, 76], OFDM receiver [23, 26–28], and OFDM transceiver [7]. Because this chapter is focused on the long-haul transmission, we will mainly discuss the real-time CO-OFDM transmission in this subsection. With increased research interest in optical OFDM, numerous publications on this topic are being produced confirming the fast pace of research. However, most of the published CO-OFDM experiments are based on off-line processing, which lags behind single-carrier counterpart, where a real-time transceiver operating at 40 Gb s^{-1} based on CMOS ASICs has already been reported [77]. More importantly, OFDM is based on symbol and frame structure, and the required DSP associated with OFDM procedures, such as window synchronization and channel estimation, remains a challenge for real-time implementation. Among many demonstrated algorithms, only a few can be practically realized due to various limitations associated with digital signal processor capability. It is thus essential to investigate efficient and realistic algorithms for real-time CO-OFDM implementation in both FPGA and ASIC platforms.

2.4.4.1 Real-Time Window Synchronization

The first DSP procedure for OFDM is symbol synchronization. Traditional offline processing uses the Schimdl approach [69], where the autocorrelation of two identical patterns inserted at the beginning of each OFDM frame gives rise to a peak indicating the starting position of the OFDM frame and symbol. The autocorrelation output is

$$P(d) = \sum_{k=0}^{L-1} r_{d+k}^{*} r_{d+k+L}. \tag{2.36}$$

and can be recursively expressed as

$$P(d+1) = P(d) + r_{d+L}{}^{*} r_{d+2L} - r_{d}{}^{*} r_{d+L}. \tag{2.37}$$

An example of DSP implementation of (2.37) can be found in Fig. 2.17, where L indicates the length of synchronization pattern, r_d indicates the complex samples, and $P(d)$ indicates the autocorrelation term whose amplitude gives peak when the synchronization is found. The relatively simple equation (2.37) and the architecture in Fig. 2.18, however, assume that the incoming signal is a serial stream, and this implementation only works if the process clock rate is the same as the sample rate.

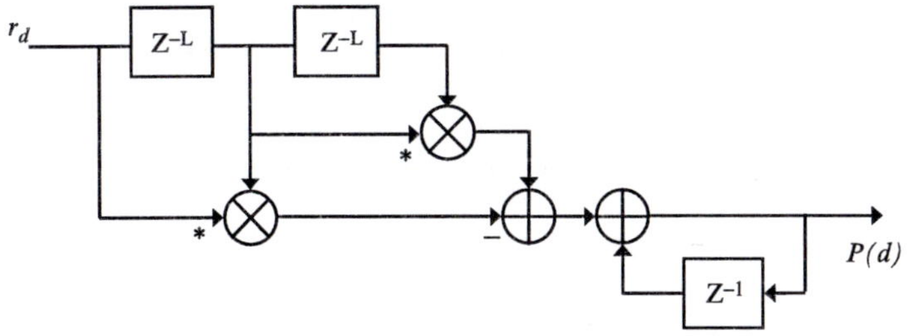

Fig. 2.17 DSP block diagram of autocorrelation for symbol synchronization based on serial processing

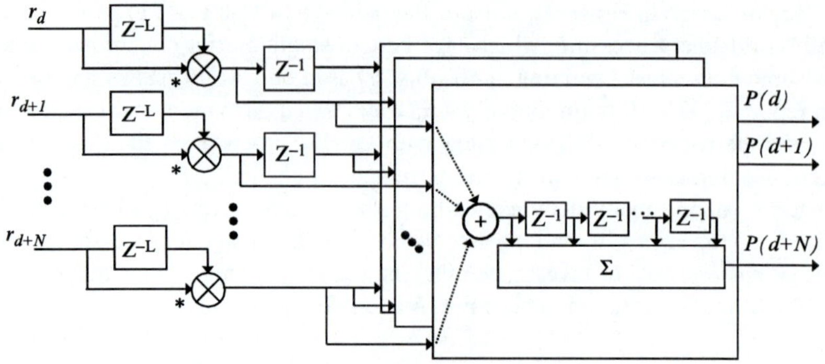

Fig. 2.18 DSP block diagram of autocorrelation for symbol synchronization based on parallel processing

This is because the moving window for autocorrelation needs to be taken sample by sample while multiple samples need to be processed simultaneously at a parallel process clock cycle. As there was no direct information available to indicate the frame starting point in the 16 parallel channels in our setup, locating the exact frame beginning would involve heavy computation that processes the data among all the channels. To illustrate this point, an implementation of the parallel autocorrelation can be constructed such that we can divide the autocorrelation of (2.36) by length N for the N parallel processing:

$$P(d) = \sum_{k=0}^{(L/N)} \sum_{m=Nk}^{N(k+1)-1} \left(r^*_{d+m} r_{d+m+L}\right), \tag{2.38}$$

which does not have an apparent recursive equation. The DSP realization is presented in Fig. 2.18. As shown in (2.38) and Fig. 2.18, by restricting the synchronization pattern length L to multiple of the number of de-multiplexed bits N, a simple implementation of autocorrelation suitable for parallel processing is realized. However, for the case of $N = 16$ and $L = 32$, the processing resource required in this parallel implementation is estimated as 16 complex multipliers and $16 \times 15 + 16 = 256$ complex adders at each clock cycle. This indicates further efficiency improvement of symbol synchronization in parallel processing is desired.

2.4.4.2 Real-Time Frequency Offset Synchronization

Frequency offset between signal laser and local lasers must be estimated and compensated before further processing. The algorithm used in this stage is the same as (2.29). In the experiment, the local laser frequency is placed within ±2 subcarrier spacings from the signal laser, which guarantees that the phase difference $\hat{\phi}$ between these two synchronization patterns remains bounded within $\pm\pi$. It can be shown that the error of multiple of the subcarrier spacing has no significance. The frequency offset can be derived as:

$$f_{\text{offset}} = \hat{\phi}/(\pi T/2). \tag{2.39}$$

The COordinate-Rotation-DIgital-Computer (CORDIC) algorithm is used to calculate the frequency offset angle and compensate input data in vectoring and rotation modes, respectively. Figure 2.19 shows the frequency offset angle output against the sampling points with the frequency offset normalized to $2/(\pi T)$. Once the timing estimate signal from window synchronization stage is detected, the current output value of (2.39) is the correct frequency offset.

Once the frequency offset is obtained, frequency-offset compensation will be started. The implementation of frequency offset compensation in real-time is to use the cumulative phase information. The DSP diagram for frequency compensation is shown in Fig. 2.20. Assuming that $\Delta\Phi$ is the phase difference between adjacent samples, which is derived from the auto-correlation, within one FPGA sampling period, N samples are distributed among the multiplexed channels. For the ith channel, the phase is cumulated as $i \times \Delta\Phi$, and then compensated for that channel.

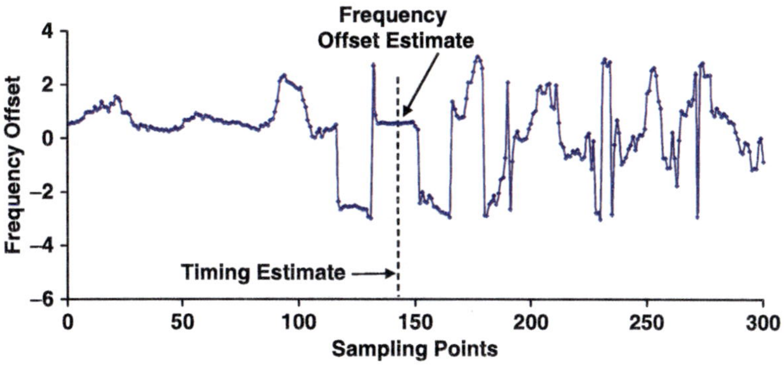

Fig. 2.19 Real-time measurement of frequency offset estimation for the OFDM signal. The frequency offset is normalized to $2/(\pi T)$

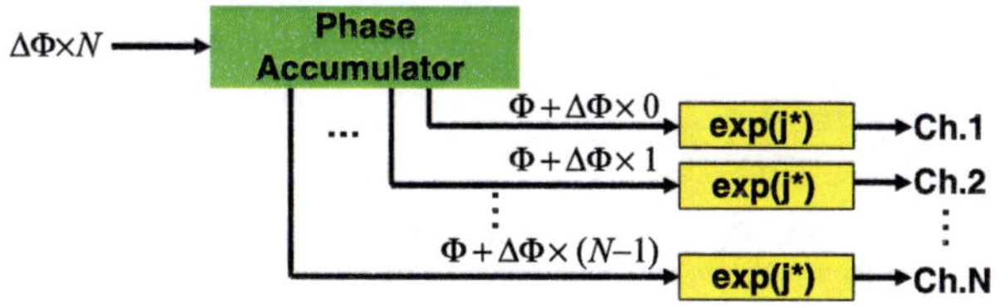

Fig. 2.20 DSP diagram for frequency offset compensation

2.4.4.3 Real-Time Channel Estimation

Figure 2.21 shows the diagram for real-time CO-OFDM channel estimation. Once the OFDM window is synchronized, an internal timer will be started, which is used to distinguish the pilot symbols and payload. Two steps are involved in this procedure, channel matrix estimation and compensation. In the time slot for pilot symbols, the received signal is multiplied with locally stored transmitted pilot symbols to estimate the channel response. The transmitted pattern typically has very simple numerical orientation. Thus, multiplication can be changed into addition/subtraction of real and imaginary parts of the complex received signal, which can give additional resource saving. Taking average of the estimated channel matrixes over time and frequency can be used to alleviate error due to the random noise. Then the averaged channel estimation will be multiplied to the rest of the received payload symbols to compensate for the channel response. It is worth pointing out that one complex multiplier can be composed of only three (instead of four) real number multipliers.

To further save the hardware resources, the realization of the channel estimation can be done in a simple lookup table when pilot subcarriers are modulated with QPSK as in Table 2.3, avoiding the use of costly multipliers.

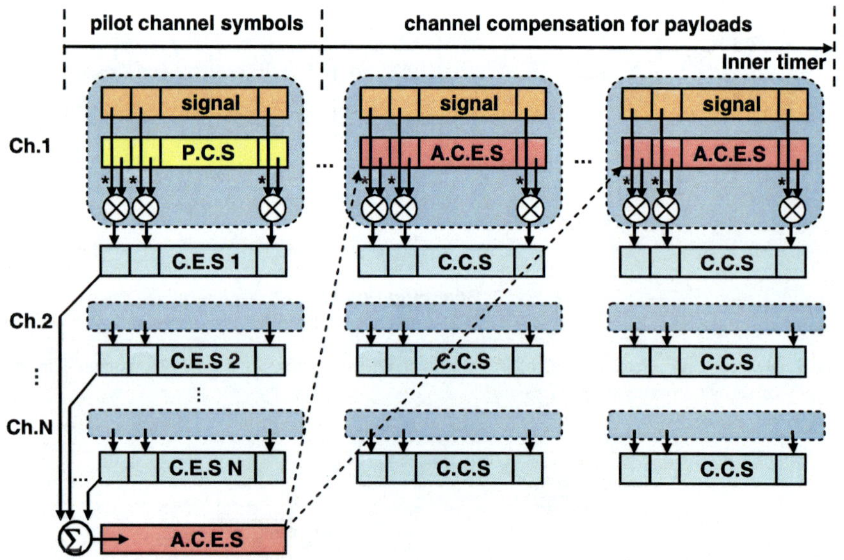

Fig. 2.21 Channel estimation diagram. *P.C.S* Pilot channel symbol; *C.E.S* Channel estimated symbol; *A.C.E.S* Averaged channel estimated symbol; *C.C.S* Compensated channel symbol

Table 2.3 Lookup table for channel and phase estimate in case of QPSK pilot subcarrier. Received signal is $R = a + jb$

Message symbols of pilot	Modulated symbols of pilot	H^{-1} or B^{-1} Real	Imaginary
0	$-1 + j$	$-a - b$	$a - b$
1	$-1 - j$	$-a + b$	$-a - b$
2	$1 + j$	$a - b$	$a + b$
3	$1 - j$	$a + b$	$-a + b$

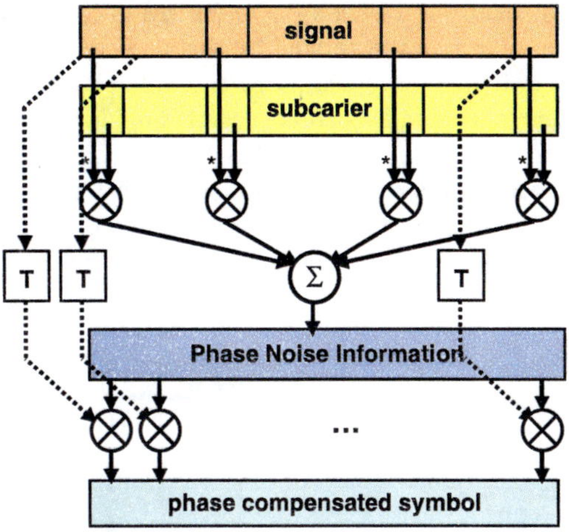

Fig. 2.22 Phase estimation diagram

2.4.4.4 Real-Time Phase Estimation

Similar to channel estimation, phase estimation procedure can also be divided into estimation and compensation parts, which is shown in Fig. 2.22. Pilot subcarriers within one symbol will be selected by the inner timer. These pilot subcarriers then are compared with local stored transmitted pattern to obtain the phase noise information. The same symbol is delayed, and then compensated with the estimated phase noise factor.

2.4.5 Experimental Demonstrations for CO-OFDM, from 100 Gb s^{-1} to 1 Tb s^{-1}, from Offline to Real-Time

Before 2008, the maximum line rate of CO-OFDM was limited to 52.5 Gb s^{-1}, insufficient to meet the requirement of 100 Gb s^{-1} Ethernet. The main limitation is the electrical RF bandwidth of off-shelf DAC/ADC components. To implement 107 Gb s^{-1} optical coherent OFDM based on QPSK, the required electrical bandwidth is about 15 GHz. The best commercial DACs/ADCs in silicon IC at that time had a bandwidth of only 6 GHz [77], so the realization of 100 Gb s^{-1} CO-OFDM in a cost-effective manner remained challenging. To overcome this electrical bandwidth bottleneck associated with DAC/ADC devices, we used the orthogonal band multiplexing to demonstrate 107 Gb s^{-1} transmission over 1,000 km [19].

At the transmitter side, the 107 Gb s^{-1} OBM-OFDM signal is generated by multiplexing 5 OFDM subbands. In each band, 21.4 Gb s^{-1} OFDM signals are transmitted in both polarizations. The multi-frequency optical source with tones spaced at 6406.25 MHz is generated by cascading two intensity modulators (IMs). The guard-band equals to just one subcarrier spacing ($m = 1$). The experimental setup for 107 Gb s^{-1} CO-OFDM is shown in Fig. 2.23. Figure 2.24 shows the multiple tones generated by this cascaded architecture using two IMs. Only the middle five tones with large and even power are used for performance evaluation. The transmitted signal is generated off-line by MATLAB program with a length of $2^{15} - 1$ PRBS and mapped to 4-QAM constellation. The digital time domain signal is formed after IFFT operation. The FFT size of OFDM is 128, and guard interval is 1/8 of the symbol window. The middle 82 subcarriers out of 128 are filled, from which four pilot subcarriers are used for phase estimation. The I and Q components

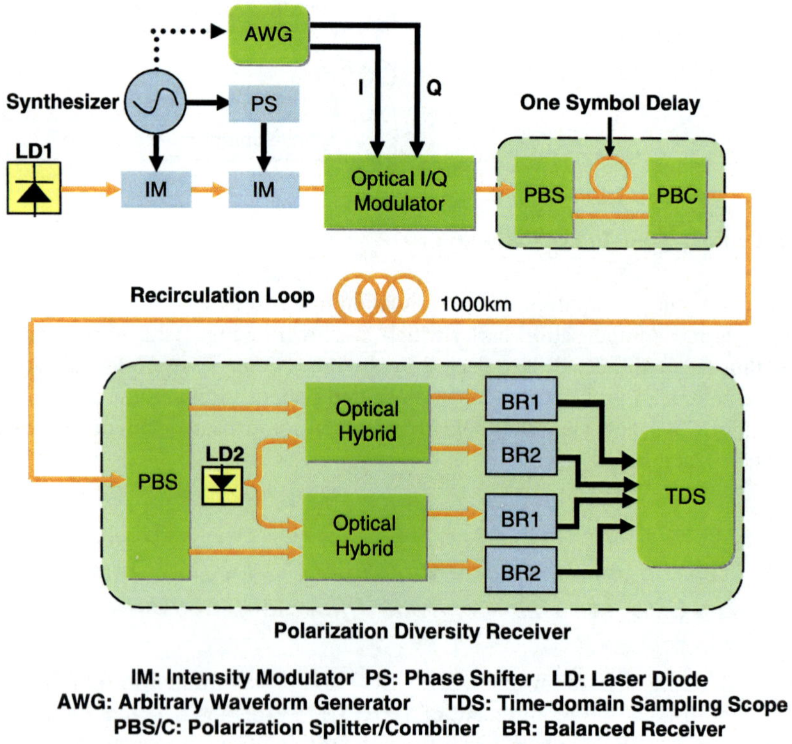

Fig. 2.23 Experimental setup for 107 Gb s^{-1} OBM-OFDM systems

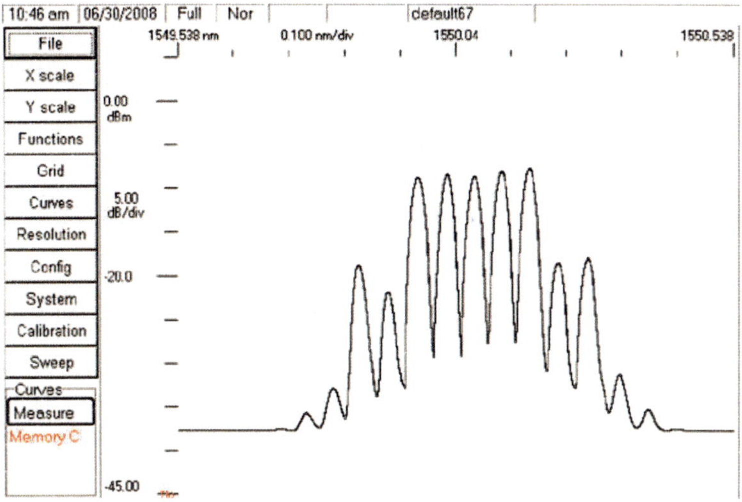

Fig. 2.24 Multiple tones generated by two cascaded intensity modulators [78]

of the time domain signal is uploaded onto a Tektronix Arbitrary Waveform Generator (AWG), which provides the analog signals at 10 GS s^{-1} for both I and Q parts. The AWG is phase locked to the synthesizer through 10 MHz reference. The optical I/Q modulator comprising two MZMs with 90° phase shift is used to directly impress the baseband OFDM signal onto five optical tones. The modulator is biased at null point to suppress the optical carrier completely and perform linear baseband-to-optical up-conversion [79]. The optical output of the I/Q modulator consists of five-band OBM-OFDM signals. Each band is filled with the same data at 10.7 Gb s^{-1} data rate and is consequently called "uniform filling" in this paper. To improve the spectrum efficiency, 2×2 MIMO-OFDM is employed, with the two OFDM transmitters being emulated by splitting the transmitted signal and recombining on orthogonal polarizations with a one OFDM symbol delay. These are then detected by two OFDM receivers, one for each polarization.

At the receiver side, the signal is coupled out of the recirculation loop and received with a polarization diversity coherent optical receiver [64, 80] comprising a polarization beam splitter, a local laser, two optical 90° hybrids, and four balanced photoreceivers. The complete OFDM spectrum comprises 5 subbands. The entire bandwidth for 107 Gb s^{-1} OFDM signal is only 32 GHz. The local laser is tuned to the center of each band, and the RF signals from the four balanced detectors are first passed through the anti-aliasing low-pass filters with a bandwidth of 3.8 GHz, such that only a small portion of the frequency components from other bands is passed through, which can be easily removed during OFDM signal processing. The performance of each band is measured independently. The detected RF signals are then sampled with a Tektronix Time Domain-sampling Scope (TDS) at 20 GS s^{-1}. The sampled data is processed with a MATLAB program to perform 2×2 MIMO-OFDM processing.

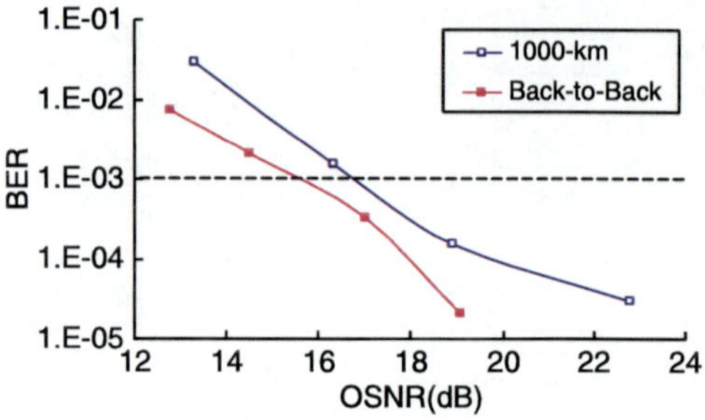

Fig. 2.25 BER sensitivity of 107 Gb s^{-1} CO-OFDM signal at the back-to-back and 1,000-km transmission

Figure 2.25 shows the BER sensitivity performance for the entire 107 Gb s^{-1} CO-OFDM signal at the back-to-back and 1,000-km transmission with the launch power of −1 dBm. The BER is counted across all five bands and two polarizations. It can be seen that the OSNR required for a BER of 10^{-3} is, respectively, 15.8 dB and 16.8 dB for back-to-back and 1,000-km transmission.

As 100-Gb s^{-1} Ethernet has almost become a commercial reality, 1-Tb s^{-1} transmission starts to receive growing attention. Some industry experts believe that the Tb/s Ethernet standard should be available in the time frame as early as 2012–2013 [74]. In the Tb/s experimental demonstrations [4, 5], we show that by using multiband structure of the proposed 1-Tb s^{-1} signal, parallel coherent receivers each working at 30-Gb s^{-1} can be used to detect 1-Tb s^{-1} signal, namely, we have an option of receiver design in 30-Gb s^{-1} granularity, a small fraction of the entire bandwidth of the wavelength channel. However, extension from current 100-Gb s^{-1} demonstration to 1-Tb s^{-1} requires tenfold bandwidth expansion, which is a significant challenge. To optically construct the multiband CO-OFDM signal using cascaded optical modulators, it entails ten times higher drive voltage, or use of the nonlinear fiber which may introduce unacceptable noise to the Tb/s signal. We here adopt a novel approach of multi-tone generation using a recirculating frequency shifter (RFS) architecture that generates 36 tones spaced at 8.9 GHz with only a single optical IQ modulator without a need for excessive high drive voltage. In this work, we extend the report of the first 1-Tb s^{-1} CO-OFDM transmission with a record reach of 600 km over SSMF fiber and a spectral efficiency of 3.3 bit s^{-1} Hz^{-1} without either Raman amplification or optical compensation [81]. Our demonstration signifies that the CO-OFDM may potentially become an attractive candidate for future 1-Tb s^{-1} Ethernet transport even with the installed fiber base.

Figure 2.26a shows the architecture of the RFS consisting of a closed fiber loop, an IQ modulator, and two optical amplifiers to compensate the frequency conversion loss. The IQ modulator is driven with two equal but 90° phase shifted RF tones through I and Q ports, to induce a frequency shifting to the input optical signal [82]. As shown in Fig. 2.26b, in the first round, an OFDM band at the center frequency of f_1 (called f1 band) is generated when the original OFDM band at the center frequency of f_0 passes through the optical IQ modulator and incurs a frequency shift equal to the drive voltage frequency of f. The f_1 band is split into two branches, one coupled out and the other recirculating back to the input of the optical IQ modulator.

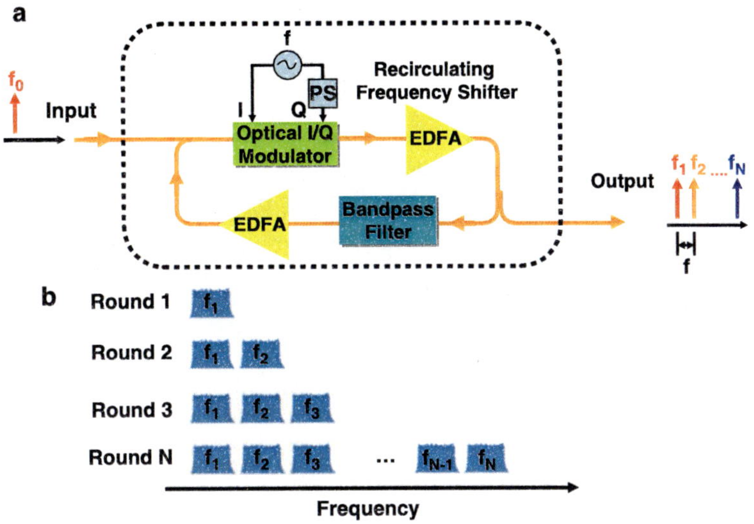

Fig. 2.26 (**a**) Schematic of the recirculating frequency shifter (RFS) as a multi-tone generator, and (**b**) illustration of replication of the OFDM bands using an RFS. Each OFDM band is synchronized but yet uncorrelated due to the delay of multiple of the OFDM symbol period. *PS* Phase shifter

In the second round, f_2 band is generated by shifting f_1 band along with a new f_1 band, which is shifted from original f_0 band. Similarly, in the Nth round, we will have f_N band shifted from the previous f_{N-1} band, and f_{N-1} shifted from previous f_{N-2}, etc. The f_{N+1} band and beyond will be filtered out by the bandpass filter placed in the loop. With this scheme, the OFDM bands f_1 to f_N are coming from different rounds and hence contain uncorrelated data pattern. In addition, such bandwidth expansion does not require excessive drive voltage for the optical modulator. Another major benefit of using the RFS is that we can adjust the delay of the recirculating loop to an integer number (30 in this experiment) of the OFDM symbol periods, and therefore the neighboring bands not only reside at the correct frequency grids, but are also synchronized in OFDM frame at the transmit. Replicating uncorrelated multiple OFDM bands using RFS is thus an extremely useful technique as it does not require duplication of the expensive test equipments including AWG and optical IQ modulators, etc. The RFS has been proposed and demonstrated for a tunable delay, but with only one tone being selected and used [82]. We here extend the application of RFS for multi-tone generation, or more precisely, for bandwidth expansion of uncorrelated multi-band OFDM signal.

Figure 2.27 shows the experimental setup for the 1-Tb s^{-1} CO-OFDM systems. The optical sources for both transmitter and local oscillators are commercially available external-cavity lasers (ECLs), which have linewidth of about 100 kHz. The first OFDM band signal is generated by using a Tektronix AWG. The time domain OFDM waveform is generated with a MATLAB program with the parameters as follows: 128 total subcarriers; guard interval 1/8 of the observation period; middle

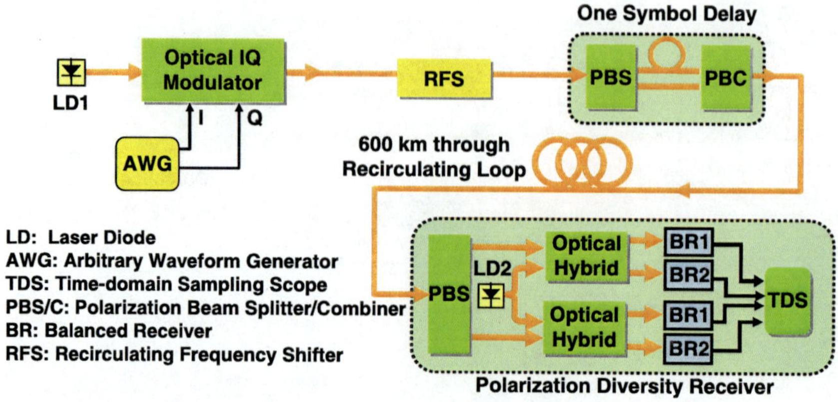

Fig. 2.27 Experimental setup for 1 Tb s^{-1} CO-OFDM transmission

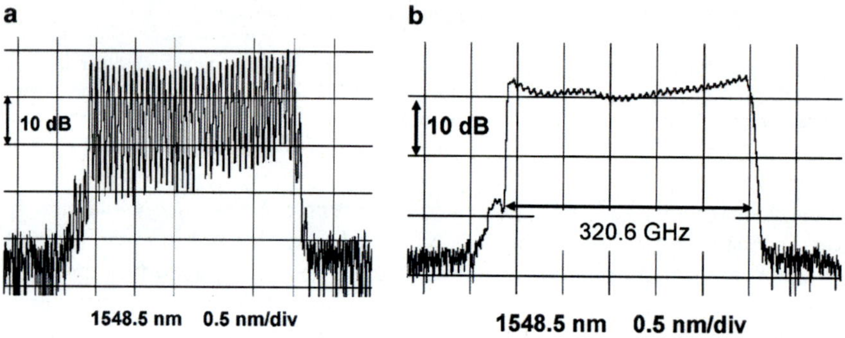

Fig. 2.28 (**a**) Multi-tone generation when the optical IQ modulator is bypassed, and (**b**) the 1.08 Tb s^{-1} CO-OFDM spectrum comprising continuous 4,104 spectrally overlapped subcarriers

114 subcarriers filled out of 128, from which four pilot subcarriers are used for phase estimation. The real and imaginary parts of the OFDM waveforms are uploaded into the AWG operated at 10 GS s^{-1} to generate IQ analog signals, and subsequently fed into I and Q ports of an optical IQ modulator, respectively. The net data rate is 15 Gb s^{-1} after excluding the overhead of cyclic prefix, pilot tones, and unused middle two subcarriers. The optical output from the optical IQ modulator is fed into the RFS, replicated 36 times in a fashion described in Fig. 2.26b, and is subsequently expanded to a 36-band CO-OFDM signal with a data rate of 540 Gb s^{-1}. The optical OFDM signal from the RFS is then inserted into a polarization beam splitter, with one branch delayed by one OFDM symbol period (14.4 ns), and then recombined with a polarization beam combiner to emulate the polarization multiplexing, resulting in a net date rate of 1.08 Tb s^{-1}.

Figure 2.28a shows the multitone generation if the optical IQ modulation in Fig. 2.27 is bypassed. It shows a successful 36-tone generation with a tone-to-noise ratio (TNR) of larger than 20 dB at a resolution bandwidth of 0.02 nm. Figure 2.28b shows the optical spectrum of 1.08 Tb s^{-1} CO-OFDM signal spanning 320.6 GHz in bandwidth consisting of 4,104 continuous spectrally overlapped subcarriers, implying a spectral efficiency of 3.3 bit s^{-1} Hz^{-1}.

Figure 2.29 shows the BER sensitivity performance for the entire 1.08 Tb s^{-1} CO-OFDM signal at the back to back. The OSNR required for a BER of 10^{-3} is 27.0 dB, which is about 11.3 dB higher than 107 Gb s^{-1} we measured in [5]. The inset shows the typical constellation diagram for the detected CO-OFDM signal. The additional 1.3 dB OSNR penalty is attributed to the degraded TNR at the right-edge of the CO-OFDM signal spectrum (see Fig. 2.28a). Figure 2.30 shows the BER performance for all the 36 bands at the reach of 600 km with a launch power of 7.5 dBm, and it can be seen that all the bands can achieve a BER better than 2×10^{-3}, the FEC threshold with 7% overhead. The inset shows the 1-Tb s^{-1} optical signal spectrum at 600-km transmission. It is noted that the reach performance for this first 1-Tb s^{-1} CO-OFDM transmission is limited by two factors: (1) the noise accumulation for

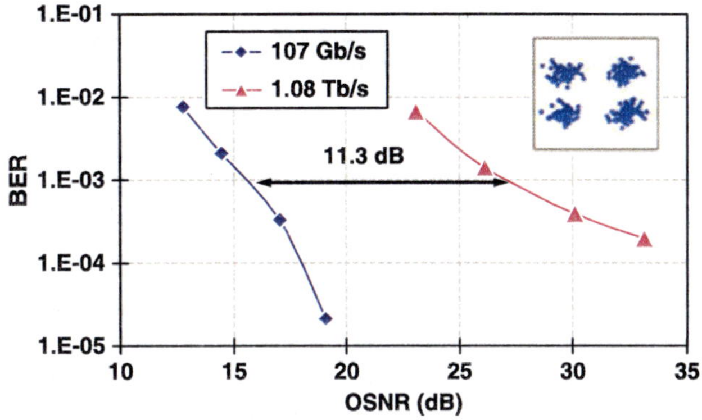

Fig. 2.29 Back-to-back OSNR sensitivity for 1 Tb s^{-1} CO-OFDM signal

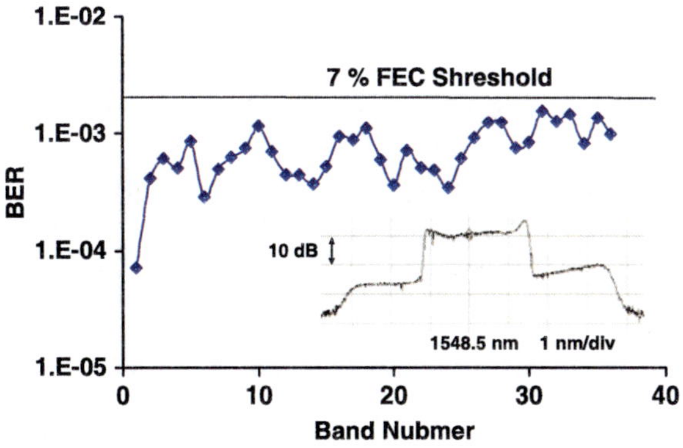

Fig. 2.30 BER performance for individual OFDM subbands at 600 km. The inset shows the optical spectrum of 1-Tb s^{-1} CO-OFDM signal after 600 km transmission

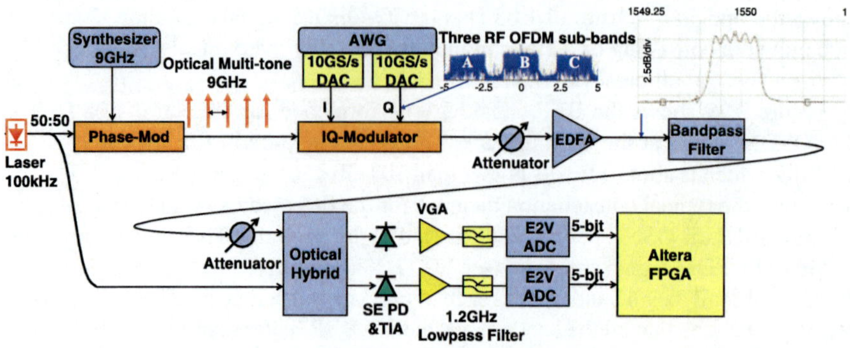

Fig. 2.31 Real-time CO-OFDM transmission experimental setup (*left*) and the DSP programming diagram of the real-time receiver (*right*). Insets: sample generated three OFDM band signal spectrums

the edge subcarriers that have gone through most of the frequency shifting, and (2) the two-stage amplifier exhibits over 9 dB noise figure because of the difficulty of tilt control in the recirculation loop. Both of the two issues can be overcome, and 1,000 km and beyond transmission at 1-Tb s^{-1} is practically reachable.

Another important development is the real-time CO-OFDM transmission. In 2009, 3.6 Gb s^{-1} per band CO-OFDM real-time OFDM reception was demonstrated by using a 54 Gb s^{-1} multi-band CO-OFDM signal [26]. Figure 2.31 shows the experimental setup and the DSP programming diagram of the real-time CO-OFDM receiver. At the transmitter, a data stream consisting of pseudo-random bit sequences (PRBSs) of length $2^{15}-1$ was first mapped onto three OFDM subbands with QPSK modulation. Three OFDM subbands were generated by an AWG at 10 GS s^{-1}. Each subband contained 115 subcarriers modulated with QPSK. Two unfilled gap bands with 62 subcarrier-spacings were placed between the three subbands, which allowed them to be evenly distributed across the AWG output bandwidth. In each OFDM subband, the filled subcarriers, together with eight pilot subcarriers and 13 adjacent unfilled subcarriers, were converted to the time domain via inverse Fourier transform (IFFT) with size of 128. The number of filled subcarriers was restricted by the 1.2 GHz RF low-pass filter, which was used to select the subband to be received. A cyclic prefix of length 16 sample point was used, resulting in an OFDM symbol size of 144. The total number of OFDM symbols in each frame was 512. The first 16 symbols were used as training symbols for channel estimation. The real and imaginary parts of the OFDM symbol sequence were converted to analog waveforms via the AWG, before being amplified and used to drive an optical I/Q modulator that was biased at null. The transmitter laser and the receiver local laser were originated from the same ECL with 100-kHz linewidth through a 3-dB coupler. By doing so, frequency offset estimation was not needed in this experiment. The maximum net data rate of the signal after the optical modulation was 3.6 Gb s^{-1} for each OFDM subband. The multifrequency optical source contained 5 optical carriers at 9-GHz spacing, and was generated by using an MZM-driven by a high-power RF sinusoidal

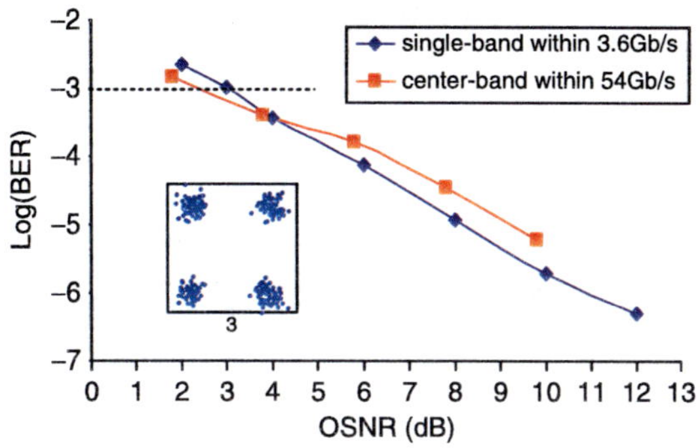

Fig. 2.32 Measured BER vs. OSNR for a single 3.6-Gb s^{-1} signal and for the center subband of the 54-Gb s^{-1} multi-band signal

wave at 9 GHz. The total number of subbands was then 15, resulting in a total net data rate of 54 Gb s^{-1}. Unlike earlier works [19], the adjacent subbands in the multi-band OFDM signal contained independent data contents, more closely emulating an actual system. At the receiver, the OFDM signal in each sub-subband was detected by a digital coherent receiver consisting of an optical hybrid and two single-ended input photodiode with a transimpedance amplifier (PIN-TIA). Two variable gain amplifiers (VGAs) amplified the signals to the optimum input amplitude before the ADCs, which were sampling at a rate of 2.5 GS s^{-1}. The five most significant bits of each ADC were fed into an Altera Stratix II GX FPGA. All the CO-OFDM DSP was performed in the FPGA. The bit error rate was measured from the defined inner registers through embedded logic analyzer SignalTap II ports in Altera FPGA.

Figure 2.32 shows the measured BER as a function of optical signal-to-noise ratio (OSNR) for two cases: (1) a single 3.6-Gb s^{-1} CO-OFDM signal; (2) the center subband of the 54-Gb s^{-1} multi-band signal. In case (1), a BER better than 1×10^{-3} can be observed at OSNR of 3 dB. The OSNR is defined as the signal power in the subband under measurement over the noise power in a 0.1-nm bandwidth. In case (2), the required OSNR for BER 1×10^{-3} is 2.5 dB. There is virtually no penalty introduced by the band-multiplexing.

2.5 Promising Research Direction and Future Expectations

In this section, we consider some of the possible future research topics and trends of optical OFDM.

1. Optical OFDM for 1 Tb s^{-1} Ethernet transport.
   As the 100 Gb s^{-1} Ethernet has increasingly become a commercial reality, the next pressing issue would be a migration path toward 1 Tb s^{-1} Ethernet transport to cope with ever-growing Internet traffic. In fact, some industry experts forecast that standardization of 1 TbE should be available in the time frame of 2012–2013 [74]. CO-OFDM may offer a promising alternative pathway toward Tb/s transport that possesses high spectral efficiency, resilience to

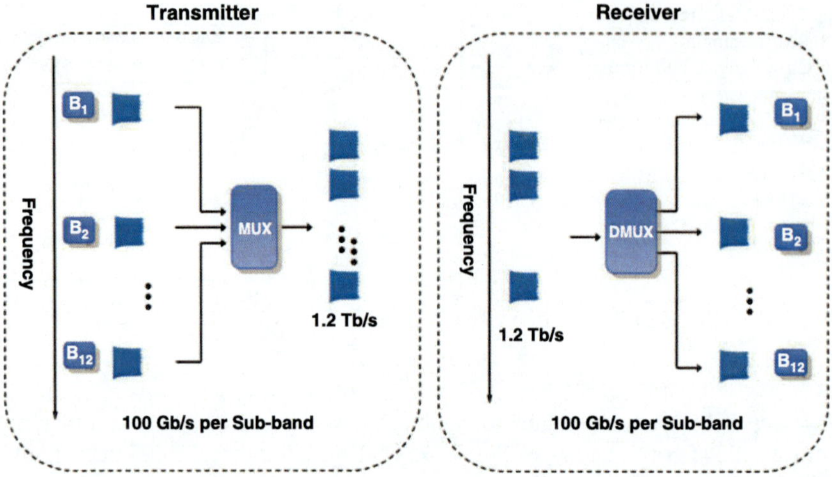

Fig. 2.33 Conceptual diagram of multiplexing and demultiplexing architecture for 1 Tb s^{-1} coherent optical orthogonal frequency-division multiplexing (CO-OFDM) systems. In particular, 1.2 Tb s^{-1} CO-OFDM signal comprising 12 bands (B1–12) is shown as an example

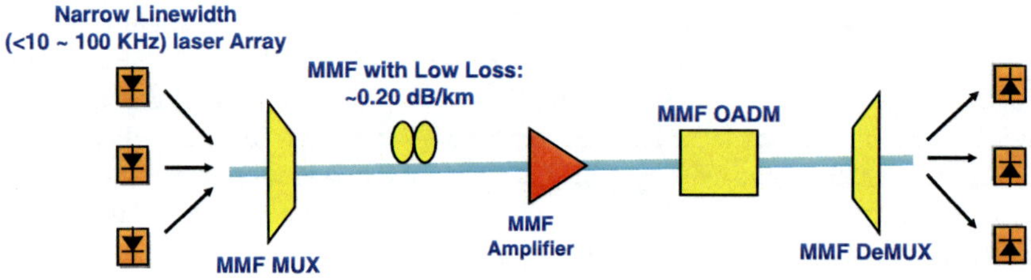

Fig. 2.34 Conceptual diagram of MMF-based long-haul communication systems

tributary timing misalignment and most important of all, the chromatic dispersion and PMD. Figure 2.33 shows the multiplexing and de-multiplexing architecture of CO-OFDM, where 1.2 Tb s^{-1} is divided into 12 frequency-domain tributaries at 100 Gb s^{-1} each. Using OBM scheme, OFDM can realize the high capacity without sacrificing spectral efficiency or increasing computational complexity [31].

2. MMF fiber for high spectral efficiency long-haul transmission.
   MMF has long been perceived as a medium that is limited to short reach systems, although it can achieve very high capacity [83, 84]. Recent experiment of 20 Gb s^{-1} CO-OFDM transmission over 200 km MMF fiber may change that stereotype and spur research interests in MMF-based long-haul transmission [85]. The ideal MMF fiber for long-haul transmission may be the few-mode MMF fiber, for instance, dual-mode fiber, where space diversity can be utilized for MIMO gain. Figure 2.34 shows the conceptual diagram of an MMF-based

long-haul system. However, the MMF-based long-haul systems not only entail massive signal processing due to higher order MIMO reception, but it also requires many critical devices that are not employed in the conventional optical communications, such as the mode multiplexer and MMF amplifier.

3. Opto-electronic integrated circuits (OEICs) for optical OFDM
The notion of the OEICs that promises to place large number of the optical and electronic devices onto a single-chip can be traced back about four decades [86]. Because of the extensive signal processing involved in optical OFDM, it is natural to expect the silicon technology as the platform of choice to integrate the electronic DSPs and photonic components onto a single chip. Figure 2.35 shows a CO-OFDM transceiver architecture that includes four functional blocks including baseband OFDM transmitter, RF-to-optical (RTO) up-converter, optical-to-RF (OTR) down-converter, and baseband OFDM Receiver. We believe that the future advances in silicon OEIC will open up new venues for the coherent optical transmission technology and these will make inroads into the broad range of optical communication applications from access to core optical networks. We anticipate that the realization of optical OFDM subsystems/systems based on compact, power-efficient OEIC will present huge challenges and rich opportunities due to potential cost saving by law of scaling.

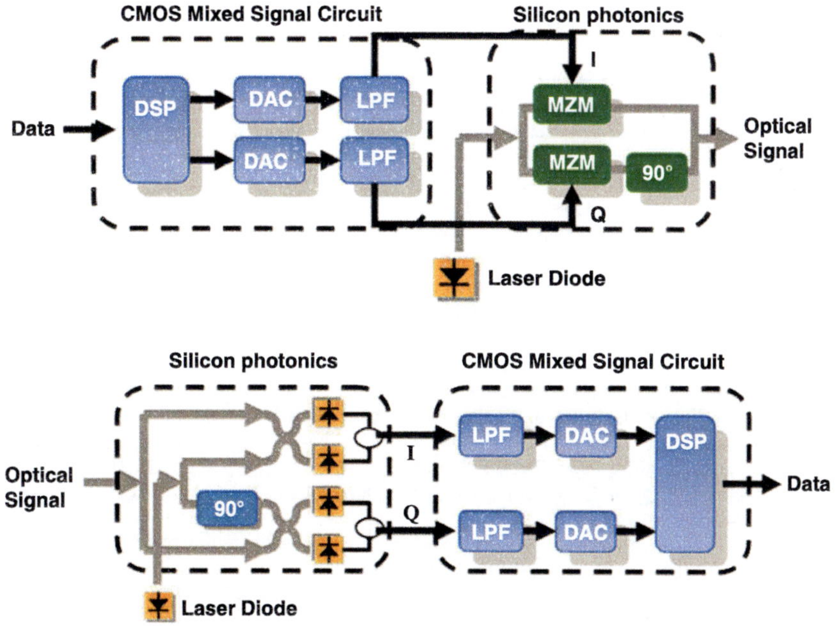

Fig. 2.35 Functional blocks of a CO-OFDM transceiver and its corresponding mapping to an integrated silicon chip

There are some other promising directions, such as adaptive coding in optical OFDM, optical OFDM-based access networks, standardizations, etc. Interested readers shall refer to [55] for more detailed discussion.

2.6 Conclusion

Optical OFDM transmission has become a fast progressing and vibrant research field in optical fiber communications. Last few years saw experimental demonstrations up to 1 Tb s^{-1} transmissions, together with rapid advance in real-time demonstrations. With the standardization of 100 GbE and prospect of emergence of the Tb/s era, much excitement is growing in the optical communications community for the application of OFDM, the modulation format of choice in RF wireless communications. The introduction of OFDM without doubt has great potential and promise in bringing about the next-generation optical networks that possess high degree of flexibility and scalability. In the meantime, the research in optical OFDM also presents tremendous challenges and opportunities in the areas of novel DSP algorithms, high-speed electronic and photonic integrated circuits.

References

1. W. Shieh, C. Athaudage, Electron. Lett. **42**, 587–589 (2006)
2. A.J. Lowery, L. Du, J. Armstrong, Orthogonal frequency division multiplexing for adaptive dispersion compensation in long haul WDM systems, Optical fiber communication conference, paper PDP, Anaheim, CA, p. 39, 2006
3. I.B. Djordjevic, B. Vasic, Opt. Express **14**, 3767–3775 (2006)
4. Y. Ma, Q. Yang, Y. Tang, S. Chen, W. Shieh, Opt. Express **17**, 9421–9427 (2009)
5. R. Dischler, F. Buchali, Transmission of 1.2 Tb/s Continuous Waveband PDM - OFDM - FDM Signal with Spectral Efficiency of 3.3 bit/s/Hz over 400 km of SSMF, Optical fiber communication conference, paper PDP C2, San Diego, USA, 2009
6. D. Hillerkuss, T. Schellinger, R. Schmogrow, M. Winter, T. Vallaitis, R. Bonk, A. Marculescu, J. Li, M. Dreschmann, J. Meyer, S. Ben Ezra, N. Narkiss, B. Nebendahl, F. Parmigiani, P. Petropoulos, B. Resan, K. Weingarten, T. Ellermeyer, J. Lutz, M. Möller, M. Hübner, J. Becker, C. Koos, W. Freude, J. Leuthold, Single source optical OFDM transmitter and optical FFT receiver demonstrated at line rates of 5.4 and 10.8 Tbit/s, Optical fiber communication conference (OFC'10), Postdeadline Paper PDPC1, San Diego, CA, USA, 21–25 March 2010
7. R.P. Giddings, X.Q. Jin, E. Hugues-Salas, E. Giacoumidis, J.L. Wei, J.M. Tang, Opt. Express **18**, 5541–5555 (2010)
8. US patent, 174,465, Improvement in Telegraphy, Alexander Graham Bell, 7 March 1876
9. R.W. Chang, Bell Sys. Tech. J. **45**, 1775–1796 (1966)
10. S.B. Weinstein, P.M. Ebert, IEEE Trans. Commun. **19**(5), 628–634 (1971)
11. A. Peled, A. Ruiz, Frequency domain data transmission using reduced computational complexity algorithms, in *Proceedings of the IEEE International Conference on Acoustics, Speech, and Signal Processing*, vol. 5, pp. 964–967, April 1980
12. L.J. Cimini, IEEE Trans. Commun. **COM-33**, 665–675 (1985)
13. Q. Pan, R.J. Green, IEEE Photon. Technol. Lett. **8**, 278–280 (1996)

14. B.J. Dixon, R.D. Pollard, S. Iezekiel, IEEE Trans. Microwave Theory Tech. **49**, 1404–1409 (2001)
15. W. Shieh, X. Yi, Y. Tang, IEEE Electron. Lett. **43**, 183–185 (2007)
16. R. You, J.M. Kahn, IEEE Trans. Commun. **49**, 2164–2171 (2001)
17. N.E. Jolley, H. Kee, R. Rickard, J. Tang, Generation and propagation of a 1550 nm 10 Gbit/s optical orthogonal frequency division multiplexed signal over 1000 m of multimode fibre using a directly modulated DFB, OFC, Paper OFP3 Proceedings, Anaheim, CA, 2005
18. A.J. Lowery, J. Armstrong, Opt. Express **14**(6), 2079–2084 (2006)
19. Q. Yang, Y. Ma, W. Shieh, 107 Gb/s coherent optical OFDM reception using orthogonal band multiplexing, in *Proceedings of the Optical Fiber Communication Conference*, PDP 7, 2008
20. S.L. Jansen et al., 10 × 121.9-Gb/s PDM-OFDM transmission with 2b/s/Hz spectral efficiency over 1,000km of SSMF, in Proceedings of OFC, paper PDP2, San Diego, USA, 2008
21. E. Yamada, A. Sano, H. Masuda, E. Yamazaki, T. Kobayashi, E. Yoshida, K. Yonenaga, Y. Miyamoto, K. Ishihara, Y. Takatori, T. Yamada, H. Yamazaki, 1Tb/s (111Gb/s/ch × 10ch) no-guard-interval COOFDM transmission over 2100 km DSF, OECC/ACOFT conference, paper PDP6, 2008)
22. S. Chandrasekhar et al., Transmission of a1.2Tb/s 24-carrier No-guard-interval Coherent OFDM Superchannel over 7200km of Ultra-large-area Fiber, ECOC'09, paper no. PD 2.6, 2009
23. S. Chen, Q. Yang, Y. Ma, W. Shieh, Multi-gigabit real-time coherent optical OFDM receiver, OFC'2009, Paper OTuO4, 2009
24. T. Pfau, S. Hoffmann, R. Peveling, S. Bhandare, S. Ibrahim, O. Adamczyk, M. Porrmann, R. Noé, Y. Achiam, IEEE Photon. Technol. Lett. **18**(18), 1907–1909 (2006)
25. A. Leven, N. Kaneda, A. Klein, U.V. Kco, Y.K. Chen, Electron. Lett. **42**(24), 1421–1422 (2006)
26. Q. Yang, N. Kaneda, X. Liu, S. Chandrasekhar, W. Shieh, Y.K. Chen, Real-time coherent optical OFDM receiver at 2.5 GS/s for receiving a 54 Gb/s multi-band signal, OFC 2009 Paper PDPC5, 2009
27. S. Chen, Y. Ma, W. Shieh, 110-Gb/s multi-band real-time coherent optical OFDM reception after 600-km transmission over SSMF fiber, in Optical fiber communication conference, OSA Technical Digest (CD) (Optical Society of America, 2010), paper OMS2, 2010
28. D. Qian, T.T. Kwok, N. Cvijetic, J. Hu, T. Wang, 41.25 Gb/s real-time OFDM receiver for variable rate WDM-OFDMA-PON transmission, in Optical fiber communication conference, OSA Technical Digest (CD) (Optical Society of America, 2010), paper PDPD9, 2010
29. R.R. Mosier, R.G. Clabaugh, AIEE Trans. **76**, 723–728 (1958)
30. M.S. Zimmerman, A.L. Kirsch, AIEE Trans. **79**, 248–255 (1960)
31. W. Shieh, Q. Yang, Y. Ma, Opt. Express **16**, 6378–6386 (2008)
32. W. Shieh, H. Bao, Y. Tang, Opt. Express **16**, 841–859 (2008)
33. X. Liu, S. Chandrasekhar, B. Zhu, D.W. Peckham, Efficient digital coherent detection of A 1.2-Tb/s 24-carrier no-guard-interval CO-OFDM signal by simultaneously detecting multiple carriers per sampling, in Optical fiber communication conference, paper OMO2, 2010
34. Q. Yang, W. Shieh, Y. Ma, Opt. Lett. **33**, 2239–2241 (2008)
35. J. Armstrong, IEEE Trans. Commun. **47**(3), 365–369 (1999)
36. P. Duhamel, H. Hollmann, IET Elect. Lett. **20**, 14–16 (1984)
37. S. Hara, R. Prasad, *Multicarrier Techniques for 4G Mobile Communications* (Artech House, Boston, 2003)
38. L. Hanzo, M. Munster, B.J. Choi, T. Keller, *OFDM and MC-CDMA for Broadband Multi-User Communications, WLANs and Broadcasting* (Wiley, New York, 2003)
39. X. Yi, W. Shieh, Y. Ma, Phase noise on coherent optical OFDM systems with 16-QAM and 64-QAM beyond 10 Gb/s, European conference on optical communication, paper 5.2.3, Berlin, Germany, 2007
40. H. Takahashi, A. Al Amin, S.L. Jansen, I. Morita, H. Tanaka, 8 × 66.8-Gbit/s Coherent PDM-OFDM Transmission over 640 km of SSMF at 5.6-bit/s/Hz Spectral Efficiency, European conference on optical communication, paper Th.3.E.4, Brussels, Belgium 2008
41. A.J. Lowery, S. Wang, M. Premaratne, Opt. Express. **15**, 13282–13287 2007

42. R. Dischler, F. Buchali, Measurement of non linear thresholds in O-OFDM systems with respect to data pattern and peak power to average ratio, Optical fiber communication conference, paper Mo.3.E.5, San Diego, CA, 2008
43. M. Nazarathy, J. Khurgin, R. Weidenfeld, Y. Meiman, P. Cho, R. Noe, I. Shpantzer, V. Karagodsky, Opt. Express **16**, 15777–15810 2008
44. R. O'Neil, L.N. Lopes, Envelope variations and spectral splatter in clipped multicarrier signals, in *Proceedings of IEEE 1995 International Symposium on Personal, Inddor and Mobile Radio Communications*, pp. 71–75, 1995
45. X. Li, L.J. Cimini Jr., IEEE Commun. Lett. **2**, 131–133 (1998)
46. J. Armstrong, IET Elect. Lett. **38**, 246–247 (2002)
47. D.J.G. Mestdagh, P.M.P. Spruyt, IEEE Trans. Commun. **44**, 1234–1238 (1996)
48. R.W. Bauml, R.F.H. Fischer, J.B. Huber, IET Electron. Lett. **32**, 2056–2057 (1996)
49. S.H. Muller, J.B. Huber, A novel peak power reduction scheme for OFDM, in *Proceedings of IEEE 1997 International Symposium on Personal. Indoor and Mobile Radio Communications*, pp. 1090–1094, 1997
50. M. Friese, OFDM signals with low crest-factor, in *Proceedings of 1997 IEEE Global Telecommunications Conference*, pp. 290–294, 1997
51. J. Tellado, J.M. Cioffi, Peak power reduction for multicarrier transmission, in *Proceedings of 1998 IEEE Global Telecommunication Conference*, pp. 219–224, 1998
52. B.S. Krongold, D.L. Jones, IEEE Trans. Broadcasting. **49**, 258–268 (2003)
53. J.M. Tang, K.A. Shore, J. Lightwave Technol. **25**, 787–798 (2007)
54. X.Q. Jin, J.M. Tang, P.S. Spencer, K.A. Shore, J. Opt. Networking, **7**, 198–214 (2008)
55. W. Shieh, I. Djordjevic, *OFDM for Optical Communications*. (Elsevier, Amsterdam, 2009)
56. W. Shieh, X. Yi, Y. Ma, Y. Tang, Opt. Express, **15**, 9936–9947 (2007)
57. S.L. Jansen, I. Morita, N. Takeda, H. Tanaka, 20-Gb/s OFDM transmission over 4,160-km SSMF enabled by RF-Pilot tone phase noise compensation, Optical fiber communication conference, paper PDP15, Anaheim, CA, USA, 2007
58. E. Yamada, A. Sano, H. Masuda, T. Kobayashi, E. Yoshida, Y. Miyamoto, Y. Hibino, K. Ishihara, Y. Takatori, K. Okada, K. Hagimoto, T. Yamada, H. Yamazaki, Novel no-guardinterval PDM CO-OFDM transmission in 4.1 Tb/s (50 88.8-Gb/s) DWDM link over 800 km SMF including 50-GHz spaced ROADM nodes. Optical fiber communication conference, paper PDP8, San Diego, CA, USA, 2008
59. R.C. Giles, K.C. Reichman, Electron. Lett. **23**, 1180–1180 (1987)
60. L.G. Kazovsky, S. Benedetto, A.E. Willner, *Optical Fiber Communication Systems* (Artech House, Boston, 1996)
61. T. Okoshi, K. Kikuchi, *Coherent Optical Fiber Communications* (Springer, Heidelberg, 1988)
62. A.H. Gnauck et al., 2.5Tb/s6442.7 Gb/s1 transmission over 40100 km NZDSF using RZ-DPSK format and all-Raman-amplified spans, in Optical fiber communication conference and exposition. Technical Digest, Optical Society of America, paper FC2, 2002
63. D.S. Ly-Gagnon, S. Tsukarnoto, K. Katoh, K. Kikuchi, J. Lightwave Technol. **24**, 12–21 (2006)
64. S.L. Jansen, I. Morita, H. Tanaka, 16 × 52.5-Gb/s, 50-GHz spaced, POLMUX-COOFDM transmission over 4,160 km of SSMF enabled by MIMO processing KDDI R&D Laboratories, Presented at the european conference on optical communications, Paper PD1.3, Berlin, Germany, 16–20 September 2007
65. J.M. Tang, P.M. Lane, K.A. Shore, 30 Gb/s transmission over 40 km directly modulated DFB laser-based SMF links without optical amplification and dispersion compensation for VSR and metro applications, in Optical fiber communication conference, Paper JThB8, Optical Society of America, 2006
66. D. Qian, J. Hu, J. Yu, P. Ji, L. Xu, T. Wang, M. Cvijetic, T. Kusano, Experimental demonstration of a novel OFDM-A based 10 Gb/s PON architecture, Presented at the european conference on optical communications, Paper 5.4.1, Berlin, Germany, 16–20 September 2007
67. M. Nazarathy, R. Weidenfeld, R. Noe, J. Khurgin, Y. Meiman, P. Cho, I. Shpantzer, Recent advances in coherent optical OFDM high-speed transmission, PhotonicsGlobal@Singapore, 2008. IPGC 2008, IEEE, pp.1–4, 8–11 December 2008

68. W. Shieh, X. Yi, Y. Ma, Q. Yang, J. Opt. Networking **7**, 234–255 (2008)
69. T.M. Schmidl, D.C. Cox, IEEE Trans. Commun. **45**, 1613–1621 (1997)
70. F. Buchali, R. Dischler, M. Mayrock, X. Xiao, Y. Tang, Improved frequency offset correction in coherent optical OFDM systems, 34th european conference on optical communication, 2008. ECOC 2008, pp. 1–2, 21–25 September 2008
71. H. Sari, G. Karam, I. Jeanclaude, IEEE Commun. Mag. **33**(2), 100–109 (1995)
72. X. Yi, W. Shieh, Y. Ma, J. Lightwave Technol. **26**, 1309–1316 (2008)
73. X. Liu, F Buchali, Opt. Express **16**, 21944–21957 (2008)
74. J. McDonough, IEEE Communications Magazine 45, Appl. Practice, 6–9 (2007)
75. F. Buchali, R. Dischler, A. Klekamp, M. Bernhard, D. Efinger, Realization of a real-time 12.1 Gb/s optical OFDM transmitter and its application in a 109 Gb/s transmission system with coherent reception, European conference on optical communication (ECOC), PD paper 2.1, Vienna, 2009
76. Y. Benlachtar, P.M. Watts, R. Bouziane, P. Milder, R. Koutsoyannis, J.C. Hoe, M. Puschel, M. Glick, R.I. Killey, 21.4 GS/s real-time DSP-based optical OFDM signal generation and transmission over 1600km of uncompensated fibre, European conference on optical communication (ECOC), PD paper 2.4, Vienna, 2009
77. H. Sun, K.T. Wu, K. Roberts, Opt. Express **16**, 873–879 (2008)
78. Q. Yang, Y. Tang, Y. Ma, W. Shieh, J. Lightwave Technol. **27**, 168–176 (2009)
79. Y. Tang, W. Shieh, X. Yi, R. Evans, IEEE Photon. Technol. Lett. **19**, 483–485 (2007)
80. W. Shieh, Coherent optical MIMO-OFDM for optical fibre communication systems, Workshop 5, European conference on optical communication, Berlin, Germany, 2007
81. Y. Ma, Q. Yang, Y. Tang, S. Chen, W. Shieh, 1-Tb/s per channel coherent optical OFDM transmission with subwavelength bandwidth access, Optical fiber communication conference, Paper PDP C1, San Diego, USA, 2009
82. T. Kawanishi, S. Oikawa, K. Higuma, M. Izutsu, Photon. Technol. Lett. **14**, 1454–1456 (2002)
83. I. Gasulla, J. Capmany, Opt. Express **16**, 8033–8038 (2008)
84. E.J. Tyler, P. Kourtessis, M. Webster, E. Rochat, T. Quinlan, S.E.M. Dudley, S.D. Walker, R.V. Penty, I.H. White, J. Lightwave Technol. **21**, 3237–3243 (2003)
85. Z. Tong, Y. Ma, W. Shieh, IET Electron. Lett. **44**, 1373–1374 (2008)
86. S.E. Miller, Bell Syst. Tech. J. **48**, 2059–2069 (1969)

Chapter 3
Nonlinear Impairments in Coherent Optical OFDM Systems and Their Mitigation

Moshe Nazarathy and Rakefet Weidenfeld

3.1 Introduction

This chapter addresses the analysis of the fiber channel Kerr-effect-induced nonlinearities as well as the synthesis of mitigation methods for these nonlinear (NL) impairments, in the specific context of multicarrier coherent optical Orthogonal Frequency-Division Multiplexing (OFDM) transmission [1–5] (Chap. 2), which imposes a particular spectral and temporal structure on the transmitted signals. We should mention that our analysis is restricted to coherent optical OFDM multicarrier OFDM transmission, making our NL scalar treatment somewhat distinct from those pursued in Chaps. 6–8 applicable to single-carrier systems. In fact, whenever we use the term "OFDM" in this chapter, we imply *coherent* (rather than direct-detection) *optical* OFDM.

An OFDM signal, as described in Chap. 2, is a special case of a multichannel signal, consisting of modulated (sub-) carriers regularly spaced in the frequency domain. A predecessor to OFDM, also featuring a regular multichannel structure, was introduced in optical communication a couple of decades ago – the dense wavelength division multiplexing (DWDM) technique. The analysis of the four-wave mixing (FWM) nonlinearity in the DWDM context was addressed in the early days of optical communication and repeatedly visited the course of the last two decades, in works such as [6–28]. Those studies are, in principle, relevant to OFDM NL [6–9] analysis, with two caveats: the OFDM subchannels are much denser than the DWDM channels; the OFDM subcarriers are usually phase or phase/amplitude modulated, whereas the DWDM channels were typically used to carry OOK modulation. Therefore, despite superficial similarities, although a common formalism may, in principle, describe OFDM and DWDM, the respective regimes of operation of these two applications are vastly different, translating into distinctly different qualities for the two system types. Moreover, the early DWDM NL analyses [6–23] provided fragmented descriptions, not carrying the model through to a unified

M. Nazarathy (✉) and R. Weidenfeld
Electrical Engineering Department, Technion, Israel Institute of Technology, Israel
e-mail: nazarat@ee.technion.ac.il; wrakefet@gmail.com

S. Kumar (ed.), *Impact of Nonlinearities on Fiber Optic Communications*, Optical and Fiber Communications Reports 7, DOI 10.1007/978-1-4419-8139-4_3,
© Springer Science+Business Media, LLC 2011

view in its full generality, e.g., stopping short of modeling both FWM and cross-phase modulation (XPM) for arbitrary position-dependent α, β_2, γ fiber parameters, missing the remarkable compact Fourier Transform theorem, which we formulated in [30]. The NL generation efficiency [*Volterra Transfer Function* (VTF) – see below] is proportional to the Fourier Transform of the spatial power gain profile for systems with constant β_2, γ [but arbitrary $\alpha(z)$ profiles and *optical amplifier* (OA) gains]. It is this simple theorem, which leads us in [29, 30] to the phased-array interpretation, providing the comprehensive analytical justification for the modern trend that coherent optical transmission links be best operated dispersion-unmanaged, i.e., with the fiber compensation modules removed [4, 31, 32].

With the emergence of OFDM transmission in the context of coherent detection, interest in NL modeling was reignited [4, 28–30, 33–43]. Among the first works in this research wave, a fundamental analysis of the FWM impairment in the absence of dispersion was carried out in [34], working out the combinatorics of triplets of OFDM subcarriers forming intermodulation (IM) products falling onto and perturbing other subcarriers. However, in the presence of dispersion, the FWM formation becomes much more complex a process, as the IM products no longer add up in phase, but rather sum up on a phasor basis, with angles depending in complicated ways on the three participating frequencies in each FWM IM triplet. In principle, a solution of the NL Schroedinger equation (NLSE) simultaneously accounting for dispersion and nonlinearity is called for. Such an approach has been pursued in [30] in terms of a perturbation solution to the NLSE, the first two relevant orders of which assume that the "pumps" (i.e., the transmitted subcarriers) remain un-depleted. In this chapter, we review elements of that approach, but proceed to supplement it with an alternative equivalent method, which is more straightforward to derive, yet provides considerable physical insight into the formation of the NL perturbation: the *optical path integral* (OPI) method, whereby the FWM nonlinearity is modeled by summing up (integrating) the FWM contributions generated by each differential length segment of the fiber, further integrating over all relevant frequency triplets. From a mathematical formalism point of view, the OPI physical approach is complemented by the rigorous Volterra NL modeling description, which is extended here from a second-order treatment in [44] to a third-order and higher order model. The OPI and Volterra methodologies provide the main analysis and synthesis tools for the modeling of the Kerr-induced nonlinearity (FWM/XPM) in the presence of chromatic dispersion (CD), as developed in the first part of this chapter, applied in the second part of this chapter to conceiving and designing efficient NL compensation (NLC) methods to counteract the detrimental effect of the nonlinearities in OFDM transmission.

In Sect. 3.2, we develop a rigorous OFDM *Transmitter* (Tx) model, including the effects of interpolation and digital up-shifting, as well as an analog-related model of the Tx (derived in Appendix A), which well approximates the rigorous Tx model, but is more amenable to the subsequent NL analysis. In Sect. 3.3, we proceed to model the optically amplified fiber-channel, both linearly and nonlinearly, analyzing the SPM/XPM/FWM channel impairment. The Volterra NL methodology is formally developed in Appendix B, however, for those less bent on rigor, the

VTF is alternatively justified in Sect. 3.3 on NL optics physical grounds. Section 3.4 addresses the linear and NL modeling of the analog and digital processing occurring in the OFDM *Receiver* (Rx) front-end, including the effects of aliasing and oversampling (further explored in Appendix C). In Sect. 3.5, we proceed to analytically derive the VTF of a general optically amplified multispan link providing the complete description of NL behavior, based on aforementioned Volterra formalism and the OPI approach, streamlining the perturbation analysis by means of the concept of virtual back-propagated fields using the *quasilinear propagation transfer function* (QLP-TF). A compact analytic expression is derived for the FWM impairment for the most general case of fiberoptic link, which is irregular and inhomogeneous, in the sense that the lengths of the fiber-spans and gains of the OAs are allowed to be arbitrary, and the $\alpha,\ \beta_2,\ \gamma$ fiber parameters are generally position dependent. As special cases of this most general description, following [30], we analyze regular multispan systems in which all fiber spans are identical, identifying the *phased-array effect* similar to that occurring in microwave antenna arrays, whereby over a dispersion-unmanaged long-haul link (i.e., without dispersion compensation modules), the NL contributions of multiple fiber spans tend to interfere destructively, resulting in substantial enhancement of the *NL tolerance* (NLT), under proper conditions. In this section, we also model an optically amplified fiber link with *dispersion compensating fiber* (DCF) modules positioned every few spans, derived as a special case of our general formalism pertaining to arbitrary position-dependent $\alpha,\ \beta_2,\ \gamma$ parameters. We then proceed to develop analytic expressions for the OFDM system performance [Q-factor and *bit error rate* (BER)], in terms of the system parameters and in particular in terms of the NLT parameter reflecting the RMS VTF over all possible IM products. Finally, the last subsection of the section develops analytic insight into the NLT parameter for broadband OFDM systems, which is shown to vary as the *bandwidth*^{2} × *length* × *GVD* product, leading to a most compact expression for the Q-factor performance over a dispersion-unmanaged link, indicating that the Q-factor varies as $N_{\text{spans}}^{-1/2}$, vs. the number of fiber spans, N_{spans}, i.e., even more favorably than if the mechanism of spans NL addition were incoherent.

Starting with Sect. 3.8, we address NL mitigation or compensation methods for OFDM transmission, first reviewing the current main NLC approaches, then proceeding to develop our own Volterra-based NLC method. In Sect. 3.9, we analyze in detail the operation of the simplest NLC conventional method [33] based on *backward NL phase rotation* (B-NLPR). We highlight an aspect not explicitly mentioned in previous works: a high oversampling factor must be applied in order to enable this NLC method – when baud-rate sampling is used, the B-NLPR compensation breaks down completely. We then modify the B-NLPR NLC to operate with baud-rate sampling, essentially attaining the same performance as in the high oversampling case. However, even with this improvement, the B-NLPR NLC merely provides limited relief, as it is frequency agnostic, ignoring the interplay between CD and nonlinearity. In Sect. 3.10, we introduce our Frequency-Shaped Volterra *decision-feedback* (DF)-based NLC, providing insight into the DF-aided principle of operation and the mitigation of the frequency-dependent nonlinearity in terms of a genie-based paradigm, The system block diagram is successively evolved over three versions, culminating in a complete system operating in three passes (iterations), as described in Sect. 3.11. The overall characteristics and performance of our OFDM link with the final version of the Volterra NL DF-based NLC, first presented in [45] are elaborated. An NLT improvement of up to ~4 dB is achieved. Section 3.12 explains, in terms of signal processing interpolation and filtering principles, how our final system manages to achieve baud-rate operation despite the substantial spectral broadening associated with the NL nature of the fiber channel and the digital compensator. Section 3.13 highlights another beneficial aspect of our NLC system, namely its low-error propagation, traced to the fact that the percentage of errored IM triplets, affected by an errored preliminary decision in the first pass, is very low. In Sect. 3.14, the role of higher order (5th, 7th,...) nonlinearities is discussed, indicating that although the balancing of the nonlinearities of the fiber link and the compensator is nominally carried out at third-order, the higher orders further mutually cancel to some extent. Section 3.15 details another key ingredient leading to improved performance of our Volterra DF NLC scheme, namely the decoupling of the XPM and FWM mitigation mechanisms by means of the "XPM undo and derotate" signal processing procedure. Section 3.16 presents the simulations of the improved performance attained by our method. Section 3.17 considers the associated computational complexity. In Sect. 3.18, the Volterra DF NLC is compared with the back-propagation (BP) method proposing a roadmap of future improvements merging the Volterra and BP NLC approaches. Finally, this chapter is concluded in Sect. 3.9, surveying the advantages and limitations of our approach, and suggesting future research directions.

3.2 Rigorous OFDM Transmission Model

The OFDM signal is a succession of symbol *blocks* (alternatively called *OFDM symbols* or *frames*), each of duration T_B. The OFDM signal complex envelope (CE) (denoted by an overtilde), is usually modeled, over the first T seconds of the block [up to the moment when the cyclic prefix (CP) extension starts], as emitting a superposition of M analog subcarriers, modulated by complex symbols $\underset{\sim}{A}_i$, which are effectively specified in the frequency domain:

$$\underset{\sim}{s}_{\text{TX}}(t) = \mathbf{1}_{[0,T]} \sum_{i=0}^{M-1} \underset{\sim}{A}_i e^{j2\pi i \Delta\nu t} . \tag{3.1}$$

Most OFDM treatments resort to this simplistic model; however in reality, the transmitted signal is not strictly an analog subcarrier multiplexed one, but it is rather generated by a digital processing chain terminated in a pair of digital-to-analog converters (DACs) driving the IQ modulator, generating a transmitted analog CE slightly different than (3.1).

We assume a generic structure for the OFDM *Transmitter* (Tx) and *Receiver* (Rx) as described in Chap. 2. We proceed to precisely model the OFDM Tx, further incorporating into the treatment aspects of interpolation and frequency shifting, which are often ignored. We show that the exact expression of the analog transmitted signal may be approximately cast in an equivalent form akin to (3.1), under certain assumptions. For simplicity, we initially provide a preliminary OFDM Tx description ignoring interpolation and frequency up/down conversion, subsequently extending the model to include these effects.

Each OFDM block is generated by superposing M subchannels, equispaced, at spectral separation $\Delta\nu$ between adjacent subcarriers; the ith tone in the subcarrier grid is at frequency $\nu_i = i\,\Delta\nu + \nu_0, i = 0, 1, \ldots, M-1$. The aggregate OFDM signal bandwidth is approximately $B_{\mathrm{T}} = M\Delta\nu = MT^{-1}$, with $T = \Delta\nu^{-1}$ the *net* duration of the OFDM block, i.e., its duration net of the CP extension, which lasts $T_{\mathrm{CP}} = T_{\mathrm{B}} - T$. The data stream to be transmitted over the block is mapped into a vector $\underset{\sim}{\mathbf{A}} = \{\underset{\sim}{A}_i\}_{i=0}^{M-1}$ of M complex symbols, each selected out of a specified complex-valued transmission constellation, e.g., Quaternary phase-shift keying (QPSK) or QAM. The symbols vector $\underset{\sim}{\mathbf{A}}$ undergoes an IDFT, $\underset{\sim}{a}_n = \sum_{i=0}^{M-1} \underset{\sim}{A}_i e^{j2\pi ni/M}$, yielding the time-domain vector $\underset{\sim}{\mathbf{a}} = \{\underset{\sim}{a}_n\}_{n=0}^{M-1}$, which is CP padded, generating an extended vector $\underset{\sim}{\mathbf{s}} = \{\underset{\sim}{s}_n\}_{n=-\mu}^{M-1}$ of length $M + \mu$, with its first μ elements duplicating the last μ elements of the IDFT output vector $\underset{\sim}{\mathbf{a}}$, i.e., $\underset{\sim}{s}_n = \underset{\sim}{a}_{n \bmod M}$, $n = -\mu, -\mu + 1, \ldots, M-1$, where the number of samples $\mu = T_{\mathrm{CP}}/T_{\mathrm{c}}$ in the CP is typically taken equal to the delay spread of the channel expressed in *chip* intervals, T_{c}, where the term chips refers to the fast modulation intervals $T_{\mathrm{c}} \equiv T_{\mathrm{B}}/(M + \mu) = T/M$ at the DAC clock rate. The complex vector $\underset{\sim}{\mathbf{s}}$ (or equivalently the pair of its real and imaginary parts) is converted into a pair of I/Q analog signals by means of a pair of DACs, modeled as generating the complex signal $\underset{\sim}{s}_{\mathrm{TX}}(t) = \sum_{n=-\mu}^{M-1} \underset{\sim}{s}_n g_{\mathrm{TX}}(t - nT_{\mathrm{c}})$, driving the IQ modulator. Here, $g_{\mathrm{TX}}(t) = g_{\mathrm{DAC}}(t) \otimes g_{\mathrm{MOD}}(t)$ represents the combined effects of the aperture of the DAC sample & hold circuitry, $g_{\mathrm{DAC}}(t)$ (the DAC reconstruction function) and the RF driver and IQ modulator impulse response $g_{\mathrm{MOD}}(t)$.

3.2.1 *Interpolation and Digital Frequency Up-Shifting*

In the description above, the DAC operates at a chip rate T_{c}^{-1}, in par with the baud-rate of a comparable single-carrier system. To the extent that a faster DAC is available (say, L_{INTP} times faster with the integer interpolation factor, L_{INTP}, typically equal to 2 or 4), such DAC enables shaping the aperture response $g_{\mathrm{DAC}}(t)$, by running the DAC at the elevated clock rate, $L_{\mathrm{INTP}}T_{\mathrm{c}}^{-1}$, with the DAC input sequence obtained by digital interpolation of the $\underset{\sim}{s}_n$ sequence, yielding an interpolated sequence $\underset{\sim}{s}_n^{\mathrm{INTP}}$, e.g., generated by filling the original sequence $\underset{\sim}{s}_n$ with $L_{\mathrm{INTP}} - 1$

zeros in between the consecutive samples, followed by a convolution with a suitable digital interpolation kernel. Note that the orders of interpolation and CP extension may be exchanged, interpolating the $\underset{\sim}{a}_n$ sequence first, yielding $\underset{\sim}{a}_n^{\text{INTP}}$ then CP extending it to $\underset{\sim}{s}_n^{\text{INTP}}$. Digital interpolation circumvents the need to carefully shape the DAC analog reconstruction function. Interpolation may also be used to model the analog transmitted signal in numerical simulation studies of OFDM transmission.

As the $\underset{\sim}{a}_n$ sequence to be interpolated is the IDFT of $\underset{\sim}{A}_i$, the time-domain interpolation may be equivalently obtained by zero-padding prior to applying a larger size IDFT [46]: The M symbols $\underset{\sim}{A}_i$ are zero-padded (ZP) to a total length $D_{\text{ZP}} = L_{\text{INTP}} M > M$, generating the vector $\underset{\sim}{A}_i^{\text{ZP}:D_{\text{ZP}}}$, onto which a D_{ZP}-sized IDFT is applied:

$$\underset{\sim}{a}_n^{\text{INTP}} = \sum_{i=0}^{D_{\text{ZP}}-1} \underset{\sim}{A}_i^{\text{ZP}:D_{\text{ZP}}} e^{j2\pi in/D_{\text{ZP}}} = \sum_{i=0}^{M-1} \underset{\sim}{A}_i e^{j2\pi in/D_{\text{ZP}}} = \underset{\sim}{a}(t)\Big|_{t\to nT/D};$$

$$\underset{\sim}{a}(t) \equiv \sum_{i=0}^{M-1} \underset{\sim}{A}_i e^{j2\pi i\,\Delta\nu\, t}, \tag{3.2}$$

(the ZP vector $\{\underset{\sim}{A}_i^{\text{ZP}:D_{\text{ZP}}}\}_{i=0}^{D_{\text{ZP}}-1}$ is defined as $\underset{\sim}{A}_i^{\text{ZP}:D_{\text{ZP}}} = \underset{\sim}{A}_i,\ k = 0, 1, \ldots, M-1$, else $\underset{\sim}{A}_i^{\text{ZP}:D_{\text{ZP}}} = 0$).

The analog function $\underset{\sim}{a}(t)$ in (3.2), which is effectively being sampled at a rate $D_{\text{ZP}} T^{-1}$ at the ZP IDFT output, is a finite Fourier series (FFS) with period T, i.e., a Fourier series (FS) with a finite number of harmonics $\{\underset{\sim}{A}_i\}_{i=0}^{M-1}$. If zero-padding were not applied, then we would have $\underset{\sim}{a}_i = \sum_{i=0}^{M-1} \underset{\sim}{A}_i e^{j2\pi in/M} = \underset{\sim}{a}(t)|_{t\to nT/M}$, to be compared with $\underset{\sim}{a}_n^{\text{INTP}}$ in (3.2). This indicates that zero-padding the input vector $\underset{\sim}{\mathbf{A}}$ to length $D_{\text{ZP}} > M$ and applying an IDFT, amounts to sampling the FFS $\underset{\sim}{a}(t)$ over a finer grid with spacing $\Delta t = T/D_{\text{ZP}}$, rather than $\Delta t = T/M$, collecting $D_{\text{ZP}} > M$ samples over the T-period of the periodic analog waveform $\underset{\sim}{a}(t)$ with harmonic coefficients $\underset{\sim}{A}_i$. We conclude that the mechanism of zero-padding the IDFT input yields an interpolated time-domain output $\underset{\sim}{a}_n^{\text{INTP}}$, L_{INTP} times more densely sampling the FFS $\underset{\sim}{a}(t)$ vs. the case of the non-ZP sequence $\underset{\sim}{a}_n$.

Note that this interpolation-by-zero-padding-the-IDFT-input technique is useful not only in actual Tx realization, but it may also be conveniently employed in simulation, digitally synthesizing an analog-like OFDM transmitted signal by selecting a large L_{INTP} factor (of the order of 10), to be subsequently propagated through the optical channel via the *split-step-Fourier* (SSF) method.

We next observe that the spectrum of the signal $\underset{\sim}{a}_n^{\text{INTP}}$ applied to the DAC is *Single Sideband* (SSB), consistent with the IDFT definition. It is advantageous to generate a more symmetrical spectrum of the transmitted CE (nearly centering the CE spectrum around DC, nearly halving the IQ modulator bandwidth). To this end, $\underset{\sim}{a}_n^{\mathrm{INTP}}$ is modulated by a discrete-time subcarrier $c_n = (-1)^n = e^{-j\pi n} = e^{j2\pi(-D_{\mathrm{ZP}}/2)n/D_{\mathrm{ZP}}}$ effecting down-conversion (D/C), shifting the CE band frequency closer to the origin:

$$\underset{\sim}{a}_n^{\mathrm{INTP-D/C}} = c_n \underset{\sim}{a}_n^{\mathrm{INTP}} = e^{j2\pi(-M/2)n/D} \sum_{i=0}^{M-1} \underset{\sim}{A}_i e^{j2\pi in/D_{\mathrm{ZP}}}$$
$$= \sum_{i=0}^{M-1} \underset{\sim}{A}_i e^{j2\pi(i-D_{\mathrm{zp}}/2)n/D_{\mathrm{ZP}}} = \sum_{i=-M/2}^{M/2-1} \underset{\sim}{A}_{i+M/2} e^{j2\pi in/D_{\mathrm{ZP}}}. \quad (3.3)$$

The D/C vector $\{\underset{\sim}{a}_n^{\mathrm{INTP-D/C}}\}_{n=0}^{D-1}$ is subsequently CP-appended, prepending its last μL_{INTP} samples at the beginning of the record, yielding D CP-extended samples, with $D = D_{\mathrm{ZP}} + \mu L_{\mathrm{INT}} = M L_{\mathrm{INTP}} + \mu L_{\mathrm{INTP}} = (M+\mu) L_{\mathrm{INTP}}$: $\underset{\sim}{s}_n = \underset{\sim}{a}_{n \bmod D_{\mathrm{ZP}}}^{\mathrm{INTP-D/C}}$, $n = -\mu L_{\mathrm{INTP}}, -\mu L_{\mathrm{INTP}} + 1, \ldots, M L_{\mathrm{INTP}} - 1$. Substituting (3.3) into the last equation yields:

$$\underset{\sim}{s}_n = \underset{\sim}{a}_{n \bmod D_{\mathrm{ZP}}}^{\mathrm{INTP-D/C}} = \sum_{i=-M/2}^{M/2-1} \underset{\sim}{A}_{i+M/2} e^{j2\pi i(n \bmod D_{\mathrm{ZP}})/D_{\mathrm{ZP}}}$$
$$= \sum_{i=-M/2}^{M/2-1} \underset{\sim}{A}_{i+M/2} e^{j2\pi in/D_{\mathrm{ZP}}}, \quad n = -\mu L_{\mathrm{INT}}, -\mu L_{\mathrm{INT}} + 1, \ldots, D_{\mathrm{ZP}} - 1, \quad (3.4)$$

where in the last equality we were able to discard the mod D_{ZP} operation in the exponent, as the mapping $n \to n + D_{\mathrm{ZP}}$, occurring over $-\mu L_{\mathrm{INT}} \le n < 0$, merely adds a 2π integer multiple to the exponent. Note that in our processing chain the D/C operation preceded the CP extension; however, the order of these two operations may be exchanged. The resulting sequence, $\underset{\sim}{s}_n$, finally drives the DAC pair, with reconstruction function $h_{\mathrm{DAC}}(t)$ and L_{INPL} times faster clock interval, $T_c \equiv T/D = T/[L_{\mathrm{INPL}}(M+\mu)]$. The analog DAC output is convolved with the IQ modulator analog E-O response $h_{\mathrm{MOD}}(t)$, yielding the transmitted CE:

$$\underset{\sim}{s}(t) = \sum_{n=-\mu L_{\mathrm{INT}}}^{D_{\mathrm{ZP}}-1} \underset{\sim}{s}_n h_{\mathrm{DAC}}(t - nT_c) \otimes h_{\mathrm{MOD}}(t) = \sum_{n=-\mu L_{\mathrm{INT}}}^{D_{\mathrm{ZP}}-1} \underset{\sim}{s}_n h_{\mathrm{TX}}(t - nT_c)$$
$$= \sum_{n=-\mu L_{\mathrm{INT}}}^{D_{\mathrm{ZP}}-1} \sum_{i=-M/2}^{M/2-1} \underset{\sim}{A}_{i+M/2} e^{j2\pi in/D_{\mathrm{ZP}}} h_{\mathrm{TX}}(t - nT_c). \quad (3.5)$$

The complete "digital OFDM + DAC" signal generation model is compactly and accurately described by the last equation, capturing the key digital processing and D/A conversion effects in the OFDM Tx. Note that this precise expression seems superficially different from the mathematical description (3.1), which is usually invoked in the literature. Nevertheless, for the purpose of NL channel propagation analysis, an "analog-like OFDM" model akin to the form (3.1) would be more convenient, but can such model be formally derived starting from (3.5), and under what assumptions would it be applicable?

3.2.2 OFDM Analog-Like Tx Model

We now show that (3.5) reduces to an expression akin to (3.1), yielding a quite accurate description provided that a relatively large number of subcarriers M is used; hence, the number of time samples in the OFDM window satisfies $D \gg 1$, and moreover the Tx analog response $H^{\mathrm{TX}}(\nu) = F\{h_{\mathrm{RX}}(t)\}$ is bandlimited to the frequency interval $[-T_c^{-1}/2, T_c^{-1}/2]$, with cutoff frequency $T_c^{-1} = D/T_{\mathrm{B}} = (M+\mu)\, L_{\mathrm{INT}} T_{\mathrm{B}}^{-1} = (M+\mu)\, L_{\mathrm{INT}} \Delta\nu = (1+\mu/M)\, L_{\mathrm{INT}} B_{\mathrm{T}}$. All we require is the bandwidth limitation of the Tx response, but $H^{\mathrm{TX}}(\nu)$ should not necessarily be flat over its pass-band, i.e., the Tx analog impulse response need not be an ideal sinc function. It is then shown in Appendix A, based on sampling theorem considerations, that the precise OFDM signal generation model (3.5) may be cast in the approximate form

$$\underset{\sim}{s}(t) = h^{\mathrm{TX}}(t) \otimes \underset{\sim}{a}(t) \cong \sum_{i=-M/2}^{M/2-1} \underset{\sim}{A}_i^{\mathrm{TX}}\, \mathrm{e}^{j2\pi i \Delta\nu t} \mathbf{1}_{[-T_{\mathrm{CP}}, -T_{\mathrm{CP}}+T_{\mathrm{B}}]}(t);$$

$$\underset{\sim}{a}(t) \cong \mathbf{1}_{[-T_{\mathrm{CP}}, -T_{\mathrm{CP}}+T_{\mathrm{B}}]}(t) \sum_{i=-M/2}^{M/2-1} \underset{\sim}{A}_i^{\mathrm{TX}}\, \mathrm{e}^{j2\pi i \Delta\nu t}, \tag{3.6}$$

where we introduced the indicator function ($\mathbf{1}_{[a,b]}(t) \equiv 1$ if $t \in [a,b]$, $\mathbf{1}_{[a,b]}(t) \equiv 0$, otherwise), relabeled the time-window as $[-T_{\mathrm{CP}}, -T_{\mathrm{CP}} + T_{\mathrm{B}}] = [-\mu L_{\mathrm{INT}} T_{\mathrm{c}}, (D_{\mathrm{ZP}}-1) T_{\mathrm{c}}]$, we denoted by $H_i^{\mathrm{TX}} \equiv H^{\mathrm{TX}}(i\,\Delta\nu)$ the frequency samples of the Tx response $H_{\mathrm{TX}}(\nu) = F\{h_{\mathrm{TX}}(t)\}$, and defined $\underset{\sim}{A}_i^{\mathrm{TX}} \equiv \underset{\sim}{A}_{i+M/2} H_i^{\mathrm{TX}}$. The ith subcarrier is represented in (3.6) as an analog harmonic tone $\mathrm{e}^{j2\pi i \Delta\nu t}$ rectangular-windowed over the OFDM block duration T_{B}. scaled by the complex symbol. This establishes the approximate equivalence between the conventional analog simplified representation of OFDM (3.6), and the precise digital–analog OFDM Tx model (3.5).

3.3 Fiber Channel Model: Third-Order Volterra Description of the FWM/XPM Impairment

3.3.1 Complex Representation

Let $u(t;z)$ be the *real-valued* scalar optical field at time t and position z along the fiber, $\underset{\sim}{u}(t;z)$ its CE, and $\underset{\sim}{u}(t;z)$ its *spatiotemporal CE* (STCE) (note the $\sqrt{2}$ normalization factor in our convention):

$$u(z,t) = \sqrt{2}\,\mathrm{Re}\left\{\underset{\sim}{u}(z,t)\mathrm{e}^{j2\pi\,\nu_0 t}\right\} = \sqrt{2}\,\mathrm{Re}\left\{\underset{\frown}{u}(z,t)\mathrm{e}^{-j(\beta_0 z-2\pi\,\nu_0 t)}\right\}. \tag{3.7}$$

The CE and STCE are related by $\underset{\sim}{u}(z,t) = \underset{\frown}{u}(z,t)\mathrm{e}^{-j\beta_0 z}$. In turn, the *analytic signal* (AS) $u_a(z,t)$ is related to the other representations by

$$u_a(z,t) = \underset{\sim}{u}(z,t)\mathrm{e}^{j\omega_0 t} = \underset{\frown}{u}(z,t)\mathrm{e}^{-j(\beta_0 z-\omega_0 t)};\quad u(z,t) = \sqrt{2}\,\mathrm{Re}\left\{u_a(z,t)\right\}. \tag{3.8}$$

Although the related quantities $u, u_a, \underset{\sim}{u}, \underset{\frown}{u}$ above share the same letter u, this is not strictly necessary; in the sequel, various representations of a given signal might involve different letters. Finally, depending on the context, spatiotemporal signals, which are functions of z, t, will be sometimes explicitly labeled just by one of the two variables z or t, with other one implicit.

3.3.2 Fiber Channel Model

We proceed to model the linear and NL propagation of the OFDM transmitted signal (3.6) over a scalar fiberoptic channel, starting with linear propagation. We express the signal launched into the fiber link, at $z = 0$, as

$$\begin{aligned}\underset{\frown}{u}(0,t) &= \underset{\sim}{s}(t) = \sum_{i=M_1}^{M_2} \underset{\sim}{A}_i^{\mathrm{TX}}\,\mathrm{e}^{j2\pi i\Delta\nu t},\quad t\in[-T_{\mathrm{CP}}, -T_{\mathrm{CP}}+T_{\mathrm{B}}];\\ M_1 &= -M/2;\quad M_2 = M/2-1.\end{aligned} \tag{3.9}$$

i.e., we consider a lone OFDM block, or equivalently consider a sequence of blocks while ignoring inter-block interference, which is effectively mitigated by the CP extension. We decompose the propagating SCTE into narrowband subchannels, $\underset{\frown}{u}(z,t) = \sum_{i=M_1}^{M_2} \underset{\frown}{u}_i(z,t)$, corresponding to the OFDM sub-carries, modeling their (not necessarily linear) propagation and interactions. These subchannels are launched at z = 0 with initial conditions as determined by the OFDM Tx model (3.6):

$$\underset{\sim}{s}(t) = \underset{\sim}{u}(0,t) = \sum_{i=M_1}^{M_2} \underset{\sim}{s}_i(t); \qquad \underset{\sim}{s}_i(t) = \underset{\sim}{u}_i(0,t)$$
$$= \underset{\sim}{A}_i^{\mathrm{TX}}\, \mathrm{e}^{j2\pi i\Delta\nu t} 1_{[-T_{\mathrm{CP}},-T_{\mathrm{CP}}+T_{\mathrm{B}}]}(t). \tag{3.10}$$

Note that unlike in [30], the subchannels SCTEs $\underset{\sim}{u}_i(z,t)$ have their frequency shifts $\mathrm{e}^{j2\pi i\Delta\nu t}$ implicitly included in the subchannel CEs; all STCEs are defined here relative to the same spatiotemporal carrier $\mathrm{e}^{-j(\beta_0 z-2\pi\nu_0 t)}$.

The launched signal (3.9) propagates along the fiber link of length L, arriving at the *receiver* (Rx), where the received CE $\underset{\sim}{r}(t) \equiv \underset{\sim}{u}(L,t)$ is extracted by the coherent optical hybrid front-end.

The fiber link typically consists of N_{span} identical spans, each of length L_{span}, i.e., the total link length is $L = N_{\mathrm{span}}L_{\mathrm{span}}$. Each span is terminated in an OA, typically perfectly compensating the power loss $\mathrm{e}^{-\alpha L_{\mathrm{span}}}$ by providing power gain $G_{\mathrm{OA}} = \mathrm{e}^{\alpha L_{\mathrm{span}}}$, possibly incorporating a DCF module, to change the balance of accumulated dispersion over the span or prior few spans. Beyond this "regular" multispan fiber configuration, we shall model in Sect. 3.5.8 a generalized inhomogeneous fiber link configuration, comprising multiple fiber segments with arbitrary linear and NL fiber parameters, in particular the linear propagation constant $\beta(z)$ and the NL parameter $\gamma(z)$ will both be taken as piecewise-constant functions of z, whereas the loss profile of the fiber will be allowed to be an arbitrary function $\alpha(z)$ of z. We allow an arbitrary differential loss function $\alpha(z)$ along the fiber link, possibly containing impulsive components, modeling the lumped gains of the OAs, which are formally described as negative spatial impulses at the fiber spans ends. The initial transmitter OA is excluded from the fiber link description as it is considered part of the optical source, but the last OA at the Rx (the Rx pre-amplifier) is included. In the particular case of a "regular" multispan system with identical spans, we have the same fixed loss, $\alpha(z) = \alpha_0$ over any span. The differential loss profile and the power gain are then given by (with $\int_0^L \alpha(z)\mathrm{d}z = 0$ consistent with $G(L) = 1$):

$$\alpha(z) = \alpha_0 - \alpha_0 L_{\mathrm{span}} \sum_{s=1}^{N_{\mathrm{span}}} \delta(z - sL_{\mathrm{span}})\; G_{\mathrm{p}}(z) \equiv \mathrm{e}^{-\int_0^z \alpha(z')\mathrm{d}z'} \mathbf{1}_{[0,L]}(z)$$
$$= \mathrm{e}^{-\alpha_0 (z \bmod L_{\mathrm{span}})} \mathbf{1}_{[0,L]}(z) = \sum_{s=0}^{N_{\mathrm{span}}-1} \delta(z - sL_{\mathrm{span}}) \otimes \mathrm{e}^{-\alpha_0 z} \mathbf{1}_{[0,L_{\mathrm{span}}]}(z). \tag{3.11}$$

The three z-dependent parameters $\alpha(z), \beta(z), \gamma(z)$ feature in the NLSE:

$$\partial_z \underset{\sim}{u}(z,t) - \frac{j}{2}\beta_2(z)\partial_t^2 \underset{\sim}{u}(z,t) + \frac{1}{2}\alpha(z)\underset{\sim}{u}(z,t) = -j\gamma(z)|\underset{\sim}{u}(z,t)|^2 \underset{\sim}{u}(z,t), \tag{3.12}$$

where t is the retarded time i.e., the substitution $t \to t - \beta' z$ is assumed, ∂_t, ∂_t^2 are the first and second derivatives with respect t, $\beta_1 \equiv \partial_\omega \beta(\omega)$ and $\beta_2 \equiv \partial_\omega^2 \beta(\omega)$.

In [30], our NL modeling approach was based on substituting (3.10) into the NLSE and deriving coupled mode equations, solved by a perturbation method. Here, we de-emphasize such differential equation-based approach, instead applying the perturbation rationale to an equivalent OPI formulation, more amenable to physical intuition (Sect. 3.5).

3.3.3 Linear + SPM/XPM Propagation of the Subcarriers

We model the propagation of the individual subchannels, initially neglecting FWM cross-NL effects among the subchannels, as well as the distortive effect of dispersion on the block-long approximately rectangular envelopes, while still accounting for the CD-induced delay of each rectangular envelope, for the SPM of each subchannel as well as for the XPM among the subchannels. As FWM coupling among the subchannels is ignored at this point, we may separately propagate each of the summand signals (subchannels), $\underset{\sim}{s}_i(t)$, in (3.10), all the way to the Rx, with each subchannel being affected by the other channels only via the XPM mechanism (and by itself via SPM):

$$\underset{\sim}{u}(L,t) = \sum_{i=M_1}^{M_2} \underset{\sim}{u}_i(L,t) = \sum_{i=M_1}^{M_2} \underset{\sim}{r}_i(t) \tag{3.13}$$

$$\begin{aligned} \underset{\sim}{r}_i(t) &\equiv \underset{\sim}{u}_i(L,t) = \underset{\sim}{u}_i(0,t) \mathrm{e}^{-j\int_0^L \beta_i^{\mathrm{T}}(z')\mathrm{d}z'} 1_{[-T_{\mathrm{CP}},-T_{\mathrm{CP}}+T_{\mathrm{B}}]}(t-\tau_i) \\ &= \underset{\sim}{s}_i(t) \mathrm{e}^{-j\int_0^L \beta_i^{\mathrm{CD}}(z')\mathrm{d}z'} \mathrm{e}^{-j\int_0^L \beta_i^{\mathrm{NL}}(z')\mathrm{d}z'} \mathrm{e}^{-\int_0^L \alpha(z')\mathrm{d}z'} 1_{[-T_{\mathrm{CP}},-T_{\mathrm{CP}}+T_{\mathrm{B}}]}(t-\tau_i) \end{aligned} \tag{3.14}$$

with total effective propagation constant

$$\beta_i^{\mathrm{T}} = \beta_i^{\mathrm{CD}}(z) + \beta_i^{\mathrm{NL}}(z) - j\alpha(z)/2, \tag{3.15}$$

where the NL propagation constant accounting for SPM and XPM is given by

$$\beta_i^{\mathrm{NL}}(z) = \gamma(z)\left[2P^{\mathrm{T}}(z) - p_i(z)\right]; P^{\mathrm{T}}(z) \equiv \sum_{i=M_1}^{M_2} |\underset{\sim}{u}_i(z)|^2; \ p_i(z) \equiv |\underset{\sim}{u}_i(z)|^2. \tag{3.16}$$

Also note that each rectangular envelope was group-delayed, due to CD, by $\tau_i = \Delta\tau \cdot i + \tau_0$, where $\Delta\tau \equiv 2\pi\beta_2 L \Delta\nu$ and τ_0 is the group delay experienced at frequency ν_0. Indeed,

$$\tau_i - \tau_0 = \tau(\nu_i) - \tau(\nu_0) = \frac{\mathrm{d}\tau}{\mathrm{d}\omega}\Delta\omega = \frac{\mathrm{d}}{\mathrm{d}\omega}(L\beta_1)\,\Delta\omega = L\beta_2 2\pi\Delta\nu\cdot i. \quad (3.17)$$

The CP duration is set equal to the delay spread – difference of the group delays at the extreme frequency indexes $M-1$ and 0:

$$\begin{aligned} T_{\mathrm{CP}} &= \tau_{M-1} - \tau_0 = \tau(\nu_{M-1}) - \tau(\nu_0) = L\beta_2 2\pi\Delta\nu\cdot(M-1) \\ &\cong 2\pi\Delta\nu\cdot M = 2\pi\beta_2 L B_{\mathrm{T}} \end{aligned} \quad (3.18)$$

We discard the fixed τ_0 delay (in effect shifting the time-origin by τ_0 at the receiver side). The ith received subcarrier CE is then

$$\underset{\sim}{r}_i(t) = \underset{\sim}{s}_i(t)\mathrm{e}^{-j\int_0^L \beta_i^{\mathrm{T}}(z')\mathrm{d}z'}1_{[-T_{\mathrm{CP}},-T_{\mathrm{CP}}+T_{\mathrm{B}}]}(t-\Delta\tau\cdot i). \quad (3.19)$$

Note that the two extreme subchannels (with indexes $i=0, M-1$) are associated with the respective time-windows $1_{[-T_{\mathrm{CP}},-T_{\mathrm{CP}}+T_{\mathrm{B}}]}(t)$ and $1_{[-T_{\mathrm{CP}},-T_{\mathrm{CP}}+T_{\mathrm{B}}]}(t-T_{\mathrm{CP}}) = 1_{[0,T_{\mathrm{B}}]}(t)$, consistent with the delay spread being equal T_{CP}. The Rx discards the CP, i.e., deletes the sampled data over the interval $[-T_{\mathrm{CP}}, 0]$, retaining just the samples over the $[0, -T_{\mathrm{CP}}+T_{\mathrm{B}}] = [0, T]$ interval, in which interval is included in the windows of both extreme subcarriers. In fact, this $[0, T]$ "net" interval is also included in the window $1_{[-T_{\mathrm{CP}},-T_{\mathrm{CP}}+T_{\mathrm{B}}]}(t-\Delta\tau\cdot i)$ of any of the subcarriers. Over the $[0, T]$ interval, the received ith subcarrier is expressed as

$$\underset{\sim}{r}_i(t) = \underset{\sim}{A}_i^{\mathrm{TX}}\mathrm{e}^{j2\pi i\Delta\nu t}\mathrm{e}^{-j\int_0^L \beta_i^{\mathrm{CD}}(z')\mathrm{d}z'}\mathrm{e}^{-j\int_0^L \beta_i^{\mathrm{NL}}(z')\mathrm{d}z'}\mathrm{e}^{-\int_0^L \alpha(z')\mathrm{d}z'};\quad t\in[0,T] \quad (3.20)$$

featuring an harmonic variation $\mathrm{e}^{j2\pi i\Delta\nu t}$ for the ith subchannel, conducive to frequency analysis by means of a DFT.

We develop a most general treatment allowing for z-varying fiber parameters, namely the (linear, CD related) propagation constant, $\beta_i^{\mathrm{CD}}(z)$, the NL constant $\gamma(z)$ and the differential loss, $\alpha(z)$. In particular, $\alpha(z)$ may contain (impulsive) negative components to describe the (lumped) gains of the OAs, as discussed above. However, we assume that $\alpha(z)$, $\gamma(z)$ are independent of frequency, whereas the frequency dependence of $\beta_i^{\mathrm{CD}}(z) \equiv \beta^{\mathrm{CD}}(\nu_i)$ (its dependence on the index i) is modeled as second-order dispersive (as reduced time is used in the equivalent NLSE description [30], the first-order dispersion term is absent). For example, for a homogeneous fiber link, with fixed β, γ along the fiber link, the frequency dependence of the propagation constant is:

$$\beta_i^{\mathrm{CD}} \equiv \beta^{\mathrm{CD}}(\nu_i) = \beta_0 + \frac{1}{2}\beta_2(2\pi\Delta\nu\cdot i)^2. \quad (3.21)$$

Assuming perfect compensation of the distributed losses by means of the lumped gains (negative impulses in $\alpha(z)$,) as in (3.11), we have $\int_0^L \alpha(z')\mathrm{d}z' = 0$, i.e., unity power gain, $G_p(L) = 1$ – the signal at the Tx optical preamp output is received with the same power as transmitted. Finally, assuming that all spans are identical, having constant loss α, and all signals are launched with identical power, we have $p_i(z) = \frac{1}{M} P^T(0)e^{-\alpha z}$, hence $2P^T(z) - p_i(z) = (2M-1)p_i(z) = \frac{2M-1}{M} P^T(0)e^{-\alpha z}$, yielding a total NL phase-shift

$$\begin{aligned}\phi_{NL} &\equiv \int_0^L \beta_i^{NL}(z')dz' = N_{span} \int_0^{L_{span}} \beta_i^{NL}(z')dz' = N_{span}\gamma \int_0^{L_{span}} [2P^T(z) - p_i(z)]dz' \\ &= N_{span}\gamma \frac{2M-1}{M} P^T(0) \int_0^{L_{span}} e^{-\alpha z}dz' \\ &= \frac{2M-1}{M} P^T(0)\gamma N_{span} L_{eff} = (2 - M^{-1}) P^T(0) g_{eff},\end{aligned} \tag{3.22}$$

where the effective NL gain factor g_{eff}, was introduced, with L_{eff} the *nonlinear effective length*:

$$g_{eff} \equiv \gamma N_{span} L_{eff}; \qquad L_{eff} = \int_0^{L_{span}} e^{-\alpha z}dz = (1 - e^{-\alpha L_{span}})/\alpha. \tag{3.23}$$

Thus, the ith received subchannel CE (3.14) is compactly expressed as

$$\begin{aligned}\underset{\sim}{r}_i(t) = \underset{\sim}{s}_i(t)e^{-j[\beta_0 L + \phi_{NL} + \frac{1}{2}\beta'' L(2\pi\Delta\nu \cdot i)^2]} &= \underset{\sim}{A}_{i+M/2} H_i^{TX} H_i^{CH} e^{j2\pi i \Delta\nu t} \\ 1_{[-T_{CP}, -T_{CP}+T_B]}(t - \Delta\tau \cdot i), &\quad -M/2 \le i \le M/2 - 1,\end{aligned} \tag{3.24}$$

where the subcarrier-spacing sampled TF is identified as $H_i^{CH} = \exp\{j\,[\beta_0 L + \phi_{NL} + \frac{1}{2}\beta'' L(2\pi\Delta\nu \cdot i)^2\,]\}$, i.e., each received subchannel CE is phase rotated relative to the transmitted $\underset{\sim}{s}_i(t)$ by an angle, $\beta_0 L + \phi_{NL}$, corresponding to the accumulated linear and XPM/SPM phase-shifts, as well as by a frequency-dependent angle proportional to the square $(\Delta\nu \cdot i)^2$ of the subchannel frequency deviation, corresponding to second-order CD. All channel-induced phase-shifts may be canceled by means of channel equalization and XPM compensation in the Rx. The total received signal (labeled by $^{(1)}$ to indicate that this is the linear, first-order component) is finally expressed as a superposition of the individual subchannels:

$$\underset{\sim}{r}^{(1)}(t) = \sum_{i=M_1}^{M_2} \underset{\sim}{A}_{i-M_1} H_i^{TX} H_i^{CH} e^{j2\pi i \Delta\nu t} 1_{[-T_{CP}, -T_{CP}+T_B]}(t - \Delta\tau \cdot i). \tag{3.25}$$

3.3.4 VTF for the FWM Among the Subcarriers

We next derive FWM coupling between the subcarriers, presenting the results in the streamlined Volterra NL formalism. Practitioners of NL optics, even if unfamiliar with the mathematical language of Volterra theory [44], as reviewed and elaborated in Appendix B, should find the VTF concept intuitively appealing, formalizing optical physics already well known to them. Reviewing FWM basics, three tones at freqs. ν_j, ν_k, ν_l generate a fourth tone at freq. $\nu_i = \nu_j + \nu_k - \nu_l$. In OFDM, the center frequencies (subcarriers) of the subchannels fall on a regularly spaced frequency grid: $\nu_i = i\Delta\nu + \nu_0,\ i = 1, 2, \ldots M$, hence it is convenient to label all the discrete tones by their integer indexes, $i \in \mathbb{Z}$, setting a one-to-one correspondence $\nu_i = \nu_j + \nu_k - \nu_l = (j + k - l)\Delta\nu$. Let between frequencies and their indexes the rotating phasors (ASs) describing the optical fields of the three input tones be given by,

$$u_a^j(t) = \underset{\sim}{A}_j e^{j2\pi\nu_j t}, u_a^k(t) = \underset{\sim}{A}_k e^{j2\pi\nu_k t}, u_a^l(t) = \underset{\sim}{A}_l e^{j2\pi\nu_l t}, \tag{3.26}$$

then, in elementary FWM analysis, we seek the mixing product generated by the third-order ideal nonlinearity corresponding to a lumped FWM generation mechanism. The NL-generated optical field contribution generated at frequency ν_i (indexed by i), in a differential length element of an NL medium, due to excitation by three tones with frequencies indexed by j,k,l, is $u_a^{i;jkl}(t) = (-j\gamma \mathrm{d}z)\, u_a^j(t) u_a^k(t) u_a^{l*}(t)$. Substituting the three phasors (3.26) into the last in-line equation, the NL output field at newly generated mixing frequency ν_i has the following AS and CE:

$$\begin{aligned} u_a^{i;jkl}(t) &= (-j\gamma \mathrm{d}z) \underset{\sim}{A}_j \underset{\sim}{A}_k \underset{\sim}{A}_l^* e^{j2\pi(\nu_j+\nu_k-\nu_l)t} = \underset{\sim}{U}_{i;jkl}^{(3)} e^{j2\pi\nu_i t}; \\ \underset{\sim}{U}_{i;jkl}^{(3)} &\equiv (-j\gamma \mathrm{d}z) \underset{\sim}{A}_j \underset{\sim}{A}_k \underset{\sim}{A}_l^*. \end{aligned} \tag{3.27}$$

So far we treated a differential NL element excited by three tones. For a more complicated distributed NL channel (e.g., an optically amplified fiber link), the factor $-j\gamma \mathrm{d}z$ in the elementary triple product expression (3.27) is to be replaced by a complex scaling factor $H_{i;jkl}^{\mathrm{CH}}$, generally depending on the three input tones j,k,l (which in turn determine the output tone $i = j + k - l$):

$$\underset{\sim}{U}_{i;jkl}^{(3)} = H_{i;jkl}^{\mathrm{CH}} \underset{\sim}{A}_j^{\mathrm{TX}} \underset{\sim}{A}_k^{\mathrm{TX}} \underset{\sim}{A}_l^{\mathrm{TX}*} \tag{3.28}$$

with $\underset{\sim}{A}_i^{\mathrm{TX}}$ the frequency domain sample of the input signal into the NL channel. For OFDM, we have $\underset{\sim}{A}_i^{\mathrm{TX}} \equiv \underset{\sim}{A}_i H_i^{\mathrm{TX}}$.

The complex scaling factor $H_{i;jkl}^{\mathrm{CH}}$ in (3.28), mapping the triple product of phasors of the three exciting tones into the phasor of the resulting tone, is defined as the VTF of the third-order NL system, describing the amplitude attenuation or gain and the phase-shift experienced by the mixing product excited by the three input tones. Relevant elements of Volterra NL theory are formally developed in Appendix A, generalizing to third-order the second-order Volterra treatment of [44]; however for more physically inclined readers, the description in this section may suffice. The VTF is a generalization of the concept of linear TF, applicable to NL systems. The conventional linear TF describing the complex gain of a single frequency tone is denoted in the current context $H_i \equiv H_{i;i} \equiv H(\nu_i)$. The CE of the ith tone linearly propagates according to $\tilde{U}^{(1)}_{i;i} = H_{i;i} \underset{\sim}{A}_i$.

Note that in FWM generation, for a specified output (target) tone i, once the two input tones j,k are also given, the third input tone, l, becomes redundant, as it is uniquely determined by the constraint $l = j + k - i$. We then discard this implied fourth index, l, introducing the abbreviated three-index VTF notation $H^{\mathrm{CH}}_{i;jk} \equiv H^{\mathrm{CH}}_{i;j,k,l}\Big|_{l \to j+k-i}$, expressing the output FWM contribution due to the three tones (j,k and the corresponding l making the mixing product fall onto i) as follows:

$$\underset{\sim}{U}^{(3)}_{i;jk} = H^{\mathrm{CH}}_{i;jk} \underset{\sim}{A}^{\mathrm{TX}}_j \underset{\sim}{A}^{\mathrm{TX}}_k \underset{\sim}{A}^{\mathrm{TX}*}_{j+k-i}; \quad u_a^{i;jk}(t) = \underset{\sim}{U}^{(3)}_{i;jk} \mathrm{e}^{j2\pi\nu_i t}. \tag{3.29}$$

When the input contains a multitude of tones, e.g., the multiple subcarriers in an OFDM signal, the mixing products i.e., IM tones, in brief referred to as intermods, from all possible tone triplets must be superposed. Let the input into the NL system be given by an FS, implying that it is either time-limited or periodic. Further assume that the input is represented as band-limited (BL) FFS with M harmonics.

$$\underset{\sim}{a}(t) = \sum_{i=M_1}^{M_2} \underset{\sim}{A}^{\mathrm{TX}}_i \mathrm{e}^{j2\pi i \Delta\nu t}; \quad \Delta\nu \equiv T^{-1}; \quad M = M_2 - M_1 + 1. \tag{3.30}$$

For the sake of generality, we used arbitrary summation limits M_1, M_2. Note that modifying the central frequency (carrier), relative to which the CE is defined, results in rigidly shifting all frequencies (and shifting the frequency index limits M_1, M_2 in the FFS accordingly). Another way to effectively shift M_1, M_2 is by active digital modulation (Sect. 3.2.1). Two cases of interest are the one-sided CE spectrum, with $M_1 = 0,\ M_2 = M - 1$ (corresponding to the IDFT generation in the OFDM Tx) and the almost symmetric CE spectrum, with $M_1 = -M/2,\ M_2 = M/2 - 1$ (for even M, which is typically the case in OFDM). A multitone signal such as (3.30) generates a superposition of IMs stemming from all possible triplets of frequencies. The total third-order NL field accruing all the IMs falling onto the ith frequency is given by

$$\underset{\sim}{u}^{(3)}_i(t) = \sum_{j=M_1}^{M_2} \sum_{k=M_1}^{M_2} \underset{\sim}{U}^{(3)}_{i;jk} \mathrm{e}^{j2\pi i \Delta\nu t}; \quad t \in [0, T], \tag{3.31}$$

where the summation is formally carried out over all index pairs in the domain $[M_1,\ M_2] \times [M_1,\ M_2]$; however, we allow for the possibility that given a target index i, then $H^{\mathrm{CH}}_{i;jk}$ (and $\underset{\sim}{U}^{(3)}_{i;jk}$) may be null for certain indexes j,k since for these index values, $l = j + k - i$ falls outside the $[M_1,\ M_2]$ range of data subcarriers, i.e., $\underset{\sim}{A}^{\mathrm{TX}}_{j+k-i} = 0$, nulling the FWM, hence some terms in the summation (3.31) are zero. Restricting the summation to nonnegative terms, given i, it suffices to sum j,k just over the set $S[i] \equiv \{[j,k] : j,k, M_1 \le j + k - i \le M_2,\ j \ne i \ne k\}$ of subchannel index pairs $[j,k]$ for which $l = j + k - i$ also falls within the transmitted subcarriers range $[M_1, \ldots, M_2]$. The third-order NL distortion (3.31) falling on the ith subchannel is expressed as

$$\underset{\sim}{u}_i^{(3)}(t) = \sum_{[j,k]\in S[i]} \underset{\sim}{U}_{i;jk}^{(3)} e^{j2\pi i\,\Delta\nu t} + 2 \cdot \sum_{\substack{k = M_1 \\ k \neq i}}^{M_2} \underset{\sim}{U}_{i;ik}^{(3)} e^{j2\pi i\,\Delta\nu t} + \underset{\sim}{U}_{i;ii}^{(3)} e^{j2\pi i\,\Delta\nu t}; \quad t \in [0, T]. \tag{3.32}$$

Note that by means of the condition $j \neq i \neq k$ within the definition of the set $S[i]$ of IMs we exclude from this set the XPM and SPM triplets for which $j = i$ *or* $k = i$ i.e., triplets of either the form $[i, k, l] = [i, k, k]$ or $[i, k, l] = [j, i, j]$ or $[i, k, l] = [i, i, i]$, for which IM field contributions are of the respective forms $H_{i;ik}^{\mathrm{CH}} \underset{\sim}{A}_i^{\mathrm{TX}} |\underset{\sim}{A}_k^{\mathrm{TX}}|^2$, $H_{i;ji}^{\mathrm{CH}} \underset{\sim}{A}_i^{\mathrm{TX}} |\underset{\sim}{A}_j^{\mathrm{TX}}|^2$, $H_{i;ii}^{\mathrm{CH}} \underset{\sim}{A}_i^{\mathrm{TX}} |\underset{\sim}{A}_i^{\mathrm{TX}}|^2$, seen to be coherent with the transmitted channel $\underset{\sim}{A}_i^{\mathrm{TX}}$ (XPM/SPM will be separately treated by introducing a power-dependent effective propagation constant β_i^{NL} for each narrowband subchannel). In contrast set $S[i]$ of pairs $[j, k]$ uniquely specifying the valid IMs $[j, k, j + k - i]$ falling onto subchannel i, solely includes "proper FWM" non-coherent terms, excluding the coherent terms of the form above. This set is illustrated in Fig. 3.1.

Finally note that for *out-of-band* (OOB) target indexes (i.e., $i < M_1$ or $i > M_2$), the summation (3.32) comprises noncoherent terms solely. So far we derived the FWM field at a single target frequency i. The total NL field over the full band is a superposition over all i tones: $\underset{\sim}{u}^{(3)}(t) = \sum_{i=2M_1-M_2}^{2M_2-M_1} \underset{\sim}{u}_i^{(3)}(t)$. This field spectrally spans the in-band region as well as two OOB regions adjacent to the in-band region from either side, wherein there are no transmitted subchannels, yet IM products

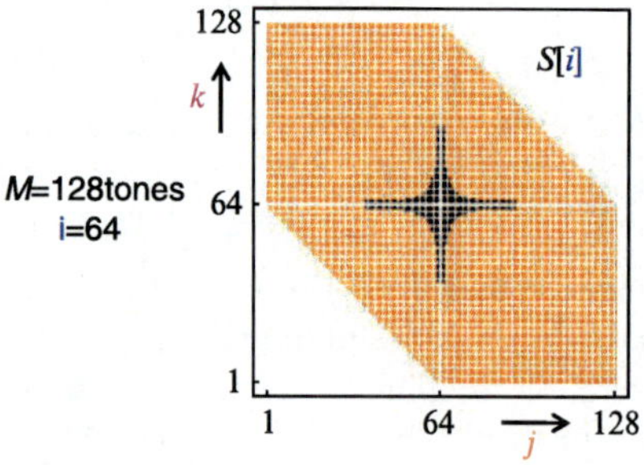

Fig. 3.1 The set of $[j, k]$ subcarrier labels in unique correspondence with the set of proper FWM triplets of subcarriers with IM falling on a given subchannel i. Adapted with permission from Fig. 1 of [30]

do fall within these OOB regions. Substituting (3.32) into the last equation yields the complete FS expansion of the NL system output over the $[0, T]$ interval, partitioned into three spectral regions (lower-out-of-band, in-band, upper-out-of-band) corresponding to the three lines in the equation below (note that the middle line, describing the in-band intermods, includes both FWM, XPM and SPM, whereas the OOB intermods – first and last line – solely comprise FWM):

$$\begin{aligned}\underset{\sim}{u}^{(3)}(t) = &\sum_{i=2M_1-M_2}^{M_1-1} \mathrm{e}^{j2\pi i\,\Delta\nu t} \sum\sum_{[j,k]\in S[i]} H_{i;jk}^{\mathrm{CH}} \underset{\sim}{A}_j^{\mathrm{TX}} \underset{\sim}{A}_k^{\mathrm{TX}} \underset{\sim}{A}_{j+k-i}^{\mathrm{TX}*} + \sum_{i=M_1}^{M_2} \mathrm{e}^{j2\pi i\,\Delta\nu t} \\ &\times \left[\sum\sum_{[j,k]\in S[i]} H_{i;jk}^{\mathrm{CH}} \underset{\sim}{A}_j^{\mathrm{TX}} \underset{\sim}{A}_k^{\mathrm{TX}} \underset{\sim}{A}_{j+k-i}^{\mathrm{TX}*} + 2\cdot \sum_{\substack{lk=M_1 \\ k\neq i}}^{M_2} H_{i;ik}^{\mathrm{CH}} \underset{\sim}{A}_i^{\mathrm{TX}} \left|\underset{\sim}{A}_k^{\mathrm{TX}}\right|^2 \right. \\ &\left. + H_{i;ii}^{\mathrm{CH}} \underset{\sim}{A}_i^{\mathrm{TX}} \left|\underset{\sim}{A}_i^{\mathrm{TX}}\right|^2 \right] + \sum_{i=M_2+1}^{2M_2-M_1} \mathrm{e}^{j2\pi i\,\Delta\nu t} \\ &\times \sum\sum_{[j,k]\in S[i]} H_{i;jk}^{\mathrm{CH}} \underset{\sim}{A}_j^{\mathrm{TX}} \underset{\sim}{A}_k^{\mathrm{TX}} \underset{\sim}{A}_{j+k-i}^{\mathrm{TX}*} = \sum_{i=D_1}^{D_2} \underset{\sim}{U}_i^{(3)} \mathrm{e}^{j2\pi i\,\Delta\nu t}. \end{aligned} \tag{3.33}$$

The summation limits are $D_1 = 2M_1 - M_2;\ D_2 = 2M_2 - M_1$. The total number of harmonics in the NL output (3.33) due to excitation in the $[M_1,\ M_2]$ range is

$$\begin{aligned} D_h &= D_2 - D_1 + 1 = (2M_2 - M_1) - (2M_1 - M_2) + 1 = 3(M_2 - M_1) + 1 \\ &= 3(M-1) + 1 = 3M - 2. \end{aligned}$$

The harmonic coefficients $\underset{\sim}{U}_i^{(3)}$ in the last expression of (3.33) are given by the sum of all IMs (mixing products) falling onto tone i, each weighted by the corresponding VTF, e.g., in-band, i.e., for $M_1 \le i \le M_2$, we have

$$\begin{aligned}\underset{\sim}{U}_i^{(3)} = &\sum\sum_{[j,k]\in S[i]} H_{i;jk}^{\mathrm{CH}} \underset{\sim}{A}_j^{\mathrm{TX}} \underset{\sim}{A}_k^{\mathrm{TX}} \underset{\sim}{A}_{j+k-i}^{\mathrm{TX}*} + 2\cdot \sum_{\substack{k=M_1 \\ k\neq i}}^{M_2} H_{i;ik}^{\mathrm{CH}} \underset{\sim}{A}_i^{\mathrm{TX}} \left|\underset{\sim}{A}_k^{\mathrm{TX}}\right|^2 \\ &+ H_{i;ii}^{\mathrm{CH}} \underset{\sim}{A}_i^{\mathrm{TX}} \left|\underset{\sim}{A}_i^{\mathrm{TX}}\right|^2,\ M_1 \leqslant i \leqslant M_2. \end{aligned} \tag{3.34}$$

Letting $M_1 = -M/2,\ M_2 = M/2 - 1$, yields $D_1 = 2M_1 - M_2 = -1.5M + 1$; $D_2 = 2M_2 - M_1 = 1.5M - 2$. The overall NL signal is then expressed as the FFS

$$\underset{\sim}{u}^{(3)}(t) = \sum_{i=-1.5M+1}^{1.5M-2} \underset{\sim}{U}_i^{(3)} \mathrm{e}^{j2\pi i\,\Delta\nu t}. \tag{3.35}$$

3.4 OFDM Receiver: Linear and Nonlinear Modeling

The OFDM receiver was modeled in [30] in terms of an equivalent analog front-end consistent with the analog-like OFDM transmitter representation (3.6). The received CE over the full block interval is given by (3.25). Upon discarding the CP, the received CE is effectively restricted to the interval $[0, -T_{\rm CP} + T_{\rm B}] = [0, T]$. The received linear signal component over this interval is

$$\underset{\sim}{r}^{(1)}(t) = \sum_{i=M_1}^{M_2} \underset{\sim}{A}_{i-M_1} H_i^{\rm TX} H_i^{\rm CH} e^{j2\pi i \Delta\nu t} 1_{[0,T]}(t). \tag{3.36}$$

3.4.1 Rx Processing

The form of the last equation suggests that a band-pass correlator bank may be used for detection of such an orthogonal PAM signal, correlating the received signal against the orthogonal basis functions $\{e^{j2\pi i\Delta\nu t} 1_{[0,T]}(t)\}_{i=M_1}^{M_2}$. In principle, this may be realized by splitting $\underset{\sim}{r}^{(1)}(t)$ into multiple identical paths, down-converting each path to baseband, in effect frequency demultiplexing $\underset{\sim}{r}^{(1)}(t)$ by demodulating each path according to its subcarrier frequency, removing the modulation factors $\exp[j2\pi i\Delta\nu t]$, then applying integrate-and-dump (I&D) filtering $y(t) = \frac{1}{T}\int_{-T/2}^{T/2} x(t){\rm d}t$ onto each of the down-converted signals. The complex-valued output of each I&D filter is sampled at the OFDM block rate T^{-1}, then one-tap-equalized (i.e., multiplied by a complex weight) canceling the linear channel distortion, i.e., realigning the received constellation axes and normalizing the magnitude. Each of the equalized subchannel constellations is input into its own decision device (slicer). Essentially, this was the Rx model used in [30].

A more precise receiver description is based on faithful representation of the actual Rx processing, as described next: The Rx front-end consists of a coherent optical hybrid, extracting the received signal CE by beating the received signal with In-Phase and Quadrature (I/Q) *local oscillators* (LO) at the carrier frequency ν_0 around which the transmitted CE is approximately situated. The coherent hybrid I/Q outputs are fed to a pair *analog-to-digital converters* (ADCs). Let $h_{\rm RX}(t)$ be the analog response of the Rx front-end, including the ADC *antialiasing* (AA) filter. Let us initially assume that the ADC samples the received CE at "baud-rate," i.e., samples are taken at the receiver chip intervals, $T_{\rm c}^{\rm RX} = T_{\rm F}/D = T/M$ ($T_{\rm c}^{\rm RX}$ may differ from the transmitter chip intervals $T_{\rm c}$, as the Tx may use DAC interpolation), yielding the following sequence of samples of the received OFDM block (ignoring NL impairments):

$$\begin{aligned}\underset{\sim}{r}_n^{(1)} &= \underset{\sim}{r}^{(1)}(t) \otimes h_{\rm RX}(t)\Big|_{t\to nT/M} \\ &= \sum_{i=M_1}^{M_2} \underset{\sim}{A}_{i-M_1} H_i^{\rm TX} H_i^{\rm CH} e^{j2\pi i\Delta\nu t}\, 1_{[0,T]}(t) \otimes h_{\rm RX}(t)\Big|_{t\to nT/M}\end{aligned}$$

$$= \sum_{i=M_1}^{M_2} \underset{\sim}{A}_{i-M_1} H_i^{\mathrm{TX}} H_i^{\mathrm{CH}} H_i^{\mathrm{RX}}(t) \mathrm{e}^{j2\pi i \Delta\nu nT/M}$$
$$= \sum_{i=-M/2}^{M/2-1} \underset{\sim}{A}_{i+M/2} H_i^{\mathrm{LINK}} \, \mathrm{e}^{j2\pi i\, n/M}; \; n = 0, 1, \ldots, M-1, \quad (3.37)$$

where $H_i^{\mathrm{RX}} \equiv H^{\mathrm{RX}}(i\,\Delta\nu)$ are frequency samples of the BL Tx response $H^{\mathrm{RX}}(\nu)$, the *link* TF is $H_{\mathrm{i}}^{\mathrm{LINK}} = H_i^{\mathrm{TX}} H_i^{\mathrm{CH}} H_i^{\mathrm{RX}}$, and in the last expression in (3.37) the generic summation limits M_1, M_2 were set to $M_1 = -M/2;\; M_2 = M/2-1$, their two-sided values, as transmitted.

Note that the third equality in (3.37) an approximation (similarly to the (3.133) at the Tx side) ignoring end-interval effects, and assuming that the duration of $h^{\mathrm{RX}}(t)$ is small relative to the $\mathbf{1}_{[-T_{\mathrm{CP}},-T_{\mathrm{CP}}+T_{\mathrm{F}}]}(t)$ window duration:

$$h_{\mathrm{RX}}(t) \otimes \left\{ \mathbf{1}_{[-T_{\mathrm{CP}},-T_{\mathrm{CP}}+T_{\mathrm{B}}]}(t) \mathrm{e}^{j2\pi i \Delta\nu t} \right\} \cong H_i^{\mathrm{RX}} \mathrm{e}^{j2\pi i \Delta\nu t} \mathbf{1}_{[-T_{\mathrm{CP}},-T_{\mathrm{CP}}+T_{\mathrm{B}}]}(t). \quad (3.38)$$

The two-sided spectrum (3.37) is up-converted (U/C) in the Rx to a one-sided spectrum (directly amenable to FFT analysis), by digitally modulating it with the same midband digital carrier $c_n = (-1)^n = \mathrm{e}^{j\pi n} = \mathrm{e}^{j2\pi(M/2)n/M}$ as used in the Tx to map the SSB spectrum to a two-sided version (note that c_n is its own inverse). This alternate-sign-flipping operation, of very low complexity, up-shifts the spectrum by $M/2$ units:

$$\underset{\sim}{r}_n^{(1)\,\mathrm{U/C}} = c_n \underset{\sim}{r}_n^{(1)} = \mathrm{e}^{j2\pi(M/2)n/M} \sum_{i=-M/2}^{M/2-1} \underset{\sim}{A}_{i+M/2} H_i^{\mathrm{LINK}} \mathrm{e}^{j2\pi i\, n/M}$$
$$= \sum_{i=-M/2}^{M/2-1} \underset{\sim}{A}_{i+M/2} H_i^{\mathrm{LINK}} \mathrm{e}^{j2\pi(i+M/2)n/M} = \sum_{i=0}^{M-1} \underset{\sim}{A}_i H_{i-M/2}^{\mathrm{LINK}} \mathrm{e}^{j2\pi in/M}. \quad (3.39)$$

The last expression in (3.39) identifies the vector of received samples at the ADC outputs as an IDFT:

$$\underset{\sim}{r}_n^{\mathrm{U/C}} = M \cdot \mathrm{IDFT}_M \{ \underset{\sim}{A}_i H_{i-M/2}^{\mathrm{LINK}} \}; \; n = 0, 1, \ldots, M-1. \quad (3.40)$$

This immediately evokes that the next Rx processing step ought to undo the IDFT by means of a DFT, yielding

$$\underset{\sim}{\rho}_i = M^{-1} \cdot \mathrm{DFT}_M \left\{ \underset{\sim}{r}_n^{\mathrm{U/C}} \right\}; \; i = 0, 1, \ldots, M-1 \quad (3.41)$$

$$\underset{\sim}{\rho}_i^{(1)} = \underset{\sim}{A}_i H_{i-M/2}^{\mathrm{LINK}} = \underset{\sim}{A}_i H_{i-M/2}^{\mathrm{TX}} H_{i-M/2}^{\mathrm{CH}} H_{i-M/2}^{\mathrm{RX}}; \; i = 0, 1, \ldots, M-1. \quad (3.42)$$

The linear distortion affecting the transmitted symbols is readily undone (equalized) by dividing each of the $\underset{\sim}{\rho}_i$ out by $H^{\text{LINK}}_{i-M/2}$ (in effect applying one complex tap to each of the subcarriers – DFT output samples), provided the overall link response H^{LINK}_i has been estimated in advance (in a practical implementation the complex taps would be adjusted adaptively).

Our receiver digitally samples, at baud-rate, the optical wave-field at the output of the NL fiber transmission channel. We next consider the impairment due to the NL fluctuation components corrupting in the receiver input, accounting for the sampling rate effects. The insights of our analysis are critical to crafting an effective NL compensation strategy.

3.4.2 Aliasing of NL Components in a Baud-Rate OFDM Receiver

The input into the channel is modeled as an FFS signal (3.9). The NL propagation of this signal through the channel generates spectral broadening – new harmonics appear in the channel output. For a third-order Volterra nonlinearity, the input frequency span (difference between extreme tones) is $(M-1)\Delta\nu$, while the output span is approximately three times larger, due to the NL broadening, $(M_h - 1)\,\Delta\nu = (3M-3)\Delta\nu$, where M_h is the total number of harmonics, including the NL-generated ones. However, accounting for the finite width of the spectral shape convolved around each of the frequency tones, the extreme subcarriers further extend out by $\Delta\nu/2$ on each side. The input *spectral span* is then $B_{\text{T}} \equiv M\Delta\nu$. A similar argument for the output spectral extent adds up twice $3{\cdot}\Delta\nu/2$ to $(3M-3)\Delta\nu$ yielding $3M\Delta\nu = 3B_{\text{T}}$, i.e., the third-order nonlinearity generates threefold spectral expansion. The same conclusion may be alternatively be obtained by convolving-correlating the analog input spectrum with itself three times. The received signal is of the form (3.33). Inspecting the summation limits in that equation corroborates the spectral broadening claim. In order to conserve transmission bandwidth, while exploiting I/Q multiplexing, the transmitted spectrum is typically centered around the carrier by applying digital D/C, such that its harmonics span the $\{-M/2,\ M/2-1\}$ range, as explained in Sect. 3.2.1, i.e., the linear component of the transmitted CE becomes two-sided over the range $[-W, W]$, with $W = B_{\text{T}}/2$. The NL components of the received envelope are then of the form (3.35).

To reconstruct the linear component in the received signal, it suffices to sample it at the Nyquist rate $f_s = B_{\text{T}}$; however at this sampling rate, the threefold spectrally wider NL component in the received signal is evidently severely undersampled. Let us develop some insight into the resulting aliasing of the time-domain third-order NL signal at the channel output, at over the $[0, T]$ interval, which signal is expressed as follows by specializing (3.33) to $z = L$:

$$\underset{\sim}{r}^{(3)}(t) \equiv \underset{\sim}{u}^{(3)}(L,t) = \sum_{i=-1.5M+1}^{1.5M-2} \underset{\sim}{R}_i^{(3)} e^{j2\pi i\,\Delta\nu t} 1_{[0,T]}(t) \quad (3.43)$$

$$\underset{\sim}{R}_i^{(3)} = \begin{cases} \sum\sum_{[j,k]\in S[i]} H^{\mathrm{CH}}_{i;jk} \underset{\sim}{A}_j^{\mathrm{TX}} \underset{\sim}{A}_k^{\mathrm{TX}} \underset{\sim}{A}_{j+k-i}^{\mathrm{TX}*}; \\ D_1 = -1.5M+1 \leqslant i \leqslant -0.5M-1 = M_1 - 1 \\ \sum\sum_{[j,k]\in S[i]} H^{\mathrm{CH}}_{i;jk} \underset{\sim}{A}_j^{\mathrm{TX}} \underset{\sim}{A}_k^{\mathrm{TX}} \underset{\sim}{A}_{j+k-i}^{\mathrm{TX}*} + 2\cdot \sum_{\substack{lk=M_1 \\ k\neq i}}^{M_2} H^{\mathrm{CH}}_{i;ik} \underset{\sim}{A}_i^{\mathrm{TX}} \\ \left|\underset{\sim}{A}_k^{\mathrm{TX}}\right|^2 + H^{\mathrm{CH}}_{i;ii} \underset{\sim}{A}_i^{\mathrm{TX}} \left|\underset{\sim}{A}_i^{\mathrm{TX}}\right|^2; \quad M_1 = -0.5M \leqslant i \leqslant 0.5M-1 \\ = M_2 \sum\sum_{[j,k]\in S[i]} H^{\mathrm{CH}}_{i;jk} \underset{\sim}{A}_j^{\mathrm{TX}} \underset{\sim}{A}_k^{\mathrm{TX}} \underset{\sim}{A}_{j+k-i}^{\mathrm{TX}*}; \\ M_2 + 1 = 0.5M \leqslant i \leqslant 1.5M-2 = D_2. \end{cases} \quad (3.44)$$

3.4.3 Oversampling the NL Output

As the output (3.44) is a T-periodic FFS with $M_h = 3M - 2$ NL-generated harmonics, which are generally nonzero, the proper Nyquist rate to sample it at, is that which would collect M_s samples over the T interval, such that $M_s \geq M_h = 3M-2$ (indeed, the FFS bandwidth – size of the spectral support – is $M_h\Delta\nu$, whereas the sampling rate may be expressed as $M_s/T = M_s\Delta\nu$, thus the sampling rate does exceed the two-sided bandwidth, satisfying the Nyquist criterion). A sampling rate $\geq 3M$ per T seconds would then avoid aliasing of the third-order nonlinearities generated in the fiber channel. However, as both $M,\ M_s$ should be powers-of-two for efficient FFT realizations, we should adopt oversampling by a factor which is a power-of-two, the lowest such factor mitigating aliasing being 4, i.e., $M_s = 4M$ samples are to be collected over the T-interval to reconstruct the full NL information. In fact as there may be some residual energy beyond three times the transmitted bandwidth, due to higher order IM products generated by higher-order nonlinearity in the fiber (e.g., fifth order, or seventh order – must be odd order due to the centro-symmetry of the fiber), then sampling at four times the transmitted signal bandwidth may somewhat alleviate the additional spectral broadening. Let us then declare the effective number of NL harmonics to be $M_h^{\mathrm{eff}} = 4M$ (even if the actual number of harmonics were $3M$, e.g., as for strictly third-order nonlinearity, we may always extend the $3M$-long vector of harmonic coefficients to length $4M$, by zero-filling). If higher-order nonlinearity is considered, the number of nonzero NL harmonics will extend beyond $4M$, and we shall just cutoff the tails of higher order harmonics at $4M$, by means of an AA filter with four times the bandwidth, assuming that the energy of the higher-order harmonics beyond $4M$ is small – if these higher-order NL harmonics are nonnegligible and they are not antialiased, then they will alias back in-band, introducing some error. As $M_h^{\mathrm{eff}} = 4M$, the proper Nyquist sampling rate for it is $M_s = M_h^{\mathrm{eff}} = 4M$. The NL coefficients $\underset{\sim}{R}_i^{(3)}$ (both in-band and

OOB) would then be precisely reconstructed. Such oversampling strategy, precisely reconstructing the NL components, enables in principle full NL compensation. Unfortunately, fourfold oversampling is practically prohibitive for ultra-high speed applications (e.g., to carry 100G OFDM with QPSK modulation of the subcarriers may require $B_T \approx 32\,\text{GHz}$, which would call for a prohibitive 128 Gsamp s^{-1} oversampling sampling rate). A Volterra NL compensation method was introduced [45, 50], not requiring oversampling, but rather sampling the OFDM signal at the baud-rate (just M rather than $4M$ samples per T interval). Nevertheless, oversampling is conceptually simpler to explain, and may also be used in simulations. A baud-rate sampled version of NL compensators is introduced in Sect. 3.12.

The effect of Nyquist sampling the linear component, which amounts to undersampling the NL component, is analyzed in Appendix C, along with the effect of the AA filtering.

3.5 Derivation of the FWM VTF: OPI Model of Third-Order NL+CD Propagation

In this section, we analytically derive the VTF of the NL impairment over a dispersive medium with $\chi^{(3)}$ nonlinearity interacting with CD, providing an analytical description of the FWM/XPM/SPM nonlinearity for an OFDM signal launched into an arbitrary fiber link, possibly with inhomogeneous fiber parameters, $\alpha(z), \beta(z), \gamma(z)$. We introduce a novel OPI formulation of the problem, which is equivalent to the perturbation-based solution of the NLSE (3.12), as carried in [30], yet is more physically insightful and intuitive.

3.5.1 OPI Approach

A differential equation solution of the NLSE for a multitone OFDM signal was pursued in [30], whereas here we develop an alternative derivation in terms of the OPI point of view, which turns out to provide the most intuitive understanding of the mechanisms of NL FWM generation in propagation along a distributed medium. The key idea is that the NL polarization current, induced in each differential length element along the fiber, acts in effect as a tiny antenna radiating an infinitesimal field contribution, which propagates forward to the end of the link. Each elemental "antenna" is in turn excited by the NL mixing of three incident pump fields. We shall evaluate the contribution of each span to the build-up of each FWM IM, by integrating over all the differential length elements along the span. Subsequently, superposing the "macro" contributions from all the spans will be seen to amount to the action of a *phased array* (PA) of spatially distributed antennas, yielding the so-called "phased-array effect" [29].

3.5.2 *Quasilinear Propagation Transfer Function*

We introduce an effective TF $H_{[z_1,z_2]}(\nu)$, referred to as QLP-TF, describing evolution of a monochromatic optical field at frequency ν from position z_1 to position z_2 along the fiber link (possibly, the segment $[z_1, z_2]$ includes multiple spans or heterogeneous fiber segments, and/or parts thereof) accounting for dispersion, loss and SPM/XPM of a narrowband STCE $\underset{\sim}{u}_i(t;z)$ centered on frequency ν, but ignoring the FWM NL interaction with similar wave-packets at other frequencies:

$$H_{[z_1,z_2]}(\nu) = F_t\{\underset{\sim}{u}_i(t;z_2)\}/F_t\{\underset{\sim}{u}_i(t;z_1)\}, \tag{3.45}$$

where the subscript t indicates that the Fourier transform is over the time variable (all relevant CE signals in this chapter are functions of time, though the time dependences are not always explicitly indicated). We shall use the shorthand notation $H^i_{[z_1,z_2]} = H_{[z_1,z_2]}(\nu_i)$ for the propagation TF sampled at the center frequency $\nu = \nu_i$ of the narrowband signal.

The index i indicates that the propagated narrowband wave-packet is centered on a point of the frequency grid, $\nu_i = i\Delta\nu + \nu_0$. Note this is not a proper TF in the linear sense (hence the terminology quasilinear), as it accounts for XPM/XPM, i.e., the QLP-TF is dependent on the power of the ith subchannel and of the neighboring subchannels.

Similarly to the derivation in (3.14), (see also [30]), the narrowband packet centered at frequency ν_i propagates as

$$\underset{\sim}{u}_i(t;z_2) = \underset{\sim}{u}_i(t;z_1)\mathrm{e}^{-j\int_{z_1}^{z_2}\beta_i^{\mathrm{T}}(z')\mathrm{d}z'} = \underset{\sim}{u}_i(t;z_1)H^i_{[z_1,z_2]}, \tag{3.46}$$

where in the second inequality we identified the QLP-TF as

$$H^i_{[z_1,z_2]} = \mathrm{e}^{-j\int_{z_1}^{z_2}\beta_i^{\mathrm{T}}(z')\mathrm{d}z'} = G_{[z_1,z_2]}\mathrm{e}^{j\angle H^i_{[z_1,z_2]}} \tag{3.47}$$

with magnitude and phase given by

$$G_{[z_1,z_2]} = \left|H^i_{[z_1,z_2]}\right| = \mathrm{e}^{-j\int_{z_1}^{z_2}\alpha(z')\mathrm{d}z'};\ \angle H^i_{[z_1,z_2]}$$
$$= -\int_{z_1}^{z_2}\beta_i^{\mathrm{CD}}(z')\mathrm{d}z' - \int_{z_1}^{z_2}\beta_i^{\mathrm{NL}}(z')\mathrm{d}z', \tag{3.48}$$

where the total effective propagation constant, β_i^{T}, includes a linear component (labeled as CD to indicate its dispersive origin), a NL (power-dependent) component, and a loss component represented as imaginary propagation constant:

$$\beta_i^{\mathrm{T}} = \beta_i^{\mathrm{CD}}(z) + \beta_i^{\mathrm{NL}}(z) - j\alpha(z)/2;\ \beta_i^{\mathrm{NL}}(z) = 2\gamma(z)\left[P^{\mathrm{T}}(z) - p_i(z)\right];$$
$$P^{\mathrm{T}}(z) \equiv \sum_{i=M_1}^{M_2}\left|\underset{\sim}{u}_i(z)\right|^2;\ p_i(z) \equiv \left|\underset{\sim}{u}_i(z)\right|^2. \tag{3.49}$$

Here, we pursue a general treatment allowing z-varying parameters: propagation constant, $\beta_i^{\mathrm{CD}}(z)$, NL constant $\gamma(z)$, and differential loss, $\alpha(z)$. However, we assume that $\alpha(z), \gamma(z)$ are independent of frequency, whereas the frequency dependence of $\beta_i^{\mathrm{CD}}(z)$ (its dependence on the index i) is modeled as in (3.21) as dispersive to second-order (as reduced time is used, the first-order dispersion term is absent):

$$\beta_i^{\mathrm{CD}}(z) = \beta_0 + \frac{1}{2}\beta_2(z)\Omega_i^2 = \beta_0 + \frac{1}{2}\beta_2\,(2\pi\Delta\nu)^2\,i^2; \qquad \Omega_i \equiv 2\pi\Delta\nu\cdot i. \tag{3.50}$$

The following transitivity property of the narrowband propagation TF readily stems from the definition (3.45) [or from (3.47)]:

$$H^i_{[z_1,z_2]}H^i_{[z_2,z_3]} = H^i_{[z_1,z_3]}. \tag{3.51}$$

3.5.3 Virtual Backpropagated Fields

A normalized version $\underset{\sim}{v}_i(t;z)$ of the STCE $\underset{\sim}{u}_i(t;z)$ was introduced in [30, (22)] leading to a simplification of the NLSE solution. The v-normalization is reformulated here as division of the u-field at point z through the QLP-TF from the input to point z:

$$\underset{\sim}{v}_i(t;z) \equiv \underset{\sim}{u}_i(t;z)/H^i_{[0,z]} = \underset{\sim}{u}_i(t;z)\mathrm{e}^{j\int_0^z \beta_i^{\mathrm{T}}(t,z')\mathrm{d}z'}. \tag{3.52}$$

The v-normalized field is essentially the u-field at z referred back to the input $z = 0$ $\left(\text{back-propagated through } \left(H^i_{[0,z]}\right)^{-1}\right)$. The v-field, $\underset{\sim}{v}_i(t;z)$, associated with a given u-field, $\underset{\sim}{u}_i(t;z)$, at position z, may be described as a *virtual field* at $z = 0$, which, after forward propagation through $H^i_{[0,z]}$ would coincide with the actual u-field at position z:

$$\underset{\sim}{u}_i(t;z) = \underset{\sim}{v}_i(t;z)H^i_{[0,z]} = \underset{\sim}{v}_i(t;z)\mathrm{e}^{-j\int_{z_1}^{z_2}\beta_i^{\mathrm{T}}(z')\mathrm{d}z'}. \tag{3.53}$$

It is readily seen that the virtual field $\underset{\sim}{v}_i^{(1)}(t;z)$ of any first-order narrowband field stays invariant along z. Indeed, as per (3.46), first-order fields evolve according to the TF $H^i_{[0,z]} : \underset{\sim}{u}_i^{(1)}(t;z) = \underset{\sim}{u}_i^{(1)}(t;0)H^i_{[0,z]}$. Substituting this into (3.52) yields

$$\breve{v}_i^{(1)}(t;z) \equiv \underset{\sim}{u}_i^{(1)}(t;z)/H^i_{[0,z]} = \underset{\sim}{u}_i^{(1)}(t;0)H^i_{[0,z]}/H^i_{[0,z]} = \underset{\sim}{u}_i^{(1)}(t;0) = \underset{\sim}{v}_i^{(1)}(t;0), \tag{3.54}$$

where the last equality was obtained by setting $z = 0$ in (3.53), and using $H^i_{[0,0]} = 1$. Thus, the v-normalized virtual first-order field is constant along z, in fact equal to the u-field initial condition: $\underset{\sim}{v}_i^{(1)}(t,z) = \underset{\sim}{u}_i^{(1)}(t;0)$. In the special case of m-ary PSK (e.g., QPSK) OFDM transmission, of interest in this paper, and assuming all subchannel powers are launched equal, we have

$$\underset{\sim}{v}_i^{(1)}(t,z) = \underset{\sim}{u}_i^{(1)}(t;0) = \sqrt{p_0(t)}\mathrm{e}^{j\phi_i(t)}. \tag{3.55}$$

The invariance along z of virtual first-order fields yields a simple description of the quasilinear (linear + XPM/SPM) propagation components. The utility of the virtual field concept (3.53) pertains to modeling higher order perturbation fields, providing the most compact description of the generation of higher perturbation orders. The virtual field concept facilitates the analysis of NL propagation by referring all fields to a common plane, $z = 0$.

3.5.4 *OPI Derivation of the VTF of a General Inhomogeneous Fiber Link*

We next work out the third-order perturbation fields without solving the differential NLSE, but rather adopting a more insightful OPI approach. The main physical idea is to propagate the three first-order subcarrier waves from the input until they reach a differential length element dz at position z; the three waves nonlinearly mix within the NL element, and the resulting IM, at a new frequency, propagates to the output; the IMs generated by all triplets of subcarriers are superposed, and the output contributions from all differential length elements are integrated along the fiber.

The superposition of the FWM IMs falling on the ith frequency, due to a differential length element at position z, is given by

$$\mathrm{d}\underline{u}_i^{(3)}(z) \equiv -j\gamma \mathrm{d}z \sum_{[j,k]\in S[i]} \underline{u}_j^{(1)}(z)\underline{u}_k^{(1)}(z)\underline{u}_{j+k-i}^{(1)*}(z), \tag{3.56}$$

where for all fields the t-dependence is not explicitly mentioned. The $^{(3)}$ superscript indicates the mixing of three "pump" fields, each of which is propagated from the input to the differential element at z, via its respective QLP-TF, e.g., $\underline{u}_j^{(1)}(z) = \underline{u}_j^{(1)}(0)H_{[0,z]}^j = \underline{v}_j^{(1)}H_{[0,z]}^j$, with similar relations for the other two terms. Substituting these QLP-TF relations into (3.56) yields (with $l = j + k - i$):

$$\mathrm{d}\underline{u}_i^{(3)}(z) \equiv -j\gamma \mathrm{d}z \sum_{[j,k]\in S[i]} \underline{v}_j^{(1)}\underline{v}_k^{(1)}\underline{v}_{j+k-i}^{(1)*}H_{[0,z]}^j H_{[0,z]}^k H_{[0,z]}^{l*}. \tag{3.57}$$

The total third-order IM at frequency i at the end of the fiber link is obtained by propagating the differential contribution from position z to the fiber end $z = L$, and integrating over all the differential contributions (we present both u- and v-versions):

$$\underline{u}_i^{(3)}(L) \equiv \int_0^L H_{[z,L]}^i \mathrm{d}\underline{u}_i^{(3)}(z) \tag{3.58}$$

$$\begin{aligned}\underline{v}_i^{(3)}(L) &= \left(H_{[0,L]}^i\right)^{-1}\underline{u}_i^{(3)}(L) = \left(H_{[0,L]}^i\right)^{-1}\int_0^L H_{[z,L]}^i \mathrm{d}\underline{u}_i^{(3)}(z)\\ &= \int_0^L \left(H_{[0,z]}^i\right)^{-1}\mathrm{d}\underline{u}_i^{(3)}(z),\end{aligned} \tag{3.59}$$

where we used $\left(H^i_{[0,L]}\right)^{-1} H^i_{[z,L]} = \left(H^i_{[0,z]}\right)^{-1} \Leftrightarrow H^i_{[0,L]} = H^i_{[0,z]} H^i_{[z,L]}$ consistent with the transitivity property (3.51).

The integrand in the last expression in (3.59) is interpreted as propagating the IM differential contribution at z back to the input plane $z = 0$. Substituting (3.57) into the last expression in (3.59) and interchanging the orders of summation and integration yields the following Volterra trilinear superposition expression:

$$\begin{aligned} \underset{\sim}{v}_i^{(3)}(L) &= \sum_{[j,k]\in S[i]} \underset{\sim}{v}_j^{(1)} \underset{\sim}{v}_k^{(1)} \underset{\sim}{v}_{j+k-i}^{(1)*} \int_0^L (-j\gamma)\, H^j_{[0,z]} H^k_{[0,z]} H^{l*}_{[0,z]} \left(H^i_{[0,z]}\right)^{-1} \mathrm{d}z \\ &= \sum_{[j,k]\in S[i]} \underset{\sim}{v}_j^{(1)} \underset{\sim}{v}_k^{(1)} \underset{\sim}{v}_{j+k-i}^{(1)*} H^{i;jk}_{[0,L]}, \end{aligned} \tag{3.60}$$

where in the last expression in (3.60) we introduced the overall fiber link VTF, $H^{i;jk}_{[0,L]}$, expressed by integrating the FWM contributions of all the differential elements in the range $[0, L]$:

$$H^{i;jk}_{[0,L]} \equiv \int_0^L (-j\gamma)\, H^j_{[0,z]} H^k_{[0,z]} H^{l*}_{[0,z]} \left(H^i_{[0,z]}\right)^{-1} \mathrm{d}z. \tag{3.61}$$

We physically account for this VTF expression as follows: The integration superposes the IM contributions (associated with each triplet of tones) from all the differential elements along the fiber, and then virtually back-propagates it to the input (effecting the v-normalization). Indeed, the first-order perturbation fields incident onto the differential element dz at z are obtained by propagating the incident v-fields from position 0 to position z, via the three respective QLP-TFs at frequencies j,k,l. The NL polarization current generated in the element dz at z, and its induced secondary field at the ith IM frequency, are proportional to the product of the three exciting fields (with the third field complex-conjugated): $-j\gamma(z)\big(H^j_{[0,z]}\underset{\sim}{v}_j^{(1)}\big)\big(H^k_{[0,z]}\underset{\sim}{v}_k^{(1)}\big)\big(H^{j+k-i}_{[0,z]}\,\underset{\sim}{v}_{j+k-i}^{(1)}\big)^*$, where $\underset{\sim}{v}_j^{(1)}$ coincides with the initial condition $\underset{\sim}{u}_j^{(1)}(0)$, and likewise for j,k. Finally, the multiplication of the last expression by the TF$(H^i_{[0,z]})^{-1}$ back-propagates the secondary field (excited at the intermod frequency i) from position z back to the input $z = 0$ (this is equivalent to propagating the secondary field from z all the way to the end of the link ($z = L$), over a distance L–z then back-propagating over a distance L to the origin, $z = 0$). It remains to evaluate the VTF integral expression (3.61). First evaluate its integrand, compactly denoted as $H^{i;jk}_{[0,z,0]}$ (the label $[0, z, 0]$ indicates propagation of the three first-order fields from $z = 0$ to the differential element at z, then back-propagating to $z = 0$):

$$H^{i;jk}_{[0,z,0]} \equiv -j\gamma(z) H^j_{[0,z]} H^k_{[0,z]} \left(H^{j+k-i}_{[0,z]}\right)^* \left(H^i_{[0,z]}\right)^{-1}; \qquad H^{i;jk}_{[0,L]} \equiv \int_0^L H^{i;jk}_{[0,z,0]} \mathrm{d}z. \tag{3.62}$$

Expressing the QLP-TFs appearing in $H^{i;jk}_{[0,z,0]}$ (3.62) in terms of magnitudes and phases, as in (3.48), yields

$$H^i_{[0,z]} = G_{[0,z]} \mathrm{e}^{j\angle H^i_{[0,z]}}; \quad \angle H^i_{[0,z]} = -\int_0^z \beta_i^{\mathrm{CD}}(z')\mathrm{d}z' - \int_0^z \beta_i^{\mathrm{NL}}(z')\mathrm{d}z', \quad (3.63)$$

where the frequency superscript i was discarded off $G^i_{[0,z]}$, as the fiber loss $\alpha(z)$ is assumed independent of frequency. Substituting (3.63) into (3.62) and algebraically simplifying finally yields

$$\begin{aligned} H^{i;jk}_{[0,z,L]} &= -j\gamma(z) G_{[0,z]} G_{[0,z]} \left(G_{[0,z]}\right)^* \left(G_{[0,z]}\right)^{-1} \mathrm{e}^{j\angle H^j_{[0,z]}} \mathrm{e}^{j\angle H^k_{[0,z]}} \\ &\quad \times \left(\mathrm{e}^{j\angle H^{j+k-i}_{[0,z]}}\right)^* \left(\mathrm{e}^{j\angle H^i_{[0,z]}}\right)^{-1} \\ &= -j\gamma(z) G^2_{[0,z]} \mathrm{e}^{j\left(\angle H^j_{[0,z]} + \angle H^k_{[0,z]} - \angle H^{j+k-i}_{[0,z]} - \angle H^i_{[0,z]}\right)} = -j\gamma(z) G^2_{[0,z]} \cdot \\ &\quad \times \exp\left\{-j\left[\int_0^z \Delta\beta^{\mathrm{CD}}_{i;jk}(z')\mathrm{d}z' + \int_0^z \Delta\beta^{\mathrm{NL}}_{i;jk}(z')\mathrm{d}z'\right]\right\}, \end{aligned} \quad (3.64)$$

where (omitting the z-dependence for brevity) the CD-induced $\Delta\beta$ mismatch is given by

$$\begin{aligned} \Delta\beta^{\mathrm{CD}}_{i;jk} &\equiv \beta_j^{\mathrm{CD}} + \beta_k^{\mathrm{CD}} - \beta^{\mathrm{CD}}_{j+k-i} - \beta_i^{\mathrm{CD}} = \frac{\beta_2}{2}\left[\left(\Omega_j^2 + \Omega_k^2 - \Omega^2_{j+k-i}\right) - \Omega_i^2\right] \\ &= -\beta_2 \left(2\pi\Delta\nu\right)^2 (j-i)(k-i) \end{aligned} \quad (3.65)$$

with the two last equalities obtained using (3.50). The NL-induced $\Delta\beta$ mismatch in (3.64) is given by

$$\Delta\beta^{\mathrm{NL}}_{i;jk}(z) \equiv \beta_j^{\mathrm{NL}}(z) + \beta_k^{\mathrm{NL}}(z) - \beta^{\mathrm{NL}}_{j+k-i}(z) - \beta_i^{\mathrm{NL}}(z) = -2\gamma(z)\Delta p^{(1)}_{i;jk}(z), \quad (3.66)$$

where

$$\Delta p^{(1)}_{i;jk}(z) \equiv p_i^{(1)}(z) + p_j^{(1)}(z) - p_k^{(1)}(z) - p^{(1)}_{j+k-i}(z) \quad (3.67)$$

is called the *power imbalance* of the IM triplet. If all OFDM subcarriers are launched with equal power (e.g., when equal power m-ary PSK constellations are used for all subchannels), then the four power terms in (3.67) evolve identically along the link, hence the four terms in the right-hand side of (3.67) are equal, and the power imbalance nulls out everywhere: $\Delta p^{(1)}_{i;jk}(z) = 0$. In this equi-power case, the NL term with integrand $\Delta\beta^{\mathrm{NL}}_{i;jk}(z')$ may be discarded in (3.64), reducing the differential VTF (3.64) to

$$H^{i;jk}_{[0,z,0]} \equiv -j\gamma(z) G_{\mathrm{p}}(z) \mathrm{e}^{-j\Delta\phi^{\mathrm{CD}}_{i;jk}[0,z]}, \quad (3.68)$$

where we introduced the cumulative $\Delta\beta$-phase between two z positions (and in the second expression in (3.65) was substituted):

$$\Delta\phi_{i;jk}^{\mathrm{CD}}[z_1, z_2] \equiv \int_{z_1}^{z_2} \Delta\beta_{i;jk}^{\mathrm{CD}}(z')\mathrm{d}z' = -(2\pi\Delta\nu)^2 (j-i)(k-i) \int_{z_1}^{z_2} \beta_2(z')\mathrm{d}z' \tag{3.69}$$

and defined the power gain from the input $z = 0$ to position z, as the square of the amplitude gain, $G_{\mathrm{p}}(z) \equiv G_{[0,z]}^2$.

Finally, substituting the compact differential VTF expression (3.68) into the VTF integral (3.62) yields the overall VTF from the input at $z = 0$ to the link output at $z = L$, for an arbitrary multispan link with inhomogeneous (z-dependent) $\gamma(z)$, $\beta_2(z)$,

$$H_{[0,L]}^{i;jk}\Big|_{\mathrm{inhom.}}^{\text{multi-span}} = \int_0^L H_{[0,z,0]}^{i;jk}\mathrm{d}z = -j\int_0^L \gamma(z)G_{\mathrm{p}}(z)\mathrm{e}^{-j\Delta\phi_{i;jk}^{\mathrm{CD}}[0,z]}\mathrm{d}z \tag{3.70}$$

compactly expressed in terms of integrating over the z-dependencies of the nonlinearity profile $\gamma(z)$, the power gain (and loss) profile $G_{\mathrm{p}}(z)$, and the cumulative $\Delta\beta$-phase (3.69). This is our new key result for the I/O VTF of a most general fiber link with equi-power subchannels. In the sequel, this general result is specialized to particular configurations.

3.5.5 *Homogeneous Fiber Link*

Let us assume the special case of a *homogeneous* multispan link with z-independent $\beta_i^{\mathrm{CD}}, \gamma$ parameters (but with possibly different span lengths and gain/loss profiles, i.e., allowing for arbitrary $G_{\mathrm{p}}(z)$). In this case, the $\Delta\beta$ phase integration (3.69) yields a linear function in z : $\Delta\phi_{i;jk}^{\mathrm{CD}}(z) = \Delta\beta_{i;jk}^{\mathrm{CD}}z$. Substitution into (3.70) yields a compact *Fourier transform* (FT) expression:

$$H_{[0,L]}^{i;jk}\Big|_{\mathrm{hom.}}^{\text{multi-span}} = -j\gamma\int_0^L G_{\mathrm{p}}(z)\mathrm{e}^{-j\Delta\beta_{i;jk}^{\mathrm{CD}}z}\mathrm{d}z = -j\gamma\cdot {}^{\Delta\beta_{i;jk}^{\mathrm{CD}}}F_z\left\{G_{\mathrm{p}}(z)\right\}, \tag{3.71}$$

where the FT was labeled by a right subscript z and left superscript w, respectively, indicating its input and output:

$$^{w}F_z\{f(z)\} = \int f(z)\mathrm{e}^{-jwz}\mathrm{d}z.$$

The VTF of the homogeneous link is seen to be expressed as the spatial FT of the power amplification/attenuation profile, evaluated at a spatial frequency equal to the $\Delta\beta$-mismatch. This result for the VTF of a homogeneous fiber with arbitrary gain and loss profile was already derived in [30] by means of a perturbation solution of the NLSE, but is rederived here by the OPI approach. Glimpses of this homogeneous case result (emergence of FT-like expressions) may be found in earlier works [6–23]; however, the current compact formulation has never been heretofore rigorously derived and stated in its full generality, as it is here. Moreover, we presently generalize this result to inhomogeneous links (3.70) for the first time. Prior to that, let us explore two special cases of the formalism.

3.5.6 Single Homogeneous Span

As a first application, we readily derive the VTF describing the FWM build-up for an OFDM signal over a single homogeneous fiber span: lossy, dispersive, with gain profile given by $G_{\mathrm{p}}^{\mathrm{span}}(z) = \mathrm{e}^{-\alpha z} 1_{[0,L_{\mathrm{span}}]}(z)$:

$$\begin{aligned} H_{[0,L_{\mathrm{span}}]}^{i;jk}\Big|_{\mathrm{hom.}}^{\text{single-span}} &= -j\gamma^{\Delta\beta_{i;jk}^{\mathrm{CD}}} F_z\left\{G_{\mathrm{p}}^{\mathrm{span}}(z)\right\} = -j\gamma^{\Delta\beta_{i;jk}^{\mathrm{CD}}} F_z\left\{\mathrm{e}^{-\alpha z} 1_{[0,L_{\mathrm{span}}]}(z)\right\} \\ &= -j\gamma \int_0^{L_{\mathrm{span}}} \mathrm{e}^{-\alpha z} \mathrm{e}^{-j\Delta\beta_{i;jk}^{\mathrm{CD}} z} \mathrm{d}z = -j\gamma \int_0^{L_{\mathrm{span}}} \mathrm{e}^{-\left(j\Delta\beta_{i;jk}^{\mathrm{CD}}+\alpha\right)z} \mathrm{d}z \\ &= -j\gamma \frac{1-\mathrm{e}^{-\left(j\Delta\beta_{i;jk}^{\mathrm{CD}}+\alpha\right)L_{\mathrm{span}}}}{j\Delta\beta_{ijk}+\alpha}. \end{aligned} \tag{3.72}$$

In particular, in the dispersion-free or $\Delta\beta$-matched case, $\Delta\beta_{i;jk}^{\mathrm{CD}} = 0$, (3.72) reduces to a constant expression proportional to the well-known *Effective Nonlinear Length* (ENL) parameter, L_{eff} (3.23):

$$H_{[0,L_{\mathrm{span}}]}^{i;jk}\Big|_{\Delta\beta_{i;jk}^{\mathrm{LN}}=0} = -j\gamma(1-\mathrm{e}^{-\alpha L_{\mathrm{span}}})/\alpha \equiv -j\gamma L_{\mathrm{eff}}. \tag{3.73}$$

More generally, the factor multiplying $-j\gamma$ in (3.72) has dimensions of length, and is designated *Effective FWM length* (generalizing the ENL concept, $L_{\mathrm{eff}} = (1 - \mathrm{e}^{-\alpha L_{\mathrm{span}}})/\alpha$, reducing to it in the absence of dispersion):

$$\begin{aligned} L_{i;jk}^{\mathrm{FWM}} &\equiv \left[1-\mathrm{e}^{-\left(j\Delta\beta_{i;jk}^{\mathrm{CD}}+\alpha\right)L_{\mathrm{span}}}\right] \Big/ \left[j\Delta\beta_{ijk}^{\mathrm{CD}}+\alpha\right];\ H_{[0,L_{\mathrm{span}}]}^{i;jk} = -j\gamma L_{i;jk}^{\mathrm{FWM}} \\ &= -j\gamma L_{\mathrm{eff}} \hat{L}_{i;jk}^{\mathrm{FWM}}, \end{aligned} \tag{3.74}$$

where in the last expression we normalized the Effective FWM length by the ENL: $\hat{L}_{i;jk}^{\mathrm{FWM}} \equiv \hat{L}_{i;jk}^{\mathrm{FWM}}/L_{\mathrm{eff}}$.

It is readily seen that $\left|\hat{L}_{i;jk}^{\mathrm{FWM}}\right| \le 1$ with equality achieved in the absence of dispersion, or when there is perfect phase matching.

3.5.7 "Regular" Multispan Link

Next consider a "regular" multispan link consisting of N_{span} identical optically amplified fiber spans, modeled by expressing the gain profile $G_{\text{p}}(z)$ as a finite periodic function with N_{span} identical periods ("regular" means identical spans):

$$G_{\text{p}}(z) = \sum_{s=0}^{N_{\text{span}}-1} G_{\text{p}}^{\text{span}}(z - sL_{\text{span}}) = G_{\text{p}}^{\text{span}}(z) \otimes \sum_{s=0}^{N_{\text{span}}-1} \delta(z - sL_{\text{span}}). \quad (3.75)$$

Substituting this gain profile into the VTF (3.71) and evaluating the FT yields

$$\begin{aligned} H_{[0,L]}^{i;jk}\Big|^{\text{reg. spans}} &= -j\gamma\ {}^{\Delta\beta_{i;jk}^{\text{CD}}} F_z\left\{G_{\text{p}}(z)\right\} \\ &= -j\gamma^{\Delta\beta_{i;jk}^{\text{CD}}} F_z\left\{G_{\text{p}}^{\text{span}}(z)\right\} \cdot{}^{\Delta\beta_{i;jk}^{\text{CD}}} F\left\{\sum_{s=0}^{N_{\text{span}}-1} \delta(z - sL_{\text{span}})\right\} \\ &= -j\gamma^{\Delta\beta_{i;jk}^{\text{CD}}} F_z\left\{G_{\text{p}}^{\text{span}}(z)\right\} \cdot \sum_{s=0}^{N_{\text{span}}-1} \mathrm{e}^{-j\Delta\beta_{i;jk}^{\text{CD}} L_{\text{span}} s}. \end{aligned} \quad (3.76)$$

The first term in the last expression ($-j\gamma$ times the FT) is identified as the VTF of a single span, as per (3.71):

$$H_{[0,L_{\text{span}}]}^{i;jk}\Big|^{\text{single-span}} = -j\gamma^{\Delta\beta_{i;jk}^{\text{CD}}} F_z\{G_{\text{p}}^{\text{span}}(z)\}. \quad (3.77)$$

Note that this single-span expression is still more general than the particular result (3.72), pertaining to a homogeneous span, as we have not yet specified the nature of $G_{\text{p}}^{\text{span}}(z)$. The summation in the last line of (3.76) is identified as $N_{\text{span}} F_{i;jk}$, where

$$\begin{aligned} F_{i;jk} &\equiv \frac{1}{N_{\text{span}}} \sum_{s=0}^{N_{\text{span}}-1} \mathrm{e}^{-j\Delta\beta_{i;jk}^{\text{CD}} L_{\text{span}} s} \\ &= \mathrm{e}^{-\frac{j}{2}\Delta\beta_{ijk} L_{\text{span}}(N_{\text{span}}-1)} \frac{\sin(\Delta\beta_{i;jk}^{\text{CD}} N_{\text{span}} L_{\text{span}}/2)}{N_{\text{span}} \sin(\Delta\beta_{i;jk}^{\text{CD}} L_{\text{span}}/2)} \\ &= \mathrm{e}^{-\frac{j}{2}\Delta\beta_{ijk}^{\text{CD}}(L-L_{\text{span}})} \mathrm{dinc}_{N_{\text{span}}}\left[\frac{L\Delta\beta_{ijk}^{\text{CD}}}{2\pi}\right] \end{aligned} \quad (3.78)$$

with $\mathrm{dinc}_N[u] \equiv \sin(\pi u)/\left[N \sin(\pi u/N)\right]$ a "digital sinc" or Dirichlet kernel. The function $F_{i;jk}$ is called the *array factor*, as it arises in the radiation pattern of antenna-PAs [47]. The array factor is dependent on the fiber spans geometry (length of each span and number of span), but not on the detail of each span. The array factor magnitude [dB] is plotted in Fig. 3.2.

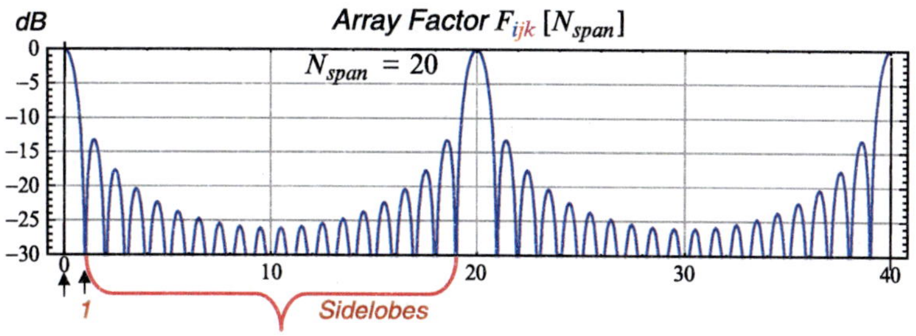

Fig. 3.2 Magnitude of normalized Array factor, $\mathrm{dinc}_N[u]$ on a dB scale. This function has period N, and its mainlobe spans the normalized argument range $|u| \le 1$

Using (3.77) for the single-span VTF and (3.78) for the array factor, the overall VTF (3.76) may be compactly expressed as

$$H_{[0,L]}^{i;jk}\Big|^{\text{reg. spans}} = N_{\text{span}}\, H_{[0,L_{\text{span}}]}^{i;jk}\Big|^{\text{single-span}} \cdot F_{i;jk}. \tag{3.79}$$

This is our main result for the VTF of a regular multispan link, irrespective of the nature of each span (which may be inhomogeneous, as long as all spans are identical). Thus, the overall VTF (3.79) of a regular link is expressed as N_{span} times the VTF of a single span (which would have corresponded to coherent addition of identical spans), scaled down by the array factor, which satisfies $|F_{i;jk}| \le 1$, reflecting partial destructive interference between the coherent, yet de-phased, contributions of the multiple spans. In the particular case that all identical spans are homogeneous (constant β, γ vs. z), we may use the particular form (3.74) for the single-span VTF, thus (3.79) reduces to the following more definite expression

$$\begin{aligned} H_{[0,L]}^{i;jk}\Big|_{\text{hom.}}^{\text{reg. spans}} &= H_{[0,L_{\text{span}}]}^{i;jk} \cdot N_{\text{span}} F_{i;jk} = -j\gamma L_{i;jk}^{\text{FWM}} \cdot N_{\text{span}} F_{i;jk} \\ &= -j\gamma L_{\text{eff}} N_{\text{span}} \hat{L}_{i;jk}^{\text{FWM}} \cdot F_{i;jk} = -j g_{\text{eff}} \hat{L}_{i;jk}^{\text{FWM}} \cdot F_{i;jk} - j g_{\text{eff}} \hat{H}_{i;jk}^{\text{FWM}}, \end{aligned} \tag{3.80}$$

where $g_{\text{eff}} \equiv \gamma L_{\text{eff}} N_{\text{span}}$ and $\hat{H}_{i;jk}^{\text{FWM}} \equiv \hat{L}_{i;jk}^{\text{FWM}} \cdot F_{i;jk}$ is a normalized form of the VTF, and all three normalized quantities have their magnitude bounded by unity: $|\hat{H}_{i;jk}^{\text{FWM}}| \le 1$; $|\hat{L}_{i;jk}^{\text{FWM}}| \le 1$; $|F_{i;jk}| \le 1$. Substituting this VTF into (3.60) yields our key result for the overall FWM contribution at the far end of a homogeneous regular link:

$$\begin{aligned} \underset{\sim}{v}_i^{(3)}(L)\Big|_{\text{hom.}}^{\text{reg. spans}} &= \sum_{[j,k]\in S[i]} \underset{\sim}{v}_j^{(1)}\, \underset{\sim}{v}_k^{(1)}\, \underset{\sim}{v}_{j+k-i}^{(1)*}\, H_{[0,L]}^{i;jk} \\ &= -j g_{\text{eff}} \sum_{[j,k]\in S[i]} \underset{\sim}{v}_j^{(1)}\, \underset{\sim}{v}_k^{(1)}\, \underset{\sim}{v}_{j+k-i}^{(1)*}\, \hat{L}_{i;jk}^{\text{FWM}} F_{i;jk}. \end{aligned} \tag{3.81}$$

For perfect phase-match (e.g., ideal CD-free link) $\hat{L}^{\text{FWM}}_{i;jk} = 1 = F_{i;jk}$, yielding $\breve{v}^{(3)}_i(L) = -jg_{\text{eff}} \sum_{[j,k]\in S[i]} \underset{\frown}{v}^{(1)}_j \underset{\frown}{v}^{(1)}_k \underset{\frown}{v}^{(1)*}_{j+k-i}$.

Note: For readers of our prior work [30], it is useful to reconcile our VTF formula (3.80) vs. the early D-notation used in [30]:

$$H^{i;jk}_{[0,L]}\Big|^{\text{reg. spans}}_{\text{hom.}} = -j\gamma D^{\text{FWM}}_{i;jk} = -j\gamma \overbrace{L_{\text{eff}} N_{\text{span}} \hat{D}^{\text{FWM}}_{i;jk}}^{D^{\text{FWM}}_{i;jk}}$$

$$= -j\gamma \overbrace{L_{\text{eff}} N_{\text{span}} \underbrace{\hat{L}^{\text{FWM}}_{i;jk} \cdot F_{i;jk}}_{\hat{H}^{\text{FWM}}_{i;jk} \equiv \hat{D}^{\text{FWM}}_{i;jk}}}^{D^{\text{FWM}}_{i;jk}}. \tag{3.82}$$

3.5.8 Irregular Inhomogenenous Links

Let us next generalize the treatment beyond [30] modeling here an irregular inhomogeneous fiber link with piecewise constant fiber parameters $\beta(z), \gamma(z)$ and with arbitrary continuous, discontinuous or even impulsive $\alpha(z)$ i.e., allowing arbitrary gain and/or loss profile, possibly different from one fiber segment to the next. This models a general fiber link configuration, allowing for concatenating diverse fiber types (including DCFs), with each fiber segment assumed uniform in its β, γ parameters, though the parameters may differ from one fiber segment to the next one (note that by concatenating a very large number of very short piecewise-constant segments, even continuously varying distributions of $\beta(z), \gamma(z)$ may be precisely approximated in the limit). In any case, the lumped or distributed gains and losses may vary within each segment and from segment to segment as reflected in the arbitrary $\alpha(z)$ profile. Let the fixed fiber parameters over the sth segment $[z_s, z_{s+1}]$ be given by $\beta^{\text{CD}}_i[s], \gamma[s]$, where $s = 0, 1, \ldots, N_{\text{seg}} - 1$.

The cumulative $\Delta\beta$-phase is readily integrated over the sth piecewise constant segment. For $z \in [z_s, z_{s+1}]$ (3.69) yields:

$$\Delta\phi^{\text{CD}}_{i;jk}(z) = \Delta\phi^{\text{CD}}_{i;jk}(z_s) + \Delta\beta^{\text{CD}}_{i;jk}[s](z - z_s). \tag{3.83}$$

The accumulated phase at the sth segment right end is

$$\Delta\phi^{\text{CD}}_{i;jk}(z_{s+1}) = \Delta\phi^{\text{CD}}_{i;jk}(z_s) + \Delta\beta^{\text{CD}}_{i;jk}[s](z_{s+1} - z_s). \tag{3.84}$$

This recursion readily yields an explicit expression for the cumulative $\Delta\beta$-phase at the right end of the $[z_{s-1}, z_s]$ segment:

$$\Delta\phi^{\text{CD}}_{i;jk}(z_s) = \sum_{s'=0}^{s-1} \Delta\beta^{\text{CD}}_{i;jk}[s'] L^{\text{seg}}_{s'}; \qquad L^{\text{seg}}_s \equiv z_{s+1} - z_s. \tag{3.85}$$

The integration in (3.70) may then be partitioned into a sum of integrals over the individual piecewise constant segments:

$$\begin{aligned} H_{[0,L]}^{i;jk} &= -j \sum_{s=0}^{N_{\text{seg}}-1} \gamma[s] \int_{z_s}^{z_s+1} G_{\text{p}}(z) \mathrm{e}^{-j\left[\Delta\phi_{i;jk}^{\text{CD}}(z_s)+\Delta\beta_{i;jk}^{\text{CD}}[s](z-z_s)\right]} \mathrm{d}z \\ &= -j \sum_{s=0}^{N_{\text{seg}}-1} \gamma[s] \mathrm{e}^{-j\Delta\phi_{i;jk}^{\text{CD}}(z_s)} \int_0^{L_s^{\text{seg}}} G_{\text{p}}(z-z_s) \mathrm{e}^{-j\Delta\beta_{i;jk}^{\text{CD}}[s]z} \mathrm{d}z \\ &= \sum_{s=0}^{N_{\text{seg}}-1} \mathrm{e}^{-j\Delta\phi_{i;jk}^{\text{CD}}(z_s)} \left[-j\gamma[s]\ {}^{\Delta\beta_{i;jk}^{\text{CD}}[s]}F_z\left\{G_{\text{p}}(z-z_s)\right\}\right]. \end{aligned} \tag{3.86}$$

Comparing the expression in square brackets in the last line with that for the VTF (3.71) of a uniform fiber, the bracketed expression is identified as the VTF of the standalone sth piecewise constant segment:

$$H_{[z_f,z_f+1]}^{i;jk} = -j\gamma[s]\ {}^{\Delta\beta_{i;jk}^{\text{CD}}[s]}F_z\left\{G_{\text{p}}(z-z_s)\right\}. \tag{3.87}$$

The VTF (3.86) is then compactly expressed as

$$\begin{aligned} H_{[0,L]}^{i;jk}\Big|_{\text{inhom.}}^{\text{irregular}} &= \sum_{s=0}^{N_{\text{seg}}-1} \mathrm{e}^{-j\Delta\phi_{i;jk}^{\text{CD}}(z_s)} H_{[z_s,z_s+1]}^{i;jk} \\ &= \sum_{s=0}^{N_{\text{seg}}-1} \mathrm{e}^{-j\sum_{s'=0}^{s-1}\Delta\beta_{i;jk}^{\text{CD}}[s']L_{s'}^{\text{seg}}} H_{[z_s,z_s+1]}^{i;jk} \\ &= H_{[0,z_1]}^{i;jk} + \mathrm{e}^{-j\Delta\beta_{i;jk}^{\text{CD}}[0]L_0^{\text{seg}}} H_{[z_1,z_2]}^{i;jk} \\ &\quad + \mathrm{e}^{-j\left\{\Delta\beta_{i;jk}^{\text{CD}}[0]L_0^{\text{seg}}+\Delta\beta_{i;jk}^{\text{CD}}[1]L_1^{\text{seg}}\right\}} H_{[z_2,z_3]}^{i;jk} \\ &\quad + \ldots \mathrm{e}^{-j\sum_{s'=0}^{N_{\text{seg}}-2}\Delta\beta_{i;jk}^{\text{CD}}[s']L_{s'}^{\text{seg}}} H_{[z_{N_{\text{seg}}-1},z_{N_{\text{seg}}}]}^{i;jk}. \end{aligned} \tag{3.88}$$

This is our main result for the VTF of an irregular, inhomogeneous fiber link with piecewise constant $\beta(z), \gamma(z)$ (constant over each segment but generally different from segment to segment), expressed as a linear combination of the VTFs of the individual segments. Graphically, we formulate the following rule for superposing the VTFs of a collection of spans or segments:

VTF dephasing rule – 1st formulation: The VTF phasor corresponding to each segment, taken standalone, is rotated by an angle equal the cumulative $\Delta\beta$-phase $-\Delta\phi_{i;jk}^{\text{CD}}(z_s)$ at the beginning of the particular segment, i.e., the phasor contribution of that span is rotated by an angle equal to the (linear) cumulative $\Delta\beta$ phase-shift from z $= 0$ up to the input of that span.

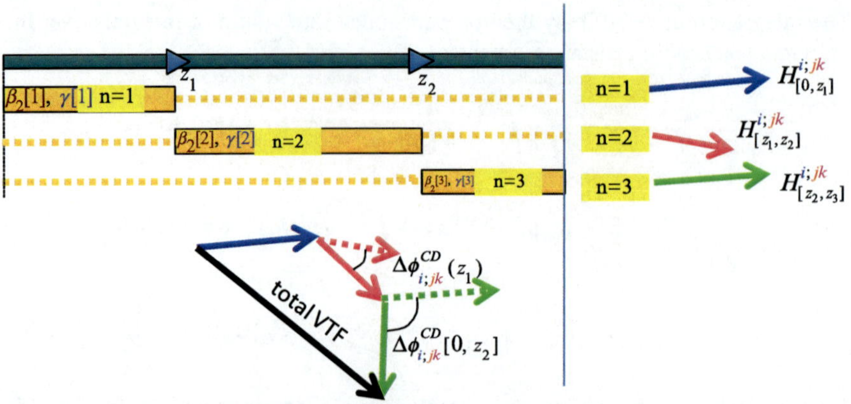

Fig. 3.3 Formation of the total VTF for an irregular inhomogeneous link. The VTFs of the individual spans are successively dephased and their phasors are added up to form the overall VTF. Just the superposition of three spans is illustrated in the figure

Second equivalent formulation: The VTF phasor of each segment is rotated relative to the previous one by an extra angle equal to the phase increment over the previous segment. The overall VTF is obtained as the sum of all the rotated phasors corresponding to all the segments.

The formation of (3.88) may be graphically visualized (Fig. 3.3) as addition of a set of phasors successively rotated by $-\Delta\phi^{\text{CD}}_{i;jk}(z_s)$ with $\Delta\phi^{\text{CD}}_{i;jk}(z_s)$ given by (3.85), i.e., the $s+1$ th phasor is clock-wise rotated by an extra angular increment $\Delta\beta^{\text{CD}}_{i;jk}[s]L^{\text{seg}}_s$ relative to the s th phasor. It follows from the triangle inequality that $\left| H^{i;jk}_{[0,L]}\Big|^{\text{irregular}}_{\text{inhom.}} \right| \leq \sum_{s=0}^{N_{\text{seg}}-1} \left| H^{i;jk}_{[z_s,z_{s+1}]} \right|$, with the maximum attained when all the phasors in (3.88) are collinear, $\Delta\phi^{\text{CD}}_{i;jk}(z_s) = 0$, i.e., under phase-matched conditions: $\Delta\beta^{\text{CD}}_{i;jk}[s] = 0, \ \forall s$.

Note that the $\Delta\beta$-phase accrued over the incremental segment has no effect on the VTF at the segment end; however, this accrued phase over the segment does contribute to the VTF of the *next* segment to be appended.

The incremental rotations of the individual span phasors are instrumental in mitigating the NL build-up by reducing the absolute value of the resulting VTF, the formation of which is visualized as summation of successively rotated phasors. The superposition of rotated phasors forming the overall VTF of a multispan system is exemplified in Fig. 3.3.

3.5.9 Dispersion-Unmanaged "Regular" Spans Revisited

It is useful to re-derive the VTF (3.79) for the regular link with homogeneous identical spans configuration, as a special case of the inhomogeneous irregular fiber VTF just derived in (3.88), setting the following special parameters in the general model:

$$\Delta\beta^{\mathrm{CD}}_{i;jk}[s] = \Delta\beta^{\mathrm{CD}}_{i;jk},\, L^{\mathrm{seg}}_{s} = L_{\mathrm{span}},\, N_{\mathrm{seg}} = N_{\mathrm{span}}$$

$$\Rightarrow H^{i;jk}_{[0,L]}\Big|^{\mathrm{reg.\,spans}} = H^{i;jk}_{[0,L]}\Big|^{\mathrm{single\ span}} \sum_{s=0}^{N_{\mathrm{span}}-1} \mathrm{e}^{-j\Delta\beta^{\mathrm{CD}}_{i;jk} L_{\mathrm{span}} s}. \quad (3.89)$$

The last sum is identified as $N_{\mathrm{span}} F_{i;jk}$ as per (3.78); hence, (3.89) reduces to the following expression, reproducing (3.79):

$$H^{i;jk}_{[0,L]}\Big|^{\mathrm{reg.\ spans}}_{\mathrm{hom.}} = H^{i;jk}_{[0,L_{\mathrm{span}}]}\Big|^{\mathrm{single\text{-}span}} \cdot N_{\mathrm{span}} F_{i;jk}. \quad (3.90)$$

3.5.10 Phased-Array Effect Tends to Reduce FWM Build-up

The array factor (3.78) captures the salient structure of the multispan configuration. The FWM contributions from the various spans may coherently interfere either constructively or destructively, much like in a *phased array* of radio antennas, i.e., a collection of antennas in which the relative phases of the respective signals feeding the antennas are shaped in such a way that the effective radiation pattern of the array is reinforced in certain directions and suppressed in other directions (or for a fixed direction, there is dependence on frequency). The *array factor* describes the geometrical structure – the relative positioning of the antennas – independent of the common radiation pattern of the individual antennas. For certain parameter combinations, we may even have $|F_{i;jk}| = 0$. The PA mechanism, tending to reduce the overall FWM, is graphically visualized by inspecting (3.88), (3.89) and constructing phasor addition diagrams for these expressions. Phasor diagrams for the "regular" case (3.89) are shown in Fig. 3.4a, b. Each phasor represents the contribution of a particular span to the total FWM IM, for a particular triplet at the fiber link output. The phasors addition forms a partial regular polygon, which tends to close upon itself for particular values of the angle by which adjacent phasors are successively rotated, as detailed in the figure caption. The irregular (inhomogeneous) case described by (3.89) is also readily visualized in terms of a distortion of regular polygon structure of Fig. 3.4a, b, as shown in Fig. 3.4c, constructing an irregular (partial) polygon with varying side lengths and vertex angles, with the side lengths corresponding to the VTF of each fiber span, and with the vertex angles determined by the $\Delta\beta$-phase accumulation over successive spans. We shall further investigate the PA effect in Sect. 3.7, in particular consider the compounding of a very large number of effective PAs, one for each IM product.

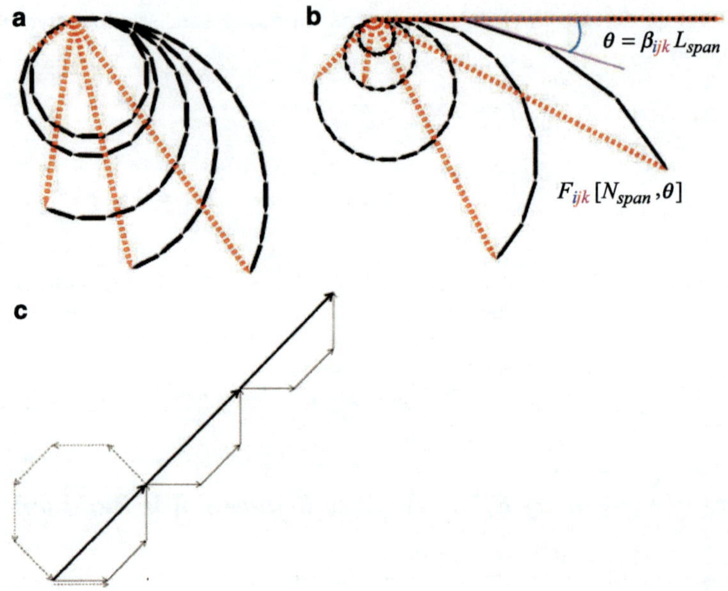

Fig. 3.4 Array factor (dinc function) graphical formation as resultant (*dotted arrow*) of adding up Nspan phasors (*continuous-line arrows*), each of length 1/Nspan, regularly dephased by an angle $\theta = \beta_{ijk} L_{\text{span}} = 2\pi/N_{\text{coh}} = 2\pi u^{ijk} N_{\text{span}}$ between successive phasors. (**a**): $N_{\text{span}} = 12$, $\theta = 10°, 14°, 18°, 26°, 30°$. In the last case, the polygon closes upon itself ($12 \cdot 30° = 360°$) corresponding to the first zero crossing of the dinc. The other five points sample the dinc in its mainlobe (**b**): $\theta = 18°$, $N_{\text{span}} = 1, 4, 8, 16, 32, 64$. In last two cases, the polygon retraces itself, making several revolutions. In fact, $N_{\text{span}} = 20$ accomplishes one full revolution. Sixty-four modulo $12 = 4$, hence the resultants for $N_{\text{span}} = 4, 64$ are parallel. The condition for making one complete revolution (which yields zero resultant) is $N_{\text{span}}\theta = 2\pi$ or $N_{\text{span}} = N_{\text{coh}}$. The condition for zero resultant (possibly making multiple complete revolutions) is that N_{coh} divide N_{span}. When $N_{\text{span}} < N_{\text{coh}}$ ($N_{\text{span}} > N_{\text{coh}}$) the dinc is sampled in its mainlobe (sidelobes). In the dinc sidelobes, the polygon curls up upon itself, completing at least one full revolution, while becoming quite small. (**c**): FWM build-up for a regular fiber link with nine identical spans, with DCF applied every three spans, assuming that $\Delta\phi^{\text{CD}}_{i;jk}[0,\ L_{\text{span}}] = \pi/4$. The VTF of each standalone span is $H_0 \equiv H^{i;jk}_{[0,L_{\text{span}}]}$ (for simplicity we assumed that $\angle H_0 = 0$, else the whole figure would need to be rotated by the angle $\angle H_0$). In the absence of DCF (*dotted arrows*), the first eight summed up phasors would curl up to form a regular octagon, i.e., interfere destructively to zero resultant, leaving a net contribution just from the ninth phasor, i.e., the total FWM VTF would be equal to that associated with the last span, H_0. With DCF, phasor addition is "reset" every three spans, recommencing the phasors addition from zero phase within each group of three spans, hence the three groups (each referred to as a "superspan") each add up in phase, to three times the vector sum $H_0 + e^{j\pi/4} H_0 + e^{j\pi/2} H_0$ of the first three phasors. The resultant has length $3(1 + \sqrt{2})|H_0|$, i.e., the FWM power tolerance in this example is significantly impaired by the dispersion-management, by a factor $[3(1 + 2^{1/2})]^2 = 17.2\,\text{dB}$. Note that this picture is associated with a particular triplet of intermodulation tones, as determined by the *i;jk* indexes. The overall NL tolerance is determined by power superposition of thousands of such IM triplets, each having different $\Delta\phi^{\text{CD}}_{i;jk}[0,\ L_{\text{span}}]$ "curl-up rate," hence the phasor addition illustrated must be modified for each triplet, and the resultant vectors must all be rms averaged. Part of the figure is reproduced from [30]

3.5.11 *The Effect of Dispersion-Compensation Fiber: Dispersion-Managed Links*

The result (3.79) or (3.90) pertains to a dispersion-unmanaged multispan link with identical spans. At the other extreme, we have a dispersion-managed-per-span link. The DCF modules terminating each provide $\Delta\beta$ dispersion of opposite sign to that accumulated over the span, and introduce some residual nonlinearity. It turns out that the modeling of links incorporating DCFs is readily addressed by the VTF dephasing rule stated in Sect. 3.5.8 for an irregular inhomogeneous multispan link.

Let us consider the effect of inserting a DCF module of length Δz at point z. Assume the DCF perfectly cancels the second-order dispersion accrued up to point z. Using (3.69), the accrued $\Delta\beta$ phase, must be zero:

$$\Delta\phi_{i;jk}^{\mathrm{CD}}[0,z+\Delta z] = \Delta\phi_{i;jk}^{\mathrm{CD}}[0,z] + \Delta\phi_{i;jk}^{\mathrm{CD}}[z,z+\Delta z] = -\,(2\pi\Delta\nu)^2\,(j-i)(k-i)$$
$$\times\left[\int_0^z \beta_2(z')\mathrm{d}z' + \int_z^{z+\Delta z} \beta_2(z')\mathrm{d}z'\right] = 0. \tag{3.91}$$

Next consider the contribution to the VTF of the remainder of the fiber, past the DCF, namely the segment $[z+\Delta z, L]$, all the way to the link end. According to the first form of the VTF dephasing rule, this segment contributes

$$H_{[z+\Delta z,L]}^{i;jk}\,\mathrm{e}^{-j\Delta\phi_{i;jk}^{\mathrm{CD}}[0,z+\Delta z]}\Big|_{\Delta\phi_{i;jk}^{\mathrm{CD}}[0,z+\Delta z]=0} = H_{[z+\Delta z,L]}^{i;jk} \tag{3.92}$$

i.e., this contribution of the fiber remainder after the DCF is "straightened up" rather than being rotated, since the cumulative phase $\Delta\phi_{i;jk}^{\mathrm{CD}}[0,z+\Delta z]$ up to the beginning of the remaining NL segment has been nulled out by the DCF. Thus, past the DCF, the VTF of the remainder of the fiber is not rotated at all, but rather its rotation angle is reset to zero, as if the NL system over the segment $[z+\Delta z, L]$ were positioned starting at $z=0$.

Equivalently, we may obtain the same conclusion by applying the second form of the VTF dephasing graphical rule, namely that *each segment VTF phasor is rotated relative to the prior one by an extra angle equal to the phase increment through the previous segment.* Consider the succession of the initial section of the link up with the DCF, the DCF segment, and the span following the DCF. The DCF is assumed ideally linear, hence it adds up an infinitesimal NL VTF, albeit in the particular direction of the cumulative phase angle at the output of the preceding section, essentially imparting a definite direction to its infinitesimal NL VTF contribution. Upon considering the current span, its prior segment is the DCF, hence the extra phase rotation to be applied to the current span VTF equals to the phase increment through the DCF, which by the definition of full dispersion compensation, equals minus the phase accrued up to the input of the DCF, i.e., the span following the DCF has its VTF phasor derotated back to zero. We conclude that generally the effect of DCF is to worsen (increase) the FWM build-up, as the summed up VTF phasors of the fiber spans are no longer allowed to continue to "curl up" (which would have reduced the length of the vector resultant), but rather at the end of each fully compensating DCF, the phasor of the next span is counter-rotated and reset to zero phase, such that the addition of subsequent span VTF phasors starts from the beginning. This effect is exemplified in Fig. 3.4c for a link with identical spans, with DCF applied every three spans assuming that the $\Delta\beta$-phase accrued in each span is $\pi/4$.

Similar considerations may be applied in order to graphically or analytically model arbitrary dispersion maps with partial DCF compensations and with residual DCFs nonlinearities.

3.5.12 Intermod Statistics: Power Propagation Over a "Regular" Spans Link

Heretofore, we have treated the VTF for a generic triplet as indexed by *i; jk*. We now work out the superposition of the multitude of IM contributions from all relevant IM triplets of tones. Assuming equi-power m-ary PSK transmission over all subcarriers, we substitute (3.55) $\left(\check{v}_i^{(1)} = p_0^{1/2} e^{j\phi_i}\right)$ into (3.81), yielding:

$$\underset{\frown}{v}_i^{(3)}(L) = -jg_{\text{eff}} \sum_{[j,k]\in S[i]} \underset{\frown}{v}_j^{(1)} \underset{\frown}{v}_k^{(1)} \underset{\frown}{v}_{j+k-i}^{(1)*} \hat{L}_{i;jk}^{\text{FWM}} F_{i;jk} = -jg_{\text{eff}}\, p_0^{3/2} \times \sum_{[j,k]\in S[i]} \left|\hat{L}_{i;jk}^{\text{FWM}} F_{i;jk}\right| e^{j\left[\phi_j+\phi_k-\phi_{j+k-i}+\angle L_{i;jk}^{\text{FWM}}(L)+\angle F_{i;jk}\right]}. \tag{3.93}$$

As the m-ary PSK angles $\phi_j, \phi_k, \phi_{j+k-i}$ are equi-probable over the m-ary PSK set, and independent over distinct indexes, the summands in the last equation are mostly i.i.d. phasors (with an exception described below), adding up on a complex amplitude basis like *fully developed speckle* [48]. For a large number of summands, the distribution of the sum of phasors tends to complex circular Gaussian, as illustrated in Fig. 3.5 for the special case of a QPSK modulation format ($m = 4$). Absolute-squaring the FWM field (3.81) yields the FWM output optical power:

$$\left|\underset{\frown}{v}_i^{(3)}(L)\right|^2 = g_{\text{eff}}^2 p_0^3 \left|\sum_{[j,k]\in S[i]} \left|\hat{L}_{i;jk}^{\text{FWM}} F_{i;jk}\right| e^{j\left[\phi_j+\phi_k-\phi_{j+k-i}+\angle L_{i;jk}^{\text{FWM}}(L)+\angle F_{i;jk}\right]}\right|^2. \tag{3.94}$$

The set of IM triplets, $S[i]$, was partitioned in [30] into two sets: a *degenerate* (DG) subset $S^{\text{DG}}[i]$ consisting of the points in the hexagonal domain along the bisector of the $[j,k]$ plane in Fig. 3.1, for which $[j,k,j+k-i]$ degenerates to $[j,j,2j-i]$, and the *nondegenerate* (NDG) subset ($j \neq k$), in turn expressed as the union of two subsets, $S_{>}^{\text{NDG}}[i]$ with all its $[j,k]$ elements satisfying $j > k$ and $S_{<}^{\text{NDG}}[i]$ with its

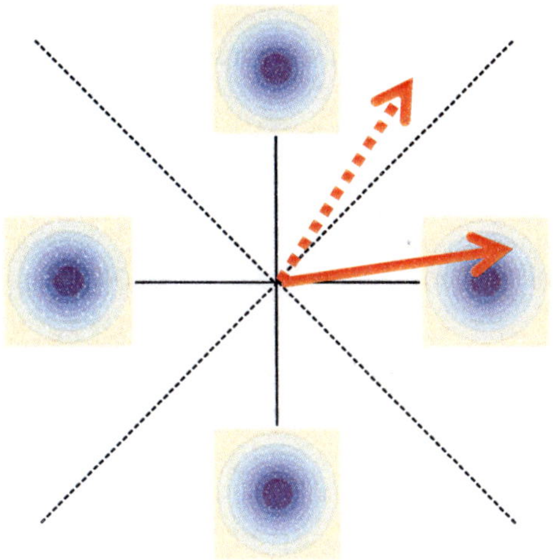

Fig. 3.5 Received constellation for QPSK OFDM transmission in the presence of the FWM impairment – each ideal constellation point is blurred into a circular Gaussian distribution due to the addition of noise and FWM fluctuations/

elements satisfying $j < k$. Note that each element of the one-sided set $S_<^{\mathrm{NDG}}[i]$ is obtained by transposition of an element in $S_>^{\mathrm{NDG}}[i]$ and vice versa. The statistical argument in [30] states that the summand terms in (3.94) are mutually incoherent, since the m-ary PSK angles $\phi_j, \phi_k, \phi_{j+k-i}$ are equi-probable over the m-ary PSK set, and independent over distinct indexes, therefore summing up on a power basis. There is, however, a notable exception: The transposed pairs $[j,k]$, $[k,j]$ are indistinguishable, yielding identical phases in their respective intermods,

$$\begin{aligned}&\exp\left[j\left(\phi_j + \phi_k - \phi_{j+k-i} + \angle L_{i;jk}^{\mathrm{FWM}}(L) + \angle F_{i;jk}\right)\right]\\&\quad = \exp\left[j\left(\phi_k + \phi_j - \phi_{k+j-i} + \angle L_{i;kj}^{\mathrm{FWM}}(L) + \angle F_{i;kj}\right)\right] \qquad (3.95)\end{aligned}$$

hence the IMs from these two pairs add up coherently, on an amplitude basis, providing a power gain double that which would have been generated if the addition were incoherent. The rigorous statistical analysis was carried out in [30]; it is also shown there that the contribution of the degenerate set is quite negligible. The total output power (3.94) is the sum of the powers of the NDG set (with an amplitude factor of 2 squared, i.e., 4, in the one-sided NDG set, or equivalently a factor of 2 in the full NDG set), and the DG set, which is neglected here (precise power expressions including the DG contribution were derived in [30]):

$$\begin{aligned}\sigma^2_{\underset{\sim}{\nu}_i(L)} &= \left\langle \left|\underset{\sim}{\nu}_i^{(3)}(L)\right|^2 \right\rangle\\&= g_{\mathrm{eff}}^2 p_0^3 \left\{ 4 \sum_{[j,k]\in S_>^{\mathrm{NDG}}[i]} \left|\hat{L}_{i;jk}^{\mathrm{FWM}} F_{i;jk}\right|^2 + \sum_{[j,j]\in S^{\mathrm{DG}}[i]} \left|\hat{L}_{i;jj}^{\mathrm{FWM}} F_{i;jj}\right|^2 \right\}\\&\cong g_{\mathrm{eff}}^2 p_0^3 \cdot 2 \sum_{[j,k]\in S[i]} \left|\hat{L}_{i;jk}^{\mathrm{FWM}} F_{i;jk}\right|^2. \qquad (3.96)\end{aligned}$$

Define $N_{\text{beats}}[i, M]$ as the cardinality of $S[i]$, i.e., number of FWM IM triplets or "beats" falling on the ith frequency,

$$N_{\text{beats}}[i, M] \equiv |S[i]| = \left(M^2 - 5M + 2\right)/2 + (M+1)i - i^2, \tag{3.97}$$

where the actual function of i, M, as given in the last equation, was derived in [30]. Further introduce a *root-mean-square* (rms) average of $\left|\hat{H}^{\text{FWM}}_{i;jk}\right| = \left|\hat{L}^{\text{FWM}}_{i;jk} F_{i;jk}\right|$ over all over all j, k pairs in $S[i]$, called NLT parameter:

$$\begin{aligned}\hat{G}^{\text{FWM}}_{\text{eff}} &\equiv \left\langle\left|\hat{H}^{\text{FWM}}_{i;jk}\right|\right\rangle_{\text{rms}} = \left\langle\left|\hat{L}^{\text{FWM}}_{i;jk} F_{i;jk}\right|\right\rangle_{\text{rms}} \\ &\equiv \sqrt{\frac{1}{N_{\text{beats}}[i, M]} \sum_{[j,k]\in S[i]} \left|\hat{L}^{\text{FWM}}_{i;jk} F_{i;jk}\right|^2}.\end{aligned} \tag{3.98}$$

As $\hat{L}^{\text{FWM}}_{i;jk}$, $F_{i;jk}$ are known in closed-form (see (3.74) and (3.78)), the rms average above is readily evaluated. Note that since $\left|\hat{L}^{\text{FWM}}_{i;jk}\right| \leq 1, \left|F_{i;jk}\right| \leq 1$, the NLT parameter is bounded by unity: $\hat{G}^{\text{FWM}}_{\text{eff}} \leq 1$. With these definitions, (3.96) leads to the following compact formula for the FWM power at the output dispersion-unmanaged "regular" link (i.e., a link with identical spans and with DCFs removed), where we denoted the *received* field at the end of the link as $\underset{\sim}{r}_i \equiv \underset{\sim}{u}_i(L)$:

$$\begin{aligned}\sigma^2_{\underset{\sim}{r}i} \equiv \sigma^2_{\underset{\sim}{u}i(L)} = \sigma^2_{\underset{\sim}{v}i(L)} = 2\left(\gamma L_{\text{eff}} N_{\text{span}} \hat{G}^{\text{FWM}}_{\text{eff}}\right)^2 N_{\text{beats}}[i, M] p_0^3 \\ \text{dispersion-unmanaged.}\end{aligned} \tag{3.99}$$

The second equality above stems from assuming a unity gain link, $\left|H^i_{[0,L]}\right| = 1$ (i.e., using amplifiers precisely offsetting the end-to-end losses):

$$\sigma^2_{\underset{\sim}{u}i(L)} = \left\langle\left|\underset{\sim}{u}^{(3)}_i(L)\right|^2\right\rangle = \left\langle\left|\underset{\sim}{v}^{(3)}_i(L) H^i_{[0,L]}\right|^2\right\rangle = \left\langle\left|\underset{\sim}{v}^{(3)}_i(L)\right|^2\right\rangle = \sigma^2_{\underset{\sim}{v}i(L)}.$$

3.5.13 The FWM Power for Dispersion-Managed Links

Finally, let us treat a link wherein DCFs are inserted every N_{interDCF} spans. We refer to each group of N_{interDCF} spans as a "super-span," the number of such super-spans being $N_{\text{super}} = N_{\text{spans}} N_{\text{interDCF}}$. As exemplified in Fig. 3.4c, the super-spans have their contributions adding up coherently; however, the N_{interDCF} spans within each super-span compound according to the phased-array effect. Hence, (3.99) applies within each super-span, which by itself would contribute FWM power $\sigma^2_{\text{superspan}} = 2\left(\gamma L_{\text{eff}} N_{\text{interDCF}} \hat{G}^{\text{FWM}}_{\text{eff}}\right)^2 N_{\text{beats}}[i, M] p_0^3$, where we labeled the array factor in the last formula by N_{interDCF} to indicate that it is evaluated over this number of spans $\Big($specifically, the array factor used in the evaluation of $\sigma^2_{\text{superspan}}$ is given by $\left|F_{i;jk}[N_{\text{interDCF}}]\right| = \left|\text{dinc}_{N_{\text{interDCF}}}\left[N_{\text{interDCF}} L_{\text{span}} \Delta\beta^{CD}_{ijk}/2\pi\right]\right|\Big)$. Finally, the N_{super} super-spans add up coherently, i.e., their combined FWM power contribution is N^2_{super} times higher than that of a single super-span. The overall power is then given by $\sigma^2_{\mathfrak{L}i} = N^2_{\text{super}} \cdot 2\left(\gamma L_{\text{eff}} N_{\text{interDCF}} \hat{G}^{\text{FWM}}_{\text{eff}}\right)^2 N_{\text{beats}}[i, M] p_0^3$. Finally, using $N_{\text{super}} = N_{\text{spans}} N_{\text{interDCF}}$, the formula for the overall FWM power reduces to

$$\sigma^2_{\mathfrak{L}i} = 2\left(\gamma L_{\text{eff}} N_{\text{span}} \hat{G}^{\text{FWM}}_{\text{eff}}\right)^2 N_{\text{beats}}[i, M] p_0^3$$
$$\text{dispersion-managed every } N_{\text{interDCF}} \text{ spans.} \quad (3.100)$$

Note that this result differs from (3.99) just in having the array factor evaluated for N_{interDCF} spans [which tends to make the array factor larger (still bounded by unity)]. The worst case is obtained for $N_{\text{interDCF}} = 1$, i.e., $N_{\text{super}} = N_{\text{span}}$ DCFs are used, one per span. In this case, the array factor becomes unity, yielding (with $\hat{G}^{\text{FWM}}_{\text{eff}} = \left\langle\left|\hat{L}^{\text{FWM}}_{i;jk}\right|\right\rangle_{\text{rms}}$):

$$\sigma^2_{\mathfrak{L}i} = 2\left(\gamma L_{\text{eff}} N_{\text{span}} \hat{G}^{\text{FWM}}_{\text{eff}}\right)^2 N_{\text{beats}}[i, M] p_0^3 \quad \text{dispersion-managed-per-span.} \quad (3.101)$$

This result is worth comparing with the single span result, formally obtained from (3.99) by setting $N_{\text{span}} = 1$:

$$\sigma^2_{\mathfrak{L}i} = 2\left(\gamma L_{\text{eff}} \hat{G}^{\text{FWM}}_{\text{eff}}\right)^2 N_{\text{beats}}[i, M] p_0^3 \quad \text{single-span.} \quad (3.102)$$

Evidently, the dispersion-managed-per-span configuration generates N^2_{span} worse FWM power than each span, as the multiple spans add up coherently (their phasors are collinear) due to the $\Delta\beta$ cumulative phase being reset at the end of each span.

3.6 OFDM Link Performance

In this section, we work out the end-to-end OFDM link performance, in the absence of an active compensation means for the FWM impairment, highlighting the beneficial role of the phased-array effect, significantly improving NLT under certain conditions, especially when DCF modules-based dispersion compensation is entirely removed or is scarcely applied (i.e., in case N_{interDCF} is large and N_{super} is small).

3.6.1 Angular Variance

Assuming m-ary PSK transmission, let us work out the variance $\sigma^2_{\angle\text{FWM}} \equiv \text{var}\{\varphi_i\}$ of the phase noise induced by FWM in the angular decision variable $\varphi_i \equiv \angle \underset{\sim}{r}_i$. Here, $\underset{\sim}{r}_i$ is a circular Gaussian random variable with equal variance of its real and imaginary parts, which point was made when we described the speckle-like formation of (3.93). We assume that the FWM-induced phase noise is small relative to the angular distance of the noiseless angle to the decision boundary, which is π/m, for m-ary PSK. In this case, the phase noise, φ_i is essentially determined by the variance of the fluctuations in the imaginary part r_i^{im} of $\underset{\sim}{r}_i$ (equal to half the variance of $\underset{\sim}{r}_i$), normalized by the signal power:

$$\sigma^2_{\angle\text{FWM}}[i,M] \approx var\left\{r_i^{im}/A\right\} = \sigma^2_{\tilde{r}_i}/(2A^2) = \sigma^2_{\underset{\sim}{r}_i}/2p_0$$
$$= \left(g_{\text{eff}}\hat{G}^{\text{FWM}}_{\text{eff}}\right)^2 N_{\text{beats}}[i,M]p_0^2, \quad (3.103)$$

where we used the fact that the end-to-end magnitude gain is unity (due to the OAs compensating the losses), setting the received power equal to the transmitted power per subchannel, $A^2 = p_0$, and in the last equality, we substituted (3.99) for $\sigma^2_{\underset{\sim}{r}i}$ and canceled a $2p_0$ factor. Next, we substitute $p_0 = P_{\text{T}}/M$ into (3.103), yielding our final result for the angular variance, and its square root, the angular standard deviation, for a *dispersive* regular multispan fiber link:

$$\sigma^2_{\angle\text{FWM}}[i,M,N_{\text{span}}] = \left(g_{\text{eff}}\hat{G}^{\text{FWM}}_{\text{eff}}\right)^2 \hat{N}_{\text{beats}}[i,M]\cdot P_{\text{T}}^2;$$
$$\sigma_{\angle\text{FWM}}[i,M,N_{\text{span}}] = \gamma L_{\text{eff}} N_{\text{span}} \hat{G}^{\text{FWM}}_{\text{eff}} \sqrt{\hat{N}_{\text{beats}}[i,M]}\cdot P_{\text{T}}, \quad (3.104)$$

where we introduced a scaled version of N_{beats} (3.97), normalized by M^2:

$$\hat{N}_{\text{beats}}[i,M] \equiv N_{\text{beats}}[i,M]/M^2 = 0.5+(i-2.5)/M+(1+i-i^2)/M^2. \quad (3.105)$$

Since $N_{\text{beats}}[i,M]$ (3.97) has a quadratic dependence on M, then, for large M, its normalized version is weakly dependent on M, as seen in (3.105). In particular, at the mid-band frequency, $i = M/2$ (assuming even M), we obtain a numerical value 0.734:

$$\hat{N}_{\text{beats}}[M/2,M] = 3/4 - 2/M + 1/M^2;$$
$$\hat{N}_{\text{beats}}[64,128] = 0.734 \approx \hat{N}_{\text{beats}}[M/2,M]. \quad (3.106)$$

We may approximate $\hat{N}_{\text{beats}} \approx 0.734$ for other values of M ($\neq 128$) as well, since $\hat{N}_{\text{beats}}$ is weakly dependent on M. Considering now the dispersion-free special case, we set $\hat{G}_{\text{eff}}^{\text{FWM}} = 1$ in (3.104) and use the approximation (3.106) for $\hat{N}_{\text{beats}}$, as well as $g_{\text{eff}} \equiv \gamma L_{\text{eff}} N_{\text{span}}$, in order to reproduce a result equivalently stated in [34]:

$$\sigma^2_{\angle\text{FWM}}[M/2, M] \approx 0.734(g_{\text{eff}} \gamma L_{\text{eff}} N_{\text{span}} P_{\text{T}})^2 \qquad \text{dispersion-free.} \tag{3.107}$$

In the absence of dispersion, the FWM-induced phase noise power is proportional to the total power of all OFDM subchannels, nearly independent of the number of subchannels. However, beyond the dispersion-free approximation (3.107), our general expression (3.104) accounts for FWM+CD effects, compactly described in terms of the key FWM NLT parameter, $\hat{G}_{\text{eff}}^{\text{FWM}} \equiv \left\langle \left| \hat{L}_{i;jk}^{\text{FWM}} F_{i;jk} \right| \right\rangle_{\text{rms}}$, which is upper bounded by unity, representing the rms-averaged FWM attenuation over all IMs. Unlike the dispersion-free result (3.107), the FWM power in the presence of dispersion may exhibit nonnegligible dependence on M (via the NTP which depends on the array factor). Finally, taking the square root of the approximate value $\hat{N}_{\text{beats}} \approx 0.734$ as numerical coefficient in the angular standard-deviation equation in (3.104) yields the approximation:

$$\sigma_{\angle\text{FWM}}[M/2, M, N_{\text{span}}] \approx 0.857 g_{\text{eff}} \hat{G}_{\text{eff}}^{\text{FWM}} P_{\text{T}} \qquad \text{dispersion-unmanaged.} \tag{3.108}$$

In the absence of CD, we set $\hat{G}_{\text{eff}}^{\text{FWM}} = 1$, yielding $0.857\, g_{\text{eff}} P_{\text{T}}$; in the presence of CD, the angular standard deviation is attenuated by the NLT parameter. In all these expressions (3.104)–(3.108), in order to get substantial suppression, the NLT factor ought to be very small. As the NLT parameter is an rms average of the two-dimensional function $\hat{L}_{i;jk}^{\text{FWM}} F_{i;jk}$ having the two indexes j, k as arguments (for given observation index i), visual inspection of this function, as plotted above the $S[i]$ set in the $[j, k]$ plane is indicative of the amount of FWM supression. For example, in the plot of Fig. 3.6, $\hat{H}_{i;jk}^{\text{FWM}} \equiv \hat{L}_{i;jk}^{\text{FWM}} F_{i;jk}$ is very small except at some "ridges," hence its RMS average gets quite small. For practical parameters, $\hat{L}_{i;jk}^{\text{FWM}}$, representing the normalized VTF of a single span hardly falls under unity, hence the variations of $\hat{H}_{i;jk}^{\text{FWM}}$, which is essentially a normalized VTF of the overall system, are dominated by the behavior of the array factor $F_{i;jk}$, which acquires a mainlobe + sidelobes structure, provided the argument of the "dinc" function (3.78) exceeds unity in absolute value. Fortunately, for intermods sampling the sidelobes of the "dinc" function, the array factor becomes very small, and the proportion of these IMs in the overall IM "population" may be very large. The formation of the array factor may be best understood via the phased-array effect, which was briefly introduced above, and is further elaborated in the next section.

We mention that the result (3.108) for the dispersion-unmanaged link is readily adapted to describe a dispersion managed link, noticing that the only difference in (3.100) relative to (3.99) is the usage of N_{span} in the dispersion-unmanaged case as argument of the array factor, vs. usage of N_{interDCF} in the dispersion-managed case. Hence, making the substitution $N_{\text{span}} \to N_{\text{interDCF}}$ within the array factor in (3.108)

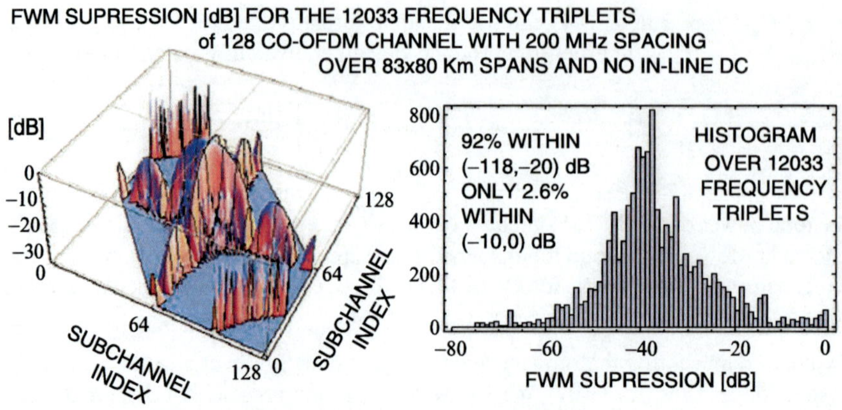

Fig. 3.6 Plot and histogram of FWM suppression for the 12,033 IM triplets for an OFDM system with M = 128 subcarriers. The 3-D plot axes are the [j,k] indexes. It is apparent that most of the triplets experience very large FWM suppression, as also verified by the histogram. Part a of the figure is reproduced from [30]

yields the corresponding formula for the angular variance in the dispersion-managed case:

$$\sigma_{\angle\mathrm{FWM}}[M/2, M, N_{\mathrm{span}}, N_{\mathrm{interDCF}}] \approx 0.857 g_{\mathrm{eff}} \hat{G}_{\mathrm{eff}}^{\mathrm{FWM}} P_{\mathrm{T}}$$

$$\text{dispersion-managed every } N_{\mathrm{interDCF}} \text{ spans.} \tag{3.109}$$

(Evidently, the more accurate formula (3.104) may also be similarly adapted, simply by using N_{interDCF} in the array factor).

At this point, we derive the overall receiver performance in the wake of FWM fluctuations and ASE noise.

3.6.2 Q-Factor, Symbol Error Rate, BER

As seen above, the FWM fluctuations are speckle-like adding up to a circular Gaussian noise-like perturbation of the ideal constellation points. The key additional mechanism of ASE noise from the OAs is also additive Gaussian; hence, the overall evaluation of BER performance is relatively straightforward, as it is governed by Gaussian statistics.

For example, for m-ary PSK, the *symbol error rate* (SER) is given by SER $\cong 2Q[q_\angle]$. The argument $q_\angle$ of the $Q[q] = (2\pi)^{-1/2} \int_{-\infty}^{q} \exp\left[-\frac{1}{2}(x/\sigma)^2\right] \mathrm{d}x$ function is called Q-factor. In particular for QPSK ($m = 2$), the BER for Gray encoding of bit pairs to QPSK symbols, is precisely given by BER $= Q[q_\angle]$. The Q-factor is given in this case by $q_\angle = \pi \Big/ \left(m\sqrt{\kappa_m \sigma_\angle^2}\right)$ with $\sigma_\angle^2 = \sigma_{\angle\mathrm{FWM}}^2 + \sigma_{\angle\mathrm{ASE}}^2$ the total variance of the decision variable due to the two independent noise sources, and κ_m a correction factor shown in [49] to provide an improved fit for the tails of the actual distribution, yielding improved accuracy of the linear phase noise model induced by circular Gaussian noise fluctuations (e.g., for QPSK $\kappa_4 = 1.11$). Introducing respective Q-factors $q_{\angle\mathrm{FWM}} = \pi \Big/ \left(m\sqrt{\kappa_m \sigma^2_{\angle\mathrm{FWM}}}\right)$, $q_{\angle\mathrm{ASE}} = \pi \Big/ \left(m\sqrt{\kappa_m \sigma^2_{\angle\mathrm{ASE}}}\right)$ for FWM and ASE acting alone (assuming the other noise source was turned off), we readily obtain the total Q-factor: $q_\angle = \left(q^{-2}_{\angle\mathrm{ASE}} + q^{-2}_{\angle\mathrm{FWM}}\right)^{-1/2}$.

It remains to evaluate the individual Q-factors. Using (3.104), the FWM-related Q-factor is

$$q_{\angle\mathrm{FWM}}[i, M, N_{\mathrm{span}}] \equiv \frac{\pi/m}{\sqrt{\kappa_m}\sigma_{\angle\mathrm{FWM}}} = \frac{\pi/m}{\gamma L_{\mathrm{eff}} N_{\mathrm{span}} \hat{G}^{\mathrm{FWM}}_{\mathrm{eff}} \sqrt{\kappa_m \hat{N}_{\mathrm{beats}}[i, M]} \cdot P_{\mathrm{T}}}. \tag{3.110}$$

The FWM Q-factor is seen to degrade, as the number of spans and the optical power are increased.

The Q^2-factor (in electrical dB units, $20\log_{10}(q_{\angle\mathrm{FWM}})$) decreases 6 dB per octave (doubling) of the spans number, and the optical power. In the presence of dispersion, the NLT parameter $\hat{G}^{\mathrm{FWM}}_{\mathrm{eff}} = \left\langle \left| \hat{L}^{\mathrm{FWM}}_{i;jk} F_{i;jk} \right| \right\rangle_{\mathrm{rms}} \leq 1$ acts to improve the Q^2-factor by the positive increment $-20\log_{10}\hat{G}^{\mathrm{FWM}}_{\mathrm{eff}}$, referred to as *FWM suppression*. The ASE Q^2-factor was evaluated in [30], consistent with [3], seen to be proportional to the PSD $P_{\mathrm{T}}/B_{\mathrm{T}}$ [Watt/Hz] of the OFDM signal, and inversely proportional to the number of OAs, $N_{\mathrm{span}} + 1$ (F_N is the OA noise figure):

$$q^2_{\angle\mathrm{ASE}} \equiv \frac{(\pi/m)^2}{\kappa_m}\sigma^{-2}_{\angle\mathrm{NL}} = \frac{(2\pi/m)^2}{\kappa_m}\left[F_N(G_{\mathrm{OA}} - 1)h\nu_0(N_{\mathrm{span}} + 1)\right]^{-1}\frac{P_{\mathrm{T}}}{B_{\mathrm{T}}}. \tag{3.111}$$

Evidently, there is an optimum optical power P_{T} balancing the opposing trends of the $q_{\angle\mathrm{FWM}}$ and $q_{\angle\mathrm{ASE}}$ vs. P_{T}.

3.7 PA Effect for Dispersion-Unmanaged Regular Multispan Links

In this section, we revisit the PA effect introduced in Sect. 3.5.7, where we established the formal equivalence between the compounding of FWM from multiple spans and the radiation build-up from an analogous PA of antennas.

The FWM problem is far more complex than analyzing a single effective PA and deriving its array factor $F_{i;jk}$. In fact, one must average a very large number (typically thousands) of effective PAs, one for each frequency triplet associated onto the observation subchannel, i. At first sight, this averaging process seems intractable. In this section, we derive simple approximate analytic rules for the NL tolerance of the FWM impairment over a regular multispan homogeneous link.

3.7.1 Compounding Multiple PAs

The number of superposed PAs to work out the statistics of, equals the cardinality of the set $S[i]$ of intermods (e.g., for $M = 128$ subcarriers and $i = 64$, there are 12,033 IMs, each of which has a different array factor). The statistics of power superposition of the multiple PAs is captured in the NLT parameter $\hat{G}_{\text{eff}}^{\text{FWM}} = \left\langle \left| \hat{L}_{i;jk}^{\text{FWM}} F_{i;jk} \right| \right\rangle_{\text{rms}}$, which is substantially reduced by having most of the PAs satisfy $\left|F_{i;jk}\right| \ll 1$ (allowing just a small fraction of the IMs to have their array factor close to unity), in which case a large amount of FWM suppression is attained by virtue of the PA effect. In [30], we investigated the conditions under which $\left|F_{i;jk}\right| \ll 1$: the intermod corresponding to $i; jk$ must sample the dinc[u] function in its sidelobes, which requires that the argument of the dinc function satisfy $|u| = L\left|\Delta\beta_{ijk}\right|/2\pi > 1$. Now, using (3.65) the last stated condition amounts to $L\beta_2(2\pi\Delta\nu)^2|j-i||k-i| > 1 \Leftrightarrow \left[2\pi(\Delta\nu)^2 L\beta_2\right]^{-1} < |j-i| \cdot |k-i|$. We may arrange for this condition to hold for the vast majority of frequency triplets, provided the product $(\Delta\nu)^2\ L\beta_2$ is made large (the LSH of the last inequality is made small).

3.7.2 The NLT is set by Bandwidth^2 × Length × GVD

It is remarkable that the NLT parameter turns out to be nearly independent of M, only depending on the product $B_{\text{T}} = M\Delta\nu$ rather than on $\Delta\nu$, M individually. It is shown in [30] that the total amount of FWM suppression attained via the PA effect actually varies as the *bandwidth*^{2} × *length* × *GVD* product, $B_{\text{T}}^2 L\beta_2$. The effect at work here, as detailed in [30], is that in the $[j,k]$ plane (wherein each discrete point corresponds to an IM), the array factor mainlobe area of a two-dimensional map of the VTF power in the $[j,k]$ turns out to be linear in M^2, as is the total area of the set of triplets. Thus, upon evaluating the ratio of the number of points belonging the mainlobe, vs. the overall number N_{beats} of points in the $S[i]$ domain, M^2 cancels out and it turns out that the resulting ratio is inversely proportional to $B_{\text{T}}^2 L\beta_2 :\ N_{\text{mainlobe}}/N_{\text{beats}} \propto \left(2\pi L B_{\text{T}}^2 \beta_2\right)^{-1}$.

Now approximating all points within the sidelobes as having zero array factor, while all points within the mainlobe are set to unity array factor, it is apparent that $\left\langle\left|F_{i;jk}\right|\right\rangle_{\text{rms}}^2 \cong N_{\text{mainlobe}}/N_{\text{beats}}$ hence $\left\langle\left|F_{i;jk}\right|\right\rangle_{\text{rms}}^2 \propto \left(2\pi L B_{\text{T}}^2 \beta_2\right)^{-1}$. Recalling that $\hat{L}_{i;jk}^{\text{FWM}}$ is very close to unity, we then also have $\left(\hat{G}_{\text{eff}}^{\text{FWM}}\right)^2 = \left\langle\left|\hat{L}_{i;jk}^{\text{FWM}} F_{i;jk}\right|\right\rangle_{\text{rms}}^2 \approx \left\langle\left|F_{i;jk}\right|\right\rangle_{\text{rms}}^2 \propto \left(2\pi L B_{\text{T}}^2 \beta_2\right)^{-1}$ or $\hat{G}_{\text{eff}}^{\text{FWM}} \propto \left(2\pi L B_{\text{T}}^2 \beta_2\right)^{-1/2}$.

The longer and more dispersive the fiber is, and the wider band the OFDM system is, the better its FWM NLT, which is plausible, as bandwidth, length and GVD are measures of increased dispersion, tending to mitigate FWM by enhancing the phase mismatch which tends to reduce the NL build-up.

3.7.3 A Simple Q-Factor Performance Lower Bound for Dispersion Unmanaged Links

We would now like to more precisely assess the general behavior of the NLT parameter, $\hat{G}_{\text{eff}}^{\text{FWM}}$ over the $[N_{\text{span}}, B_{\text{T}}, P_{\text{T}}]$ space of performance variables for a given fiber. Note that for a regular fiber link with given type of fiber (specified β_2), the fiber length, L, is proportional to the number of spans, hence the parameters $[N_{\text{span}}, B_{\text{T}}]$ (and β_2) uniquely determine the *bandwidth*^{2} × *length* × *GVD* combination, which was just seen to essentially determine the NLT. Moreover, it is the total power, P_{T}, rather than the power per subchannel, $p_0 = P_{\text{T}}/M$ that determines the Q-factor (along with the NLT), as borne out in the formulas (3.110), (3.111), which were seen to be very mildly dependent on M (note that (3.111) does not depend on M, whereas (3.110) depends M on just via the $\hat{N}_{\text{beats}}[i, M] \equiv N_{\text{beats}}[i, M]/M^2 = 0.5 + (i - 2.5)/M + (1 + i - i^2)/M^2$ term, which hardly varies with M, for large M). It is our objective to compress the apparent numerical complexity of description, distilling the ASE + dispersive FWM statistics into a very compact analytic model for the Q-factor, which no longer involves complicated averaging of array factors as reflected in the NLT parameter. Rather our target Q-factor formula should be uniquely determined by the $[N_{\text{span}}, B_{\text{T}}, P_{\text{T}}]$ parameters, at least asymptotically (as M and B_{T} becomes large, as typical for long-haul high-speed OFDM).

Let us define the NLT suppression as the reciprocal of the squared NLT parameter, $(\hat{G}_{\text{eff}}^{\text{FWM}})^2$, i.e., on a dB scale the NLT suppression is given by $-\text{NLT}_{\text{dB}} \equiv -20\log_{10}\hat{G}_{\text{eff}}^{\text{FWM}}$. From the insightful geometric argument made in [30] regarding the distribution of (tens of) thousands of FWM mixing products, as reviewed in Sect. 3.3, the NLT over an optically amplified PDM-OFDM link of length $L = N_{\text{span}}L_{\text{span}}$, containing N_{span} identical homogeneous spans, is essentially determined by the *bandwidth*^{2} × *length* × *GVD* product:

$$\left(\hat{G}_{\text{eff}}^{\text{FWM}}\right)^2 = C/\left(N_{\text{span}}B_{\text{T}}^2\beta_2\right); \quad \text{NLT}_{\text{dB}} = -10\log_{10}\left(\hat{G}_{\text{eff}}^{\text{FWM}}\right)^2$$
$$= 10\log_{10}(\beta_2/C) + 20\log_{10}B_{\text{T}} + 10\log_{10}N_{\text{span}}. \tag{3.112}$$

The NLT suppression is plotted in Fig. 3.7b against the total bandwidth, allowing to extract the proportionality coefficient C of the bound (3.112), as described next. The numerical results of Fig. 3.7 indicate substantial attainable FWM suppression (>15 dB for large aggregate bandwidth $B_{\text{T}} = M\Delta\nu$). Note that for large M (number of OFDM subchannels), the NLT measure tends to be nearly independent of M, as illustrated by the flattening of the curves in Fig. 3.7a.

For definiteness, the coefficients in all ensuing formulas are taken numeric rather than symbolic, assuming specific numerical values for the system parameters as follows: G.652 standard fiber ($\beta_2 = -21.7\,\text{psec}^2/\text{Km}$); fiber loss $\alpha_0 = 0.22\,\text{dB/Km}$; NL coefficient $\gamma = 1.3/\text{W/Km}$; fiber spans of $L_{\text{span}} = 80\,\text{Km}$; OAs gain $G_0 = e^{\alpha_0 L_{\text{span}}} = 17.6\,\text{dB}$; noise figure $F_N = 6.5\,\text{dB}$.

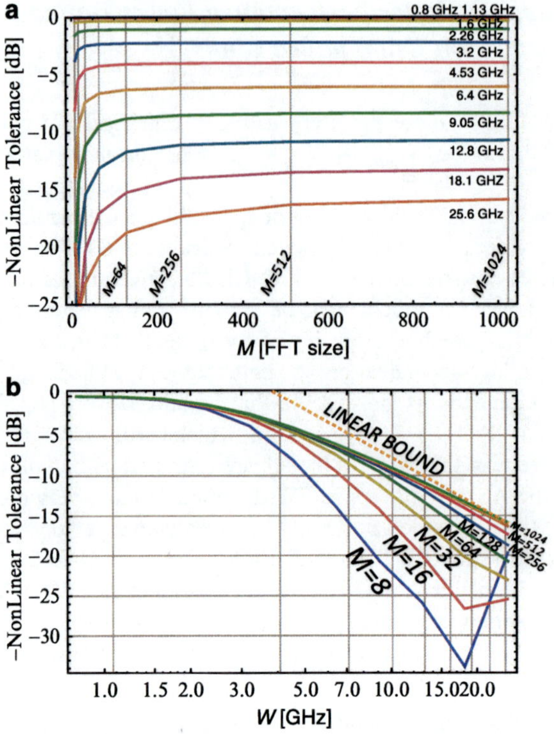

Fig. 3.7 Nonlinear tolerance (−NLT [dB]) for dispersion-unmanaged OFDM transmission over an 87 spans link: **(a)** plotted vs. the number of subchannels (FFT size) M, parameterized by total bandwidth $W \equiv B_T$ per OFDM channel, in half-octave steps. **(b)** NLT plotted vs. B_T (log scale), parameterized by M, in octave steps. Substantial FWM suppression is attained for large bandwidth, and the NLT is nearly independent of M, for large M. The upper linear bound (*dotted line* in (**b**)) is essential for developing the simple analytic Q-factor limit. Note: the bound in Fig. 3.7a assumes a different power optimization P_T^{opt} at each distance (N_{span} value); however, the dependence of P_T^{opt} on N_{span} is weak anyway, e.g., as N_{span} ranges from 10 to 74, P_T^{opt} varies just by 2.7%, hence we might as well optimize the power to attain a target BER $= 10^{-3}$ right at the end of the link (attained for 74 spans), then use bound (3.113) with this fixed power instead. The (3.113) bound would differ imperceptibly on the scale of Fig. 3.7b if power-optimized at the link end. This indicates the feasibility of inserting multiple add-drops along dispersion-unmanaged OFDM links that have been optimized for best performance at the far end

The $-\text{NLT}_{\text{dB}}$ formula (3.112) is linear in $\log_{10} B_T$ (i.e., should appear as a straight line sloped −20 dB/decade when using log-dB scales) as plotted in the dotted straight line bound at the top of Fig. 3.7b. This leads to a remarkably simple new lower bound as derived here for the Q-factor of dispersive OFDM transmission, accounting for the main FWM and ASE impairments. In Fig. 3.7b, this bound corresponds to a linear asymptote approaching the top $M = 1,024$ numerically generated curve, for large B_T. From this linear curve-fit, we extract

$C/\beta_2 = 1477.36$. Substituting (3.112) along with this coefficient into (3.110) yields $q_{\angle \mathrm{FWM}} = 8.64 \cdot 10^{-13} \cdot B_\mathrm{T}/N_\mathrm{span}^{1/2} P_\mathrm{T}$. Substituting the system parameters into (3.111) yields the ASE partial Q-factor $q_{\angle \mathrm{ASE}} = 1637.03/\left[P_\mathrm{T}/(N_\mathrm{span}+1)\right]^{1/2}$ As noise powers are additive, the two partial Q-factors compound according to $q_\mathrm{T} = \left(q_{\angle \mathrm{FWM}}^{-2} + q_{\angle \mathrm{ASE}}^{-2}\right)^{-1/2}$, yielding a total Q-factor bound:

$$q_\mathrm{T}[N_\mathrm{span}, B_\mathrm{T}, P_\mathrm{T}] \geq \left(1.34 \cdot 10^{24} N_\mathrm{span}(P_\mathrm{T}/B_\mathrm{T})^2 + 1.46 \cdot 10^{-17}(1+N_\mathrm{span})(P_\mathrm{T}/B_\mathrm{T})^{-1}\right)^{-1/2}. \tag{3.113}$$

Note the opposite dependences of the FWM and ASE contributions on the transmitted PSD $P_\mathrm{T}/B_\mathrm{T}$ [Watts/Hz]. Maximizing (3.113) by differentiating over P_T yields the optimal launch power $P_\mathrm{T}^\mathrm{opt} = 1.76 \cdot 10^{-14}\left(1 + N_\mathrm{span}^{-1}\right)^{1/3} B_\mathrm{T}$. Plugging $P_\mathrm{T}^\mathrm{opt}$ into (3.113), the B_T dependence is seen to cancel out, leaving a sole dependence of the total Q-factor on transmission range:

$$q_\mathrm{T}[N_\mathrm{span}, P_\mathrm{T}^\mathrm{opt}] \geq 28.36\, N_\mathrm{span}^{-1/6}\left(N_\mathrm{span}+1\right)^{-1/3} \approx 28.36 N_\mathrm{span}^{-1/2}. \tag{3.114}$$

Consistent with Fig. 3.7b, the lower bound on Q-factor is tight whenever B_T, M are large, which is the case of interest in ultra-broadband OFDM systems (the Q-factor for low B_T, M, may be substantially better than the bound we derived). It is remarkable that upon compounding a very large number of FWM mixing products, the power-optimized Q-factor bound comes out bandwidth-independent (provided the bandwidth is sufficiently high).

The dependence of the overall Q-factor bound on the number of spans is quite remarkable: The Q-factor degrades neither as N_span^{-2} (coherently) nor as N_span^{-1} (incoherently) but rather declines even more slowly over distance, approximately as $N_\mathrm{span}^{-1/2}$ (decreasing even slower than an incoherent build-up of FWM power with the number of spans). This is indicative of very favorable NLT characteristics for dispersion-unmanaged OFDM transmission, by virtue of the PA effect. The numerical coefficients would become even more favorable for higher GVD coefficient β_2 (raising the Q-factor lower bound while retaining its $N_\mathrm{span}^{-1/2}$ dependence).

Finally, note that the dispersion unmanaged system described here attains quite a large range, almost 6,000 Km (74 spans times 80 Km/span) for 10^{-3} BER. However, this simplistic model excludes multiple additional impairment factors, e.g., ADC and DAC quantization noise and distortion, IQ modulator distortion, laser source and LO phase noise, accuracy of the timing and carrier recovery circuits, etc., which will eventually further limit ultimate performance. Hence, the model derived here provides a Q-factor performance upper bound summarized in Fig. 3.8, reducing the numerical complexity of treating thousands of FWM mixing products, distilling it into a compact all-analytic model.

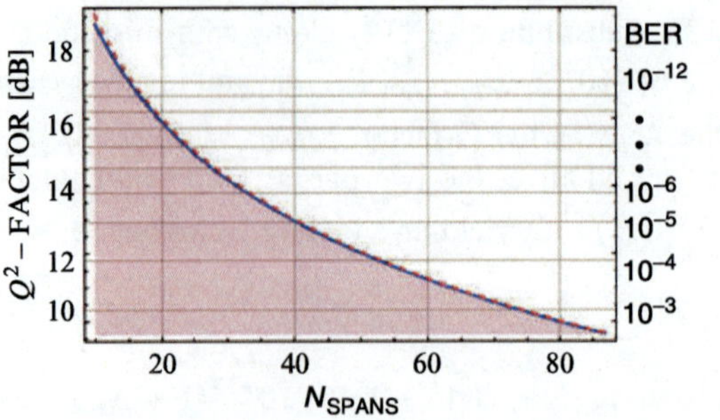

Fig. 3.8 Dispersion-unmanaged OFDM performance bounds: Q-factor bound ($20\log_{10} Q$) vs. link reach (expressed in span length units). This is a lower (conservative) bound, quite tight for large *W,M* (ultrabroadband transmission). The horizontal grid lines correspond to BER levels in decade steps. The dotted line is the $N_{\mathrm{span}}^{-1/2}$ approximation in (3.3), barely differing from the solid one (the precise expression)

3.8 Overview of NLC Methods

Heretofore, we have developed simple, insightful, yet precise analytic models of the NL impairment generated in optical OFDM. We now address the mitigation of this NL impairment by means of a *NL compensator* (NLC) in the OFDM receiver. We start by briefly reviewing prior NLC approaches, then introduce our own Volterra-based improved OFDM NLC method [45, 50].

Let us first review the first OFDM NLC scheme introduced by Lowery [33], referred to here as *Backward NonLinear Phase Rotator* B-NLPR.

This technique may be applied both at the Tx (as a NL predistorter) or at the Rx, or be distributed between the Tx and the Rx. Here, we focus on Rx-based NLC techniques. As shown in the simplified model of Fig. 3.9, M symbols are to be transmitted over an OFDM link. The symbols are IFFT-ed in the Tx, then propagated through the fiber link. In a simplified description of the Rx, the received sampled signal is passed through a memoryless nonlinearity referred to as B-NLPR, then FFT-ed and sliced to obtain decisions, which are improved relative to what would be obtained if the NLPR were not inserted. The B-NLPR NL operation consists of multiplication of its input by the quadratic phase factor $\exp[\left[jg_{\mathrm{eff}}\,|\bullet|^2\right]$, where $\bullet$ denotes the input, in this case the reconstructed complex field samples. This operation is the inverse of the field transformation $\exp[\left[-jg_{\mathrm{eff}}\,|\bullet|^2\right]\times$ that would occur along the fiber link in the absence of CD, i.e., just accounting for SPM in the propagation process. We thus refer to this NLC method as B-NLPR. We note that there have been polarization-vectorial extensions of this NLC method [36–38]; however, we focus here on the scalar version. Simulations of the scalar B-NLPR performance are shown in Fig. 3.10. Evidently, the performance is better under low dispersion conditions, as this memoryless NLC method is frequency agnostic, ignoring the interaction between CD and NL, solely accounting for the SPM NL. Also note that the

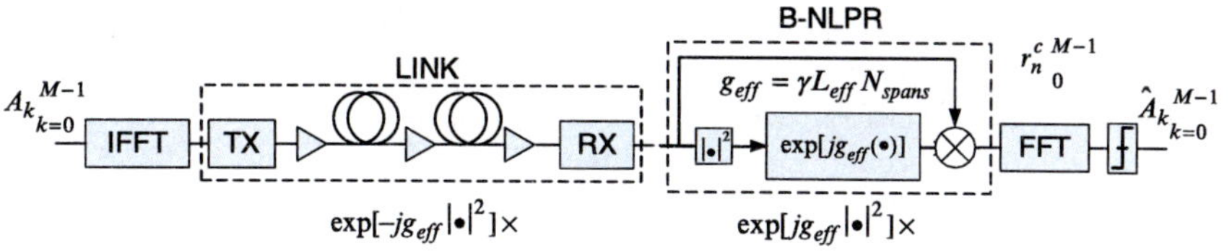

Fig. 3.9 The backward nonlinear phase rotation (B-NLPR) nonlinear compensation (NLC) method

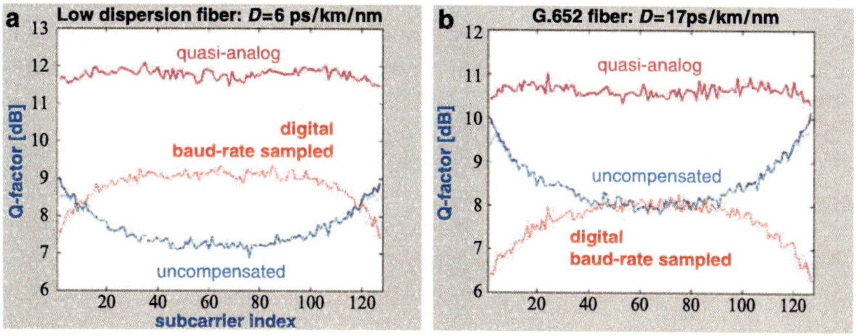

Fig. 3.10 Performance of the B-NLPR vs. uncompensated: Q-factor vs. subcarrier index (frequency). Two B-NLPR versions are considered: quasianalog (with 12× oversampling), and baud-rate sampled. (**a**): Low-dispersion fiber. (**b**): standard fiber. The B-NLPR fares worse in higher dispersion (**b**). Moreover, the performance of the baud-rate sampled version is deteriorated to the extent of becoming unusable. The parameters assumed in the simulations are: 112 Gb s^{-1} OFDM system with $M = 128$ subcarriers over $B_T = 32$ GHz, ~10% pilot tones, cyclic prefix overhead 8.7%, $\gamma = 1.3$ /W/Km, $\alpha = 0.2$ dB/Km, 25 spans of 80 Km each, optical amplifier gain 17.6 dB fully balancing the loss, noise figure 6.5 dB

improvement deteriorates at the band-edges – for standard fiber the improvement is ~2 dB at the mid-band subchannel. A frequency filtered extension of this method has been investigated by [35]. Here, we shall adopt a Volterra-based systematic approach striving to introduce frequency dependence in the VTF, and optimizing performance.

The top curves in Fig. 3.10 are actually quasi-analog – we used very large (12×) oversampling in the simulation, in order to avoid aliasing of the NL spectrum. For practical realization, it would be desirable to operate this scheme with baud-rate sampling. However, in this case, the performance deteriorates considerably (see the "digital baud-rate sampled" curves in Fig. 3.10), in fact breaking down completely for standard fiber (Fig. 3.10b), for which the usage of the baud-rate sampled NLC actually worsens performance, rather than improving it. We shall revisit this baud-rate sampling issue in the next section, motivated by baud-rate operation being extremely desirable at ultra-high speeds. Moreover, B-NLPR is actually a building block in our own NLC scheme.

At this point, let us briefly review the BP method [2, 28, 51–58], which has been extensively investigated in recent years. The underlying concept is that the NLSE is mathematically invertible (even in the presence of loss), simply by propagating the received signal through a version of the NLSE with the signs of its α, β_2, γ parameters all inverted. This may be accomplished at the receiver, in the digital domain, by simulating the NLSE inversion by means of an SSF algorithm (with the appropriate inverted parameters). In the absence of noise, over a scalar channel, this method is evidently optimal. Polarization-vectorial extensions of the method have also been pursued [36–38]. If the PMD dynamics along the fiber were known, the vectorial polarization-aware NLSE would be strictly invertible just as the scalar version is. As information on the PDM instantaneous evolution is not practically retrievable, one resorts to working with average values – the Manakov equation is used and inverted [28].

While in principle providing optimal or near-optimal performance, BP methods suffer from a key deficiency: prohibitive computational complexity incurred in evaluating a large number of stages of the split-step Fourier method, with each stage comprising a pair of FFTs. Here, we restrict attention, for simplicity, to scalar BP methods, which are evidently less demanding than vector methods but still pose a prohibitive computational load. The NL tolerance performance vs. complexity may evidently be traded off, by taking fewer stages, at the expense of the attained NL tolerance, but even with several stages the complexity is still prohibitive. Moreover, we conjecture that by using our DF-based Volterra NLC instead of the BP algorithm, a better performance-complexity tradeoff is obtained (Sect. 3.18). In addition to using the Volterra NL representation, our NLC approach also differs from the conventional BP methods, in that it is DF based, operating in multiple iterations, using the slicer preliminary decisions in order to synthesize an approximation of the NL signal component accounting for the interplay between dispersion and NL, then subtracting this synthesized nonlinearity from received signal. In contrast, current BP methods are invariably based on *feed-forward* (FF) NL equalization, rather than using DF.

3.9 Baud-Rate Sampled Version of the B-NLPR NLC

The performance degradation incurred by the B-NLPR method upon attempting baud-rate sampling was highlighted in Fig. 3.10. The source of this degradation is the spectral broadening due to NL propagation, which generates both in-band and OOB distortion. For a third-order NL mechanism, the spectrum would be broadened by a factor of 3; however, higher-order NL components (predominantly fifth-order) are non-negligible, such that the spectrum is more than three times broader. We shall assume that the spectrum is essentially broadened by a factor of four, neglecting the spectral energy beyond four times the baud-rate.

Note that the OOB sidebands in the received signal can be removed by means of an AA filter inserted prior to the sampler in the Rx; however, the removal of the OOB distortion generated in the fiber by means of AA filtering does not solve the

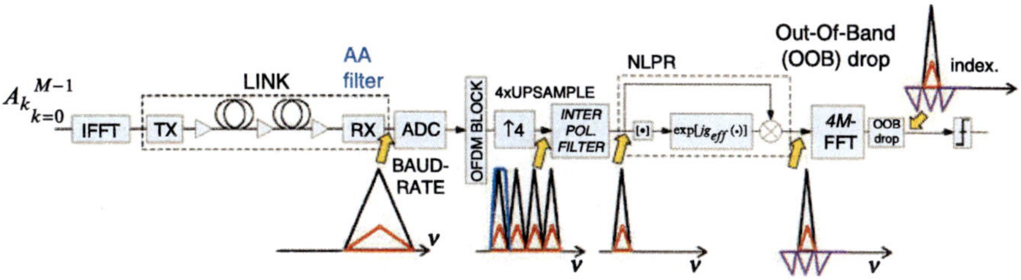

Fig. 3.11 Baud-rate sampled version of the B-NLPR NL compensator, showing the spectra at various points in the Rx

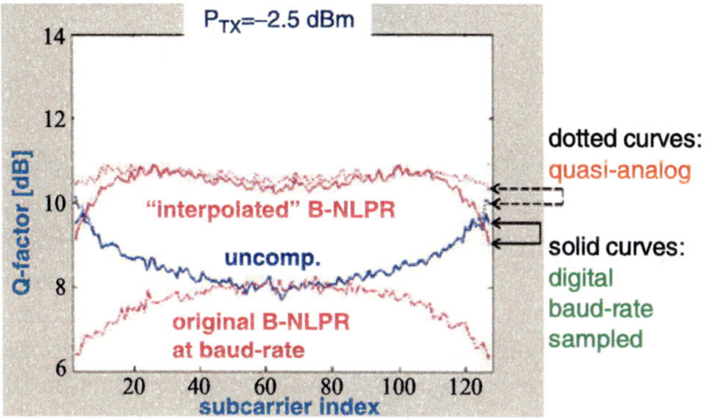

Fig. 3.12 Performance of two B-NLPR versions, both at baud-rate with and without the baud-rate signal processing procedure proposed in Fig. 3.11, for the same conditions as in Fig. 3.10b (standard fiber). The uncompensated performance is also shown for comparison. Evidently, the proposed signal processing scheme enables baud-rate operation of the nonlinear compensator

problem, since the OOB distortion is regenerated upon propagation through the digital nonlinearity, and aliases back in-band due to the digital processing operations. Thus, we may attain some degree of cancelation of the in-band original NL components; however, the new digitally generated OOB products get aliased and reappear back in-band, once an M-point FFT of a signal with M harmonics is taken. These OOB components, which are aliased back in-band, account for the degradation experienced by the B-NLPR NLC, when simplistically operated at baud-rate.

A baud-rate version of B-NLPR was introduced in [50] (Fig. 3.11). The ADC is preceded by a relatively sharp AA filter, blocking the OOB analog components generated in the fiber, then in the digital domain 4× up-sampling is applied onto the ADC output, followed by a 4× interpolation filter, then followed by the B-NLPR NL module, then followed by a $4M$-FFT, the output of which is digitally filtered by an "OOB drop" filter, essentially retaining just the M in-band samples out of the 4M output samples, while discarding the OOB components. The performance attained by this system is presented in Fig. 3.12. The "interpolated B-NLPR" scheme again exceeds (by $\sim$2 dB mid-band) the uncompensated link performance, in fact the baud-rate sampled B-NLPR performance is almost as good as that of the quasi-analog B-NLPR.

The principle of operation of this baud-rate sampled ADC is inferred in the frequency domain, by inspecting the spectral plots in Fig. 3.11: The 4× upsampling generates four spectral images of the input in each spectral period of the sampled signal. The interpolating filter selects the first image and blocks the three remaining images, vacating three times as much spectral room (previously occupied by the other three spectral images) allowing for subsequent expansion of the NL spectrum. At the NLPR output, the spectrum does get broadened – acquiring OOB components. However, at the FFT output we simply block the OOB components, essentially retaining the M in-band samples, which now have their nonlinearity reduced.

3.10 Volterra DF-Based NLC: Principle of Operation

In this section, we introduce the principle of operation of our main Volterra NL DF-based NLC for an OFDM link (Fig. 3.13)

We recall that each triplet of subcarriers out of the OFDM spectrum mixes nonlinearly in the fiber, generating an FWM product, which may fall back in-band and perturb one of the subcarriers. The total FWM NL component, falling on subchannel i, is given by a sum of triple products of the complex amplitudes of all relevant triplets of subcarriers, with coefficients $H_{i;jk}$, which depend on the three participating frequency indexes, which coefficients form the NL VTF:

$$\underset{\sim}{R}_i^{\mathrm{NL}} = \sum_j \sum_k H_{i;jk} \underset{\sim}{A}_j \underset{\sim}{A}_k \underset{\sim}{A}^*_{j+k-i}. \qquad (3.115)$$

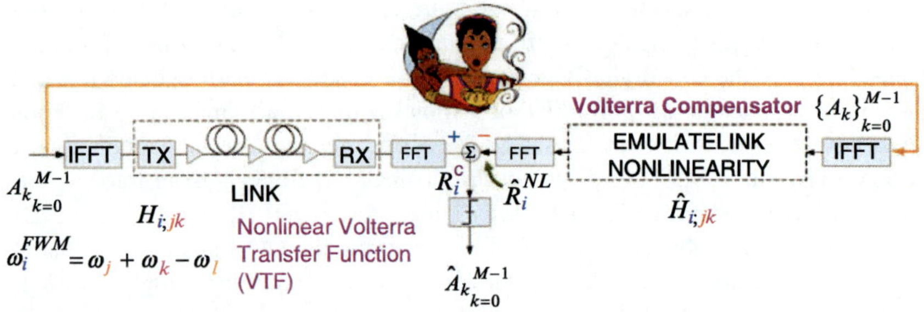

Fig. 3.13 An OFDM link aided by a genie who informs the Rx what the Tx symbols were, yet forbids the Rx to use that info for its decisions. However, the genie allows using the Tx symbols info for emulating propagation along the link, in order to obtain an estimate of the nonlinearity in the received signal and subtract that estimate from the received signal, improving the nonlinear tolerance

To intuitively explain the DF operation of our NLC, we invoke the services of a genie, who magically conveys to the receiver what the transmitted symbols $\underset{\sim}{A}_k$ were. The contract with the genie precludes the Rx from directly using the Tx information for its decisions. Instead, the genie graciously allows using the Tx info in order to reduce the nonlinearity prior to detection. In order to accomplish that, since the Rx has been informed of what has been transmitted, the Rx can simply emulate the link propagation digitally, by passing the $\underset{\sim}{A}_k$ symbols through an IFFT (emulating the Tx), then through a *Volterra filter* (VF) emulating the fiber nonlinearity, and finally taking an FFT (identical to that of the Rx FFT). This way the receiver generates an estimate of the nonlinearity in the received vector, $\hat{\underset{\sim}{R}}_i^{\mathrm{NL}}$, which is subtracted (in the frequency domain) off the received vector, $\underset{\sim}{R}_i$, generating a cleaner signal with reduced nonlinearity, $\hat{\underset{\sim}{R}}_i$, which is finally sliced, obtaining improved decisions. This NL emulated component is also expanded as a sum of triple products, similarly to (3.115), albeit with coefficients $\hat{H}_{i;jk}$ representing the VTF of the compensating NL filter, approximating $H_{i;jk}$. The residual NL components falling on subchannel i, after the NL compensation (subtraction) are expressed as

$$\begin{aligned}\hat{\underset{\sim}{R}}_i^{C} &= \underset{\sim}{A}_i + \sum_j \sum_k H_{i;jk} \underset{\sim}{A}_j \underset{\sim}{A}_k \underset{\sim}{A}^*_{j+k-i} - \sum_j \sum_k \hat{H}_{i;jk} \underset{\sim}{A}_j \underset{\sim}{A}_k \underset{\sim}{A}^*_{j+k-i} \\ &= \underset{\sim}{A}_i + \sum_j \sum_k \left(H_{i;jk} - \hat{H}_{i;jk}\right) \underset{\sim}{A}_j \underset{\sim}{A}_k \underset{\sim}{A}^*_{j+k-i}. \end{aligned} \tag{3.116}$$

(with the superscript C meaning "compensated"). To the extent that the $\hat{H}_{i;jk}$ VTF well approximates the $H_{i;jk}$ VTF, then the coefficients in the last sum are small, and the overall nonlinearity is substantially reduced.

It remains to mechanize our genie (Fig. 3.14). The idea is to use multiple iterations or passes (at least two). In the initial pass, designated pass-0, we use the best FF scheme at our disposal, recording the 'preliminary' decisions $\left\{\hat{\underset{\sim}{A}}_k^{(0)}\right\}_{k=0}^{M-1}$ made in this initial pass, which are declared to be the genie info, i.e., it is assumed that the preliminary decision symbols equal the actually transmitted symbols (the possibility of error is ignored): $\hat{\underset{\sim}{A}}_k^{(0)} = \hat{\underset{\sim}{A}}_k$. We shall later consider the impact of pass-0 errors, i.e., the so-called error propagation effect, showing that the degradation is negligible in high OSNR. In pass-1, the preliminary decisions are IFFT-ed, then propagated through a VF (to be specified below) emulating the link nonlinearity, the output $\hat{r}_k^{\mathrm{NL}}$ of which estimates the time-domain nonlinearity generated in the link, which quantity is subtracted off the received signal vector, yielding the compensated coefficients $\{\hat{r}_k^C\}_{k=0}^{M-1}$, which are then OFDM detected as usual, i.e., are FFT-ed and sliced.

The compensating VF is implemented as the cascade of a linear (LIN) and NL filter. The NL part is implemented as a memoryless nonlinearity, an NLPR similar to the one in the forward path (except for a subtraction by −1, as this NLPR only generates NL components, blocking the linear part of the signal). The LIN filter is in

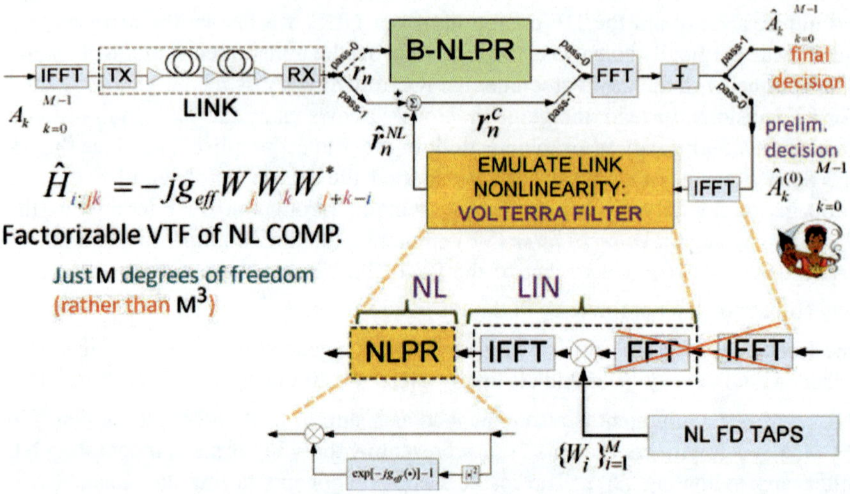

Fig. 3.14 The OFDM link of Fig. 3.13 with the mythical genie replaced by realistic decision feedback, exhibiting an NL-LIN structure for the Volterra filter emulating the link nonlinearity. The LIN part is a frequency-domain equalizer (the cascade of an FFT, complex taps, W, in the frequency domain, and an IFFT), whereas the NLPR is memoryless nonlinearity corresponding having SPM alone (no CD) in the fiber. The frequency-dependent impact of CD is approximated by the interplay of the frequency shaping by the W-coefficients and the time-domain nonlinearity. Finally, the IFFT in the DF loop, and the FFT of the LIN section of the Volterra filter mutually cancel out, yielding the block diagram of Fig. 3.15

turn implemented as a *frequency domain equalizer* (FDE), i.e., a "sandwich" of an FFT and IFFT with multiplicative frequency-domain complex taps $\mathbf{W} \equiv \{W_i\}_{k=0}^{M-1}$ applied in the middle, one such coefficient for each subcarrier, i.e., implementing the VF by means of M rather than M^3 degrees of freedom (DOFs), keeping the compensating VTF evaluation complexity relatively low. This amounts to resorting to a factorizable VTF,

$$\hat{H}_{i;jk}(\mathbf{W}) \propto W_j W_k W_{j+k-i}^* + \text{higher orders} \tag{3.117}$$

for the NL compensator (it remains to show that sufficient cancellation may still be obtained, once we give up on the full complexity). We finally note that the IFFT and the FFT in the DF path cancel out in Fig. 3.14, thus we progress to the block diagram of Fig. 3.15. The extra complexity incurred in this scheme, relative to an uncompensated Rx, is essentially M multipliers for the W-coefficients, the extra NLPR (essentially $3M$ multipliers and a lookup table) and an extra IFFT. The frequency shaping W-coefficients are evaluated offline at this point, by solving the following minimization problem (with I a set of target indexes to minimize the total distortion energy at):

$$P_{\text{opt}}^{c(\text{FWM})} = \min_{\mathbf{W}} \sum_{i \in I} \sum_{[j,k] \in S[i]} \left| H_{i;jk} - \hat{H}_{i;jk}(\mathbf{W}) \right|^2. \tag{3.118}$$

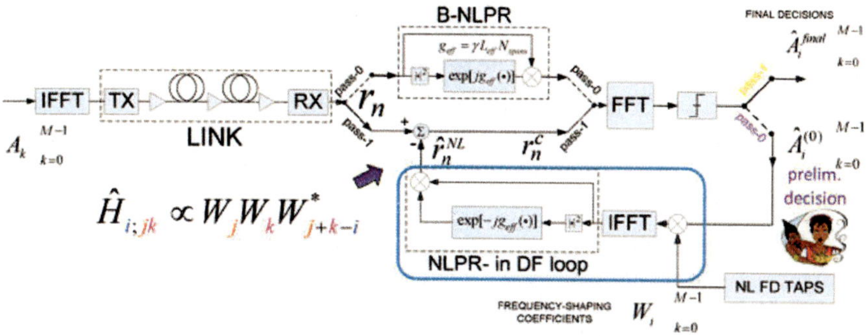

Fig. 3.15 An OFDM link, showing the Rx resulting from Fig. 3.14, detailing the top level functions required for two-pass operation of the Volterra NL DF-based NLC. In pass-0, the received time-domain signal is passed through a B-NLPR then FFT-ed and sliced, yielding preliminary decisions, which, in pass-1, are frequency-shaped, IFFT-ed, nonlinearly distorted through the NLPR in the DF loop, in effect implementing a separable VTF with M rather than M^3 degrees of freedom, yielding an estimate of the nonlinearity in the received signal, to be subtracted off the received signal. The corrected signal is then FFT-ed and sliced, yielding improved final decisions

Note that in the first part of this chapter we developed analytic solutions for the link VTF $H_{i;jk}$ under various conditions [e.g., (3.88) or (3.79)], whereas $\hat{H}_{i;jk}\,(\mathbf{W})$ is given by the factorizable expression (3.117) above. The optimization problem is reduced to a related problem, which is apparently nonoptimal yet simpler and quite close to optimal: The key idea is to convert the NL optimization problem (3.118), which appears nonconvex, into two linear least-mean-square (LMS) problems providing a nonoptimal yet close-to-optimal solution by reasoning that the requirement $H_{i;jk} \approx W_j W_k W^*_{j+k-i}$ amounts to requiring that the phases of both sides of the approximate equality be close, and likewise the log-magnitudes be close:

$$\angle\left\{H_{i;jk}\right\} \approx \angle\left\{W_j W_k W^*_{j+k-i}\right\};\quad \log\left|H_{i;jk}\right| \approx \log\left|W_j W_k W^*_{j+k-i}\right| \tag{3.119}$$

or equivalently,

$$\begin{aligned}\angle H_{i;jk} &\approx \angle W_j + \angle W_k - \angle W_{j+k-i};\\ \log\left|H_{i;jk}\right| &\approx \log\left|W_j\right| + \log\left|W_j\right| + \log\left|W_{j+k-i}\right|.\end{aligned} \tag{3.120}$$

Fortunately, the modified suboptimal problems correspond to minimizing quadratic target functions, tractable by computationally efficient linear projection methods, using pseudoinverse matrices. It turns out that the resulting weights provide very good NLC performance. In a practical system, it would be desired to use automatic coefficients adaptation. This is not pursued here; however, the convexity of the optimization problem (3.120) indicates that such an objective is attainable.

3.11 Volterra DF NLC: Complete Block Diagram, Overall Characteristics and Performance

The final Rx block diagram for an OFDM system with Volterra NL DF NLC is presented in Fig. 3.16 and is detailed in the figure caption. This system attains several desirable features and characteristics:

1. Baud-rate sampling, which is highly desirable feature at ultra-high-speed, given that analog-to-digital conversion continues to pose a major bottleneck for coherent optical transmission. Baud-rate operation is achieved as an extension of the baud-rate sampling approach introduced in Sect. 3.19 for the simpler B-NLPR method, also based on smart DSP comprising four-fold oversampling and interpolation, applying more parallelism and/or faster operations in the ASIC DSP. We shall elaborate on the baud-rate sampling principles in Sect. 3.12.

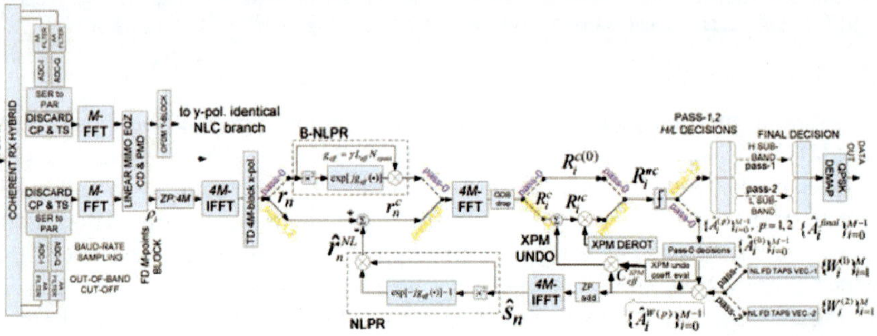

Fig. 3.16 Complete block diagram of an Rx for QPSK OFDM transmission, incorporating the Volterra NL DF NLC. The Rx front-end is a conventional dual polarization coherent OFDM one. Following M-FFTs of the x and y polarization signals, linear frequency domain (FD) MIMO processing is applied to mitigate CD and PMD, generating two separate x and y time-domain (TD) OFDM blocks (records of M points), to be processed in three passes, during the block duration T (before the next block of M samples arrives). The x-polarization processing sequence is as follows: pass-0 comprises a B-NLPR, $4M$-FFT, OOB drop retaining the M in-band points, then slicing to generate the preliminary pass-0 decisions $\{\hat{A}_i^{(0)}\}_{i=0}^{M-1}$, which are kept in a register. In each of the passes $p = 1, 2$, the pass-0 decisions are sample-by-sample multiplied by the frequency taps $\{W_i^{(p)}\}_{i=0}^{M-1}$, yielding the frequency shaped symbols $\{\hat{A}_i^{W(p)}\}_{i=0}^{M-1}$, which are passed through the NL DF loop to generate an estimate $\hat{r}_n^{\mathrm{NL}}$ of the nonlinearity in the received signal. The NL DF loop includes zero-padding to $4M$ points, a $4\,M$ –IFFT, an NLPR performing the NL memoryless operation $(\cdot)\{\exp[-jg_{\mathrm{eff}}|\cdot|^2]-1\}$, with $\cdot$ denoting the input into the NLPR. The NL estimate $\hat{r}_n^{\mathrm{NL}}$ (comprising $4M$ samples) is subtracted from the time domain (TD) 4M-point block $\hat{r}_n$, yielding the corrected signal $\underline{r}_n^C = \underline{r}_n - \hat{\underline{r}}_n^{\mathrm{NL}}$, which is $4M$-FFTed and low-pass filtered to M-points length by the OOB drop, then passed through the XPM UNDO and XPM DEROT respective additive and multiplicative operations, as described in Sect. 3.15, the output $R_i^{\prime\prime C}$ of which is presented to the slicer for the partial $p = 1, 2$ decisions. The final decisions $\{\hat{A}_i^{\mathrm{final}}\}_{i=0}^{M-1}$ are obtained by combining the upper half decisions from pass-1 and the lower half decisions from pass-2: $\hat{A}_i^{\mathrm{final}} = \hat{A}_i^{(1)},\ 1 \le i \le M/2;\ \hat{A}_i^{\mathrm{final}} = \hat{A}_i^{(2)},\ M/2 < i \le M$

Let us enumerate the additional features of the overall system of Fig. 3.16, significantly improving the NL tolerance by adopting a number of measures, the next one in line having already been discussed in the last section:

2. Frequency shaping (usage of the optimized W-coefficients) to synthesize a VTF better tracking CD + NL.
3. Low error propagation in the NL DF process (Sect. 3.13). This is a key enabler of the DF-based method.
4. An "XPM UNDO" original technique intended to decouple the XPM and FWM cancellation strategies, significantly boosting performance, as elaborated in Sect. 3.15.
5. Three passes extension (rather than the two passes implied in Fig. 3.15): The H/L subbands (high/low i.e., upper/lower halves) are separately acquired in passes 1, 2 (in pass-0 preliminary "genie" decisions are generated, as before). Such multipass approach is enabled by the block processing employed in OFDM, as M raw received samples are recorded every T seconds, and processed at a time, with the processing entailing multiple DF-based iterations completed during each of the successive T seconds intervals. The performance impact of splitting the NLC processing in two passes 1, 2 (further to pass-0) is shown in Fig. 3.17, which indicates the piecewise optimization of the two halves in parts (a) and (b), and illustrates in part (c) how the two H/L subbands are stitched together, attaining high Q-factor performance throughout. In pass-1, we use one set of W-coefficients (M of them, as shown in the block diagram of Fig. 3.16), aiming to optimize just the upper (H) subband in terms of Q-factor (Fig. 3.17a), while ignoring the lower subband performance, which makes it easier to attain improved optimization results, albeit just for the upper subband subchannels, as fewer constraints are imposed in the optimization of the compensated VTF. Similarly, in pass-2 we use a different set of W-coefficients (also M of them), aiming to optimize just the lower (L) subband Q-factor performance (Fig. 3.17b), while ignoring the lower subband performance. It turns out that the resulting performance is significantly improved relative to the initial approach of the last section,

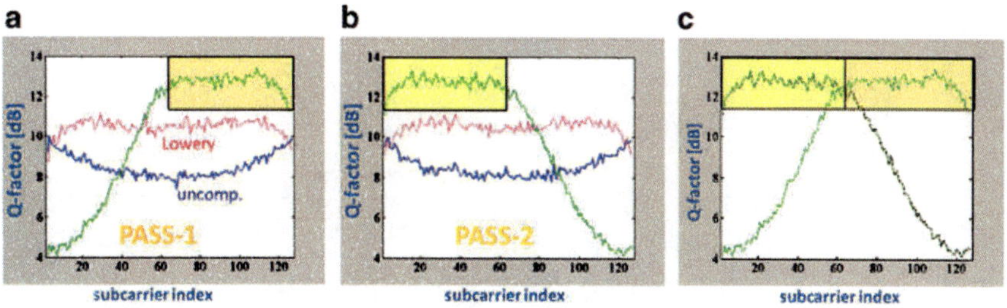

Fig. 3.17 Q-factor vs. subcarrier index in passes 1, 2, separately optimizing the lower and higher subbands performance, then stitching the two halves into the final decision for all subchannels. **(a)** Pass-1 performance optimizes performance in the upper half subband ($64 < i \leq 128$). **(b)** Pass-2 performance optimizes performance in the lower half subband ($1 \leq i \leq 64$). **(c)** Final performance of the two subbands stitched together

which aimed to achieve suppression for all subchannels at once. The price to be paid for the improved performance is that during the T seconds (at the end of which a final decision must be made on all M samples), we must accommodate two iterations rather than a single one, i.e., all processing (W-coefficients modulations, NLPRs, IFFTs) must be doubled up, enhancing the overall complexity of the scheme.

3.12 Baud-Rate Sampling Principles for the Volterra DF NLC

The key DSP concept enabling baud-rate operation is to allow the NL sidebands (generated in the B-NLPR of the DF loop) spectral room to grow without aliasing. This is accomplished by zero-padding M-point records to $4M$ prior to IFFT, and also by low-pass filtering (OOB-drop) of $4M$-point outputs, just retaining the M in-band points. In order to explain how the DSP structure of Fig. 3.16 enables baud-rate ADC, the system is probed at a dozen points and the relevant signals or spectra, tagged (a),(b),...,(k), are shown in Fig. 3.18. The spectral signal $\underset{\sim}{\rho}_i$ (a) contains M harmonic samples corrupted by FWM, XPM/SPM, and noise. The in-band NL distortion in the received signal is illustrated as a small triangle inscribed within the much higher triangle representing the spectrum of the in-band signal. The (a) signal is ZP to total length $4M$, then IFFT-ed, yielding the time-domain signal $\underset{\sim}{r}_n$, the

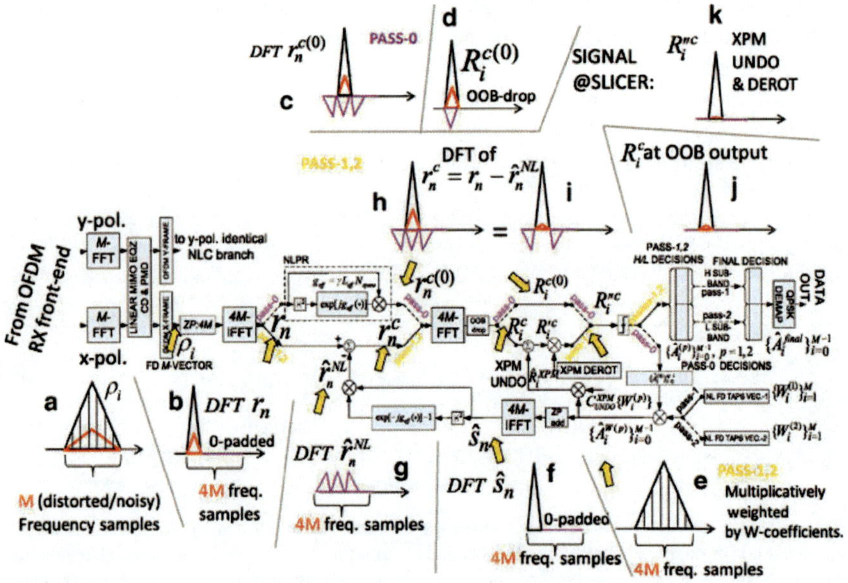

Fig. 3.18 Signal and spectral analysis of the operation of the Volterra NL DF NLC of Fig. 3.16, highlighting that the system functions with baud-rate sampling

spectrum (DFT) of which, shown in (b), is evidently sparse, with support M, out of its $4M$ points. In pass-0 (the upper path, with the switches flipped up), the B-NLPR broadens the spectrum (see also Fig. 3.11 where we analyzed B-NLPR operation), however the OOB components are filtered out by the OOB drop at the $4M$-FFT output. In detail, there are three spectral components generated at the B-NLPR output, one in-band and two OOB, shown as three small inverted triangles in spectral signal (c) representing the DFT of $\underset{\sim}{r}_n^{c(0)}$. The spectral signal $\underset{\sim}{R}_i^{c(0)}$ at the OOB-drop output is shown in (d). The in-band NL distortion has been much (but not sufficiently) suppressed in pass-0, as indicated by the two little in-band triangles in (d), representing the link and B-NLPR distortions, which approximately cancel each other. Based on this pass-0 signal with somewhat reduced distortion, the slicer makes its preliminary decisions, $\hat{\underset{\sim}{A}}_n^{(0)}$, which are subsequently multiplicatively shaped by W-coefficients in each of the passes-1, 2, yielding an M-point spectral signal $\hat{\underset{\sim}{A}}_i^{W(p)} = \hat{\underset{\sim}{A}}_i^{(0)} W_i$ (e) [note that for simplicity, the pass index 1, 2 is not explicitly attached, and the spectral distortion is not graphically illustrated in the triangular spectral shape plotted in (e)]. Further progressing through the DF loop, $\hat{\underset{\sim}{A}}_i^{W(p)}$ is ZP from length M to length $4M$, and a $4M$-IFFT is applied. The spectrum of the time-domain signal at the $4M$-IFFT output is shown in (f). This is a sparse ZP signal with in-band spectral support of M points out of the $4M$ points, making room for the spectral broadening which is about to occur upon traversing the DF-loop NLPR, the DFT of the output $\hat{\underset{\sim}{r}}_n^{\mathrm{NL}}$ of which, is shown in (g), seen to contain three NL components, one in-band and two OOB. Note that a linear term is absent in this signal, as the DF NLPR differs from the one used in pass-0 (the B-NLPR) by a -1 additive term, which suppresses the linear component. The signal $\hat{\underset{\sim}{r}}_n^{\mathrm{NL}}$ represents a synthetically generated time-domain estimate of the nonlinearity in the received signal, $\hat{\underset{\sim}{r}}_n$ (the output of the $4M$-IFFT). This estimate, $\hat{\underset{\sim}{r}}_n^{\mathrm{NL}}$, is subtracted off $\hat{\underset{\sim}{r}}_n$. The DFT of the output $\hat{\underset{\sim}{r}}_n^{c}$ of the subtractor is shown in (h), seen to contain the in-band signal (the tall triangle) and its in-band NL distortion (the smaller upward pointing triangle), as well as the three distortion terms generated in the DF loop, shown as downward pointing little triangles, two of them two OOB, and one in-band nearly canceling the upward pointing in-band small triangle, i.e., just small net residual distortion is left in band, as shown in (i). As for the two OOB side-bands also present in (i), those are blocked by the OOB drop at the output of the 4M-FFT, as shown in the R_i^c spectral signal in (j), which features the in-band signal component, with its very small in-band residual distortion. In principle, this signal could be sliced to yield final decisions for passes 1, 2; however, it turns out that even better performance may be obtained by applying the XPM UNDO and DEROT processing, essentially decoupling the FWM and XPM mitigation strategies, as detailed in Sect. 3.15. The final FWM and XPM corrected signal $R_i^{\prime\prime c}$ is illustrated in (k), featuring an even tinier in-band distortion, graphically suggestive of the improved suppression of distortion. It is this type of signal which is presented to the slicer in each of the passes 1, 2 generating improved decisions for the upper and lower subbands, stitched together to form the final decisions.

3.13 Low Error Propagation for the Volterra DF NLC

Our Monte-Carlo simulations (Fig. 3.19) counted the errors generated in pass-0 (referred to as "B-NLPR" errors) and at the end of passes 1–2 (referred to as "Volterra errors"). This was done for various levels of optical power and for various numbers of repetitions, typically several thousands. For example, at −3.5 dBm (the optimal power where best BER is attained) and over 4,000 repetitions (each repetition making decisions on each of the $M = 128$ OFDM subchannels), we collected 2,355 uncompensated errors over all subchannels. The B-NLPR cuts the number of errors down to 169, whereas the number of errors left after Volterra is 5 – in fact just one of the 169 B-NLPR errors still stands as a Volterra error; however, the Volterra procedure introduces four new errors. This dramatic reduction in the error rate (2,355 down to 169 then down to 5) is indicative of very low error propagation.

We next provide a simple theoretical analysis justifying why the Volterra NL DF method benefits from low error propagation.

In the absence of a genie, we resort to imperfect pass-0 decisions in the DF loop, replacing (3.116) by

$$\begin{aligned}\hat{\underset{\sim}{R}}{}_k^C &= \underset{\sim}{A}_i + \sum_j \sum_k H_{i;jk} \underset{\sim}{A}_j \underset{\sim}{A}_k \underset{\sim}{A}{}^*_{j+k-1} - \sum_j \sum_k \hat{H}_{i;jk} \hat{\underset{\sim}{A}}_j \hat{\underset{\sim}{A}}_k \hat{\underset{\sim}{A}}{}^*_{j+k-i} \\ &\approx \sum_j \sum_k H_{i;jk} \left(\underset{\sim}{A}_j \underset{\sim}{A}_k \underset{\sim}{A}{}^*_{j+k-i} - \hat{\underset{\sim}{A}}_j \hat{\underset{\sim}{A}}_k \hat{\underset{\sim}{A}}{}^*_{j+k-i} \right),\end{aligned} \tag{3.121}$$

where in the last expression we assumed for simplicity that the approximation $H_{i;jk} \approx \hat{H}_{i;jk}$ is actually a strict equality.

The residual variance of the compensated signal is expressed as

$$\left\langle \left| \hat{\underset{\sim}{R}}{}_k^C - \underset{\sim}{A}_i \right|^2 \right\rangle = \sum_j \sum_k \left| H_{i;jk} \right|^2 \left| \underset{\sim}{A}_j \underset{\sim}{A}_k \underset{\sim}{A}{}^*_{j+k-i} - \hat{\underset{\sim}{A}}_j \hat{\underset{\sim}{A}}_k \hat{\underset{\sim}{A}}{}^*_{j+k-i} \right|^2, \tag{3.122}$$

where we used the property that distinct triplets add up on a power basis, as they are mutually incoherent whenever the transmitted sequence is white. In this case, the only imperfection in the distortion cancelation process is due to pass-0 slicer errors, causing $\underset{\sim}{A}_j \underset{\sim}{A}_k \underset{\sim}{A}{}^*_{j+k-i} - \hat{\underset{\sim}{A}}_j \hat{\underset{\sim}{A}}_k \hat{\underset{\sim}{A}}{}^*_{j-k-i} \neq 0$. In QPSK transmission, given that an error was committed, we most likely ventured into a neighboring quadrant, such that A-phasor gets rotated by ±90°, causing the triple product $\hat{\underset{\sim}{A}}_j \hat{\underset{\sim}{A}}_k \hat{\underset{\sim}{A}}{}^*_{j+k-i}$ to also get rotated by ±90° relative to $\underset{\sim}{A}_j \underset{\sim}{A}_k \underset{\sim}{A}{}^*_{j+k-i}$, thus we have $\hat{\underset{\sim}{A}}_j \hat{\underset{\sim}{A}}_k \hat{\underset{\sim}{A}}{}^*_{j+k-i} = (\pm j) \underset{\sim}{A}_j \underset{\sim}{A}_k \underset{\sim}{A}{}^*_{j+k-i}$ (assuming a single error occurs in the three A phasors, as probability of more than one error is negligible). It follows that $\left| \underset{\sim}{A}_j \underset{\sim}{A}_k \underset{\sim}{A}{}^*_{j+k-i} - \hat{\underset{\sim}{A}}_j \hat{\underset{\sim}{A}}_k \hat{\underset{\sim}{A}}{}^*_{j+k-i} \right|^2 = 2 \left| \underset{\sim}{A}_j \underset{\sim}{A}_k \underset{\sim}{A}{}^*_{j+k-i} \right|^2$, i.e., the errored triplets are not compensated at all but are rather spoiled, having their FWM power doubled,

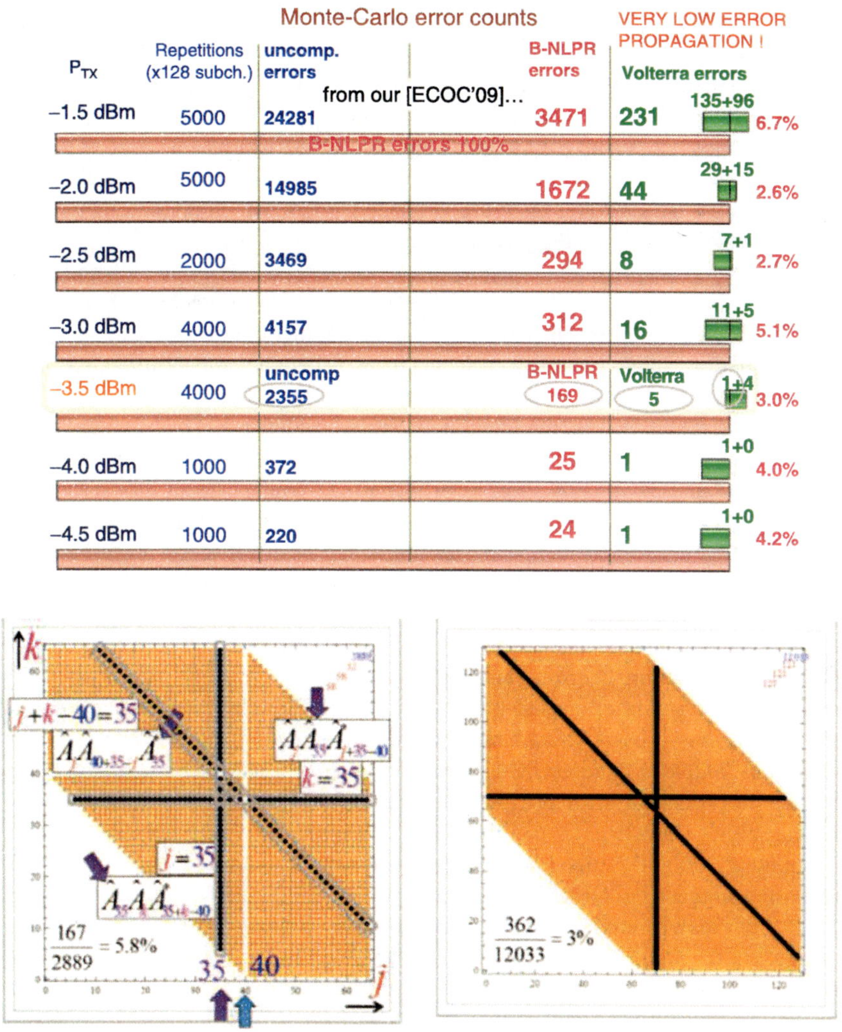

Fig. 3.19 Error propagation properties of the Volterra NL DF NLC: (*left*) Monte Carlo error counts: The B-NLPR errors in preliminary pass-0 were normalized to 100%, such that the green little bars represent the final Volterra errors, labeled and graphically scaled according to their percentage relative to of the B-NLPR errors. The simulations were run for various optical powers and numbers of repetitions, as listed. We split the Volterra errors into two types of errors – those which occur within the B-NLPR errors, which represent error propagation and new Volterra errors occurring when B-NLPR is correct. It is seen that the proportion of Volterra errors is quite low, i.e., Volterra is much more efficient than NLPR alone (middle and right): Graphical displays of the total number of triplets vs. errored triplets for $M = 64$ subcarriers (*middle*) and $M = 128$ subcarriers (*right*). The error triplets (*black points*) are arrayed along three lines in the $[j, k]$ plane, corresponding to an error in the first, second and third index

detracting from the overall cancelation for the "good" triplets. The question is how many such errored triplets are there. If it is just a small number of triplets that are in error, then although their FWM power is doubled, their percentage relative to the vast majority of triplets (whose FWM has been canceled or vastly reduced) is still negligible, thus the overall FWM cancelation is still substantial.

A rough order of magnitude of the percentage of errored triplets is obtained as follows: For M subcarriers, there are $O\left[M^3\right]$ triplets, which divided by M sub-channels, yields $O\left[M^2\right]$ falling on each subchannel. Now, when an index is in error, there are $O\left[M^2\right]$ triplets involving that index (the errored index with each one of the $M-1$ other indexes, twice), hence, dividing by the number of subchannels, there are $O\left[M\right]$ errored triplets per subchannel. Thus, the number of errored triplets over the total number of triplets falling on each subchannel (i.e., the probability to get an errored triplet fall on any given subchannel) is given by $O\left[M\right]/O\left[M^2\right] = O\left[M^{-1}\right]$. For example, for $M = 128$, the fraction of errored triplets is $O\left[1\%\right]$.

Two numerical examples of the errored triplet counts are shown in Fig. 3.19, for $M = 64$ and $M = 128$, respectively. The diagrams represent the $[j, k]$ plane of index pairs labeling each FWM triplet. For $M = 64$, we assume observation index $= 40$ and errored index $= 35$. Actually, the error can occur in three ways, either in the first, second, or third A term, respectively, corresponding to the vertical, horizontal, and slanted black lines, each black point in these lines representing an errored triplet. There are 167 errored (black) triplets out of 2,889 total triplets, i.e., 5.8% of the triplets are in error. The chart on the right, for $M = 128$ displays similar traits, but the fraction of errored triplets is reduced. The observation index is now $M = 64$ (also mid-band where most distortion is generated), and the errored index is taken as 70. Now, there are 12,033 FWM triplets, out of which 362 are in error, i.e., the proportion of errored triplets dropped to 3% (consistent with the $O\left[1\%\right]$ rough analysis above).

Suppose we got 10 dB FWM suppression, barring error propagation for a dispersion-unmanaged OFDM system with $M = 128$ (actually in excess of 15 dB suppression may be attained). Thus, for 97% of the triplets, those which are not in error, we get -10 dB i.e., a factor of 0.1 FWM suppression, whereas for 3% of the triplets, those which are in error, we actually get a doubling of the FWM power. In this example, compounding those two effects we have $97\%{*}0.1 + 3\%{*}2 = -8$ dB, rather than the original -10 dB assumed without the error propagation effect. We conclude that despite the doubling of FWM for the errored triplets, the small proportion of error triplets leads to the error propagation effect being fairly small. The simulations shown in Sect. 3.16 actually incorporate the effect of error propagation, demonstrating that excellent NL tolerance improvement is attainable.

3.14 The Role of Higher-Order (5th, 7th, . . .) Nonlinearities

Considering the "undepleted pumps" perturbation approach, it turns out that the modeling must be extended up to fifth or even seventh order to achieve sufficient accuracy. The question is why higher orders would be needed to describe FWM, which is solely a third-order effect. In the perturbation method, each triplet of subchannels ("pumps") is linearly propagated while neglecting changes in their complex amplitude due to FWM "back-reaction." The third-order FWM due to the "undepleted pumps" must be first evaluated. The fifth order is generated by two "pumps" and a third-order product, all three mixing again through the third-order FWM nonlinearity. The perturbation series may be continued, yielding a multi-wave mixing (MWM) series description of the FWM effect, albeit expanded in terms of the original excitation of the "undepleted pumps":

$$R_i^{\mathrm{NL}} = \sum_{j,k} H_{i;jk} A_j A_k A_{j+k-i}^* + \sum_{j,k,m,n} H_{i;jkmn} A_j A_k A_m A_n^* A_{j+k+m-n-i}^* + \ldots. \tag{3.123}$$

In our NLC realization, we balance the third-order mixing products of the compensator against the third-order mixing products of the fiber. However, the MWM expansion indicates that we must also contend with the effect of the higher order terms.

Note that the memoryless part of our NLC (the NLPR) is not purely a third-order nonlinearity, but has been patterned to correspond to the SPM effect in the fiber link, purposely designed to include NL orders higher than the third in its Taylor expansion (only odd-order terms appear in the expansion: 5th, 7th, ... order; typically up to fifth order suffices):

$$\begin{aligned} \underset{\sim}{u} \exp\left[-j g_{\mathrm{eff}} |\underset{\sim}{u}|^2\right] - 1 &= u \left(-\phi - \frac{1}{2}\phi^2 - \frac{1}{3!}\phi^3 - \ldots \bigg|_{\phi \to j g_{\mathrm{eff}} |\underset{\sim}{u}|^2} \right) \\ &= -j g_{\mathrm{eff}} |\underset{\sim}{u}|^2 u + \frac{1}{2} g_{\mathrm{eff}}^2 |\underset{\sim}{u}|^4 \underset{\sim}{u} + \frac{j}{3!} g_{\mathrm{eff}}^3 |\underset{\sim}{u}|^6 \underset{\sim}{u}. \end{aligned} \tag{3.124}$$

Inferring from the improved NLT attained with our Volterra NLC, the higher orders of its DF NLPR appear to cancel the corresponding higher orders of the fiber fairly well (once third-orders are mutually balanced). In the next section, treating the XPM analysis and mitigation, we shall see that MWM modeling up to fifth order becomes important in the XPM context as well.

3.15 "XPM Undo and Derotate" Decoupling XPM and FWM Mitigation in the Volterra DF NLC

The FWM and XPM respective contributions in the received signal are given by:

$$\text{FWM\_R: } \sum\sum_{[j,k]\in S[i]} H_{i;jk} \underset{\sim}{A}_j \underset{\sim}{A}_k \underset{\sim}{A}_{j+k-i}^* ; \qquad \text{XPM\_R: } 2\tilde{A}_i \sum_{k\neq i} H_{i;ik} |\tilde{A}_k|^2. \tag{3.125}$$

The XPM generated by the NLPR in the NL DF loop is given by XPM_DF: $\tilde{A}_i \cdot 2W_i^{(p)} \sum_{k \neq i} H_{i;ik} \left|W_k^{(p)}\right|^2 \left|\tilde{A}_k\right|^2$, where $W_i^{(p)}$ are the frequency shaping coefficients in the pth pass $(p = 1, 2)$.

It is apparent that XPM_DF (i.e., the XPM generated in the compensator) is not a good canceller for XPM_R, unlike the FWM component generated in the compensator, which does provide an excellent canceller for the received FWM. The way it stands now while correcting FWM we actually spoil XPM. It is thus desirable to decouple the FWM and XPM mitigation processes, performing each one individually in an optimized way, eliminating the tradeoff between the two effects. As it is inevitable that XPM be generated within the DF loop NLPR, alongside FWM, the proposed strategy for decoupling the two processes is to subtract (or rather to add with opposite sign) the XPM_DF component out of the compensated signal in the frequency domain, in effect "undoing" the XPM correction of the NLPR by means of the XPM UNDO adder indicated in Fig. 3.16. Once the XPM has been "undone," i.e., removed from R_i^c, yielding the output $R_i'^c$, then XPM remains present in full strength in the signal $R_i'^c$ at the output of the XPM UNDO adder, and it must be somehow mitigated. XPM is known to be an impairment consisting of an overall rotation of the complex-plane received constellation, with the amount of rotation determined by the power of all subcarriers (we assume SPM to be included as a special case, with half the power efficiency). Its mitigation is then readily accomplished by means of an XPM DEROT multiplier, which simply derotates the constellation back to its original position: $R_i''^c = R_i'^c e^{j g_{\text{eff}}(2M-1)P_0}$, $i = 0, 1, 2, \ldots, M$, where $P_0 \equiv \sum_{k=0}^{M-1} \left|\hat{A}_k\right|^2$ is the total received power. It is the XPM-"undone" and derotated spectral signal $R_i''^c$ that is presented to the slicer in each of the passes 1, 2.

It remains to describe the novel XPM UNDO procedure. For this method to be effective, it must cancel out of the NL DF loop output all mechanisms of higher-order XPM generation, beyond the third order, at least up to the fifth order. This is in the spirit of the higher-order perturbation approach of the last section, whereby triplets of subcarriers generate NL products (the third order), and in turn third-order XPM experiences XPM itself, interacting with the power of the other subcarriers to generate a fifth-order XPM product. The mathematical description of this process of XPM generation in the DF loop, up to the fifth order, is given by the following expression of the XPM component at the output of the NLPR in the DF-loop:

$$\hat{\underset{\sim}{r}}_n^{\text{XPM}} = C_{\text{eff}}^{\text{XPM}} \sum_{i=0}^{M} \hat{A}_i^W e^{j\omega_i n} = C_{\text{eff}}^{\text{XPM}} \hat{s}_n^W;$$

$$C_{\text{eff}}^{\text{XPM}} \equiv \frac{1}{1!}(-j g_{\text{eff}})^1 C^{(3)} + \frac{1}{2!}(-j g_{\text{eff}})^2 C^{(5)} + \ldots$$

$$C^{(3)} = -2\hat{P}^W = -2 \sum_{k=0}^{M-1} \left|\hat{A}_k^W\right|^2; \quad C^{(5)} \approx (12M-9) \sum_{j=0}^{M-1} \left|\hat{A}_j^W\right|^4. \quad (3.126)$$

The frequency-domain XPM UNDO procedure is actually very simple: $R_i^{\prime c} = R_i^c + C_{\text{eff}}^{\text{XPM}} \hat{A}_i^W$, with $C_{\text{eff}}^{\text{XPM}}$ generated by the "XPM undo coeffs eval" module of Fig. 3.16, according to (3.126). The complexity involved in generating $R_i^{\prime c}$ is low, just $2M + 2$ complex multipliers (CMs) per OFDM block (we count two multiplications in evaluating $\left|\hat{A}_j^W\right|^4 = \left(\hat{A}_j^W \hat{A}_j^{W*}\right)^2$ for each of the M j-indexes, and two extra multiplications by prescribed Taylor coefficients, which are functions of g_{eff}). The block diagram of Fig. 3.16, including the XPM UNDO and DEROT procedure, yields large improvement in NL tolerance, as detailed next.

3.16 Volterra DF NLC Performance Simulations (Q-Factor and BER)

In this section, we compare the Volterra NL DF-based NLC, with the B-NLPR system, and with an uncompensated OFDM system. The parameters used in our performance simulations are identical to those stated in the caption of Fig. 3.10, which described the performance of a B-NLPR NLC system.

We start with the ASE turned off (Fig. 3.20) to assess how well the FWM and XPM nonlinearities are suppressed, without getting the NL performance obscured by the noise. It is apparent that from the viewpoint of FWM suppression, we attain 3 to 4 db improvement above the B-NLPR and 2–7 db above an uncompensated system.

The performance with both NL and ASE noise is shown in Fig. 3.21, presenting the Q-factor vs. subcarrier index (Fig. 3.21-left), and BER vs. launched optical power (Fig. 3.21-right). It is apparent that the Volterra NLC is a ~2 dB above the B-NLPR. In turn, the B-NLPR is ~2 dB on top of an uncompensated system (at mid band), i.e., the Volterra system is about 4 dB above the uncomp system. Moreover, some decent margin above the uncompensated system is retained by the Volterra system even at the band edges. From Fig. 3.21-right, it is apparent that we can turn

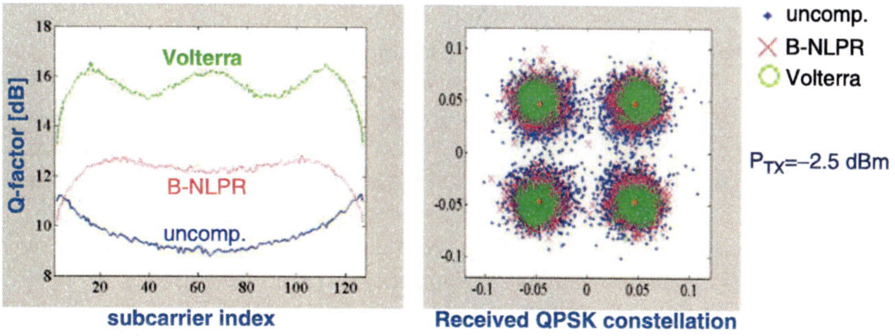

Fig. 3.20 FWM and XPM alone, turning the ASE off in the simulation. (*Left*): Q-factor vs. subcarrier index. (*Right*): Received constellation

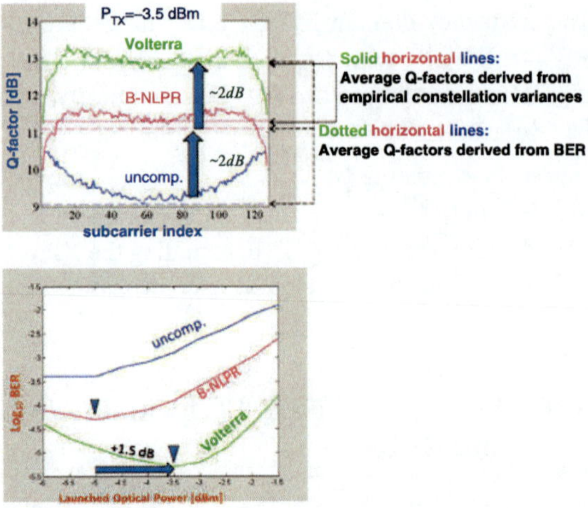

Fig. 3.21 Volterra vs. B-NLPR NLC vs. uncompensated performance. (*Left*): Q-factor vs. subcarrier index. (*Right*): BER vs. optical power

up the power by 1.5 dB and still attain more than two orders of magnitude improvement in BER, indicative of the highly improved NL tolerance of the Volterra NLC.

3.17 Computational Complexity vs. NL Tolerance Performance Trade-Offs

We now consider the complexity price to be paid in exchange for the improved NLT. In the plot of Fig. 3.22-left, the horizontal axis is the number of subcarriers, M, and the vertical axis is the number $\kappa(M) = C(M)/(T \cdot B_T)$ of CMs per OFDM block, further normalized by T, the block duration, and by B_T, the total OFDM bandwidth. Thus, the units of the complexity measure along the vertical axis are CM per sec per Hz. Since $T \cdot B_T = T \cdot \Delta\nu M = M$, then our complexity measure is alternatively expressed as $\kappa(M) = C(M)/(T \cdot B_T) = C(M)/M$, i.e., CM per subcarrier. Another interpretation is that for a given modulation format of each subcarrier, the total data rate is $R_T = \eta B_T$, where η is the spectral efficiency in units of b/s/Hz, thus, $T \cdot B_T = T \cdot R_T/\eta = b_T/\eta$, where b_T is the total number of bits conveyed during an OFDM block (T sec duration). Therefore, our measure of complexity is re-expressed as $\kappa(M) = \eta C(M)/b_T$, i.e., it is proportional to the number of CMs per bit of conveyed information (irrespective of the rate). However, for evaluation purposes, we prefer the $\kappa(M) = C(M)/M$ form. The number of CMs per frame, $C(M)$, is evaluated for our Volterra NL DF system (referred to as "OUR"), for the B-NLPR system as well as for an uncompensated system, by itemized counting all the DSP operations (FFT, CD + XPM, PMD derotation, interpolation, frequency shaping, IFFT, XPM undo, yielding the counts:

$$C_{\mathrm{OUR}}(M) \equiv 73M + 12.5M \cdot \log M; \quad C_{\mathrm{B-NLPR}}(M) \equiv 23M + 4.5M \cdot \log M;$$
$$C_{\mathrm{UNCOMP}}(M) \equiv 3M + \frac{1}{2}M \cdot \log M. \tag{3.127}$$

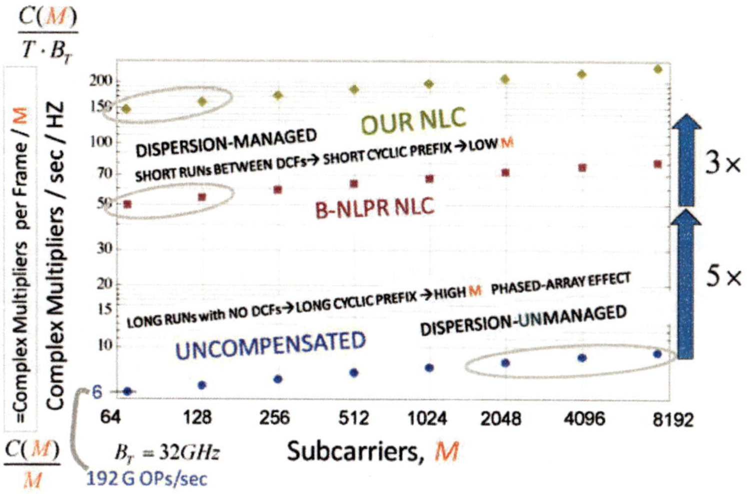

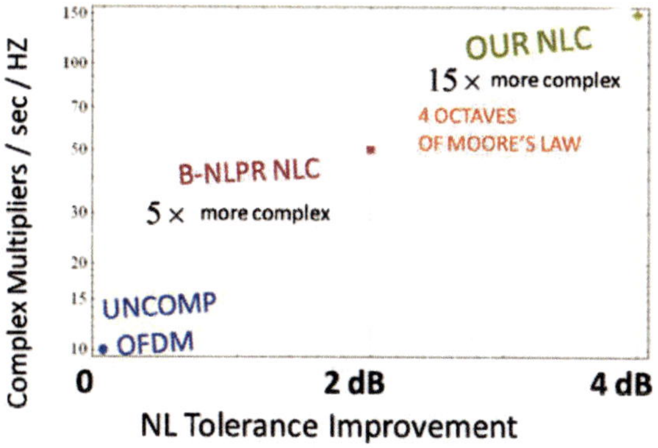

Fig. 3.22 Complexity-performance trade-offs. (*Left*): Complexity-measure (per-bit or per-second-per-Hz), vs. number of subcarriers. (*Right*): Complexity measure vs. NL tolerance improvement, for the B-NLPR and Volterra (our) NLC, with an uncompensated system used as a baseline

Once we divide these counts by M, we obtain the following formulas for the respective complexity measures:

$$\kappa_{\text{OUR}}(M) \equiv 73 + 12.5 \cdot \log M; \quad \kappa_{\text{B-NLPR}}(M) \equiv 23 + 4.5 \cdot \log M;$$
$$\kappa_{\text{UNCOMP}}(M) \equiv 3 + \frac{1}{2}\log M. \tag{3.128}$$

These complexities may be all described as $O(\log M)$. Intuitively, the FFT, which is one of the heaviest computational resources in the overall DSP chain, has complexity $\frac{M}{2}\log M$; however for larger M, the FFT duration is proportionally extended, hence the rate (ops/s) tends is scaled back by a factor of M, thus the final complexity measure of an FFT merely grows as $\frac{1}{2}\log M$.

However, besides the $O(\log M)$ order trend, the actual numerical factors in (3.128) are important, as they weigh heavily on the computational burden. For example, for a 32 GHz total bandwidth OFDM system, required to carry 112 Gb s^{-1} each point on the vertical axis represents 32 G multipliers per sec, e.g., the 6 multipliers per sec per Hz required for an uncompensated system with $M = 64$ map into an actual complexity of 192 G Ops s.

Note that a dispersion unmanaged link would be typically used without compensation, relying on the PA effect to suppress FWM, taking large M values in order to keep down the CP overhead. In contrast, in the dispersion-managed case, NL compensation would be applied to counteract the nonlinearity in each span, which adds coherently from span to span, and since the dispersion is low, one can adopt low M values without incurring substantial overhead. Assessing the required complexities in Fig. 3.22-left, the good news is that our scheme is just a factor of 3 more complex than that the baud-rate version of the B-NLPR basic NLC scheme; however, the bad news is of the (baud-rate) B-NLPR is already a factor of 5 more complex relative to an uncompensated system. Thus, altogether, in exchange for its ~4 dB NL tolerance improvement, our NLC is 15 times worse in complexity than an uncompensated system.

Evidently, complexity should not be considered alone, but in be assessed conjunction with the performance improvement benefit it brings about. Figure 3.22-right shows the performance-complexity plane, with the horizontal axis being the amount of NL tolerance improvement (FWM suppression) in dB, while the vertical axis is the complexity measure, normalized by that of an uncompensated system. Thus, with the uncompensated case taken as baseline, the B-NLPR is 5 times more complex while it improves NL tolerance performance by 2 dB, and finally our NLC is 15 times more complex but improves performance by 4 dB. It is suggested that the performance of all competitive NLC schemes be pegged on such complexity vs. performance chart, carefully counting the normalized numbers of operations (per bit or per sec per Hz) relative to an uncompensated system, vs. the achieved NL tolerance improvement.

3.18 Discussion: Volterra DF NLC vs. BP – Suggested Roadmap for Future NLC

The BP NLC method was reviewed in Sect. 3.8. BP is intuitively appealing to those used to physical thinking, as it precisely emulates the physics of propagation, albeit in reverse. If unlimited computing power were available, i.e., a very large number of SSF sections could be realized, and in the absence of noise, BP would be an optimal method in the scalar (single polarization) case. In the vector case accounting for both two polarizations, and in the absence of knowledge of the PMD dynamics, a form of the BP based on inverting the Manakov equation would be optimal [28].

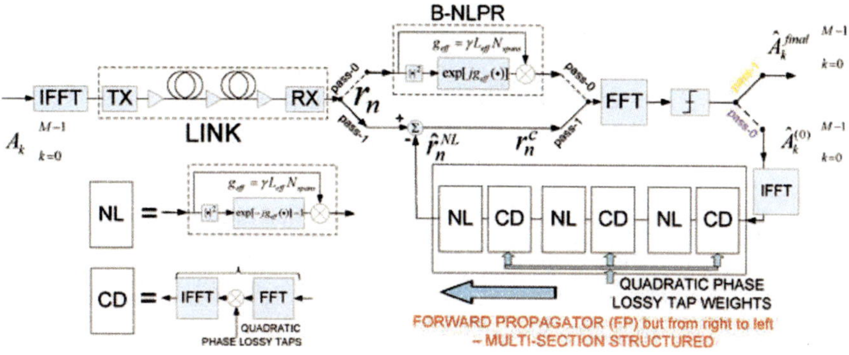

Fig. 3.23 A decision-feedback-based version of the BP NLC, best referred to as DF forward propagation (FP) NLC. The preliminary pass-0 decisions are IFFTed then used to emulate forward propagation through the fiber through an SSF structure (rather than back-propagation)

However, when computing power is constrained, e.g., if just several SSF sections may be afforded, we conjecture that BP ceases to be optimal, and an optimized VF of the same computational complexity might provide better performance.

To justify this, note that BP is a form of FF NL equalization. It is well known that DF equalizers are preferred to FF equalizers, thus we conjecture that this rule extends to the NL case as well. We then propose to introduce a DF-based version of BP, as shown in Fig. 3.23. Such DF BP system would have better performance than the corresponding FF BP system using the same number and complexity of elementary NL-CD sections.

However, we conjecture that the optimality of BP in the complexity unconstrained case is misleading, and does not necessarily project to the finite computing power case. In this case, allocating the available operations to elementary CD-NL, CD-NL, CD-NL,... sections may not be the optimal way to organize the DF NLC. We may exemplify this in the special case that the DF loop contains a single elementary CD-NL section. As the fiber emulator is fed by an IFFT of the pass-0 decisions, and the CD consists of the cascade of an FFT a multiplication by quadratic phase taps and an FFT, then it is apparent that the IFFT and the FFT cancel out, and we are left with the multiplication by quadratic phase taps followed by the NL section, which amounts to an NLPR, mimicking a dispersion-free NL fiber, i.e., the SPM NL. But this structure is almost the same as that of our Volterra DF NLC, with the exception of using here quadratic phase taps rather than optimized general W-taps used there. Yet, we know that our optimization of the frequency domain weights does not yield a quadratic phase dependence! So, we have just exemplified in the case of DF with a single section, that the BP-based version fares worse than a fully optimized VF in the DF loop. The resemblance of our Volterra DF NLC to a single section DF BP NLC, suggests an extended Volterra DF structure (Fig. 3.24), based on multiple sections (LIN-NL) (LIN-NL) (LIN-NL)....rather than a single LIN-NL section (Fig. 3.14) in the DF loop. This novel structure is inspired by physical intuition in its NL realization, using the $\exp\left[-j\gamma N_{\text{span}} L_{\text{eff}} |\underline{u}|^2\right] - 1$ memoryless

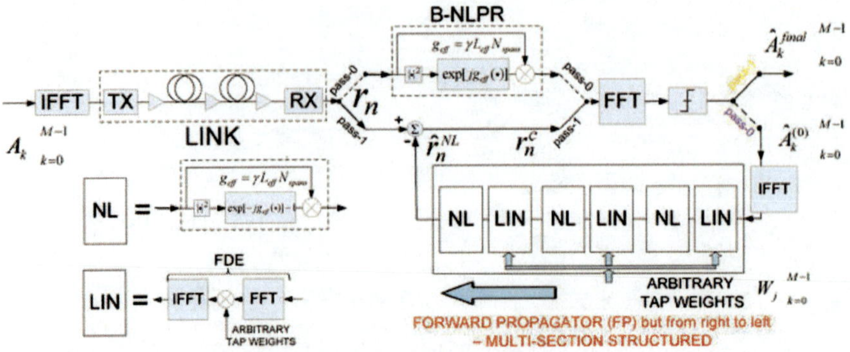

Fig. 3.24 A decision-feedback based version with improved multisection filter inspired by the DF NLC system of Fig. 3.23. The preliminary pass-0 decisions are IFFTed, then used to emulate forward propagation through the fiber through a multisection Volterra filter generalizing the forward propagating SSF structure. The multisection Volterra filter consists of an alternation of LIN and NL sections as shown. The LIN sections are more general than the CD sections of Fig. 3.23, thus the whole NLC structure includes the one in Fig. 3.24 as a special case, indicating that upon optimizing the tap weights in the LIN sections here, we may obtain better performance than in the decision-feedback based system of Fig. 3.23, which in turn would yield better performance than the BP method which is a form of feedforward NL equalization. Also note that this structure generalizes the one in Fig. 3.14, which amounts to taking a single LIN-NL section rather than multiple ones

nonlinearity corresponding to a CD-free fiber (SPM), however, unlike in Fig. 3.23, the LIN sections of the structure of Fig. 3.24 are detached from CD physical meaning, allowing for arbitrary linear taps (W-coefficients) to be used in each of the LIN sections, which enables improved optimization over those taps. The modeling of the two DF structures proposed in Figs. 3.23 and 3.24, and the assessment of their relative performance, are relegated to future work.

3.19 Conclusions

In this chapter, we derived a fully analytic model for the NL impairments within a single OFDM channel. The mathematical Volterra formalism the physical OPI perturbation approach provides the most suitable tools for treating the Kerr-induced nonlinearity. Based on these analytical tools, as developed in the first half of the chapter, we proceeded in the second half of the chapter beyond analysis, to synthesis of efficient NL compensators for CO-OFDM.

It turns out that the relative amounts of CD vs. NL and the extent of dispersion management adopted for the fiber-link, set one of three operational regimes:

(1) CD $\gg$ NL: If the dispersion dominates over the nonlinearity, and the link is dispersion unmanaged (no DCFs), efficient PA cancelation of NL [30], may occur even without requiring an NLC, providing the most high-performance

solution. The removal of DCFs, however, may not be always possible (e.g., on certain legacy links, especially submarine ones).

(2) CD $\ll$ NL: For dispersion-managed links using low-dispersion fiber, a simple memoryless B-NLPR NLC [33, 35], modified to enable baud-rate operation as outlined in this chapter would suffice, roughly requiring 5$\times$ higher complexity relative to an uncompensated OFDM system.

(3) CD $\sim$ NL: If the CD and the NL interact on equal footing, e.g., for regular dispersion fiber with DCF in every span or nearly every span, a frequency-shaped NLC, based on the Volterra DF structure, may provide up to $\sim$4 dB NLT improvement. Unfortunately, the signal required signal processing load (15$\times$ higher) still currently poses a challenge, requiring a few more octaves of Moore's law evolution in terms of the DSP capabilities of Silicon ASICs.

Note that throughout this chapter we analyzed (and synthesized NLC for) just a single OFDM channel, e.g., as carried over a single DWDM 50 GHz band. We essentially modeled the "intrachannel" FWM mutually generated among the subcarriers of a single OFDM channel, which may be alternatively viewed as the SPM of the composite OFDM signal (it all depends whether our vantage point is the distinct OFDM subchannels or the composite OFDM channel). Here, we ignored the NL interaction among multiple OFDM channels, i.e., the NL impact on an OFDM channel due to the OFDM channels at the neighboring wavelengths, which impact may be alternatively described either as XPM between the composite OFDM channels or as "inter-channel FWM" among the subcarriers of one OFDM channel and the subcarriers of neighboring OFDM channels. For modern broadband OFDM systems, with the OFDM spectra extending to cover most of the WDM band slots, the interaction with neighboring OFDM channels turns out to be substantial. Studies of the "inter-channel" effect [28] indicate that the "interchannel" effect, ignored in this chapter, has about the same magnitude as the "intrachannel" effect addressed here. Unfortunately, there is no mitigation method available yet for mitigating inter-channel effects.

Therefore, despite the high performance of our Volterra mitigation method, providing $\sim$4 dB suppression of the "intrachannel" nonlinearity, in the absence of an XPM mitigation method the final NLT improvement is likely to be reduced down to $\sim$2 dB.

Back to considering NL analysis, an interesting point of view is that even a "single-carrier" communication signal may be effectively viewed as superposition of a multitude of "subcarriers" – the key idea is that a continuous spectrum of a long block of single-carrier symbols, may always be approximated in terms of a finite yet very large number of "frequency components" (amounting to the approximation of the FT by a DFT). Each of these "frequency components" amounts to a narrowband wave-packet, viewed as an effective "subcarrier." Thus, our derivation is actually independent of modulation format (not necessarily restricted to OFDM), in principle applicable to the propagation of any optical signal over any distributed dispersive optical medium with Kerr-induced third-order nonlinearity, with the broadband signal decomposed into a stack of equi-spaced narrowband frequency components, for the sake of analysis, even if not explicitly synthesized as such, unlike in OFDM.

By this token, the analysis pursued in this section equally applies to OFDM and non-OFDM signals. This leads to the interesting insight that the NL impairments in single-carrier and multicarrier may fundamentally described by an identical formalism (though actual behaviors of the two types may diverge due to different parameter values and different time scales), in principle facilitating a comparison between single-carrier and multicarrier systems, though we have not attempted such a comparison here, focusing in this chapter on deriving the modeling tools, and applying them to the OFDM case.

Future research directions to be considered are: (1) The application of pre-emphasis of the transmitted subchannel amplitudes, to even out frequency-dependent performance. (2) Vector (polarization) extending the scalar single-polarization treatment combining the approach of [36–38] FF NLC with the current frequency-shaped DF NLC. (3) The Volterra frequency shaping coefficients, W, are currently evaluated offline. It is imperative to work adaptation algorithms for the compensator coefficients, as the amount of link nonlinearity is unknown. (4) Combine DF with Forward Propagators/VFs, either or both at the Tx or at the Rx. (5) Evaluate and optimize multisection Volterra DF NLC performance, as outlined in Sect. 3.18 (6) Port the current method to single-carrier transmission using the frequency domain equalization (FDE) approach. (7) Further investigate the trade-offs between complexity and performance in systems which adapt their performance to varying conditions of the photonic network.

3.20 Appendix A: Derivation of the Analog-Like OFDM Transmitter Model

The derivation of (3.6) invokes the assumption $h^{\mathrm{TX}}(t) = \mathrm{sinc}\,(t/T_{\mathrm{c}}) \otimes h^{\mathrm{TX}}(t)$, amounting to a band-limitation specification for $h_{\mathrm{TX}}(t)$, as readily verified in the frequency domain. We may then rewrite (3.5) in the form:

$$\begin{aligned}
\underset{\sim}{s}(t) &= \sum_{n=-\mu L_{\mathrm{INT}}}^{D_{\mathrm{ZP}}-1} \sum_{i=-M/2}^{M/2-1} \underset{\sim}{A}_i \mathrm{e}^{j2\pi i n/D_{\mathrm{ZP}}} \mathrm{sinc}\,[(t-nT_{\mathrm{c}})/T_{\mathrm{c}}] \otimes h^{\mathrm{TX}}(t) \\
&= h^{\mathrm{TX}}(t) \otimes \sum_{i=-M/2}^{M/2-1} \underset{\sim}{A}_i \sum_{n=-\mu L_{\mathrm{INT}}}^{D_{\mathrm{ZP}}-1} \mathrm{e}^{j2\pi i n/D_{\mathrm{ZP}}} \mathrm{sinc}\,(t/T_{\mathrm{c}}-n) \\
&\cong h^{\mathrm{TX}}(t) \otimes \sum_{i=-M/2}^{M/2-1} \underset{\sim}{A}_i\, \mathrm{e}^{j2\pi i \Delta\nu t} \mathbf{1}_{[-\mu L_{\mathrm{INT}} T_{\mathrm{c}}, (D_{\mathrm{ZP}}-1)T_{\mathrm{c}}]}(t).
\end{aligned} \tag{3.129}$$

The last equation is compactly expressed as

$$\underset{\sim}{s}(t) = h^{\mathrm{TX}}(t) \otimes \underset{\sim}{a}(t); \qquad \underset{\sim}{a}(t) \equiv \mathbf{1}_{[-T_{\mathrm{CP}}, -T_{\mathrm{CP}}+T_{\mathrm{F}}]}(t) \sum_{i=-M/2}^{M/2-1} \underset{\sim}{A}_i \mathrm{e}^{j2\pi i \Delta\nu t}, \tag{3.130}$$

where we relabeled the time-window in the last expression in (3.129) as $[-T_{CP}, -T_{CP} + T_F] = [-\mu L_{INT} T_c, (D_{ZP} - 1)T_c]$, and the sampling theorem was applied in order to express the CT harmonic tones $e^{j2\pi in/D_{ZP}}$ in terms of their DT samples,

$$e^{j2\pi in/D_{ZP}} = e^{j2\pi i \Delta\nu t}\Big|_{t\to nT_c} : e^{j2\pi i\Delta\nu t}\mathbf{1}_{[-T_{CP},-T_{CP}+T_F]}(t)$$

$$\cong \sum_{n=-\mu L_{INT}}^{D_{ZP}-1} e^{j2\pi in/D_{ZP}}\text{sinc}\,(t/T_c - n). \tag{3.131}$$

For this interpolation relation to be strictly correct, the band-pass analog signal in the LHS must be BL to a spectral support T_c^{-1}. Evidently, this can only approximately hold, as the spectral support of the shifted sinc in LSH of (3.131) is infinite: the time-domain rectangular window, of duration $T_F = DT_c = (M + \mu) L_{INT} T_c$ (the OFDM block duration) has an FT with magnitude given by $\left|\text{sinc}\left(\nu/T^{-1}\right)\right|$ i.e., has approximate bandwidth $T_F^{-1} = T_c^{-1}/D$. The LHS waveform is actually oversampled at a rate $T_c^{-1} = DT_F^{-1}$ (its samples $e^{j2\pi in/D_{ZP}} = e^{j2\pi i\Delta\nu t}\big|_{t\to nT_c}$ are taken at intervals T_c apart). The sampling rate is then D times larger than the approximate spectral extent of the sinc (the position T_F^{-1} of its first zero-crossing), hence for large D (implying large number of subcarriers M) the sinc function is indeed BL to $T_c^{-1} = DT_F^{-1}$, to a very good approximation, establishing the accuracy of (3.131). Our result (3.130) may be finally expressed in the form:

$$\underset{\sim}{s}(t) \cong h^{TX}(t) \otimes \underset{\sim}{a}(t) = h^{TX}(t) \otimes \left\{\mathbf{1}_{[-T_{CP},-T_{CP}+T_F]}(t) \sum_{i=-M/2}^{M/2-1} \underset{\sim}{A}_i H_i^{TX} e^{j2\pi i\Delta\nu t}\right\}$$

$$\cong \sum_{i=-M/2}^{M/2-1} \underset{\sim}{A}_i H_i^{TX} e^{j2\pi i\Delta\nu t}\mathbf{1}_{[-T_{CP},-T_{CP}+T_F]}(t), \tag{3.132}$$

where $\underset{\sim}{a}(t)$ is given by (130), $H_k^{TX} \equiv H^{TX}(k\Delta\nu)$ are frequency samples of the BL Tx response $H^{TX}(\nu)$, i.e., the transmitted symbols are scaled by the transmitter TF. In the last equality of (3.132), we further made the approximation

$$h^{TX}(t) \otimes \left\{\mathbf{1}_{[-T_{CP},-T_{CP}+T_F]}(t)e^{j2\pi i\Delta\nu t}\right\} \cong H_i^{TX}e^{j2\pi i\Delta\nu t}\mathbf{1}_{[-T_{CP},-T_{CP}+T_F]}(t) \tag{3.133}$$

ignoring end-interval effects, and assuming that the duration of $h^{TX}(t)$ is small relative to the duration of the window $\mathbf{1}_{[-\mu L_{INT}T_c,(D_{ZP}-1)T_c]}(t)$ (the ratio of the two durations is $1/D$, with D assumed large).

3.21 Appendix B: Volterra NL Systems Formalism Extending [44] to Third-Order

Here, we develop some NL systems theory background, extending the second-order treatment in [44] to third-order NL (*trilinear*) systems. The resulting formalism mathematically streamlines our physical description of Kerr-induced nonlinearities in the main text of this chapter. The main concepts and derivations extend those of [44], wherein a second-order NL Volterra theory was developed; here, the analysis is extended to third order. A similar extension may be carried out to higher-orders.

Trilinearity: Let $\underset{\sim}{r}^{(3)}(t) = T^{(3)}\{\underset{\sim}{a}(t), \underset{\sim}{b}(t), \underset{\sim}{c}(t)\}$ be the response of a *trilinear* system to a triplet of periodic excitations. A trilinear system is additive and homogeneous (i.e., linear) in each of its three inputs *separately* (while the other two inputs are held constant), e.g., for the first slot (argument) we have:

$$\underset{\sim}{r}^{(3)}(t) = T^{(3)}\left\{\sum_j \underset{\sim}{a}_i(t), \underset{\sim}{b}(t), \underset{\sim}{c}(t)\right\} = \sum_j T^{(3)}\left\{\underset{\sim}{a}_j(t), \underset{\sim}{b}(t), \underset{\sim}{c}(t)\right\};$$
$$T^{(3)}\{\alpha\underset{\sim}{a}(t), \underset{\sim}{b}(t), \gamma\underset{\sim}{c}(t)\} = \alpha T^{(3)}\{\underset{\sim}{a}(t), \underset{\sim}{b}(t), \underset{\sim}{c}(t)\} \tag{3.134}$$

For lightwave systems modeling purposes, it is convenient to introduce a complex-valued form of trilinearity, satisfying a modified tri-homogeneity property with the third coefficient conjugated: $T^{(3)}\{\underset{\sim}{a}(t), \underset{\sim}{b}(t), \gamma\underset{\sim}{c}(t)\} = \gamma^* T^{(3)}\{\underset{\sim}{a}(t), \underset{\sim}{b}(t), \underset{\sim}{c}(t)\}$.

By repeated application of tri-aditivity and (conjugate) tri-homogeneity, the following *trilinear superposition* property is shown:

$$T^{(3)}\left\{\sum_j \alpha_j \underset{\sim}{a}_j(t), \sum_k \beta_k \underset{\sim}{b}_k(t), \sum_l \gamma_l \underset{\sim}{c}_l(t)\right\}$$
$$= \sum_j \sum_k \sum_l \alpha_j \beta_k \gamma_l^* T^{(3)}\left\{\underset{\sim}{a}_j(t), \underset{\sim}{b}_k(t), \underset{\sim}{c}_l(t)\right\}. \tag{3.135}$$

This is a generalization of the bilinear superposition property introduced in [44]. A single-input single-output (SISO) third-order NL system is obtained from the trilinear system by setting the three inputs equal: $T^{(3)}\left[\underset{\sim}{a}(t)\right] = T^{(3)}\{\underset{\sim}{a}(t), \underset{\sim}{a}(t), \underset{\sim}{a}(t)\}$. In our methodology, the full NL Volterra theory is developed starting from these trilinearity definitions and properties. This is a different approach than the usual exposition of the topic [59], starting from time-domain Volterra series kernels, $h(t_1, t_2, t_3)$ (generalizations of the concept of impulse response) featuring in Volterra series forms formulated as NL convolutions:

$$\underset{\sim}{r}^{(3)}(t) \equiv \int_{-\infty}^{\infty}\int_{-\infty}^{\infty}\int_{-\infty}^{\infty} h(t-t_1, t-t_2, t-t_3)\underset{\sim}{a}(t_1)\,\underset{\sim}{b}(t_2)\,\underset{\sim}{c}_3^*(t_3)\mathrm{d}t_1\mathrm{d}t_2\mathrm{d}t_3. \tag{3.136}$$

Finite Fourier Series: Let $\underset{\sim}{a}(t), \underset{\sim}{b}(t), \underset{\sim}{c}(t)$ be time-limited complex-valued signals, with support over a time-window T, expressible as FS with finite number $M = M_2 - M_1 + 1$ of not-necessarily-zero harmonic coefficients of coefficients:

$$\underset{\sim}{a}(t) = \sum_{j=M_1}^{M_2} \underset{\sim}{A}_j e^{j2\pi j\Delta\nu t}; \quad \underset{\sim}{b}(t) = \sum_{k=M_1}^{M_2-1} \underset{\sim}{B}_k e^{j2\pi k\Delta\nu t};$$

$$\underset{\sim}{c}(t) = \sum_{l=M_1}^{M_2-1} \underset{\sim}{C}_l e^{j2\pi l\Delta\nu t}, \quad t \in [0, T]. \tag{3.137}$$

When periodically extended over all t, these are in fact *Finite FS* expansions – defined as BL FS, i.e., FS with finite numbers of harmonics. In practice, the BL condition may be approximately satisfied by neglecting weak higher-order harmonics. The total band-limitation bandwidth is related to the number of harmonic coefficients by $M \equiv WT + 1 = W/\Delta\nu + 1$ with $\Delta\nu \equiv T^{-1}$ the fundamental frequency ($\Delta\nu$ is also the spectral separation between adjacent harmonics). FFS are also referred to as *trigonometric polynomials* in signal analysis. In particular, the CE of an OFDM composite signal is identified as an FFS.

When T-periodic FFS are input into a time-invariant third-order NL system, the third-order NL output component $\underset{\sim}{r}^{(3)}(t)$ is also T-periodic (as may be proven from the time-invariance), hence may also be represented by an FS with coefficients denoted $\underset{\sim}{R}_i^{(3)}$, which we set out to derive. Substituting the FFS expansions (3.137) of the inputs and applying trilinearity (3.135) yields:

$$\begin{aligned}
\underset{\sim}{r}^{(3)}(t) &= T^{(3)}\{\underset{\sim}{a}(t), \underset{\sim}{b}(t), \underset{\sim}{c}(t)\} \\
&= T^{(3)}\left\{\sum_{j=M_1}^{M_2-1} \underset{\sim}{A}_j e^{j2\pi j\Delta\nu t}, \sum_{k=M_1}^{M_2-1} \underset{\sim}{B}_k e^{j2\pi k\Delta\nu t}, \sum_{l=M_1}^{M_2-1} \underset{\sim}{C}_l e^{j2\pi l\Delta\nu t}\right\} \\
&= \sum_{j=0}^{M-1}\sum_{k=0}^{M-1}\sum_{l=0}^{M-1} \underset{\sim}{A}_j \underset{\sim}{B}_k \underset{\sim}{C}_l^* M\,[t; j\Delta\nu, k\Delta\nu, l\Delta\nu] \\
&= \sum_{j=M_1}^{M_2-1}\sum_{k=M_1}^{M_2-1}\sum_{l=M_1}^{M_2-1} \underset{\sim}{A}_j \underset{\sim}{B}_k \underset{\sim}{C}_l^* H(j\,\Delta\nu, k\,\Delta\nu, l\,\Delta\nu) e^{j2\pi(j+k-l)\,\Delta\nu\, t},
\end{aligned} \tag{3.138}$$

where we introduced the *IM frequency response* (IFR) (the time response of the NL system to a three-tone-test):

$$\begin{aligned}
\mathrm{M}\,[t; j\Delta\nu, k\Delta\nu, l\Delta\nu] &\equiv T^{(3)}\left\{e^{j2\pi j\,\Delta\nu\, t}, e^{j2\pi k\,\Delta\nu\, t}, e^{j2\pi l\,\Delta\nu\, t}\right\} \\
&= e^{j2\pi(j+k-l)\,\Delta\nu\, t} H(j\,\Delta\nu, k\,\Delta\nu, l\,\Delta\nu),
\end{aligned} \tag{3.139}$$

with the "analog" VTF $H\left[\nu_1,\nu_2,\nu_3\right]$ defined as a triple FT of the Volterra kernel $h(t_1,t_2,t_3)$ appearing in (3.136) (with the last sign flipped in the FT definition, corresponding to the conjugated slot):

$$H\left[\nu_1\,,\nu_2\,,\nu_3\right]\equiv\int_{-\infty}^{\infty}\int_{-\infty}^{\infty}\int_{-\infty}^{\infty}h(t_1,t_2,t_3)\mathrm{e}^{-j2\pi(\nu_1 t_1+\nu_2 t_2-\nu_3 t_3)}\mathrm{d}t_1\mathrm{d}t_2\mathrm{d}t_3. \tag{3.140}$$

We now perform a change of variables in the trilinear summation (3.138), from *j,k,l* to *j,k,i*, with $i=j+k-l$ being the IM frequency. Substituting $l=j+k-i$ into (3.138), the summation over l is replaced by a summation over i:

$$\underset{\sim}{r}^{(3)}(t)=\sum_{i=2M_1-M_2}^{2M_2-M_1}\sum_{j=M_1}^{M_2}\sum_{k=M_1}^{M_2}H^{(3)}_{i;jk}\underset{\sim}{A}_j\underset{\sim}{B}_k\underset{\sim}{C}^*_{j+k-i}\mathrm{e}^{j2\pi\nu_i t};\ t\in[0,T] \tag{3.141}$$

with $H^{(3)}_{i;jk}\equiv H^{(3)}\left[j\,\Delta\nu,k\,\Delta\nu,(j+k-i)\Delta\nu\right]$ a sampled version of the VTF $H^{(3)}\left[\nu_1,\nu_2,\nu_3\right]$, and with the upper (lower) limit in the i summation obtained by taking the max (min) of *j,k* i.e., M_2 (M_1) and the min (max) of l i.e.,M_1 (M_2). The outer summation in (3.141) is identified as an FFS, with harmonic coefficients $\tilde{R}^{(3)}_i$ as specified:

$$\underset{\sim}{r}^{(3)}(t)=\sum_{i=2M_1-M_2}^{2M_2-M_1}\underset{\sim}{R}^{(3)}_i\mathrm{e}^{j2\pi i\,\Delta\nu\,t};$$

$$\underset{\sim}{R}^{(3)}_i=\sum_{j=M_1}^{M-1}\sum_{k=M_1}^{M-1}\underset{\sim}{A}_j\underset{\sim}{B}_k\underset{\sim}{C}^*_{j+k-i}H^{(3)}_{i;jk},\ -M+1\le i\le 2M-2. \tag{3.142}$$

The expression for $\underset{\sim}{R}^{(3)}_i$ yields the discrete-frequency spectral propagation rule, mapping the Fourier coefficients of the periodic excitations to those of the NL response. The double summation yielding the output FS coefficient, $\underset{\sim}{R}^{(3)}_i$ is referred to as weighted cross-correlation-convolution (WCCC), with this terminology motivated by observing that when the VTF is unity, $H^{(3)}_{i;jk}=1$, then $\underset{\sim}{R}^{(3)}_i$ reduces to a convolution and a cross-correlation of the coefficients:

$$\underset{\sim}{R}^{(3)}_i=\sum_{j=0}^{M-1}\sum_{k=0}^{M-1}\underset{\sim}{A}_j\underset{\sim}{B}_k\underset{\sim}{C}^*_{j+k-i}=\underset{\sim}{A}_i\otimes\underset{\sim}{B}_i\otimes\underset{\sim}{C}^*_{-i}. \tag{3.143}$$

In the time-domain, a system with unity VTF is described by the memoryless (conjugate) multiplication relation $\underset{\sim}{y}(t)=\underset{\sim}{a}(t)\underset{\sim}{b}(t)\underset{\sim}{c}^*(t)$, transforming to (3.143) in the frequency domain. More generally, for a third-order NL system with memory, the correlation-convolution (3.143) is generalized to (3.142) by incorporating the index-dependent weighting $H_{i;jk}$ in the double summation. Note that spectral width of a WCCC coincides with that of a conventional correlation-convolution, namely it is the sum of the three input spectral widths. For W-bandlimited FS inputs, the spectral width of (3.142) is then $3W$. Finally, to convert from a third-order trilinear system to a third-order SISO system, we must set $\underset{\sim}{b}(t) = \underset{\sim}{a}(t); \underset{\sim}{c}(t) = \underset{\sim}{a}(t)$, or in the frequency domain $\underset{\sim}{B}_k = \underset{\sim}{A}_k; \underset{\sim}{C}_l^* = \underset{\sim}{A}_l$, i.e., the triple products $\underset{\sim}{A}_j \underset{\sim}{B}_k \underset{\sim}{C}_{j+k-i}^*$ above are replaced by $\underset{\sim}{A}_j \underset{\sim}{A}_k \underset{\sim}{A}_{j+k-i}^*$ everywhere, in particular (3.142) reduces to

$$\underset{\sim}{r}^{(3)}(t) = \sum_{i=-M+1}^{2M-2} \underset{\sim}{R}_i^{(3)} e^{j2\pi i\,\Delta\nu\, t};$$

$$\underset{\sim}{R}_i^{(3)} = \sum_{j=M_1}^{M-1} \sum_{k=M_1}^{M-1} \underset{\sim}{A}_j \underset{\sim}{A}_k \underset{\sim}{A}_{j+k-i}^* H_{i;jk}^{(3)}, -M+1 \le i \le 2M-2 \tag{3.144}$$

Linear–nonlinear (LN–NL) and linear–nonlinear-linear (LNL) structures: A structured NL system acting as a prototype for more complex ones, consists of the LN–NL cascade of two modules: An LTI filter with TF $H^{\text{in}}(\nu)$, followed by a memoryless nonlinearity. By running a 3-tone test, the third-order VTF of this structure is obtained as $H(\nu_1, \nu_2, \nu_3) = H^{\text{in}}(\nu_1) H^{\text{in}}(\nu_2) H^{\text{in}^*}(\nu_3)$.

Accordingly, the sampled VTF is given by $H_{i;jk}^{\text{LN-NL}} = H_j^{\text{in}} H_k^{\text{in}} H_{j+k-i}^{\text{in}^*}$. A slightly more complicated overall nonlinearity is the so-called LNL structure [60], consisting of an ideal third-order memoryless nonlinearity, specified for our purposes as $\underset{\sim}{y} = \underset{\sim}{x}^2 \underset{\sim}{x}^* = \underset{\sim}{x} \left|\underset{\sim}{x}\right|^2$, "sandwitched" in between two linear filters, $H^{\text{in}}(\nu), H^{\text{out}}(\nu)$. By propagating three test tones through this compound NL structure, the VTF of the LNL structure is obtained $H(\nu_1, \nu_2, \nu_3) = H^{\text{in}}(\nu_1) H^{\text{in}}(\nu_2) H^{\text{in}^*}(\nu_3) H^{\text{out}}(\nu_1 + \nu_2 - \nu_3)$, whereas the sampled VTF is given by $H_{i;jk}^{\text{LNL}} = H_j^{\text{in}} H_k^{\text{in}} H_{j+k-i}^{\text{in}^*} H_i^{\text{out}}$. These results may be proven by extending the analogous derivations in [44] for the VTF of a second-order LNL structure with detector-like nonlinearity $\underset{\sim}{y} = \underset{\sim}{x}\underset{\sim}{x}^* = \left|\underset{\sim}{x}\right|^2$, yielding in the current notation $H_{i;j}^{\text{LNL}} = H_j^{\text{in}} H_{i-j}^{\text{in}^*} H_i^{\text{out}}$, for the VTF of the second-order LNL structure. An even more general formulation replaces the memoryless nonlinearity in the LNL model by a general VTF, $H_{i;jk}^{\text{inner}}$, "sandwitched" in between two linear filters, $H_i^{\text{in}}, H_i^{\text{out}}$. The overall VTF of the Generalized LNL (G-LNL) system, is given by

$$H_{i;jk}^{\text{LNL}} = H_j^{\text{in}} H_k^{\text{in}} H_{j+k-i}^{\text{in}^*} H_{i;jk}^{\text{inner}} H_i^{\text{out}}. \tag{3.145}$$

3.22 Appendix C: Sampling and Nonlinearity Effects in the OFDM Receiver

3.22.1 *Nyquist Sampling the Linear Component Under Samples the NL Component*

Following the notation of Sect. 3.4, we mathematically analyze the effect of under-sampling the received NL component, $\underset{\sim}{r}^{(3)}(t)$, which would occur if the Rx sampled the received signal, $\underset{\sim}{r}(t) = \underset{\sim}{r}^{(1)}(t) + \underset{\sim}{r}^{(3)}(t)$, at the Nyquist rate corresponding the linear component, $\underset{\sim}{r}^{(1)}(t)$ ($\underset{\sim}{r}^{(1)}(t)$ is given by (3.36) and is $\underset{\sim}{r}^{(3)}(t)$ given by (3.44)). Let the Rx collect $M_s = M$ samples per T interval at the instants $t \to nT/M$ (this is the Nyquist rate for the linear component, which has spectral support $M\Delta\nu = M/T$). The sampled third-order received signal is expressed as

$$\begin{aligned}
\underset{\sim}{r}_n^{(3)} &= \underset{\sim}{r}^{(3)}(t) \otimes h_{\mathrm{RX}}(t)\Big|_{t\to nT/M} \\
&= \sum_{i=-1.5M+1}^{1.5M-2} \underset{\sim}{R}_i^{(3)} \mathrm{e}^{j2\pi i\,\Delta\nu t} 1_{[0,T]}(t) \otimes h_{\mathrm{RX}}(t)\Big|_{t\to nT/M} \\
&\cong \sum_{i=-1.5M+1}^{1.5M-2} \underset{\sim}{R}_i^{(3)} H_i^{\mathrm{RX}} \mathrm{e}^{j2\pi i\,\Delta\nu t}\Big|_{t\to nT/M} \\
&= \sum_{i=-1.5M+1}^{1.5M-2} \underset{\sim}{R}_i^{(3)} H_i^{\mathrm{RX}} \mathrm{e}^{j2\pi i\,\Delta\nu\, nT/M} \\
&= \sum_{i=-1.5M+1}^{1.5M-2} \underset{\sim}{R}_i^{(3)} H_i^{\mathrm{RX}} \mathrm{e}^{j2\pi i n/M};\quad n = 0, 1, \ldots, M-1.
\end{aligned} \tag{3.146}$$

The U/C operation (3.39) is next applied, up-shifting the spectrum of sampled signal by $M/2$ units, which makes the received linear spectrum properly one-sided. However, following the U/C operation, the spectrum of the third-order received NL signal, spanning the index range $-1.5M + 1 \le i \le 1.5M - 2$, becomes skewed with respect to the origin ($M + 1 \le i \le 2M - 2$):

$$\begin{aligned}
\underset{\sim}{r}_n^{(3)\,\mathrm{U/C}} &= c_n \underset{\sim}{r}_n^{(3)} = \mathrm{e}^{j2\pi(M/2)n/M} \sum_{i=-1.5M+1}^{1.5M-2} \underset{\sim}{R}_i^{(3)} H_i^{\mathrm{RX}} \mathrm{e}^{j2\pi i n/M} \\
&= \sum_{i=-1.5M+1}^{1.5M-2} \underset{\sim}{R}_i^{(3)} H_i^{\mathrm{RX}} \mathrm{e}^{j2\pi(i+M/2)n/M} \\
&= \sum_{i=-M+1}^{2M-2} \underset{\sim}{R}_{i-M/2}^{(3)} H_{i-M/2}^{\mathrm{RX}} \mathrm{e}^{j2\pi i n/M}.
\end{aligned} \tag{3.147}$$

We partition the last summation into three sums over the three index sets $-M+1 \le i \le -1;\ 0 \le i \le M-1;\ M \le i \le 2M-2$, corresponding to the lower-out-of-band, in-band and upper-out-of-band spectral regions, respectively:

$$\begin{aligned}
\underset{\sim}{r}_n^{(3)\,\mathrm{U/C}} &= \sum_{i=-M+1}^{-1} \underset{\sim}{R}_{i-M/2}^{(3)} H_{i-M/2}^{\mathrm{RX}} \mathrm{e}^{j2\pi i n/M} \\
&\quad + \sum_{i=0}^{M-1} \underset{\sim}{R}_{i-M/2}^{(3)} H_{i-M/2}^{\mathrm{RX}} \mathrm{e}^{j2\pi i n/M} \\
&\quad + \sum_{i=M}^{2M-2} \underset{\sim}{R}_{i-M/2}^{(3)} H_{i-M/2}^{\mathrm{RX}} \mathrm{e}^{j2\pi i n/M} \\
&= \sum_{i=-M+1}^{-1} \underset{\sim}{R}_{i-M/2}^{(3)\mathrm{RX}} \mathrm{e}^{j2\pi i n/M} + \sum_{i=0}^{M-1} \underset{\sim}{R}_{i-M/2}^{(3)\mathrm{RX}} \mathrm{e}^{j2\pi i n/M} \\
&\quad + \sum_{i=M}^{2M-2} \underset{\sim}{R}_{i-M/2}^{(3)\mathrm{RX}} \mathrm{e}^{j2\pi i n/M},
\end{aligned} \tag{3.148}$$

where in the last line we introduced the shorthand $\underset{\sim}{R}_i^{(3)\mathrm{RX}} \equiv \underset{\sim}{R}_i^{(3)} H_i^{\mathrm{RX}}$. Next, (3.148) is algebraically manipulated as follows:

$$\begin{aligned}
\underset{\sim}{r}_n^{(3)\,\mathrm{U/C}} &= \sum_{i'=0}^{M-2} \underset{\sim}{R}_{i'-1.5M+1}^{(3)\mathrm{RX}} \mathrm{e}^{j2\pi(i'-M+1)n/M} + \sum_{i=0}^{M-1} \underset{\sim}{R}_{i-M/2}^{(3)\mathrm{RX}} \mathrm{e}^{j2\pi i n/M} \\
&\quad + \sum_{i''=0}^{M-2} \underset{\sim}{R}_{i''+M/2}^{(3)\mathrm{RX}} \mathrm{e}^{j2\pi(i''+M)n/M} \\
&= \sum_{i'=0}^{M-1} \underset{\sim}{R}_{i'-1.5M+1}^{(3)\mathrm{RX\,ZP}:M} \mathrm{e}^{j2\pi(i'-M+1)n/M} + \sum_{i=0}^{M-1} \underset{\sim}{R}_{i-M/2}^{(3)\mathrm{RX}} \mathrm{e}^{j2\pi i n/M} \\
&\quad + \sum_{i''=0}^{M-1} \underset{\sim}{R}_{i''+M/2}^{(3)\mathrm{RX\,ZP}:M} \mathrm{e}^{j2\pi(i''+M)n/M} \\
&= \sum_{i'=0}^{M-1} \mathrm{e}^{j2\pi n/M} \underset{\sim}{R}_{i'-1.5M+1}^{(3)\mathrm{RX\,ZP}:M} \mathrm{e}^{j2\pi i' n/M} + \sum_{i=0}^{M-1} \underset{\sim}{R}_{i-m/2}^{(3)\mathrm{RX}} \mathrm{e}^{j2\pi i n/M} \\
&\quad + \sum_{i''=0}^{M-1} \underset{\sim}{R}_{i''+M/2}^{(3)\mathrm{RX\,ZP}:M} \mathrm{e}^{j2\pi i'' n/M}
\end{aligned}$$

$$= \sum_{i=0}^{M-1} \left[e^{j2\pi n/M} \underset{\sim}{R}_{i-1.5M+1}^{(3)\text{RX ZP:}M} + \underset{\sim}{R}_{i-1.5M+1}^{(3)\text{RX}} + \underset{\sim}{R}_{i+M/2}^{(3)\text{RX ZP:}M} \right] e^{j2\pi - in/M}$$

$$= M \cdot \text{IDFT}_M \left\{ e^{j2\pi n/M} \underset{\sim}{R}_{i-1.5M+1}^{(3)\text{RX ZP:}M} + \underset{\sim}{R}_{i-M/2}^{(3)\text{RX}} + \underset{\sim}{R}_{i+M/2}^{(3)\text{RX ZP:}M} \right\}, \tag{3.149}$$

where in the first line of the last equation, change-of-summation-variable transformations were applied, making the summation limits one-sided; in the second line the first and last summands were ZP from length $M-1$ to length M (just appending a zero at the end), extending all summations over the in-band range $0 \le i \le M-1$; in the third line a $e^{j2\pi n/M}$ factor was extracted from the first summand, such that the $e^{j2\pi - in/M}$ IDFT kernel appeared; in the last line the three sums were combined into a single sum over the $0 \le i \le M-1$ in-band range, which was identified as an IDFT.

It is apparent that the effect of undersampling the received NL components, is to shift the (lower and upper) OOB segments of the spectrum, by $\pm M$, respectively, aliasing them into the in-band interval $0 \le i \le M-1$. If more harmonics were present further out (e.g., due to higher-order nonlinearity), then these harmonics would also alias back into the $[0, M-1]$ range. In the last line of (3.149), the received sampled signal was expressed as an M-point IDFT of the superposition of these aliased bands. The final step (3.41) in the receiver processing chain, namely taking the scaled DFT of $\tilde{r}_n^{\text{U/C}}$ extracts the aliased superposition of spectral bands:

$$\underset{\sim}{\rho}_i^{(3)} = M^{-1} \cdot \text{DFT}_M \left\{ \underset{\sim}{r}_n^{(3)\text{U/C}} \right\} = e^{j2\pi n/M} \underset{\sim}{R}_{i-1.5M+1}^{(3)\text{RX ZP:}M} + \underset{\sim}{R}_{i-M/2}^{(3)\text{RX}} + \underset{\sim}{R}_{i+M/2}^{(3)\text{RX ZP:}M}. \tag{3.150}$$

3.22.2 AA Filtering

An enabling strategy for baud-rate sampling is to apply a high quality analog AA filter, band-limiting the overall received signal to the in-band spectral range $[-B_T/2, B_T/2]$, passing through the linear and the in-band NL component while blocking out the OOB NL components. We show in Sect. 3.12 that such AA measure is essential for achieving NL compensation at baud-rate sampling. The reason AA filtering is effective is that the useful signal for our purposes does not reside in the OOB regions, but it is rather the in-band components that are of interest – this includes the in-band NL signal, to which we must have access in order to have it canceled by the NL compensation procedure – however it suffices that this cancelation just occur in-band, as there is no useful information signal residing OOB, hence the effort to recover and cancel out OOB NL components would be futile (note: the OOB signal is actually a form of XPM affecting the adjacent WDM-OFDM channel – it would be useful if the receivers of the adjacent WDM channels were able to cancel the XPM, but this does not seem possible without multiple cooperating receivers).

The receiver then filters out the OOB spectral regions in the analog domain prior to sampling at the baud-rate. In detail, the received signal $\underset{\sim}{r}^{(3)}(t)$ is passed through an AA filter with a sharp pass-band, blocking as much of the OOB signal as possible while distorting as little of the in-band signal as possible (ideally the AA filter response is $\mathbf{1}_{[-B_T/2,B_T/2]}(\nu)$). The linear and NL in-band harmonics with indexes $-0.5M \le i \le 0.5M - 1$ are passed through, whereas the NL harmonics in the lower and upper out-of-band regions are blocked out. This means that the OOB images are suppressed – only the middle sum is retained in (3.149) (after U/C shifting the frequency indexes up by $M/2$). To work this out formally, we recall our definition $\underset{\sim}{R}_i^{(3)\mathrm{RX}} \equiv \underset{\sim}{R}_i^{(3)} H_i^{\mathrm{RX}}$, where $H_i^{\mathrm{RX}} \equiv H^{\mathrm{RX}}(i\Delta\nu)$. The edges of the analog passband of the AA are at $\pm B_T/2 = \pm M\Delta\nu/2$, i.e., upon sampling the frequency domain at $\Delta\nu$ intervals, the edges of the AA passband occur at $\pm M/2$. We then model the AA filter in the discrete frequency domain as $1_{[-M/2+1,M/2-1]}[i]$, with $1_{[M_1,M_2]}[i]$ a discrete-time indicator function assuming unity value in the range $M_1 \le i \le M_2$ zero otherwise. The overall receiver response, including the AA filter is then modeled as $H_i^{\mathrm{RX}} = H_i^{\mathrm{RX}} 1_{[-M/2+1,M/2-1]}[i]$. This condition does not imply that the receiver sampled analog frequency response is flat, but rather that it is BL (there might be other sources of roll-off in the receiver front-end). Substituting $H_i^{\mathrm{RX}}\mathbf{1}_{[-M/2+1,M/2-1]}[i]$ for H_i^{RX} in the first line of (3.148) amounts to replacing $H_{i-M/2}^{\mathrm{RX}}$ by $H_{i-M/2}^{\mathrm{RX}}\mathbf{1}_{[-M/2+1,M/2-1]}[i-M/2] = H_{i-M/2}^{\mathrm{RX}}\mathbf{1}_{[1,M-1]}[i]$, yielding

$$\begin{aligned}\underset{\sim}{r}_n^{(3)\,\mathrm{U/C}} &= \sum_{i=-M+1}^{-1} \underset{\sim}{R}_{i-M/2}^{(3)} H_{i-M/2}^{\mathrm{RX}}\mathbf{1}_{[1,M-1]}[i]\mathrm{e}^{j2\pi in/M} \\ &\quad + \sum_{i=0}^{M-1} \underset{\sim}{R}_{i-M/2}^{(3)} H_{i-M/2}^{\mathrm{RX}}\mathbf{1}_{[1,M-1]}[i]\mathrm{e}^{j2\pi in/M} \\ &\quad + \sum_{i=M}^{2M-2} \underset{\sim}{R}_{i-M/2}^{(3)} H_{i-M/2}^{\mathrm{RX}}\mathbf{1}_{[1,M-1]}[i]\mathrm{e}^{j2\pi in/M}. \end{aligned} \tag{3.151}$$

The presence of the $\mathbf{1}_{[1,M-1]}[i]$ indicator in the summands of the first and last sum nulls these sums out, since the indicator is zero in the index ranges of these two sums. Discarding the first and last sums in (3.151) (or equivalently, discarding the first and last sums in (3.149)), yields

$$\begin{aligned}\underset{\sim}{r}_n^{(3)\,\mathrm{U/C}} &= \sum_{i=0}^{M-1} \underset{\sim}{R}_{i-M/2}^{(3)} H_{i-M/2}^{\mathrm{RX}}\mathbf{1}_{[1,M-1]}[i]\mathrm{e}^{j2\pi in/M} \\ &= M \cdot \mathrm{IDFT}_M\left\{\underset{\sim}{R}_{i-M/2}^{(3)} H_{i-M/2}^{\mathrm{RX}}\mathbf{1}_{[1,M-1]}[i]\right\}. \end{aligned} \tag{3.152}$$

The indicator nulls out the $i = 0$ term $\underset{\sim}{R}^{(3)}_{-M/2} H^{\mathrm{RX}}_{-M/2}$, i.e., last summation is actually restricted to $\sum_{i=1}^{M-1} \underset{\sim}{R}^{(3)}_{i-M/2} H^{\mathrm{RX}}_{i-M/2} e^{j2\pi in/M}$, which does not pose a problem; however, the corresponding linear component of the subcarrier is also blocked by the AA (the blocked subcarrier has index $i = -M/2$ prior to Rx U/C, i.e., at the Rx input, corresponding at the Tx side, prior to D/C, to the "DC" term $i = 0$ in the IDFT input in the Tx). To model this formally, we start from the AA band-limitation condition $H_i^{\mathrm{RX}} = H_i^{\mathrm{RX}} \mathbf{1}_{[-M/2+1, M/2-1]}[i]$. The DFT-ed linear response (3.42) is then expressed (using $\mathbf{1}_{[-M/2+1,M/2-1]}[i - M/2] = \mathbf{1}_{[1,M-1]}[i]$) as:

$$\begin{aligned}\underset{\sim}{\rho}_i^{(1)} &= \underset{\sim}{A}_i H^{\mathrm{TX}}_{i-M/2} H^{\mathrm{CH}}_{i-M/2} H^{\mathrm{RX}}_{i-M/2} \mathbf{1}_{[-M/2+1,M/2-1]}[i - M/2] \\ &= \underset{\sim}{A}_i H^{\mathrm{TX}}_{i-M/2} H^{\mathrm{CH}}_{i-M/2} H^{\mathrm{RX}}_{i-M/2} \mathbf{1}_{[1,M-1]}[i]; \quad i = 0, 1, \ldots, M - 1\end{aligned}$$

i.e., $\underset{\sim}{\rho}_0^{(1)} = 0$, therefore the lowest frequency subcarrier should not be modulated with useful information, hence the symbol $\underset{\sim}{A}_0$ is not to be used to map information bits in the Tx, as its corresponding subcarrier would be blocked by the AA filter.

Returning to consider the OOB NL components, aliasing of these components is mitigated by ideal AA filtering. Finally, applying a scaled DFT onto the up-converted signal retrieves just the in-band NL component:

$$\underset{\sim}{\rho}_i^{(3)} = M^{-1} \cdot \mathrm{DFT}_M \left\{ \underset{\sim}{r}_n^{(3)\mathrm{U/C}} \right\} = \underset{\sim}{R}^{(3)}_{i-M/2} H^{\mathrm{RX}}_{i-M/2} 1_{[1,M-1]}[i]. \tag{3.153}$$

To the extent, the AA filter stop-band is not ideal, there will be some residual OOB components, aliasing in-band. Such residual effect may also be modeled by the formalism above. The total DFT output is given by the sum of the linear and NL components (3.42) and (3.153), which were separately propagated through the receiver:

$$\begin{aligned}\underset{\sim}{\rho}_i &= M^{-1} \cdot \mathrm{DFT}_M \left\{ \tilde{r}_n^{\mathrm{U/C}} \right\} = \underset{\sim}{\rho}_i^{(1)} + \underset{\sim}{\rho}_i^{(3)} \\ &= \underset{\sim}{A}_i H^{\mathrm{LINK}}_{i-M/2} + \underset{\sim}{R}^{(3)}_{i-M/2} H^{\mathrm{RX}}_{i-M/2}, \quad 1 \le i \le M - 1.\end{aligned} \tag{3.154}$$

It remains to incorporate the received in-band NL components $\left\{ \underset{\sim}{R}_i^{(3)} \right\}_{i=M/2}^{M/2\ 1}$ in the last expression. This NL sequence at the channel output is given by the middle line of the expression for $\tilde{R}_i^{(3)}$ in (3.44). Recalling that XPM and SPM are already included in the modeling of the "linear" (1) term (mislabeled as "linear," being actually linear + XPM/SPM), we may discard the terms involving $H_{i;ik}$, $H_{i;ii}$ in (3.44), making the substitution $\underset{\sim}{R}_i^{(3)} \to \sum\sum_{[j,k]\in S[i]} H^{\mathrm{CH}}_{i;jk} \underset{\sim}{A}_j^{\mathrm{TX}} \underset{\sim}{A}_k^{\mathrm{TX}} \underset{\sim}{A}_{j+k-i}^{\mathrm{TX}*}$ in (3.154), yielding

$$\underset{\sim}{\rho}_i = \underset{\sim}{A}_i H^{\text{LINK}}_{i-M/2} + \left\{ \sum\sum_{[j,k]\in S[i]} H^{\text{CH}}_{i;jk} \underset{\sim}{A}^{\text{TX}}_j \underset{\sim}{A}^{\text{TX}}_k \underset{\sim}{A}^{\text{TX}*}_{j+k-i} H^{\text{RX}}_i \Big|_{i\to i-M/2} \right\},$$
$$1 \le i \le M-1. \tag{3.155}$$

Recalling that $\underset{\sim}{A}^{\text{TX}}_i \equiv \underset{\sim}{A}_{i+M/2} H^{\text{TX}}_i$, the summand in the last equation reduces to

$$\begin{aligned} H^{\text{CH}}_{i;jk} \underset{\sim}{A}^{\text{TX}}_j \underset{\sim}{A}^{\text{TX}}_k \underset{\sim}{A}^{\text{TX}*}_{j+k-i} H^{\text{RX}}_i &= H^{\text{TX}}_i H^{\text{TX}}_j H^{\text{TX}}_k H^{\text{TX}*}_{j+k-i} H^{\text{CH}}_{i;jk} H^{\text{RX}}_i \underset{\sim}{A}_j \underset{\sim}{A}_k \underset{\sim}{A}^*_{j+k-i} \\ &= H^{\text{LINK}}_{i;jk} \underset{\sim}{A}_j \underset{\sim}{A}_k \underset{\sim}{A}^*_{j+k-i}, \end{aligned} \tag{3.156}$$

where $H^{\text{LINK}}_{i;jk}$ is the VTF of the overall link (Tx + CH + Rx – for completeness we also repeated the linear TF of the overall link):

$$H^{\text{LINK}}_{i;jk} = H^{\text{TX}}_i H^{\text{TX}}_j H^{\text{TX}}_k H^{\text{TX}*}_{j+k-i} H^{\text{CH}}_{i;jk} H^{\text{RX}}_i; \quad H^{\text{LINK}}_i = H^{\text{TX}}_i H^{\text{CH}}_i H^{\text{RX}}_i. \tag{3.157}$$

The expression just derived for the VTF of the NL cascade of the Tx, CH, Rx is consistent with the result derived in Appendix A, for the VTF of a linear-NL-linear cascade. Substituting (3.156) into (3.155) yields our final result

$$\underset{\sim}{\rho}_i = \underset{\sim}{\rho}^{(1)}_i + \underset{\sim}{\rho}^{(3)}_i = \underset{\sim}{A}_i H^{\text{LINK}}_{i-M/2} + \left\{ \sum\sum_{[j,k]\in S[i]} H^{\text{LINK}}_{i;jk} \underset{\sim}{A}_j \underset{\sim}{A}_k \underset{\sim}{A}^*_{j+k-i} \Big|_{i\to i-M/2} \right\},$$
$$1 \leqslant i \leqslant M-1. \tag{3.158}$$

This is our final expression for the signal at the DFT output in the receiver. The NL distortion term is given in braces, with the i index appearing in the double sum ranging over the two-sided transmitted frequencies range $-M/2+1 \le i \le M/2-1$. In a simple linear receiver, there is no mitigation of NL distortion, and the $\{\underset{\sim}{\rho}_i\}_{i=1}^{M-1}$ signal is equalized by dividing its ith sample through $H^{\text{LINK}}_{i-M/2}$, recovering the $\underset{\sim}{A}_i$ symbols (corrupted by noise and distortion). In a more sophisticated receiver with NL compensation, the NL distortion term is mitigated as described in Sect. 3.12.

In order to model the NL distortion and the resulting performance, the next step is to evaluate the VTF of the link, $H^{\text{LINK}}_{i;jk}$.

Glossary

AA	Antialiasing
ADC	Analog-to-digital converter
AS	Analytic Signal
ASE	Amplified Spontaneous Emission

BER	Bit error ratio
BL	Band-Limited
B-NLPR	Backward Nonlinear Phase Rotation (or Rotator)
BP	Back-Propagation
CD	Chromatic dispersion
CE	Complex Envelope
CP	Cyclic Prefix
DAC	Digital-to-analog converter
DCF	Dispersion Compensating Fiber
DF	Decision Feedback
DFT	Discrete Fourier Transform
DWDM	Dense wavelength-division multiplexing
ENL	Effective Nonlinear Length
FDE	Frequency Domain Equalizer
FFT	Fast Fourier Transform
FFS	Finite Fourier Series
FS	Fourier Series
FT	Fourier Transform
FWM	Four-Wave-Mixing
IFFT	Inverse FFT
IFR	Intermodulation Frequency Response
IM	Intermodulation tone (intermod)
IM	(intermod) – InterModulation product
LO	Local Oscillator
NL	NonLinear
NLC	NonLinear Compensation
NLSE	Nonlinear Schroedinger Equation
NLT	NonLinear Tolerance
OA	Optical amplifier
OA	Optical Amplifier
OFDM	Orthogonal Frequency-Division Multiplexing
OOB	Out-of-band
OPI	Optical Path Integral
PA	Phased Array
PAM	Pulse Amplitude Modulation
PSD	Power Spectral Density
QLP-TF	Quasi Linear Propagation Transfer Function
QPSK	Quaternary phase-shift keying
Rx	Receiver
Rx	Receiver
SER	Symbol Error Rate
SPM	Self phase modulation
SSB	Single Side Band
STCE	Spatiotemporal Complex Envelope
TF	Transfer Function

Tx	Transmitter
VTF	Volterra Transfer Function
WCCC	Weighted Cross-Correlation Convolution
XPM	Cross phase modulation

References

1. W. Shieh, C. Athaudage, Electron. Lett. **42**, 587–589 (2006)
2. W. Shieh, H. Bao, Y. Tang, Opt. Express **16**, 841–859 (2008)
3. A.J. Lowery, Opt. Express **16**, 860–865 (2008)
4. S.L. Jansen, Application scenarios for optical OFDM, SPPCom – Signal processing in photonic communications – OSA Technical Digest, Optical Society of America, p. SPThB1, 2010
5. E. Forestieri, G. Colavolpe, T. Foggi, G. Bruno, Signal processing for 100Gb/s: OFDM vs. single carrier – OSA Technical Digest (CD), SPPCom – Signal processing in photonic communications – OSA Technical Digest, Optical Society of America, p. SPThC2, 2010
6. D. Schadt, Electron. Lett. **27**, 1805 (1991)
7. D. Schadt, T. Stephens, J. Lightwave Technol. **10**, 1715–1721 (1992)
8. K. Inoue, Opt. Lett. **17**, 801 (1992)
9. D. Marcuse, A. Chraplyvy, R. Tkach, J. Lightwave Technol. **12**, 885–890 (1994)
10. H. Kagi, T. Chian, T. Fong, M. Imarhic, L. Kazovsky, Electron. Lett. **30**, 1878–1879 (1994)
11. N. Kagi, T. Chiang, T. Fong, M. Marhic, L. Kazovsky, Cross phase modulation in fiber links with optical amplifiers, in *Proceedings of LEOS'94*, pp. 188–189, 1994
12. W. Zeiler, F. Di Pasquale, P. Bayvel, J. Midwinter, J. Lightwave Technol. **14**, 1933–1942 (1996)
13. W. Szczesny, M. Marciniak, Results of numerical simulation of wavelength multiplexed transmission in non-linear optical fibre telecommunication systems, *MMET conference proceedings. 1998 international conference on mathematical methods in electromagnetic theory. MMET 98 (Cat. No.98EX114)*, IEEE, pp. 923–926, 1998
14. H. Thiele, R. Killey, P. Bayvel, Electron. Lett. **34**, 2050–2051 (1998)
15. S. Song, C. Allen, K. Demarest, R. Hui, J. Lightwave Technol. **17**, 2285–2290 (1999)
16. M. Eiselt, J. Lightwave Technol. **17**, 2261–2267 (1999)
17. F. Matera, A. Mecozzi, M. Settembre, M. Tamburrini, M. Joindot, M. Midrio, Reduction of the cross-phase modulation impairment in wavelength division multipled systems with dispersion management, Opt. Soc. America, 1999
18. A.V. Cartaxo, J. Lightwave Technol. **17**, 178–190 (1999)
19. E. Neddam, S. Wabnitz, IEEE Photon. Technol. Lett. **12**, 798–800 (2000)
20. G. Bellotti, S. Bigo, IEEE Photon. Technol. Lett. **12**, 726–728 (2000)
21. F. Yang, M. Marhic, L. Kazovsky, J. Lightwave Technol. **18**, 512–520 (2000)
22. M. Premaratne, IEEE Photon. Technol. Lett. **12**, 1630–1632 (2000)
23. H. Kim, J. Lightwave Technol. **21**, 1770–1774 (2003)
24. H. Bao, W. Shieh, Opt. Express **15**, 4410–4418 (2007)
25. N.M. Costa, A.V. Cartaxo, J. Lightwave Technol. **26**, 3640–3649 (2008)
26. M.S. Islam, A. Dewanjee, M.S. Monjur, S. Majumder, Dependency of cross-phase and self-phase modulation on different link parameters for a multispan WDM system, *2009 IEEE 9th Malaysia international conference on communications (MICC)*, IEEE, pp. 280–284, 2009
27. A. Dewanjee, M.S. Islam, M.S. Monjur, S. Majumder, Impact of cross-phase and self-phase modulation on the performance of a multispan WDM system, *2009 IEEE 9th Malaysia international conference on communications (MICC)*, IEEE, pp. 285–289, 2009
28. G. Li, F. Yaman, X. Xie, E. Mateo, Signal processing for polarization multiplexed coherent WDM transmission – OSA Technical Digest (CD), *SPPCom – Signal Processing in Photonic Communications – OSA Technical Digest*, Optical Society of America, 2010, p. SPTuB1

29. M. Nazarathy, J. Khurgin, R. Weidenfeld, Y. Meiman, P. Cho, R. Noe, I. Shpantzer, The FWM impairment in coherent OFDM compounds on a phased-array basis over dispersive multi-span links, *Coherent Optical Technologies and Applications (COTA)*, Optical Society of America, 2008, p. CWA4
30. M. Nazarathy, J. Khurgin, R. Weidenfeld, Y. Meiman, P. Cho, R. Noe, I. Shpantzer, V. Karagodsky, Phased-Array Cancellation of Nonlinear FWM in Coherent OFDM Dispersive Multi-Span Links, Opt. Express. **16**, 15777–15810 (2008)
31. K. Forozesh, S.L. Jansen, S. Randel, The influence of the dispersion map in coherent optical OFDM transmission systems, *2008 digest of the IEEE/LEOS summer topical meetings*, IEEE, pp. 135–136, 2008
32. S. Adhikari, S.L. Jansen, V.A. Sleiffer, W. Rosenkranz, On the nonlinear tolerance of 42.8-Gb/s DPSK with co-propagating OFDM neighbors, *LEOS – IEEE lasers and electro-optics society annual meeting conference proceedings*, IEEE, pp. 40–41, 2009
33. A.J. Lowery, Opt. Express. **15**, 12965–12970 (2007)
34. A.J. Lowery, S. Wang, M. Premaratne, Opt. Express. **15**, 13282–13287 (2007)
35. L.B. Du, A.J. Lowery, Opt. Express. **16**, 19920–19925 (2008)
36. X. Liu, F. Buchali, R.W. Tkach, J. Lightwave Technol. **27**, 3632–3640 (2009)
37. X. Liu, S. Chandrasekhar, A. Gnauck, R. Tkach, Experimental demonstration of joint SPM compensation in 44-Gb/s PDM-OFDM transmission with 16-QAM subcarrier modulation, Vienna, Paper 2.3.4, 2009
38. X. Liu, R.W. Tkach, Joint SPM compensation for inline-dispersion- compensated 112-Gb/s PDM-OFDM transmission, *OFC/NFOEC – Conference on optical fiber communication and the national fiber optic engineers conference*, Paper OTuO5, 2009
39. W. Qiu, S. Yu, J. Zhang, J. Shen, W. Li, H. Guo, W. Gu, J. Lightwave Technol. **27**, 5321–5326 (2009)
40. Y. Tang, Y. Ma, W. Shieh, IEEE Photon. Technol. Lett. **21**, 1042–1044 (2009)
41. X. Liu, Fiber nonlinear impairments and their mitigation in coherent optical OFDM transmission – technical digest (CD), *Asia communications and photonics conference and exhibition*, Optical Society of America, p. ThF1, 2009
42. M. Nazarathy, Nonlinear impairments in coherent optical OFDM systems and their mitigation – OSA Technical Digest (CD), *SPPCom – Signal processing in photonic communications – OSA Technical Digest*, Optical Society of America, p. SPThC1, 2010
43. J. Leibrich, A. Ali, W. Rosenkranz, Single polarization direct detection optical OFDM with 100 Gb/s throughput: A concept taking into account higher order modulation formats – OSA Technical Digest (CD), *SPPCom – Signal Processing In Photonic Communications – OSA Technical Digest*, Optical Society of America, p. SPThC4, 2010
44. M. Nazarathy, B. Livshitz, Y. Atzmon, M. Secondini, E. Forestieri, J. Lightwave Technol. Optically Amplified Direct Detection with Pre- and Post- Filtering: A Volterra series approach, **26**, 3677–3693 (2008)
45. R. Weidenfeld, M. Nazarathy, R. Noe, I. Shpantzer, Volterra nonlinear compensation of 112 Gb/s ultra-long-haul coherent optical OFDM based on frequency-shaped decision feedback, *European conference of optical communication (ECOC)*, pp. 1–2 (2009)
46. B. Porat, *A Course in Digital Signal Processing* (Wiley, NY, 1996)
47. R. Feynman, R. Leighton, M. Sands, *The Feynman Lectures on Physics* (Addison Wesley, MA, 1965)
48. J. Goodman, *Speckle Phenomena in Optics: Theory and Applications* (Roberts and Company, CO, 2007)
49. Y. Atzmon, M. Nazarathy, J. Lightwave Technol. **27**, 4650–4659 (2009)
50. R. Weidenfeld, M. Nazarathy, R. Noe, I. Shpantzer, Volterra nonlinear compensation of 100G coherent OFDM with Baud-rate ADC, tolerable complexity and low intra-channel FWM/XPM error propagation, *OFC/NFOEC – Conference on optical fiber communication and the national fiber optic engineers conference*, Paper OTuE3, 2010
51. G. Goldfarb, M.G. Taylor, G. Li, Experimental demonstration of distributed impairment compensation for high-spectral efficiency transmission, *Coherent optical technologies and applications (COTA)*, Optical Society of America, p. CWB3, 2008

52. X. Li, X. Chen, G. Goldfarb, E. Mateo, I. Kim, F. Yaman, G. Li, Opt. Express **16**, 880–888 (2008)
53. E. Ip, A.P. Lau, D.J. Barros, J.M. Kahn, *Compensation of Dispersion and Nonlinearity in WDM Transmission Using Simplified Digital Backpropagation*, IEEE, 2008
54. E. Ip, J.M. Kahn, J. Lightwave Technol. **26**, 3416–3425 (2008)
55. G. Goldfarb, M.G. Taylor, G. Li, IEEE Photon. Technol. Lett. **20**, 1887–1889 (2008)
56. G. Goldfarb, G. Li, *Wavelet Split-Step Backward-Propagation for Efficient Post-Compensation of WDM Transmission Impairments*, 2009
57. E. Ip, J. Lightwave Technol. **28**, 939–951 (2010)
58. E. Ip, J.M. Kahn, J. Lightwave Technol. **28**, 502–519 (2010)
59. M. Schetzen, *The Volterra and Wiener Theories of Nonlinear Systems* (Wiley, NY, 1980)
60. G. Mathews, V.J. Sicuranza, *Polynomial Signal Processing* (Wiley, NY, 2000)

Chapter 4
Systems with Higher-Order Modulation

Matthias Seimetz

4.1 Introduction

With the objective of reducing costs per information bit in optical communication networks, per fibre capacities and optical transparent transmission lengths have been stepped up by the introduction of new technology in recent years. The innovation of the erbium-doped fibre amplifier (EDFA) at the beginning of the nineties facilitated long distances to be bridged without electro-optical conversion. Wavelength division multiplexing (WDM) technology allowed a lot of wavelength channels to be simultaneously transmitted over one fibre and to be amplified by one EDFA with high bandwidth, offering a huge network capacity. At this time, the modulation format of choice was the simple "on-off keying" (OOK), and there was no need for increasing spectral efficiency. The internet traffic growth during the nineties required increasing transmission rates. In that context, the transmission impairments of the optical fibre had to be counteracted and the application of differential binary phased shift keying (DBPSK) became an issue, providing for a higher robustness against nonlinear effects [1]. Moreover, the transmission behaviour of binary intensity modulation was optimized by using alternative optical pulse shapes such as return to zero (RZ) and by employing schemes with auxiliary phase coding, such as optical duobinary, which exhibits a higher tolerance against chromatic dispersion (CD). The capacity-distance product was further enhanced by applying optical dispersion compensation, Raman amplification and advanced optical fibres, as well as through electronic means, such as forward error correction (FEC) and the adaptive compensation of CD and polarization mode dispersion (PMD).

Driven by the immense need for transmission capacity expected in future optical fibre networks, transmission formats with increased spectral efficiency became more and more an important issue of research in the last years. To be able to fulfil the enormous future bandwidth requirements, higher-order modulation formats and

M. Seimetz (✉)
Beuth Hochschule für Technik Berlin, FB VII: Elektrotechnik und Feinwerktechnik, Luxemburger Str. 10, 13353 Berlin, Germany
e-mail: matthias.seimetz@beuth-hochschule.de

S. Kumar (ed.), *Impact of Nonlinearities on Fiber Optic Communications*, Optical and Fiber Communications Reports 7, DOI 10.1007/978-1-4419-8139-4_4,
© Springer Science+Business Media, LLC 2011

orthogonal frequency division multiplexing (OFDM) – both emerging technologies on the way to highest spectral efficiency and 100 Gbit s^{-1} line rates – are now pending to be deployed in optical fibre networks.

In this chapter, optical "single-carrier systems" where optical carriers are higher-order phase and quadrature amplitude modulated by a complex electrical base-band signal are described ("multi-carrier systems" with several electrical sub-carriers such as OFDM are not considered). In those systems with higher-order modulation, $m = \log_2 M$ data bits are encoded on M symbols and transmission can be accomplished at a symbol rate, which is reduced by m compared with the data rate. This allows upgrading to higher channel data rates by using existing lower-speed equipment and thus exceeding the limits of present high-speed electronics and digital signal processing (DSP). From another point of view, when assuming a given channel data rate, the transmission with lower symbol rates in WDM networks leads to a reduction in spectral width of a WDM channel, and thus to higher spectral efficiency, which is defined as the ratio of data rate per channel to WDM channel spacing.

Recently, successful practical implementation of optical systems with higher-order modulation is greatly facilitated by the availability of high-speed DSP technology. This allows for performing the necessary coding functions and generating multi-level electrical driving signals at the transmitter side by digital means. Moreover, it enables demodulation, synchronization and decoding to be implemented digitally within the receiver. Higher-order modulation formats can be detected using direct detection receivers as well as coherent receivers. Due to linear detection of the optical field, the latter allow for detecting arbitrary modulation formats and very efficient compensation of optical transmission impairments. Since the entire optical field parameters (amplitude, phase, frequency and polarization) are available in the electrical domain, especially coherent receivers benefit from DSP, and critical operations such as phase locking, frequency synchronization and polarization control can now be implemented in the electronic domain.

4.2 Higher-Order Modulation Formats

Through the deployment of optical *higher-order modulation formats*, symbol rate is reduced by m compared to the data rate by encoding $m = \log_2 M$ data bits on M symbols, as mentioned above, and higher spectral efficiencies can be obtained due to spectral narrowing. One of the $M = 2^m$ symbols is assigned to each symbol interval of length $T_S = m \cdot T_B$, where $r_B = 1/T_B$ is the data rate. The assignment of appropriate combinations of m bits to symbols with particular amplitude and phase states (bit mapping) is defined in a so-called *constellation diagram*. For the best noise performance, bit mapping should be arranged in such a way that only one bit per symbol differs from a neighbouring symbol (Gray coding). The symbols are transmitted on the reduced symbol rate $r_S = 1/T_S = r_B/m$.

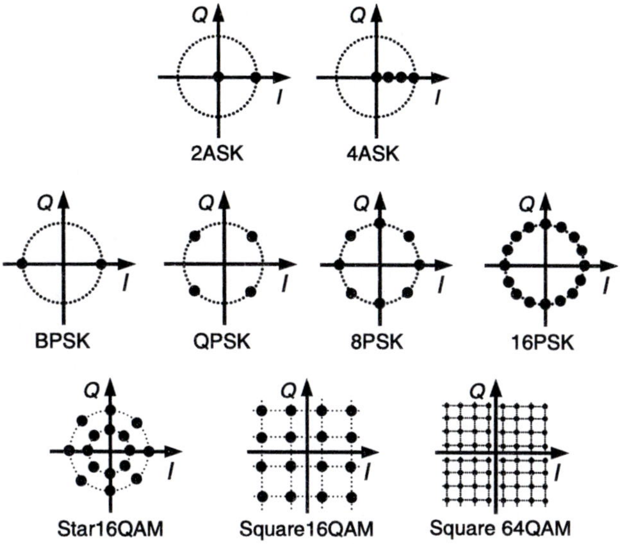

Fig. 4.1 Constellation diagrams of selected modulation formats applicable in future optical fibre networks

Figure 4.1 illustrates the constellation diagrams of selected higher-order modulation formats, which are possible candidates for future application in optical fibre networks.

A simple optical multi-level signalling scheme is M-ary amplitude shift keying (MASK). The constellation diagrams of binary ASK (2ASK) and quaternary ASK (4ASK) are shown in the upper part of Fig. 4.1. Information is encoded here at several amplitude levels. The 2ASK, usually denoted as OOK, is the standard modulation format in currently deployed optical transmission systems and defines only two symbol points (just one bit is assigned to each symbol). MASK was shown in [2, 3] to require high optical signal-to-noise ratios (OSNRs) for direct detection, especially in optically amplified links, due to the intensity dependence of the signal-spontaneous beat noise. For instance, a 2.5 times higher dispersion tolerance compared to OOK can be achieved by 4ASK, but only at the expense of a 5 dB power penalty due to noise. MASK formats will not be considered further in this chapter.

The constellation diagrams of different *phase modulation formats* are illustrated in the second row of Fig. 4.1. In the case of phase modulation, all constellation points lie in one circle and all symbols exhibit the same amplitude level, but different phase states. The first optical multi-level phase modulation format, whose transmission characteristics were intensively examined, for instance in [4], is the quadrature phase shift keying (QPSK). Because it features good transmission performance and doubled spectral efficiency at only a relatively moderate increase of complexity, it is already used for 40 Gbit s^{-1} networks. Optionally, spectral efficiency of QPSK can be further doubled by the use of polarization division multiplexing (PDM-QPSK).

The differential version of the QPSK – differential quadrature phase shift keying (DQPSK) – on the one hand, is typically detected by a direct detection receiver with lower complexity, which, on the other hand, does not provide for equally effective equalization.

Encouraged by the current trends and today's progress in high-speed electronics and DSP technology, even higher-order modulation formats have been investigated in various research groups in recent years. With direct detection, 8-ary differential phase shift keying (8DPSK) has been theoretically examined by Ohm [5] and Yoon et al. [6], and experimentally demonstrated by Serbay et al. [7]. By using coherent detection, 8-ary PSK has been experimentally reported by Tsukamoto et al. [8], Seimetz et al. [9], Freund et al. [10], Zhou et al. [11] and Yu et al. [12]. The 16PSK/16DPSK formats, which exhibit relatively poor OSNR performance have been so far investigated by computer simulations only [13, 14].

By combining intensity and phase modulation [quadrature amplitude modulation (QAM)], the number of phase states can be reduced for the same number of symbols, leading to modulation formats with larger Euclidean distances between the symbols. As shown in the lower part of Fig. 4.1, the symbols can be arranged in different circles (Star QAM) or can be positioned in a square (Square QAM). In *Star QAM* constellations, first suggested by Cahn in 1960 [15], the same number of symbols is placed on different concentric circles. The phases can be arranged with equal spacing, as shown in Fig. 4.1, for Star 16QAM (which can also be denoted as 2ASK-8PSK or 2ASK-8DPSK, respectively), so the phase difference of any two symbols corresponds to a phase state defined in the constellation diagram and phase information can be differentially encoded as for DPSK formats. Thus, on the one hand, Star QAM signals with differentially encoded phases can be detected by receivers with differential detection. In contrast, Star QAM constellations are not optimal with respect to noise performance because symbols on the inner ring are closer together than symbols on the outer ring. In order to improve noise performance, Hancock and Lucky suggested placing more symbols on the outer ring than on the inner ring [16], leading to constellations with more balanced Euclidean distances. But they came to the conclusion that such systems are more complicated to implement. For optical transmission, Star QAM experiments have been reported so far with four phase levels in [17] and [18, 19] for 2ASK-DQPSK and 4ASK-DQPSK, respectively. The Star 16QAM format shown in Fig. 4.1 has been investigated by computer simulations [13, 14, 20] and experimentally as well [21]. Moreover, the 8QAM format with two rings – each of them containing four symbols – that are shifted by 45° against each other has been experimentally demonstrated in [22].

Formats widely used in electrical communication systems are the *Square QAM formats*, where the symbols are arranged in a square, leading to larger Euclidean distances between the symbols and thus to an improvement of noise performance. Square QAM constellations, shown in Fig. 4.1 for Square 16QAM and Square 64QAM, were introduced for the first time in 1962 by Campopiano and Glazer [23]. Square QAM signals are conveniently detected by coherent synchronous receivers, although they can also be detected by differential detection when phase pre-integration is employed at the transmitter [24]. Thinking in terms of two quadrature carriers, relatively simple modulation and demodulation schemes are possible due to the regular structure of the constellation projected onto the in-phase and quadrature axes. Recently, Square QAM has been successfully demonstrated also for optical fibre transmission: Square 16QAM signals were transmitted over large distances of more than 1000 kilometres for single-channel transmission [25, 26], as well as with a high baud rate of 28 Gbaud and a high spectral efficiency of 6.2 bit s^{-1} Hz^{-1} for WDM transmission [27]. Even very high-order Square 256QAM transmission has already been performed at a lower baud rate of 4 Gbaud [28].

4.3 Signal Generation

Optical higher-order modulation signals can be generated using various transmitter configurations. Generally, the migration to higher-order formats brings about an increase in transmitter complexity. The upgrade can be performed by adding optical modulators and accordingly creating more elaborate optical modulator structures or by providing more complex electrical level generators for the generation of multi-level electrical driving signals. For this purpose, analogue or digital technology can be employed. An analogue creation of multi-level electrical driving signals with sufficient high power for the modulator inputs is quite challenging since overlapping different binary electrical signals to generate a multi-level signal leads to increased eye spreading and thus to a degradation of system performance [29]. In contrast, when looking at digital solutions, high-speed digital-to-analogue (D/A) converters just start appearing. Figure 4.2 illustrates that the overall complexity of transmitters for higher-order modulation can be traded off between the optical and electrical parts. Optical complexity can be reduced through increased electrical complexity and vice versa.

4.3.1 *External Optical Modulators*

The optical part of higher-order modulation transmitters is typically composed of one or more fundamental external optical modulator structures: the phase modulator (PM), the Mach–Zehnder modulator (MZM) and the optical IQ modulator (IQM).

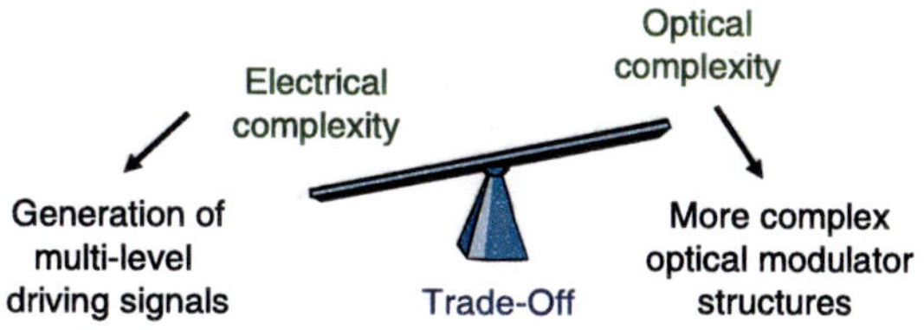

Fig. 4.2 Transmitter complexity: trade-off between the optical and electrical parts

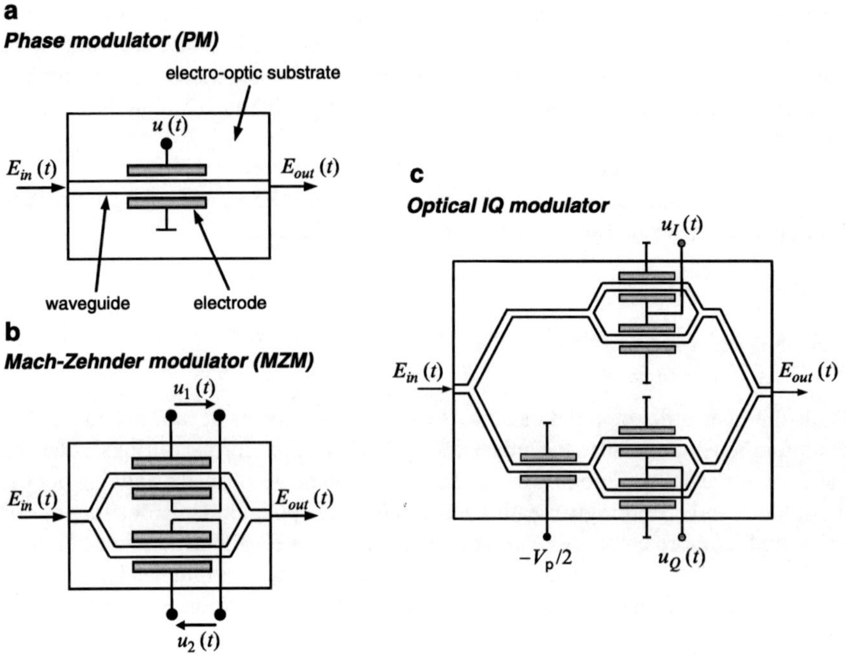

Fig. 4.3 Fundamental optical modulator structures

An optical PM can be fabricated as an integrated optical device by embedding an optical waveguide in an electro-optical substrate (mostly $LiNbO_3$), see Fig. 4.3a. By utilizing the fact that the refractive index of a material, and thus the effective refractive of the waveguide, can be changed by applying an external voltage $u(t)$ through a coated electrode, the electrical field of the incoming optical carrier $E_{in}(t)$ can be modulated in phase. When solely considering the Pockels effect, the change of the refractive index can be assumed to be linear to the applied external voltage.

By utilizing the principle of interference, the process of phase modulation can also be used to cause an intensity modulation of the optical lightwave when the interferometric structure shown in Fig. 4.3b is employed. This represents a dual-drive MZM. The incoming light is split into two paths, both equipped with PMs. After acquiring some phase differences relative to each other, the two optical fields are recombined. The interference varies from constructive to destructive, depending on the relative phase shift.

The field and power transfer functions of an MZM are shown in Fig. 4.4, illustrating two different operation principles, where the operation points (OPs) of the MZM are chosen differently. For achieving modulation in intensity, the MZM can be operated at the quadrature point, with a DC bias of $-V_\pi/2$ and a peak-to-peak modulation of V_π (see Fig. 4.4, left), assuming V_π to be the voltage inducing a phase shift of π in the power transfer function of the MZM. When the MZM is operated

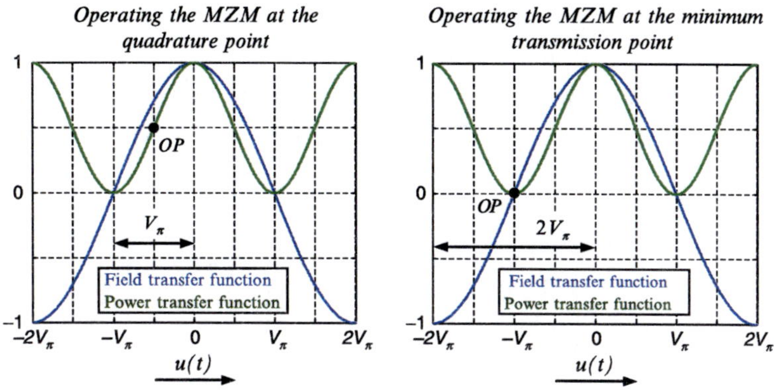

Fig. 4.4 Operating the MZM at the quadrature point (*left*) and the minimum transmission point (*right*)

at the minimum transmission point (see Fig. 4.4, right), with a DC bias of V_π and a peak-to-peak modulation of $2V_\pi$, a phase skip of π occurs when crossing the minimum transmission point. This becomes apparent from the field transfer function. This way, the MZM can be used for binary phase modulation and for modulation of the field amplitude in each branch of an IQM.

A third fundamental optical modulator structure is the IQM, which can be composed of a PM and two MZMs. It is commercially available in an integrated form. As illustrated in Fig. 4.3c, the incoming light is equally split into two arms, the in-phase and the quadrature arm. In both paths, a field amplitude modulation is performed by operating the MZMs at the minimum transmission point. Moreover, a relative phase shift of $\pi/2$ is adjusted in one arm, for instance by an additional PM. This way, any constellation point can be reached in the complex IQ-plane after recombining the light of both branches.

4.3.2 *Higher-Order PSK/DPSK and QAM Transmitters*

For generation of PSK/DPSK, Star QAM and Square QAM formats, transmitter configurations with multi-level electrical driving signals (moderate optical complexity) or binary driving signals (higher optical complexity) are possible. Some of them are discussed in the following two sections.

4.3.2.1 Transmitters Based on Multi-Level Driving Signals

Theoretically, a single dual-drive MZM in the optical transmitter part is sufficient to generate arbitrary higher-order PSK/DPSK and QAM signals [29]. However,

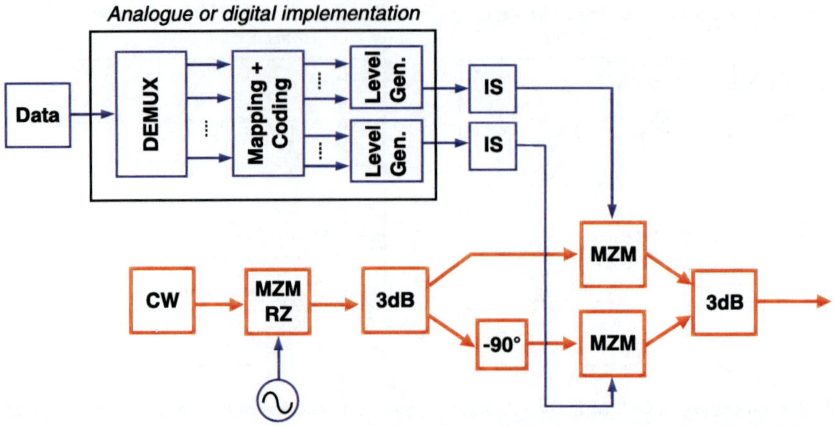

Fig. 4.5 Higher-order modulation transmitter suitable for generating arbitrary PSK/DPSK and QAM formats based on an optical IQM and multi-level electrical driving signals; *CW* Continuous wave laser, *IS* Impulse shaper, *DEMUX* Demultiplexer, *MZM* Mach-Zehnder modulator, *RZ* Return-to-zero

generation of multi-level electrical driving signals with a very high number of levels is then required for generating formats with high order. To give an example, 16-level driving signals are needed for Square 16QAM.

Another option suitable for generating arbitrary higher-order PSK and QAM signals is to use a single IQM in the optical transmitter part. Figure 4.5 shows this transmitter including its electrical part, where the data signal is first parallelized with a demultiplexer. Parallelized data bits are fed into a module performing mapping and coding – for instance, a differential encoding which allows for differential detection at the receiver side or to resolve phase ambiguity within the carrier synchronization [optical phase-locked loop (OPLL) or digital phase estimation] when a receiver with coherent synchronous detection is applied. Otherwise, the differential encoding can be omitted. Afterwards, multi-level in-phase and quadrature driving signals are generated, either by analogue level-generators or by digital means using D/A-converters. The necessary number of levels of the driving signals depends on the respective modulation format and corresponds to the number of projections of the symbols onto the in-phase and the quadrature axes. The driving signals can be formed by an impulse shaper (IS) filter before being fed into the both MZMs of the IQM. In the optical domain, an MZM can optionally be used behind the continuous wave (CW) laser for carving RZ pulses.

The shown transmitter based on a single IQM in the optical part may not be the best choice for the generation of higher-order PSK and Star QAM signals because the in-phase and quadrature driving signals have a high number of signal states and the distances between these signal states are small. Nevertheless, due to the regular structure of the Square QAM constellation projected onto the in-phase and quadrature axes, this transmitter is a suitable device for generating Square QAM signals. However, the generation of multi-level driving signals is required here as well (quaternary driving signals must be generated for Square 16QAM, and 8-ary driving signals for Square 64QAM). Because multi-level electrical driving signals are currently hard to generate at high data rates, transmitter configurations are attractive, which require solely binary electrical driving signals. However, this increases the necessary number of optical modulators and thus the complexity of the optical transmitter part. As will be shown in the next section, transmitters with binary electrical driving signals are possible for arbitrary PSK/DPSK, Star QAM and Square QAM formats.

4.3.2.2 Transmitters Based on Binary Driving Signals

A simple way of generating *optical PSK/DPSK signals* with binary electrical driving signals is to use several consecutive PMs with phase shifts of $\pi/2^{n-1}$ ($n = 1, \ldots, m$). After the first PM (phase shift π), a signal with binary phase modulation is obtained, after the second PM (phase shift $\pi/2$) a signal with quaternary phase modulation, and so on. Figure 4.6 illustrates this kind of transmitter, including the electrical transmitter part, which is shown here with differential encoding.

The complexity and configuration of the differential encoder in the electrical part of the transmitter depend on the order of the DPSK modulation [30]. In the optical domain, the first PM accomplishing the phase modulation by π can also be replaced by an MZM driven at the minimum transmission point, as done in the experiment reported in [11]. This leads to higher phase accuracy and to a better transmission performance in the case of NRZ pulse shape. From a practical point of view, phase modulation using PMs necessitates high accuracy of the electrical driving signals, since the optical phase changes linearly with the applied voltage. Any variation in the amplitude of the driving voltage will appear as phase noise in the optical signal.

Another transmitter configuration suitable for generating arbitrary PSK/DPSK signals, which has been employed in recent experiments with higher-order phase

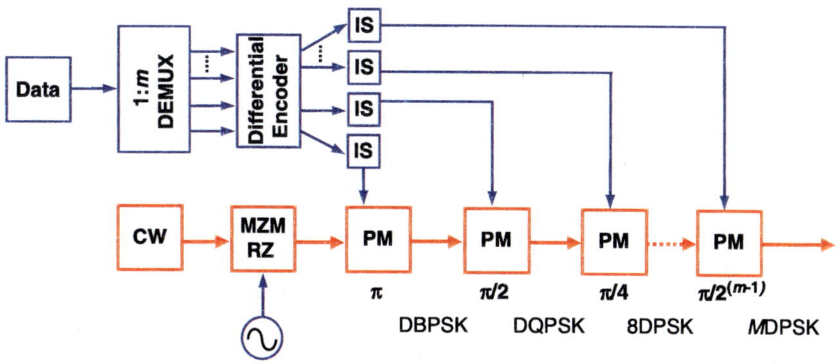

Fig. 4.6 Higher-order DPSK transmitter composed of consecutive phase modulators (PM)

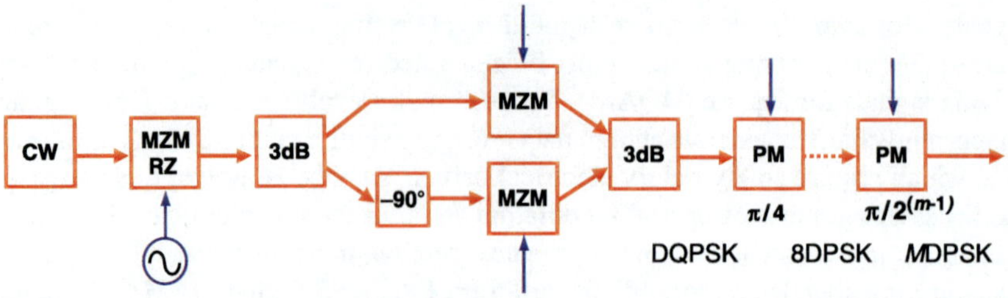

Fig. 4.7 Optical part of a higher-order PSK/DPSK transmitter composed of an optical IQM and consecutive phase modulators (PM)

modulation [9], uses also binary electrical driving signals and is composed of a combination of an IQM and consecutive PMs, as depicted in Fig. 4.7. The IQM, whose MZMs are driven at the minimum transmission point, accomplishes a quaternary phase modulation, and higher-order phase modulation signals are generated by the consecutive PMs. The electrical transmitter part (not shown in Fig. 4.7) is identical to the one for the transmitter composed of consecutive PMs, with the exception of the internal setup of the differential encoder [30].

For generation of *Star QAM signals* using binary driving signals, almost the same transmitter structures as described for PSK/DPSK can be employed. The PSK/DPSK transmitters described above have to be extended only by an additional intensity modulator, usually an MZM. This modulator allows for placing symbols at different intensity levels. For instance, a transmitter for Star 16QAM (2ASK-8PSK/2ASK-8DPSK) can be composed of an 8PSK/8DPSK transmitter extended by an additional MZM. In the case of Star QAM constellations with only two intensity rings, the driving signal of the MZM is binary. Otherwise, in the case of more than two rings, the driving signal of the MZM is multi-level. To differentially encode the phases of Star QAM signals, the same differential encoders can be used as for the respective DPSK format with the same number of phase states. An important parameter, which can optimize the OSNR performance for Star QAM formats with only two amplitude states, is the ring ratio $RR = r_2/r_1$, where r_1 and r_2 are the amplitudes of the inner and outer circles, respectively. It can be adjusted by changing the driving and bias voltages of the MZM.

In the case of *Square QAM*, various options exist for signal generation. Due to the regular structure of the constellation projected on the in-phase and quadrature axes, the use of the transmitter based on a single optical IQM described above is a beneficial solution for Square QAM. However, if generation of multi-level driving signals shall be avoided, transmitter configurations with binary driving signals become attractive. In contrast to Star QAM, the phases are arranged unequally spaced in Square QAM constellations. For this reason, it is not possible to adjust all the phase states of the symbols by driving consecutive PMs with binary electrical signals. Nevertheless, several options exist for generating square-shaped constellations using

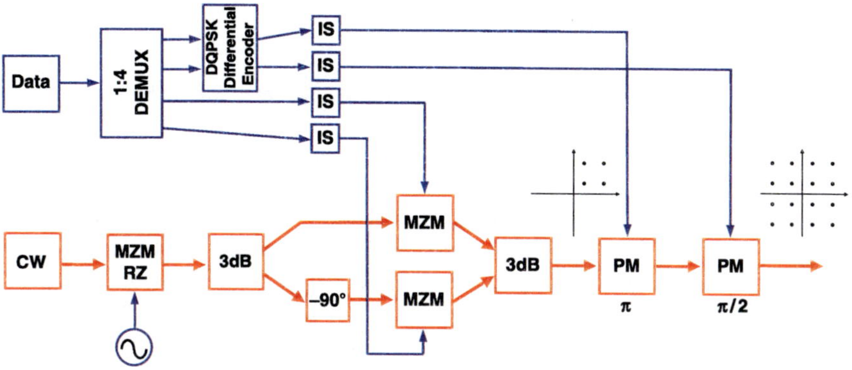

Fig. 4.8 "Tandem-QPSK transmitter" for generating optical Square 16QAM signals with binary driving signals

binary driving signals. Different transmitter configurations, denoted as "Enhanced IQ transmitter," "Tandem-QPSK transmitter" and "Multi-parallel MZM transmitter" are described in detail in [30]. Exemplarily, Fig. 4.8 shows the Tandem-QPSK transmitter for generation of Square 16QAM signals. A first IQM is employed to generate a constellation with four symbols in the first quadrant. This is achieved by using the MZMs of the IQM as intensity modulators and operating them in the quadrature point. Using a consecutive QPSK modulator, which can be realized using another IQM or by using two PMs as shown in Fig. 4.8, for instance, the four-symbol constellation in the first quadrant can be shifted into the other three quadrants, thereby generating a complete Square 16QAM constellation.

If a quadrant ambiguity must be resolved in the carrier synchronization of the receiver, a DQPSK differential encoding has to be performed on two of the bits in the transmitter's electrical part. Moreover, it must be ensured that the chosen bit mapping is symmetric in rotation with respect to the remaining bits. It is a beneficial side effect of the transmitter shown in Fig. 4.8 that – initiated through signal generation – the resulting constellation is inherently symmetric in rotation and additional coding can be avoided. For the other transmitters, a mapping symmetric in rotation can be achieved by an additional coder. More details regarding the electrical part of the transmitters can be found in [30].

Choosing a particular transmitter structure is not only a matter of trading off complexities and looking at the transmitter's practical feasibilities, but the respective transmitters can also be rated by considering the influence of their individual signal properties such as intensity shape, symbol transitions and chirp characteristics on the overall system performance. Especially in the case of NRZ pulse shape, different transmitter configurations exhibit a different system performance, whereas the differences are only small for RZ [30].

4.4 Signal Detection

An overview about receiver schemes applicable for the detection of optical higher-order modulation signals is given in Fig. 4.9. They can be roughly divided into two basic groups: Direct detection and coherent detection. In the latter case, two fundamental coherent detection principles can be distinguished: homodyne and heterodyne detection. In the case of homodyne detection, the carrier frequencies of the signal laser and the LO laser aspire to be identical and the optical spectrum is directly converted to the electrical baseband. In the case of heterodyne detection, the frequencies of the signal laser and the LO are chosen to be different, so that the field information of the optical signal wave is transferred to an electrical carrier at an intermediate frequency, which corresponds to the frequency difference of the signal laser and the LO. On the one hand, heterodyne detection permits simple demodulation schemes and enables carrier synchronization with an electrical phase locked loop. On the other, the occupied electrical bandwidth for heterodyne detection is more than twice as high as for homodyne detection, and image-rejection techniques are required to allow for acceptable spectral efficiencies for WDM. For this reason, only direct detection and homodyne detection will be discussed in the following subsections.

4.4.1 Direct Detection Receivers

Although only the intensity of the optical field can be detected by a simple photodiode, the information encoded in the optical phase can also be obtained when employing additional optics. By using an optical interferometer, the phase difference information of two consecutive symbols can be converted into intensity information,

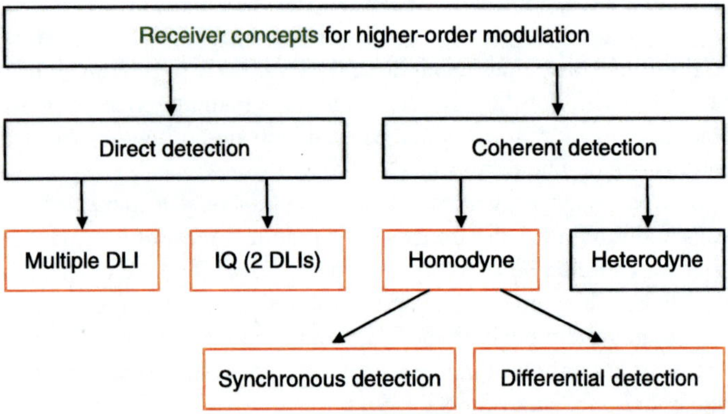

Fig. 4.9 Overview about detection schemes applicable for detection of optical higher-order modulation signals

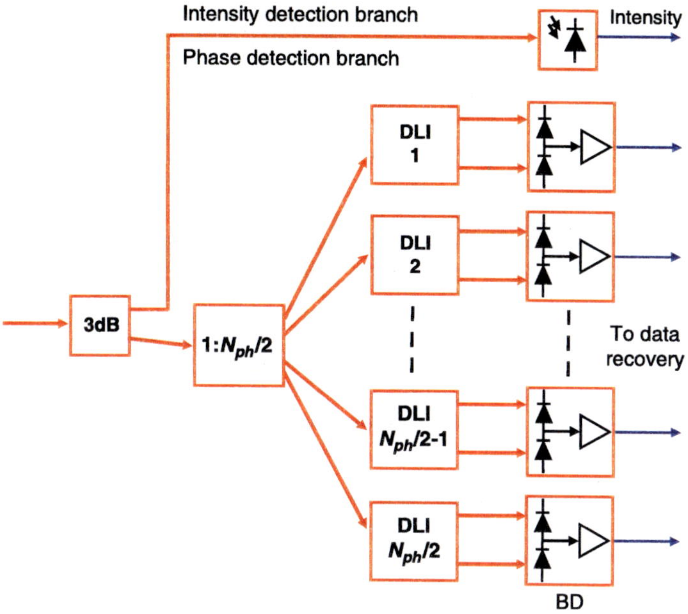

Fig. 4.10 Optical part of a Star QAM direct detection receiver composed of an array of delay line interferometers (DLIs); *BD* Balanced detector

which can then be detected by a photodiode. This allows for the detection of arbitrary DPSK signals. With a separate intensity detection branch, arbitrary Star QAM signals with differentially encoded phases can also be received when appropriate data recovery methods are employed [30, 31]. Square QAM signals have recently been detected by differential detection using an additional phase pre-integration at the transmitter [24].

The usual way for constructing direct detection receivers is employing delay line interferometers (DLIs) to convert differential phase modulation into intensity modulation before photodiode square-law detection. One receiver option – whose optical part is shown in Fig. 4.10 – is to use $N_{\mathrm{ph}}/2$ DLIs with appropriate phase shifts, where N_{ph} represents the number of phase states ($N_{\mathrm{ph}} = M$ for an MDPSK signal). For the detection of DPSK signals, only the branch with the DLIs (phase detection branch) is needed. Another branch (intensity detection branch) must be provided for a separate evaluation of the intensity when detecting Star QAM signals. Phase information can finally be demodulated by performing bi-level decisions on the resulting $N_{\mathrm{ph}}/2$ electrical photocurrents. This receiver concept with multiple DLIs was investigated for 8DPSK in [6]. Unfortunately, the optical effort becomes quite high for modulation formats with a high number of phase states. Four DLIs are needed for 8DPSK, and as many as eight DLIs for 16DPSK.

The complexity of the optical receiver part can be reduced by employing a receiver structure with only two DLIs, which is sufficient to obtain the phase

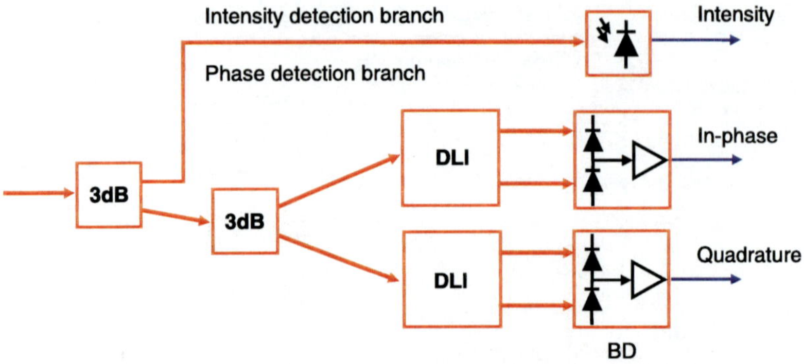

Fig. 4.11 Optical part of a direct detection IQ receiver composed of two delay line interferometers (DLI) and two balanced detectors (BD) and comprising an intensity detection branch for Star QAM

difference information of arbitrary DPSK and Star QAM signals by detecting their in-phase and quadrature components (direct detection IQ receiver). However, a more complex data recovery with decisions on electrical multi-level signals and multiple thresholds becomes necessary in that case for modulation formats with $N_{\text{ph}} > 4$. Moreover, decision thresholds are then no longer located at zero. Figure 4.11 shows the optical part of a direct detection IQ receiver comprising a separate intensity detection branch for Star QAM.

To enhance the sensitivity, an optical pre-amplifier, commonly followed by an optical filter, is typically placed in front of the receiver (not shown in Fig. 4.11). Looking at the internal setup of the DLIs, the phase shifts of the upper and lower DLI in the phase detection branch should be set to $-45°$ and $-135°$ in the case of the detection of DQPSK signals, for instance, so that information retrieval can be accomplished based upon binary signals in the in-phase and quadrature arms. More general, the in-phase and quadrature components of arbitrary DPSK constellations can be obtained by choosing the phase shifts of the DLIs as $0°$ and $-90°$. Principles of electrical data recovery from the in-phase and quadrature photocurrents for arbitrary DPSK and Star QAM formats are described in [30].

Direct detection receivers feature a relatively simple setup (no phase, frequency or polarization control is necessary) and lower laser linewidth requirements in comparison with coherent receivers. However, receiver sensitivities attainable are not as high as for coherent receivers and electronic equalization cannot be carried out as efficiently.

4.4.2 Homodyne Receivers

Since laser linewidth requirements have relaxed with increasing data rates (enabling the use of commercial communication lasers) and high-speed DSP technology provides now for an easier implementation, coherent receivers have reappeared as an area of interest in the last years and are even now deployed by carrier companies. In modern homodyne receivers based on DSP, a free running LO which does not have to be phase locked by an OPLL can be used. Due to the linear detection of all optical field parameters, demodulation schemes are not limited to the detection of phase differences as for direct detection, but arbitrary modulation formats and modulation constellations can be received. Compensation of transmission impairments such as CD and fibre nonlinearities can be accomplished efficiently using DSP. Moreover, WDM channel separation can be accomplished by highly selective electrical filtering. Nevertheless, when being compared to direct detection receivers, additional effort must be spent in coherent receivers on tasks such as carrier synchronization and polarization control. However, these tasks can all be accomplished using signal processing. Demodulation concepts in homodyne receivers can be based on synchronous or differential detection. Both detection schemes are briefly discussed in the following two subsections.

4.4.2.1 Receivers with Homodyne Synchronous Detection

Figure 4.12 shows the basic setup of a typical digital coherent receiver with homodyne synchronous detection and polarization division de-multiplexing. The signal launched into the receiver is split by a polarization beam splitter (PBS) first. Afterwards, both polarization components are interfered with the LO light in two 2×4 90°-hybrids. The splitting of the LO light by another PBS in Fig. 4.12 has to be understood schematically. In practice, both separated polarization components of the information signal at the PBS outputs exhibit the same linear polarization state, and it suffices when the LO light, whose polarization must then be aligned to the polarization of the signal at the two PBS outputs, is equally split with a 3 dB coupler.

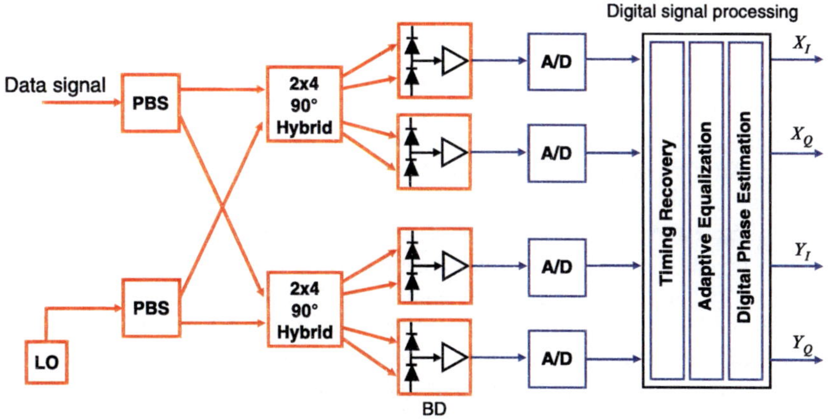

Fig. 4.12 Digital coherent receiver with homodyne synchronous detection employing timing recovery, adaptive equalization, polarization de-multiplexing and digital phase estimation

Since carrier synchronization is performed by digital means in the electrical part of the receiver, a free running LO, which does not have to be phase-locked can be used in modern homodyne receivers based on DSP. The output signals of the two 2×4 90°-hybrids are detected by two pairs of balanced detectors, which provide the in-phase and quadrature photocurrents of both polarization components at the outputs of the optical receiver front-end.

In the electrical receiver part, the in-phase and quadrature signals are sampled by A/D-converters and then further processed by elaborate DSP. Typically, the first functional block in the DSP part is a non-adaptive time or frequency domain equalizer (not shown in Fig. 4.12), which compensates for the main part of CD having accumulated along the fibre link [32, 33]. Afterwards, a timing recovery is accomplished in order to synchronize the sample rate with the signal's symbol rate. Widely used algorithms are the Gardner [34] and the square timing recovery [35] here. Timing recovery is typically followed by an adaptive time domain equalizer, which compensates for degradation effects and performs the polarization de-multiplexing. The equalizer is usually implemented as an FIR butterfly equalizer [36], whose coefficients are adapted using the constant modulus algorithm (CMA) or the decision-directed least mean square (LMS) algorithm. In order to ensure a proper operation of the equalizer, a sample rate of at least twice the symbol rate is mostly chosen (fractionally spaced equalizer). For digital phase estimation – the functional block behind the adaptive equalizer – just one sample per symbol is required which must be properly selected for the case that more than one sample per symbol is utilized for equalization. Phase estimation can be performed by treating both polarizations independently (selected algorithms are described in [30, 37, 38]) or by using a joint-polarization approach [39]. After carrier synchronization, the constellation diagrams are appropriately aligned and data can be recovered from the received symbols by evaluating their amplitudes and absolute phase states (synchronous detection), as in detail described for PSK, Star QAM and Square QAM formats in [30]. In the case of single-polarization systems, the optical effort is approximately half (the PBS, one 2×4 90°-hybrid and two balanced detectors can be saved). Moreover, the DSP becomes less complex.

To give a better insight into the signal processing block, the following paragraphs describe – as an example – a possible algorithm for digital phase estimation for the single-polarization case, which is denoted as feed forward Mth power block scheme. After timing recovery and equalization, the signal is typically resampled to one sample per symbol and then processed by digital phase estimation. The Mth power phase estimation procedure for MPSK signals is illustrated in Fig. 4.13.

After demultiplexing the incoming stream of received complex samples X_k into blocks of length N, the N parallel samples are first raised to the Mth power to wipe off the M-ary phase modulation. To more accurately estimate the phase error out of the shot-noise/optical amplifier (OA) noise, an averaging is performed by adding the raised samples of a block of length N. Afterwards, a common phase error estimate φ_{est} for all symbols of the block is obtained by calculating the argument of the complex sum vector and dividing it by M. On the one hand, averaging lowers the influence of the shot-noise/amplifier noise on the phase error estimate. On the other,

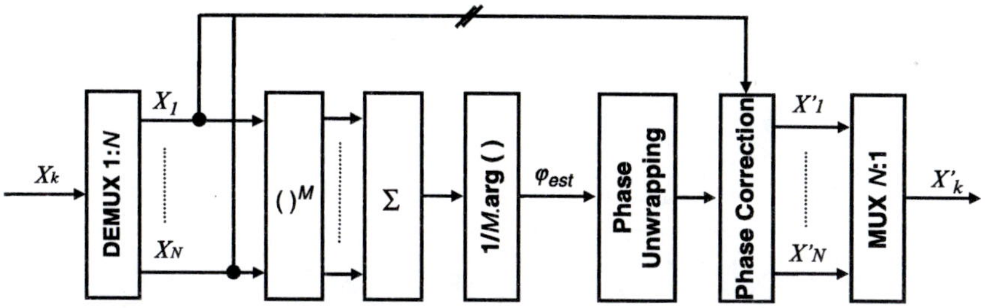

Fig. 4.13 Digital phase estimation according to the *M*th power feed forward block scheme for *M*PSK formats

an inherent error is introduced since an average phase error estimate is calculated, commonly used for the phase correction of all symbols in the block. An optimal block length can be found as a trade-off between the shot-noise/amplifier noise and the phase noise effects. Alternatively to the scheme shown here, a particular phase error estimate can be calculated for each symbol, corresponding to a sliding window technique. This can lead to a higher tolerance against phase noise, but leads to a higher implementation complexity.

When the random walk of the phase noise is passing one of the boundaries between two segments at $n \cdot 2\pi/M,\ n \in \{0, 1, \ldots, M-1\}$, the phase error estimate performs phase jumps (cycle slips) and does not follow the trajectory of the physical phase [40]. These phase jumps must be corrected by performing a phase unwrapping. Moreover, since the angle values calculated by the arg-operation are limited to the interval $[-\pi,\ \pi]$, the phase error estimate takes only values between π/M and π/M and an M-fold phase ambiguity of $n \cdot 2\pi/M$ is induced. This problem can be overcome by periodically sending synchronization sequences, or better still by the use of differential decoding.

The described phase estimation scheme yields phase error estimate inaccuracies when the summed phasors are not of the same length. Thus, it might not be appropriate for phase estimation of highly distorted *M*PSK signals and cannot be used for QAM formats without further modification. The scheme can be improved and made usable for carrier phase estimation of Star QAM signals by normalizing the phasors to a common amplitude before being summed [30,41]. In the case of Square QAM formats which have constellations with non-equidistantly spaced phases, the scheme can be modified by partitioning the symbols into two subgroups as shown in Fig. 4.14 for Square 16QAM and Square 64QAM (showing only one quadrant). The Class I symbols (solid points) have in common that they exhibit modulation angles of $\pi/4 + n \cdot \pi/2 (n = 0, \ldots, 3)$, so that the modulation can be wiped off as for QPSK modulation by raising to the fourth power when selecting only these symbols for determination of the phase error estimate within each block. The selection between Class I and Class II symbols can be accomplished by performing amplitude decisions on the received complex signal samples. More details can be found in [30].

The digital phase estimation scheme described in the last paragraphs represents only one of a large number of possible algorithms. There are various alternative

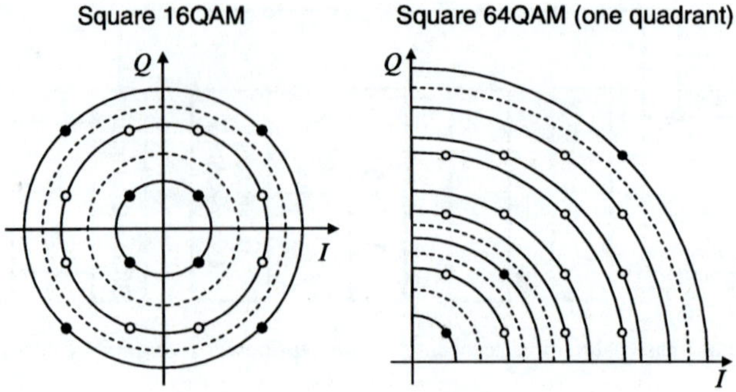

Fig. 4.14 Class partitioning for Square 16QAM (*left*) and Square 64QAM (*right*)

schemes which can be based on symbol-to-symbol phase correction, enhanced filtering (Wiener filtering, for instance) or decision-directed techniques. Moreover, enhanced phase estimation algorithms for Square QAM have been recently proposed, which can significantly reduce the requirements on laser phase noise [42,43].

On the one hand, homodyne receivers with synchronous detection feature various advantages. First, the same optical frontend can be used for the detection of any modulation format. The digital algorithms, however, as well as the data recovery, must be adapted in accordance with the particular received format. Second, receiver sensitivity is increased in comparison with receivers based on differential detection schemes. Furthermore, the availability of the optical phase information in the electrical domain enables an efficient digital equalization to compensate for transmission impairments. On the other hand, homodyne receivers with synchronous detection show the disadvantage of more stringent laser linewidth requirements in comparison with receivers based on differential detection [30].

4.4.2.2 Receivers with Homodyne Differential Detection

If laser linewidth requirements cannot be fulfilled using a homodyne receiver with synchronous detection but the advantage of an efficient equalization shall still be exploited, receivers with homodyne differential detection are an interesting option. Differential detection in the electrical part of the receiver can be accomplished by analogue means or by applying DSP, as shown for the single-polarization receiver in Fig. 4.15.

After sampling the in-phase and quadrature signals at the outputs of the optical front-end by A/D-converters and optionally performing digital equalization, an arg-operation is performed on the in-phase and quadrature samples to calculate the instantaneous phase of the current symbol. Subsequently, the current phase difference can be determined by subtracting the phase sample delayed by one symbol time from the current phase sample. In practice, these steps necessitate only a table-lookup for phase determination and a subtraction operation for phase differentiation, so the signal processing part is by far less complex than for homodyne synchronous

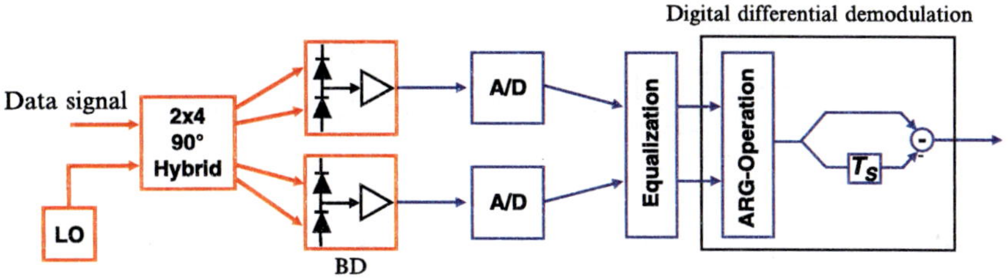

Fig. 4.15 Homodyne receiver with digital differential demodulation, illustrated here for the reception of arbitrary DPSK signals

detection. This way, phase information of arbitrary DPSK signals and Star QAM signals with differentially encoded phases can be differentially demodulated. Moreover, the amplitude of Star QAM signals can be easily calculated by squaring and adding the in-phase and quadrature samples. It should be noted that digital equalization of transmission impairments can be performed in the same manner as for synchronous detection.

Due to the differential demodulation, laser phase noise becomes not critical until the phase noise-induced phase change takes considerable values within the symbol duration – same as for direct detection. Thus, linewith requirements are relaxed in comparison with homodyne synchronous detection. In comparison with direct detection, requirements are doubled when the same linewidth are assumed for the signal laser and the LO [30]. Frequency offsets and frequency offset drifts, which lead to corresponding fixed phase rotations and to slow varying rotations of the constellation diagram, respectively, can be compensated for by an AFC loop or digital frequency offset estimation [38]. Moreover, a polarization control must be implemented to align polarizations of the signal laser and the LO. The drawback of homodyne differential detection scheme in comparison with synchronous detection is the lower receiver sensitivity, being only in the range of direct detection receivers [30].

4.4.2.3 2 × 4 90° Hybrid Optical Front-end

Whereas DSP represents a key technology for coherent receivers in the electrical domain, the optical front-end – comprising one optical 2 × 4 90°-hybrid and two balanced detectors (single polarization case) or two optical 2 × 4 90°-hybrids and four balanced detectors (polarization multiplexing case) – is the key component in the receiver's optical part. Fortunately, this optical front-end has become commercially available from several companies in recent years. Hybrids and balanced detectors can be obtained separately or integrated in a single component.

The 2 × 4 90°-hybrid is a key component in optical coherent receivers allowing the in-phase and quadrature components of the complex optical field to be detected [30,44], and can be realized by different implementation options, which are depicted in Fig. 4.16.

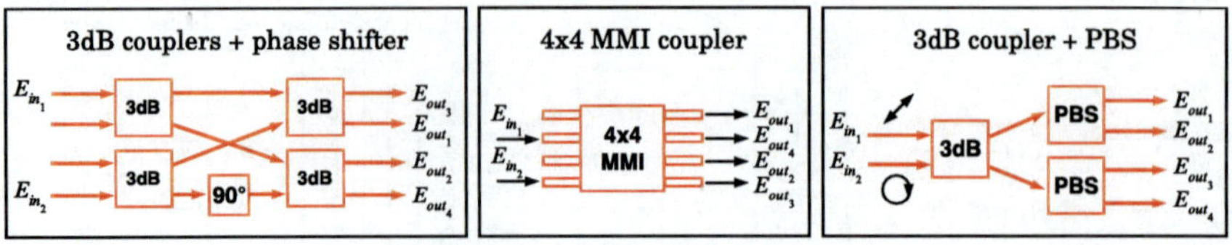

Fig. 4.16 Implementation options for 2 × 4 90°-hybrids; *left*: four 3dB-couplers and phase shifter, *middle*: 4 × 4-multimode interference (MMI) coupler, *right*: 3dB-coupler and polarization beam splitters (PBS)

One possibility is to construct the hybrid with four 3dB-couplers and an additional phase shifter in one branch (see Fig. 4.16, left). This configuration has to be implemented in an integrated form to achieve sufficient IQ-balance. A version fabricated on $LiNbO_3$ was analyzed and discussed in [45]. This device is commercially available at present, and can be adjusted with six different electrodes. Four electrodes control the uniformity of the 3dB-couplers. With the remaining two electrodes, the phase shifts in the upper and the lower branches can be set [46]. To ensure orthogonality, the relative phase shift between two branches has to be tuned to 90°. Imprecise relative phase shifts lead to a degradation of the IQ balance, whereas the asymmetries of the 3 dB couplers affect the power symmetry of the hybrid output signals and thus the symmetry of the subsequent balanced detection processes. For commercial application, accuracy of the phase shift should be stabilized by an external control loop [44]. Alternatively, I-Q phase errors can be compensated by the DSP engine [47]. A nice feature of the commercially available device, which offers adjustable phase shifts in two branches, is its usability also for other applications. For example, with an additional time delay of one symbol period in front of one input, and phase shifts of $\pi/4$ in both branches, it can be used for optical demodulation of DQPSK signals.

A very promising alternative for fabricating a phase-stable 2 × 4 90°-hybrid component without the requirement of an additional phase control is to exploit the properties of a 4 × 4 MMI coupler (Fig. 4.16, middle). Using the right inputs and for proper waveguide dimensioning, this component inherently exhibits the desired phase relations. Furthermore, MMI couplers are broadband, which make them suitable for WDM application. In addition, the balanced detectors of the receiver can be integrated on the chip, possibly with polarization diversity. The device has to be carefully designed to achieve equal splitting ratios together with the appropriate phase relations, as it was shown using simulations in [44].

A third option to implement the 2 × 4 90°-hybrid, which has been realized with discrete components [48] as well as in an integrated form [49], is a configuration with a 3 dB-coupler and two PBSs (see Fig. 4.16, right). This arrangement, however, requires specific polarization states from the signals feeding into the hybrid inputs. One input signal must be linearly polarized at 45° with respect to the PBS reference directions, and the other one must be circularly polarized.

4.5 Trends in System Performance

The migration from traditionally used binary modulation formats to higher-order formats with more bits per symbol leads to a reduction in symbol rate and spectral width. Therefore, higher spectral efficiencies and per fibre capacities can be achieved. At the same time, migration to higher-order modulation strongly influences system performance. This section discusses the basic trends in system performance resulting from migration to higher-order modulation formats, regarding relevant parameters such as noise, laser linewidth requirements, CD tolerance and self-phase modulation (SPM) tolerance. The discussion presented here is based on computer simulations for 40 Gbit s^{-1} systems employing homodyne receivers with synchronous detection.

Figure 4.17a shows the *back-to-back OSNR requirements* of various modulation formats in the case of using a homodyne receiver with synchronous detection – assuming an ideal carrier synchronization, a data rate of 40 Gbit s^{-1} as well as the use of second-order Gaussian optical bandpass and fifth-order electrical Bessel filters within the receiver, with 3 dB bandwidths of 2.5 and 0.75 times the symbol rate, respectively. It can be observed that – when assuming a fixed data rate – the noise performance degrades when going up to higher-order formats since the Euclidean distances between the symbols become smaller with increasing number of bits per symbol. Higher-order QAM formats exhibit a significantly better noise performance than higher-order phase modulation formats for a certain number of bits per symbol, in particular Square QAM formats. In comparison with 16PSK, Square 16QAM has an OSNR performance gain of about 4 dB, for instance.

Another important system parameter, which can become critical in systems with higher-order modulation, is the *laser linewidth*. As illustrated in Fig. 4.17b for systems employing receivers with homodyne synchronous detection based on *M*th power feed-forward phase estimation, requirements on laser linewidth increase with an increasing number of phase states, since a certain level of laser phase noise is

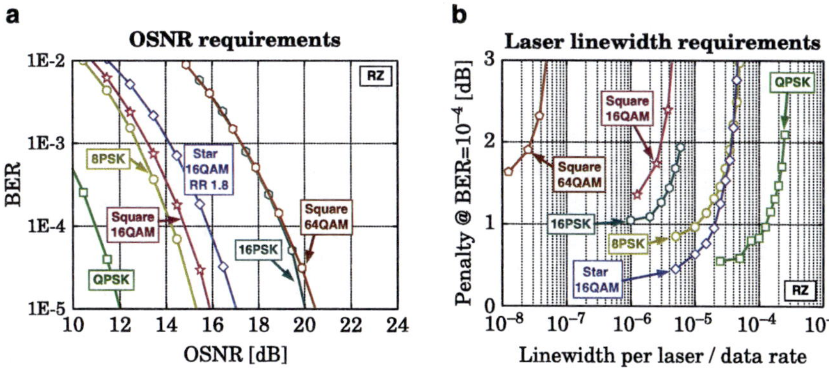

Fig. 4.17 OSNR requirements at 40 Gbit/s (**a**) and laser linewidth requirements with *M*th power feed forward phase estimation (**b**) of various modulation formats when using homodyne receivers with synchronous detection

more problematic for closer phase distances. In addition – if the different formats are compared at the same data rate – the reduction in the symbol rate makes the laser phase noise more critical for modulation formats with a higher number of bits per symbol. When the *M*th power feed forward scheme described in Sect. 4.4.2.1 is employed, requirements on laser phase noise become stringent for higher-order formats such as 16PSK, Square 16QAM and Square 64QAM, although this carrier recovery scheme is not impaired by processing delay. The required linewidths at 40 Gbit s^{-1} are then in the range of 240 kHz, 120 kHz and 1 kHz for 16PSK, Square 16QAM and Square 64QAM, respectively [41]. These requirements cannot be fulfilled with currently available low-cost lasers. As a consequence, a commercial application of those modulation formats in systems with homodyne synchronous detection necessitates the development of low-cost lasers with very low linewidths. Moreover, the application of improved phase estimation schemes offers a way of further relaxing the requirements on laser linewidth [42, 43]. In comparison with systems with homodyne synchronous detection, the linewidth requirements are relatively relaxed in systems with direct detection. Even 16DPSK can tolerate a linewidth of about 1 MHz at 40 Gbit s^{-1} [30]. In the case of homodyne differential detection, the effective phase noise, which affects the electrical differential demodulation process, is determined by the beat-linewidth. The linewidth requirements on each laser are approximately doubled in comparison with direct detection when the same linewidths are assumed for the signal laser and the LO.

In the following paragraphs, the tolerance of different modulation formats regarding two important fibre transmission effects is outlined: CD and SPM. Due to the reduced symbol rates and the longer symbol durations therewith aligned, modulation formats of higher order feature an improved tolerance against CD. The same is true for tolerance against PMD. Figure 4.18a illustrates the *CD tolerance* of a wide range of modulation formats at 40 Gbit s^{-1} for RZ pulse shape when homodyne synchronous receivers without digital equalization are used. Results were obtained by Monte Carlo simulations. It can be observed that – at a fixed data rate – CD tolerances improve when the order of the modulation format is increased.

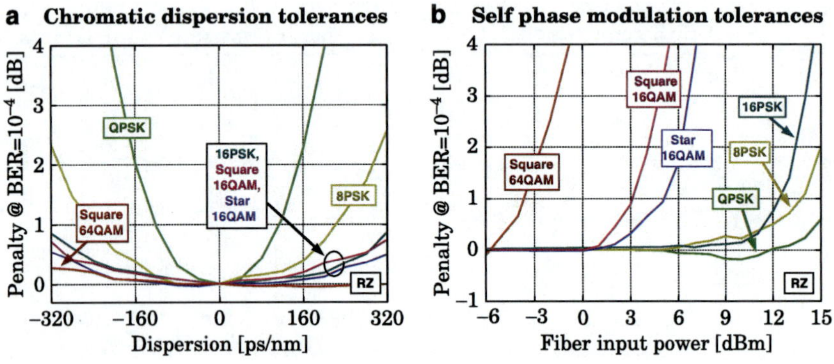

Fig. 4.18 Chromatic dispersion tolerance (**a**) and self-phase modulation tolerance (**b**) of various modulation formats for 40 Gbit s^{-1}; parameters: RZ pulse shape, homodyne synchronous detection with *M*th power feed forward digital phase estimation

Figure 4.18b illustrates the *SPM tolerance* of various modulation formats, which was determined by transmitting the signals over a single dispersive and nonlinear fibre link [standard single mode fibre (SSMF)] with a length of 80 km. The CD is completely compensated for after the link and the average fibre input power is varied. SPM induces a power-dependent phase shift on a signal propagating through the fibre [50]. Generally, SPM tolerances tend to become worse as the number of phase states increases in modulation formats and phase distances between symbols become smaller. Each symbol of an idealized phase modulated signal with constant power would be affected by the same nonlinear phase shift during fibre propagation if there was no other effect than SPM. In this case, the received constellation would be rotated, but not distorted. However, CD and SPM interact during propagation. Power fluctuations induced by CD cause the nonlinear phase shifts experienced by the symbols to become different so that the received constellation diagrams become distorted. Since phase distances are getting smaller, the robustness against SPM decreases with an increasing order of the PSK/DPSK format.

When QAM signals have been propagated through the fibre, the constellation diagrams are deformed even in the absence of CD since symbols with different power levels are affected by different mean nonlinear phase shifts. This effect is shown in Fig. 4.19 showing 16PSK, Star 16QAM and Square 16QAM. In the case of phase modulation, all symbols are located on one intensity ring and the nonlinear phase shift induces only a phase rotation common to all symbols. In the case of QAM formats, however, constellations become not only rotated due to SPM but also distorted. This phenomenon constitutes an inherent problem of optical QAM transmission and is the reason for the poor SPM performance of all QAM formats (see Fig. 4.18b).

The SPM-induced distortions of the signal constellations cannot be compensated for by phase estimation solely, which just rotates back the entire constellation by the phase error, but must be compensated for by an additional nonlinear phase shift compensator to enable further use of simple decision techniques. In the case of Square 16QAM, for instance, the optimal decision boundaries are spiral-like when not

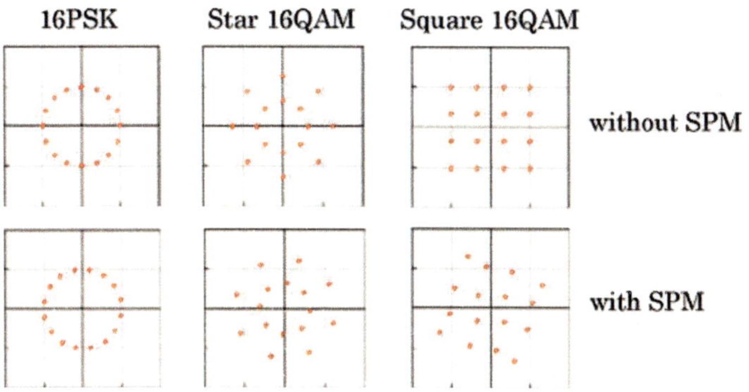

Fig. 4.19 Deformation of the signal constellations of 16PSK (*left*), Star 16QAM (*centre*) and Square 16QAM (*right*) caused by the SPM-induced nonlinear phase shift

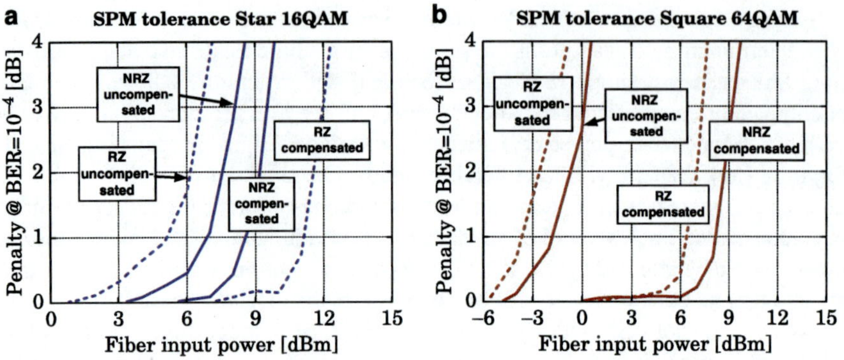

Fig. 4.20 Enhancement of the SPM tolerance by compensation of the SPM-induced mean nonlinear phase shift for Star 16QAM (**a**) and Square 64QAM (**b**)

employing compensation, whereas the usual straight-line decision boundaries can be used after nonlinear phase shift compensation. A simple compensation scheme is modulating the signal phase in front of the receiver according to the power of the received signal, as it is done with the compensator shown in the right part of Fig. 4.31 [30]. The improvements of the SPM tolerance for Star 16QAM and Square 64QAM attained with this simple scheme are illustrated in Fig. 4.20, assuming the same scenario as for the determination of the SPM tolerances described above (80 km fibre link, 100% CD post-compensation, varied fibre input powers) and showing results for NRZ and RZ pulse shapes.

It can be seen that SPM tolerance can be greatly enhanced for both formats shown. However, although this simple compensation scheme turns out to be quite effective for the single-span system configuration discussed here, it may be less efficient in multi-span transmission systems, where signals propagating along the fibre are highly distorted due to CD. The interaction between CD and SPM prevents a complete compensation of the nonlinear phase shift. Alternatively or additionally to the compensation scheme described here, signal distortions through the SPM-induced mean nonlinear phase shift can potentially be reduced by means of digital equalization in the electrical part of the receiver, for instance using decision-directed adaptive equalization schemes, or by applying pre-distortion techniques on the transmitter side. Compensation of the nonlinear phase shift in multi-span long-haul transmission systems will be briefly discussed later on in Sect. 4.6.3.

4.6 Long-Haul Transmission

Concluding the performance trends discussed in the last section, migration to modulation formats with more bits per symbol leads to higher spectral efficiencies and higher CD and PMD tolerances. At the same time, laser linewidth requirements get more stringent, noise performance deteriorates and self phase modulation tolerances go down. Optical multi-span long-haul transmission systems, which are typically composed of multiple transmission sections each containing a fibre – usually with a length of about 80 km – and OAs compensating for fibre attenuation are mainly limited by amplifier noise and fibre nonlinearities. Thus, systems applying higher-order modulation formats show a reduced transmission reach. CD can be compensated for within each span (optical inline dispersion compensation) or electrically at the receiver. Already installed long-haul fibre transmission systems are mainly based on OOK and differential binary phase shift keying. QPSK systems are also starting to be commercially deployed. Even higher-order formats are not yet adopted in commercially deployed systems. But the imminent need for optical data transmission capacity feeds the interest in system concepts allowing for high spectrally efficient transmission by the use of higher-order modulation formats and motivates the current research activities in this field. However, to be applicable for long-haul fibre links, transmission formats must also exhibit an attractive transparent transmission reach. In this paragraph, some simulative and experimental work identifying performance and distances attainable in optical multi-span transmission systems with higher-order modulation is presented, which has been performed in the former research group of the author at the Fraunhofer Institute for Telecommunications, Heinrich-Hertz-Institute, Berlin.

4.6.1 System Experiments with Optical Inline CD Compensation

This section presents some experimental results, which have been published in [9, 21] investigating transmission distances attainable with RZ-QPSK, RZ-8PSK and RZ-Star 16QAM at a common symbol rate of 10 Gbaud for multi-span transmission with optical inline CD compensation and homodyne synchronous detection.

In Fig. 4.21, the schematic of the experimental system setup with optical inline CD compensation used is shown. The transmitter consists of an external cavity laser (ECL) with a linewidth specified as 100 kHz. For RZ pulse carving an MZM is used. Afterwards, an optical RZ-QPSK signal is generated by an optical IQM. With the consecutive PM, an additional $\pi/4$ phase modulation is accomplished to obtain an RZ-8PSK signal. A further MZM is used for Star 16QAM signal generation. By changing the driving and bias voltages of this MZM, different ring ratios (RRs) can be adjusted. The underlying data signal is a 2^{11} de Bruijn sequence, which is given to the modulator inputs with different delays. Moreover, polarization multiplexed transmission is investigated by splitting the signal at the MZM output with a PBS, delaying one polarization component, and afterwards adding both polarization components in a polarization beam combiner (PBC).

The transmission link is based on a re-circulating fibre loop with adjustable number of sections. Each section consists of 80 km SSMF and about 13 km dispersion compensating fibre (DCF) which fully compensates for the SSMF CD. EDFAs are used to compensate for the fibre loss and control the launch powers into the SSMF

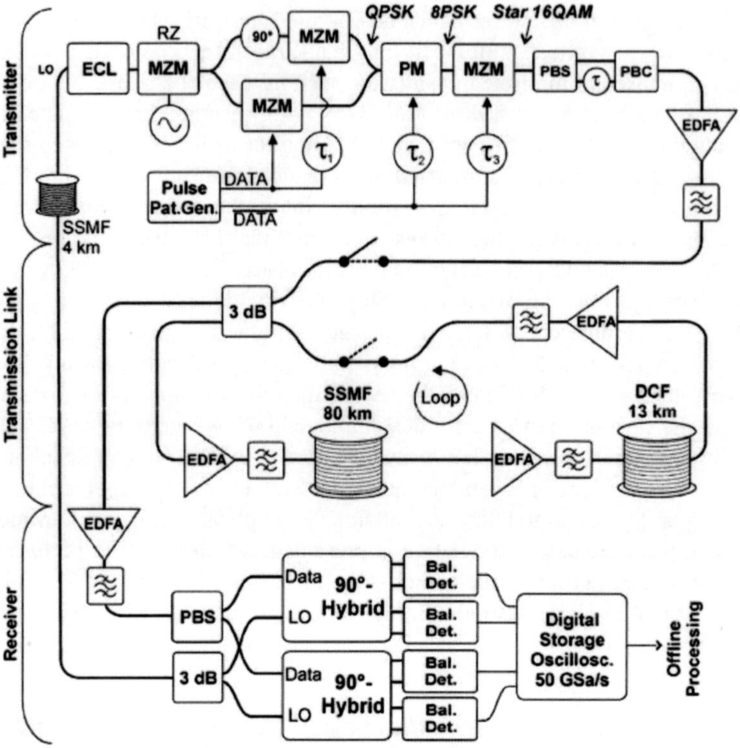

Fig. 4.21 Experimental system setup for the coherent multi-span long-haul transmission experiments with inline chromatic dispersion compensation performed in [21]

and DCF. The noise power of the OAs outside the signal band is reduced by optical band-pass filters. The signal can be sent to the receiver after being transmitted over a desired number of cascaded sections by the use of acousto-optical switches.

At the receiver end, the received signal is split by a PBS first in case of polarization de-multiplexing. Signal polarization is controlled manually in front of the PBS. Afterwards, both polarization components are interfered with the light of a local oscillator (LO) in two 2×4 90°-hybrids. For experimental simplicity, the LO light is taken here from the transmitter laser to avoid an automatic frequency control loop. In the back-to-back (BtB) case where the transmitter is directly connected to the receiver, the received information signal and the LO signal are de-correlated by a 4 km long SSMF. The hybrid output signals are detected by four balanced detectors and the photocurrents are digitized using a 50 GSa s^{-1} digital storage oscilloscope. Finally, data is recovered offline by applying digital phase estimation (using a feed-forward block scheme with rectangular time domain filtering and averaging over eight symbols) and appropriate data recovery. Further electrical equalization of transmission impairments is not performed. In the single-polarization case, the PBS, one hybrid and two balanced detectors can be saved. The optical part of the receiver

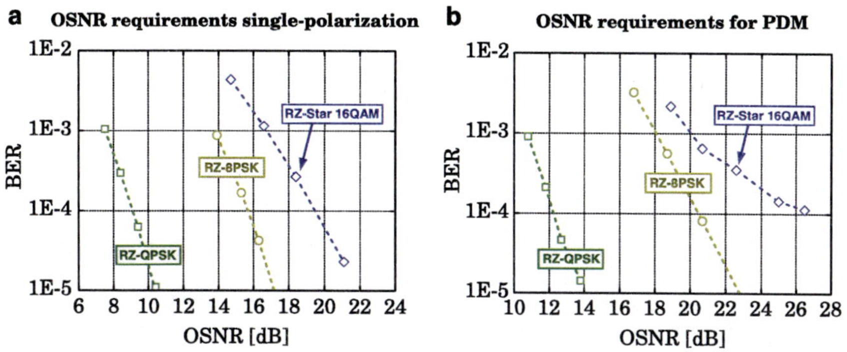

Fig. 4.22 Back-to-back OSNR requirements of RZ-QPSK, RZ-8PSK and RZ-Star 16QAM measured in [21] for single-polarization (**a**) and polarization division multiplexing (**b**), assuming a common symbol rate of 10 Gbaud

is identical for all modulation formats examined here. For offline calculation of bit error rates, DSP and data recovery algorithms must be adapted in accordance with the investigated modulation format.

A first indicator for the transmission length achievable with a particular modulation format is the back-to-back noise performance. In Fig. 4.22, the back-to-back OSNR requirements measured for RZ-QPSK, RZ-8PSK and RZ-Star 16QAM are compared for single-polarization and PDM

To obtain a BER of 10^{-3}, an OSNR of about 16.5 dB and 20.0 dB was required for RZ-Star 16QAM in the case of single-polarization and PDM, respectively. The measured OSNR penalty at BER $= 10^{-3}$ is 2–3 dB and 9 dB compared with RZ-8PSK and RZ-QPSK, respectively. The differences in the required OSNR between these formats are larger than expected from numerical simulation (1.5 dB and 5 dB, see Sect. 4.5), since in the practical transmitter setup every new modulation stage led to higher inter-symbol interference caused by pattern effects of the electrical driving signals and thus to higher implementation penalties. OSNR requirements increase by about 3 dB when upgrading from single-polarization to PDM.

Transmission distances achieved with RZ-QPSK, RZ-8PSK and RZ-Star 16QAM are compared in Fig. 4.23 for single-polarization (a) and PDM (b), assuming a common symbol rate of 10 Gbaud for all formats.

The experimental results presented in Fig. 4.23 assume optimized launch powers into the SSMF and DCF and demonstrate that the attainable transmission distances are considerably reduced when migrating from QPSK to 8PSK, and even more when applying Star 16QAM. This is primarily caused by the more stringent OSNR requirements of the higher-order formats, as well as by their reduced tolerance against nonlinear effects. However, it should be noted that the curves for RZ-Star 16QAM in Fig. 4.23 are shown without compensation of the SPM-induced mean nonlinear phase shift. This effect was already discussed in Sect. 4.5 and causes a relative rotation of the symbols located on the inner and outer rings. It can also be seen from the experimentally obtained constellation diagrams shown in the left part of

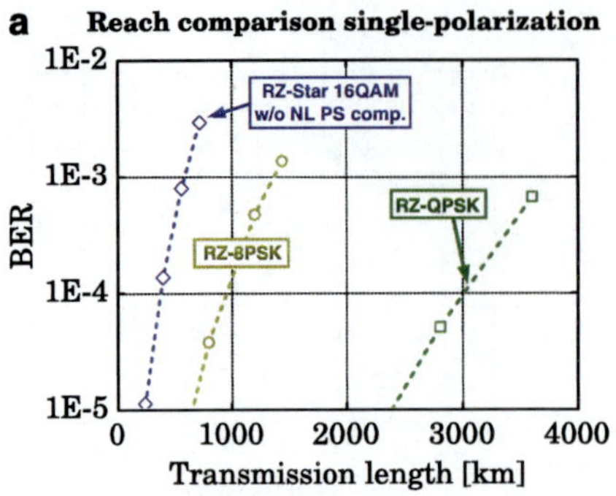

Fig. 4.23 Transmission distances achieved in [21] with RZ-QPSK, RZ-8PSK and RZ-Star 16QAM for multi-span transmission with optical inline CD compensation for single-polarization (**a**) and PDM (**b**), assuming a common symbol rate of 10 Gbaud

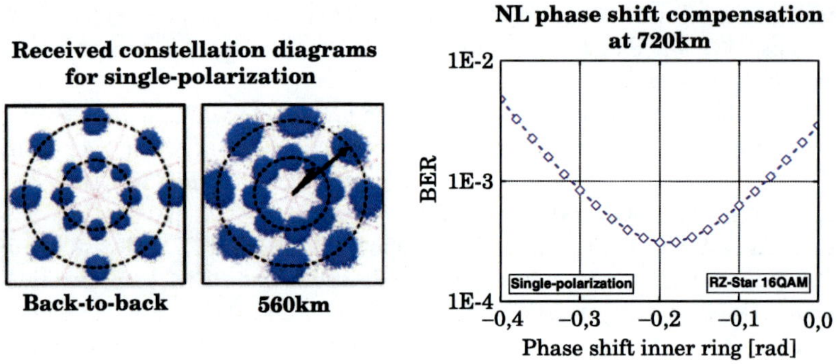

Fig. 4.24 Received constellation plots for single-polarization for BtB/after 560 km (*left*); BER improvement through nonlinear phase shift compensation at 720 km for single-polarization RZ-Star 16QAM (*right*) [21]

Fig. 4.24 – after 560 km symbols on the inner and outer rings have experienced different nonlinear phase shifts. Transmission distances for Star 16QAM can be increased when the relative nonlinearity-induced phase difference of both rings is compensated for. As it becomes apparent from the right diagram in Fig. 4.24, an optimum BER performance at 720 km is obtained when symbols on the inner ring are rotated by about −0.19 rad for single-polarization case. In the experiments performed in [21], this relative phase shift has been compensated for electrically and transmission reach for Star 16QAM could be increased to about 1,000 km.

It should be noted that the comparison of transmission distances made in this section is based upon a common symbol rate for all the modulation formats. The differences between the maximum transmission distances would be smaller if the comparison were made at the same data rate.

4.6.2 System Experiments with Electrical CD Compensation

The traditional way of compensating for CD in optical fibre networks is by applying optical inline compensation. However, since coherent receivers are now on the way towards a commercial deployment, a pure electrical CD compensation at the receiver becomes a promising option. A great advantage of a coherent receiver in comparison with a direct detection receiver is its capability to efficiently compensate for transmission impairments in the electrical domain. All information parameters of the optical signal are accessible after detection. The accumulated CD can simply be compensated for by convolution of the received complex signal with the inverse fibre impulse response. Removing the DCFs from the link leads not only to a lower system complexity, but also transmission reach can be increased – as has been shown in recent experiments [10]. This can be explained by an improvement of OSNR, caused by the noise reduction through the removal of the EDFA-amplified DCF, as well as by a decrease of accumulated fibre nonlinearities due to the removal of the DCF and a more constant signal envelope along the fibre (symbol power levels become indistinguishable after certain transmission distances due to CD). In the following paragraphs, some system experiments with pure electrical CD compensation at the receiver employing RZ-QPSK, RZ-8PSK, RZ-Star 16QAM and RZ-Square 16QAM are described, which have been published in [10, 25, 30, 51].

To discover the attainable transmission lengths with RZ-QPSK and RZ-8PSK without optical inline CD compensation, a similar experimental setup has been used as in the experiments with inline CD compensation (whose setup is shown in Fig. 4.21). However, the DCF and the EDFA in front of the DCF are removed from the transmission section and the raw data is electrically equalized off-line by an ideal FIR filter before digital phase estimation within the receiver. For practical filter implementation, the equalization performance will be limited, for instance by the number of taps. As can be seen from Fig. 4.25, transmission lengths could be increased for both formats by replacing optical inline CD compensation with an electrical CD compensation at the receiver. In this way, transmission distances of >6,000 km/2,800 km for a target BER of 10^{-3} could be attained at 10 Gbaud for single-polarization RZ-QPSK/RZ-8PSK.

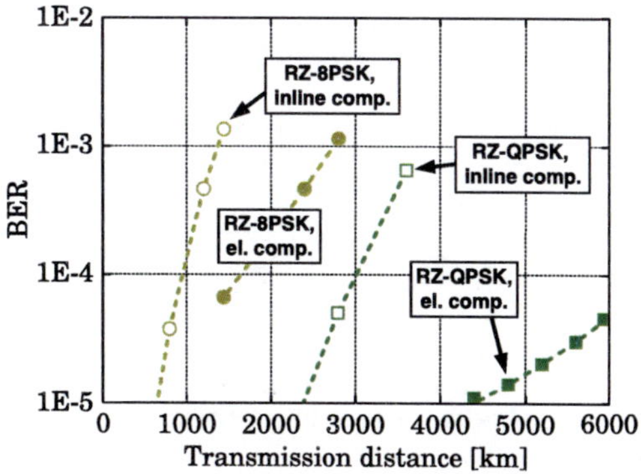

Fig. 4.25 Comparison of transmission distances obtained with RZ-QPSK and RZ-8PSK with inline CD compensation and electrical CD compensation at 10 Gbaud [10, 30]

In another experiment with electrical dispersion compensation performed in [51], five 10 Gbaud RZ-Star 16QAM WDM channels were transmitted over 800 km/1,400 km SSMF with/without PDM on a 50 GHz frequency grid centred at 1550.92 nm. The central channel is demodulated with the aid of electronic equalization at the receiver. The system setup employed in the experiment is very similar to the single-channel experimental setup shown in Fig. 4.21 but upgraded to WDM at some locations. The transmitter consists of five ECLs, which are coupled by a set of 3 dB couplers. The transmission section within the re-circulating fibre loop is composed of three 80 km SSMF spans without inline dispersion compensating modules. EDFAs are used to compensate for the loop loss and control the launch powers into the fibre spans. The noise power of the amplifiers is reduced by optical bandpass filters and a gain equalizer controls the gain of each WDM channel. In the electrical part of the receiver, the in-phase and quadrature photocurrents of both polarizations are digitized and processed by digital equalization. CD is compensated for by using a filter that is implemented in the frequency domain. Thereafter, an enhanced CMA – denoted as multiple moduli algorithm (MMA) and published in [42], where it was applied to Square 16QAM – is used to compensate for residual transmission impairments such as nonlinearities and polarization crosstalk. After equalization, feed forward Mth power phase estimation is used to compensate for laser phase noise. Finally, decoding and error counting are performed. Figure 4.26a shows the measured WDM spectra at the output of the transmitter and after transmission over 1,200 km for a fibre launch power of -1 dBm/channel.

BER values of the central channel were measured after different numbers of loop round trips. Figure 4.26b depicts BER vs. transmission distance when using an adaptive MMA equalizer with 9 taps. Applying this approach, transmission distances of about 800 km and 1,400 km for 80 Gbit s^{-1} and 40 Gbit s^{-1} RZ-Star 16QAM with and without PDM could be achieved, respectively.

Square QAM formats have also been investigated in recent experiments. In [25], single channel Square 16QAM transmission has been demonstrated with a symbol

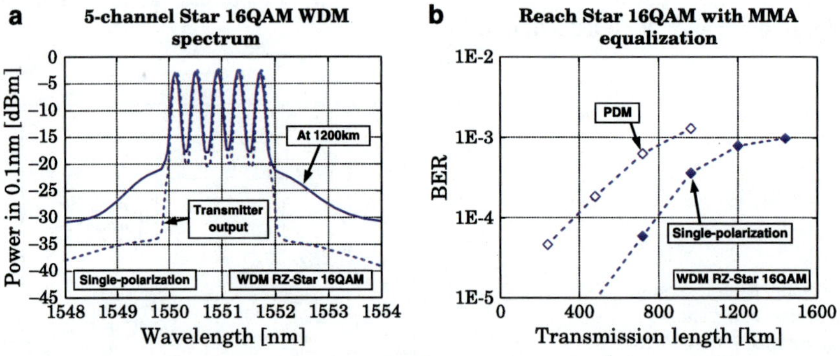

Fig. 4.26 A 5-channel 10 Gbaud RZ-Star 16QAM WDM spectrum (**a**); transmission reach obtained in [51] with 10 Gbaud WDM RZ-Star 16QAM for single-polarization and PDM using MMA equalization (**b**)

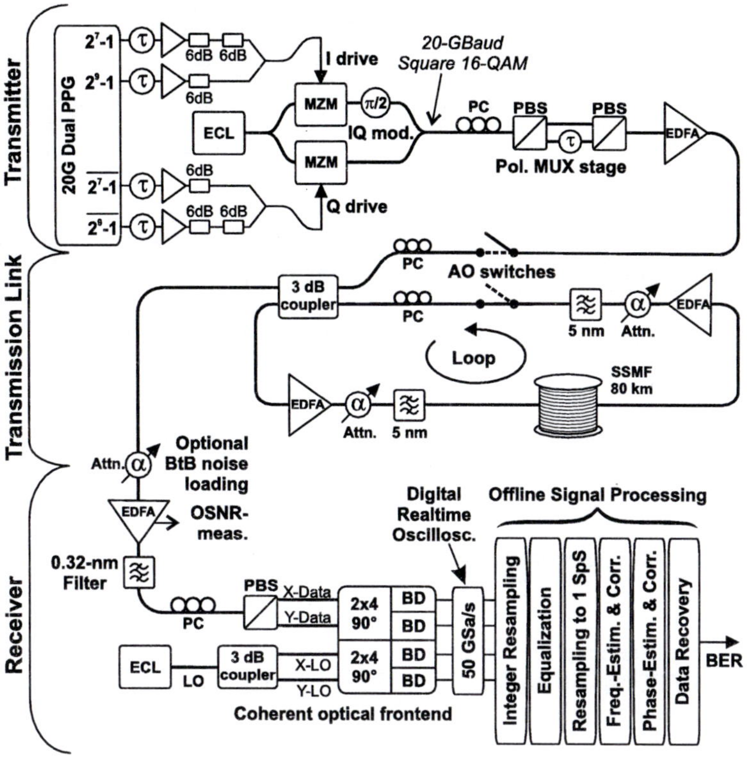

Fig. 4.27 Experimental setup used in [25] for investigation of 20 Gbaud PDM-Square 16QAM

rate of 20 Gbaud corresponding to bit rates of 80 Gbit s^{-1} and 160 Gbit s^{-1} in the single-polarization and polarization multiplexed case, respectively. Figure 4.27 shows the experimental setup which has been used.

Within the transmitter, a single optical IQM in nested Mach–Zehnder configuration is used. However, choosing this relatively simple optical configuration is accompanied by the need for generating high-quality electrical quaternary signals for driving the modulator. As shown in Fig. 4.27, the in-phase and quadrature driving signals were created by passively combining appropriately levelled binary data signals carrying $2^7 - 1$ and $2^9 - 1$ PRBS sequences. Electrical attenuators were used to adjust the voltage levels and to reduce amplifier interactions. All binary data streams are delayed with respect to each other for decorrelation by several symbol durations. An ECL with a linewidth of 100 kHz is used as transmitter laser. Polarization multiplexing is done by splitting the output light of the IQM with a PBS, delaying one component by several tens of symbol durations and orthogonally adding the two paths by another PBS. The 20 Gbaud quaternary modulator driving signal as well as the optical envelope of the Square 16QAM signal at the modulator output are shown in the left part of Fig. 4.28.

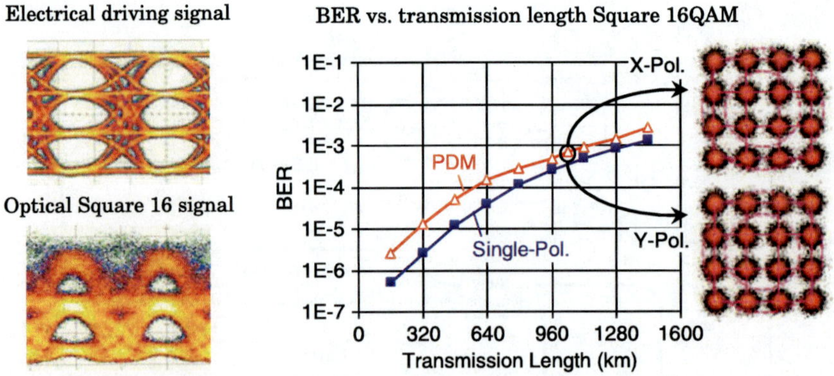

Fig. 4.28 Experimental results obtained in [25]: Quaternary electrical driving signal and optical Square 16QAM transmitter output signal (*left*); BER vs. transmission length for 20 Gbaud Square 16QAM (*right*)

As in the other experiments described before, the transmission link is built as a recirculating fibre loop. In the Square 16QAM experiment performed here, the loop contains one EDFA amplified section of 80 km SSMF. The launch power into the SSMF and the loop unity gain are controlled by optical attenuators. The noise power of the EDFAs (noise figure about 5 dB) outside the channel bandwidth is reduced by optical band-pass filters with 5 nm bandwidth. After transmission over a desired number of cascaded sections, the signal is coupled out to a digital coherent receiver. The received signal is split into its two polarizations, which are then combined with the light of a free running LO in two optical quadrature front-ends. The four photocurrents within the receiver are digitized by synchronous sampling with 2.5 samples per symbol and 8-bit resolution using a commercial digital real-time oscilloscope with 16 GHz bandwidth. Data is then recovered offline by an extensive signal processing block comprising the following sub-functions: Resampling to an integer number of samples per symbol, FFT-based CD compensation, equalization of the system's frequency response, IQ gain equalization, resampling to one sample per symbol at the optimum sample time, frequency estimation and correction based on the phase differential algorithm [38], two-stage phase estimation and correction using a Viterbi–Viterbi algorithm with class partitioning based on the QPSK partitioning scheme [38] and data recovery based on a rectilinear decision grid and Gray-decoding. Bit error counting was performed on averaged blocks of up to 2 million samples corresponding to 1.6 million bits per polarization tributary. The right diagram in Fig. 4.28 illustrates the BER vs. transmission length for single polarization and PDM transmission. Transmission distances of more than 1,300 km and 1,100 km were achieved with a BER smaller than 10^{-3} for single-polarization and PDM, respectively. Received constellation diagrams for both tributaries after PDM transmission over 1,040 km are also depicted in Fig. 4.28.

In recent years, investigation of long-haul transmission systems employing higher-order modulation has become an important field of research. The system experiments described in this section are only a small sample of the whole set of experiments, which have been performed in several research groups in the last years. Various modulation formats known from electrical and wireless transmission have been transmitted over fibre, employing PDM and WDM. For instance, impressive spectral efficiencies of 4.2 bit s^{-1} Hz^{-1} for PDM-8PSK [12], 6.4 bit s^{-1} Hz^{-1} for PDM-Square 16QAM [27, 52] and 11.8 bit s^{-1} Hz^{-1} for PDM-Square 256QAM [28] have been demonstrated, the latter still at a lower baud rate of 4 Gbaud. Moreover, transmission distances are aimed to be increased by employing fibres with lower loss and larger nonlinear effective area, distributed Raman amplification and receiver-sided nonlinear equalization [53]. Using Raman amplified 80 km ultra large area fibre spans, a transmission distance of 1,200 km for 28 Gbaud PDM-Square 16QAM with a spectral efficiency of 4.2 bit s^{-1} Hz^{-1} has recently be achieved in a 10 channel WDM environment [27]. Even 3,123 km could be bridged with 20 Gbaud Square 16QAM for single-channel transmission [26]. Looking at practical systems with higher-order modulation, one of the main challenges is real-time implementation of the digital parts of the transmitter and the receiver. FPGA-based implementations of transmitters and receivers are currently being developed for baud rates up to 32 Gbaud [54].

4.6.3 System Simulations with Nonlinear Phase Shift Compensation

This section presents simulation results obtained in [20] examining the influence of the SPM-induced mean nonlinear phase shift on Star 16QAM signals in optical multi-span transmission systems. Differences between system configurations with optical inline CD compensation and electrical CD compensation at the receiver are pointed out, and possible compensation schemes are discussed. Figure 4.29

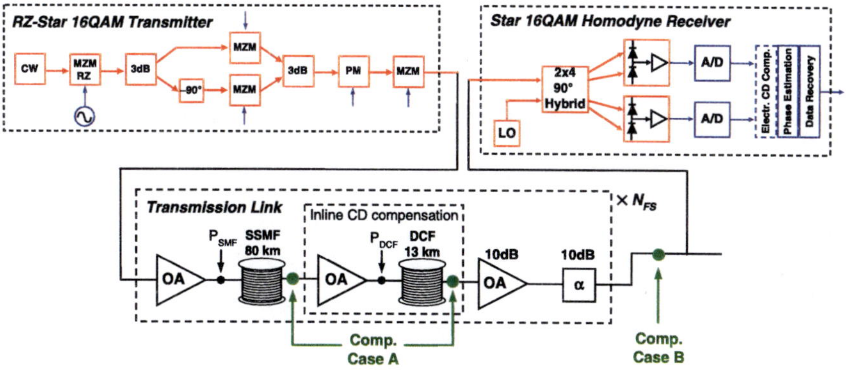

Fig. 4.29 Single-polarization RZ-Star 16QAM multi-span system setup used in [20] to investigate different schemes for compensation of the SPM-induced mean nonlinear phase shift

shows the RZ-Star16QAM multi-span system setup with optical inline CD compensation employed in [20] for simulative investigation of the single-polarization case. The RZ-Star 16QAM signal is generated by using an RZ-Star 16QAM transmitter composed of an IQM followed by a PM and an MZM performing intensity modulation. The transmission link consists of N_{FS} sections, each being composed of 80 km SSMF, 13 km DCF (fully compensating for the CD of the SSMF) and OAs with a noise figure of 5.6 dB. An additional attenuation of 10 dB is used in each section to better emulate the behaviour of an experimental re-circulating fibre loop test bed. At the receiver side, the signal is detected by a digital homodyne receiver, which is performing digital CD compensation (optionally) and phase estimation.

In the case of PDM, the transmitter is doubled and both polarizations are multiplexed in a PBC before the PDM signal is launched into the fibre. Moreover, the receiver frontend is enhanced as shown in Fig. 4.21. SPM-induced signal distortions are different for single-polarization and PDM systems, as illustrated in Fig. 4.28 for RZ-Star 16QAM transmission over a single non-dispersive noise-free transmission section with nonlinear propagation coefficients of the SSMF and DCF given by $\gamma_{\mathrm{SMF}} = 1.43\,\mathrm{W}^{-1}\mathrm{km}^{-1}$ and $\gamma_{\mathrm{DCF}} = 5.84\,\mathrm{W}^{-1}\mathrm{km}^{-1}$, respectively, and for fibre input powers into the SSMF and DCF of 6 dBm and 1 dBm, respectively. It can be observed from the single-polarization case (Fig. 4.30, left) that symbols with different power levels undergo different degrees of phase rotation. In the case of PDM, distortions are different due to nonlinear cross-polarization effects (see Fig. 4.30, right).

As already discussed in Sect. 4.5 for single-span transmission systems, the resulting distortions of the signal constellations must be compensated for by a nonlinear phase shift compensator. Without compensation, attainable transmission lengths for multi-span QAM transmission are strongly limited. This was already demonstrated in the experiments described in Sect. 4.6.1. For comparison, some results for PDM systems at 10 Gbaud determined by computer simulations in [20] are illustrated in the left part of Fig. 4.31. These are valid for optimized fibre input powers and indicate that the transmission distances achieved experimentally for 8PSK in [9] and Star 16QAM in [21] can potentially be increased by further practical system optimization. Nevertheless, attainable transmission distances for RZ-Star 16QAM are limited to about 800 km at BER $= 10^{-3}$ due to the SPM-induced mean nonlinear phase shift and significantly reduced in comparison with RZ-8PSK.

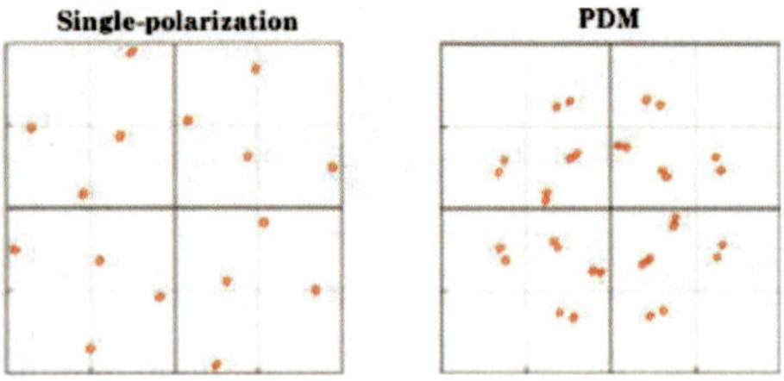

Fig. 4.30 Effect of the SPM-induced mean nonlinear phase shift on RZ-Star 16QAM signals in single polarization systems (*left*) and for PDM (*right*)

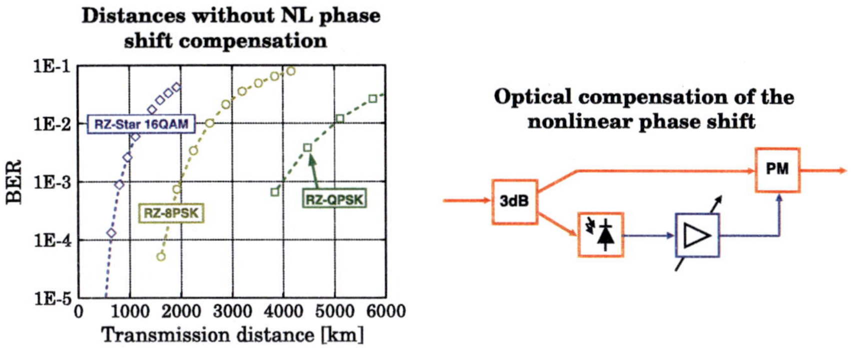

Fig. 4.31 Attainable transmission distances for PDM systems at 10 Gbaud determined by computer simulations in [20] (*left*); simple optical compensator of the nonlinear phase shift (*right*)

The distortions caused by the SPM-induced mean nonlinear phase shift can be partly compensated for using the simple optical compensator depicted in the right part of Fig. 4.31. The optical phase is rotated back by $\varphi(t) = c \cdot \alpha_{NL} \cdot P_{in}(t)$, proportionally to the instantaneous power at the compensator input $P_{in}(t)$. The proportionality factor c depends on the link parameters and the location, where the compensator is placed within the system. In systems with optical inline CD compensation, the compensator could principally be placed behind each fibre in each span (denoted here as "Case A"). Another, more practical option is to place only one compensator directly in front of the coherent receiver (denoted here as "Case B"). Both compensation schemes are indicated in Fig. 4.29. It should be noted that in both cases compensation is not ideal since the intensity shape of the propagating signal changes along the fibre and interaction between CD and SPM prevents a complete compensation of the mean nonlinear phase shift. Moreover, the simple compensator depicted in Fig. 4.31 does not work ideally for PDM where distortions due to cross-polarization effects necessitate a more complex compensator for achieving best performance. Furthermore, the nonlinear phase noise should be considered additionally in practical systems and an appropriate scaling factor α_{NL} should be found to reduce the variance of the nonlinear phase shift [55]. Nevertheless, both compensation schemes presented here lead to a significant transmission reach enhancement. This is illustrated in the case of RZ-Star 16QAM transmission for single-polarization in Fig. 4.32a and for PDM in Fig. 4.32b, assuming optimized launched powers into the SSMF and DCF.

In single-polarization systems, the transmission lengths attainable with RZ-Star 16QAM at 10 Gbaud can be increased from 900 km to about 1,500 km when placing the compensator only at the receiver (Case B) and almost doubled to 1,750 km when using a compensator behind each fibre (Case A). However, compensation with this simple optical compensator does not work equally effective for PDM, where transmission distances are increased to 1,100 km and 1,200 km for Case B with scaling factors of $\alpha_{NL} = 1$ and $\alpha_{NL} = 0.85$, respectively, and to 1,400 km for Case A

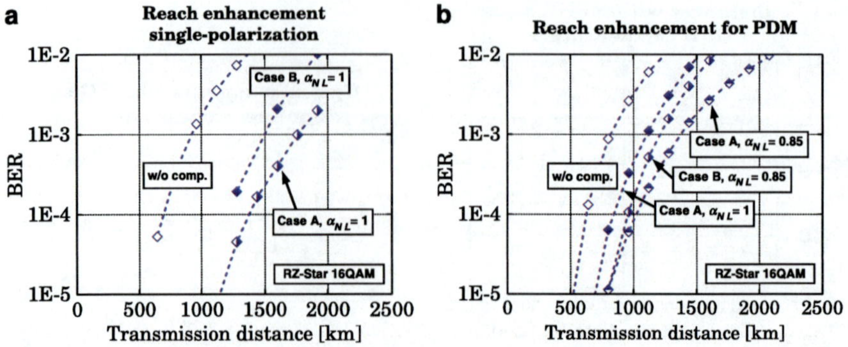

Fig. 4.32 Enhancement of transmission reach for RZ-Star 16QAM at 10 Gbaud for single polarization (**a**) and PDM (**b**) using different schemes of nonlinear phase shift compensation based on the optical compensator depicted in the right part of Fig. 4.31 [20]

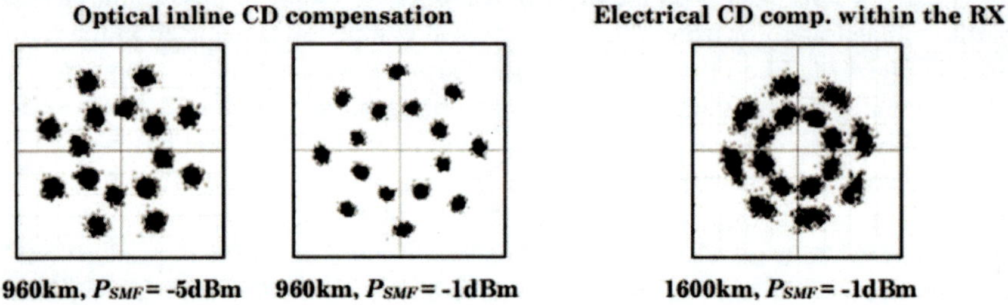

Fig. 4.33 RZ-Star 16QAM constellation diagrams received in systems with optical inline CD compensation and electrical CD compensation at the receiver for selected transmission distances and fibre input powers

(with $\alpha_{NL} = 0.85$). It can be observed from Fig. 4.32b that scaling factors not equal to one are optimal for PDM due to nonlinear cross-polarization effects. Nonlinear phase noise was neglected in these investigations.

When CD is not compensated for periodically in each transmission section but solely by an electrical CD compensation module within the receiver (see Fig. 4.29; the DCF and the OA in front of the DCF are then removed from the transmission link), the difference of the mean nonlinear phase shifts experienced by symbols with different power levels is smaller because the symbol power levels become indistinguishable after certain transmission distances due to CD. The two left plots in Fig. 4.33 show the received constellation diagrams before digital phase estimation within the receiver in systems with inline CD compensation after 960 km for SSMF input powers of −5 dBm (optimal) and −1 dBm, respectively. The mean nonlinear phase shift difference between symbols of the different intensity rings can be clearly seen as the limiting degradation effect. On the contrary, the relative nonlinearity-induced phase difference of both rings is smaller in systems without optical inline

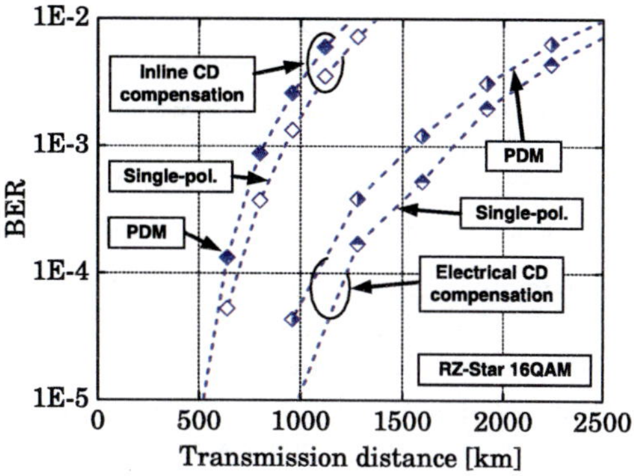

Fig. 4.34 Distances attainable for RZ-Star 16QAM in systems with optical inline CD compensation and electrical CD compensation at the receiver for single-polarization and PDM at 10 Gbaud without nonlinear phase shift compensation, determined in [20]

CD compensation. This becomes apparent from the constellation diagram depicted in the right part of Fig. 4.33 which is received at 1,600 km after electrical CD compensation when an optimal SSMF input power of −1 dBm is chosen.

Due to the reduced relative phase shift between the symbols of both rings, transmission distances of 1,700 km (single-polarization) and 1,500 km (PDM) can be bridged in systems with electrical CD compensation at the receiver even without nonlinear phase shift compensation, as illustrated in Fig. 4.34. These distances are similar to or even greater than in systems with optical inline CD compensation, which additionally use nonlinear phase shift compensation.

Transmission distances in systems with electrical CD compensation at the receiver can be further increased by compensating for the small relative phase difference of both rings observable in the right diagram in Fig. 4.33. Generally, in systems with optical inline or electrical CD compensation, signal distortions through the SPM-induced mean nonlinear phase shift can be reduced additionally by means of adaptive digital equalization within the receiver or by applying pre-distortion techniques at the transmitter side. Both techniques have not been employed in the investigations described here.

4.7 Issues of Future Research

A continuous extension of network capacities is of high relevance, and it can be achieved by applying higher-order modulation formats, which provide a higher spectral efficiency. However, at the same time, it is important that systems maintain an attractive system reach. The reduction of transmission distances aligned with the application of higher-order modulation formats can be mitigated by optimization and high-quality fabrication of the system components required for generating and detecting optical signals with higher-order modulation, and especially by reducing transmission impairments, such as noise and fibre nonlinearities using low-noise optical amplification and Kerr effect compensation. Future research should cover the following areas:

Transmission distances achievable with higher-order modulation formats: Analysis of multi-span fibre transmission systems with higher-order modulation is still at an early stage. Further investigations are here indispensable. Link configurations must be optimized for optimal fibre input powers and dispersion maps in systems with optical inline CD compensation. Moreover, a key issue is the development of techniques, which will efficiently compensate for fibre nonlinearities.

Behaviour of higher-order modulation formats in WDM systems: The transmission lengths and channel spacings achievable with higher-order modulation formats in WDM systems are a matter of particular interest. Attention must be paid to channel filtering, crosstalk and inter-channel nonlinearities. The channel spacing attainable depends on the signal bandwidth and on how narrowly optical signals can be filtered. Narrower channel spacing induces higher penalties due to cross-phase modulation and four-wave mixing. Thus, the system penalty induced by narrow optical filtering and the impact of linear and nonlinear inter-channel crosstalk must be determined for the various modulation formats.

Capacity, spectral efficiency and capacity-distance product attainable in WDM systems: If the fibre were linear and there were no system degradation through fibre nonlinearities, spectral efficiency could theoretically be increased to infinity by applying modulation formats of higher and higher order. Thereby, the expected increase of spectral efficiency would be about the ratio of the data rate to the symbol rate. More demanding noise requirements of the higher-order formats could then be met by simply launching more and more power into the fibre. However, in real transmission systems, performance degrades due to fibre nonlinearities when the fibre input power is increased. It is an open question whether the capacity-distance product can be improved through the application of higher-order modulation formats as a consequence of the reduced transmission distances. For instance, the capacity-distance product of 16.58 Pbit s^{-1} km reported so far with Square 16QAM [52] is more than six times smaller than the record product of 111.6 Pbit s^{-1} km obtained with QPSK [56]. However, there is potential for further optimization of systems applying higher-order modulation.

Practical system optimization: One key challenge on the way towards widespread deployment of systems using higher-order modulation is the optimization of system components at low cost. At the transmitter, distortions of the electrical driving signals accumulate in multiple modulator stages and can lead to implementation penalties. Moreover, multi-level electrical driving signals are not being generated easily. Thus, to realize transmitters performing close to the theoretical performance limits, high-speed integrated optical modulator structures and fast analogue-to-digital converters for generating multi-level driving signals of high quality are currently being developed. At the receiver end, developments aim at integrating the whole optical receiver frontend in a single chip, and to exploit DSP technology to compensate for performance degradation effects and facilitate the recovery of information.

Utilization of polarization: Polarization information provides an additional degree of freedom in optical fibre transmission systems and by utilizing PDM the spectral efficiency of any modulation format can be doubled. The extent to which crosstalk between the multiplexed channels degrades the performance of systems applying higher-order modulation will be a topic of future research. In addition, modulation formats exploiting all the parameters of the electrical field and encoding information additionally into the polarization are available for optical transmission and should also be considered in future investigations.

References

1. M. Rohde, C. Caspar, N. Heimes, M. Konitzer, E.J. Bachus, N. Hanik, Electron. Lett. **36**, 1483–1484 (1999)
2. S. Walklin, J. Conradi, J. Lightwave Technol. **17**(11), 2235–2248 (1999)
3. J. Zhao, L. Huo, C. Chan, L. Chen, C. Lin, Analytical investigation of optimization, performance bound, and chromatic dispersion tolerance of 4-amplitude-shifted-keying format, in *Proceedings of OFC-2006*, p. JThB15, 2006
4. C. Wree, J. Leibrich, W. Rosenkranz, Differential quadrature phase-shift keying for cost-effective doubling of the capacity in existing WDM systems, in *Proceedings of the 4th Conference on Photonic Networks*, pp. 161–168, 2003
5. M. Ohm, Optical 8-DPSK and receiver with direct detection and multilevel electrical signals, IEEE/LEOS workshop on advanced modulation formats, pp. 45–46, 2004
6. H. Yoon, D. Lee, N. Park, Opt. Express **13**(2), 371–376 (2005)
7. M. Serbay, C. Wree, W. Rosenkranz, Experimental investigation of RZ-8DPSK at 3 (10.7Gb/s, The 18th annual meeting of the IEEE lasers and electro-optics society, Sydney, p. WE3, 2005
8. S. Tsukamoto, K. Katoh, K. Kikuchi, Coherent demodulation of optical 8-phase shift-keying signals using homodyne detection and digital signal processing, in *Proceedings of OFC-2006*, p. OThR5, 2006
9. M. Seimetz, L. Molle, D.D. Gross, B. Auth, R. Freund, Coherent RZ-8PSK transmission at 30Gbit/s over 1200km employing homodyne detection with digital carrier phase estimation, in *Proceedings of ECOC-2007*, p. We834, 2007
10. R. Freund, D.D. Groß, M. Seimetz, L. Molle, C. Caspar, 30 Gbit/s RZ-8-PSK transmission over 2800 km standard single mode fibre without inline dispersion compensation, in *Proceedings of OFC-2008*, p. OMI5, 2008
11. X. Zhou, J. Yu, D. Qian, T. Wang, G. Zhang, P. Magil, 8 (114Gb/s, 25-GHz-spaced, PolMux-RZ-8PSK transmission over 640km of SSMF employing digital coherent detection and EDFA-only amplification, in *Proceedings of OFC-2008*, p. PDP1, 2008
12. J. Yu, X. Zhou, M.F. Huang, Y. Shao, D. Qian, T. Wang, M. Cvijetic, P. Magill, L. Nelson, M. Birk, S. Ten, H.B. Matthew, S.K. Mishra, 17 Tb/s (161 × 114 Gb/s) PolMux-RZ-8PSK transmission over 662 km of ultra-low loss fiber using C-band EDFA amplification and digital coherent detection, in *Proceedings of ECOC-2008*, p. Th3E2, 2008
13. M. Seimetz, M. Noelle, E. Patzak, J. Lightwave Technol. **25**(6), 1515–1530 (2007)
14. M. Seimetz, Optical fiber transmission systems with high-order phase and quadrature amplitude modulation, Dissertation, Technical University of Berlin, Germany, 2008
15. C.R. Cahn, IRE Trans. Commun. **CS-8**, 150–155 (1960)
16. J.C. Hancock, R.W. Lucky, IRE Trans. Commun. **CS-8**, 232–237 (1960)
17. M. Ohm, J. Speidel, Receiver sensitivity, chromatic dispersion tolerance and optimal receiver bandwidths for 40 Gbit/s 8-level optical ASK-DQPSK and optical 8-DPSK, in *Proceedings of 6th Conference on Photonic Networks*, Leipzig, Germany, pp. 211–217, 2005

18. K. Sekine, N. Kikuchi, S. Sasaki, S. Hayase, C. Hasegawa, T. Sugawara, Proposal and demonstration of 10-Gsymbol/sec 16-ary (40 Gbit/s) optical modulation/demodulation scheme, in *Proceedings of ECOC-2004*, p. We345, 2004
19. M. Serbay, T. Tokle, P. Jeppesen, W. Rosenkranz, 42.8 Gbit/s, 4 Bits per symbol 16-ary inverse-RZ-QASK-DQPSK transmission experiment without Polmux, in *Proceedings of OFC-2007*, p. OThL2, 2007
20. M. Seimetz, System degradation by the SPM-induced mean nonlinear phase shift in optical QAM transmission, in *Proceedings of OFC-2009*, p. JWA38, 2009
21. M. Seimetz, L. Molle, M. Gruner, R. Freund, Transmission reach attainable for single-polarization and PolMux coherent star 16QAM systems in comparison to 8PSK and QPSK at 10Gbaud, in *Proceedings of OFC-2009*, p. OTuN2, 2009
22. X. Zhou, J. Yu, M.F. Huang, Y. Shao, T. Wang, P. Magill, M. Cvijetic, L. Nelson, M. Birk, G. Zhang, S. Ten. H.B. Matthew, S.K. Mishra, 32Tb/s (320 (114 Gb/s) PDM-RZ-8QAM transmission over 580 km of SMF-28 ultra-low-loss fiber, in *Proceedings of OFC-2009*, p. PDPB4, 2009
23. C.N. Campopiano, B.G. Glazer, IRE Trans. Commun. **CS-10**, 90–95 (1962)
24. N. Kikuchi, S. Sasaki, Optical dispersion-compensation free incoherent multilevel signal transmission over single-mode fiber with digital pre-distortion and phase pre-integration techniques, in *Proceedings of ECOC-2008*, Tu1E2, 2008
25. L. Molle, M. Seimetz, D.D. Gross, R. Freund, M. Rohde, Polarization multiplexed 20 Gbaud Square 16QAM long-haul transmission over 1120 km using EDFA amplification, in *Proceedings of ECOC-2009*, p. 8.4.4, 2009
26. T. Kobayashi, A. Sano, H. Masuda, K. Ishihara, E. Yoshida, Y. Miyamoto, H. Yamazaki, T. Yamada, 160-Gb/s polarization-multiplexed 16-QAM long-haul transmission over 3,123 km using digital coherent receiver with digital PLL based frequency offset compensator, in *Proceedings of OFC-2010*, p. OTuD1, 2010
27. A.H. Gnauck, P.J. Winzer, S. Chandrasekhar, X. Liu, B. Zhu, D.W. Peckham, 10 (224-Gb/s WDM transmission of 28-Gbaud PDM 16-QAM On A 50-GHz grid over 1,200 Km of fiber, in *Proceedings of OFC-2010*, p. PDPB8, 2010
28. M. Nakazawa, S. Okamoto, T. Omiya, K. Kasai, M. Yoshida, 256 QAM (64 Gbit/s) coherent optical transmission over 160 km with an optical bandwidth of 5.4 GHz, in *Proceedings of OFC-2010*, p. OMJ5, 2010
29. K.P. Ho, H.W. Cuei, J. Lightwave Technol. **23**(2), 764–770 (2005)
30. M. Seimetz, *High-Order Modulation for Optical Fiber Transmission*, Springer Series in Optical Sciences, vol. 143, ISBN 978–3–540–93770–8 (Springer, Berlin, 2009)
31. M. Seimetz, Optical receiver for reception of M-ary star-shaped quadrature amplitude modulation with differentially encoded phases and its application, Patent DE 10 2006 030 915.4, German Patent and Trade Mark Office, 2006
32. M. Kuschnerov, F.N. Hauske, K. Piyawanno, B. Spinnler, E.D. Schmidt, B. Lankl, Joint equalization and timing recovery for coherent fiber optic receivers, in *Proceedings of ECOC-2008*, p. Mo3D3, 2008
33. S.J. Savory, Compensation of fibre impairments in digital coherent systems, in *Proceedings of ECOC-2008*, p. Mo3D1, 2008
34. F.M. Gardner, IEEE Trans. Commun. **COM-34**(5), 423–429 (1986)
35. M. Oerder, H. Meyr, IEEE Trans. Commun. **36**(5), 605–612 (1988)
36. S.J. Savory, G. Gavioli, R.I. Killey, P. Bayvel, Transmission of 42.8 Gbit/s polarization multiplexed NRZ-QPSK over 6400 km of standard fiber with no optical dispersion compensation, in *Proceedings of OFC-2007*, p. OTuA1, 2007
37. J.G. Proakis, *Digital Communications*, ISBN 978–0071263788 (McGraw-Hill, NY, 2008)
38. F. Rice, Bounds and Algorithms for Carrier Frequency and Phase Estimation, Dissertation, University of South Australia, 2002
39. M. Kuschnerov, D. van den Borne, K. Piyawanno, F.N. Hauske, C.R.S. Fludger, T. Duthel, T. Wuth, J.C. Geyer, C. Schulien, B. Spinnler, E.-D. Schmidt, B. Lankl, Joint-polarization carrier phase estimation for XPM-limited coherent polarization-multiplexed QPSK transmission with OOK-neighbors, in *Proceedings of ECOC-2008*, p. Mo4D2, 2008

40. R. Noé, IEEE Photon. Technol. Lett. **17**(4), 887–889 (2005)
41. M. Seimetz, Laser linewidth limitations for optical systems with high-order modulation employing feed forward digital carrier phase estimation, in *Proceedings of OFC-2008*, p. OTuM2, 2008
42. H. Louchet, K. Kuzmin, A. Richter, Improved DSP algorithms for coherent 16-QAM transmission, in *Proceedings of ECOC-2008*, p. Tu1E6, 2008
43. T. Pfau, S. Hoffmann, R. Noé, J. Lightwave Technol. **27**(8), 989–999 (2009)
44. M. Seimetz, C.-M. Weinert, J. Lightwave Technol. **24**(3), 1317–1322 (2006)
45. D. Hoffmann, H. Heidrich, G. Wenke, R. Langenhorst, E. Dietrich J. Lightwave Technol. **6**(5), 794–798 (1989)
46. A. Kaplan, K. Achiam, $LiNbO_3$ integrated optical QPSK modulator and coherent receiver, in *Proceedings of ECIO-2003*, pp. 79–82, 2003
47. I. Fatadin, S.J. Savory, D. Ives, IEEE Photon. Technol. Lett. **20**(20), 1733–1735 (2008)
48. W.R. Leeb, Electron. Lett. **26**, 1431–1432 (1990)
49. R. Langenhorst, Optische Koppelelemente für den kohärent optischen Mehrtorempfänger, Dissertation, Technical University of Berlin, Germany, 1992
50. G.P. Agrawal, *Nonlinear Fiber Optics*, ISBN 978–0123695161 (Academic, NY, 2006)
51. R. Freund, H. Louchet, M. Gruner, L. Molle, M. Seimetz, A. Richter, 80 Gbit/s/λ polarization multiplexed star-16QAM WDM transmission over 720 km SSMF with electronic distortion equalization, in *Proceedings of Optoelectronics and Communications Conference*, OECC-2009, Hong Kong, 2009
52. A. Sano, H. Masuda, T. Kobayashi, M. Fujiwara, K. Horikoshi, E. Yoshida, Y. Miyamoto, M. Matsui, M. Mizoguchi, H. Yamazaki, Y. Sakamaki, H. Ishii, 69.1-Tb/s (432 (171-Gb/s) C- and extended L-band transmission over 240 Km using PDM-16-QAM modulation and digital coherent detection, in *Proceedings OFC-2010*, p. PDPB7, 2010
53. S. Makovejs, D.S. Millar, V. Mikhailov, G. Gavioli, R.I. Killey, S.J. Savory, P. Bayvel, Experimental investigation of PDM-QAM16 transmission at 112 Gbit/s over 2400 km, in *Proceedings of OFC-2010*, p. OMJ6, 2010
54. J. Hilt, M. Nölle, L. Molle, M. Seimetz, R. Freund, 32 Gbaud real-time FPGA-based multi-format transmitter for generation of higher-order modulation formats, 9th Conference on Optical Internet (COIN 2010), Korea, 2010
55. K.P. Ho, *Phase Modulated Optical Communication Systems*, ISBN 0–387–24392–5 (Springer, Berlin, 2009)
56. M. Salsi, H. Mardoyan, P. Tran, C. Koebele, E. Dutisseuil, G. Charlet, S. Bigo, 155 (100 Gbit/s coherent PDM-QPSK transmission over 7,200 km, in *Proceedings of ECOC-2009*, p. PD2.5, 2009

Chapter 5
Power-Efficient Modulation Schemes

Magnus Karlsson and Erik Agrell

5.1 Introduction

Coherent optical fiber communications had a brief period of popularity in the early 1990s, mainly because the optical links of that day were significantly power limited. Coherent detection provided a possibility of optically amplifying the signal to a power level that, after photodetection, made the thermal noise negligible. Two things, however, caused those coherent systems to be abandoned. The first was the sheer technical difficulties: a coherent receiver requires a local oscillator laser that is to be phase- and polarization-locked to the received signal. This gave rise to significant technical obstacles, and only a few limited and expensive coherent receiver solutions were demonstrated [17,27]. The second was the development of the Erbium-doper fiber amplifier (EDFA) that provided an elegant and practical solution to the problem of the thermal noise. By 1995, the EDFA was a commodity in fiber communication systems, simple on-off keying modulation worked well enough, and coherent communication was forgotten.

However, coherent transmission systems got renewed attention around 2005 [12, 34]. This time the motivation was entirely different. A coherent receiver gives access to both the optical phase and the amplitude, which provides two important benefits; (1) advanced multilevel modulation formats can be used, which can improve the spectral efficiency; and (2) electronic distortion mitigation can be used, as the optical field is directly mapped to the electrical signal. Moreover, the practical problems with the coherent detection could now be solved by performing the phase- and polarization tracking by fast digital signal processing. This enabled a third significant benefit: (3) a practical use of both polarization components for data

M. Karlsson (✉)
Photonics Laboratory, Department of Microtechnology and Nanoscience,
Chalmers University of Technology, SE-412 96 Göteborg, Sweden
e-mail: magnus.karlsson@chalmers.se

E. Agrell
Communication Systems Group, Department of Signals and Systems,
Chalmers University of Technology, SE-412 96 Göteborg, Sweden
e-mail: agrell@chalmers.se

S. Kumar (ed.), *Impact of Nonlinearities on Fiber Optic Communications*, Optical and Fiber Communications Reports 7, DOI 10.1007/978-1-4419-8139-4_5,
© Springer Science+Business Media, LLC 2011

transmission. By 2008, a landmark development was reported by Sun et al. [51]: the first 10 Gbaud coherent transmission system, with a working coherent receiver based on digital signal processing. In this work, we will investigate modulation formats for such links, which have the peculiarity that the signaling space is four-dimensional.

5.1.1 Optical Coherent Modulation: Background

An electromagnetic carrier wave offers essentially four degrees of freedom (DOFs) in which data can be independently modulated; the I and Q quadratures (or the real and imaginary parts) of each of the x and y polarization components. These four DOFs can also be interpreted as the amplitude, absolute phase, and polarization state of the wave. We will refer to the number of available DOFs in a transmission system as the *dimensionality*, N, of the constellation space. Binary phase-shift keying (BPSK) requires a one-dimensional constellation space and its higher-dimensional generalizations, quaternary phase-shift keying (QPSK) and dual-polarization QPSK (DP-QPSK), have $N = 2$ and $N = 4$, respectively. These constellations form an N-dimensional cube in their respective constellation spaces.

The polarization state is used for information transmission in fixed microwave communication links, e.g., the Ericsson Mini-Link system, and similar methods have also been considered for mobile radio communications, although impairments such as fading and polarization interference pose severe difficulties in the latter case [58].

In coherent optical systems, however, all four DOFs can be readily detected and used for signaling. And indeed, in recent coherent transmission research, this is precisely what is done: a binary modulation in each of the four quadratures, enabling four parallel binary data streams that produce a signal with a data rate that is four times the symbol rate [13, 40, 51, 54]. This modulation format is often referred to as DP-QPSK. It is a 16-level modulation format formed by the vertices of a cube in a four-dimensional (4d) constellation space.

Coherent fiber systems using optical amplifiers can, to a good approximation, be modeled as additive white Gaussian noise (AWGN) channels [23–26, 30], which is important since all fundamental theorems and results of AWGN channels will apply [43]. To compare the performance of different modulation formats, we will use the receiver *sensitivity*, which is defined as the signal-to-noise ratio (SNR) required to reach a bit error rate (BER) or symbol error rate (SER) of 10^{-9}, or, which is increasingly common, 10^{-3}. BPSK is often chosen as a reference format, and is (at least in the optical research community) often believed to have the best sensitivity among all possible modulation formats at a given bit rate. Since the DP-QPSK format is four parallel and independent BPSK channels, its sensitivity is the same as that for BPSK. However, as we will show in this chapter, thanks to the geometrical properties of four-dimensional constellation space, there exist modulation formats that have better sensitivities than BPSK [1, 28]. The improvement comes from jointly optimizing the constellations over all four DOFs, rather than applying independent modulation in each polarization.

In this paper, we will analyze some of those formats, and quantify their sensitivities within the AWGN model. Besides being of fundamental interest, such power-efficient modulation formats may be of practical relevance as they provide means to reduce nonlinear fiber transmission impairments [28], by allowing reduced transmitter power for the same BER. We will here extend previous studies of modulation formats based on average-energy minimization to peak-energy minimization. As will be discussed in Sect. 5.5, the peak energy may be more critical than the average in systems limited by fiber nonlinearities, such as self- and cross-phase modulation (SPM, XPM). We will give several examples of optimized constellations and present their coordinate representations.

Error correction coding is a way of increasing the dimensionality by introducing more DOFs in the transmitted signal space, however at the price of increased system complexity. In this work, we will limit the discussion to the constellation space of the uncoded modulated signal, which is four-dimensional.

Modulation in a four-dimensional constellation space has been investigated previously in the communication theory literature, e.g., [8, 32, 42, 53, 56, 62]. In [56], constellations with more than 12 levels were analyzed in terms of SER. Some simpler formats, including 5-, 8- and 16-level systems, were analyzed in [62]. For reasons that will be apparent later on in this article, the 5-, 8-, 16-, and 24-level schemes are of most interest.

In the optical communication context, 4d modulation was investigated in the early 1990s [5–7, 16], when coherent systems were popular. These papers demonstrated theoretically how optical transmission systems could benefit from 4d modulation techniques, by showing how transmitters and receivers could be realized. Some fundamental sensitivity limits were given in [5, 6]. However, it is not entirely clear from these works under what circumstances the constellations were optimized (for example, under an average or maximum symbol energy constraint). Nor do they point out that sensitivity improvements over BPSK could be achieved, which in our opinion is a most important, and not widely known, observation.

We will give a number of examples of modulation formats (e.g., based on 5, 8, 16, and 24 levels) that have *improved* receiver sensitivities over BPSK and DP-QPSK. Two of these (the 8- and 24-level formats) have a reasonable complexity and, contrary to the 5-level system, the transmitter and the bit-to-symbol mapping problem can be solved without too much loss of performance, so we will describe those modulation formats and their implementations in more detail. It should be noted that we are not the first to point out that multilevel formats with sensitivities better than BPSK exist. Rather, their asymptotic sensitivity gains were originally given in [8, 42, 53]. However, that context was different, as they considered increasing the dimensionality of the signal by using two carrier waves, rather than the two polarization components that can be used in fiber communications.

This chapter is structured as follows: In Sect. 5.2, we lay out the basic definitions and notation, discuss the relation between polarization states and signals in four-dimensional space, and explain the relation between dense sphere packings and power-efficient constellations. In Sect. 5.3, we review sphere packing in two and four dimensions, and present two different optimization principles (minimization of average and maximum symbol energy, respectively) that we use. Then we present optimum constellations and compare them in terms of sensitivity and spectral efficiency. In Sect. 5.4, we compute and discuss symbol- and bit-error rates for some of the most promising constellations. In Sect. 5.5, we present fundamental sensitivity limits for the coherent (four-dimensional) channel, and discuss the influence of fiber nonlinearities on the results. We also compare and discuss the two families of optimal constellations we have found in more detail. Finally in Sect. 5.6, we summarize this chapter.

5.2 Definitions and System Model

This section describes the basic properties of the electromagnetic field and how we interpret it as a four-dimensional signal. Then we will go on to describe how this relates to digital signal transmission, and finally show how sphere packings can be used to find power-efficient formats. Much of the material in this section is standard textbook material, but as it is scattered over different texts we wish to include it for completeness.

5.2.1 The Four-Dimensional Optical Signal

As mentioned in the introduction, the electromagnetic field has two quadratures in two polarization components, thus in total four DOFs, which span a 4d signal space. The electric field amplitude of the optical wave can be written as a complex, 2-component vector

$$\mathbf{E} = \begin{pmatrix} E_{x,r} + iE_{x,i} \\ E_{y,r} + iE_{y,i} \end{pmatrix} = \begin{pmatrix} |E_x| \exp(i\varphi_x) \\ |E_y| \exp(i\varphi_y) \end{pmatrix}, \tag{5.1}$$

where indices x and y denote the polarization components, and r and i the real and imaginary parts, resp., of the field. The coordinate directions x and y are orthogonal to the propagation direction z. The phases φ_x and φ_y are by definition in the interval $(-\pi, \pi]$.

The electric field may be equivalently described in terms of its phase, amplitude and polarization state (the latter being the *relative* phase and amplitude between the x and y field components) as

$$\mathbf{E} = \|\mathbf{E}\| \exp(i\varphi_a)J = \|\mathbf{E}\| \exp(i\varphi_a) \begin{pmatrix} \cos\theta \exp(i\varphi_r) \\ \sin\theta \exp(-i\varphi_r) \end{pmatrix}, \tag{5.2}$$

where $\|\mathbf{E}\|^2 = |E_x|^2 + |E_y|^2$ and $\theta = \sin^{-1}(|E_y|/\|\mathbf{E}\|)$. J denotes the Jones vector, which is usually normalized to unity, i.e., $J^+J = |J|^2 = 1$. Note the distinction between the absolute phase $\varphi_a = (\varphi_x + \varphi_y)/2$ of the field and the relative phase $\varphi_r = (\varphi_x - \varphi_y)/2$ between the field vector components. The relative phase $\varphi_r \in (-\pi, \pi]$ describes the *ellipticity* of the polarization state, with the special cases $\varphi_r = 0, \pm\pi/2, \pi$ for linear polarization and $\varphi_r = \pm\pi/4, \pm 3\pi/4$ for circular polarization, and all other cases are called elliptical states of polarization. The angle $\theta \in [0, \pi/2]$ is usually called the *azimuth* as it describes the orientation in the xy plane of the linear polarization states, or, more generally, the major axis of the polarization ellipse.

A final way of expressing the signal is as a four-dimensional vector $\mathbf{s}$ with real components

$$\mathbf{s} = \begin{pmatrix} E_{x,r} \\ E_{x,i} \\ E_{y,r} \\ E_{y,i} \end{pmatrix} = \begin{pmatrix} \|\mathbf{E}\| \cos\varphi_x \sin\theta \\ \|\mathbf{E}\| \sin\varphi_x \sin\theta \\ \|\mathbf{E}\| \cos\varphi_y \cos\theta \\ \|\mathbf{E}\| \sin\varphi_y \cos\theta \end{pmatrix}. \tag{5.3}$$

The transmitted optical power is $P = \|\mathbf{s}\|^2 = \|\mathbf{E}\|^2 = E_{x,r}^2 + E_{x,i}^2 + E_{y,r}^2 + E_{y,i}^2$. Note that this four-dimensional vector should not be confused with the Stokes vector description of polarization states, which is defined in a completely different way and proportional to the intensity rather than being linear in the field. The three-dimensional Stokes space was used as a signal space for so-called polarization shift keying modulation in the 1990s [4]. However, the lack of an absolute phase description makes constellation points with different absolute phase but same polarization coincide in Stokes space, and it is therefore less useful as a signal space in a coherent communication system with additive noise (see Sect. 5.2.2). Yet, the Stokes space description of the optical field is useful when discussing the polarization properties of the different modulation formats.

As an example, we consider the DP-QPSK modulation format, which uses independent QPSK modulation in both polarization components, i.e., $\varphi_x = m\pi/4$ and $\varphi_y = n\pi/4$ where $m, n \in \{-3, -1, 1, 3\}$, while $|E_x|$ and $|E_y|$ remain the same for all phases. In the notation of (5.2), the absolute and relative phases φ_a and φ_r are both multiples of $\pi/4$. The 16 possible combinations are schematically shown in Fig. 5.1, along with the polarization states they correspond to. Thus, the polarization of DP-QPSK varies between four states; linear in the +45° direction for $\varphi_r = 0$, linear in the –45° direction for $\varphi_r = \pm\pi/2$, left-hand circular (LHC) for $\varphi_r = \pi/4$ or $\varphi_r = -3\pi/4$, and right-hand circular (RHC) for $\varphi_r = -\pi/4$ or $\varphi_r = 3\pi/4$.

5.2.2 Digital Transmission Over a Noisy Channel

In general, all entities in (5.3) vary continuously with time. For the purpose of digital communications, $\mathbf{s}(t)$ is designed to transmit a sequence of information symbols $(\mathbf{s}_0, \mathbf{s}_1, \mathbf{s}_2, \ldots)$, one symbol every T seconds. The symbol $\mathbf{s}_n$ is taken from a finite set, or *constellation*, $\mathscr{C} = \{\mathbf{c}_1, \ldots, \mathbf{c}_M\}$ of N-dimensional vectors. We assume all constellation vectors to be equally likely. Thus, $\log_2 M$ information bits are transmitted every T seconds, yielding an information bit rate of $R_{\mathrm{B}} = \log_2 M/T$ bits/s.

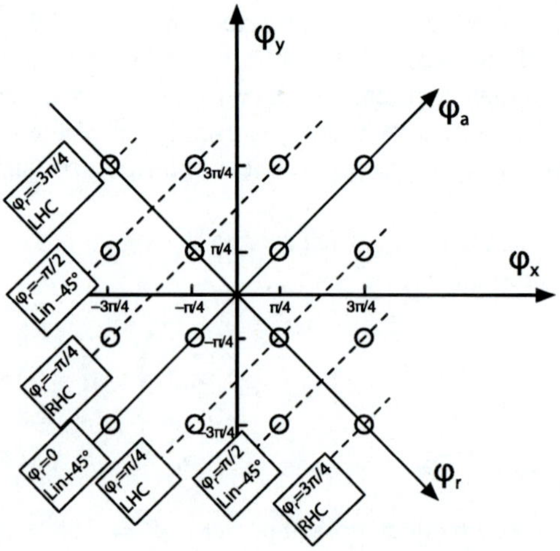

Fig. 5.1 The phase values used for DP-QPSK modulation. The diagonal axes show the φ_r and φ_a phases. For the φ_r levels, the corresponding states of polarization are denoted as linear ±45°, LHC, or RHC

With linear modulation, $\mathbf{s}(t)$ is generated as

$$\mathbf{s}(t) = \sum_n \mathbf{s}_n p(t - nT), \tag{5.4}$$

where $p(t)$ is a pulse-shaping function. It may, e.g., be taken as a rectangular pulse of duration T to provide perfect constant-intensity modulation, or a narrower function for RZ pulse shaping. Without loss of generality, we normalize $p(t)$ to unit energy, so that $\int_{-\infty}^{\infty} p^2(t)\mathrm{d}t = 1$.

The signal $\mathbf{s}(t)$ is now transmitted over a noisy channel. In the coherent optical systems of today, the dominating noise source is usually either amplified spontaneous emission (ASE) noise from in-line optical amplifiers or shot noise from the local oscillator in the receiver [23, 24, 31]. Both these noise sources are accurately modeled by the AWGN channel, for which the received N-dimensional signal is $\mathbf{r}(t) = \mathbf{s}(t) + \mathbf{z}(t)$, where $\mathbf{z}(t)$ is a vector of N independent, white, and Gaussian noise processes, each with a double-sided spectral density of $N_0/2$ (which is the standard notation in communications literature).

The purpose of the receiver is to recover the sequence $(\mathbf{s}_0, \mathbf{s}_1, \ldots)$ as reliably as possible, given an observation of the signal $\mathbf{r}(t)$. It is well known (see [3, Sect. 2.6] or [39, Sect. 5.1]) that in the absence of inter-symbol interference, the optimal receiver operates by filtering $\mathbf{r}(t)$ and sampling, creating a sequence of so-called *received vectors* $(\mathbf{r}_0, \mathbf{r}_1, \ldots)$, where

$$\mathbf{r}_n = \int_{-\infty}^{\infty} \mathbf{r}(t + nT)p(t)\mathrm{d}t. \tag{5.5}$$

It can be shown that $\mathbf{r}_n = \mathbf{s}_n + \mathbf{z}_n$, where $\mathbf{z}_n$ are independent, Gaussian random vectors with variance $N_0/2$ in each dimension. This equation is a *discrete-time channel model,* which includes modulation, optical transmission, and demodulation. It should not be confused with its continuous-time counterpart $\mathbf{r}(t) = \mathbf{s}(t) + \mathbf{z}(t)$. For instance, the average of the squared field amplitude $\|\mathbf{s}(t)\|^2$ is the optical transmitted power P, while the average of $\|\mathbf{s}_n\|^2$ equals the *average energy per symbol*

$$E_s = \frac{1}{M} \sum_{k=1}^{M} \|\mathbf{c}_k\|^2 = PT \tag{5.6}$$

assuming that each symbol in the set is transmitted with the same probability. We also find it useful to define the *maximum energy per symbol* as

$$E_{s,\max} = \max\left\{\|\mathbf{c}_1\|^2, \ldots, \|\mathbf{c}_M\|^2\right\}. \tag{5.7}$$

Similarly, while the optical noise power $\|\mathbf{z}(t)\|^2$ is (in theory) infinite, the discrete-time noise energy $\|\mathbf{z}_n\|^2$ is finite and equals on average $NN_0/2$, because each of the N components of $\mathbf{z}_n$ has variance $N_0/2$.

The spectral efficiency, SE, is generally defined either as the information bitrate per bandwidth (in bits/s/Hz) or as information bits per channel use, where a "channel use" refers to the transmission of two (or sometimes one) real vectors over the discrete-time channel, i.e., to two (or one) dimensions in signal space [3, p. 219]. We follow the latter approach, defining the spectral efficiency as the number of transmitted bits per polarization, where each polarization represents a dimension pair. Formally,

$$\mathrm{SE} = \frac{\log_2 M}{N/2} \ [\text{bits}/(\text{symbol} \cdot \text{polarization})]. \tag{5.8}$$

With this definition, BPSK, QPSK, and DP-QPSK all have the same spectral efficiency of 2 bits/sym/pol, which actually makes sense, since BPSK uses only one quadrature, i.e., 1/2 polarization.

5.2.3 Symbol Error Rates and Sphere Packing

If the pulse $p(t)$ is suitably chosen, there is no inter-symbol interference and $\mathbf{s}_n$ can be optimally estimated from the single received vector $\mathbf{r}_n$. The AWGN model means that the received vector $\mathbf{r}_n$ has an isotropic distribution around $\mathbf{s}_n$ in an N-dimensional space, and for a maximum likelihood receiver, the symbol decision is based on which signal in the constellation set is closest (in the Euclidian sense) to the received vector. To put this on more solid mathematical grounds, consider the constellation $\mathscr{C} = \{\mathbf{c}_1, \ldots, \mathbf{c}_M\}$ of M signaling points, or symbols. Each symbol $\mathbf{c}_k$ is surrounded by a decision region, also known as a *Voronoi region,* defined as all points in the N-dimensional Euclidean space that are closer to $\mathbf{c}_k$ than to any

$\mathbf{c}_j \neq \mathbf{c}_k$. The probability of receiving symbol $\mathbf{c}_k$ in error is then the probability for a Gaussian variable centered at $\mathbf{c}_k$ to be outside its Voronoi region. For constellations in many dimensions, this probability in general cannot be calculated exactly, since the Voronoi regions may have very complex shapes.

However, a simple, yet useful, approximation to the SER is the *union bound*. It builds on the fact that the *pairwise* error probability of confusing the symbols $\mathbf{c}_k$ and $\mathbf{c}_j$ is easy to calculate – it is simply a function of the distance $d_{kj} = \|\mathbf{c}_k - \mathbf{c}_j\|$. The overall SER of a symbol $\mathbf{c}_k$ is then upperbounded by the sum of these pairwise error probabilities over all $j \neq k$. Finally, averaging over all equiprobable symbols $\mathbf{c}_k$, the union bound on the SER can be expressed as [3, p. 191]

$$\text{SER} \leq \frac{1}{M} \sum_{k=1}^{M} \sum_{\substack{j=1 \\ j \neq k}}^{M} \frac{1}{2} \operatorname{erfc}\left(\frac{d_{kj}}{2\sqrt{N_0}} \right), \tag{5.9}$$

where erfc denotes the complementary error function. This bound is in most cases sufficiently accurate at large SNR, and it approaches the true SER asymptotically. We will show numerically later on that it, in our cases, agrees well with exact results for SERs less than 10^{-3}.

We may see directly from (5.9) that in the limit of high SNR (and low SER), the errors will be dominated by the signals in the set that are closest together, i.e., the term containing $\operatorname{erfc}(d_{\min}/2\sqrt{N_0})$, where $d_{\min} = \min_{j \neq k}\{d_{kj}\}$ is the *minimum distance* of the constellation. Therefore, a judicious selection of signaling levels $\mathbf{c}_k$ that minimizes the average energy per symbol E_s without decreasing $d_{\min}$ is crucial for a modulation format to perform well. This selection is equivalent to the problem of packing M N-dimensional spheres so that E_s (which is equal to the average second moment of $\mathbf{c}_k$) is minimized. In fact, at a more fundamental level, most coding and modulation problems for AWGN-limited systems may, in the high-SNR regime, be reformulated as sphere-packing problems. Unfortunately, while such sphere packing problems are often easy to formulate, they are notoriously difficult to solve analytically, and one must often resort to numerical optimization techniques to find the best constellations.

We now wish to compare the performance of constellations with different numbers of levels M at a fixed bit rate R_B. We therefore rewrite the dominant term in (5.9) as

$$\operatorname{erfc}\left(\sqrt{\gamma \frac{P}{R_B N_0}} \right) = \operatorname{erfc}\left(\sqrt{\gamma \frac{E_b}{N_0}} \right), \tag{5.10}$$

where

$$\gamma = \frac{d_{\min}^2}{4E_b} \tag{5.11}$$

and $E_b = P/R_B = E_s/\log_2 M$ is the average *energy per bit*. In the following, we will refer to both E_s/N_0 and E_b/N_0 as the SNR, depending on the context. The parameter γ, which captures the constellation's influence on the SER and is usually given in dB, is called the *asymptotic power efficiency* [3, p. 220], because the power needed for a certain required SER, still at asymptotically high SNR, is proportional to $1/\gamma$. Another interpretation of γ is as the *sensitivity gain* over BPSK to transmit the same data rate, since $\gamma = 0$ dB for BPSK, QPSK, and DP-QPSK.

In fact, most common modulation formats have a penalty with respect to BPSK; for example, M-PSK and M-QAM have [3, pp. 226, 234]

$$\gamma_{M\text{-PSK}} = \sin^2(\pi/M)\log_2 M, \tag{5.12}$$

$$\gamma_{M\text{-QAM}} = \frac{3\log_2 M}{2(M-1)}, \tag{5.13}$$

where (5.13) is valid for M being a power of 4. We can show from these expressions that both M-PSK and M-QAM have efficiencies $\gamma \leq 0$ dB for all values of M (with the notable exception of 3-PSK, which will be discussed in the next section).

The first general investigation on how the SER depends on the dimensionality N, the constellation size M, and the SNR was done by Shannon in 1959 [44]. By using geometrical sphere-packing arguments, he managed to obtain upper and lower bounds on the SER under rather general conditions. While Shannon's objective was to quantify the performance of capacity-approaching coded systems, our focus in this paper is on uncoded transmission, i.e., low-dimensional constellations, in particular $N = 2$ and 4.

Specifically, we will consider the question: At a given dimension N, and constellation size M, and asymptotic SNR, which modulation format (constellation) has the highest asymptotic power efficiency γ? Quite surprisingly, this issue was not addressed until recently by us [1, 28] and then only when minimizing the average symbol energy E_s. As noted earlier [44], minimizing the maximum energy $E_{s,\max}$ is also a relevant problem. In the next section, we will therefore present results for both average-energy and maximum-energy minimization.

5.3 N-Dimensional Sphere Packing Results

Before presenting the main results, we will give a brief historical background and introduction to the area of sphere packing.

5.3.1 *Sphere Packings: Background*

As we noted in Sect. 5.2.3, the problem of finding the constellation with maximum asymptotic power efficiency is equivalent to finding the densest packing of M N-dimensional spheres. Here, "densest" can be interpreted either as a minimization of the maximum distance from the origin, or as a minimization of the average squared distance from the origin, as mentioned above. In this chapter, we will refer to a sphere-packing constellation designed to minimize the average squared distance as a *cluster* and one designed to minimize the maximum distance as a *ball.*[1] It is actually challenging enough to find the best constellations for a fixed number of levels M in a given dimension N. In general, no formal mathematical proof that a certain constellation is the densest is known, and conclusions are rather supported by empirical evidence in the sense that "no better constellations have been found." In reality, sphere packing optimization often involves the creation of thousands of dense constellations (and various efficient algorithms for this have been proposed), and then selecting the best among these. For high dimensionality and constellation sizes, this can be quite demanding.

For planar clusters, some conjectured optimal constellations were originally presented by Foschini et al. [20] for selected values of M up to 16. They are typically hexagonal packings of M circles centered around the origin. This was further demonstrated by Graham et al. [22], who numerically computed conjectured optimum packings up to $M \leq 100$ in the plane and even larger constellations ($M \leq 500$) with a suboptimal, greedy technique. In $N = 3$ dimensions, the best known sphere packings, including images of the cases $M \leq 20$, were originally reported by Sloane et al. in [46]. Their work has been updated and extended to tables of the best known packings for $N = 3$, $M \leq 99$ and $N = 4$, $M \leq 32$, which are available online [47]. Some early work on ball optimization were reported by Lachs [33], but limited to 10 points in 3 and 4 dimensions. Also, other tables based on numerical optimization have been reported, e.g., in [38], but it is noteworthy that some of the constellations reported there are inferior to those of [47] (one such example is the case $M = 8$, $N = 4$ which is of particular interest to us). We performed our own sphere-packing optimizations for $N = 2, 3, 4$ and $M \leq 16$ that verified the reported values from [47]. For higher dimensions, not much is known about good constellations of finite sizes M. Much more is known about the densest infinite-size packings, particularly *lattices,* for higher dimensions, and most of this work can be found in the extensive review by Conway and Sloane [14].

If the target is to design balls instead of clusters, i.e., to minimize $E_{s,max}$ instead of E_s, the optimization problem can be interpreted as packing M unit-size spheres into a larger sphere, which should be as small as possible. In two dimensions, this problem and its variants have received a lot of attention, as evidenced by Stephenson's extensive bibliography [50]. The best known balls are tabulated by E. Specht for $M \leq 900$ [49]. We are not aware of any published results for $N \geq 3$, but we can derive presumably optimal constellations of moderate sizes based on available results for *spherical codes.*

In a spherical code, all constellation points are required to have the same distance to the origin, and a good spherical code is one where this distance is as low as possible. It is known since the days of Shannon that spherical codes are good for

[1] Mathematically, a "ball" is defined as the set of points in Euclidean space whose distance to a given point is upperbounded by a given constant, i.e., the region bounded by a sphere. "Although physicists often use the term 'sphere' to mean the solid ball, mathematicians definitely do not" states Weisstein [55].

communication over the AWGN channel in very high dimensions [43,44], but this is generally not the case in the low-dimensional applications considered in this chapter. The best known spherical codes are tabulated for $M \leq 130$ and dimensions up to 5 [48]. In this work, we derive balls of size $M \leq K_N + 1$ from spherical codes, where the *kissing number* K_N is the maximum number of nonoverlapping spheres in N-dimensional space that can touch a given sphere with the same size. For two and three dimensions, one has $K_2 = 6$ and $K_3 = 12$, respectively [14], and in four dimensions one has $K_4 = 24$. Like many sphere-packing problems, rigorous proofs of these values are very difficult, and although $K_4 = 24$ was long conjectured [14], it was only recently proven formally [35].

It can be shown that the optimal N-dimensional ball is identical to the optimal spherical code if $M \leq K_N$. Furthermore, if $M = K_N + 1$, we conjecture that the optimal ball is constructed as a spherical code of size K_N with the addition of an extra constellation point at the origin.

As an example of the difference between the maximum and average symbol energy minimization, two-dimensional balls and clusters of size $M = 5$ are shown in Fig. 5.4. This case is further discussed in Sect. 5.3.3.1.

5.3.2 Results: Sensitivity vs. Spectral Efficiency

A common way to compare modulation formats [3, 39] is to represent each format as a point in the spectral efficiency vs. sensitivity plane. These sensitivities can be obtained by using the union bound (5.9) to plot SER vs. SNR as shown for example in Fig. 5.9 in Sect. 5.4, and then finding the E_b/N_0 required to get a certain SER. This is convenient as it directly shows the SE–sensitivity trade-off, and in addition it can be compared to the Shannon capacity limit, which relates the SNR and spectral efficiency as

$$\frac{E_b}{N_0} = \frac{2^{SE} - 1}{SE}. \tag{5.14}$$

The results are shown in Fig. 5.2, plotting the optimized constellations for $SER = 10^{-3}$ and $SER = 10^{-9}$. The balls are marked with circles and the clusters with triangles in this graph. One can clearly see the required extra SNR as the SER demand increases to 10^{-9}. Also, the difference in sensitivity between the balls and the clusters increases at 10^{-9}, as does the difference between the two- and four-dimensional constellations. It should be noted that the balls will always have a sensitivity penalty relative to the clusters, as we choose to define sensitivity in terms of average energy per bit, E_b. In Sect. 5.5.2, we will show the difference when we use maximum energy per bit, $E_{b,max} = E_{s,max}/\log_2 M$, as a sensitivity measure instead.

Asymptotically, for very low required SERs, the relative difference in sensitivities between the formats approach constant values, although the absolute sensitivity in E_b/N_0 will approach infinity. This situation can be shown by plotting the for-

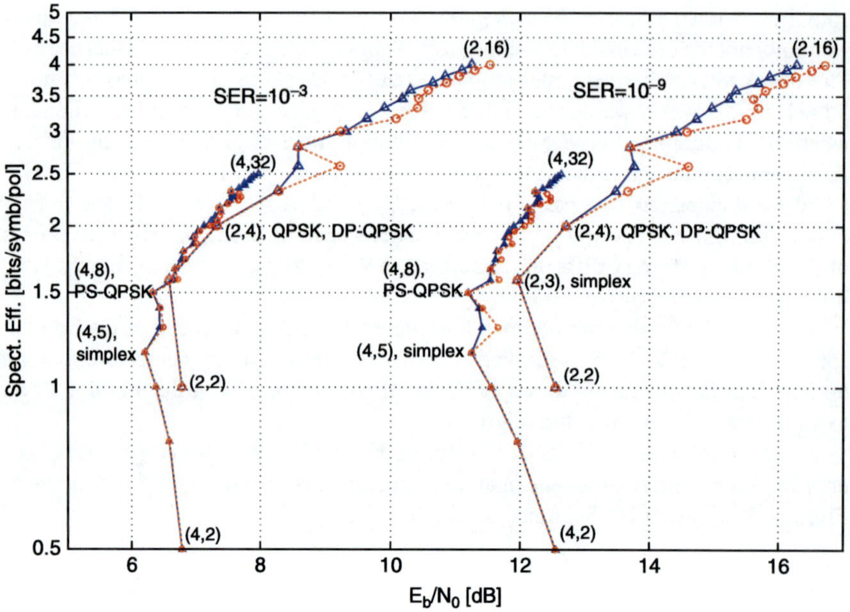

Fig. 5.2 Spectral efficiency vs. required E_b/N_0 for $SER = 10^{-3}$ and $SER = 10^{-9}$. The optimum constellations are referred to as (N, M), where N is the number of dimensions and M is the number of points in the constellation. We plot constellations in $N = 2$ up to $M = 16$. In $N = 4$ dimensions, we plot balls (shown as *circles* connected with *dashed lines*) up to $M = 25$ as well as clusters (shown as *triangles* connected with *solid lines*) up to $M = 32$. Some common modulation formats (QPSK, DP-QPSK) are identical with the optimized (2,4)-constellation. The PS-QPSK format (4,8) is also shown, as are the simplices

mats as in Fig. 5.3 with the (inverse) asymptotic power efficiency on the x-axis. This facilitates a direct comparison between the constellations, as the relative sensitivity differences are approximately the same as in the absolute sensitivity scale of Fig. 5.2, but the Shannon limit cannot, for example, be included. In this plot, we removed the balls from simplicity, but have included some other known formats such as M-PSK, and rectangular 8- and 16-QAM for comparison. We also indicate the *kissing configurations*, i.e., the configurations involving the K_N spheres touching a central sphere, which emerge as local minima for the power efficiency at $M = K_N + 1$ for $N = 2$ and $N = 4$ (but not, e.g., $N = 3$).

As M increases for a given (low) dimension N, the best (densest) packings are known to approach a regular structure called a *lattice*. In two dimensions, the best lattice is generated by placing three circles in a regular triangle (simplex) and extending the pattern indefinitely in all directions. This generates the well-known honeycomb, or hexagonal lattice, usually denoted A_2. Its density is $\Delta(2) = \pi/(2\sqrt{3}) = 0.91$, which means that the circles cover 91% of the plane. The three-dimensional analogy is the face-centered cubic lattice A_3, obtained by extending a regular tetrahedron (three-dimensional simplex), with the density $\Delta(3) =$

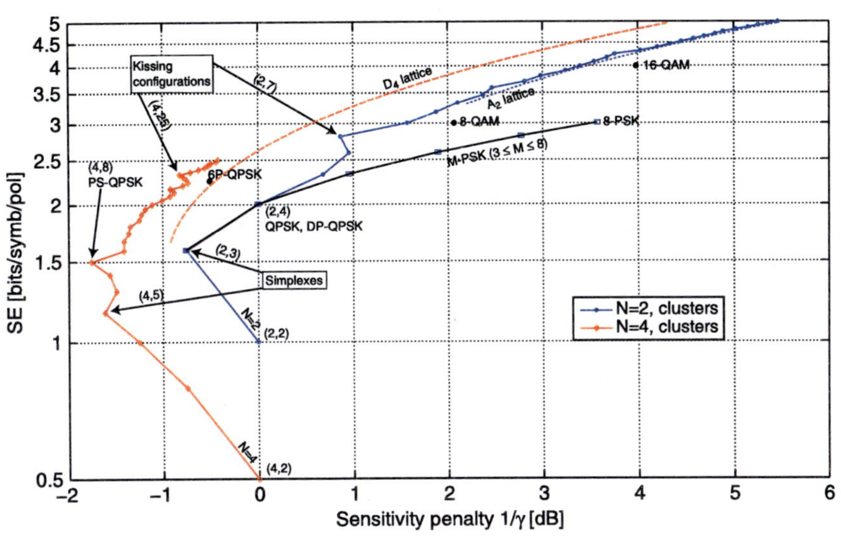

Fig. 5.3 Spectral efficiency vs. asymptotic power efficiency for SER= 10^{-3}. We plot optimized clusters in $N = 2$ and $N = 4$ dimensions. For comparison, we also plot the M-PSK, 8- and 16-QAM, and 6P-QPSK formats, and the best lattice packings in 2 and 4 dimensions (*dashed lines*). The optimum constellations have in some cases been marked by (N, M), indicating dimensionality and number of points

$\pi/(3\sqrt{2}) = 0.74$. In four dimensions, however, something unexpected happens. Even though a four-dimensional lattice, A_4, can be generated from a 4d simplex in perfect analogy with A_2 and A_3, it is not the densest lattice possible. The densest lattice in four dimensions is denoted D_4 [14], and can be seen as a 4d analogy of the checkerboard pattern. It can be represented by all integer coordinate points such that the coordinates sum to an even integer, and it has the density $\Delta(4) = \pi^2/16 = 0.62$.

The asymptotic power efficiency of a lattice is [14, (32)]

$$\gamma_{\text{lat}} = \log_2(M)\left(1 + \frac{2}{N}\right)\left(\frac{\Delta(N)}{M}\right)^{2/N}, \tag{5.15}$$

where the densities $\Delta(N)$ are tabulated in [14, Table 1.2]. The performance of the densest lattices, A_2 and D_4, are included as dashed-line asymptotes in Fig. 5.3.

5.3.3 *Specific Formats*

In this section, we will discuss some of the optimized constellations from Figs. 5.2 and 5.3, and present their coordinates when known. We denote the optimized constellations for M points in N dimensions with $\mathscr{C}_{N,M}$ for clusters and $\mathscr{B}_{N,M}$ for balls. When the coordinates of the constellations are presented, they have been normalized to make the minimum distance between points $d_{\min} = 2$, which corresponds to the packing of unit-radius spheres. We will present both balls and clusters for selected sizes, and emphasize when they are equal, which occurs, we believe, only in a finite number of cases. We will discuss each dimension in turn.

We use the following sources for the best known constellations.

- $\mathscr{C}_{2,M}$ and $\mathscr{B}_{2,M}$ for $N = 2, 4$ and $M = 2, 3, 4$ are M-PSK constellations.
- $\mathscr{C}_{2,M}$ for $M \geq 5$ were designed by Graham and Sloane [22], but the obtained constellations were not reported, only their average second moments. We have reconstructed these constellations based on the conjecture in [22] that they are all subsets of the lattice A_2.
- $\mathscr{C}_{4,M}$ for $M \geq 5$ were taken from Sloane's website [47].
- $\mathscr{B}_{2,M}$ for $M \geq 5$ were taken from Specht's website [49].
- $\mathscr{B}_{4,M}$ for $M \geq 5$ were constructed from the spherical codes in [48] using the methods described in Sect. 5.3.1.

5.3.3.1 Two-Dimensional Constellations, $N = 2$

On the one hand, the two-dimensional clusters are always subsets of the hexagonal lattice, as pointed out in [22]. The two-dimensional balls, on the other hand, have more irregular structures, and the best known are listed in [49] for $M \leq 900$ (with pictures for $M \leq 804$). The only cases we have found where the balls and clusters are identical are for $M = 2, 3, 4, 7, 31, 55$. We believe these are the only such cases in two dimensions. A property of some balls (but no clusters) is the presence of "loose points," which are constellation points that are further than the minimum distance from all neighbors and the surrounding circle. Such points can move freely without affecting $E_{\mathrm{s,max}}$, which makes the ball nonunique, and having a continuum of possible average powers E_{s}. The first loose point arises for $M = 8$ and such points become increasingly common as the constellation size increases. The largest known balls without loose points are $M = 37, 61, 91$. We will below briefly discuss a few two-dimensional balls and clusters of particular interest.

$M = 2, 3, 4$

These modulation formats are the well-known binary, ternary, and quaternary PSK. The clusters and balls coincide for these. The smallest sensitivity over all sizes M is obtained for $M = 3$, and the optimal constellation is the triangle, or *simplex*. It was suggested for modulation in [18, 37] under the name ternary phase-shift keying (3-PSK), and it has a $\gamma = (3/4)\log_2 3 = 0.75\,\mathrm{dB}$ asymptotic sensitivity gain over BPSK. Due to the moderate gain as well as the difficulty of mapping bits to three levels, this format has gained little attention, however. The other constellation points are given by $\mathscr{C}_{2,2} = \mathscr{B}_{2,2} = \{(\pm 1, 0)\}$ for BPSK and $\mathscr{C}_{2,4} = \mathscr{B}_{2,4} = \{(\pm 1, \pm 1)\}$ for QPSK. It is noteworthy that $\mathscr{C}_{2,4}$ is not unique; the constellation points can

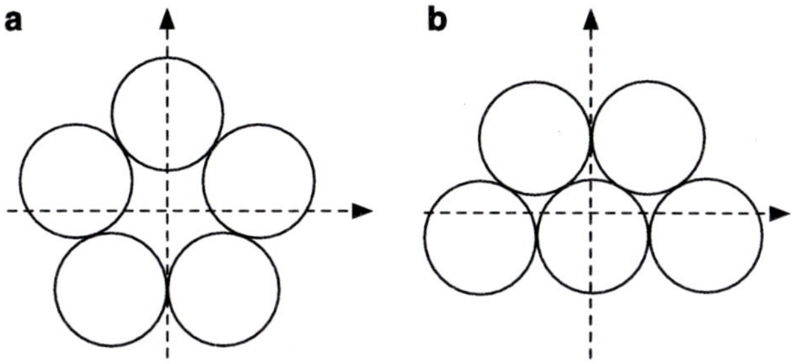

Fig. 5.4 Optimum five-point constellations in the plane, $(N, M) = (2, 5)$. Minimizing the maximum energy gives the *ball* $\mathscr{B}_{2,5}$ shown in (**a**) where all symbols lie on a regular pentagon, and minimizing the average energy gives the *cluster* $\mathscr{C}_{2,5}$ in (**b**) which is a subset of the hexagonal packing

be continuously deformed to $\mathscr{C}_{2,4} = \{(0, \pm 2/\sqrt{3}), (\pm 1, -1/\sqrt{3})\}$, which is an extension of $\mathscr{C}_{2,3}$ with one point. This constellation is also a cluster, since it has the same E_s [22]. Note also that both BPSK and QPSK have the same power efficiency, 0 dB.

$M = 5$

This is the first case for which the cluster and the ball are not identical. The two cases are shown in Fig. 5.4. The pentagonal structure, Fig. 5.4a, has the same maximum and average energy, $E_s = E_{s,max} = 8/(5 - \sqrt{5}) \approx 2.89$, whereas the hexagonal structure, Fig. 5.4b, has average energy $E_s = 68/25 = 2.72$ and maximum energy $E_{s,max} = 112/25 = 4.48$.

$M = 6, 7$

The $M = 7$ constellation is the kissing configuration in two dimensions: six circles touching a unit circle at the origin. The ball and the cluster are identical to this kissing configuration, i.e., $\mathscr{B}_{2,7} = \mathscr{C}_{2,7} = \{(0, 0), (\pm\sqrt{3}, \pm 1), (0, \pm 2)\}$, for all sign combinations. The maximum energy is $E_{s,max} = 4$ and the average energy is $E_s = 24/7 = 3.43$. The asymptotic power efficiency is $\gamma = \log_2(7)/E_s = -0.87$ dB.

The cluster $\mathscr{C}_{2,6}$ is obtained by removing an edge point from $\mathscr{B}_{2,7}$ and recentering the constellation, which gives $E_s = 29/9 = 3.22$. The ball $\mathscr{B}_{2,6}$ is obtained by removing an edge point or a center point, since $E_{s,max} = 4$ irrespective of which point is removed. The average energy will be larger and equals $E_s = 4$ if the center point is removed, which is the choice used in [49] and in the results presented here.

$M = 8, 9$

These balls have both $M - 1$ points in a circle of radius $1/\sin(\pi/(M-1))$ and a loose point inside this circle.

$M = 15$

This ball consists of a regular structure with 5 inner points in a pentagon and an outer ring of 10 points, arranged so that two outer points touch each inner point.

$M = 19$

The ball and the cluster are different, but very close in structure. Both have hexagonal symmetry, with a $\mathscr{B}_{2,7}$ ball of 7 points in the center, surrounded by 12 outer points. The cluster $\mathscr{C}_{2,19}$ is formed when the outer points form a large hexagon, while in $\mathscr{B}_{2,19}$, the outer points form a circle, as shown in Fig. 5.5.

$M = 31, 55$

The two largest known constellations for which the cluster is also a ball occurs for $M = 31$ and $M = 55$. They are shown in Fig. 5.6. For $M = 55$, the ball has six loose points (black) that can be moved without changing $E_{s,\max}$. The cluster forces these loose points to lie in the hexagonal lattice.

a b

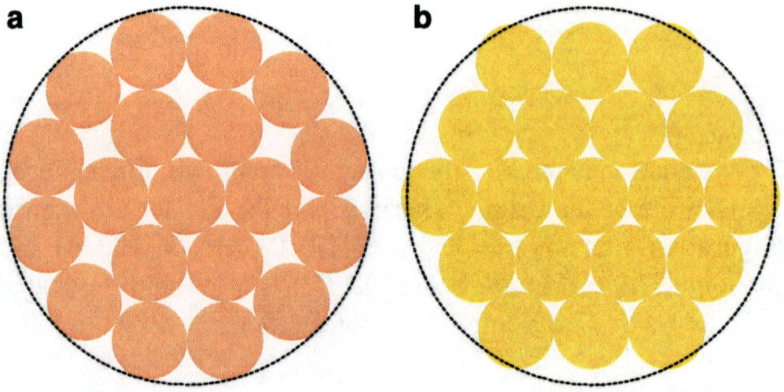

Fig. 5.5 The ball $\mathscr{B}_{2,19}$ (**a**) and the cluster $\mathscr{C}_{2,19}$ (**b**) can be obtained from each other by shifting the outer ring of disks. The *dashed circles* have the same size, showing that $E_{s,\max}$ of the cluster is higher

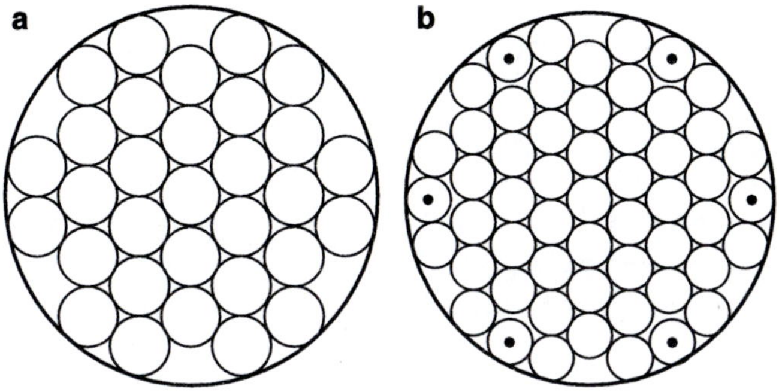

Fig. 5.6 The constellations $\mathscr{B}_{2,31} = \mathscr{C}_{2,31}$ (**a**) and $\mathscr{B}_{2,55}$ (**b**), with coordinates taken from [49]. The cluster $\mathscr{C}_{2,55}$ is obtained by moving the loose points (denoted with *black dots*) closer to the center, which does not change $E_{s,\max}$

5.3.3.2 Four-Dimensional Constellations, $N = 4$

In four dimensions, the constellations are a bit more difficult to visualize. For $M = 2$ and 4, the clusters and balls are all $(M - 1)$-dimensional simplices, i.e., 3-PSK and the tetrahedron constellation. We will present some interesting special cases of clusters and balls below, referring to them with the number of points.

$M = 5$

The four-dimensional simplex has 5 points, and is called the *pentachoron*, or *pentatope*, or *5-cell*. It is both cluster and ball. It was discussed in several papers analyzing four-dimensional modulation [6, 8, 32, 53, 56, 62]. Its coordinates can be compactly expressed as

$$\mathscr{C}_{4,5} = \mathscr{B}_{4,5} = \left\{ \sqrt{\frac{2}{5}}(1,1,1,1),\ -\frac{1}{2\sqrt{10}}\left(1-3\sqrt{5}, 1+\sqrt{5}, 1+\sqrt{5}, 1+\sqrt{5}\right) \right\}, \tag{5.16}$$

where the second vector should be repeated with all four coordinate permutations [63]. Asymptotically, the pentachoron has a $\gamma = (5/8)\log_2 5 = 1.62\,\text{dB}$ gain over BPSK. As for most constellations in this section, the difficulty of using it for transmission lies partly in its generation and partly in the difficulty to map bits to five constellation levels.

$M = 6$

This is the first instance for which the cluster and the ball differ. The cluster, which is the pentachoron plus an extra point, has the coordinates

$$\mathscr{C}_{4,6} = \left\{ \pm\sqrt{\frac{5}{8}}(1,1,1,1),\ \frac{1}{\sqrt{8}}(-3,1,1,1) \right\} \tag{5.17}$$

with both signs for the first vector and all four permutations of the second.

The ball is not unique. We use the constellation from [48], whose coordinates can be obtained by rescaling the first vector of (5.17). After renormalization, this yields

$$\mathscr{B}_{4,6} = \left\{ \pm\frac{1}{\sqrt{2}}(1,1,1,1),\ \frac{1}{\sqrt{6}}(-3,1,1,1) \right\}. \tag{5.18}$$

Other, equally good, balls can be obtained by removing any two points from the cross-polytope constellation $\mathscr{B}_{4,8}$ described below.

$M = 7$

Again, the ball is not unique. The constellation in [48] can be identified as

$$\mathscr{B}_{4,7} = \left\{ (\pm 1, \pm 1, 0, 0), \left(0, 0, \sqrt{2}, 0\right), \left(0, 0, -\frac{1}{\sqrt{2}}, \pm\sqrt{\frac{3}{2}}\right) \right\} \tag{5.19}$$

with all signs. Thus, it consists of four points forming a square in one plane, and three points forming an equilateral triangle in the orthogonal plane. Other versions of the ball can be obtained from $\mathscr{B}_{4,8}$ by removing an arbitrary point.

The cluster $\mathscr{C}_{4,7}$ is obtained from $\mathscr{B}_{4,8}$ by removing any point and shifting the resulting constellation to have zero mean.

$M = 8$

In terms of average bit energy requirements, the cluster $\mathscr{C}_{4,8}$ is the best 4d constellation of any size M, as can be seen from Figs. 5.2 and 5.3. A projection of the constellation is shown in Fig. 5.7a. All its points lie on the 4d sphere, and thus $\mathscr{B}_{4,8} = \mathscr{C}_{4,8}$. Its eight points follow from the biorthogonal representation, which is given by all signs and all permutations of

$$\mathscr{C}_{4,8} = \mathscr{B}_{4,8} = \left\{ \left(\pm\sqrt{2}, 0, 0, 0\right) \right\}. \tag{5.20}$$

The structure is known as the *cross-polytope*, and it is invariant under a number of symmetries, which simplifies its implementation in a transmission system. A 45° absolute phase rotation will bring it into the modified representation

$$\mathscr{C}'_{4,8} = \{(\pm 1, \pm 1, 0, 0), (0, 0, \pm 1, \pm 1)\}. \tag{5.21}$$

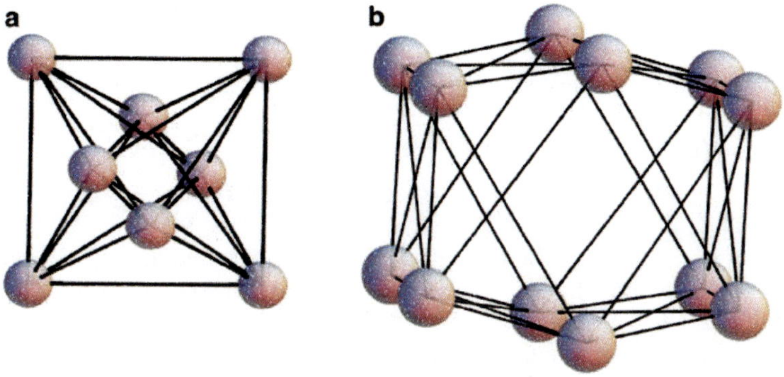

Fig. 5.7 Projections of the constellations $\mathscr{B}_{4,8} = \mathscr{C}_{4,8}$ (**a**) and $\mathscr{B}_{4,12}$ (**b**). The *black lines* connect nearest neighbors, and they have all the same length in four-dimensional space

This shows that a modulator based on the cross-polytope can be implemented as QPSK transmission in either x or y polarization, but *not* both simultaneously as in DP-QPSK [28]. Therefore, we call this modulation format *polarization-switched QPSK* (PS-QPSK).

A third representation is possible as half of the points (e.g., those whose coordinates sum to an even integer) of the cubic (DP-QPSK) constellation. It was described in more detail (including transmitter configurations) in [28]. Since it has only eight levels, its spectral efficiency is reduced to 3 bits per symbol (1.5 bits per polarization), but this is more than compensated for by the minimum distance increasing by a factor of $\sqrt{2}$. Thus, the asymptotic power efficiency becomes $\gamma = 3/2 = 1.76\,\text{dB}$ better than DP-QPSK.

$M = 10$

The cluster and ball are identical also here, and this constellation is known as the *rectified 5-cell,* which is formed by the ten points that lie midway between all pairs of points in the 4d simplex. After normalizing, the coordinates can be expressed as

$$\mathscr{C}_{4,10} = \mathscr{B}_{4,10} = \left\{ \frac{1}{2\sqrt{10}} \left(3 + 3\sqrt{5}, 3 - \sqrt{5}, 3 - \sqrt{5}, 3 - \sqrt{5}\right), \right.$$
$$\left. -\frac{1}{\sqrt{10}} \left(1 - \sqrt{5}, 1 - \sqrt{5}, 1 + \sqrt{5}, 1 + \sqrt{5}\right) \right\} \quad (5.22)$$

where the first vector should be taken with its four coordinate permutations and the second vector with its six permutations. This is a rather regular structure, where each point has 6 nearest neighbors at an angular distance of $\cos^{-1}(1/6)$, and the three furthest points all lie at an angular distance of $\cos^{-1}(-2/3)$. The asymptotic power efficiency of this constellation is $\gamma = 1.41\,\text{dB}$. This structure was originally identified as the optimum by Lachs [33].

$M = 12$

The ball is given by the neat structure

$$\mathscr{B}_{4,12} = \{(\pm a, -b, -b, -b), (\pm a, b, b, -b), (\pm a, b, -b, b), (\pm a, -b, b, b), \\ (0, c, c, c), (0, c, -c, -c), (0, -c, c, -c), (0, -c, -c, c)\}, \quad (5.23)$$

where $a = \sqrt{7/6}$, $b = 1/\sqrt{2}$, and $c = 2\sqrt{2}/3$. As illustrated in Fig. 5.7b, the ball consists of three tetrahedra, uniformly spread along the first coordinate.

The cluster $\mathscr{C}_{4,12}$ is obtained by stretching the middle tetrahedron by about 4% and then pushing the two outer tetrahedra closer together along the first dimension until all three touch each other. Thus, the ball and the cluster have the same symmetries. Graphically, $\mathscr{C}_{4,12}$ looks almost exactly as Fig. 5.7b, with the addition of four more lines representing nearest neighbors. Its coordinates are also given by (5.23), where in this case $a = 1$, $b = 1/\sqrt{2}$, and $c = (2\sqrt{5} + \sqrt{2})/6$.

$M = 16$

We denote the cubic constellation DP-QPSK with $\mathscr{D}_{4\text{cube}} = \{(\pm 1, \pm 1, \pm 1, \pm 1)\}$, with all possible sign selections. This is the most common modulation format in coherent systems, as it is easy to generate and detect. However, it is not a very optimized configuration, either in an average-energy or maximum-energy sense. The optimum cluster $\mathscr{C}_{4,16}$ is instead a remarkable structure comprising two subsets of the D_4-lattice, with 7 and 9 points, rotated and translated with respect to each other. Its coordinates can be given as

$$\mathscr{C}_{4,16} = \Big\{\big(a + \sqrt{2}, 0, 0, 0\big), \big(a, \pm\sqrt{2}, 0, 0\big), \big(a, 0, \pm\sqrt{2}, 0\big), \big(a, 0, 0, \pm\sqrt{2}\big), \\ (a - c, \pm 1, \pm 1, \pm 1), (a - c - 1, 0, 0, 0)\Big\} \quad (5.24)$$

with all combinations of signs, where $a = (1 - \sqrt{2} + 9c)/16$ and $c = \sqrt{2\sqrt{2} - 1}$. With this representation, which is illustrated in Fig. 5.8a, the cluster can be regarded as four three-dimensional constellations stacked on top of each other along the first dimension: a single point, an octahedron, a cube, and finally another single point. The average symbol energy of this constellation can be expressed as $E_s = (279 + 64\sqrt{2} + (7 + 9\sqrt{2})c)/128 = 3.09$, which can be compared to $E_s = 4$ for $\mathscr{D}_{4\text{cube}}$, which makes the sensitivity of $\mathscr{C}_{4,16}$ 1.11 dB better than DP-QPSK. A comparison between these two formats with and without coding was performed in [64].

The ball $\mathscr{B}_{4,16}$ has no apparent useful symmetries facilitating a nice coordinate representation. Another constant-energy constellation was given in [32] with almost as good performance as $\mathscr{B}_{4,16}$ (having about 0.1% higher $E_{s,\max}$), but the two constellations are geometrically different. This illustrates the occurrence of multiple local minima in numerical constellation optimization.

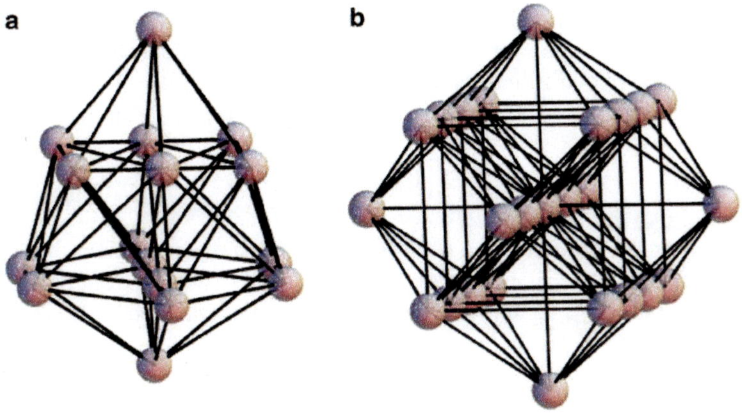

Fig. 5.8 Projections of the constellations $\mathscr{C}_{4,16}$ (**a**) and $\mathscr{B}_{4,25} = \mathscr{C}_{4,25}$ (**b**). The *black lines* connect nearest neighbors, and they have all the same length in four-dimensional space

$M = 23, \ldots, 27$

All clusters, and some balls, in the range $M = 23, \ldots, 27$ can be derived from the kissing configuration $\mathscr{B}_{4,25} = \mathscr{C}_{4,25}$, which is the four-dimensional analogy of $\mathscr{B}_{2,7} = \mathscr{C}_{2,7}$. It consists of a sphere at the origin and 24 spheres touching this sphere. There is a unique way to arrange 25 spheres in this manner, illustrated in Fig. 5.8b. It forms a subset of the D_4 lattice and is a very symmetrical and dense constellation. It can be formally defined as $\mathscr{B}_{4,25} = \mathscr{C}_{4,25} = \mathscr{B}_{4,24} \cup \{(0,0,0,0)\}$, where $\mathscr{B}_{4,24}$ represents the 24-cell defined below. The constellation $\mathscr{B}_{4,25}$ was discussed in [56] and it has an asymptotic power efficiency of $\gamma = 0.83$ dB.

The ball for $M = 24$ is obtained by removing any point from $\mathscr{B}_{4,25}$. The choice of point to remove does not influence the performance (in perfect analogy with $\mathscr{B}_{4,6}$) and we choose $(0,0,0,0)$ to preserve the symmetry. The ball $\mathscr{B}_{4,24}$ thus defined consists of the 24 vertices of the 4d regular polytope sometimes referred to as the *24-cell*. All five regular Platonic solids in three dimensions (tetrahedron, cube, octahedron, dodecahedron, and icosahedron) have extensions to four dimensions. The 24-cell, however, is the only regular 4d polytope, that, according to Coxeter, is unique: "...having no analogue [in dimensions] above or below." [15, p. 289]. The 24-cell was considered for communications in [8, 32, 53, 56, 62]. Its coordinates can be expressed in two distinct ways. The first is as the union of the 16 levels of the 4d cube (DP-QPSK) and the 8 levels of a cross-polytope:

$$\mathscr{B}_{4,24} = \mathscr{D}_{4\text{cube}} \cup \sqrt{2}\mathscr{B}_{4,8} = \{(\pm 1, \pm 1, \pm 1, \pm 1), (\pm 2, 0, 0, 0)\}, \tag{5.25}$$

again including all signs and permutations. This demonstrates how the DP-QPSK format can be extended to 24 points without increasing the average symbol energy or reducing the minimum distance. These additional modulation levels were also recently suggested by Bülow [11] to be utilized for forward error correction overhead. The modulation format can be seen as using four absolute phase levels for each of the six polarization states (x, y, $\pm 45°$, LHC, RHC).

The second and more compact description of the 24-cell is

$$\mathscr{B}'_{4,24} = \left\{\sqrt{2}(\pm 1, \pm 1, 0, 0)\right\}, \quad (5.26)$$

again allowing for arbitrary sign choices and coordinate permutations. This is an equally common representation of the 24-cell. A point $\mathbf{c}'$ in $\mathscr{B}'_{4,24}$ can be obtained from a point $\mathbf{c}$ in $\mathscr{B}_{4,24}$ by applying the coordinate transformation [14]

$$\mathbf{c}' = \frac{1}{\sqrt{2}} \begin{pmatrix} 1 & 1 & 0 & 0 \\ 1 & -1 & 0 & 0 \\ 0 & 0 & 1 & 1 \\ 0 & 0 & 1 & -1 \end{pmatrix} \mathbf{c}. \quad (5.27)$$

In fiber-optics language, a similar transformation that can be used to transform $\mathbf{c}$ to $\mathbf{c}'$ is $\mathbf{E}' = \mathbf{E}\exp(i\pi/4)$.[2]

By using the set $\mathscr{B}_{4,24}$, the sensitivity of the DP-QPSK format can be improved by $\log_2(24)/\log_2(16) = 0.59\,\text{dB}$, but mapping bits to 24 symbols is nontrivial. In [1] we introduced a modulation format called 6P-QPSK by mapping nine information bits to two sequential points in $\mathscr{B}_{4,24}$, which enables 4.5 bits per symbol to be transmitted. This gives an improvement of $\gamma = 9/8 = 0.51\,\text{dB}$ over DP-QPSK.

The cluster $\mathscr{C}_{4,24}$ is obtained by removing an outer point (i.e., not $(0,0,0,0)$) from $\mathscr{B}_{4,25}$ and shifting the resulting constellation to zero mean. It improves on DP-QPSK by 0.79 dB.

For $M = 23$, a ball is obtained by removing two arbitrary points from $\mathscr{B}_{4,25}$. The remaining balls can be shifted around in many ways without changing the maximum energy. The cluster $\mathscr{C}_{4,23}$ is however unique, and it is obtained by removing two adjacent outer points from $\mathscr{B}_{4,25}$ (say, $(1,1,1,\pm 1)$) and recentering the constellation.

Clusters for $M = 26$ and $M = 27$ are obtained by adding points from the next layer of D_4 to $\mathscr{B}_{4,25}$. Specifically, $\mathscr{C}_{4,26}$ is obtained by centering $\mathscr{B}_{4,25} \cup \{(2,2,0,0)\}$ and $\mathscr{C}_{4,27}$ is obtained by centering $\mathscr{B}_{4,25} \cup \{(2,2,0,0),(2,0,2,0)\}$. These two clusters are, however, very weak in terms of maximum power, as will be shown in Sect. 5.5.2.4. The balls for $M = 26$ and $M = 27$ have no apparent relation to the kissing configuration $\mathscr{B}_{4,25}$ or the D_4 lattice.

[2] It was erroneously stated in [1] that the transformation (5.27) is equivalent to a 45° rotation of the carrier phase of the electric field. It is, if one interchanges row 1 with 2 and row 3 with 4 of the matrix in (5.27).

$M > 27$

There are several regular 4d constellations with more points. For example, a 48-point constellation can be formed as $\mathscr{B}_{4,24} \cup \mathscr{B}'_{4,24}$, which was discussed in [8, 62]. There are also the regular 600-cell (for $M = 120$) and 120-cell (for $M = 600$) [8, 32, 56, 62], of which the former is good in terms of both average and maximum energy and the second is not good, in analogy with the icosahedron and dodecahedron, resp., in three dimensions [2]. At asymptotically high M, optimal constellations in both senses can be constructed as circular subsets of the D_4 lattice.

5.4 Symbol- and Bit-Error Rates

In this section, we will discuss SER for some of the common modulation formats, and also discuss the difference between maximum-energy and average-energy SNR. We will start with this latter point.

Based on the union bound (5.9), we can now plot SER vs. SNR for all constellations we known with coordinates. In general, the union bound agrees well with the exact SER for SER $< 10^{-3}$. Note, however, that the SNR can be defined in two different ways: either (which is most common) as E_b/N_0, i.e., with respect to the average energy per bit, or as $E_{b,max}/N_0$, i.e., with respect to the maximum energy per bit. Figures 5.9 and 5.10 show the SER for the same group of constellations plotted vs. these two SNR definitions. For formats where the average and peak symbol energies are the same (e.g., BPSK, QPSK, and PS-QPSK), there will be no difference. However, for formats where the peak and symbol energy differ (as for $\mathscr{C}_{4,25}$), the x-axis will be rescaled when plotting vs. $E_{b,max}$. A more dramatic difference can be seen when comparing clusters and balls that are nonidentical. As a simple example of this, we plotted the SER for $\mathscr{C}_{2,6}$ (solid lines, triangles) and $\mathscr{B}_{2,6}$ (dashed lines, triangles) in Figs. 5.9 and 5.10. Quite obviously, a constellation that has been optimized with respect to averge energy (a cluster) will perform better than a ball when plotted vs. average energy (in Fig. 5.9). The situation is reversed when plotting the SER vs. maximum energy (Fig. 5.10); here, the ball performs better than the cluster.

We will now go beyond the union bounds and present exact SER for three of the most interesting formats, which are:

- the cubic constellation $\mathscr{D}_{4\text{cube}}$, which corresponds to the DP-QPSK format,
- the cross-polytope $\mathscr{C}_{4,8}$, which corresponds to the PS-QPSK format, and
- the 24-cell constellation, $\mathscr{B}_{4,24}$, which is used for the 6P-QPSK format.

The exact SER expressions for these constellations are, resp.,

$$\text{SER}_{4\text{cube}} = 1 - \left[1 - \frac{1}{2}\,\text{erfc}\left(\sqrt{\frac{E_s}{4N_0}}\right)\right]^4 \tag{5.28}$$

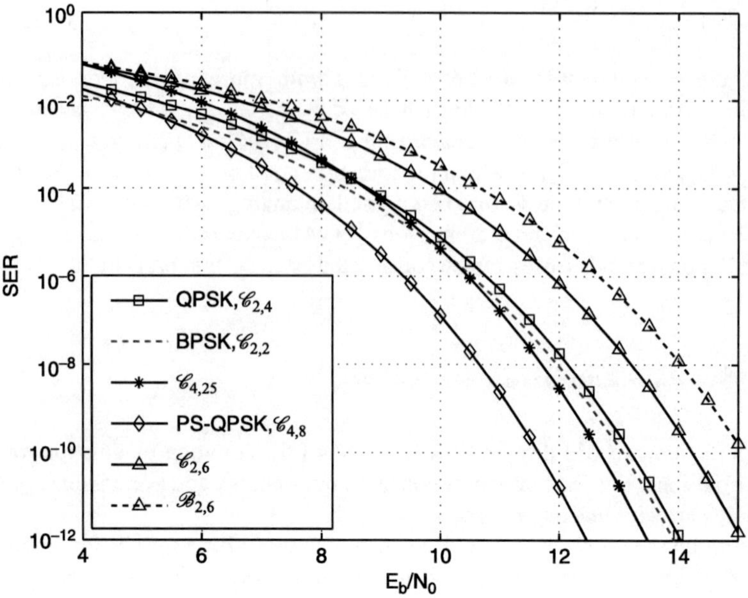

Fig. 5.9 SER vs. E_b/N_0 (average-energy SNR) for a number of constellations, including QPSK and BPSK

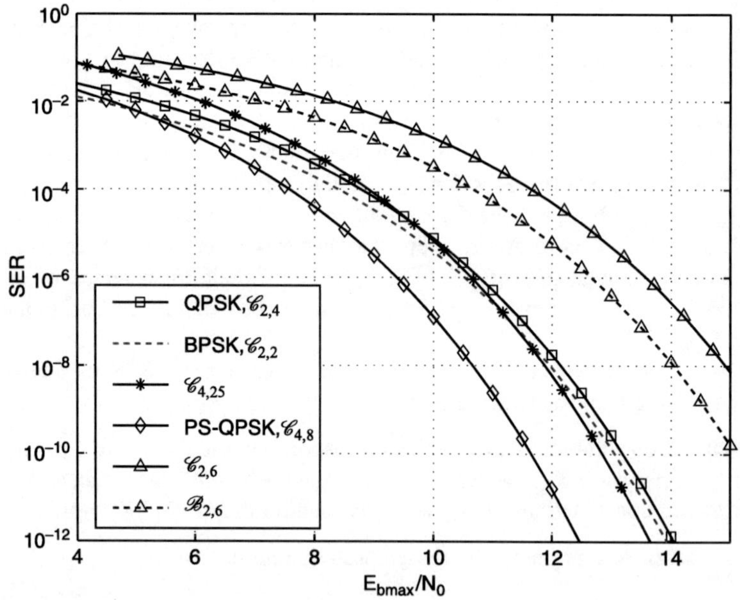

Fig. 5.10 SER vs. $E_{b,max}/N_0$ (maximum-energy SNR) for a number of constellations, including QPSK and BPSK

$$\text{SER}_{4,8} = 1 - \frac{1}{\sqrt{\pi}} \int_0^\infty (1 - \operatorname{erfc} x)^3 e^{-\left(x - \sqrt{\frac{E_s}{N_0}}\right)^2} \mathrm{d}x \tag{5.29}$$

$$\text{SER}_{4,24} = 1 - \frac{1}{\sqrt{\pi}} \int_0^\infty (1 - \operatorname{erfc} x)^2 \operatorname{erfc}\left(x - \sqrt{\frac{E_s}{2N_0}}\right) \cdot e^{-\left(x - \sqrt{\frac{E_s}{2N_0}}\right)^2} \mathrm{d}x. \tag{5.30}$$

Equation (5.28) is straightforward to derive due to the simple geometry of the cubic constellations. The $\text{SER}_{4,8}$ expression (5.29) can be found in standard textbooks [3, p. 210], [45, p. 201] by recognizing $\mathscr{C}_{4,8}$ as an 8-ary biorthogonal constellation. The derivation of the $\text{SER}_{4,24}$-expression (5.30) is more cumbersome and reported in [2].

We do not recommend (5.28)–(5.30) for numerical evaluation at high E_s/N_0, as cancellation occurs when subtracting two almost equal numbers. As observed in [59] for the case of $\mathscr{C}_{4,8}$, expanding the polynomials in erfc x and integrating out the constant term yields

$$\begin{aligned} \text{SER}_{4\text{cube}} &= \frac{1}{16} \operatorname{erfc}\left(\sqrt{\frac{E_s}{4N_0}}\right) \left[4 - \operatorname{erfc}\left(\sqrt{\frac{E_s}{4N_0}}\right)\right] \\ &\quad \times \left[8 - 4 \operatorname{erfc}\left(\sqrt{\frac{E_s}{4N_0}}\right) + \operatorname{erfc}^2\left(\sqrt{\frac{E_s}{4N_0}}\right)\right] \end{aligned} \tag{5.31}$$

$$\begin{aligned} \text{SER}_{4,8} &= \frac{1}{2} \operatorname{erfc}\left(\sqrt{\frac{E_s}{N_0}}\right) + \frac{1}{\sqrt{\pi}} \int_0^\infty \operatorname{erfc} x \\ &\quad \times (3 - 3 \operatorname{erfc} x + \operatorname{erfc}^2 x) \mathrm{e}^{-\left(x - \sqrt{\frac{E_s}{N_0}}\right)^2} \mathrm{d}x \end{aligned} \tag{5.32}$$

$$\begin{aligned} \text{SER}_{4,24} &= \operatorname{erfc}\left(\sqrt{\frac{E_s}{2N_0}}\right) \left[1 - \frac{1}{4} \operatorname{erfc}\left(\sqrt{\frac{E_s}{2N_0}}\right)\right] \\ &\quad + \frac{1}{\sqrt{\pi}} \int_0^\infty \operatorname{erfc} x (2 - \operatorname{erfc} x) \operatorname{erfc}\left(x - \sqrt{\frac{E_s}{2N_0}}\right) \\ &\quad \times \mathrm{e}^{-\left(x - \sqrt{\frac{E_s}{2N_0}}\right)^2} \mathrm{d}x. \end{aligned} \tag{5.33}$$

In Fig. 5.11, we plot the SER as a function of E_b/N_0 by using these expressions. Union bounds from (5.9) are also shown. It is noteworthy that the union bound becomes indistinguishable from the exact values when the SER is less than 10^{-3}.

The BER performance depends on the mapping from information bits to symbols, which in turn depends on the modulator (and demodulator) implementation. If M is not a power of two, all constellation points cannot be used for binary data transmission, but the excess points can be used for framing and control purposes, as in, e.g., Fast Ethernet and Gigabit Ethernet, where 3- and 5-level modulation formats are standardized [52, pp. 285–289]. The amount of excess points can be controlled by mapping bits to a block of symbols rather than to independent symbols. The

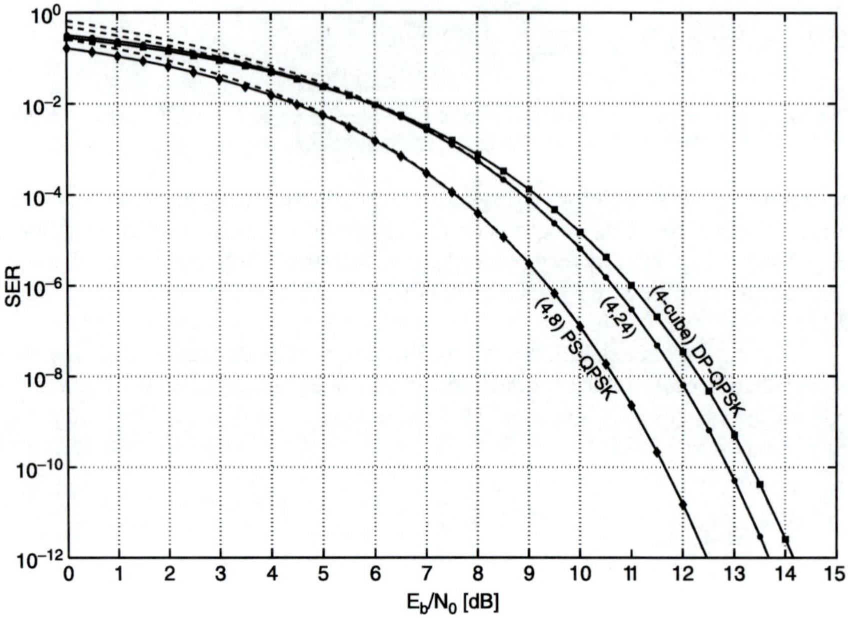

Fig. 5.11 SER vs. E_b/N_0 for $\mathscr{C}_{4,8}$ (PS-QPSK), $\mathscr{B}_{4,24}$, and $\mathscr{D}_{4\text{cube}}$ (DP-QPSK). The *dashed lines* are union bound calculations, whereas the *solid lines* are exact calculations from (5.28)–(5.30). The expected asymptotic improvements are 1.76 dB for PS-QPSK and 0.59 dB for $\mathscr{B}_{4,24}$

BER performance of DP-QPSK (or, equivalently, BPSK), PS-QPSK (exact), and 6P-QPSK (approximation) are compared in Fig. 5.12. We omit these details, which are discussed in [1], and give the results only.

For the DP-QPSK format, the BER performance is equivalent to that of the BPSK channel, which is $(1/2)\,\mathrm{erfc}(\sqrt{E_b/N_0})$. This property holds for any N-dimensional cubic modulation format, such as BPSK, QPSK, or DP-QPSK. For the PS-QPSK format, we map the bits so that opposite points in the constellation have opposite bit patterns and find that $\mathrm{BER}_{\text{PS-QPSK}} \approx \mathrm{SER}_{4,8}/2$. For the 6P-QPSK format, we map nine bits to two consecutive symbols, and then it is possible to obtain $\mathrm{BER}_{\text{6P-QPSK}} \approx (5/18)\mathrm{SER}_{4,24}$.

5.5 Sensitivities and Nonlinearities

We will now discuss how these power-efficient modulation formats will improve the fundamental quantum-limited sensitivities of optical systems, and also discuss the role of fiber nonlinearities.

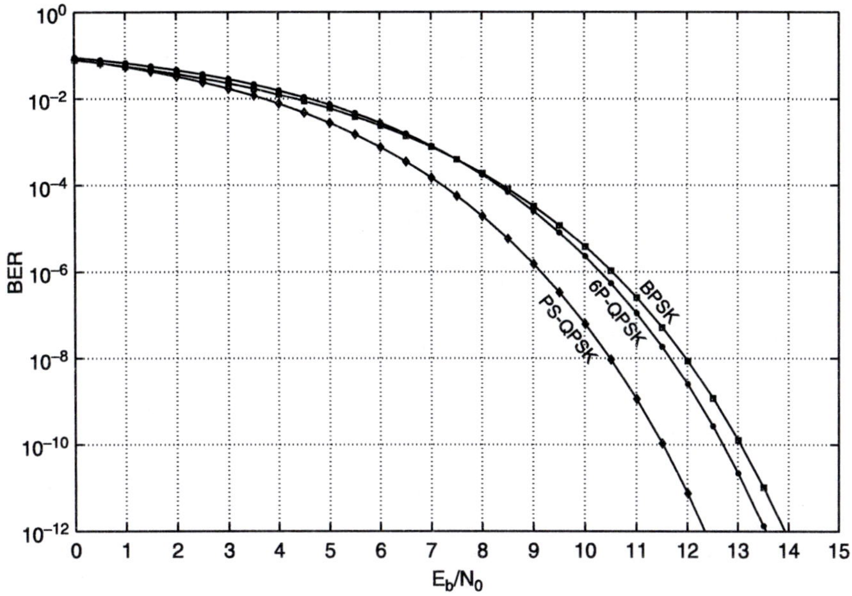

Fig. 5.12 BER vs. E_b/N_0 for PS-QPSK, 6P-QPSK, and BPSK. QPSK and DP-QPSK have the same BER performance as BPSK. The improvement of PS-QPSK over BPSK is 0.97 dB at a BER of 10^{-3} and 1.51 dB at 10^{-9}. The asymptotic gains are again 1.76 dB for PS-QPSK but only 0.51 dB for 6P-QPSK

5.5.1 Fundamental Sensitivity Limits

Under the reasonable assumption that coherent links will use optical amplifiers, the main limiting noise source will be ASE noise from the amplifiers. It has been shown [21] that ASE noise is additive and Gaussian in nature, i.e., that the AWGN model applies to such a system. The optical noise at the receiver has a power spectral density of

$$N_0 = N_a n_{sp} h\nu \frac{G-1}{G} \approx N_a n_{sp} h\nu \tag{5.34}$$

per polarization [24, 30]. Here, N_a denotes the number of in-line amplifiers, G the gain, n_{sp} the spontaneous emission factor of the amplifiers, and $h\nu$ the photon energy. In a polarization diversity homodyne coherent receiver, the optical amplitude is directly mapped to the electrical signal, so our AWGN results can be interpreted by using $E_b/N_0 = n_b/N_a n_{sp}$, where n_b is the average number of photons per bit. In the limit of a single amplifier with 3 dB noise figure ($N_a = n_{sp} = 1$), this implies that E_b/N_0 has a physically appealing interpretation as the number of photons per bit of the received signal. This can be used to translate the results from Fig. 5.12 to sensitivities (i.e., the number of photons per bit required to get BER $= 10^{-9}$). For BPSK, we get the well-known result $E_b/N_0 = 12.5$ dB $= 18$ photons per bit

Table 5.1 The properties of some common modulation formats, including the ones presented by us. The QAM formats are square grids; the 8-QAM being a 3×3 grid with the center point removed

Name	Nbr. of pts. M	Nbr. of dims. N	Pow. eff. γ (dB)	Spectral eff. (bits/symb/pol)	Sens. at BER $= 10^{-3}$ E_b/N_0 (dB)
BPSK	2	1	0	2	6.8
QPSK	4	2	0	2	6.8
8-PSK	8	2	–3.57	3	10.0
8-QAM	8	2	–3.01	3	9.0
16-QAM	16	2	–3.98	4	10.5
DP-QPSK = $\mathscr{D}_{4\text{cube}}$	16	4	0	2	6.8
PS-QPSK = $\mathscr{C}_{4,8}$	8	4	1.76	1.5	5.8
6P-QPSK	$2^{9/2} = 22.6$	4	0.51	2.25	6.9

[26, 30]. The most sensitive format, PS-QPSK, improves this with 1.5 dB to 13 photons per bit [28]. The 6P-QPSK format is with 17 photons per bit slightly better than BPSK. All sensitivities (including some other formats discussed in [28] are found in Table 5.1.

We believe that these relative improvements of PS-QPSK and 6P-QPSK over BPSK will translate also to other coherent optical channels where the AWGN model applies, such as the shot-noise limit [23, 24]. Neglecting pulse position modulation (which has been shown to provide unbounded capacity but is impractical in high-speed links [36]), we can thus conclude that the PS-QPSK modulation format gives the best sensitivity in uncoded optical links [28].

To get some real numbers into these sensitivities, we may note that at a bit rate of $1/T = 10$ Gbit/s, one photon per bit equals a received optical power of –59 dBm, and the sensitivity for BPSK in the ASE limit is then 12.5 dB above this, at –46.5 dBm. Recent experiments, based on offline synchronization algorithms, have succeeded in reaching remarkably close, within 4 dB, of this limit [31]. At higher rates, e.g., 100 Gbit/s, the sensitivity power levels become 10 dB higher in absolute power terms. Eventually, at this and higher rates, the nonlinear distortions of optical fibers will limit the BER, and power-efficient modulation formats such as those outlined in this paper may play an important role in improving the performance.

5.5.2 Nonlinear Effects

The widespread deployment of EDFAs, and the development of high-power optical amplifiers have made the available optical power less of a problem than in the pre-EDFA days. Instead, fiber nonlinearities such as SPM and XPM are becoming increasingly important as limiting factors of fiber capacity [9, 10, 19, 60, 61]. The influence of nonlinearities is complicated by the fact that they are more or less impossible to discuss without also considering the dispersion. Different dispersion management schemes will lead to different impacts of the nonlinearities.

For example, links with dispersion compensating fiber inserted periodically will not influence the signal in the same way as links that compensate all accumulated dispersion in the receiver (which is becoming more and more common in coherent systems) [41, 61]. The latter situation is significantly more difficult to analyze; to our knowledge, no analytic approaches are available and one usually has to resort to tedious simulations [10, 61].

The case when the accumulated dispersion is not allowed to grow significantly (by, e.g., in-line compensation) is easier to analyze. The simplest approach is to just neglect dispersion, or only account for the walk-off effects in WDM systems. Then it is simpler to investigate how the SPM or XPM alone, or together with ASE noise, distorts the signal. Such links are mainly penalized by, to first order, the SPM/XPM-induced nonlinear phase shift, and to second order, nonlinear phase noise (NLPN). On the one hand, SPM is usually less relevant for equal-amplitude formats, since all constellation points will get the same nonlinear phase shift. On the other hand, it acts over all high-power sections in the system. In absence of dispersion and noise, SPM can be completely cancelled in the receiver by rotating the phase back in proportion to the detected amplitude.

XPM, in contrast, induces phase shifts in proportion to the instantaneous power in all WDM channels, but acts mainly over the walk-off-length between the two WDM channels considered. It cannot be compensated, unless all WDM channels are simultaneously received and post-processed, which seems very challenging in today's systems. In general, XPM acts in two ways, one is direct phase modulation and the other is polarization changes, sometimes referred to as *cross-polarization modulation*, XPolM [29, 57].

NLPN comes from the simultaneous action of ASE-induced intensity noise and SPM (or XPM). It will make the channel differ from the AWGN model by causing the phase noise to be larger than the amplitude noise.

There are three different aspects of the nonlinear influence on modulation formats that we shall briefly discuss here. They are (1) the role of the format's power efficiency, (2) the format's robustness against nonlinear impairments and (3) the format's influence on other wavelengths via XPM. In general, all these three items will be relevant, but which one is most limiting may likely vary between different system configurations, and would require full WDM system simulations to analyze, which is beyond the scope of this paper.

5.5.2.1 Power Efficiency

Obviously, power-efficient formats allow the transmitted power to be reduced, and as a result, the induced nonlinearities will decrease. Thus, for example, we can expect the PS-QPSK format to have 1.76 dB less power than DP-QPSK when transmitting at the same data rate, and naturally, this will be beneficial in links that are affected by nonlinearities.

5.5.2.2 Nonlinear Robustness

The power efficiency is not the whole truth when it comes to nonlinear robustness. We must also consider the robustness to SPM/XPM of the formats. For example, the multilevel pulse-amplitude modulation (PAM) format may tolerate more NLPN than QPSK, since the NLPN will move the points in the phase rather than amplitude direction, and hence not closer to a decision boundary. Thus, from this point of view, amplitude modulation might be beneficial in NLPN-limited links. However, amplitude-modulated formats will get more distorted from SPM, so it may not necessarily be a benefit.

Only scattered work has been done on comparing the nonlinear robustness of different formats in coherent links, so this is a rather open field for research. Recent simulation work on PS-QPSK have shown an improved robustness to XPM nonlinearities over DP-QPSK [65, 66].

5.5.2.3 XPM-Induced Crosstalk

Even if, as we saw above, a PAM format may be more robust to nonlinear phase rotation in itself, amplitude-modulated formats are much worse when it comes to their influence on other WDM channels via XPM. This means that the amount of XPM-induced phase shift will depend on which symbols in the WDM channels overlap at a specific instance of time. Therefore, from this point of view, one would prefer equal-amplitude formats. For example, it has been shown that coherent DP-QPSK channels are more severely affected by on-off keying WDM channels than other DP-QPSK channels [10, 41].

However, in the presence of dispersion, also initially equal-amplitude formats will become amplitude-varying, so how large this effect is will depend on the details of the link and its dispersion management. There is, for example, work indicating that no optical dispersion compensation reduces the XPM influence [41, 61].

5.5.2.4 Relevance of Maximum Energy Optimization

In general, all these three items will be relevant, but which one is most limiting may likely vary between different system configurations, and would require full WDM system simulations to analyze, which is beyond the scope of this paper.

It should thus be evident from the above discussion that nonlinear limitations are complex, and depend strongly on link design parameters such as dispersion map, amplifier spacing, WDM channel powers and separation, and, last but not least, modulation formats. As we know that SPM and XPM are determined by instantaneous rather than average power levels, we believe that minimization of maximum symbol energy power is preferred over average energy minimization in situations where nonlinearities are significant. There is thus reason to compare the two optimization schemes in more detail, and it would be interesting to show the formats also on a maximum-energy scale rather than the average bit-energy scale that is usually chosen. This is done in Figs. 5.13 and 5.14, which shows the

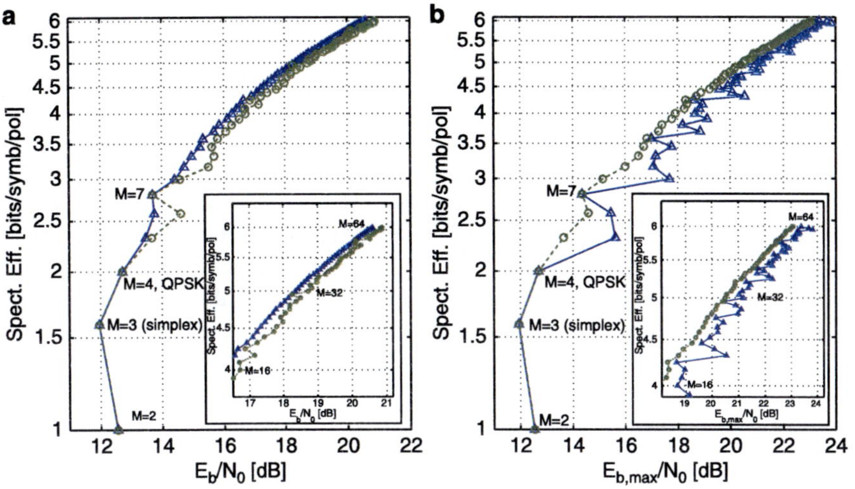

Fig. 5.13 SE vs. sensitivity for two-dimensional balls (*circles, dashed lines*) $\mathcal{B}_{2,M}$ and clusters (*triangles, solid lines*) $\mathcal{C}_{2,M}$, at a sensitivity defined at SER $= 10^{-9}$. The two plots show average (**a**) and maximum (**b**) SNR, and the insets are magnifications of the last points up to $M = 64$

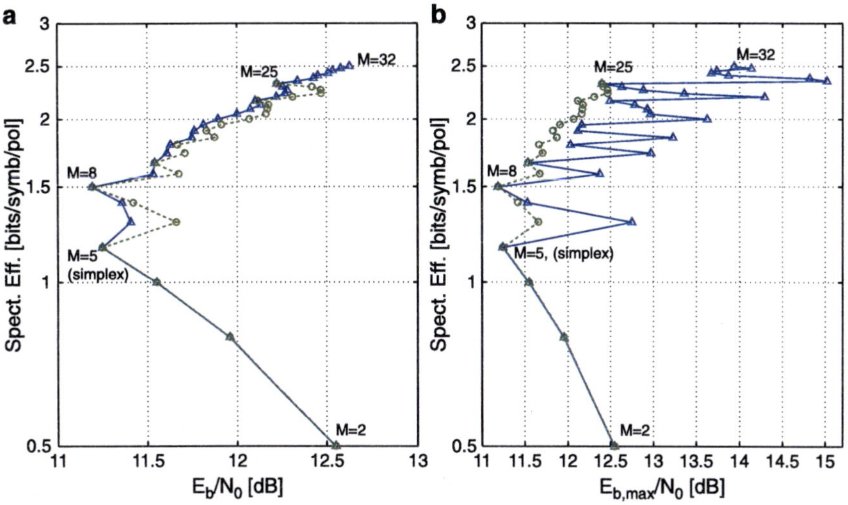

Fig. 5.14 SE vs. sensitivity for four-dimensional balls (*circles, dashed lines*) $\mathcal{B}_{4,M}$ and clusters (*triangles, solid lines*) $\mathcal{C}_{4,M}$, at a sensitivity defined as SER $= 10^{-9}$. The two plots show the same constellations vs. average (**a**) and maximum (**b**) SNR, for clusters up to $M = 32$ and balls up to $M = 25$

performance of the clusters and balls of Sect. 5.3 in terms of average bit energy E_b and maximum bit energy $E_{b,max} = E_{s,max}/\log_2 M$. Obviously, the clusters outperform the balls in terms of average energy, and the balls are better in terms of maximum energy. It is, however, interesting to see that many clusters are very bad in terms of maximum energy (the (b)-plots), whereas the balls perform fairly well for both measures. The cases in which the cluster and the ball coincide seem, however, to be very good constellations in general. In two dimensions, this occurs for $M = 2, 3, 4, 7, 31, 55$, which we believe are the only cases. In four dimensions, it occurs for $M = 2, 3, 4, 5, 8, 10, 25$, and although this list may not be conclusive as we have not analyzed balls beyond $M = 25$, we believe there are only a finite number of coinciding cases.

A next step in the research of these optimized constellations will be to make full simulations, including nonlinearities and thereby judging the nonlinear robustness of these formats. Their practical realization may in some cases be complicated by the number of symbols in a constellation not being a power of 2. The transmitters and receivers for nonrectangular constellations are more complex as well, and those are also problems to look into. Nevertheless, a format such as PS-QPSK has none of these problems [28], and to investigate its nonlinear robustness and performance relative to, e.g., DP-QPSK appears to be quite interesting.

5.6 Summary and Outlook

By using numerically optimized sphere constellations, we computed the best sensitivities of four-dimensional modulation formats up to 32 levels, which resulted in the conclusion that PS-QPSK is the format with the overall best sensitivity, 1.76 dB better than BPSK. We have shown that this is the most power-efficient modulation format when using four-dimensional constellations, unless the dimension is somehow increased. This can be done, for example, by using error-correcting codes, wavelength/space/time division multiplexing, or different modes in multimode fibers.

We also studied constellations that were optimized with respect to peak power, which we believe are relevant in nonlinearly limited systems. Our comparisons show that the mismatch penalty when using a format optimized for peak power in a scenario, where the average power is critical, is much less than vice versa. Hence, formats optimized for peak power are more robust and should be preferred in applications where both average and peak power are relevant, which is the case for most nonlinear impairments. Analyzing the performance of these modulation formats in nonlinear situations is an open area for future research.

Acknowledgements We wish to acknowledge funding from Vinnova within the IKT grant, and the Swedish strategic research foundation (SSF). We also acknowledge numerous stimulating discussions with all the researchers within the Chalmers fiber-optic communications research center FORCE. Dr. Seb Savory is gratefully acknowledged for a useful discussion, help with the $\mathscr{C}_{4,16}$ cluster, and for providing a few previously overlooked references.

References

1. E. Agrell, M. Karlsson, J. Lightwave Technol. **27**(22), 5115–5126 (2009)
2. E. Agrell, M. Karlsson, On the symbol error rate of regular polyhedra (2010). IEEE Trans. Inform. Theor., to appear, 2011
3. S. Benedetto, E. Biglieri, *Principles of Digital Transmission: With Wireless Applications* (Kluwer, New York, 1999)
4. S. Benedetto, P. Poggiolini, IEEE Trans. Commun. **40**(4), 708–721 (1992)
5. S. Betti, F. Curti, G. De Marchis, E. Iannone, Electron. Lett. **26**(14), 992–993 (1990).
6. S. Betti, F. Curti, G. De Marchis, E. Iannone, J. Lightwave Technol. **9**(4), 514–523 (1991).
7. S. Betti, G. De Marchis, E. Iannone, P. Lazzaro, J. Lightwave Technol. **9**(10), 1314–1320 (1991).
8. E. Biglieri, *Advanced Modulation Formats for Satellite Communications*, ed. by J. Hagenauer. Advanced Methods for Satellite and Deep Space Communications (Springer, Berlin, 1992) pp. 61–80
9. A. Bononi, M. Bertolini, P. Serena, G. Bellotti, J. Lightwave Technol. **27**(18), 3974–3983 (2009).
10. A. Bononi, P. Serena, N. Rossi, Opt. Fiber Technol. **16**, 73–85 (2010)
11. H. Bülow, Polarization QAM modulation (POL-QAM) for coherent detection schemes. *Proceedings of optical fiber communication and national fiber optic engineers conference, OFC/NFOEC'09*. Paper OWG2, 2009
12. G. Charlet, N. Maaref, J. Renaudier, H. Mardoyan, P. Tran, S. Bigo, Transmission of 40 Gb/s QPSK with coherent detection over ultra-long distance improved by nonlinearity mitigation. *Proceedings of European conference on optical communications, ECOC'06*. Paper PDP Th.4.3.6, 2006
13. G. Charlet, M. Salsi, J. Renaudier, O. Pardo, H. Mardoyan, S. Bigo, Electron. Lett. **43**(20), 1109–1111 (2007).
14. J.H. Conway, N.J.A. Sloane, *Sphere Packings, Lattices and Groups*, 3rd edn. (Springer, New York, 1999)
15. H.S.M. Coxeter, *Regular Polytopes* (Dover Publications, New York, 1973)
16. R. Cusani, E. Iannone, A. Salonico, M. Todaro, J. Lightwave Technol. **10**(6), 777–786 (1992)
17. F. Derr, Electron. Lett. **26**(6), 401–403 (1990)
18. N. Ekanayake, T. Tjhung, IEEE Trans. Inform. Theor. **IT-28**(4), 658–660 (1982)
19. R. Essiambre, G. Kramer, P. Winzer, G. Foschini, B. Goebel, J. Lightwave Technol. **28**(4), 662–701 (2010)
20. G. Foschini, R. Gitlin, S. Weinstein, IEEE Trans. Commun. **22**(1), 28–38 (1974)
21. J.P. Gordon, L.R. Walker, W.H. Louisell, Phys. Rev. **130**(2), 806–812 (1963).
22. R.L. Graham, N.J.A. Sloane, Discrete Comput. Geom. **5**(1), 1–11 (1990)
23. K.-P. Ho, *Phase-Modulated Optical Communication Systems* (Springer, New York, 2005)
24. E. Ip, A.P.T. Lau, D.J.F. Barros, J.M. Kahn, Opt. Express **16**(2), 753–791 (2008); Opt. Express **16**(26), 21943 (2008)
25. G. Jacobsen, *Noise in Digital Optical Transmission Systems* (Artech House Publishers, Boston, 1994)
26. J.M. Kahn, K.-P. Ho, IEEE J. Select. Top. Quant. Electron. **10**(2), 259–272 (2004).
27. J.M. Kahn, A.H. Gnauck, J.J. Veselka, S.K. Korotky, B.L. Kasper, IEEE Photon. Technol. Lett. **2**(4), 285–287 (1990).
28. M. Karlsson, E. Agrell, Opt. Express **17**(13), 10814–10819 (2009)
29. M. Karlsson, H. Sunnerud, J. Lightwave Technol. **24**(11), 4127–4137 (2006)
30. L. Kazovsky, S. Benedetto, A. Willner, *Optical Fiber Communication Systems* (Artech House Publishers, Boston, 1996)
31. K. Kikuchi, S. Tsukamoto, J. Lightwave Technol. **26**(13), 1817–1822 (2008)
32. H.G. Kim, 4-dimensional modulation for a bandlimited channel using Q^2PSK. *IEEE wireless communications and networking conference, WCNC*, vol. 3, pp. 1144–1147, 1999
33. G. Lachs, IEEE Trans. Inform. Theor. **9**(2), 95–97 (1963)

34. D. Ly-Gagnon, K. Katoh, K. Kikuchi, Electron. Lett. **41**(4), 206–207 (2005)
35. O. Musin, Ann. Math. **168**, 1–32 (2008)
36. J.R. Pierce, IEEE Trans. Commun. **26**(12), 1819–1821 (1978)
37. J.R. Pierce, IEEE Trans. Commun. **COM-28**(7), 1098–1099 (1980)
38. J.-E. Porath, T. Aulin, IEE Proc. Commun. **150**(5), 317–323 (2003).
39. J. Proakis, *Digital Communications*, 4th edn. (McGraw-Hill, Boston, 2001)
40. J. Renaudier, G. Charlet, M. Salsi, O. Pardo, H. Mardoyan, P. Tran, S. Bigo, J. Lightwave Technol. **26**(1), 36–42 (2008)
41. K. Roberts, M. O'Sullivan, K.T. Wu, H. Sun, A. Awadalla, D.J. Krause, C. Laperle, J. Lightwave Technol. **27**(16), 3546–3559 (2009).
42. D. Saha, T. Birdsall, IEEE Trans. Commun. **37**(5), 437–448 (1989).
43. C.E. Shannon, Proc. IRE **37**(1), 10–21 (1949)
44. C.E. Shannon, Bell Syst. Tech. J. **38**(3), 611–656 (1959)
45. M. Simon, S. Hinedi, W. Lindsey, *Digital Communication Techniques: Signal Design and Detection*. (PTR, Prentice Hall, 1995)
46. N.J.A. Sloane, R.H. Hardin, T.S. Duff, J.H. Conway, Discrete Comput. Geom. **14**(3), 237–259 (1995)
47. N.J.A. Sloane, R.H. Hardin, T.S. Duff, J.H. Conway, Minimal-energy clusters, library of 3-d clusters, library of 4-d clusters (1997). http://www.research.att.com/~njas/cluster/
48. N.J.A. Sloane, R.H. Hardin, T.S. Duff, J.H. Conway, Spherical codes, part 1 (2000). http://www.research.att.com/~njas/packings/
49. E. Specht, The best known packings of equal circles in the unit circle (2009). http://hydra.nat.uni-magdeburg.de/packing/cci/cci.html
50. K. Stephenson, Circle packing bibliography as of September 2005 (2005). http://www.math.utk.edu/~kens/CP-bib.pdf
51. H. Sun, K. Wu, K. Roberts, Opt. Express **16**(2), 873–879 (2008)
52. A.S. Tanenbaum, *Computer Networks*, 4th edn. (Pearson, Upper Saddle River, 2003)
53. G. Taricco, E. Biglieri, V. Castellani, Applicability of four-dimensional modulations to digital satellites: A simulation study. *Proceedings of IEEE global telecommunications conference*, vol. 4, pp. 28–34, 1993
54. S. Tsukamoto, D. Ly-Gagnon, K. Katoh, K. Kikuchi, Coherent demodulation of 40-Gbit/s polarization-multiplexed QPSK signals with 16-GHz spacing after 200-km transmission. *Proceedings of optical fiber communication and national fiber optic engineers conference, OFC/NFOEC*, vol. 6. Paper PDP 29, 2005
55. E.W. Weisstein, Ball, From Mathworld – a Wolfram Web Resource (2010). http://mathworld.wolfram.com/Ball.html
56. G. Welti, J. Lee, IEEE Trans. Inform. Theor. **20**(4), 497–502 (1974)
57. M. Winter, C.A. Bunge, D. Setti, K. Petermann, J. Lightwave Technol. **27**(17), 3739–3751 (2009)
58. J. Wu, M.C. Wu, IEEE Trans. Vehicular Technol. **49**(6), 2244–2256 (2000)
59. L. Xiao, X. Dong, IEEE Trans. Wireless Commun. **4**(4), 1418–1424 (2005)
60. C. Xie, IEEE Photon. Technol. Lett. **21**(5), 274 (2009)
61. C. Xie, Opt. Express **17**(6), 4815–4823 (2009)
62. L. Zetterberg, H. Brändström, IEEE Trans. Commun. **25**(9), 943–950 (1977)
63. H.Y. Song, S.W. Golomb, IEEE Trans. Inform. Theor. **40**(2), 504–507 (1994)
64. M. Karlsson, E. Agrell, Four-dimensional optimized constellations for coherent optical transmission systems. *Proceedings of the 36th European conference on Optical Communication, ECOC'*10. Paper We.8.C.3, 2010
65. P. Serena, A. Vanucci, A. Bononi, The performance of polarization-wwitched QPSK (PS-QPSK) in dispersion managed WDM transmissions. *Proceedings of the 36th European conference on Optical Communication, ECOC'*10. Paper Th.10.E.2, 2010
66. P. Poggiolini, Opt. Express. **18**(11), 11360–11371 (2010)

Chapter 6
A Unified Theory of Intrachannel Nonlinearity in Pseudolinear Transmission

Antonio Mecozzi

6.1 Introduction

The material of this chapter originates from a visit of the author the AT&T Laboratory in Red Bank, NJ in the summer of 2000. During that visit, the author was exposed to some experimental work on transmission using short pulses, which spread very rapidly upon propagation and for this reason were dubbed by Jay Wiesenfeld into "Tedons" from "to ted" which, according to Merriam-Webster's Collegiate Dictionary, means "to spread or turn from the swath and scatter (as new-mown grass) for drying." Tedons minimize the effects of nonlinearity by a quick spread, unlike solitons that instead resist to nonlinearity by balancing nonlinearity with dispersion, so that their shape does not change. He teamed up with Carl Clausen and Mark Shtaif and developed a perturbative theory, whose results were presented in a series of three papers [1–3]. The details of that theory and of its derivations were, however, never published in the open literature. The presentation of these details, together with some later improvements, is the purpose of this chapter.

The theory was originally developed for the only practical scheme at the time, namely on-off keying (OOK) intensity-modulation direct-detection (IMDD) transmission, a scheme that exploit only one of the four degrees of freedom (two quadratures for each polarization) of a single-mode optical field [4]. Ten years, however, did not pass in vain. It is the purpose of this chapter to extend the kind of modulations that are becoming relevant today, differential phase-shift keying (DPSK) and differential quadrature phase-shift keying (DQPSK) [5].

The maximum information rate (the capacity) that can be transmitted in a communication channel is limited by channel nonidealities. In amplified fiber-based systems, like those in the backbone of the information infrastructure, a ubiquitous nonideality is the noise of the in-line amplifiers that are used to compensate for fiber loss. Amplified spontaneous emission (ASE) is inevitably present because basic quantum mechanical principles, and namely the Heisenberg uncertainty principle,

A. Mecozzi (✉)
University of L'Aquila, 67100 L'Aquila, Italy
e-mail: antonio.mecozzi@univaq.it

S. Kumar (ed.), *Impact of Nonlinearities on Fiber Optic Communications*, Optical and Fiber Communications Reports 7, DOI 10.1007/978-1-4419-8139-4_6,
© Springer Science+Business Media, LLC 2011

would otherwise be violated [6]. It generates white Gaussian noise in the optical domain. When ASE noise is the only impairment, the channel capacity is given by the celebrated Shannon formula [7]

$$C = 2 \times \frac{1}{2T} \log_2 \left(1 + \frac{S}{N}\right), \tag{6.1}$$

where C is units of bits per time, T is the symbol duration, S is the average signal power, and N is the average noise power per degree of freedom. This formula assumes that transmitter and channel have no memory, and it is achieved when the transmitted signal has an infinite number of Gaussian distributed levels. Equation (6.1) directly applies to optical transmission as well when it is based on a coherent receiver, which is capable of recovering both quadratures of the optical signal. The coherent detection case is characterized by two independent degrees of freedom, the two quadratures of the optical field; this is the reason for the factor 2 in (6.1) [4]. In [8], it has been shown that the the spectral efficiency achieved in recent "hero" experiments over practical distances lies well below the level given by (6.1), the main reason for this being that optical transmission systems are far from being linear. High bit-rate transmission over practical distance is in fact impaired by the optical nonlinearity of the fiber, mainly Kerr nonlinearity. So, pumping up the signal power to increase the information rate, as suggested by the Shannon formula, is a successful strategy only until the fiber nonlinearity kicks in, causing signal distortion. The capacity of a realistic channel is therefore limited by both amplifier noise and fiber nonlinearity and, of course, by their interaction.

A series of recent papers [9–12] has quantified to what extent the actual channel capacity is limited by nonlinearity. For a given amount of ASE noise, increasing the power above a given level results in a reduction of the capacity because of the nonlinear impairments. Thus, for a given transmission distance, the capacity cannot exceed a maximum value. This maximum value, however, depends on the system design. Because of the large number of control parameters available in every system design, it is not obvious that the maximum capacity, estimated with a numerical optimization of the system design as in [9–12], be the actual maximum. It was indeed already shown that a careful design of the line dispersion can strongly reduce the impairments caused by the nonlinearity of the fiber [13]. Any analytical tools that may serve as a guidance for the optimization of the system design is therefore highly desired. The presentation of a first attempt toward the development of such analytical tools is given in this chapter.

6.2 Basic Formalism

Let us start with the nonlinear Schrödinger equation for the scalar electric field amplitude ψ, averaged to account for the small-scale polarization evolution (no polarization-dependent effects are considered in this chapter)

$$\frac{\partial \psi}{\partial z} = \frac{g(z) - \alpha}{2}\psi - i\frac{\beta''}{2}\frac{\partial^2 \psi}{\partial t^2} + i\gamma|\psi|^2\psi, \tag{6.2}$$

where $g(z)$ is the local power gain coefficient within the fiber (lumped with Erbium amplifiers or distributed with Raman), α is the power attenuation coefficient, β'' (negative in the anomalous dispersion region) is the group velocity dispersion, $\gamma = 2\pi n_2/(\lambda A_{\rm eff})$ is the fiber nonlinear coefficient, n_2 is the nonlinear refractive index, and $A_{\rm eff}$ is the effective area of the fiber. If we substitute into (6.2)

$$\psi(z) = \Psi(z)u(z) \tag{6.3}$$

with

$$\frac{\mathrm{d}}{\mathrm{d}z}\Psi(z) = \frac{g(z) - \alpha}{2}\Psi(z), \tag{6.4}$$

we obtain

$$\frac{\partial u}{\partial z} = -i\frac{\beta''}{2}\frac{\partial^2 u}{\partial t^2} + i\gamma f(z)|u|^2 u, \tag{6.5}$$

where $f(z) = \Psi^2(z)$ rescales the fiber nonlinearity to include the effects of a nonunform power profile. It assumes that if equally spaced Erbium amplifiers are used, that exactly compensate for the attenuation of the preceding fiber span, the expression

$$f(z) = \exp\left[-\alpha \operatorname{mod}(z, z_s)\right], \quad 0 \le z < L, \tag{6.6}$$

where mode is the modulus function, z_s is the span length, and L is the fiber length.

6.3 First-Order Perturbation Theory

It might be convenient Fourier transforming (6.5) to obtain

$$\begin{aligned}\frac{\partial \tilde{u}(z,\omega)}{\partial z} &= i\frac{\beta''}{2}\omega^2\tilde{u}(z,\omega) + i\gamma f(z)\\ &\times \int \frac{\mathrm{d}\omega'}{2\pi}\int \frac{\mathrm{d}\omega''}{2\pi}\tilde{u}(z,\omega+\omega')\tilde{u}^*(z,\omega'+\omega'')\tilde{u}(z,\omega').\end{aligned} \tag{6.7}$$

We may at this point treat the nonlinear term perturbatively, defining $\tilde{u}(z,\omega) = \tilde{u}_0(z,\omega) + \Delta u(z,\omega)$. Let us assume that the dispersion is always constant, except for lumped locations where dispersion is added linearly to the field (dispersion compensating locations). We assume that at the line input, the field is linearly predispersed by some fixed amount of dispersion (usually opposite to that of the line), transmitted through the dispersive nonlinear fiber, and the total accumulated dispersion of the field (predispersion + line dispersion) is fully compensated by a linear dispersion compensating device. In other words, we assume that the initial and final point of the first span between dispersion compensating stations are always points where the field experiences zero-accumulated dispersion. Then, in the second span between dispersion compensating stations, the field is predispersed, transmitted again through the fiber, and the total accumulated dispersion is linearly compensated. The spans after the second are treated in the same way. Using this trick, we may analyze the concatenation of more than one span between dispersion compensating stations as the concatenation of spans where the initial and final point have zero-accumulated dispersion. Then, within linear perturbation theory, the perturbation at the end of the line will be the sum of the perturbation of these zero-accumulated dispersion sections between compensating stations.

We treat the effect of nonlinearity using first-order perturbation theory, using $\tilde{u}(z,\omega) = \tilde{u}_0(z,\omega) + \Delta u(z,\omega)$ into (6.7) and preserving only terms up to first-order in $\Delta u(z,\omega)$. This approximation is well founded in the case of transmission of short pulses because of the large phase-mismatch of the different frequency components of the transmitted field. It is also a good approximation if the local dispersion is high, and the pulses weak enough. The regime of operation where first-order perturbation theory is valid is known as quasi-linear transmission. The validity of the theory will be checked self-consistently at the end.

If $\tilde{u}_0(z,\omega)$ is the Fourier transform of the field injected in the fiber, the field after precompensation and propagation up to z, at zeroth order, that is without nonlinearity or $\gamma = 0$, is

$$\tilde{u}_0(z,\omega) = \tilde{v}(\omega) \exp\left[i \frac{\beta''}{2} \omega^2 (z - z^*) \right], \tag{6.8}$$

where $\tilde{v}(\omega) = \tilde{u}(0,\omega)$ for short. Here, we have assumed that the precompensation is translated into an equivalent fiber length. Namely, if the amount of precompensation is β_{pre}, then $z^* = -\beta_{\mathrm{pre}}/\beta''$ is the point down the fiber where the accumulated linear dispersion of the fiber exactly counteracts the precompensation dispersion so that the field under linear propagation is the same as at the input, unchirped if the input field was such.

Inserting $\tilde{u}(z,\omega) = \tilde{u}_0(z,\omega) + \Delta u(z,\omega)$ into (6.7), using $\tilde{u}(z,\omega) \simeq \tilde{u}_0(z,\omega)$ within the term proportional to γ, and integrating with $\Delta\tilde{u}(0,\omega) = 0$, we obtain

$$\Delta\tilde{u}(z,\omega) = i\gamma \exp\left[i \frac{\beta''}{2} \omega^2 (z - z^*) \right] \int_0^z \mathrm{d}z' f(z') \int \frac{\mathrm{d}\omega'}{2\pi} \int \frac{\mathrm{d}\omega''}{2\pi} \\ \times \tilde{v}(\omega + \omega') \tilde{v}^*(\omega' + \omega'') \tilde{v}(\omega'') \exp\left[i\varphi(\omega,\omega',\omega'')(z' - z^*) \right], \tag{6.9}$$

where the exponent is

$$\varphi(\omega,\omega',\omega'') = \frac{\beta''}{2} \left[(\omega + \omega')^2 - (\omega' + \omega'')^2 + \omega''^2 - \omega^2 \right] = \beta'' \omega' (\omega - \omega''). \tag{6.10}$$

Let us now assume that at $z = L$ a linear dispersion compensating device adds to the optical field the total accumulated dispersion from $z = 0$ to $z = L - \mathrm{d}z = L^-$, including the predispersion. After dispersion compensation, the perturbation term becomes

$$\Delta\tilde{u}(L,\omega) = \Delta\tilde{u}(z \to L^-,\omega)\exp\left[-i\frac{\beta''}{2}\omega^2(L-z^*)\right]. \tag{6.11}$$

Equation (6.9) evaluated at $z = L$ becomes

$$\Delta\tilde{u}(L,\omega) = i\int_0^L \mathrm{d}z f(z)\int\frac{\mathrm{d}\omega'}{2\pi}\int\frac{\mathrm{d}\omega''}{2\pi}\tilde{v}(\omega+\omega')\tilde{v}^*(\omega'+\omega'')\tilde{v}(\omega'') \times \exp\left[i\beta''(z-z^*)\omega'(\omega-\omega'')\right]. \tag{6.12}$$

If we now substitute $\omega_1 = \omega'$ and $\omega_2 = \omega'' - \omega$, we arrive at

$$\Delta\tilde{u}(L,\omega) = i\gamma\int_0^L \mathrm{d}z f(z)\int\frac{\mathrm{d}\omega_1}{2\pi}\int\frac{\mathrm{d}\omega_2}{2\pi}\tilde{v}(\omega_1+\omega) \times\tilde{v}^*(\omega_1+\omega_2+\omega)\,\tilde{v}(\omega_2+\omega)\exp\left[-i\beta''(z-z^*)\omega_1\omega_2\right]. \tag{6.13}$$

Equation (6.13) shows that our first-order perturbation theory is equivalent to approximating the nonlinear interaction as a four-wave mixing interaction with undepleted-pump, namely the interaction by which three wavelengths affects a fourth or, alternatively, two photons are annihilated and two are created preserving both energy and momentum (phase matching) in the interaction.

If we assume that the input field is made of a sequence of pulses,

$$u_0(0,t) = \sum_j v_j(t-T_j), \quad \tilde{v}(\omega) = \sum_j \tilde{v}_j(\omega)\exp(i\omega T_j), \tag{6.14}$$

the perturbation becomes $\Delta\tilde{u}(L,\omega) = \sum_j\sum_k\sum_l \Delta\tilde{u}_{j,k,l}(L,\omega)$, where

$$\Delta\tilde{u}_{j,k,l}(L,\omega) = i\gamma\exp\left[i\omega(T_j-T_k+T_l)\right]\int_0^L \mathrm{d}z f(z)\int\frac{\mathrm{d}\omega_1}{2\pi}\int\frac{\mathrm{d}\omega_2}{2\pi} \times\exp\left[-i\beta''(z-z^*)\omega_1\omega_2 - i\omega_1(T_k-T_j) - i\omega_2(T_k-T_l)\right] \times\tilde{v}_k^*(\omega_1+\omega_2+\omega)\,\tilde{v}_l(\omega_2+\omega)\,\tilde{v}_j(\omega_1+\omega). \tag{6.15}$$

Transforming (6.15) back into time domain, we obtain

$$\Delta u(L,t) = \sum_{j,k,l}\Delta u_{j,k,l}(L,t), \tag{6.16}$$

where

$$\Delta u_{j,k,l}(L,t) = i\gamma\int_0^L \mathrm{d}z f(z)\int\frac{\mathrm{d}\omega}{2\pi}\int\frac{\mathrm{d}\omega_1}{2\pi}\int\frac{\mathrm{d}\omega_2}{2\pi}\exp\left[-i\beta''(z-z^*)\omega_1\omega_2\right] \times\exp\left[-i\omega(t-T_j+T_k-T_l) - i\omega_1(T_k-T_j) - i\omega_2(T_k-T_l)\right] \times\tilde{v}_j(\omega_1+\omega)\,\tilde{v}_k^*(\omega_1+\omega_2+\omega)\,\tilde{v}_l(\omega_2+\omega). \tag{6.17}$$

This is a general result within first-order perturbation theory. In the following section, it is specialized to the case of Gaussian pulses at input.

6.4 Sequence of Gaussian Pulses

The analysis is highly facilitated if we assume un-chirped Gaussian pulses with the same pulse width and possibly different complex amplitudes at input

$$v_j(t) = A_j \exp[-t^2/(2\tau^2)]. \tag{6.18}$$

The Fourier spectrum is

$$\tilde{v}_j(\omega) = A_j \tau \sqrt{2\pi} \exp(-\omega^2\tau^2/2). \tag{6.19}$$

In the Fourier domain, predispersion and linear dispersive evolution have a simple effect

$$\tilde{v}_j(\omega, z) = A_j \tau \sqrt{2\pi} \exp\left[-\omega^2 \frac{\tau^2}{2} + i \frac{\beta''}{2} \omega^2 (z - z^*)\right]. \tag{6.20}$$

If we define the dispersion length as

$$z_d = \frac{\beta''}{\tau^2}, \tag{6.21}$$

Equation (6.20) can be set in the form

$$\tilde{v}_j(\omega, z) = A_j \tau \sqrt{2\pi} \exp\left[i \frac{\omega^2\tau^2}{2} \left(\frac{z - z^*}{z_d} + i\right)\right]. \tag{6.22}$$

Entering (6.19) into (6.17) we obtain

$$\begin{aligned} \Delta u_{j,k,l}(L,t) = {} & i\gamma A_j A_k^* A_l \tau^3 (2\pi)^{3/2} \int_0^L \mathrm{d}z f(z) \int \frac{\mathrm{d}\omega}{2\pi} \int \frac{\mathrm{d}\omega_1}{2\pi} \int \frac{\mathrm{d}\omega_2}{2\pi} \\ & \times \exp\left[-i\omega(t - T_{j,k,l}) - i\omega_1(T_k - T_j) - i\omega_2(T_k - T_l)\right] \\ & \times \exp\left\{ -\frac{\tau^2}{2} \left[(\omega_1 + \omega)^2 + (\omega_1 + \omega_2 + \omega)^2 + (\omega_2 + \omega)^2\right] \right. \\ & \left. - i\beta''(z - z^*)\omega_1\omega_2 \right\}, \end{aligned} \tag{6.23}$$

where

$$T_{j,k,l} = T_j - T_k + T_l. \tag{6.24}$$

Performing the triple integral in frequency, we obtain after shifting the propagation axis $z = z' + z^*$ into the integral over z

$$\Delta u_{j,k,l}(t + T_{j,k,l}) = i\gamma A_j A_k^* A_l \mathcal{U}_{j,k,l}(L, t + T_{j,k,l}), \tag{6.25}$$

where

$$\mathcal{U}_{j,k,l}(t+T_{j,k,l}) = \exp\left(-\frac{t^2}{6\tau^2}\right)\int_{-z^*}^{L-z^*}\frac{f(z'+z^*)\mathrm{d}z'}{\sqrt{3q^*\,(q+2i/3)}}$$
$$\times\exp\left\{i\frac{[2t/3+(T_j-T_k)]\,[2t/3+(T_l-T_k)]}{\tau^2\,(q+2i/3)}-\frac{(T_j-T_l)^2}{3\tau^2 q^*\,(q+2i/3)}\right\}, \quad (6.26)$$

and the complex parameter q is defined as

$$q=\frac{z}{z_d}-i. \quad (6.27)$$

Note that the dispersion length is positive or negative depending upon the sign of β''. Equation (6.25) shows that the perturbation field does not in general overlap with the generating pulses, but is centered at the time $T_{j,k,l}$ given by (6.24). Asymptotically, the integral over z' becomes virtually independent of t, hence $\mathcal{U}_{j,k,l}(t+T_{j,k,l})\propto\exp(-t^2/6\tau^2)$. Consequently, the perturbation appears as a pulse centered at $T_{j,k,l}$ of width $\sqrt{3}$ times larger than the generating pulses. If a pulse was originally present at position $T_{j,k,l}$, the perturbation coherently overlaps with this pulse. If instead there were no pulses at time $T_{j,k,l}$, the perturbation shows up as a stretched copy of the generating pulses in a position where no pulse was originally present. This process is similar to the generation of echo pulses that show up in repetitive photon echo experiments such as those described is [14, 15].

For N spans of fiber (that is, N positions where partial dispersion compensation is performed) of length L_n, the result is

$$\Delta u_{j,k,l}(L,t+T_{j,k,l}) = i\gamma A_j A_k^* A_l \mathcal{U}_{j,k,l}(L,t+T_{j,k,l}), \quad (6.28)$$

where

$$\mathcal{U}_{j,k,l}(t+T_{j,k,l}) = \exp\left(-\frac{t^2}{6\tau^2}\right)\sum_{n=1}^{N}\int_{-z_n^*}^{L_n-z_n^*}\frac{f_n(z'+z_n^*)\mathrm{d}z'}{\sqrt{3q^*\,(q+2i/3)}}$$
$$\times\exp\left\{i\frac{[2t/3+(T_j-T_k)]\,[2t/3+(T_l-T_k)]}{\tau^2\,(q+2i/3)}-\frac{(T_j-T_l)^2}{3\tau^2 q^*\,(q+2i/3)}\right\}, \quad (6.29)$$

where we use in each span the origin of the z axis at the input of each span, and z_n^* is the zero dispersion point of the span (which can be also less than zero or larger than L_n, in which case there is no point of zero dispersion within that span).

6.5 Coherent and Direct Detection

Next step is to consider a sequence of modulated pulses. We will restrict ourselves to the case of a sequence of Gaussian pulses with the same pulse-width and complex amplitudes A_j, spaced by the symbol time T_s. The amplitudes A_j are used to define the message in a set of N possible values. In OOK-IMDD, they will be either a fixed amplitude $A_j = A$ when a logical one is transmitted, or $A_j = 0$ when a logical zero is transmitted. In DPSK, the amplitudes are constant in modulus, and with a phase either φ_0 or $\varphi_0 + \pi$. In DQPSK, the modulus is still constant, but the values of the phase are now 4 spaced by $\pi/2$. In a coherent quadrature-amplitude modulation (QAM), the modulus and the phase are both varied, following a specific constellation of symbols in the complex plane.

Let us now define our model parameters. Let us define as A a *real* parameter equal to the maximum amplitude of the transmitted pulses, $A = \max(|A_j|, j = 1, \ldots, N)$, and N normalized complex amplitudes a_j such that

$$A_j = a_j A. \tag{6.30}$$

We have of course $0 \le |a_j| \le 1$, with $|a_j| = 1$ for at least one value of j. We will assume that each amplitude occurs with probability p_j, normalized such that $\sum_{j=1}^{N} P_j = 1$.

Let us focus our analysis on coherent differential detection first. With differential detection, any pulse is let to overlap with the following pulse of the stream, possibly phase shifted by φ_d, and the real part of the beat term is detected by a differential receiver. The complex amplitude of the detected photocurrent is proportional to

$$\mathcal{I}_D = \exp(-i\varphi_d) \int dt \left[a_1 A \exp\left(-\frac{t^2}{2\tau^2}\right) + \sum_{j',k',l'} \Delta u_{j',k',l'} \right]^* \times \left[a_0 A \exp\left(-\frac{t^2}{2\tau^2}\right) + \sum_{j,k,l} \Delta u_{j,k,l} \right], \tag{6.31}$$

where $\Delta u_{j,k,l} = \Delta u_{j,k,l}(L,t)$ for short. The first sum is extended to all combinations $T_{j,k,l} = T_j - T_k + T_l = 0$ and the second to all combinations $T_{j',k',l'} = T_{j'} - T_{k'} + T_{l'} = T_s$. Using this condition, the triple sums collapse into a double one because the first implies that $j - k + l = 0$ and hence that $k = j + l$, the second that $j' - k' + l' = 1$, hence that $k = j + l - 1$. The zeroth order term is

$$\mathcal{I}_D = \exp(-i\varphi_d) a_1^* a_0 \int dt A^2 \exp\left(-\frac{t^2}{\tau^2}\right) \simeq \exp(-i\varphi_d) a_1^* a_0 \sqrt{\pi} A^2 \tau, \tag{6.32}$$

where, although the integral is extended to the symbol time T_s, we have used the good approximation of replacing the integration interval with the whole time axis. Both pulses are perturbed by the nonlinear interaction. The perturbation of the complex amplitude of the photocurrent is

$$\Delta\mathcal{I}_D = \exp(-i\varphi_d) \int dt A \exp\left(-\frac{t^2}{2\tau^2}\right) \times \left[\sum_{j,k=j+l,l} a_1^* \Delta u_{j,k,l} + \sum_{j',k'=j'+l'-1,l'} a_0 \Delta u^*_{j',k',l'} \right]. \tag{6.33}$$

Defining in the second sum $j' = j - 1$, $l' = l - 1$, and $k' = k - 1$, condition $k' = j' + l' - 1$ becomes, adding -1 at both sides, $k = j + l$. Inserting in (6.33) the expression given by (6.25), we obtain

$$\Delta\mathcal{I}_D = \exp(-i\varphi_d)\left(\Delta I_{D,1} + \Delta I^*_{D,0}\right), \tag{6.34}$$

where

$$\Delta I_{D,1} = \sum_{j,k=j+l,l} a^*_1 a_j a^*_k a_l \mathcal{J}_{j,k,l}, \tag{6.35}$$

$$\Delta I_{D,0} = \sum_{j,k=j+l,l} a^*_0 a_{j-1} a^*_{k-1} a_{l-1} \mathcal{J}_{j-1,k-1,l-1}, \tag{6.36}$$

and

$$\mathcal{J}_{j,k,l} = i\gamma A^4 \sum_{j,k=j+l,l} \int \mathrm{d}t \exp\left(-\frac{2t^2}{3\tau^2}\right) \int_{-z^*}^{L-z^*} \frac{f(z' + z^*)\mathrm{d}z'}{\sqrt{3q^*\,(q + 2i/3)}}$$
$$\times \exp\left\{ i\frac{\left[2t/3 + (T_j - T_k)\right]\left[2t/3 + (T_l - T_k)\right]}{\tau^2\,(q + 2i/3)} - \frac{(T_j - T_l)^2}{3\tau^2 q^*\,(q + 2i/3)} \right\}. \tag{6.37}$$

The photocurrent detected with a balanced detector will be proportional to the real part of $\mathcal{I}_D$,

$$I_r = \mathrm{Re}(\mathcal{I}_D), \tag{6.38}$$

and the nonlinear contribution will be

$$\langle \Delta I_r^2 \rangle = \frac{1}{4}\langle (\Delta\mathcal{I}_D + \Delta\mathcal{I}_D^*)^2 \rangle - \frac{1}{4}\langle \Delta\mathcal{I}_D + \Delta\mathcal{I}_D^* \rangle^2. \tag{6.39}$$

With OOK-IMDD transmission, the directly detected photocurrent when a "one" is detected is

$$\mathcal{I}_{\mathrm{IMDD}} = \left| \int \mathrm{d}t \left[A\exp\left(-\frac{t^2}{2\tau^2}\right) + \sum_{j,k,l} \Delta u_{j,k,l} \right] \right|^2. \tag{6.40}$$

In this case, the detected photocurrent $\mathcal{I}_{\mathrm{IMDD}}$ is proportional $\mathcal{I}_{\mathrm{IMDD}}$ itself,

$$I_{\mathrm{IMDD}} = \mathcal{I}_{\mathrm{IMDD}}, \tag{6.41}$$

and the nonlinear displacement becomes

$$\Delta I_{\mathrm{IMDD}} = 2\,\mathrm{Re}\left(\sum_{j,k=j+l,l} a_j a^*_k a_l \mathcal{J}_{j,k,l} \right). \tag{6.42}$$

The transmission formats that we have considered, employing differential detection or IMDD, project at the receiver the signal onto a temporal profile with the conjugated temporal profile of the signal itself. In these cases, the nonlinear noise depends on an integral such as $\mathcal{J}_{j,k,l}$ given by (6.37). Our findings are, however, more general. It may be shown that the nonlinear noise depends on integrals like $\mathcal{J}_{j,k,l}$ also in coherent transmission systems employing a continuous wave local oscillator and a matched optical filter [16]. Giving a compact and handy expression of this quantity is therefore a useful task, which may be accomplished by inverting the integrals over t and z into (6.37), and integrating over t. After some algebra, $\mathcal{I}_{j,k,l}$ acquires the remarkably simple expression,

$$\mathcal{J}_{j,k,l} = i\gamma\sqrt{2\pi^3}A^4\tau^3 \int_{-z^*}^{L-z^*} f(z+z^*)G\left(T_j - T_k, T_l - T_k; z\right) \mathrm{d}z, \tag{6.43}$$

having introduced the complex bivariate Gaussian distribution

$$G(T_1, T_2; z) = \frac{1}{2\pi\tau^2\sqrt{z^2/z_d^2+1}} \exp\left[-\frac{T_1^2 + T_2^2 - 2i\left(z/z_d\right)T_1T_2}{2\tau^2(z^2/z_d^2+1)}\right]. \tag{6.44}$$

If this expression is used for $T_j - T_k + T_l = 0$ hence for $T_k = T_j + T_l$, this expression can be further simplified into

$$\mathcal{J}_{j,k=j+l,l} = i\gamma\sqrt{2\pi^3}A^4\tau^3 \int_{-z^*}^{L-z^*} f(z+z^*)G(T_l, T_j; z)\mathrm{d}z, \tag{6.45}$$

where we used that $G(-T_l, -T_j; z) = G(T_l, T_j; z)$. Again, in the case of N dispersion compensation stations, we have

$$\mathcal{J}_{j,k=j+l,l} = i\gamma\tau\sqrt{2\pi^3}A^4\tau^2 \sum_{n=1}^{N} \int_{-z_n^*}^{L-z_n^*} f(z+z_n^*)G(T_j, T_l; z)\mathrm{d}z, \tag{6.46}$$

where z_n^* is the zero dispersion point within the span, or the extrapolated zero dispersion point if the accumulated dispersion does not change sign within the span, in which case z_n^* is less than zero or larger than L_n.

A few words on the physical meaning of the integral $\mathcal{J}_{j,k,l}$ are now in order. Let us refer to the relevant case of equally spaced pulses, when this quantity is given by (6.44) and (6.45). This quantity is the modulus of the time-integrated fluctuations induced on the pulse centered at $T_0 = 0$ by the annihilation of two photons belonging to pulses of amplitude A centered at $T_j = jT_s$ and $T_l = lT_s$ and the creation of two photons on pulses of the same amplitude and centered at $T_k = kT_s$ and $T_0 = 0$ (four-wave mixing interaction). The phase of this fluctuation term is the sum of the phases of the pulses at $T_j = jT_s$ and $T_l = lT_s$ minus the phases of the pulses at $T_k = kT_s$ and $T_0 = 0$. The optical nonlinearity contributes to the fluctuations at the detector, to first-order, with the sum of all these interactions and their conjugates (which correspond to the inverse process where annihilation and creation are interchanged). In the special case of direct detection, (6.44) and (6.45) give a surprisingly simple expression to the intensity fluctuations induced on a Gaussian pulse by three identical pulses interacting with the first by a Kerr effect-mediated four-wave mixing process. The simplicity of this expression should be compared with the more involved form of $\Delta u_{j,k,l}$, (6.25).

The expressions given by (6.34) and (6.41) are useful because they suggest that a bit-dependent preemphasis, in both amplitude and phase, at the transmitter is a way for compensating nonlinear effects to first-order. Although in principle the sum is extended to all pulses in the message, the only non-negligible terms are, in practice, those corresponding to pulses that overlap along the path. The other pulses give negligible $\mathcal{J}_{j,k=j+l,l}$, so that their contribution to the sum is negligible.

6.6 Effect of the Symmetry of the Dispersion Profile

When the number of overlapping pulses are very large, preemphasis may be impractical. In these cases, minimization of the linear impairments may be the only practical way to cope with nonlinear effects. In some cases, the nonlinear impairments can be ideally suppressed.

To understand how and when this result can be achieved, let us first notice that with IMDD the pulses are all in phase and with DPSK their phase is multiple of 180 degrees. We may assume, without loss of generality, that the phase of the pulses is either 0 or 180°. This implies that the perturbation added by the other pulses on a pulse centered at $T_0 = 0$, proportional to $a_j a_k^* a_l \mathcal{J}_{j,k=j+k,l}$, is in quadrature with the pulse itself if $\text{Im}(\mathcal{J}_{j,k=j+l,l}) = 0$. When condition $\text{Im}(\mathcal{J}_{j,k=j+l,l}) = 0$ is met, the amplitude fluctuations of the pulses, hence the fluctuations of the detected eye, becomes zero to first-order, because the only component contributing, to first-order, to the eye fluctuations is that in-phase with the pulses. The condition $\text{Im}(\mathcal{J}_{j,k=j+l,l}) = 0$ may be achieved if $z^* = L/2$ and $f(z)$ is a symmetric function about $z = L/2$, because $\text{Im}(\mathcal{J}_{j,k=j+l,l})$ becomes in this case an antisymmetric function of z integrated over a symmetric interval. While condition $z^* = L/2$ can be easily met evenly dividing the dispersion compensation between the input and the output of the span, a symmetric $f(z)$ is more difficult to obtain. The power profile $f(z)$ can be made approximately symmetric if loss is locally compensated by Raman gain with a counterpropagating pump, so that the power profile (the integrated loss profile) becomes approximately symmetric about the center of the span.

The minimization of the in-phase component of the fluctuation is the key objective of the design of IMDD and DPSK systems even if $f(z)$ is not symmetric, for instance when lumped amplification is used. In this case, however, the in-phase component of the nonlinear displacement cannot be made zero, and in general the in-phase component is minimized for an uneven amount of pre- and postdispersion compensation. This preliminary discussion suggests furthermore that the minimization of the in-phase component is not an effective strategy in DQPSK, because on the one hand the phase distribution of the signal is such that the field does not have a preferential orientation in the complex plane and on the other, the detection scheme is sensitive to both in-phase and out-of-phase components.

6.7 Pseudo-Random Sequence in DPSK and DQPSK

In DPSK and DQPSK, the nonlinear impairments are minimized when the fluctuations of the detected photocurrent $I_r = \mathrm{Re}(\mathcal{I}_D)$ are minimized. The variance of the fluctuations is $\langle \Delta I_r^2 \rangle$ is given by (6.39). A significant simplification arises because phase-modulated signals are proportional to $a_j = \exp(i\varphi_j)$, with $\varphi_j = 0, \pi$ for DPSK and $\varphi_n = 0, \pi/2, \pi, 3/2\pi$ for DQPSK, all symbols being transmitted with equal probability. We have therefore $\langle a_j \rangle = 0$, hence $\langle \Delta \mathcal{I}_D \rangle = 0$. Using this condition the variance of ΔI_r becomes

$$\langle \Delta I_r^2 \rangle = \langle |\Delta I_1|^2 \rangle + \mathrm{Re}\left[\cos(2\varphi_d)\langle \Delta I_1^2 \rangle \right. \\ \left. + \exp(-2i\varphi_d)\langle \Delta I_1 \Delta I_0^* \rangle + \langle \Delta I_1 \Delta I_0 \rangle\right], \tag{6.47}$$

where $\varphi_d = 0$ for DPSK and $\varphi_d = \pm\pi/4$ for DQPSK. We used that the terms ΔI_1 and ΔI_0 are statistically equivalent, so that $\langle \Delta I_1^2 \rangle = \langle \Delta I_0^2 \rangle$ and $\langle |\Delta I_1|^2 \rangle = \langle |\Delta I_0|^2 \rangle$, and we allowed for non-zero correlations between the terms ΔI_1 and ΔI_0 [17]. The expressions of the various terms are

$$\langle |\Delta I_1|^2 \rangle = \sum_{j,l,j',l'} \langle a_1^* a_j a_{j+l}^* a_l a_1 a_{j'}^* a_{j'+l'} a_{l'}^* \rangle \mathcal{J}_{l,0,j} \mathcal{J}_{l',0,j'}^*, \tag{6.48}$$

$$\langle \Delta I_1^2 \rangle = \sum_{j,l,j',l'} \langle a_1^* a_j a_{j+l}^* a_l a_1^* a_{j'} a_{j'+l'}^* a_{l'} \rangle \mathcal{J}_{l,0,j} \mathcal{J}_{l',0,j'}, \tag{6.49}$$

$$\langle \Delta I_1 \Delta I_0^* \rangle = \sum_{j,l,j',l'} \langle a_1^* a_j a_{j+l}^* a_l a_0 a_{j'}^* a_{j'+l'-1} a_{l'}^* \rangle \mathcal{J}_{l,0,j} \mathcal{J}_{l'-1,0,j'-1}^*, \tag{6.50}$$

$$\langle \Delta I_1 \Delta I_0 \rangle = \sum_{j,l,j',l'} \langle a_1^* a_j a_{j+l}^* a_l a_0^* a_{j'} a_{j'+l'-1}^* a_{l'} \rangle \mathcal{J}_{l,0,j} \mathcal{J}_{l'-1,0,j'-1}, \tag{6.51}$$

where we used that $\mathcal{J}_{j,k,l} = \mathcal{J}_{j-k,0,l-k}$. First of all, let us note that all expressions have the exchange symmetry $j \leftrightarrow l$ and $j' \leftrightarrow l'$. Condition $\langle a_j \rangle = 0$ implies that nonzero average is obtained when the terms in the averages are equal in couples.

Let us first consider (6.48) and (6.49). The average is nonzero if (a) $j = j'$ and $l = l'$, or if $j = l'$ and $l = j'$, this second condition being fully equivalent to the first by exchange symmetry. It is convenient to group these two cases into a single, twofold degenerate, one. The only exception is the case $j = j'$ where the two conditions coincide, hence there is no degeneracy. The average is also nonzero if (b) $j = 0$ or $l = 0$, and $j' = 0$ or $l' = 0$, and the other two nonzero indices arbitrary. This case corresponds to the average of FWM terms where the pulses acting on pulse 0 collapse into a single one, hence to the average of cross-phase modulation (XPM) terms. Because any combination of a zero primed index with a zero unprimed index is allowed, this case is a fourfold degenerate one. Also in this case, there are exceptions to the four-fold degeneracy. If two primed indices are simultaneously zero or two of the unprimed indices are simultaneously zero, there is only a twofold degeneracy, and there is no degeneracy when all indices are simultaneously zero. If conditions (a) or (b) are not met, the average is zero.

Let us now consider (6.51) and (6.50). The average is nonzero if (c) $j'=1$ or $l'=1$ and $j=0$ or $l=0$, and the other two indices arbitrary, (d) if $l=1$, $l'=0$ and $j'=j+1$, with again all four combinations, and finally if (e) $j=j'$, $l=l'$ and $j=1-l$. The cases (c) and (d) are fourfold degenerate, the case (e) twofold degenerate. Again, there are exceptions. In the case (d), there is a twofold degeneracy if $j=1,-1$. In the case (c), there is a twofold degeneracy if the two primed indices are simultaneously one, or if the two unprimed indices are simultaneously zero, and no degeneracy for the single case $j'=1$ $l'=1$ $j=0$ and $l=0$. Physically, the case (c) is caused by nondegenerate FWM terms where one of the pulses is the interfering pulse at the detector. This result makes of course good sense, because XPM affects two consecutive pulses in a highly correlated way. Cases (d) and (e) are instead caused by correlated FWM terms. Gathering together all these findings, we may obtain

$$\langle|\Delta I_1|^2\rangle = \sum_{j,l} f_{j,l}\langle|a_1|^2|a_j|^2|a_{j+l}|^2|a_l|^2\rangle|\mathcal{J}_{l,0,j}|^2 + \sum_{j\neq j'} g_{j,j'}\langle|a_1|^2|a_0|^2|a_j|^2|a_{j'}|^2\rangle\mathcal{J}_{0,0,j}\mathcal{J}^*_{0,0,j'}, \tag{6.52}$$

$$\langle\Delta I_1^2\rangle = \sum_{j,l} f_{j,l}\langle a_1^{*2}a_j^2a_{j+l}^{*2}a_l^2\rangle\mathcal{J}^2_{l,0,j} + \sum_{j\neq j'} g_{j,j'}\langle a_1^{*2}a_0^2|a_j|^2|a_{j'}|^2\rangle\mathcal{J}_{0,0,j}\mathcal{J}_{0,j'}, \tag{6.53}$$

$$\langle\Delta I_1\Delta I_0^*\rangle = \sum_{j,j'} h_{j,j'}\langle a_1^{*2}|a_j|^2a_0^2|a_{j'}|^2\rangle\mathcal{J}_{0,0,j}\mathcal{J}^*_{0,j'-1} + \sum_{j\neq 0} q_j\langle|a_1|^2a_{j+1}^{*2}|a_0|^2a_j^2\rangle\mathcal{J}_{1,0,j}\mathcal{J}^*_{-1,j} + \sum_{j\neq 0,1}\langle a_1^{*2}|a_j|^2a_0^2|a_{1-j}|^2\rangle|\mathcal{J}_{j,0,1-j}|^2, \tag{6.54}$$

$$\langle\Delta I_1\Delta I_0\rangle = \sum_{j,j'} h_{j,j'}\langle|a_1|^2|a_j|^2|a_0|^2|a_{j'}|^2\rangle\mathcal{J}_{0,0,j}\mathcal{J}_{0,0,j'-1} + \sum_{j\neq 0} q_j\langle|a_1|^2|a_{j+1}|^2|a_0|^2|a_j|^2\rangle\mathcal{J}_{1,0,j}\mathcal{J}_{-1,0,j} + \sum_{j\neq 0,1}\langle a_1^{*2}a_j^2a_0^{*2}a_{1-j}^2\rangle\mathcal{J}^2_{j,0,1-j}, \tag{6.55}$$

where we defined the degeneracy functions

$$f_{j,l} = \begin{cases} 1 & j=l, \\ 2 & \text{elsewhere,} \end{cases} \tag{6.56}$$

$$g_{j,j'} = \begin{cases} 2 & j=0 \quad \text{or } j'=0, \\ 4 & \text{elsewhere,} \end{cases} \tag{6.57}$$

$$h_{j,j'} = \begin{cases} 1 & j = 0 \quad \text{and } j' = 1, \\ 2 & j = 0 \quad \text{or } \ j' = 1, \\ 4 \text{ elsewhere}, \end{cases} \tag{6.58}$$

$$q_j = \begin{cases} 2 & j = 1, \quad \text{or } j = -1 \\ 4 \text{ elsewhere}. \end{cases} \tag{6.59}$$

Some indices are excluded to avoid including twice individual terms of the sums in (6.48)–(6.50). For instance, $j = j'$ has been excluded in the last sum of (6.52) and (6.53), because this case coincides, with its degeneracy factor 4, with the two double degenerate cases $l = 0$ and $j = 0$ of the first term of the same equations.

Let us now consider separately the cases of DPSK and DQPSK. For DPSK, $|a_j|^2 = 1$ and $a_j^2 = 1$, for every j. After using these properties, we obtain

$$\langle |\Delta I_1|^2 \rangle = \mathcal{A}_{\text{fwm}} + \mathcal{A}_{\text{xpm}}, \tag{6.60}$$

$$\langle \Delta I_1^2 \rangle = \mathcal{B}_{\text{fwm}} + \mathcal{B}_{\text{xpm}}, \tag{6.61}$$

$$\langle \Delta I_1 \Delta I_0^* \rangle = \mathcal{A}_{\text{corr,xpm}} + \mathcal{A}_{\text{corr,fwm},1} + \mathcal{A}_{\text{corr,fwm},2}, \tag{6.62}$$

$$\langle \Delta I_1 \Delta I_0 \rangle = \mathcal{B}_{\text{corr,xpm}} + \mathcal{B}_{\text{corr,fwm},1} + \mathcal{B}_{\text{corr,fwm},2}, \tag{6.63}$$

where we defined the quantities related to the average square of ΔI_0 and ΔI_1,

$$\mathcal{A}_{\text{fwm}} = \sum_{j,l} f_{j,l} |\mathcal{J}_{l,0,j}|^2, \qquad \mathcal{B}_{\text{fwm}} = \sum_{j,l} f_{j,l} \mathcal{J}_{l,0,j}^2, \tag{6.64}$$

$$\mathcal{A}_{\text{xpm}} = \sum_{j \neq j'} g_{j,j'} \mathcal{J}_{0,0,j} \mathcal{J}_{0,0,j'}^*, \qquad \mathcal{B}_{\text{xpm}} = \sum_{j \neq j'} g_{j,j'} \mathcal{J}_{0,0,j} \mathcal{J}_{0,0,j'}, \tag{6.65}$$

and those related to their correlations

$$\mathcal{A}_{\text{corr,xpm}} = \sum_{j,j'} h_{j,j'} \mathcal{J}_{0,0,j} \mathcal{J}_{0,0,j'-1}^*, \qquad \mathcal{B}_{\text{corr,xpm}} = \sum_{j,j'} h_{j,j'} \mathcal{J}_{0,0,j} \mathcal{J}_{0,0,j'-1}, \tag{6.66}$$

$$\mathcal{A}_{\text{corr,fwm},1} = \sum_{j \neq 0} q_j \mathcal{J}_{1,0,j} \mathcal{J}_{-1,0,j}^*, \qquad \mathcal{B}_{\text{corr,fwm},1} = \sum_{j \neq 0} q_j \mathcal{J}_{1,0,j} \mathcal{J}_{-1,0,j} \tag{6.67}$$

$$\mathcal{A}_{\text{corr,fwm},2} = 2 \sum_{j \neq 0,1} |\mathcal{J}_{j,0,1-j}|^2, \qquad \mathcal{B}_{\text{corr,fwm},2} = 2 \sum_{j \neq 0,1} \mathcal{J}_{j,0,1-j}^2. \tag{6.68}$$

Inserting (6.63)–(6.68) into (6.47), one may obtain

$$\langle \Delta I_{\text{DPSK}}^2 \rangle = \mathcal{A}_{\text{fwm}} + \text{Re}\,(\mathcal{B}_{\text{fwm}}) + \sum_{s=1}^{2} (\mathcal{A}_{\text{corr,fwm},s} + \mathcal{B}_{\text{corr,fwm},s}). \tag{6.69}$$

We used that $\mathcal{B}_{\text{xpm}}$ is real and such that $\mathcal{B}_{\text{xpm}} = -\mathcal{A}_{\text{xpm}}$, and that $\mathcal{A}_{\text{corr,xpm}}$ and $\mathcal{B}_{\text{corr,xpm}}$ are also real and that $\mathcal{B}_{\text{corr,xpm}} = -\mathcal{A}_{\text{corr,xpm}}$. The terms related to XPM correlations disappear.

For DQPSK, also for more dense formats such as eight-ary differential phase-shift keying (D8PSK), we have $|a_j|^2 = 1$ and $\langle a_j^2 \rangle = 0$. This means that, in all averages, terms such as a_j^2 average to zero unless they have a partner such as a_j^{*2}, or a_j^2 being $a_j^4 = 1$, to saturate with. Using again (6.47), one may obtain

$$\langle \Delta I^2_{\mathrm{DQPSK}} \rangle = \mathcal{A}_{\mathrm{fwm}} + \mathcal{A}_{\mathrm{xpm}} + \mathcal{B}_{\mathrm{corr,xpm}} + \mathcal{B}_{\mathrm{corr,fwm,1}}. \tag{6.70}$$

In DQPSK, the correlation of XPM terms (the term $\mathcal{B}_{\mathrm{corr,xpm}}$) do affect the photocurrent fluctuations. Let me now comment on the above results by analyzing the physical meaning of each term.

6.7.1 *FWM Terms $\mathcal{A}_{\mathrm{fwm}}$ and $\mathcal{B}_{\mathrm{fwm}}$, and Correlation Terms $\mathcal{A}_{\mathrm{corr,fwm}}$ and $\mathcal{B}_{\mathrm{corr,fwm}}$*

These terms are related to nondegenerate FWM interactions and their correlation. They appear in the expression of the photocurrent fluctuations for DPSK, and only $\mathcal{A}_{\mathrm{fwm}}$ and $\mathcal{B}_{\mathrm{corr,fwm,1}}$ in that for DQPSK because the others average out. When $f(z)$ is a symmetric function about $z = L/2$, a condition that, as mentioned, can be approximated by Raman amplification with a counter-propagating pump, and $z^* = L/2$, the photocurrent fluctuations for DPSK are zero. This result, exact within first-order perturbation theory, may be simply shown by observing that when this symmetric condition is met, if the pulses of the sequence are all in-phase, or if their phases are multiple of 180 degrees, the time-integrated fluctuations $\mathcal{J}_{j,j+l,l}$ are in quadrature with the pulse, as it may be shown by the change of variable $z' = z - L/2$ in the integral in (6.44). The amplitude fluctuations of the pulses, hence the fluctuations of the detected eye, are therefore nulled to first-order. With DQPSK, instead, this mechanism is not effective because on one side the interacting pulses are not antipodal hence the fluctuations under symmetric conditions are not in quadrature any longer with the pulse itself. On the other, in DQPSK the signal is contained in both quadratures of the field, hence to extract the signal a projection onto two axis at 45° to the symbol constellation is required. In this case, phase fluctuations are not orthogonal to the axis where the signal is projected, hence they do contribute to the fluctuations of the detected photocurrent.

6.7.2 *Cross-Phase Modulation Term $\mathcal{A}_{\mathrm{xpm}}$ and Correlation Term $\mathcal{B}_{\mathrm{corr,xpm}}$*

These terms are related to the contribution to the photocurrent fluctuations by the phase noise induced by the XPM terms, $\mathcal{A}_{\mathrm{xpm}}$, and by their correlations, $\mathcal{B}_{\mathrm{corr,xpm}}$. They appear in the expression of the photocurrent fluctuations for DQPSK, not in that of DPSK. This fact should not be surprising. Phase fluctuations do not contribute to first-order to the noise of DPSK because the receiver is sensitive only to the in-phase component of the fluctuations, hence their correlations do not affect the performance of a DPSK system to first-order either. The correlations are due the fact that phase fluctuations induced, on the two pulses overlapping at the receiver, by the same pulses through XPM are almost the same. Correlations are beneficial for DQPSK, because fully correlated fluctuations cancel at the differential receiver.

In the design of a line, the goal is therefore increasing the (negative) contributions of $\mathcal{B}_{\mathrm{fwm}}$ in DPSK and of $\mathcal{B}_{\mathrm{corr,xpm}}$ in DQPSK, to reduce the photocurrent fluctuations. It happens that both functions are minimized by very similar dispersion profiles. The amount of predispersion is in both cases one half the total line dispersion in the power symmetric case, less than one half when lumped in-line amplifiers are used, because pulse attenuation reduces the effective nonlinearity of the final part of the span. The above analysis, however, suggests that predispersion will always significantly affect DPSK performance, whereas it affects DQPSK performance only when the correlations at the receiver are significant.

6.8 Pseudo-Random Sequence in IMDD

The analysis of an IMDD system depends on the phase distribution of the pulses. If the phases are random, which occurs when the launched pulse stream originates from more than one laser source as in the case of optical time-division multiplexing (OTDM), then the analysis is not very different from that of phase modulation, and it will not be detailed here for brevity. We will assume here instead that all pulses have the same phase, which will be chosen as zero without loss of generality. This applies generally to electrical time-division multiplexing (ETDM). In this case, (6.40)–(6.42) give the photocurrent when a "one" is detected and its perturbation. The perturbation is not of zero average in this case. Using the property that $\mathrm{Re}\left(\mathcal{J}_{l,0,j}\right)$ is antisymmetric for exchanges $j \mapsto -j$ and $l \mapsto -l$ and symmetric for exchange $j \to l$, we may write

$$\Delta I_{\mathrm{IMDD}} = 2 \sum_{j>0,l>0} C_{j,l} \mathrm{Re}\left(\mathcal{J}_{l,0,j}\right), \tag{6.71}$$

where we used that $\mathcal{J}_{0,0,j}$ and $\mathcal{J}_{l,0,0}$ are real, and defined

$$C_{j,l} = a_j\, a_{j+l}\, a_l + a_{-j}\, a_{-j-l}\, a_{-l} - a_j\, a_{j-l}\, a_{-l} - a_{-j}\, a_{-j+l}\, a_l. \tag{6.72}$$

The variance of the photocurrent fluctuations has mean square

$$\langle \delta I_{\mathrm{IMDD}}^2 \rangle = \langle \Delta I_{\mathrm{IMDD}}^2 \rangle - \langle \Delta I_{\mathrm{IMDD}} \rangle^2, \tag{6.73}$$

where we used a small-case δ to denote the displacement from the (nonzero) average value of ΔI_{IMDD}, and

$$\langle \Delta I^2_{\mathrm{IMDD}} \rangle = 4 \sum_{j>0,l>0} \sum_{j',l'} \langle C_{j,l} C_{j',l'} \rangle \mathrm{Re}\left(\mathcal{J}_{l,0,j}\right) \mathrm{Re}\left(\mathcal{J}_{l',0,j'}\right), \quad (6.74)$$

$$\langle \Delta I_{\mathrm{IMDD}} \rangle = 2 \sum_{j,l} \langle C_{j,l} \rangle \mathrm{Re}\left(\mathcal{J}_{l,0,j}\right). \quad (6.75)$$

In the averages $\langle C_{j,l} C_{j',l'} \rangle$, one should use that

$$\langle a_j a_{j+l} a_l a_{j'} a_{j'+l'} a_{l'} \rangle = 1/2^m, \quad (6.76)$$

$$\langle a_j a_{j+l} a_l \rangle = 1/2^n, \quad (6.77)$$

with m the number of distinct indices in $\{j, l, j', l'\}$, and n the number of distinct indices in $\{j, l\}$. A numerical analysis has shown that the dominant terms in the averages are those with $j = j'$ and $l = l'$, degenerate with those $j = l'$ and $l = j'$. Being for $j \neq l$ $\langle C^2_{j,l} \rangle = 5/16$ and $\langle C_{j,l} \rangle = 0$ double degenerate, and for $j = l$ $\langle C^2_{j,j} \rangle = 5/8$ and $\langle C_{j,l} \rangle = 1/4$ nondegenerate, we obtain the approximation

$$\langle \delta I^2_{\mathrm{IMDD}} \rangle \simeq \frac{5}{2} \sum_{j>0,l>0} \left(\mathrm{Re}\mathcal{J}_{l,0,j}\right)^2 - \frac{1}{4} \sum_{j>0} \left(\mathrm{Re}\mathcal{J}_{j,0,j}\right)^2. \quad (6.78)$$

This approximation will be checked below against the exact expressions given in (6.73)–(6.75).

6.9 Continuous Approximation

If we consider a continuous version of $\mathcal{J}_{j,k=j+l,l}$, that is $\mathcal{J}(T_1, T_2)$, by setting $T_1 = jT_s$ and $T_2 = lT_s$

$$\mathcal{J}(T_1, T_2) = i\gamma\sqrt{2\pi^3} A^4 \tau^3 \int_{-z^*}^{L-z^*} f(z + z^*)\, G(T_1, T_2; z)\, \mathrm{d}z, \quad (6.79)$$

and approximate the sums with integrals

$$\sum_j \mapsto \int \frac{\mathrm{d}T_1}{T_s}, \quad \sum_l \mapsto \int \frac{\mathrm{d}T_2}{T_s}, \quad (6.80)$$

obtaining

$$\mathcal{A}_{\mathrm{fwm}} \simeq \int \frac{\mathrm{d}T_1}{T_s} \int \frac{\mathrm{d}T_2}{T_s} \left[2 - T_s \delta\left(T_1 - T_2\right)\right] \left|\mathcal{J}(T_1, T_2)\right|^2, \quad (6.81)$$

$$\mathcal{B}_{\mathrm{fwm}} \simeq \int \frac{\mathrm{d}T_1}{T_s} \int \frac{\mathrm{d}T_2}{T_s} \left[2 - T_s \delta\left(T_1 - T_2\right)\right] \mathcal{J}(T_1, T_2)^2, \quad (6.82)$$

where the Dirac delta function accounts for the degeneracy factor $f_{j,l}$. The integral over T_1 and T_2 can be analytically performed, yielding the compact result

$$\mathcal{A}_{\text{fwm}} = \frac{\gamma^2\pi^2 A^8\tau^4 z_d^2}{T_s^2}\mathcal{A}'_{\text{fwm}} - \frac{\gamma^2\sqrt{\pi^3}A^8\tau^3 z_d^2}{2T_s}\mathcal{A}''_{\text{fwm}}, \tag{6.83}$$

$$\mathcal{B}_{\text{fwm}} = \frac{\gamma^2\pi^2 A^8\tau^4 z_d^2}{T_s^2}\mathcal{B}'_{\text{fwm}} - \frac{\gamma^2\sqrt{\pi^3}A^8\tau^3 z_d^2}{2T_s}\mathcal{B}''_{\text{fwm}}, \tag{6.84}$$

where we defined the dimensionless constants

$$\mathcal{A}'_{\text{fwm}} = \frac{1}{z_d^2}\int_0^L\int_0^L \frac{f(z)\mathrm{d}z\, f(z')\mathrm{d}z'}{\sqrt{4+(Z-Z')^2}}, \tag{6.85}$$

$$\mathcal{B}'_{\text{fwm}} = -\frac{1}{z_d^2}\int_{-z^*}^{L-z^*}\int_{-z^*}^{L-z^*} \frac{f(z+z^*)\mathrm{d}z\, f(z'+z^*)\mathrm{d}z'}{\sqrt{4+(Z+Z')^2}}, \tag{6.86}$$

$$\mathcal{A}''_{\text{fwm}} = \frac{1}{z_d^2}\int_{-z^*}^{L-z^*}\int_{-z^*}^{L-z^*} \frac{f(z+z^*)\mathrm{d}z\, f(z'+z^*)\mathrm{d}z'}{\sqrt{(1+Z^2)(1+iZ')+(1+Z'^2)(1-iZ)}}, \tag{6.87}$$

$$\mathcal{B}''_{\text{fwm}} = -\frac{1}{z_d^2}\int_{-z^*}^{L-z^*}\int_{-z^*}^{L-z^*} \frac{f(z+z^*)\mathrm{d}z\, f(z'+z^*)\mathrm{d}z'}{\sqrt{(1+Z^2)(1-iZ')+(1+Z'^2)(1-iZ)}}, \tag{6.88}$$

where we used, for short, the dimensionless distance

$$Z = \frac{z}{z_d}. \tag{6.89}$$

This procedure, applied also to the other terms, give expression that are valid in the limit of a large number of interacting pulses, the "tedon" limit, which can be further approximated to give the results of [2]. We will not follow this route here, rather we will use the complete expressions to investigate the behavior also of system where the number of overlapping pulses is moderate, for instance when full compensation is applied at each amplifier span, which cannot be analyzed with the asymptotic expressions.

From the above equations, however, a lesson can be learned. The term $\mathcal{A}'_{\text{fwm}}$, which is the dominant one in $\mathcal{A}_{\text{fwm}}$, is surprisingly independent of the predispersion. The term $\mathcal{A}_{\text{fwm}}$ is in turn the dominant one in the expression for $\langle \Delta I^2_{\text{DQPSK}}\rangle$. This suggests that the nonlinear fluctuations at a DQPSK receiver are almost independent of the predispersion. This property will be verified below using the exact expression for the first-order fluctuations.

6.10 Numerical Examples

To illustrate these results, let us plot the Q factor at the receiver estimated by our first-order perturbation theory. We use the definition of the Q factor at the receiver

$$Q = \frac{\langle I_1 \rangle + \langle I_0 \rangle}{\sqrt{\langle \Delta I_1^2 \rangle} + \sqrt{\langle \Delta I_0^2 \rangle}}, \tag{6.90}$$

where $\langle I_0 \rangle$ and $\langle I_1 \rangle$ are the average signal of zeros and ones, and $\langle \Delta I_0^2 \rangle$ and $\langle \Delta I_1^2 \rangle$ are the variance of the fluctuations of zeros and ones. For DPSK and DQPSK, the averages and the variances of zeros and ones are equal and $\langle I_0 \rangle = -\langle I_1 \rangle$, so that the expression for Q becomes

$$Q_{\mathrm{D(Q)PSK}} = \frac{\langle I_{\mathrm{D(Q)PSK}} \rangle}{\sqrt{\langle \Delta I_{\mathrm{D(Q)PSK}}^2 \rangle}}. \tag{6.91}$$

For IMDD, the average signal and the variance of the fluctuations of the signal is in general negligible, so that a good approximation is

$$Q_{\mathrm{IMDD}} \simeq \frac{\langle I_{\mathrm{IMDD}} \rangle}{\sqrt{\langle \delta I_{\mathrm{IMDD}}^2 \rangle}}. \tag{6.92}$$

Let us first concentrate on the nonlinear impairments only, considering DPSK first. The average signal square at detection in this case is $\langle I_{\mathrm{DPSK}} \rangle = \sqrt{\pi} A^2 \tau = T_s P_{\mathrm{av}}$, where $P_{\mathrm{av}} = \sqrt{\pi} A^2 \tau / T_s$ is the average transmitted signal power. The nonlinear Q factor at the receiver is therefore inversely proportional to $1/P_{\mathrm{av}}$. With DQPSK, the average signal square at detection is $\langle I_{\mathrm{DQPSK}} \rangle = \mathrm{Re}[\exp(-i\pi/4)\sqrt{\pi} A^2 \tau] = T_s P_{\mathrm{av}}/\sqrt{2}$. With IMDD, the average signal square is $\langle I_{\mathrm{IMDD}} \rangle = \sqrt{\pi} A^2 \tau = 2 T_s P_{\mathrm{av}}$, where the extra factor 2 compared to the phase-modulated case is due the fact that the duty cycle in this case is one half, and nonzero power is transmitted only when ones are transmitted. The root-mean square of the fluctuations are in all cases proportional to $\gamma A^4 \tau^3$ hence to P_{av}^2. The nonlinear Q factor is therefore, in all cases, inversely proportional to the transmitted power.

Let us now plot the above expressions for a system with the parameters listed in Table 6.1. We will assume first that full dispersion compensation is applied at every span. Being the analysis based on linearization, and being the unperturbed evolution identical after every span, which includes precompensation, fiber propagation, and postcompensation, the perturbation is N times the perturbation of a single span. Consequently, the nonlinear Q factor will be N times lower than the Q factor of the individual span. Of course, also in this case the variance of the noise will possibly be determined by the amount of precompensation of the first span (the inline compensation is complete but, conceptually, divided into a postcompensation of the previous span and precompensation of the following one). The analysis will

Table 6.1 Numerical parameters (*FWHM* Full-width at half maximum)

Quantity	Symbol	Value	Units
Fiber loss	α	0.25	dB km^{-1}
Fiber dispersion	β''	−20.4	ps^2 km^{-1}
Nonlinear coefficient	γ	1.3	W^{-1}km^{-1}
Pulse-width (FWHM)	τ_{FWHM}	5	ps
Bit time	T_s	25	ps
Input power	P_{dBm}	3	dBm
Number of spans	N	7	
Span length	z_s	100	km
Wavelength	λ	1.55	µm
Noise figure	F	6	dB

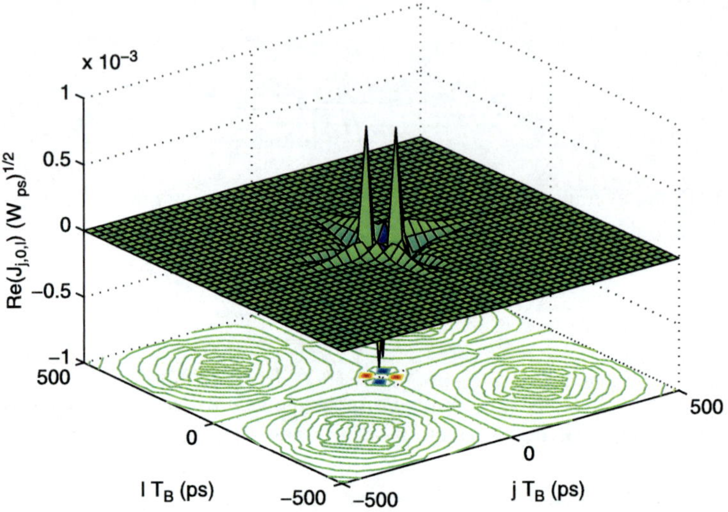

Fig. 6.1 Surface plot of the real part of $\mathcal{J}_{j,0,l}$ in (W ps)$^{1/2}$ vs. $T_j = jT_s$ and $T_l = lT_s$ in ps

be based on the numerical evaluation of the integrals $\mathcal{J}_{j,0,l}$ given by (6.45) using a Matlab code based on the Matlab command "quadv" that performs integrals that depend on matrices, in our case that containing T_j and T_l, simultaneously and efficiently. In Figs. 6.1 and 6.2, we show the real and imaginary parts of $\mathcal{J}_{j,0,l}$ for $z^* = 0$. Such curves, which can be obtained in fractions of seconds, may give an immediate visual idea on the range of the nonlinear interaction. The evaluation of $\mathcal{J}_{j,0,l}$ is the basis for the evaluation of the nonlinear Q factor. In Fig. 6.3, we show the nonlinear Q factor in a DPSK system where full dispersion compensation is performed at every span, whereas in Fig. 6.4 the same quantity in a DQPSK system, vs. the amount of precompensation quantified by the zero dispersion length z^*. In Fig. 6.5, the same quantities are given for an IMDD system. Here, with a solid blue line we show the exact expressions in equations (6.73)–(6.75), whereas with a dashed red line, the approximate expression in equation (6.78). Note that we did

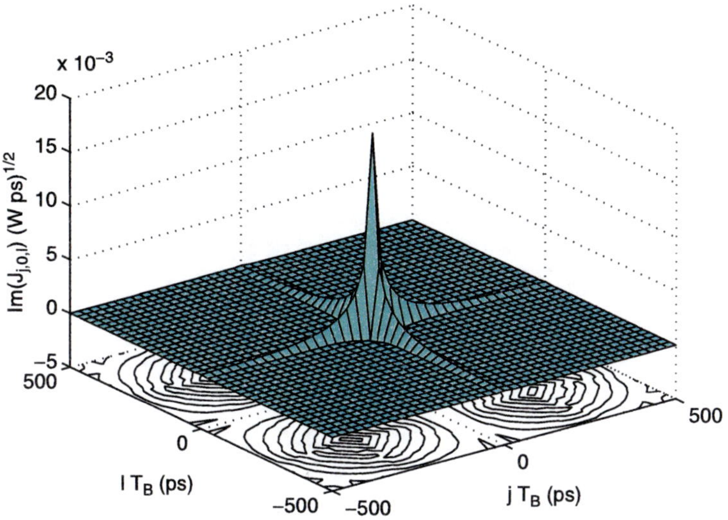

Fig. 6.2 Surface plot of the imaginary part of $\mathcal{J}_{j,0,l}$ in (W ps)$^{1/2}$ vs. $T_j = jT_s$ and $T_l = lT_s$ in ps

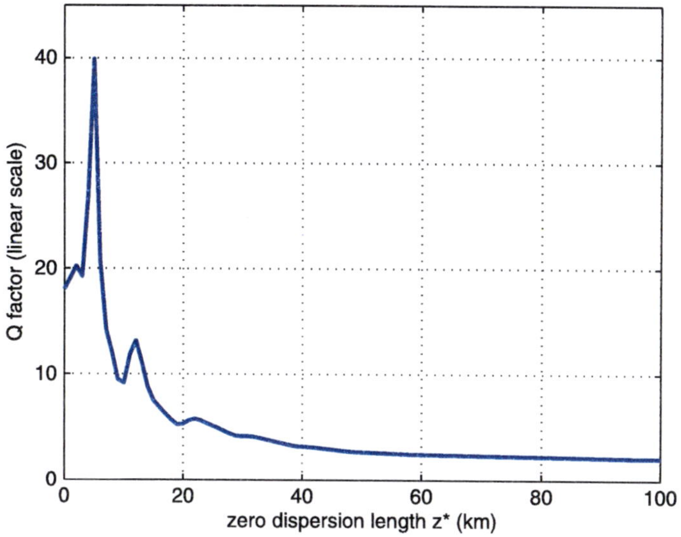

Fig. 6.3 Nonlinear Q factor Q_{DPSK} vs. the zero dispersion length z^* for DPSK transmission, with the parameters listed in Table 6.1, when dispersion compensation is complete at each span

not include here the nonlinear noise on zeros. The higher tolerance to nonlinear impairments of DQPSK over DPSK and IMDD shows up quite clearly.

Let us now compare the above examples with the case in which no inline compensation is used, but dispersion compensation is divided between both fiber ends.

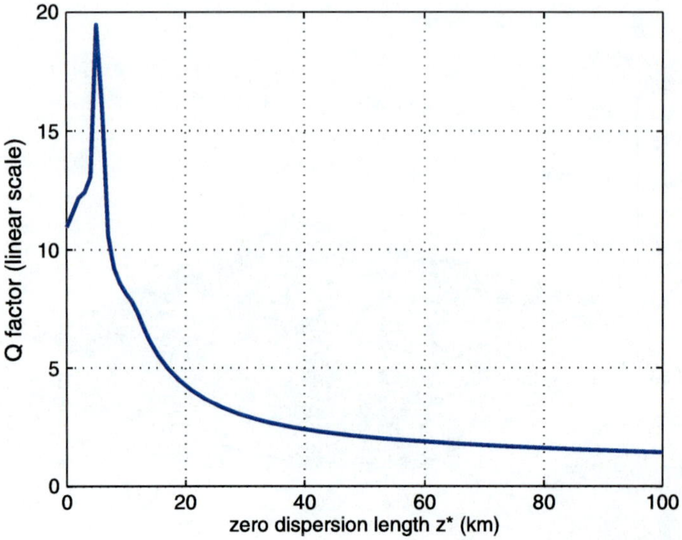

Fig. 6.4 Nonlinear Q factor Q_{DQPSK} vs. the zero dispersion length z^* for DPSK transmission, with the parameters listed in Table 6.1, when dispersion compensation is complete at each span

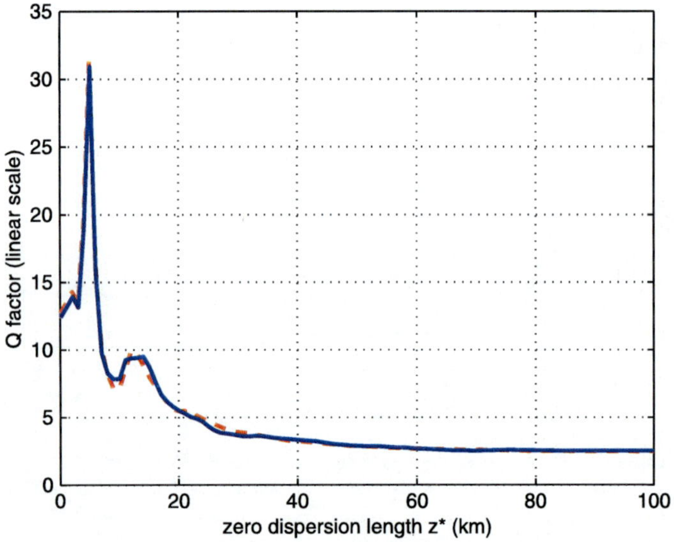

Fig. 6.5 Nonlinear Q factor Q_{IMDD} vs. the zero dispersion length z^* for IMDD transmission, with the parameters listed in Table 6.1, when dispersion compensation is complete at each span. Again, no noise on zeros has been considered. *Solid blue line*, exact expressions equations (6.73)–(6.75). *Dashed red line*, approximate expression equation (6.78)

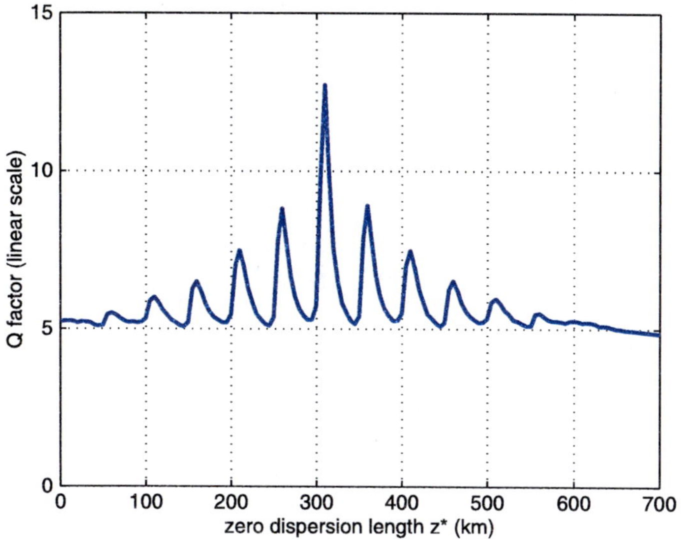

Fig. 6.6 Nonlinear Q factor Q_{DPSK} vs. the zero dispersion length z^* for DPSK transmission, with the parameters listed in Table 6.1. No inline dispersion compensation is used

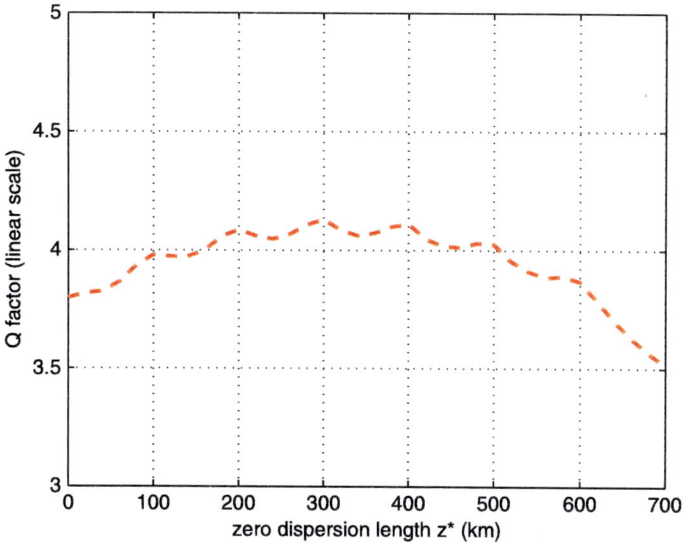

Fig. 6.7 Nonlinear Q factor Q_{DQPSK} vs. the zero dispersion length z^* for DQPSK transmission, with the parameters listed in Table 6.1. No inline dispersion compensation is used

In Fig. 6.6, we show the Q factor for DPSK Q_{DPSK}, whereas in Fig. 6.7 the Q factor for DQPSK Q_{DQPSK}, vs. the zero dispersion length z^*.

In Fig. 6.8, we show the nonlinear Q factor vs. z^* for an IMDD transmission where no inline dispersion compensation is used. The plot has been obtained

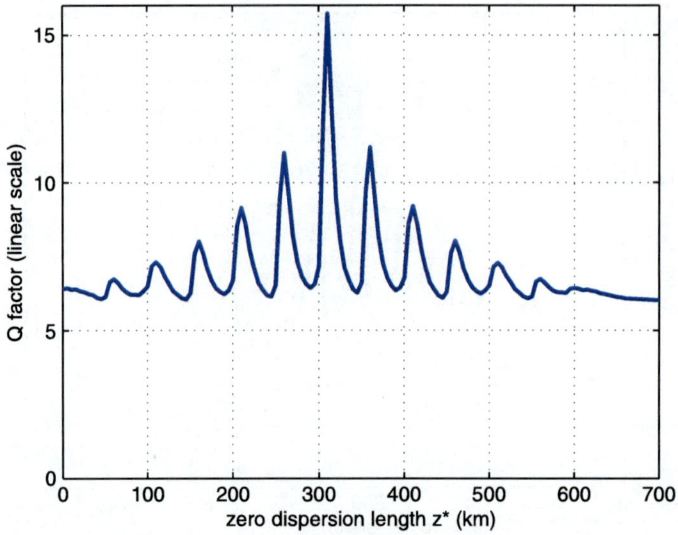

Fig. 6.8 Nonlinear Q factor Q_{IMDD} vs. the zero dispersion length z^* for IMDD transmission, with the parameters listed in Table 6.1. No inline dispersion compensation is used. Only the fluctuations of ones have been considered

by using the approximate expression given by (6.78). It is evident that, for the pulse-width considered, when dispersion compensation is applied at the fiber ends only the Q factor is lower than when complete dispersion compensation is applied at every span.

6.11 Total Receiver Noise

The nonlinear noise adds to the linear ASE noise of the amplifiers. The Q factor square with the phase-modulated schemes is

$$Q^2_{\mathrm{ASE,DPSK}} = \frac{\langle I_{\mathrm{DPSK}}\rangle^2}{\langle \Delta I^2_{\mathrm{ASE,DPSK}}\rangle} = \frac{P_{\mathrm{av}} T_s}{\hbar\omega_0 n_{\mathrm{sp}}(G-1)}, \tag{6.93}$$

$$Q^2_{\mathrm{ASE,DQPSK}} = \frac{\langle I_{\mathrm{DQPSK}}\rangle^2}{\langle \Delta I^2_{\mathrm{ASE,DQPSK}}\rangle} = \frac{P_{\mathrm{av}} T_s}{2\hbar\omega_0 n_{\mathrm{sp}}(G-1)}. \tag{6.94}$$

In the above equations, we have used that $\langle \Delta I^2_{\mathrm{ASE,D(Q)PSK}}\rangle = \hbar\omega_0 P_{\mathrm{av}} T_s n_{\mathrm{sp}}(G-1)$ and that, for the same optical power, $\langle I_{\mathrm{DQPSK}}\rangle = \langle I_{\mathrm{DPSK}}\rangle/\sqrt{2}$. With IMDD, if we assume a matched filter in the optical domain, the detected photocurrent of the ASE noise on zeros has a negative exponential distribution, with variance equal to the average squared. The Q factor is in this case, for high values of the optical signal-to-noise ratio, virtually independent of the noise on zeros. The variance of the noise on ones is instead $\langle \Delta I_{1,\mathrm{ASE,IMDD}}^2 \rangle = 2\hbar\omega_0 n_{\mathrm{sp}}(G-1)$. There is an extra factor 2 when this value is compared with that of the phase-modulated schemes. This is because, with a differential detection, the ASE noise comes from two consecutive pulses, hence it adds up incoherently, $\mathrm{Re}[(E_1+n_1)^*(E_2+n_2)] \simeq \mathrm{Re}(E_1^* E_2) + \mathrm{Re}(n_1^* E_2) + \mathrm{Re}(n_2 E_1^*)$, whereas with direct detection it comes from the beat of the pulse with itself $|E_1+n_1|^2 \simeq |E_1|^2 + 2\mathrm{Re}(n_1^* E_1)$, hence it adds coherently to itself, giving an extra factor 2 in the variance. The Q factor becomes in this case

$$Q_{\mathrm{ASE,IMDD}}^2 = \frac{\langle I_{\mathrm{IMDD}} \rangle^2}{\langle \Delta I_{\mathrm{ASE,IMDD}}^2 \rangle} = \frac{P_{\mathrm{av}} T_s}{\hbar\omega_0 n_{\mathrm{sp}}(G-1)}, \tag{6.95}$$

equal to that of DPSK. The factor 2 increase caused by the double amplitude of the detected eye of DPSK is exactly compensated by the double amplitude of the ones in IMDD for the same average power, and the factor 2 increase of the fluctuations of ones in IMDD caused by the coherent beat is compensated by the negligible contribution of the fluctuations on zeros. This fact appears in contradiction with the frequently claimed 3 dB advantage of DPSK over IMDD. Note, however, that we assumed a matched optical filter, hence $M = 1$, where $M = 2BT_s$, where B is the bandwidth of the optical filter in front of the receiver, so that neglecting the noise on zero is a good approximation. Also note that the analysis of the often quoted [18] compares IMDD with a DPSK scheme where (top of page 1,580) "as in FSK, one of the signal energies is 0 and the other is $\mathcal{E}$, depending on the data bit," so it does not seem to apply to balanced DPSK detection that we analyze here, where the noise on ones and zeros are symmetric. In addition, the results of the analysis of [18] reported in Fig. 6.5 there shows that the Gaussian approximation (the only one implying a one-to-one correspondence between the Q factor as defined here and the error probability) gives, for $M \simeq 1$, the same signal-to-noise requirements for IMDD and DPSK to achieve 10^{-9} error probability. Let us also note that with phase shift keying (PSK) employing a matched local oscillator with no noise, the noise is one half, hence the Q factor is 3 dB higher than DPSK.

As a final comment, we would like to mention that the above expressions for the Q factor assume an ideal integrate-and-dump receiver, and neglect the ASE-ASE beat noise. With a realistic receiver, a penalty is expected that depends on the electrical bandwidth of the receiver itself [19].

Being ASE and nonlinear noise independent processes, the variance add up when they act together. It is therefore useful to define the quantity $\mathcal{N} = Q^{-2}$, which is the variance of the noise normalized to the signal square. For the three schemes, the inverse of the Q factors squared when ASE and nonlinearity act alone add up to give the inverse of the overall Q factor square

$$\mathcal{N}_{\mathrm{tot,DPSK}}^2 = Q_{\mathrm{tot,DPSK}}^{-2} = Q_{\mathrm{nl,DPSK}}^{-2} + Q_{\mathrm{ASE,DPSK}}^{-2} \tag{6.96}$$

$$\mathcal{N}_{\mathrm{tot,DQPSK}}^2 = Q_{\mathrm{tot,DQPSK}}^{-2} = Q_{\mathrm{nl,DQPSK}}^{-2} + Q_{\mathrm{ASE,DQPSK}}^{-2} \tag{6.97}$$

$$\mathcal{N}_{\mathrm{tot,IMDD}}^2 = Q_{\mathrm{tot,IMDD}}^{-2} = Q_{\mathrm{nl,IMDD}}^{-2} + Q_{\mathrm{ASE,IMDD}}^{-2}, \tag{6.98}$$

Table 6.2 Minimum noise for compensation at every span. Precompensation is equivalent to $z^* = 5\,\text{km}$ of propagation, the optimum value

	Power for minimum noise (mW)	Minimum noise $\mathcal{N}$
DPSK	13.7	0.059
DQPSK	11.4	0.092
IMDD	11.5	0.065

Table 6.3 Minimum noise $\mathcal{N}$ for compensation only at the line ends. For DPSK and IMDD, precompensation is equivalent to $z^* = 370\,\text{km}$ of propagation, the optimal value, whereas for DQPSK, virtually insensitive to precompensation, $z^* = 0$

	Power for minimum noise (mW)	Minimum noise $\mathcal{N}$
DPSK	8.1	0.086
DQPSK	3.1	0.14
IMDD	7.3	0.081

where we have added the subscript "nl" to the nonlinear contribution to the Q. Being, as already mentioned, $Q_{\text{nl}}^{-2} = \kappa_1 P_{\text{av}}^2$ and $Q_{\text{ASE}}^{-2} = \kappa_2/P_{\text{av}}$, Q_{tot} is maximum for $2\kappa_1 P_{\text{av,max}} - \kappa_2/P_{\text{av,max}}^2 = 0$, that is for $P_{\text{av,max}}^3 = \kappa_2/(2\kappa_1)$. For this value of P_{av}, $Q_{\text{nl}}^2/Q_{\text{ASE}}^2 = 2$. This means that when Q is maximum the variance of the fluctuations induced by the nonlinearity, normalized to the average signal square $\mathcal{N}^2$ is one half the normalized variance square of the ASE fluctuations, and one third of the total. This property is a consequence of the quadratic dependence with power of the nonlinear contribution to $\mathcal{N}$ and the inverse proportionality of the ASE contribution to Q. In Tables 6.2 and 6.3, we give the numerical values of the optimal power, that is the power corresponding to the minimum noise, and the value of the minimum noise $\mathcal{N}$ for the cases of the two numerical examples that we considered, that is, the case of dispersion compensation at the fiber ends only, and that of dispersion compensation span by span. We have chosen the values of dispersion precompensation insuring the minimum noise. In all cases, for the system parameters assumed, the minimum noise does not exceed 15%.

In Fig. 6.9, we show the Q factor vs. the input power in dB for a DPSK transmission in which a complete compensation is performed at each span. Once again, the parameters are listed in Table 6.1 with the exception of the input power, which is used as a parameter. The blue dashed line is the $Q_{\text{ASE,DPSK}}$, that is the Q factor with no nonlinearity. The dot-dashed lines refer to the case of no ASE and only nonlinearity, and in particular the blue dot-dashed line is $Q_{\text{nl,DQPSK}}$ when $z^* = 0$, whereas the red dot-dashed line refers to the case $z^* = 5\,\text{km}$. The solid lines refer to both ASE and nonlinearity present, namely the blue solid line is the Q for $z^* = 0$ and the red solid line for $z^* = 5$ km. The Q for the other transmission schemes show a similar behavior. Remember that our analysis lies within the boundary of first-order perturbation theory. We assume that the fluctuations induced by both ASE noise and nonlinearity are small compared to the average power, and consequently their coupling is of the order of their product, hence it is of second order and can legitimately

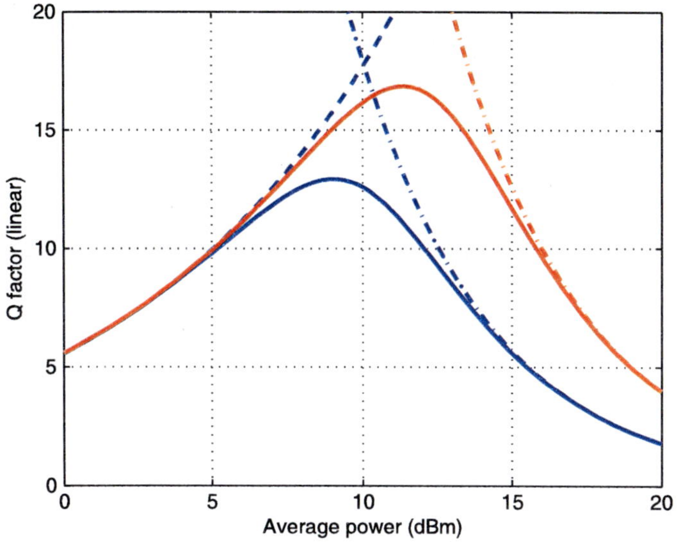

Fig. 6.9 Q factor vs. the input power P_{av} in dBm for a DPSK transmission when complete dispersion compensation is applied at every span. The *blue dashed line* is $Q_{ASE,DPSK}$ (no nonlinearity, ASE noise only). The *blue dot-dashed line* is $Q_{nl,DQPSK}$ when $z^* = 0$, the *red dot-dashed line* $Q_{nl,DQPSK}$ when $z^* = 5$ km (no ASE noise, nonlinearity only). The *blue solid line* is the Q for $z^* = 0$ and the *red solid line* the Q for $z^* = 5$ km, when both nonlinearity and ASE noise are present

be neglected. In addition, this coupling produces essentially the enhancement of phase noise (the Gordon-Mollenauer effect [20]), hence it is, again to first-order, negligible per se in DPSK. Finally, the (normalized) variances of the linear noise, nonlinear noise, and noise enhancement due to nonlinear noise coupling, are proportional to $1/P_{av}$, P_{av}^2 and P_{av} [20], so that we expect that nonlinear noise be important in a region of injected powers bounded from below and from above. The validity of our theory therefore requires that $Q \gg 1$ at the power where linear and nonlinear fluctuations are of the same order, corresponding to the point where the overall Q factor is maximum.

6.12 Discussion

The above results give a solid foundation to the common wisdom that DPSK and IMDD are more tolerant to nonlinearity than DQPSK. In addition, they show that it is very important both in simulations and in experiments that the pseudorandom bit sequence (PRBS) used is chosen with all symbols appearing with equal occurrence. If, for instance, in DQPSK a PRBS is used with a bias that gives a higher occurrence for a given symbol, then the experimentally measured, or simulated, variance of nonlinear noise will be evaluated incorrectly. This is because in this case the average $\langle a_j \rangle$ becomes artificially nonzero and therefore the variance of nonlinear noise will be affected by predispersion like with DPSK. One would then predict a dependence of the system performance by predispersion, which is instead absent in real systems where the code used is a symmetric one.

6.13 Information Rate for DPSK and DQPSK Transmission

The above analysis may lead to the conclusion that DPSK overperforms DQPSK. We will show that this is not the case, at least for practical values of signal-to-noise ratio (SNR). Let us consider first the linear case. In a DPSK system employing a balanced receiver the transmitted binary symbol $\{-\sqrt{S}, \sqrt{S}\}$ is corrupted by an additive Gaussian noise n of variance $\sigma^2 = N$, so that the detected signal is $y = x + n$. With hard decoding, the optimal threshold is $y_{\text{th}} = 0$, and the error probability is for both symbols

$$p = \frac{1}{2}\left[1 - \text{erf}\left(\sqrt{\frac{2S}{N}}\right)\right]. \tag{6.99}$$

The information rate for such a binary symmetric channel is

$$I_{\text{hard}} = \frac{1}{T_s}\left[1 - h(p)\right], \tag{6.100}$$

where $1/T$ is the symbol rate, and h is the binary entropy function

$$h(p) = -p\log_2 p - (1-p)\log_2(1-p). \tag{6.101}$$

The information rate above refers to the case of hard decoding of a DPSK signal, where the decision on the detected symbol is taken after comparing with a fixed threshold, and no further information is used. With soft decoding, where the values of the detected signal y are used to estimate the reliability of the data, the information rate is slightly higher, and can be upper-bounded by the information rate as defined by Shannon [4, 7]. After some algebra, we obtain

$$I_{\text{soft}} = \frac{1}{T_s}\left\{\log_2 2 - \int \mathrm{d}y p\left(y - \sqrt{\frac{S}{N}}\right)\log_2\left[1 + \exp\left(-2y\sqrt{\frac{S}{N}}\right)\right]\right\}, \tag{6.102}$$

where

$$p(y) = \frac{1}{\sqrt{2\pi}}\exp\left(-\frac{y^2}{2}\right). \tag{6.103}$$

For large S/N, we have $I_{\text{soft}} \to 1/T_s$ bit/symbol/s, whereas for small SNR we have

$$I_{\text{soft}} \simeq \frac{S}{2T_s N}, \quad \frac{S}{N} \ll 1. \tag{6.104}$$

A DQPSK system is equivalent to two DPSK systems, so that the information rate is exactly double. For a given total power, however, the projection on the real and imaginary axis of the electric field of the DQPSK constellation is $1/\sqrt{2}$ the projection of DPSK. If the only source of noise is ASE, this means that $I_{\text{DQPSK}}(S) = 2I_{\text{DPSK}}(S/2)$, where the two information rate are for the same noise N. This is an obvious capacity advantage of DQPSK over DPSK for realistic values of SNRs. However, for very small values of SNR, it is not, because for $S/N \ll 1$ the asymptotic formula above gives for both schemes $I_{\text{DQPSK}}(S) \simeq 2I_{\text{DPSK}}(S/2) \simeq S/(2T_s N)$. This is an indication that, in general, increasing the number of degrees of freedom for the same optical power gives a capacity advantage that reduces for small values of the SNR. This is a general result, which is valid also for the Shannon capacity limit. The capacity of a channel with additive Gaussian noise, obtained with a continuous Gaussian distribution of levels. With our notations, the capacity is

$$C = \frac{d}{2T_s} \log_2 \left(1 + \frac{S/d}{N}\right), \tag{6.105}$$

where d is the number of degrees of freedom used for transmission over which the *same* optical signal power S is divided ($d = 1$ when a single quadrature of a single-mode electric field is used like in DPSK, and $d = 2$ when the two quadrature of a single mode electric field is used, like in DQPSK). Of course, using more degrees of freedom is beneficial at high levels of the SNR S/N, because of the linear dependence of the capacity on d and the logarithmic dependence on $1/d$. For small S/N, instead, distributing the signal, for the same power, over more than one degree of freedom does not help, because asymptotically for $S/(dN) \ll 1$ we have $C \simeq d/(2T_s)S/(dN) = S/(2T_s N)$, independent of d. In addition, multilevel modulation does not help either, binary modulation already approaches the Shannon limit.

These results are illustrated in Fig. 6.10, where we show the information rate for a DPSK and a DQPSK system vs. the SNR, S/N, where the SNR is defined in terms of the *total* transmitted power. The corresponding values of the Shannon capacity limits are also given as dashed lines for comparison. The dot-dashed lines are the information rate when hard decision is used at the receiver, so that the channel is a binary symmetric one.

Let us now consider the nonlinear propagation case. With a large number of overlapping pulses, the amplitude jitter can be approximated as a Gaussian noise. In this case, the nonlinear noise can be analyzed with the theory that we have just described. In practical cases, at least in those that can be analyzed within our perturbation theory, the total noise for the optimal value of input power is small. The SNR that we have defined is related to the normalized noise power by $S/(dN) = \mathcal{N}^{-2}$, where d are the number of degrees of freedom used in the transmission. Even with the largest values of the noise in Table 6.3, the value of S/N is such that the information rate

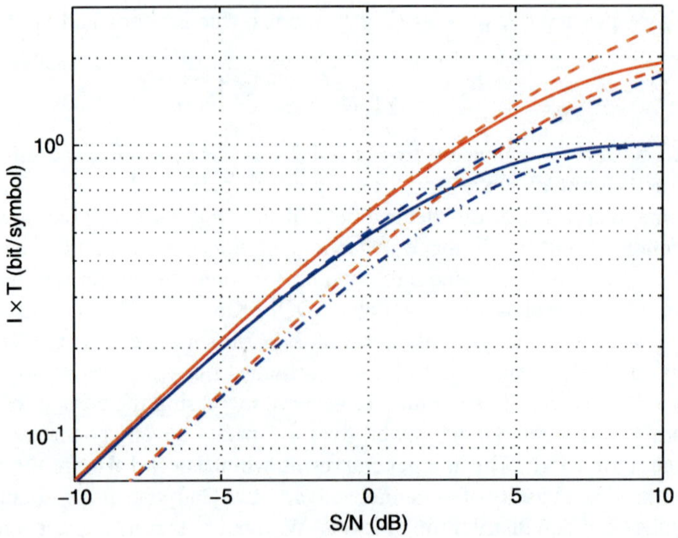

Fig. 6.10 Information rate for a system using DPSK (*solid curve below, blue*) and DQPSK (*solid curve above, red*) vs. the SNR, where the signal is the *total* transmitted power. The Shannon limits are also reported for comparison as *dashed curves*, again with the *total* transmitted power held fixed. The *dot-dashed lines* below is the information rate when hard decision is used at the receiver. The *blue line below* is for DPSK, the *red above* for DQPSK

is always 1 dB/symbol for DPSK and 2 dB/symbol for DQPSK, so that the capacity advantage of DQPSK is evident. For higher values of the optical power, however, because of the larger nonlinear noise of DQPSK, one may have at least in principle cases in which the information rate of DQPSK is lower than DPSK. These conditions occur, however, for unrealistically small values of the SNR.

6.14 Timing Jitter Between Two Pulses

Perturbations that are not symmetric in time are responsible for timing shift of the pulses. If the pulses are equally spaced in time, this occurs only for the coherent terms and the XPM term. To analyze this case, let us consider two pulses only, $u(0,t) = v_1(t) + v_2(t-T)$. In this case

$$\Delta u(L,t) = \sum_{j=1}^{2}\sum_{k=1}^{2}\sum_{l=1}^{2} \Delta u_{j,k,l}(L,t), \tag{6.106}$$

where of the 8 terms of the sum, only four are centered over the position of the two generating pulses. Let us concentrate on the two terms overlapping with pulse 1. The electric field in the neighbor of pulse 1 is then $v_1(t) + \Delta u_{122}(L,t) + \Delta u_{221}(L,t) = v_1(t) + 2\Delta u_{122}(L,t)$, where we have used the fact that the coherent and the XPM terms are equal $\Delta u_{122}(L,t) = \Delta u_{221}(L,t)$, and that $\Delta u_{122}(L,t)$ is centered around $t = 0$, see (6.25) and (6.26). Defining the timing of a pulse as the first moment of the pulse normalized intensity, the timing shift caused by the perturbation is to first-order

$$\delta T_1 = \frac{4}{\int \mathrm{d}t |v_1(t)|^2} \int t \,\mathrm{Re}\left[v_1^*(t) \Delta u_{122}(L,t)\right] \mathrm{d}t. \tag{6.107}$$

Assuming Gaussian pulses, we have $\int \mathrm{d}t |v_1(t)|^2 = \sqrt{\pi}|A_1|^2\tau$. Let us insert (6.25) and (6.26) into the expression of δT_1

$$\delta T_1 = \frac{4\gamma|A_1|^2|A_2|^2}{\sqrt{\pi}|A_1|^2\tau} \mathrm{Re}\left\{ i \int_{-z^*}^{L-z^*} \frac{f(z+z^*)\mathrm{d}z}{\sqrt{3q^*\,(q+2i/3)}} \right.$$
$$\left. \times \int \mathrm{d}t\, t\, \exp\left[-\frac{2t^2}{3\tau^2} + i\frac{2t(2t/3+T)}{\tau^2 3(q+2i/3)} - \frac{T^2}{\tau^2 3q^*\,(q+2i/3)} \right] \right\}. \tag{6.108}$$

After integrating over time, we obtain after some algebra

$$\delta T_1 = -\sqrt{2}\gamma|A_2|^2 T \int_{-z^*}^{L-z^*} \frac{(z/z_d) f(z+z^*)\mathrm{d}z}{(z^2/z_d^2+1)^{3/2}} \exp\left[-\frac{T^2}{2\tau^2(z^2/z_d^2+1)^2} \right]. \tag{6.109}$$

In the special case of lossless fiber $f(z) = 1$, the integral over z can be performed analytically, obtaining

$$\delta T_1 = \gamma|A_2|^2 \tau z_d \sqrt{\pi} \left\{ \mathrm{erf}\left[-\frac{T/(\sqrt{2}\,\tau)}{\sqrt{1+(L-z^*)^2/z_d^2}} \right] - \mathrm{erf}\left[-\frac{T/(\sqrt{2}\,\tau)}{\sqrt{1+z^{*2}/z_d^2}} \right] \right\}. \tag{6.110}$$

Note that the jitter is that of the leading one of the two pulses. It is zero if $z^* = L/2$.

Timing jitter comes from cross-gain modulation induced by intra-channel pulse collision. The above derivation does not make this point clear enough. It is therefore useful to give an alternate derivation of the timing jitter, which has the additional advantage of being suited for the analysis of pulse shapes different from Gaussian. Let us consider a pulse centered at $t = 0$ and another pulse centered at $t = T$, where T is much greater of the width of both pulses. The total field will be $u(z,t) = v_1(z,t) + v_2(z,t-T)$. If we define

$$U_1 = \int \mathrm{d}t |v_1|^2 = |A_1|^2 \tau \sqrt{\pi}, \tag{6.111}$$

$$\delta\Omega_1 = U_1^{-1} \int \mathrm{d}t v_1^* \left(i \frac{\partial}{\partial t} v_1 \right), \tag{6.112}$$

$$\delta T_1 = U_1^{-1} \int \mathrm{d}t v_1^* t v_1, \tag{6.113}$$

we may show using (6.5) and via integration by parts that the timing shift is related to the frequency shift acquired during propagation in the nonlinear fiber by

$$\frac{\partial}{\partial z} \delta T_1 = -\beta'' \delta\Omega_1, \tag{6.114}$$

integrating, we have

$$\delta T_1 = -\beta'' \int_0^z \mathrm{d}z' \delta\Omega_1(z') = -\beta'' \int_0^z \mathrm{d}z' (z - z') \frac{\partial}{\partial z'} \delta\Omega_1(z'), \tag{6.115}$$

where the last equality can be proven by integration by parts of the last integral and using the condition $\delta\omega(0) = 0$. After recompression at the dispersion compensating element of total dispersion $-\beta''(L+z^*)$, which compensate for the dispersion of the fiber plus the predispersion. If we assume the dispersion compensating fiber as linear (no conceptual problems to include the nonlinearity of the dispersion compensating element, however), the timing shift will be

$$\begin{aligned} \delta T_1(L) = -\beta'' \int_0^L \mathrm{d}z' (L - z') \frac{\partial}{\partial z'} \delta\Omega_1(z') \\ + \beta''(L - z^*) \delta\Omega_1(L) = \beta'' \int_0^L \mathrm{d}z' (z' - z^*) \frac{\partial}{\partial z'} \delta\Omega_1(z'). \end{aligned} \tag{6.116}$$

The equation for the frequency shift of pulse 1 is

$$\begin{aligned} \frac{\partial}{\partial z} \delta\Omega_1 &= 2U_1^{-1} \mathrm{Re} \int \mathrm{d}t \left(\frac{\partial v_1}{\partial t} \right)^* \left[2\gamma f(z) |v_2(z, t - T)|^2 v_1 \right] \\ &= \frac{2\gamma f(z)}{|A_1|^2 \tau \sqrt{\pi}} \int \mathrm{d}t \left[\frac{\partial}{\partial t} |v_1(z, t)|^2 \right] |v_2(z, t - T)|^2. \end{aligned} \tag{6.117}$$

Here, we have treated the effect of the pulse $v_2(t - T, z)$ on v_1 as a perturbation, by using

$$\frac{\partial v_1}{\partial z} \simeq -i \frac{\beta''}{2} \frac{\partial^2 v_1}{\partial t^2} + i\gamma f(z) \left[|v_1|^2 + 2|v_2(z, t - T)|^2 \right] v_1. \tag{6.118}$$

Substituting (6.117) with the expression for the timing shift (6.116), we obtain

$$\delta T_1(L) = \beta'' \frac{2\gamma}{|A_1|^2 \tau \sqrt{\pi}} \int_0^L \mathrm{d}z' f(z') z' \int \mathrm{d}t \left[\frac{\partial}{\partial t} |v_1(z', t)|^2 \right] |v_2(z', t - T)|^2. \tag{6.119}$$

So far, the $v_j(z,t)$ are unknown. However, in the spirit of first-order perturbation theory we may treat the effect of the XPM induced by the second pulse on the first as a perturbation. We know that without nonlinearity, we have

$$v_{j,0}(z,t) = \frac{A_j}{\sqrt{\tau^2 - i(z-z^*)/z_d}} \exp\left\{-\frac{t^2}{2\tau^2[1-i(z-z^*)/z_d]}\right\}, \quad (6.120)$$

hence the intensity is

$$|v_{j,0}(z,t)|^2 = \frac{|A_j|^2}{\sqrt{1+(z-z^*)^2/z_d^2}} \exp\left\{-\frac{t^2}{\tau^2\left[1+(z-z^*)^2/z_d^2\right]}\right\}. \quad (6.121)$$

Replacing the above expressions with (6.119), the integral over t can be analytically performed. The result is

$$\delta T_1(L) = -\beta'' \int_0^L \mathrm{d}z' f(z')(z'-z^*) \frac{\sqrt{2}\gamma|A_2|^2 T}{[1+(z'-z^*)^2/z_d^2]^{3/2}} \times \exp\left\{-\frac{T^2}{2\tau^2[1+(z'-z^*)^2/z_d^2]}\right\} \quad (6.122)$$

identical, after due changes, to the expression already obtained.

For later convenience, let us rewrite the expression for the timing jitter as

$$\delta T_1 = \tau\gamma z_d |A_2|^2 J(L,T), \quad (6.123)$$

where

$$J(L,T) = -\frac{\sqrt{2}(T/\tau)}{z_d} \int_{-z^*}^{L-z^*} \frac{(z/z_d)f(z+z^*)\mathrm{d}z}{(z^2/z_d^2+1)^{3/2}} \exp\left[-\frac{T^2}{2\tau^2(z^2/z_d^2+1)}\right]. \quad (6.124)$$

Note that if, once again, $f(z)$ is symmetric about the center of the span $z = L/2$ and $z^* = L/2$, then $J(L,T)$ is proportional to an integral of an antisymmetric function integrated over a symmetric interval, hence it is zero. This means that timing jitter induced by intra-cannel collision is in this case zero. Also in this case, it is possible to reduce for a nonsymmetric $f(z)$ the timing jitter to a minimum by a careful choice of the predispersion z^*.

6.15 Timing Jitter in a Pseudo-Random Sequence

Let $P[\Delta T, (n-1)T_s]$ be the probability distribution of the total timing jitter of a given pulse ΔT caused by a random sequence of $2(n-1)$ equally spaced pulses, $n-1$ on each side of it, each encoding one the j symbol of an alphabet of N

symbols occurring with probability p_j. If two pulses are added simultaneously at the edges of both sides, the sequence becomes of n pulses on each side. The pdf evolves according to

$$P(\Delta T, nT_s) = \sum_{j=1}^{N}\sum_{k=1}^{N} p_j p_k P[\Delta T - \delta T(a_j, n) + \delta T(a_k, n), (n-1)T_s], \quad (6.125)$$

where $\delta T(a_j, n) = \tau\gamma|a_j|^2 A^2 z_d J(L, nT_s)$ is the timing jitter if the jth symbol is added on one side. The above has been obtained using Bayes theorem and the fact that the timing jitter becomes ΔT with a sequence n pulses long at each side if the timing jitter was $\Delta T - \delta T(a_j, n) + \delta T(a_k, n)$ with a sequence of $(n-1)$ pulses and if a pulse of normalized amplitude a_j centered at timing nT_s is added at one edge, contributing the timing jitter $\delta T(a_j, n)$, and a pulse of normalized amplitude a_k centered at timing $-nT_s$ is added at the other edge, producing a timing jitter $-\delta T(a_k, n)$. Each of this case should be weighted with the corresponding probability of occurrence. Let us now use the expansions

$$P(\Delta T, nT_s) = P[\Delta T, (n-1)T_s] + T_s \left.\frac{\partial P(\Delta T, T)}{\partial T}\right|_{T=(n-1)T_s}, \quad (6.126)$$

$$\begin{aligned} &P[\Delta T - \delta T(a_j, n) + \delta T(a_k, n), (n-1)T_s] \\ &\quad = P[\Delta T, (n-1)T_s] + \frac{\partial P(\Delta T, (n-1)T_s)}{\partial \Delta T}\left[\delta T(a_k, n) - \delta T(a_j, n)\right] \\ &\quad + \frac{1}{2}\frac{\partial^2 P(\Delta T, (n-1)T_s)}{\partial \Delta T^2}\left[\delta T(a_k, n) - \delta T(a_j, n)\right]^2. \end{aligned} \quad (6.127)$$

After introducing the above into the expression for $P(\Delta T, nT_s)$ (6.125), we obtain

$$\left.\frac{\partial P(\Delta T, T)}{\partial T}\right|_{T=(n-1)T_s} = \frac{D[(n-1)T_s]}{2}\frac{\partial^2 P(\Delta T, (n-1)T_s)}{\partial \Delta T^2}, \quad (6.128)$$

where

$$\begin{aligned} D[(n-1)T_s] &= \frac{1}{T_s}\sum_{j=1}^{N}\sum_{k=1}^{N} p_j p_k \left[\delta T(a_k, n) - \delta T(a_j, n)\right]^2 \\ &= \frac{2}{T_s}\left\{\sum_{j=1}^{N} p_j \delta T^2(a_j, n) - \left[\sum_{j=1}^{N} p_j \delta T(a_j, n)\right]^2\right\}. \end{aligned} \quad (6.129)$$

Using the expression for $\delta T(a_j, n)$ now, we have

$$D[(n-1)T_s] = \frac{2\tau^2 A^4 \gamma^2 z_d^2}{T_s} \left[\sum_{j=1}^{N} p_j |a_j|^4 - \left(\sum_{j=1}^{N} p_j |a_j|^2 \right)^2 \right] |J(L, nT_s)^2. \tag{6.130}$$

It is convenient to relate the amplitude A to the average transmitted power by

$$\sqrt{\pi} A^2 \tau \sum_j p_j |a_j|^2 = P_{\mathrm{av}} T_s. \tag{6.131}$$

If we use the notation

$$\langle |a|^n \rangle = \sum_{j=1}^{N} p_j |a_j|^n, \tag{6.132}$$

we obtain

$$A^2 = \frac{P_{\mathrm{av}} T_s}{\sqrt{\pi} \tau \langle |a|^2 \rangle}. \tag{6.133}$$

Approximating now the variable nT_s with a continuous variable, we get

$$\frac{\partial P(\Delta T, T)}{\partial T} = \frac{D(T)}{2} \frac{\partial^2 P(\Delta T, T)}{\partial \Delta T^2}, \tag{6.134}$$

where

$$D(T) = \frac{2 P_{\mathrm{av}}^2 T_s^2 \gamma^2 z_d^2}{\pi T_s} M J(L, T)^2, \tag{6.135}$$

and we defined the modulation-specific parameter

$$M = \frac{\langle |a|^4 \rangle}{\langle |a|^2 \rangle^2} - 1. \tag{6.136}$$

Equation (6.134) is a diffusion equation of a particle with a nonconstant diffusion coefficient, of the kind

$$\frac{\partial}{\partial t} f(x,t) = \frac{D(t)}{2} \frac{\partial^2}{\partial x^2} f(x,t). \tag{6.137}$$

If the initial pdf is a Dirac delta centered at zero (the particle has a fixed position, which corresponds to a negligible jitter of the input pulse stream), the solution is a Gaussian, of variance

$$\sigma^2(t) = \langle x^2 \rangle - \langle x \rangle^2 = \langle x^2 \rangle = \int_0^t \mathrm{d}t' D(t'). \tag{6.138}$$

In our case, the variance is

$$\sigma^2(\Delta T) = \frac{2P_{\rm av}^2 T_s^2 \gamma^2 z_d^2}{\pi} M \int_{T_s}^{\infty} \frac{{\rm d}T}{T_s} J(L,T)^2, \tag{6.139}$$

where the upper limit is justified by the fact that a pulse experiences, in principle, the interaction with all pulses in the stream. We may at this point turn the integral back to a discrete sum,

$$\frac{\sigma^2(\Delta T)}{T_s^2} = \frac{2P_{\rm av}^2 \gamma^2 z_d^2}{\pi} M \sum_{j>0} J(L, jT_s)^2. \tag{6.140}$$

This expression, similar to those obtained for the amplitude noise, is more accurate than the integral one (6.139) and gives reliable results in all cases, including those where the interaction is effective only with a few adjacent pulses of the sequence, for instance, when dispersion compensation is applied at every span.

If the number of interacting pulses is instead large, for instance when no inline dispersion compensation is used, we may use the integral expression which, after replacing the lower limit of the integral with 0 and integrating over T, becomes

$$\frac{\sigma^2(\Delta T)}{T_s^2} = \frac{2\sqrt{2}P_{\rm av}^2 \gamma^2 z_d^2 \tau}{\sqrt{\pi} T_s} M \mathcal{T}, \tag{6.141}$$

where

$$\mathcal{T} = \int_{-z^*}^{L-z^*} \frac{{\rm d}z}{z_d} \int_{-z^*}^{L-z^*} \frac{{\rm d}z'}{z_d} \frac{(zz'/z_d^2) f(z+z^*) f(z'+z^*)}{[(z'^2+z^2)/z_d^2+2]^{3/2}}. \tag{6.142}$$

The double integral in (6.142) is computationally heavier than the sum of simple integrals in (6.140), unless $f(z) = 1$, in which case the double integral over z can be done analytically, giving the result [2]

$$\mathcal{T} = 2\sqrt{[(L-z^*)^2 + z^{*2}]/z_d^2 + 2} - \sqrt{2[(L-z^*)^2/z_d^2 + 1]} - \sqrt{2(z^{*2}/z_d^2 + 1)}. \tag{6.143}$$

With the parameters of Table 6.1, no loss and no inline compensation, (6.142) and (6.143) overlap with the exact expression given by (6.140). Note the asymptotic linear dependence on L, which replaces the asymptotic independence on L of the two pulse case. With $z^* = L/2$, we have $\mathcal{T} = 0$ and zero timing jitter. This property was anticipated above when we showed that in this case $J(L,T) = 0$ for every T.

Even for with $f(z) \neq 1$, the integral $\mathcal{T}$ is practically independent on $z_d = \tau^2/\beta''$ for large $L/|z_d|$. Being $\mathcal{T}$ virtually independent of dispersion and depending only on the link parameter, we note the cubic dependence of timing jitter on τ for constant energy pulse streams, the inverse dependence on $|\beta''|$, and the proportionality with the bit rate $1/T_s$. We may therefore infer that longer pulses propagating in low dispersion fibers are more affected by timing jitter than shorter pulses in high dispersion fibers.

Being timing jitter a phase-independent process, timing jitter is always zero for phase-modulated pulses of equal amplitudes. This is reflected by the fact that, for a pure phase-modulated signal, $M = 0$. For a symmetric OOK, we have $N = 2$, with $a_1 = 0$ and $a_2 = 1$ occurring with equal probability. In this case, $M = 1$. For a generic signal modulated in phase and amplitude, like when QAM is used, the values of M are always $0 \leq M \leq 1$ (OOK is the worst case, as obvious), and of course modulation-specific.

In Fig. 6.11, we show the ratio $\sigma(\Delta T)/T_s$ vs. the zero dispersion length z^* in km for the parameters of Table 6.1, for OOK transmission ($M = 1$) when complete compensation is performed at every span. As before, we have used that, within first-order perturbation theory, the timing jitter hence $\sigma(\Delta T)$ is N times the timing jitter of a single span if N are the number of spans. In Fig. 6.12, we show the ratio $\sigma(\Delta T)/T_s$ vs. the zero dispersion length z^* in km for the parameters of Table 6.1, for OOK transmission ($M = 1$) when no inline dispersion compensation is performed. It is interesting to notice that in this case timing jitter is less than when dispersion compensation is performed at every span. This behavior is opposite than that shown by amplitude jitter, which is less if dispersion compensation is applied at every span. The reason is that timing jitter is a two-pulse interaction, that grows linearly with the root-mean square pulse spreading. Amplitude jitter

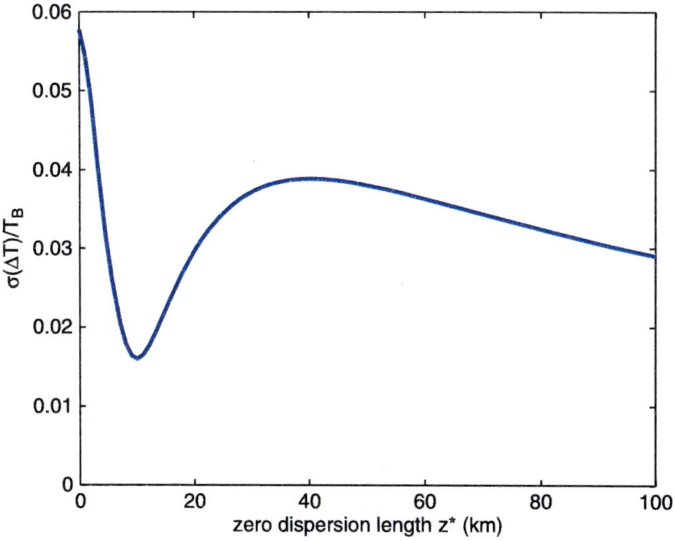

Fig. 6.11 Standard deviation of the timing jitter normalized to the bit period, $\sigma(\Delta T)/T_s$, for OOK transmission, when complete dispersion compensation is applied at every span

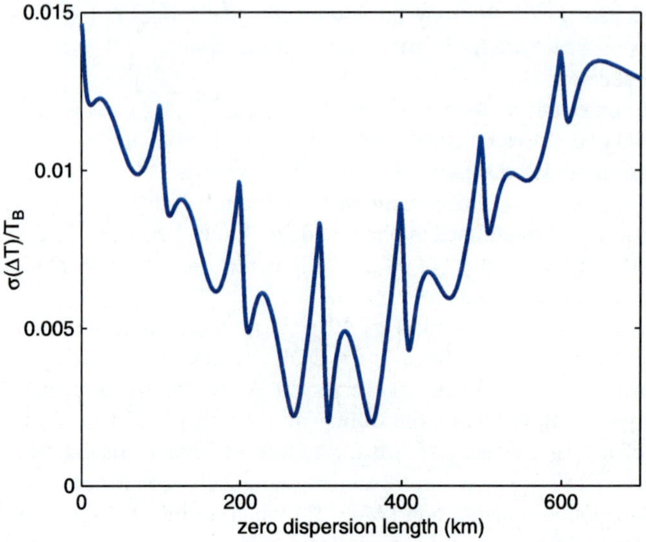

Fig. 6.12 Standard deviation of the timing jitter normalized to the bit period, $\sigma(\Delta T)/T_s$, for OOK transmission, when no inline dispersion compensation is applied

is instead dominated by FWM interaction, with the number of interacting pulses growing quadratically with the pulse spreading. This property may be important for quadrature amplitude-modulated systems if they are limited by timing jitter.

6.16 Conclusions

We have given a comprehensive analysis of the transmission of a signal under highly dispersive conditions. A significant difference between the nonlinear tolerance of the different transmission formats, and a different effect of predispersion on transmission performance are predicted and explained within a first-order perturbation theory.

References

1. A. Mecozzi, C.B. Clausen, M. Shtaif, IEEE Photon. Technol. Lett. **12**, 392–394 (2000)
2. A. Mecozzi, C.B. Clausen, M. Shtaif, IEEE Photon. Technol. Lett. **12**, 1633–1635 (2000)
3. A. Mecozzi, C.B. Clausen, M. Shtaif, P. Sang-Gyu, A.H. Gnauck, IEEE Photon. Technol. Lett. **13**, 445–447 (2001)
4. A. Mecozzi, M. Shtaif, IEEE Photon. Technol. Lett. **14**, 1029–1031 (2001)
5. P.J. Winzer, R.-J. Essiambre, Proc. IEEE **94**, 952–985 (2006)
6. H.A. Haus, J.A. Mullen, Phys. Rev. **128**, 2407–2413 (1962)

7. C.E. Shannon, Bell. Syst. Tech. J. **27**, 379–423 (1948)
8. R.J. Essiambre, G. Kramer, P.J. Winzer, G.J. Foschini, B. Goebel, J. Lightwave Technol. **28**, 662–701 (2010)
9. P.P. Mitra, J.B. Stark, Nature **411**, 1027–1030 (2001)
10. K.S. Turitsyn, S.A. Derevyanko, I.V. Yurkevich, S.K. Turitsyn, Phys. Rev. Lett. **91**, 203901 (2003)
11. I. Djordjevic, B. Vasic, M. Ivkovic, I. Gabitov, J. Lightwave Technol. **24**, 3755–3763 (2005)
12. R.-J. Essiambre, G.J. Foschini, G. Kramer, P.J. Winzer, Phys. Rev. Lett. **101**, 163901 (2008)
13. R.I. Killey, H.J. Thiele, V. Mikhailov, P. Bayvel, IEEE Photon. Technol. Lett. **13**, 1624–1626 (2000)
14. V.L. da Silva, Y. Silberberg, J.P. Heritage, E.W. Chase, M.A. Saifi, M.J. Andrejco, Opt. Lett. **16**, 1340–1342 (1991)
15. V.L. da Silva, Y. Silberberg, J.P. Heritage, Opt. Lett. **18**, 580–582 (1993)
16. D. Yang, S. Kumar, J. Lightwave Technol. **27**, 2916–2923 (2009)
17. X. Wei, X. Liu, Opt. Lett. **18** 2300–2302 (2003)
18. P.A. Humblet, M. Azizoglu, J. Lightwave Technol. **9**, 1576–1582 (1991)
19. M. Pfennigbauer, M.M. Strasser, M. Pauer, P.J. Winzer, IEEE Photon. Technol. Lett. **14**, 831–833 (2002)
20. J.P. Gordon, L.F. Mollenauer, Opt. Lett. **15**, 1351–1353 (1990)

Chapter 7
Analysis of Nonlinear Phase Noise in Single-Carrier and OFDM Systems

Shiva Kumar and Xianming Zhu

7.1 Introduction

The amplified spontaneous emission (ASE) of inline amplifiers gives rise to amplitude fluctuations of the optical field envelope and the fiber nonlinearity translates them into phase fluctuations. This is known as nonlinear phase noise. This type of noise is first studied by Gordon and Mollenauer [1] and hence, this noise is also called "Gordon–Mollenauer phase noise." The nonlinear phase noise leads to performance degradation in fiberoptic systems based on phase-shift keying (PSK) or differential phase-shift keying (DPSK) [1–4]. Gordon and Molleneuer pointed out that two degrees of freedoms (DOFs) of the noise field are of importance [1]. These noise components have the same form as the signal pulse. One of the noise components is in phase with the signal and the other in quadrature. The in-phase component of the noise changes the amplitude of the signal pulse and hence, leads to energy change while the quadrature component leads to a linear phase shift. The energy change is translated into an additional phase shift due to fiber nonlinearity. Gordon and Mollenauer argued that the noise components other than the above-mentioned modes have less significant effects if the optical bandwidth is not too large and they derived a simple analytical expression for the variance of nonlinear phase noise by ignoring fiber dispersion. When the receiver filter bandwidth is larger than the signal bandwidth, it has been found that two DOFs are not sufficient to describe the noise process [5]. Analytical expressions for the probability density function of nonlinear phase noise have been derived in [6–8] by ignoring fiber dispersion. The interaction between the nonlinearity and ASE is the strongest when the

S. Kumar (✉)
Electrical and Computer Engineering, McMaster University, ITBA 322,
1280 Main St. West, Hamilton, ON-L8S 4K1, Canada
e-mail: kumars@mail.ece.mcmaster.ca

X. Zhu
Science and Technology, Corning Incorporated, SP-TD-01-1,
Science Center Drive, Corning, NY 14831, USA
e-mail: zhux@corning.com

S. Kumar (ed.), *Impact of Nonlinearities on Fiber Optic Communications*, Optical and Fiber Communications Reports 7, DOI 10.1007/978-1-4419-8139-4_7,
© Springer Science+Business Media, LLC 2011

dispersion is zero because of phase matching and therefore, the analyses of [1, 5–8] over estimate the impact of nonlinear phase noise. Attempts have been made to calculate the impact of nonlinearphase noise in the presence of dispersion [9–23]. By assuming that the signal is CW and using the approach typically used in the study of modulational instability, it has been found that the variance of nonlinear phase noise becomes quite small in dispersion-managed transmission lines when the absolute dispersion of the transmission fiber becomes large [9]. Later in [10], the variance of nonlinear phase noise is calculated for a Gaussian pulse in a dispersion-managed transmission line and results showed that variance of nonlinear phase noise due to self-phase modulation (SPM) is quite small as compared to the case of no dispersion.

Recently, coherent optical orthogonal frequency division multiplexing (OFDM) has drawn significant attention in optical communications due to its high spectral efficiency and its robustness to fiber chromatic dispersion and polarization mode dispersion [24–28]. However, due to the large number of subcarriers, OFDM is believed to suffer from high peak-to-average power ratio leading to higher nonlinear impairments, which makes it less suitable for legacy optical communication systems with periodic inline chromatic dispersion compensation fibers [29]. In [30], a simple formula for estimating the deterministic distortions caused by four-wave mixing (FWM) is developed, and it is found that the nonlinear limit in OFDM systems is independent on the number of OFDM subcarriers in the absence of dispersion. Reference [31] analytically studied the combined effect of dispersion and FWM in OFDM multi-span systems and concluded that dispersion can significantly reduce the amount of FWM. Recently, significant research effort has been put in nonlinear compensation for coherent OFDM systems [32–39]. Of particular interest is the digital backward propagation [37–39], a technique in which the signal is propagated backward in distance using digital signal processing (DSP) so that the deterministic linear and nonlinear impairments can be compensated. However, the nonlinear phase noise caused by the interaction between ASEs noise and fiber Kerr nonlinearity cannot be compensated using digital backward propagation [37–39] or digital phase conjugation [36]. In wavelength division multiplexed (WDM) systems, nonlinear phase noise due to ASE–SPM and ASE-cross-phase modulation (XPM) interactions are important, but typically the phase noise resulting from the coupling between ASE and four-wave mixing (FWM) is negligible. But in OFDM systems, it has been found that the dominant contribution to nonlinear phase noise comes from ASE–FWM interaction [40].

This book chapter is based on a series of three papers [10, 22], and [40] on the study of nonlinear phase noise in single carrier and OFDM systems. In Sect. 7.2, the concept of DOF is reviewed and analytical expression for the linear phase noise is developed. In Sect. 7.3, analysis of nonlinear phase noise in dispersion-free fiberoptic system is carried out and the analysis is extended to a dispersive system in Sect. 7.4. In Sect. 7.5, analytical expressions for the variance of nonlinear phase noise due to ASE–SPM, ASE–XPM, and ASE–FWM interactions in OFDM systems are derived.

7.2 Linear Phase Noise

Consider the output of the optical transmitter, $s_{\text{in}}(t)$ which is confined to the bit interval $-T_b/2 < t < T_b/2$. Let

$$s_{\text{in}}(t) = a_0\sqrt{E}F(t), \tag{7.1}$$

where a_0 is the symbol in the interval, $-T_b/2 < t < T_b/2$, $F(t)$ is the pulse shape, E is the energy of the pulse, and

$$\int_{-\infty}^{\infty} |F(t)|^2 \mathrm{d}t = 1. \tag{7.2}$$

For binary phase shift keying (BPSK), a_0 takes values 1 and -1 with equal probability. In this section, we ignore the fiber dispersion and nonlinearity and include only fiber loss. To compensate for fiber loss, amplifiers are introduced periodically along the transmission line with a spacing of L_a. The amplifier compensates for the loss exactly and introduces ASE noise. In this section, let us assume that there is only one amplifier in the system and the output of the fiberoptic link can be written as

$$s_{\text{out}}(t) = s_{\text{in}}(t) + n(t), \tag{7.3}$$

where $n(t)$ is the ASE noise, which can be treated as white,

$$\langle n(t) \rangle = 0, \tag{7.4}$$

$$\left\langle n(t)n^{\star}(t') \right\rangle = \rho\delta(t - t'), \tag{7.5}$$

$$\left\langle n(t)n(t') \right\rangle = 0, \tag{7.6}$$

where ρ is the ASE power spectral density per polarization given by

$$\rho = n_{\text{sp}} h\bar{\nu}(G - 1). \tag{7.7}$$

Here, G is the gain of the amplifier, n_{sp} is spontaneous noise factor, h is Planck's constant, and $\bar{\nu}$ is the mean optical carrier frequency.

A signal of bandwidth B and duration T_b has $2J = 2BT_b$ DOF [1]. From the Nyquist sampling theorem, it follows that if the highest frequency component of a signal is $B/2$, the signal is completely described by specifying the values of the signal at instants of time separated by $1/B$. Therefore, in the interval T_b, there are BT_b complex samples which fully describe the signal. Equivalently, the signal can be described by J complex coefficients of the expansion in a set of orthonormal basis functions. Let us represent the signal and noise fields using a orthonormal set of basis functions as

$$s_{\text{in}}(t) = \sum_{j=0}^{J-1} s_j F_j(t) \tag{7.8}$$

$$n(t) = \sum_{j=0}^{J-1} n_j F_j(t), \tag{7.9}$$

where $\{F_j(t)\}$ is a set of orthonormal functions,

$$\begin{aligned} \int_{-\infty}^{\infty} F_j(t) F_k^{\star}(t) \mathrm{d}t &= 1 \text{ if } j = k \\ &= 0 \text{ otherwise.} \end{aligned} \tag{7.10}$$

Because of the orthogonality of the basis functions, it follows that

$$n_j(t) = \int_{-\infty}^{\infty} n(t) F_j^{\star}(t) \mathrm{d}t. \tag{7.11}$$

Using (7.11) and (7.4)–(7.6), we obtain

$$\langle n_j \rangle = 0, \tag{7.12}$$

$$\begin{aligned} \langle n_j n_k^{\star} \rangle &= \rho \text{ if } j = k \\ &= 0 \text{ otherwise} \end{aligned} \tag{7.13}$$

$$\langle n_j n_k \rangle = 0. \tag{7.14}$$

Using (7.8) and (7.9) in (7.3), we find

$$s_{\text{out}}(t) = \sum_{j=0}^{J-1} (s_j + n_j) F_j(t). \tag{7.15}$$

Suppose **1** is transmitted ($a_0 = 1$) we choose $F_0(t) = F(t)$ so that

$$\begin{aligned} s_j &= \sqrt{E} \text{ if } j = 0 \\ &= 0 \text{ otherwise} \end{aligned} \tag{7.16}$$

Equation (7.15) can be written as

$$s_{\text{out}}(t) = \left(\sqrt{E} + n_0\right) F(t) + \sum_{j=1}^{J-1} n_j F_j(t). \tag{7.17}$$

Let us assume that signal power is much larger than the noise power and $s_{\text{in}}(t)$ is real. Let

$$n(t) = n_r(t) + in_i(t), \tag{7.18}$$

where $n_r = \text{Re}\{n(t)\}$ and $n_i = \text{Im}\{n(t)\}$. Equation (7.3) can be written as

$$s_{\text{out}}(t) = A(t)\exp[i\phi(t)], \tag{7.19}$$

where

$$A(t) = \left\{[s_{\text{in}}(t) + n_r(t)]^2 + n_i^2(t)\right\}^{1/2} \tag{7.20}$$

$$\begin{aligned}\phi(t) &= \tan^{-1}\left\{\frac{n_i(t)}{s_{\text{in}}(t) + n_r(t)}\right\} \\ &\approx \frac{n_i(t)}{s_{\text{in}}(t)}.\end{aligned} \tag{7.21}$$

In (7.21), we have ignored the higher order terms such as n_i^2 and n_r^2. Using (7.8),(7.9),(7.16), and (7.17) in (7.21), we obtain

$$\phi(t) = \frac{n_{0i}}{\sqrt{E}} + \sum_{j=1}^{J-1} \frac{n_{ji} F_j(t)}{F(t)\sqrt{E}}, \tag{7.22}$$

where $n_{jr} = \text{Re}\{n_j\}$ and $n_{ji} = \text{Im}\{n_j\}$. From (7.22) and (7.12), it follows that

$$\langle \phi(t) \rangle = 0. \tag{7.23}$$

Squaring and averaging (7.22) and using (7.13) and (7.14), we obtain the variance of phase noise as

$$\sigma_{\text{lin}}^2 = \langle \phi^2 \rangle = \frac{\rho}{2E} + \frac{\rho}{2E} \sum_{j=1}^{J-1} \frac{F_m^2(t)}{F^2(t)}. \tag{7.24}$$

Next, let us consider the impact of a matched filter on the phase noise. When a matched filter is used, the received signal is

$$r = \int_{-\infty}^{\infty} s_{\text{out}}(t) F^{\star}(t)\text{d}t. \tag{7.25}$$

Substituting (7.17) in (7.25) and using (7.10), we obtain

$$r = \left(\sqrt{E} + n_0\right). \tag{7.26}$$

Note that the higher-order noise components given by the second term on the right-hand side of (7.17) do not contribute because of the orthogonality of basis functions. Now, (7.24) reduces to

$$\sigma_{\text{lin}}^2 = \frac{\langle n_{0i}^2 \rangle}{E} = \frac{\rho}{2E}. \tag{7.27}$$

From (7.26), we see that when a matched filter is used, the noise field is fully described by two DOFs, namely, the in-phase component n_{0r} and the quadrature component n_{0i}. The other DOFs are orthogonal to the signal and do not contribute after the matched filter. From (7.27), we see that the quadrature component n_{0i} is responsible for the linear phase noise.

7.3 Gordon–Mollenauer Phase Noise

The optical field envelope in a fiberoptic transmission system can be described by the nonlinear Schrodinger (NLS) equation,

$$i\frac{\partial q}{\partial z} - \frac{\beta_2(z)}{2}\frac{\partial^2 q}{\partial t^2} = -\gamma|q|^2 q - i\frac{\alpha(z)}{2}q, \tag{7.28}$$

where $\alpha(z)$ is the loss/gain profile, which includes fiber loss as well as amplifier gain, $\beta_2(z)$ is the dispersion profile, and γ is the fiber nonlinear coefficient. To separate the fast variation of the optical power due to fiber loss/gain, we use the following transformation [41]

$$q(z,t) = a(z)u(z,t), \tag{7.29}$$

$$\frac{\partial q}{\partial z} = u\frac{\mathrm{d}a}{\mathrm{d}z} + a\frac{\partial u}{\partial z}. \tag{7.30}$$

Let

$$\frac{\mathrm{d}a}{\mathrm{d}z} = -\frac{\alpha(z)a}{2}. \tag{7.31}$$

Substituting (7.31) and (7.30) in (7.28), we obtain the NLS equation in the loss less form,

$$i\frac{\partial u}{\partial z} - \frac{\beta_2(z)}{2}\frac{\partial^2 u}{\partial t^2} = -\gamma a^2(z)|u|^2 u. \tag{7.32}$$

Solving (7.31) with the initial condition $a(0) = 1$, we obtain

$$a(z) = \exp\left[-\frac{1}{2}\int_0^z \alpha(s)\mathrm{d}s\right]. \tag{7.33}$$

Between amplifiers, if the fiber loss is constant, (7.33) becomes

$$a(z) = \exp\left[-\alpha_0 Z/2\right], \tag{7.34}$$

where α_0 is the fiber loss coefficient, $Z = \mathrm{mod}(z, L_a)$ and L_a is the amplifier spacing. The mean optical power $\langle|q|^2\rangle$ fluctuates as a function of distance due to fiber loss and amplifier gain, but $\langle|u|^2\rangle$ is independent of distance since the variations due to loss/gain is separated out using (7.29). Note that the nonlinear coefficient is constant in (7.28), but the effective nonlinear coefficient $\gamma a^2(z)$ changes as a function of distance in (7.32). Amplifier noise effects can be introduced to (7.32) by adding a source term on the right-hand side, which leads to

$$i\frac{\partial u}{\partial z} - \frac{\beta_2(z)}{2}\frac{\partial^2 u}{\partial t^2} = -\gamma a^2(z)|q|^2 q + iR(z,t), \tag{7.35}$$

where

$$R(z,t) = \sum_{m=1}^{N_a} \delta(z - mL_a)n(t). \tag{7.36}$$

Here, N_a is the number of amplifiers and $n(t)$ is the noise field due to ASE with statistical properties defined in Sect. 7.2.

In this section, we assume that the fiber dispersion is zero. Let us first consider the solution of (7.35) in the absence of noise. Let

$$u(z,t) = A(z,t)\exp[i\phi(z,t)], \tag{7.37}$$

and

$$u(0,t) = \sqrt{E}F(t). \tag{7.38}$$

Substituting (7.37) in (7.32), we find

$$\frac{\mathrm{d}A}{\mathrm{d}z} = 0 \to A(z,t) = A(0,t) = \sqrt{E}|F(t)|, \tag{7.39}$$

$$\begin{aligned}\frac{\mathrm{d}\phi}{\mathrm{d}z} &= \gamma a^2(z)|u(0,t)|^2,\\ &= \gamma a^2(z)E|F(t)|^2.\end{aligned} \tag{7.40}$$

Solving (7.40), we find

$$\phi(z,t) = \gamma E|F(t)|^2 \int_0^z a^2(s)\mathrm{d}s, \tag{7.41}$$

$$u(z,t) = u(0,t)\exp\left[i\gamma|u(0,t)|^2 \int_0^z a^2(s)\mathrm{d}s\right], \tag{7.42}$$

We assume that the signal pulse shape is rectangular with pulse width T_b. From (7.2), it follows that $|F(t)|^2 = 1/T_b$. Since $a^2(z) = \exp(-\alpha_0 Z)$ between amplifiers, it follows that

$$\int_0^{mL_a-} a^2(z)\mathrm{d}z = mL_{\mathrm{eff}}, \tag{7.43}$$

where

$$L_{\mathrm{eff}} = \frac{1-\exp(-\alpha_0 L_a)}{\alpha_0}. \tag{7.44}$$

Substituting (7.43) in (7.41) and (7.42), we find

$$\phi(mL_a-) = \frac{\gamma E m L_{\text{eff}}}{T_b}, \tag{7.45}$$

$$u(mL_a-, t) = \sqrt{E}F(t)\exp[i\phi(mL_a-)]. \tag{7.46}$$

Next, let us consider the case when there is only one amplifier located at mL_a that introduces ASE noise. The optical field envelope after the amplifier is

$$u(mL_a+, t) = u(mL_a-, t) + n(t). \tag{7.47}$$

We assume that two DOFs of the noise field are of importance. They are in-phase component n_{0r} and quadrature component n_{0i} and ignore other noise components. In Sect. 7.2, we have seen that noise field is fully described by these two DOFs for a linear system. Gordon and Mollenauer [1] assumed that these two DOFs are adequate to describe the noise field even for a nonlinear system. Using (7.46) and (7.9) in (7.47), we find

$$\begin{aligned} u(mL_a+, t) &= \sqrt{E}F(t)\exp[i\phi(mL_a-)] + n_0 F(t) \\ &= \left(\sqrt{E} + n_0'\right) F(t)\exp[i\phi(mL_a-)], \end{aligned} \tag{7.48}$$

where

$$n_0' = n_0 \exp[-i\phi(mL_a-)] \tag{7.49}$$

n_0' is same as n_0 except for a deterministic phase shift, which does not alter the statistical properties, i.e.,

$$\langle n_0' \rangle = 0, \tag{7.50}$$

$$\langle n_0' n_0'^{*} \rangle = \rho, \tag{7.51}$$

$$\langle n_0' n_0' \rangle = 0. \tag{7.52}$$

From (7.48), we see that the complex amplitude of the field envelope has changed because of the amplifier noise. Using $u(mL_a+, t)$ as the initial condition, the NLS equation (7.32) is solved to obtain the field at the end of the transmission line as

$$\begin{aligned} u(L_{\text{tot}}, t) &= u(mL_a+, t)\exp\left\{ i\gamma |u(mL_a+, t)|^2 \int_{mL_a+}^{L_{\text{tot}}} a^2(z)\mathrm{d}z \right\} \\ &= \left(\sqrt{E} + n_0'\right) F(t)\exp\left[i\phi(mL_a-) + i\gamma |\sqrt{E} + n_0'|^2 \, (N_a - m)L_{\text{eff}}/T_b \right], \end{aligned} \tag{7.53}$$

where $L_{\mathrm{tot}} = N_a L_a$ is the total transmission distance. The phase at L_{tot} is

$$\phi = \tan^{-1}\left\{\frac{n'_{0i}}{\sqrt{E}+n'_{0r}}\right\} + \frac{\gamma|\sqrt{E}+n'_0|^2\,(N_a-m)L_{\mathrm{eff}}}{T_b} + \frac{\gamma E m L_{\mathrm{eff}}}{T_b}$$
$$\approx \frac{n'_{0i}}{\sqrt{E}} + \gamma(E + 2\sqrt{E}n'_{0r})(N_a-m)L_{\mathrm{eff}}/T_b + \gamma E m L_{\mathrm{eff}}/T_b. \quad (7.54)$$

The total phase given by (7.54) can be separated into two parts.

$$\phi = \phi_d + \delta\phi, \quad (7.55)$$

where ϕ_d is the deterministic nonlinear phase shift given by

$$\phi_d = \gamma E N_a L_{\mathrm{eff}}/T_b \quad (7.56)$$

and $\delta\phi$ represents the phase noise,

$$\delta\phi = \frac{n'_{0i}}{\sqrt{E}} + \frac{2\gamma\sqrt{E}n'_{0r}(N_a-m)L_{\mathrm{eff}}}{T_b}. \quad (7.57)$$

The first and second terms in (7.57) represent the linear and nonlinear phase noise, respectively. As can be seen, the in-phase component n'_{0r} and the quadrature component, n'_{0i} are responsible for nonlinear and linear phase noise, respectively. From (7.50), it follows that

$$\langle\delta\phi\rangle = 0. \quad (7.58)$$

Squaring and averaging (7.57) and using (7.51) and (7.52), we find the variance of the phase noise as

$$\sigma_m^2 = \frac{\rho}{2E} + 2\rho E\left[\frac{\gamma(N_a-m)L_{\mathrm{eff}}}{T_b}\right]^2. \quad (7.59)$$

So far we ignored the impact of ASE due to other amplifiers. In the presence of ASE due to other amplifiers, the expression for the optical field envelope at mL_a- given by (7.46) is inaccurate since it ignores the noise field added by the amplifiers preceding the mth amplifier. However, when the signal power is much larger than the noise power, the second order terms such as n_{0r}^2 and n_{0i}^2 can be ignored. At the end of the transmission line, the dominant contribution would come from the linear terms n_{0i} and n_{0i} of each amplifiers. Since the noise fields of amplifiers are statistically independent, total variance is the sum of variance due to each amplifier,

$$\sigma^2 = \sum_{m=1}^{N_a} \sigma_m^2$$
$$= \frac{\rho N_a}{2E} + 2\rho E \left[\frac{\gamma L_{\text{eff}}}{T_b}\right]^2 \sum_{m=1}^{N_a-1} (N_a - m)^2$$
$$= \frac{\rho N_a}{2E} + \frac{(N_a - 1)N_a(2N_a - 1)\rho E \gamma^2 L_{\text{eff}}^2}{3T_b^2}. \tag{7.60}$$

References [5–8] provide a more rigorous treatment of the nonlinear phase noise without ignoring the higher-order noise terms. From (7.60), we see that the variance of the linear phase noise (the first term on the right-hand side) increases linearly with the number of amplifiers, whereas the the variance of nonlinear phase noise (the second term) increases cubically with the number of amplifiers when N_a is large indicating that nonlinear phase noise could be the dominant penalty for ultra long haul fiberoptic transmission systems. In addition, the variance of linear phase noise is inversely proportional to the energy of the pulse, whereas the variance of nonlinear phase noise is directly proportional to the energy. This implies that there exists an optimum energy at which the total phase variance is minimum. By setting $\mathrm{d}\sigma^2/\mathrm{d}E$ to zero, the optimum energy is calculated as

$$E_{\text{opt}} = \frac{T_b}{\gamma L_{\text{eff}}} \sqrt{\frac{3}{2(N_a - 1)(2N_a - 1)}}. \tag{7.61}$$

When N_a is large, $(N_a - 1)(2N_a - 1) \approx 2N_a^2$ and using (7.56), we find that the phase variance is minimum when the deterministic nonlinear phase shift $\phi_d \approx 0.87$ rad.

7.4 Phase Noise in Dispersive Nonlinear Fiberoptic Single Carrier System

In this section, we consider a more general case in which the dispersion coefficient is not zero and the amplifier spacing is arbitrary. In this case, the noise term $R(z, t)$ of (7.35) is modified as

$$R(z, t) = \sum_{m=1}^{N_a} \delta(z - L_m) n^{(m)}(t), \tag{7.62}$$

where L_m is the location of an amplifier, N_a is the number of amplifiers, and $n^{(m)}(t)$ is the noise field due to an amplifier located at L_m. The statistical properties of $n^{(m)}(t)$ is same as that of $n(t)$. In Sect. 7.3, we assumed that pulse shape is rectangular. In a dispersive system, the pulse broadening of a rectangular pulse is hard to treat analytically. So, we assume that the launched pulse is Gaussian. In the absence of nonlinear effects and amplifier noise, if a Gaussian pulse is launched to the fiber, its propagation is given by [42]

$$u_{\text{lin}}(z,t) = \sqrt{E}F(z,t), \tag{7.63}$$

$$F(z,t) = \left[\frac{p(z)}{\sqrt{\pi}}\right]^{1/2} \exp\left\{-\frac{[p^2(z) + iC(z)]t^2}{2} + i\theta_0(z)\right\}, \tag{7.64}$$

where E is the pulse energy, $p(z)$, $C(z)$, and $\theta_0(z)$ are the inverse pulse width, chirp and phase factors, respectively, given by

$$p(z) = \frac{T_0}{\sqrt{T_0^4 + S^2(z)}}, \quad C(z) = \frac{S(z)p^2(z)}{T_0^2}, \tag{7.65}$$

$$\theta_0(z) = \frac{1}{2}\tan^{-1}\left[S(z)/T_0^2\right]. \tag{7.66}$$

Here, T_0 is the half-width at 1/e- intensity point, and $S(z)$ is the accumulated dispersion

$$S(z) = \int_0^z \beta_2(s)\mathrm{d}s. \tag{7.67}$$

The peak power, P and energy, E are related by

$$P = \frac{E}{T_{\text{eff}}}, \tag{7.68}$$

where $T_{\text{eff}} = \sqrt{\pi}T_0$ and $F(z,t)$ is normalized such that

$$\int_{-\infty}^{\infty} |F(z,t)|^2 \mathrm{d}t = 1. \tag{7.69}$$

Expanding the optical field in a series, we have

$$u(z,t) = u^{(0)}(z,t) + \gamma u^{(1)}(z,t) + \gamma^2 u^{(2)}(z,t) + \ldots \tag{7.70}$$

where $u^{(j)}(z,t)$, $j \neq 0$ is the jth order correction due to fiber nonlinearity, and $u^{(0)}(z,t)$ is the zeroth order linear solution, as given by (7.63). Here, we focus only up to the first-order correction to the optical field envelope. Substituting (7.70) in (7.32) and collecting the terms proportional to γ, we obtain

$$i\frac{\partial u^{(1)}}{\partial z} - \frac{\beta_2(z)}{2}\frac{\partial^2 u^{(1)}}{\partial t^2} = -a^2(z)|u^{(0)}|^2 u^{(0)}. \tag{7.71}$$

We will use (7.71) to calculate the impact of SPM on the signal and noise fields.

Consider the optical field envelope immediately after an amplifier located at L_m. Focusing only on the impact of the noise added by this amplifier, the linear part of the optical field envelope at $z = L_m+$ is

$$u_{\text{lin}}(L_m+, t) = u_{\text{lin}}(L_m, t) + n(t), \tag{7.72}$$

where $n(t) \equiv n^{(m)}(t)$ is the noise field added by the amplifier at L_m. As in the previous section, we first assume that two DOFs of the noise field are sufficient to describe the noise process. Similar to (7.48), the linear part of the optical field envelope immediately after the mth amplifier is

$$u^{(0)}(L_m+, t) = \left(\sqrt{E} + n_0\right) F(L_m, t). \tag{7.73}$$

Treating (7.73) as the initial condition, the zeroth order optical field envelope is described by

$$u^{(0)}(z, t) = \left(\sqrt{E} + n_0) F(z, t\right), \quad z > L_m. \tag{7.74}$$

Substituting (7.74) in (7.71), the first-order correction due to SPM can be written as

$$\begin{aligned} i\frac{\partial u^{(1)}}{\partial z} - \frac{\beta_2(z)}{2}\frac{\partial^2 u^{(1)}}{\partial t^2} &= -a^2(z)\left|\left(\sqrt{E} + n_0\right) F(z, t)\right|^2 \left(\sqrt{E} + n_0)F(z, t\right) \\ &\approx \left(\sqrt{E} + n_0\right)\left(E + 2\sqrt{E}n_{0r}\right)|F(z, t)|^2 F(z, t) \end{aligned} \tag{7.75}$$

for $z > L_m$. In (7.75), we have ignored the higher-order terms such as n_{0r}^2 and n_{0i}^2 under the assumption that the noise power is much smaller than the signal power. In practical systems operating in the psuedolinear regime, the dispersion of the transmission fibers is fully compensated at the receiver either in optical or in electrical domain, i.e., $S(L_{\text{tot}}) = 0$, where L_{tot} is the total transmission distance. Solving (7.75) with the condition, $S(L_{\text{tot}}) = 0$, we find [43–45]

$$u^{(1)}(L_{\text{tot}}, t) = i\left(\sqrt{E} + n_0\right) F(0, t)(E + \delta E)g(L_m, t), \tag{7.76}$$

where

$$\delta E = 2\sqrt{E}n_{0r} \tag{7.77}$$

$$g(z, t) = \frac{T_0}{\sqrt{\pi}}\int_z^{L_{\text{tot}}} \frac{a^2(s)\exp[-\Delta(s)t^2]\mathrm{d}s}{\sqrt{T_0^4 + 3S^2(s) + 2iT_0^2 S(s)}}, \tag{7.78}$$

$$\Delta(s) = \frac{T_0^2 - iS(s)}{T_0^2[T_0^2 + i3S(s)]}. \tag{7.79}$$

Since $S(L_{\text{tot}}) = 0$, it follows that $F(L_{\text{tot}}, t) = F(0, t)$. Combining the first-order and zeroth-order solutions ((7.74) and (7.76)), total field envelope at the end of the transmission line is

$$u(L_{\text{tot}}, t) = \left(\sqrt{E} + n_0\right) F(0, t)[1 + i\gamma(E + \delta E)g(L_m, t)]. \tag{7.80}$$

From (7.77) and (7.80), we see that the in-phase noise component n_{0r} is responsible for energy shift and the consequent nonlinear phase shift. When a matched filter is used, the received signal is

$$r = \int_{-\infty}^{\infty} u(L_{\text{tot}}, t) F^{\star}(0, t) \mathrm{d}t. \tag{7.81}$$

Substituting (7.64) and (7.80) in (7.81), we find

$$r = (\sqrt{E} + n_0)[1 + i\gamma(E + \delta E) g_f(L_m)], \tag{7.82}$$

where

$$g_f(L_m) = \frac{T_0}{\sqrt{\pi}} \int_{L_m}^{L_{\text{tot}}} G(s) \mathrm{d}s, \tag{7.83}$$

$$G(s) = \frac{a^2(s)}{\sqrt{[1 + T_0^2 \Delta(s)][T_0^4 + 3S^2(s) + 2iT_0^2 S(s)]}}. \tag{7.84}$$

The phase of the matched filter output is

$$\begin{aligned} \phi &= \tan^{-1} \frac{\mathrm{Im}[r]}{\mathrm{Re}[r]}, \\ &\approx \gamma E g_{\text{fr}}(L_m) \\ &\quad + \gamma \delta E g_{\text{fr}}(L_m) \\ &\quad + \frac{n_{0i}}{\sqrt{E}}, \end{aligned} \tag{7.85}$$

where $g_{\text{fr}}(L_m) = \mathrm{Re}[g_f(L_m)]$. In (7.85), we have ignored the terms proportional to γ^2, n_{0r}^2, n_{0i}^2, and $n_{0r} n_{0i}$. The first, second, and the last terms on the right-hand side of (7.85) represent the deterministic nonlinear phase change, nonlinear and the linear phase changes due to ASE of the amplifier located at L_m, respectively. Therefore, the phase changes due to ASE of the amplifier located at L_m are

$$\delta\phi_m = \gamma \delta E g_{\text{fr}}(L_m) + \frac{n_{0i}}{\sqrt{E}}. \tag{7.86}$$

Variance of energy shift is related to the variance of n_{0r}. From (7.5), (7.6), and (7.77) , we have

$$\langle n_{0r}^2 \rangle = \langle n_{0i}^2 \rangle = \rho_m / 2 \tag{7.87}$$

$$\langle \delta E^2 \rangle = 2\rho_m E. \tag{7.88}$$

Squaring and averaging (7.86), and using (7.87) and (7.88), we obtain

$$\langle \delta\phi_m^2 \rangle = 2\rho_m E [\gamma g_{\text{fr}}(L_m)]^2 + \frac{\rho_m}{2E}. \tag{7.89}$$

The first and the second terms in (7.89) represent the variance of nonlinear phase noise and linear phase noise, respectively, due to the amplifier located at L_m.

As in Sect. 7.3, variance of phase noise due to all the amplifiers is

$$\langle \delta\phi^2 \rangle = \sum_{m=1}^{N_a} \langle \delta\phi_m^2 \rangle. \tag{7.90}$$

To simplify (7.90) further and also to make a direct comparison with [1] and [10], we consider a transmission fiber consisting of two segments of equal lengths within an amplifier spacing. The dispersion of the first segment is anomalous, whereas that of the second segment is equal in magnitude but opposite in sign. We assume that there is no pre- and post-compensation of dispersion. Since the amplifier spans are identical, $L_m = mL_a, m = 1, 2, \ldots N_a$, where L_a is the amplifier spacing, we can write

$$g_f(L_m) = (N_a - m)h_f, \tag{7.91}$$

where

$$h_f = \frac{T_0}{\sqrt{\pi}} \int_0^{L_a} G(s)\mathrm{d}s, \tag{7.92}$$

and (7.89) is modified as

$$\langle \delta\phi_m^2 \rangle = 2\rho E[\gamma(N_a - m)h_{\mathrm{fr}}]^2 + \frac{\rho}{2E}, \tag{7.93}$$

where $h_{\mathrm{fr}} = \mathrm{Re}[h_f]$ and $\rho_m = \rho$. Adding contributions to the phase variance from all the amplifiers, we obtain the total variance as

$$\langle \delta\phi^2 \rangle = \frac{N_a(N_a - 1)(2N_a - 1)\rho E(\gamma h_{\mathrm{fr}})^2}{3} + \frac{\rho N_a}{2E}. \tag{7.94}$$

Comparing (7.60) and (7.94), we see that these two expressions are the same except that L_{eff}/T_b is replaced by h_{fr}. For a highly dispersive system, h_{fr} is much smaller than L_{eff}/T_b and hence, the variance of nonlinear phase noise due to SPM is much smaller in a highly dispersive system as compared to dispersion-free system. When $N_a \ll 1$, (7.94) can be approximated as

$$\langle \delta\phi^2 \rangle \approx \frac{2\rho E(\gamma h_{\mathrm{fr}})^2 N_a^3}{3} + \frac{\rho N_a}{2E}. \tag{7.95}$$

The optimum energy is calculated by differentiating $\langle \delta\phi^2 \rangle$ with respect to E and setting it to zero. We find the optimum energy as

$$E_{\mathrm{opt}} = \frac{1}{\gamma h_{\mathrm{fr}}} \sqrt{\frac{3}{2(N_a - 1)(2N_a - 1)}}. \tag{7.96}$$

So far we have considered only two DOFs of the noise fields. In [22], analysis has been carried out for arbitrary DOFs and the variance of phase noise is

$$\langle \delta\phi_m^2 \rangle = \frac{\rho_m E \gamma^2}{Z_0^2} \left[2g_{\mathrm{fr}}^2(L_m) + \sum_{j=1}^{J} \frac{P_j'^2 + Q_j'^2}{2} \right] + \frac{\rho_m}{2E} \left[1 + \sum_{j=1}^{J} \frac{Z_j^2}{Z_0^2} \right] + \sum_{j=1}^{J} \frac{\gamma \rho_m Q'_j Z_j}{Z_0^2}, \tag{7.97}$$

where the variables P'_j, Q'_j, and Z_j are defined in [22].The first term ($\propto \gamma^2$) on the right-hand side of (7.97) represents the nonlinear phase noise, the second term represents linear phase noise, and the last term represents the correlation between linear and nonlinear phase noise, which is absent when the DOF = 2. The variance of phase noise due to all the amplifier is given by (7.90). In the following subsection, we will use (7.95), (7.97), and (7.90) to calculate the variance of phase noise.

7.4.1 *Results and Discussion*

To test the validity of the approximations done in obtaining (7.94),(7.97), and (7.90), numerical simulations of the NLS equation by the split-step Fourier technique are carried out. We assume the following parameters throughout this section: nonlinear coefficient = 2.43 $\mathrm{W^{-1}km^{-1}}$, fiber loss coefficient = 0.2 dB/km, bit rate = 40 Gb $\mathrm{s^{-1}}$, $n_{\mathrm{sp}} = 1$, which corresponds to a noise figure of 3 dB, and spacing between inline amplifiers = 80 km. We assume that a Gaussian pulse with full width half-maximum (FWHM) of 12.5 ps is launched to the fiber link so that $T_0 = 7.5$ ps. The computational bandwidth is 320 GHz and ASE is propagated over the entire computational bandwidth. A Gaussian filter of arbitrary bandwidth is used in electrical domain and no optical filter is used. Four thousand runs of NLS equation are carried out and the phase variance of the decision variable is calculated. In Fig. 7.1, the matched filter is used at the end of the transmission line with $f_0 = 1/(2\pi T_0)$. For Figs. 7.1–7.4, two types of fibers are used between inline amplifiers, the first one is an anomalous dispersion fiber of length 40 km and the second one is the normal dispersion fiber of the same absolute dispersion and the same length. The "+" marks in Fig. 7.1 shows the numerical simulation results and the solid line shows the analytical results calculated using (7.97) with DOF = 14. As the dispersion increases, the variance of nonlinear phase noise due to SPM decreases consistent with the results of [9] and [10]. The nonlinear phase variance grows cubically with distance and therefore, the difference between the variances for the case of $|D| = 4$ ps $\mathrm{nm^{-1}}$ $\mathrm{km^{-1}}$ and $|D| = 10$ ps $\mathrm{nm^{-1}}$ $\mathrm{km^{-1}}$ increases significantly for longer transmission lengths.

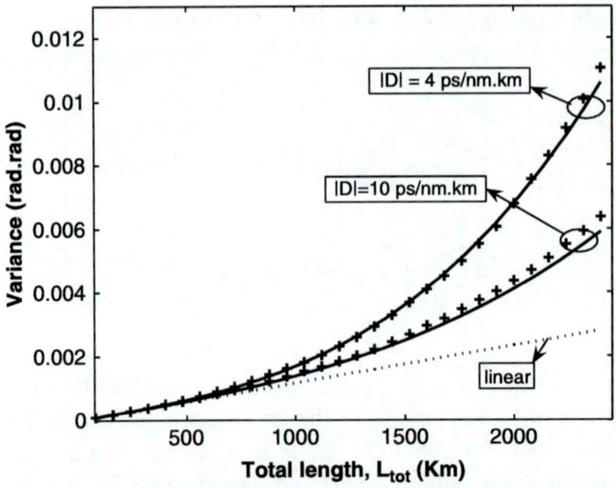

Fig. 7.1 The phase variance dependence on the total length of the transmission line. Peak power = 2 mW. *Solid line* and + marks show the analytical and numerical simulation results, respectively. The *dotted line* shows the analytical results when fiber nonlinearity is absent, which is independent of dispersion. DOF = 14 is used for analytical results. After [22] Copyright 2009 IEEE

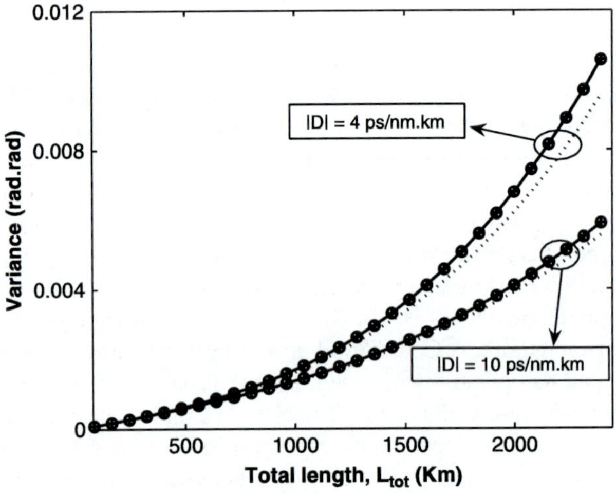

Fig. 7.2 Dependence of variance on the DOFs with a matched filter. *Dotted line*, *circles*, +, and *solid line* show the analytical results with DOF 2, 6, 10, and 14, respectively. Other parameters are same as that of Fig. 7.1. After [22] Copyright 2009 IEEE

To estimate the number of DOFs required when a matched filter (f_0 = 21.19 GHz) is used, in Fig. 7.2, we have plotted the phase variance as a function of length of transmission line for various DOFs using (7.97). From Fig. 7.2, we see that the phase variance does not change as the number of DOFs is changed

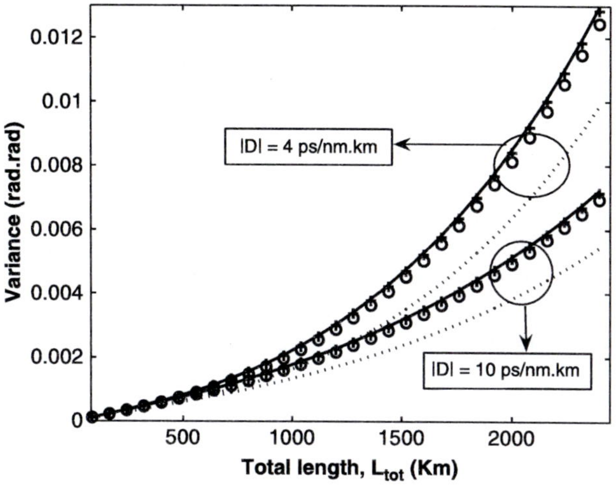

Fig. 7.3 Dependence of variance on the DOFs with a Gaussian filter with $f_0 = 42.38$ GHz. *Dotted line*, *circles*, +, and *solid line* show the analytical results with DOF 2, 6, 10, and 14, respectively. Other parameters are same as that of Fig. 7.1. After [22] Copyright 2009 IEEE

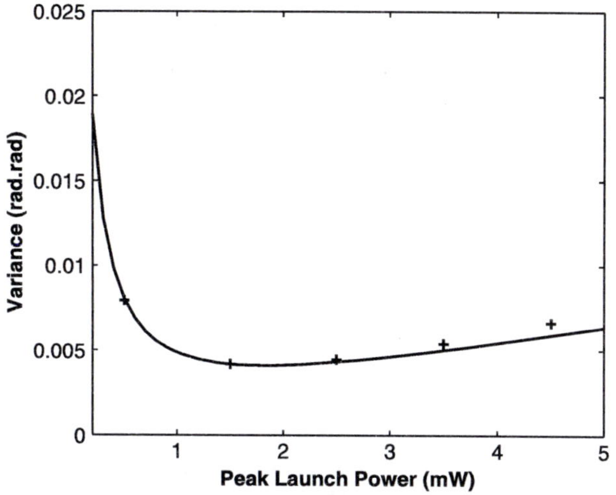

Fig. 7.4 Dependence of phase variance on peak launch power. Matched filter is used. *Solid* and "+" show the analytical and numerical simulation results, respectively. $L_{\text{tot}} = 2{,}400$ Km, and $|D| = 4\,\text{ps}\,\text{nm}^{-1}\,\text{km}^{-1}$. DOF = 14 is used for analytical results. After [22] Copyright 2009 IEEE

from 6 to 14. However, there is about 10% change in variance as the number of DOFs is changed from 2 to 6 when $|D| = 4\,\text{ps}\,\text{nm}^{-1}\,\text{km}^{-1}$ and $L_{\text{tot}} = 2,400\,\text{Km}$, and the corresponding change in variance when $|D| = 10\,\text{ps}\,\text{nm}^{-1}\,\text{km}^{-1}$ is 6%. In Fig. 7.3, a Gaussian filter with $f_0 = 42.38\,\text{GHz}$, which has a bandwidth twice that of a matched filter is used at the receiver. In this case, we see that two DOFs are not sufficient to describe the impact of noise on the phase variance. The errors introduced by using 2, 6, and 10 DOFs are 30%, 4%, and 1%, respectively, for $|D| = 4\,\mathrm{ps\,nm^{-1}\,km^{-1}}$ and $L_{\mathrm{tot}} = 2,400\,\mathrm{Km}$. As the filter bandwidth increases, higher-order noise components and noise fields due to nonlinear mixing of the signal and higher-order noise components occupy the pass band of the filter. Therefore, as the filter bandwidth increases, the variance of linear phase noise as well as nonlinear phase noise increases.

Figure 7.4 shows the dependence of phase variance on the launch power. When the launch power is low, the linear phase noise dominates (because of $1/E$ dependence in (7.94)). At high launch power, nonlinear phase noise becomes significant (because of E dependence in (7.94)). The optimum launch power is calculated to be 1.8 mW using (7.96), which is in agreement with numerical simulations. At high launch powers (>4 mW), there is a small discrepancy between the analytical results and simulation results, which is because we have ignored the terms containing γ^2 and higher. The first-order perturbation theory is known to become inaccurate at large launch powers and/or longer transmission distance. It may be possible to increase the accuracy of the calculations using the multiple-scale approaches of [46–48] when the dispersion map is periodic. Alternatively, a second-order perturbation theory [45], which is shown to be quite accurate for the description of SPM and XPM for the range of launch powers and transmission distances of practical interest could be used.

Next, we consider a dispersion map with two types of transmission fibers within an amplifier spacing. Let D_1 and D_2 be the dispersion parameters of these fibers and, l_1 and l_2 be their respective lengths. The average dispersion of these fibers is

$$D_{\mathrm{av}} = (D_1 l_1 + D_2 l_2)/(l_1 + l_2). \tag{7.98}$$

The dispersion of the transmission fibers is compensated by pre- and postcompensating fibers. The dispersion coefficients and lengths of pre- and postcompensating fibers are so selected that the total accumulated dispersion before decision is zero. The following parameters are used to obtain Fig. 7.5. The dispersion parameter of the pre- and postcompensating fiber, $D_{\mathrm{pre}} = D_{\mathrm{post}} = -100\,\mathrm{ps\,nm^{-1}\,km^{-1}}$, $l_1 = l_2 = 40\,\mathrm{Km}$, inline amplifier spacing $= l_1 + l_2 = 80\,\mathrm{Km}$, transmission distance (excluding lengths of pre- and postcompensation fibers), $L_{tr} = 2,400\,\mathrm{Km}$ and launched peak power $= 2\,\mathrm{mW}$. Approximately 50% of the total accumulated dispersion of the transmission link is compensated using the precompensating fiber. Solid line in Fig. 7.5 shows the phase variance calculated from (7.97) and (7.90) and "+" shows the numerical simulation results. As can be seen, the phase variance decreases as D_{av} or $|D_1|$ increases. As D_{av} and/or $|D_1|$ increases, the nonlinear contribution to the phase variance becomes quite small. However, in this case, pulses significantly broaden and overlap with neighboring pulses and it is likely that the ASE-induced nonlinear phase noise due to intrachannel cross-phase modulation (IXPM) could become important, which is not considered here.

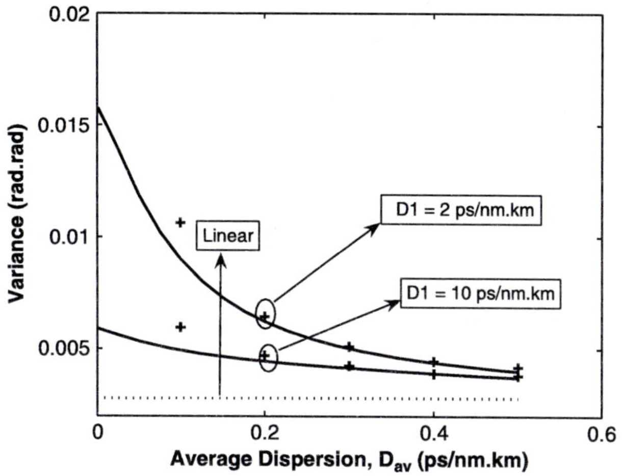

Fig. 7.5 Dependence of phase variance on the average dispersion, D_{av} and the local dispersion D_1. *Solid line* and "+" show the analytical (with J = 6) and numerical simulation results, respectively. *Dotted line* shows the analytical results for the case of $\gamma = 0$. Matched filter is used. Total transmission distance, L_{tr} (excluding pre- and post-compensation fiber) $= 2,400$ Km, peak power $= 2$ mW, location of the first inline amplifier, $L_1 = 0.5 D_{av} L_{tr} / D_{pre}$. After [22] Copyright 2009 IEEE

7.5 Phase Noise in OFDM Systems

In OFDM systems, the nonlinear interaction among subcarriers leads to performance degradation [30–32]. In this book chapter, we primarily focus on the nonlinear interaction between the signal and ASE. Typically, there are large numbers of subcarriers in OFDM systems, making each subcarrier a quasi-cw wave due to low bit rate information on each subcarrier. The OFDM signal can be described as [31]

$$u(t, z) = \sum_{l=-N/2}^{N/2-1} u_l(t, z) \exp(i \omega_l t), \tag{7.99}$$

where N is the total number of subcarriers, $u_l(t, z)$ is the slowly varying field envelope, and $\omega_l = 2\pi l / T_{\text{block}}$ is the frequency offset from a reference and T_{block} is the OFDM symbol time. First, we derive the analytical formula for the variance of nonlinear phase noise including the interaction of ASE noise with SPM and XPM. Next, we extend the analysis to include the impact of FWM.

7.5.1 SPM and XPM Induced Nonlinear Phase Noise

Inserting (7.99) into (7.32) and considering the effects of SPM and XPM only, we obtain

$$i\left(\frac{\partial u_l}{\partial z} - \beta_2\omega_l\frac{\partial u_l}{\partial t}\right) - \frac{\beta_2}{2}\omega_l\frac{\partial^2 u_l}{\partial t^2} + \frac{\beta_2}{2}\omega_l^2 u_l = -\gamma a^2(z)\left(|u_l|^2 + 2\sum_{k\neq l}|u_k|^2\right)u_l. \tag{7.100}$$

For simplicity, we assume that β_2 is constant, amplifiers are periodically spaced with a spacing of L_a, and dispersion compensation is done in the electrical domain. Within each OFDM block, u_l is constant; therefore, the first- and second- order derivatives of u_l with respect of time, appearing in (7.100) can be ignored. Now the exact solution of (7.100) can be written as

$$u_l(z) = u_l(0)\exp[i\phi(z)], \tag{7.101}$$

where

$$\phi(z) = \frac{\beta_2}{2}\omega_l^2 z + \gamma L_e(z)\left(|u_l|^2 + 2\sum_{k\neq l}|u_k|^2\right), \tag{7.102}$$

and

$$L_e(z) = \int_0^z a^2(s)\mathrm{d}s. \tag{7.103}$$

As in Sect. 7.3,we assume that two DOFs per subcarrier are sufficient to describe the noise process. Therefore, the noise field can be written as

$$n(t) = \sum_{l=-N/2}^{N/2-1} n_l \exp(i\omega_l t). \tag{7.104}$$

In (7.104), the noise field is described by $2N$ DOFs or 2 DOFs per subcarrier. The total field immediately after the amplifier located at mL_a is

$$u(t, mL_a+) = \sum_{l=-N/2}^{N/2-1} [u_l(mL_a) + n_l]\exp(i\omega_l t). \tag{7.105}$$

Let

$$u_l(mL_a+) = u_l(mL_a) + n_l = [u_l(0) + n_l']\exp[i\phi(mL_a)], \tag{7.106}$$

where

$$n_l' = n_l\exp[-i\phi(mL_a)] \tag{7.107}$$

with

$$\langle n_l' n_k'^{\star} \rangle = \langle n_l n_k^{\star} \rangle = \frac{\rho_{\text{ASE}}}{T_{\text{block}}} \delta_{lk},$$
$$\langle n_l' n_k' \rangle = 0, \tag{7.108}$$

where δ_{lk} is the Kronecker delta function. Now treating $u_l(mL_a+)$ as the initial field, (7.100) is solved to obtain the field at the end of the optical system, located at $z = N_a L_a \equiv L_{\text{tot}}$, as

$$u_l(L_{\text{tot}}) = [u_l + n_l'] \exp \left\{ i\Phi_D + i\gamma(N_a - m)L_{\text{eff}} \left[(u_l n_l'^{\star} + u_l^{\star} n_l') + 2\sum_{k \neq l} (u_k n_k'^{\star} + u_k^{\star} n_k') \right] \right\}, \tag{7.109}$$

where Φ_D is the deterministic phase shift caused by dispersion, SPM, and XPM, which has no impact on the nonlinear phase noise, and is expressed as

$$\Phi_D = \beta_2 \omega_l^2 N_a L_a / 2 + \gamma N_a L_{\text{eff}} \left(|u_l|^2 + 2\sum_{k \neq l} |u_k|^2 \right), \tag{7.110}$$

and $L_{\text{eff}} = L_e(L_a)$. The linear phase noise is embedded in the term $u_l + n_l'$, and the nonlinear phase noise of the lth subcarrier caused by SPM and XPM due to the amplifier located at $z = mL_a$ is

$$\delta\Phi_{\text{SPM+XPM},m,l} = \gamma(N_a - m)L_{\text{eff}} \left[(u_l n_l'^{\star} + u_l^{\star} n_l') + 2\sum_{k \neq l} (u_k n_k'^{\star} + u_k^{\star} n_k') \right]. \tag{7.111}$$

Squaring (7.111) and making use of (7.108), we obtain the variance of the nonlinear phase noise caused by SPM and XPM

$$\langle \delta\Phi_{\text{SPM+XPM},m,l}^2 \rangle = \frac{2\gamma^2 (N_a - m)^2 L_{\text{eff}}^2 \rho_{\text{ASE}}}{T_{\text{block}}} \left(|u_l|^2 + 2\sum_{k \neq l} |u_k|^2 \right). \tag{7.112}$$

Assuming that the number of subcarriers carrying data is N_e (equivalently the over-sampling factor is N/N_e) and each subcarrier has equal power, and summing (7.112) over all amplifiers, we obtain the nonlinear phase noise variance of the lth subcarrier caused by SPM and XPM as

$$\langle \delta\Phi_{\text{SPM+XPM},l}^2 \rangle = \frac{\rho_{\text{ASE}} N_a (N_a - 1)(2N_a - 1)\gamma^2 L_{\text{eff}}^2 (2N_e - 1) P_{\text{sc}}}{3T_{\text{block}}}, \tag{7.113}$$

where P_{sc} is the power per subcarrier. Equation (7.113) is our final expression for the nonlinear phase noise variance taking into account the interaction of ASE with SPM and XPM.

7.5.2 FWM-Induced Nonlinear Phase Noise

Substituting (7.99) into (7.32), and considering only the FWM effect, we obtain the following equation with the quasi-cw assumption

$$\frac{\partial u_l}{\partial z} - i\frac{\beta_2}{2}\omega_l^2 u_l = i\gamma a^2(z) \sum_{p+q-r=l}^{p\neq l, q\neq r} u_p u_q u_r^{\star} \exp\left[-i\left(\omega_p^2 + \omega_q^2 - \omega_r^2\right)\frac{\beta_2 z}{2}\right]. \tag{7.114}$$

The solution of (7.114) with $S(L_a N_a) = 0$ is

$$\begin{aligned} u_l(N_a L_a) &= u'_{l,z_0} \\ &\quad + i \sum_{p+q-r=l}^{p\neq l, q\neq r} u_{p,z_0} u_{q,z_0} u^{\star}_{r,z_0} \int_{z_0}^{N_a L_a} \gamma a^2(z') \exp[-i\Delta\beta_{p,q,r,l}(z')]\mathrm{d}z' \\ &= u'_{l,z_0} + i \sum_{p+q-r=l}^{p\neq l, q\neq r} u_{p,z_0} u_{q,z_0} u^{\star}_{r,z_0} Y_{p,q,r,l}(z_0, N_a L_a), \end{aligned} \tag{7.115}$$

where

$$u'_{l,z_0} = u_{l,z_0} \exp\left(-i\frac{\beta_2}{2}\omega_l^2 z_0\right), \tag{7.116}$$

with $u_{l,z_0} = u_l(z_0)$. $\Delta\beta_{p,q,r,l}(z)$ is the phase mismatch factor given by

$$\Delta\beta_{p,q,r,l}(z) = \left(\omega_p^2 + \omega_q^2 - \omega_r^2 - \omega_l^2\right)\frac{\beta_2 z}{2}, \tag{7.117}$$

and

$$Y_{p,q,r,l}(z_0, N_a L_a) = \int_{z_0}^{N_a L_a} \gamma a^2(z') \exp[-i\Delta\beta_{p,q,r,l}(z')]\mathrm{d}z'. \tag{7.118}$$

To obtain (7.115), we have ignored the depletion of FWM pumps appearing on the right-hand side (RHS) of (7.114), which is known as the undepleted pump approximation [49].

Now consider the noise added by the amplifier located at mL_a. The optical field immediately after the amplifier is given by (7.105). Equation (7.115) is solved using the initial condition of (7.105). Replacing u_{l,z_0} in (7.116) with $u_l(mL_a+)$, we obtain the optical field at the end of the fiber span as

$$
\begin{aligned}
u_l(N_a L_a) &= u^+_{l,m} \exp(-i\frac{\beta_2}{2}\omega_l^2 m L_a) \\
&\quad + i \sum_{p+q-r=l}^{p\neq l, q\neq r} u^+_{p,m} u^+_{q,m} u^{+\star}_{r,m} Y_{p,q,r,l}(mL_a, N_a L_a) \\
&= (u_{l,m} + n_l) \exp\left(-i\frac{\beta_2}{2}\omega_l^2 m L_a\right) \\
&\quad + i \sum_{p+q-r=l}^{p\neq l, q\neq r} (u_{p,m} + n_p)(u_{q,m} + n_q)(u^\star_{r,m} + n^\star_r) \\
&\quad \times Y_{p,q,r,l}(mL_a, N_a L_a),
\end{aligned}
\tag{7.119}
$$

where $u_l(mL_a+) \equiv u^+_{l,m}$. Ignoring the higher-order term of n_l, we have

$$
\begin{aligned}
u_l(N_a L_a) &\approx (u_{l,m} + n_l) \exp\left(-i\frac{\beta_2}{2}\omega_l^2 m L_a\right) + i \sum_{p+q-r=l}^{p\neq l, q\neq r} \big(u_{p,m} u_{q,m} u^\star_{r,m} \\
&\quad + n_p u_{q,m} u^\star_{r,m} + n_q u_{p,m} u^\star_{r,m} + n^\star_r u_{p,m} u_{q,m}\big) Y_{p,q,r,l}(mL_a, N_a L_a).
\end{aligned}
\tag{7.120}
$$

From (7.120), we have

$$
u_l(N_a L_a) = u_{l,m} \exp\left(-i\frac{\beta_2}{2}\omega_l^2 m L_a\right) + u_{\mathrm{FWM},l,m} + \delta u_l(N_a L_a, m), \tag{7.121}
$$

where $u_{\mathrm{FWM},l,m}$ is the deterministic distortion caused by FWM, expressed as

$$
u_{\mathrm{FWM},l,m} = i \sum_{p+q-r=l}^{p\neq l, q\neq r} u_{p,m} u_{q,m} u^\star_{r,m} Y_{p,q,r,l}(mL_a, N_a L_a). \tag{7.122}
$$

This distortion can be compensated using the digital phase conjugation, and thus, has no impact on the nonlinear phase noise. The third term on the RHS of (7.121) $\delta u_l(N_a L_a, m)$ describes the ASE–FWM interaction as well as the linear ASE noise, and can be written as

$$
\delta u_l(N_a L_a, m) = n_l \exp\left(-i\frac{\beta_2}{2}\omega_l^2 m L_a\right) + i \sum_{q=-N/2}^{N/2-1} \left(n_q A_{q,l} + n^\star_q B_{q,l}\right), \tag{7.123}
$$

where

$$A_{q,l} = 2 \sum_{p=-N/2}^{N/2-1} u_{p+l-q,m} u_{p,m}^{\star} Y_{q,p+l-q,p,l}(mL_a, N_a L_a), p \neq q, l \neq p+l-q \tag{7.124}$$

$$B_{q,l} = \sum_{p=-N/2}^{N/2-1} u_{q+l-p,m} u_{p,m} Y_{q+l-p,p,q,l}(mL_a, N_a L_a), p \neq q, l \neq p+l-q \tag{7.125}$$

From (7.123), we have

$$\langle |\delta u_l|^2 \rangle = \langle |n_l|^2 \rangle + \sum_{q=-N/2}^{N/2-1} \langle |n_q|^2 \rangle (|A_{q,l}|^2 + |B_{q,l}|^2), \tag{7.126}$$

$$\langle \delta u_l^2 \rangle = i \langle |n_l|^2 \rangle 2B_{l,l} \exp\left(-i\frac{\beta_2}{2}\omega_l^2 m L_a\right) - \sum_{q=-N/2}^{N/2-1} \langle |n_q|^2 \rangle 2A_{q,l} B_{q,l}. \tag{7.127}$$

After the digital phase conjugation removes the deterministic distortions, the phase noise of the received field due to the amplifier located at mL_a is

$$\delta \Phi_{l,m} \approx \frac{\mathrm{Im}(\delta u_l)}{|u_l|} = \frac{\delta u_l - \delta u_l^{\star}}{2i|u_l|}. \tag{7.128}$$

Since $\langle \delta \Phi_{l,m} \rangle = 0$, we can calculate the variance of the phase noise as

$$\langle \delta \Phi_{l,m}^2 \rangle \approx \left\langle -\frac{(\delta u_l - \delta u_l^{\star})^2}{2|u_l|^2} \right\rangle = \frac{2\langle |\delta u_l|^2 \rangle - (\langle \delta u_l^2 \rangle + \langle \delta u_l^{\star 2} \rangle)}{4|u_l|^2}. \tag{7.129}$$

Inserting (7.126) and (7.127) into (7.129) and using (7.108), we obtain

$$\begin{aligned} \left\langle \delta \Phi_{l,m}^2 \right\rangle \approx & \frac{\rho_{\mathrm{ASE}}}{2P_{\mathrm{sc}} T_{\mathrm{block}}} + \frac{\rho_{\mathrm{ASE}}}{2P_{\mathrm{sc}} T_{\mathrm{block}}} \sum_{q=-N/2}^{N/2-1} |A_{q,l}^{\star} + B_{q,l}|^2 \\ & + \frac{\rho_{\mathrm{ASE}}}{P_{\mathrm{sc}} T_{\mathrm{block}}} \mathrm{Im}\left[B_{l,l} \exp\left(-i\frac{\beta_2}{2}\omega_l^2 m L_a\right)\right]. \end{aligned} \tag{7.130}$$

The first term on the RHS of (7.130) is the variance of the linear phase noise, the second and third terms on the RHS of (7.130) describe the variance of the nonlinear phase noise related to FWM. Summing (7.130) over all amplifiers in the fiber system, we obtain the phase noise variance for the lth subcarrier caused by linear phase noise and FWM as follows

$$\langle \delta\Phi^2_{\text{linear},l} \rangle = \sum_{m=1}^{N_a} \left\langle \delta\Phi^2_{\text{linear},l,m} \right\rangle = \frac{\rho_{\text{ASE}} N_a}{2P_{\text{sc}} T_{\text{block}}}. \tag{7.131}$$

$$\left\langle \delta\Phi^2_{\text{FWM},l} \right\rangle = \sum_{m=1}^{N_a} \left\langle \delta\Phi^2_{\text{FWM},l,m} \right\rangle = \frac{\rho_{\text{ASE}}}{2P_{\text{sc}} T_{\text{block}}} \sum_{m=1}^{N_a} \sum_{q=-N/2}^{N/2-1} |A^{\star}_{q,l} + B_{q,l}|^2$$
$$+ \frac{\rho_{\text{ASE}}}{P_{\text{sc}} T_{\text{block}}} \sum_{m=1}^{N_a} \text{Im}\left[B_{l,l} \exp\left(-i \frac{\beta_2}{2} \omega_l^2 m L_a \right) \right]. \tag{7.132}$$

The first term on the RHS of (7.132) is the nonlinear phase noise induced by FWM, and the second term on the RHS of (7.132) is the interaction between the linear and nonlinear phase noise.

7.5.3 *Total Phase Noise*

The total phase noise for the lth subcarrier in an OFDM system including the linear phase noise and nonlinear phase noise (induced by interaction between ASE and SPM, XPM, and FWM) is as follows

$$\langle \delta\Phi_l^2 \rangle = \langle \delta\Phi^2_{\text{linear},l} \rangle + \langle \delta\Phi^2_{\text{SPM+XPM},l} \rangle + \langle \delta\Phi^2_{\text{FWM},l} \rangle, \tag{7.133}$$

where the first, second, and third terms on the RHS of (7.133) are given by (7.131), (7.113), and (7.132), respectively.

7.5.4 *Results and Discussions*

In this section, the analytical model for the variance of the total phase noise in OFDM systems given by (7.133) is validated by numerical simulations. The following parameters are used throughout this section unless otherwise specified: the bit rate is 10 Gb s^{-1}, the amplifier spacing is 100 km, and the noise figure (NF) is 6 dB. A single type of fiber is used between amplifiers. To separate the deterministic (although bit pattern dependent) distortions due to nonlinear effects from the ASE-induced nonlinear noise effects, we use digital phase conjugation [36]. Since digital phase conjugation compensates for both dispersion and deterministic nonlinear effects, we do not use the cyclic prefix. Approximately 2,048 OFDM frames are used to get a good Monte Carlo statistics. Each OFDM subcarrier is modulated with binary-phase-shift-keying (BPSK) data. Figure 7.6 shows the coherent OFDM system structure in our simulation.

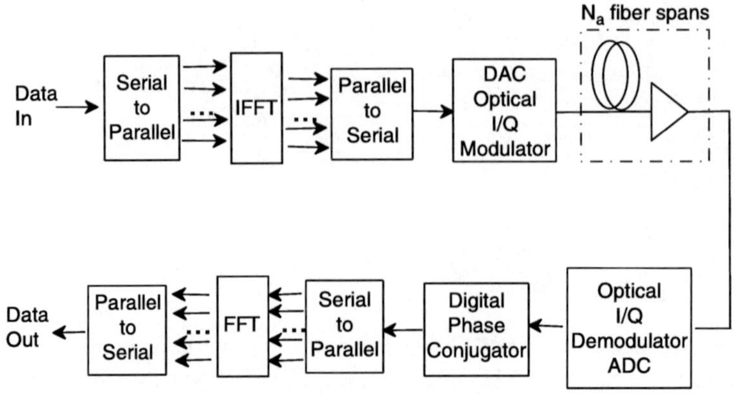

Fig. 7.6 Structure of coherent OFDM transmission systems

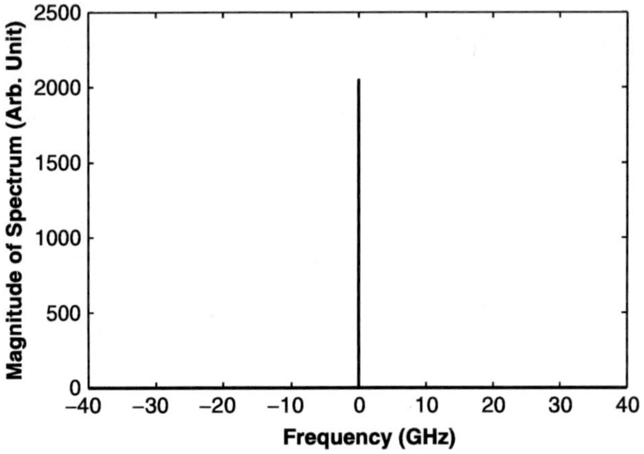

Fig. 7.7 OFDM signal spectrum before entering into fiber spans. Total number of subcarriers is 8, with one subcarrier carrying data

For Figs. 7.7 and 7.8, we choose a fiber dispersion D of 1 ps nm^{-1} km^{-1} and a total launch power of 0 dBm. Here, we use only one subcarrier ($N_e = 1$) to carry data while the total number of subcarriers is 8 (eighth-folder oversampling), so that the nonlinear phase noise model that includes SPM effects alone can be validated. The subcarrier carrying data is located at the central of the OFDM spectrum. The signal spectrum before entering into the fiber span is shown in Fig. 7.7. And in Fig. 7.8, the solid lines show the analytical linear phase noise and nonlinear phase noise variance induced by SPM only, the dashed line with triangulars show the numerical simulation results for the variance of linear phase noise and SPM-induced nonlinear phase noise, as a function of fiber propagation distance. As can be seen, the agreement is quite good.

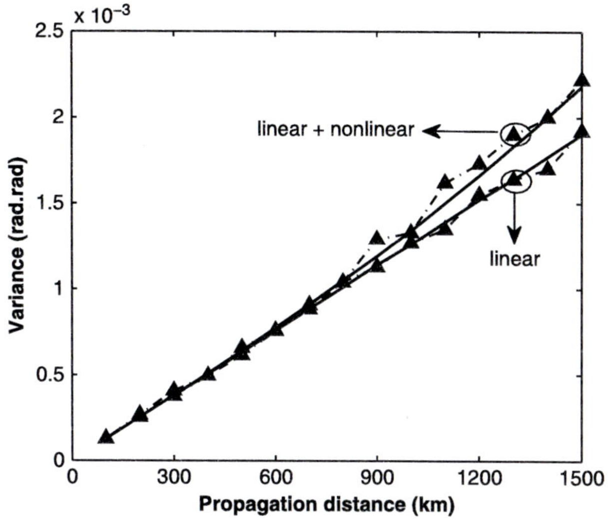

Fig. 7.8 Variance of the total phase noise as a function of propagation distance for SPM effect only. Total number of subcarrier is 8 with only one subcarrier carrying data. *Solid line* and *dashed line* with triangular show the analytical and numerical simulation results, respectively. After [40]

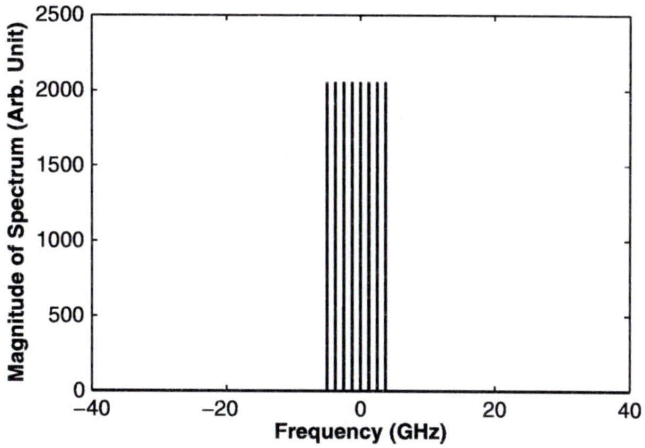

Fig. 7.9 OFDM signal spectrum before entering into fiber spans. Total number of subcarriers is 64, with 8 subcarriers carrying data. After [40]

In order to validate the nonlinear phase noise model including the ASE interaction with SPM, XPM, and FWM effects in (7.133), we turn on 8 subcarriers of an OFDM system with 64 subcarriers. The subcarrier carrying data is located at the center of the OFDM spectrum. Figure 7.9 shows the OFDM signal spectrum, and Fig. 7.10 shows the variance of the linear phase noise and nonlinear

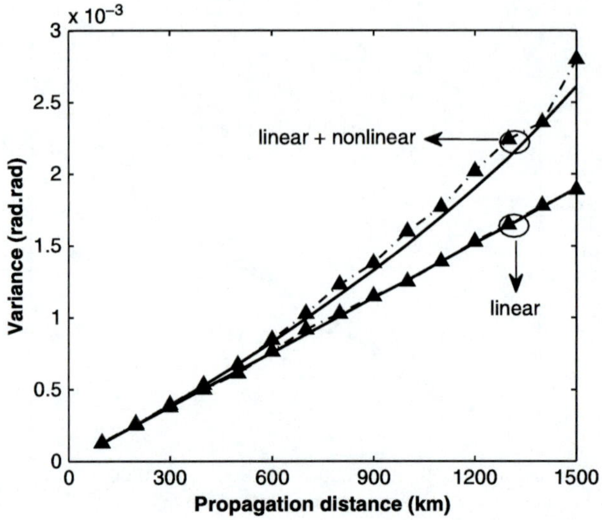

Fig. 7.10 Variance of the total phase noise as a function of propagation distance considering the ASE interaction with SPM, XPM and FWM effects. Total number of subcarriers is 64 with 8 subcarriers carrying data. *Solid line* and *dashed line* with triangular show the analytical and numerical simulation results, respectively. After [40]

phase noise from numerical simulation (dashed line with triangulars) and analytical calculation (solid line), respectively. We see that the good agreement is achieved, which validates our model for the nonlinear phase noise considering SPM, XPM, and FWM effects.

In [30], the authors showed that the nonlinear degradation due to FWM effects in OFDM systems is nearly independent of the number of ODFM subcarriers used in the system in the absence of chromatic dispersion. In [31], the authors studied the chromatic dispersion effects on the FWM and showed that chromatic dispersion could decrease the FWM effects significantly. However, both of these analyses focused on the deterministic nonlinear effects. In this section, we will study the dependence of the nonlinear phase noise effects on fiber dispersion and bit rate in an OFDM system with digital phase conjugation.

In Fig. 7.11, we fix the transmission distance to be 1,000 km, the total number of subcarriers is 128 with 64 subcarriers carrying data (twofold oversampling). We show the impact of the bit rate on the total phase noise for a transmission fiber with $D = 17\,\mathrm{ps\,nm^{-1}\,km^{-1}}$ and $D = 0\,\mathrm{ps\,nm^{-1}\,km^{-1}}$. The total launch power is -3 dBm. Solid lines and solid circles show the analytical and the numerical simulation results, respectively. From Fig. 7.11, we note that the variance of the total phase noise scales linearly with the bit rate. This could be explained by the fact that with the increase of the bit rate, the OFDM symbol time T_{block} decreases, which leads to the increase of the total phase noise as described in (7.113), (7.131), and (7.132). The qualitative explanation for the increase in phase noise when the bit rate

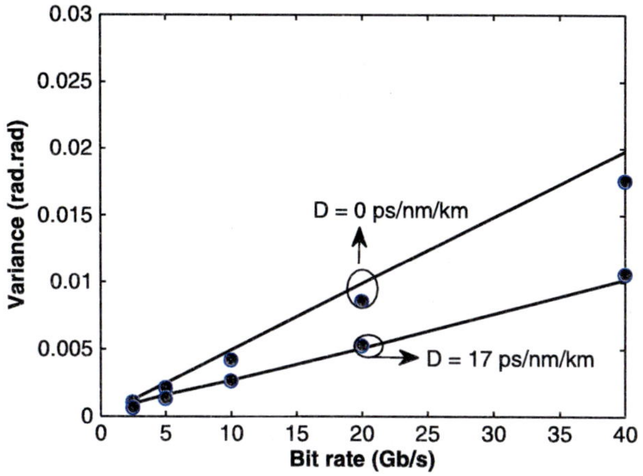

Fig. 7.11 Variance of the total phase noise as a function of bit rate in Gb/s. The total number of subcarriers is 128 with twofold oversampling, total channel power is −3 dBm, and transmission distance is 1,000 km. *Solid line* and *solid circles* show the analytical and numerical simulation results, respectively. After [40]

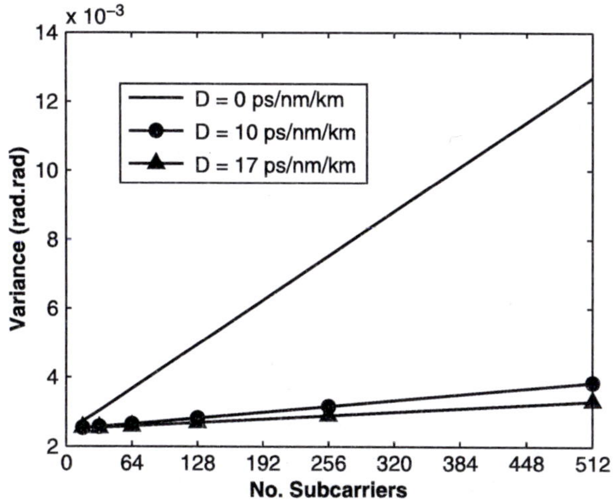

Fig. 7.12 Variance of the total phase noise as a function of number of subcarriers, obtained analytically. Two-folder oversampling is used in the simulation. Bit rate is 10 Gb s^{-1}, total channel power is −3 dBm, and transmission distance is 1,000 km. After [40]

increases is as follows: as the bit rate increases, OSNR requirement for a given BER increases. This is because the receiver filter bandwidth scales with bit rate, which leads to the increase of the total noise within the receiver bandwidth. Similarly, the variance of phase noise also scales directly with the receiver bandwidth.

In Fig. 7.12, we show the impact of the number of subcarriers on the variance of total phase noise, obtained analytically using (7.133). Twofold oversampling is

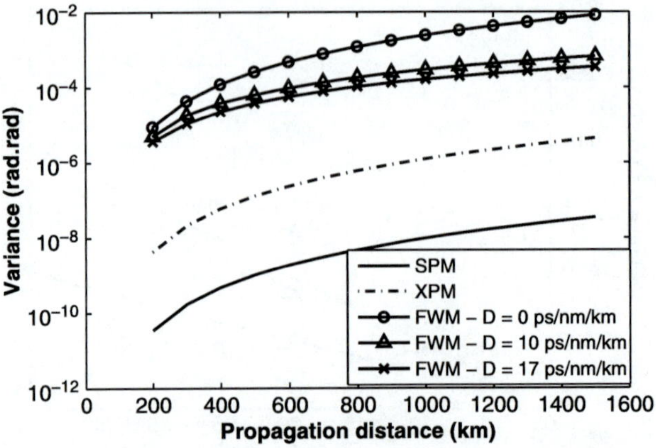

Fig. 7.13 Variance of the nonlinear phase noise due to separate effects of SPM, XPM, and FWM, as a function of propagation distance, obtained analytically. Total number of subcarriers is 128 with two-folder oversampling. Bit rate is 10 Gb s^{-1} with −3 dBm launch power. After [40]

used in the simulation. The total launch power is −3 dBm, the bit rate is 10 Gb s^{-1}. Figure 7.12 shows that in the absence of dispersion, the variance of total phase noise scales linearly with the number of subcarriers, while with moderate levels of dispersion, the variance of total phase noise is almost constant because the linear phase noise is dominant for such systems.

Finally, Fig. 7.13 shows the variance of the nonlinear phase noise as a function of propagation distance for SPM-induced nonlinear phase noise alone (solid line), XPM-induced nonlinear phase noise alone (dashed line), and FWM-induced nonlinear phase noise alone for $D = 0\,\text{ps}\,\text{nm}^{-1}\,\text{km}^{-1}$ (solid line with circles), $D = 10$ ps/nm/km (solid line with triangles) and $D = 17\,\text{ps}\,\text{nm}^{-1}\,\text{km}^{-1}$ (solid line with "x"), obtained analytically using (7.113) and (7.132). From Fig. 7.13, we note that for an OFDM system with large number of subcarriers, nonlinear phase noise induced by FWM is significantly larger than that induced by SPM and XPM. This is in contrast to the results of [50] for WDM systems, in which it is found that ASE–FWM interaction is negligible in quasilinear systems. This difference is likely due to the fact that the subcarriers of OFDM system are derived from the same laser source and interact coherently. We also note that with moderate levels of fiber chromatic dispersion, the nonlinear phase noise induced by FWM decreases since the phase matching becomes more difficult.

7.6 Conclusions

We have reviewed the interaction of the signal and noise leading to nonlinear phase noise in single carrier and OFDM systems. Although two DOFs of noise accurately describe the noise process for a linear system with matched filters, it is an approximation for the nonlinear systems. This is because the higher-order noise components interact with the signal leading to new noise components within the pass band of the matched filter. The variance of the nonlinear phase noise due to SPM decreases significantly as the fiber dispersion increases. For OFDM systems, the variance of the phase noise increases slightly with the number of subcarriers. In WDM systems, the nonlinear phase noise due to the ASE–FWM is much smaller than that due to ASE–XPM. However, for OFDM system the nonlinear phase noise due to ASE–FWM is the dominant one. This is because the subcarriers of OFDM system originate from the same laser source and interact coherently. In contrast, for WDM systems, the optical carriers are derived from different lasers with arbitrary phases.

References

1. J.P. Gordon, L.F. Mollenauer, Opt. Lett. **15**(23), 1351–1353 (1990)
2. H. Kim, A.H. Gnauck, IEEE Photon. Technol. Lett. **15**, 320–322 (2003)
3. P.J. Winzer, R.-J. Essiambre, J. Lightwave Technol. **24**(12), 4711–4728 (2006)
4. S.L. Jansen, D. van den Borne, B. Spinnler, S. Calabro, H. Suche, P.M. Krummrich, W. Sohler, G.-D.Khoe, H. de Waardt, IEEE J. Lightwave Technol. **24**, 54–64 (2006)
5. A. Mecozzi, J. Lightwave Technol. **12**(11), 1993–2000 (1994)
6. K-P. Ho, J. Opt. Soc. Am. B **20**(9), 1875–1879 (2003)
7. K-P. Ho, Opt. Lett. **28**(15), 1350–1352 (2003)
8. Mecozzi, Opt. Lett. **29**(7), 673–675 (2004)
9. A.G. Green, P.P. Mitra, L.G.L. Wegener, Opt.Lett. **28**, 2455–2457 (2003)
10. S. Kumar, Opt. Lett. **30**, 3278–3280 (2005)
11. C.J. McKinstrie, C. Xie, T. Lakoba, Opt. Lett. **27**, 1887–1889 (2002)
12. C.J. McKinstrie, C. Xie, IEEE J. Sel. Top. Quant. Electron. **8**, 616–625 (2002)
13. M. Hanna, D. Boivin, P.-A. Lacourt, J.-P. Goedgebuer, J. Opt. Soc. Am. B **21**, 24–28 (2004)
14. K.-P. Ho, H.-C. Wang, IEEE Photon. Technol. Lett. **17**, 1426–1428 (2005)
15. K.-P. Ho, H.-C.Wang, Opt. Lett. **31**, 2109–2111 (2006)
16. F. Zhang, C.-A. Bunge, K. Petermann, Opt. Lett. **31**(8), 1038–1040 (2006)
17. P. Serena, A. Orlandini, A. Bononi, J. Lightwave Technol. **24**(5), 2026–2037 (2006)
18. X. Zhu, S. Kumar, X. Li, App. Opt. **45**, 6812–6822 (2006)
19. A. Demir, J. Lightwave Technol. **25**(8) 2002–2032 (2007)
20. S. Kumar, L. Liu, Opt. Exp. **15**, 2166–2177 (2007)
21. M. Faisal, A. Maruta, Opt. Comm. **282**, 1893–1901 (2009)
22. S. Kumar, J. Lightwave Technol. **27**(21), 4722–4733 (2009)
23. A. Bononi, P. Serena, N. Rossi, Optic. Fiber Tech. **16**, 73–85 (2010)
24. W. Shieh, C. Athaudage, Electron. Lett. **42**(10), 587–588 (2006)
25. A. Lowery, L. Du, J. Armstrong, J. Lightwave. Technol. **25**(1), 131–138 (2007)
26. J. Armstrong, J. Lightwave Technol. **27**(3), 189–204 (2009)
27. A. Sano, E. Yamada, H. Masuda, E. Yamazaki, T. Kobayashi, E. Yoshida, Y. Miyamoto, R. Kudo, K. Ishihara, Y. Takatori, J. Lightwave Technol. **27**(16), 3705–3713 (2009)
28. S. Jansen, I. Morita, T. Schenk, H. Tanaka, J. Lightwave Technol. **27**(3), 177–188 (2009)
29. Y. Yang, Y. Ma, W. Shieh, IEEE Photon. Technol. Lett. **21**(15), 1042–1044 (2009)
30. A. Lowery, S. Wang, M. Premaratne, Opt. Express **15**, 13282–13287 (2007)
31. M. Nazarathy, J. Khurgin, R. Weidenfeld, Y. Meiman, P. Cho, R. Noe, I. Shpantzer, V. Karagodsky, Opt. Express **16**, 15777–15810 (2008)
32. A. Lowery, Opt. Express **15**(20), 12965–12970 (2007)

33. L. Du, A. Lowery, Opt. Express **16**(24), 19920–19925 (2008)
34. X. Liu, F. Buchali, Opt. Express **16**(26), 21944–21957 (2008)
35. X. Liu, F. Buchali, R. Tkach, J. Lightwave Technol. **27**(16), 3632–3640 (2009)
36. W. Shieh, H. Bao, Y. Tang, Opt. Express **16**(2), 841–859 (2008)
37. X. Li, X. Chen, G. Goldfarb, E. Mateo, I. Kim, F. Yaman, G. Li, Opt. Express **16**, 880–889 (2008)
38. E. Ip, J. Kahn, J. Lightwave Technol. **26**(20), 3416–3425 (2008)
39. E. Yamazaki, H. Masuda, A. Sano, T. Yoshimatsu, T. Kobayashi, E. Yoshida, Y. Miyamoto, R. Kudo, K. Ishihara, M. Matsui, Y. Takatori, Multi-staged nonlinear compensation in coherent receiver for 16,340-km transmission of 111-Gb/s no-guard-interval co-OFDM, ECOC 2009, Paper 9.4.6, 2009
40. X. Zhu, S. Kumar, Opt. Express **18**(7), 7347–7360 (2010)
41. A. Hasegawa, Y. Kodama, Phys. Rev. Lett. **66**(2), 161–164 (1991)
42. G.P. Agrawal, *Nonlinear Fiber Optics*, chap. 3 (Academic, San Diego, 2007)
43. A. Mecozzi, C.B. Clausen, M. Shtaif, IEEE Photon. Technol. Lett. **12**, 392–394 (2000)
44. R.-J.Essiambre, G. Raybon, B. Mikkelsen, in *Psuedo-Linear Transmission of High Speed TDM Signals:40 and 160 Gb/s*, chap. 6, ed. by I.P. Kaminow, T. Li. Optical Fiber Telecommunications IV B (Academic, San Diego, 2002), pp. 232–304
45. S. Kumar, D. Yang, J. Lightwave Technol. **23**(6), pp. 2073–2080 (2005)
46. J. Li, E. Spiller, G. Biondini, Phys. Rev. A **75**(5), 053818-1–053818-13 (2007)
47. S.K. Turitsyn, V.K. Mezentsev, JETP Lett. **67**(9) 616–621 (1998)
48. T.I. Lakoba, D.J. Kaup, Phys. Rev. E **58**(5), 6728–6741 (1998)
49. K. Inoue, Opt. Lett. **17**, 801–803 (1992)
50. M. Hanna, D. Boivin, P. Lacourt, J. Goedgebuer, J. Opt. Soc. Amer. B **21**, 24–28 (2004)

Chapter 8
Cross-Phase Modulation-Induced Nonlinear Phase Noise for Quadriphase-Shift-Keying Signals

Keang-Po Ho

8.1 Introduction

Recently, phase-modulated optical communication systems are used for long-haul lightwave communication systems [1–4]. With good receiver sensitivity, both quadri-phase-shift keying (QPSK) and differential QPSK (DQPSK) signals are suitable for spectrally efficient long-haul lightwave communication systems. Unlike coherent optical communications in the 1980s [5, 6], contemporary lightwave systems use optical amplifiers with high launched power per span. The system performance is dominated by optical amplifier noise and fiber nonlinearities. The optical amplifiers also have a wide bandwidth to boost all wavelength-division-multiplexed (WDM) channels together. With high launched power, signal and noise interaction is important and the nonlinear interaction between WDM channels also degrades the system performance. For optical fiber with nonzero chromatic dispersion coefficient, the interchannel nonlinearities between WDM channels are typically due to cross-phase modulation (XPM) arising from Kerr effect.

Laser phase noise was used to be the major impairments for coherent optical communications [5, 6] because of the low data rate and poor laser. Contemporary coherent systems with high-speed data rate are less likely to be degraded by phase noise from an improved laser. Self-phase modulation (SPM)-induced nonlinear phase noise [2, 7–10] is a fundamental degradation for phase-modulated signals to add phase noise directly to the signals. SPM-induced nonlinear phase noise has been studied in Chaps. 6 and 7 of this book and will not be repeated here.

XPM-induced nonlinear phase variations modulate the phase of both QPSK and DQPSK signals, giving nonlinear phase noise. XPM-induced nonlinear phase noise was studied by [11–13] for binary differential phase-shift keying (DPSK)

K.-P. Ho (✉)
SiBEAM, Sunnyvale, CA 94085, USA
e-mail: kpho@ieee.org

S. Kumar (ed.), *Impact of Nonlinearities on Fiber Optic Communications*, Optical and Fiber Communications Reports 7, DOI 10.1007/978-1-4419-8139-4_8,
© Springer Science+Business Media, LLC 2011

signal. Adjacent on-off keying (OOK) channels also give nonlinear phase noise via XPM. In practice, OOK channels induce larger nonlinear phase noise than constant-intensity phase-modulated channels. The effect of adjacent OOK channels to DPSK signal was studied in [14–19]. The effect of adjacent OOK channels to QPSK signal was studied in [19–24].

Simulation was conducted in [20, 21] to find the effect of OOK signals to QPSK signal. The simulation did not seem to include or optimize carrier recovery that may filter out part of the nonlinear phase noise, and thus improving the system performance. The measurement of [22, 23] just took constellation over a period of time, effectively ignoring the effect of carrier recovery or just rotating the signal to compensate for constant phase shift. Carrier recovery was included in [24] with a simple averaging filter. The averaging filter of [24] is not optimal as shown in [25]. Here, for QPSK signals, the optimal filter is designed for the popular feedforward-based phase tracking techniques [25, 26].

In later parts of this chapter, the effect of Gaussian-distributed phase error is first studied for both QPSK and DQPSK signals based on series expansion. The phase error standard deviation (STD) should be less than 4–6° for a raw bit-error-rate (BER) between 10^{-5} and 10^{-3} before forward error correction (FEC). The transfer function from amplitude-modulation from one WDM channel to the phase modulation of another WDM channel is then derived based on the pump-probe model for a multi-span amplified fiber link. The phase error of XPM-induced nonlinear phase noise is then calculated for both DQPSK and QPSK signals. A WDM system with pure DQPSK signals does not affect by XPM-induced nonlinear phase noise. For hybrid DQPSK and OOK WDM systems with mean nonlinear phase shift up to 0.5 rad, the SNR penalty is less than 0.5 dB due to the XPM-induced nonlinear phase noise. For QPSK signal using feedforward carrier recovery, the optimal Wiener filter is derived to reduce the XPM-induced nonlinear phase noise. With the optimal Wiener filter, QPSK signal can be operated with adjacent OOK WDM channels without guard-band, providing a great improvement compared with prior design without the optimal filter [21–23].

8.2 Gaussian-Distributed Phase Error

For both QPSK and DQPSK signals, the signals can be represented as $(\pm 1 \pm j)/\sqrt{2}$ or $s_k = \exp[j(2k+1)\pi/4]$ with $k = 0, 1, 2, 3$. With phase error and additive Gaussian noise, the received signal can be modeled as $r_k = s_k e^{j\phi_e} + n_k$, where ϕ_e is assumed to be Gaussian-distributed phase noise. Here, the impact of Gaussian-distributed phase noise is studied for QPSK and DQPSK signals.

8.2.1 DQPSK Signals

For DQPSK signal with a given phase error of ϕ_e, the bit-error probability is [27]

$$\begin{aligned} p_e(\phi_e) &= \frac{1}{2}\Big\{ Q_1(a_+, b_+) - \frac{1}{2}\mathrm{e}^{-(a_+^2+b_+^2)/2} I_0(a_+b_+) \\ &\qquad + Q_1(a_-, b_-) - \frac{1}{2}\mathrm{e}^{-(a_-^2+b_-^2)/2} I_0(a_-b_-) \Big\}, \\ a_\pm &= \sqrt{\rho_s\left[1-\cos\left(\frac{\pi}{4}\pm\phi_e\right)\right]}, \\ b_\pm &= \sqrt{\rho_s\left[1+\cos\left(\frac{\pi}{4}\pm\phi_e\right)\right]}, \end{aligned} \tag{8.1}$$

where $Q_1(\cdot,\cdot)$ is the Marcum Q function and $I_k(\cdot)$ is the kth order modified Bessel function of the first kind. If the phase error of ϕ_e is Gaussian distributed, the error probability of DQPSK signal becomes

$$p_e = \int_{-\infty}^{+\infty} p_e(\phi_e) p_{\phi_e}(\phi_e) \mathrm{d}\phi_e, \tag{8.2}$$

where $p_{\phi_e}(\phi_e)$ is the Gaussian-distributed phase error.

However, the formula of (8.2) requires numerical integration. If the phase distribution of Gaussian random variable is expressed as a Fourier series [2, App. 4.A], the bit-error probability becomes

$$p_e = \frac{3}{8} - \frac{\rho_s \mathrm{e}^{-\rho_s}}{4} \sum_{m=1}^{\infty} \frac{\exp\left(-\frac{1}{2}m^2\sigma_{\phi_e}^2\right)}{m} \sin\left(\frac{m\pi}{4}\right)\left[I_{\frac{m-1}{2}}\left(\frac{\rho_s}{2}\right) + I_{\frac{m+1}{2}}\left(\frac{\rho_s}{2}\right)\right]^2, \tag{8.3}$$

where σ_{ϕ_e} is the STD of the Gaussian-distributed phase error.

In addition to [2], the series summation of (8.3) to find error probability has very long history [28–30]. The phase distribution of a complex nonzero mean Gaussian-distributed random variable is expressed as a Fourier series to find the error probability of (8.3).

8.2.2 QPSK Signals

For QPSK signal with phase error of ϕ_e, the error probability is

$$\begin{aligned} p_e(\phi_e) &= \frac{1}{2}\mathrm{erfc}\sqrt{\rho_+} + \frac{1}{2}\mathrm{erfc}\sqrt{\rho_-} - \frac{1}{4}\mathrm{erfc}\sqrt{\rho_+}\,\mathrm{erfc}\sqrt{\rho_-} \\ \rho_\pm &= \rho_s \cos\left(\frac{\pi}{4}\pm\phi_e\right). \end{aligned} \tag{8.4}$$

Similar to (8.2), if the phase error of ϕ_e is Gaussian distributed, the error probability of QPSK signal becomes

$$p_e = \int_{-\infty}^{+\infty} p_e(\phi_e) p_{\phi_e}(\phi_e) \mathrm{d}\phi_e. \tag{8.5}$$

Similar to (8.3) using Fourier series, the bit-error probability of QPSK signal with Gaussian-distributed phase error is

$$p_e = \frac{3}{8} - \frac{\sqrt{\rho_s} e^{-\rho_s/2}}{2\sqrt{\pi}} \sum_{m=1}^{\infty} \frac{\exp\left(-\frac{1}{2}m^2\sigma_{\phi_e}^2\right)}{m} \sin\left(\frac{m\pi}{4}\right)\left[I_{\frac{m-1}{2}}\left(\frac{\rho_s}{2}\right) + I_{\frac{m+1}{2}}\left(\frac{\rho_s}{2}\right)\right]. \tag{8.6}$$

In both the series of (8.3) and (8.6), the terms of m as an integer multiple of 4 are equal to zero.

Figure 8.1 shows the signal-to-noise ratio (SNR) penalty for both QPSK and DQPSK signals as a function of the STD of the Gaussian-distributed phase noise, σ_{ϕ_e}. The raw BER for the signal is assumed to be 10^{-3}, 10^{-5}, and 10^{-9} before the application of FEC. Those three raw BERs correspond to the case with very strong, moderate, and no FEC for the signal. From Fig. 8.1, the phase noise STD should be less than 4–6° for strong-to-moderate FEC for SNR penalty less than 0.5 dB.

The required SNR for raw BER of 10^{-3}, 10^{-5}, and 10^{-9} may be found in [2, chap. 9]. Table 8.1 also lists the required SNR for those raw BER. In later parts of this chapter, the required SNR for QPSK and DQPSK signals are assumed to be 12 and 14 dB, respectively, for raw BER between 10^{-3} and 10^{-5}.

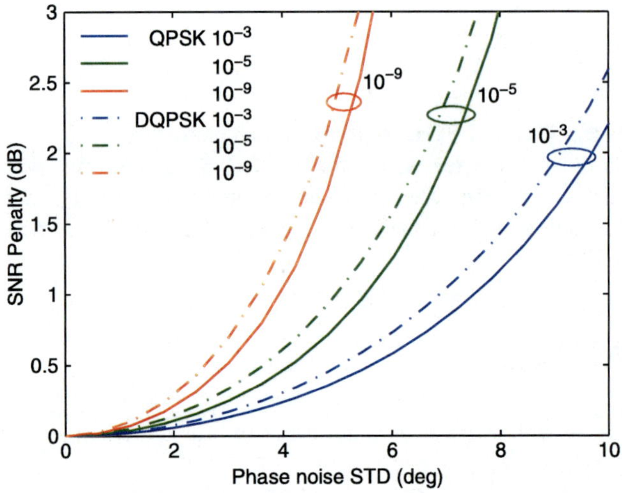

Fig. 8.1 SNR penalty as a function of the STD of Gaussian-distributed phase noise. The SNR penalties of QPSK and DQPSK signals are shown as *solid* and *dash-dot lines*, respectively, for BER of 10^{-3}, 10^{-5}, and 10^{-9}

Table 8.1 Required SNR for QPSK and DQPSK signals

BER	QPSK (dB)	DQPSK (dB)
10^{-3}	9.8	12.2
10^{-5}	12.6	15.0
10^{-9}	15.6	17.9

8.3 XPM-Induced Nonlinear Phase Noise

The phase of each WDM channel is modulated by the intensity of other WDM channels due to XPM. Even if a WDM channel has constant intensity, the amplifier noise within the signal bandwidth beats with the signal, induces intensity variations, and modulates other WDM channels. Nonlinear phase noise is a fundamental limit for phase-modulated signals [2, 7].

8.3.1 Pump-Probe Model

To study the impact of XPM from one to another WDM channel, the simplest model uses two WDM channels as the pump-probe model [23, 31–34]. The overall nonlinear phase shift to the first channel is equal to

$$\Phi_{\mathrm{NL}} = \gamma \int_0^L \left[|E_1(z)|^2 + 2|E_2(z)|^2 \right] \mathrm{d}z, \tag{8.7}$$

where E_1 and E_2 are the electric field of the first and second channels, respectively. In (8.7), the first term of the right-hand size is from SPM and the second term is from XPM. If both the first and second channels propagate in the same speed in the fiber, the contribution from XPM is the same as that from SPM other than the factor of 2. With channel walk-off due to chromatic dispersion, the XPM term is an average over an interval of time and typically smaller than the SPM term even after the factor of 2.

Based on the pump-probe model, the phase modulation of channel 1 (probe) induced by channel 2 (pump) is

$$\phi_{1,\mathrm{XPM}}(L,t) = 2\gamma \int_0^L P_2(0, t + d_{12}z)\mathrm{e}^{-\alpha z}\mathrm{d}z, \tag{8.8}$$

where $P_2(z,t)$ is the power of channel 2 as a function of position z and time t, γ is the fiber nonlinear coefficient, α is fiber attenuation coefficient, L is the fiber length, $d_{12} \approx D\Delta\lambda$ is the relative walk-off between two channels with wavelength separation of $\Delta\lambda$, and D is the dispersion coefficient of the fiber chromatic dispersion. The phase of $\phi_{1,\mathrm{XPM}}(L,t)$ assumes that the waveform of $P_2(z,t) = P(0, t - z/c_2)$ without distortion along the fiber, where c_2 is the speed of light at channel 2. When waveform distortion is ignored, the walk-off effect is included by the parameter of d_{12}. Because the impact of chromatic dispersion increases with wavelength separation, the walk-off between two channels is far larger than the chromatic dispersion within the same channel. Results from [23, 35] showed that the waveform distortion is a second-order effect.

By taking the Fourier transform of the autocorrelation function, when the power spectral density of $P_2(0,t)$ is $\Phi_{P_2}(f)$, the power spectral density of $\phi_{1,\mathrm{XPM}}(L,t)$ is

$$\Phi_{\phi_1}(f) = \Phi_{P_2}(f)|H_{12}(f)|^2, \tag{8.9}$$

where $H_{12}(f) = 2\gamma \int_0^L \mathrm{e}^{-\alpha z + j2\pi f d_{12} z} \mathrm{d}z$ or

$$H_{12}(f) = 2\gamma \frac{1 - \mathrm{e}^{-\alpha L + j2\pi f d_{12} L}}{\alpha - j2\pi f d_{12}}. \tag{8.10}$$

The transfer function of (8.10) ignores the distortion of the pump in the fiber [23, 31–33]. If the distortion of the pump is included, the denominator of (8.10) may be modified to $\alpha - j\omega d_{12} - j\beta_2\omega^2/2$ with $\omega = 2\pi f$ [24, 36, 37]. Numerical results show that the distortion of the pump may be ignored for the systems studied here.

For a system with many fiber spans, the transfer function is similar to (8.10). After K spans, the transfer function becomes

$$2\gamma \frac{1 - \mathrm{e}^{-\alpha L + j2\pi f d_{12} L}}{\alpha - j2\pi f d_{12}} \sum_{k=0}^{K-1} \mathrm{e}^{-j2\pi k f(1-\kappa) d_{12} L} \tag{8.11}$$

or

$$H_{12}^{(K)}(f) = 2\gamma \frac{1 - \mathrm{e}^{-\alpha L + j2\pi f d_{12} L}}{\alpha - j2\pi f d_{12}} \times \frac{1 - \mathrm{e}^{-j2\pi f(1-\kappa) d_{12} K L}}{1 - \mathrm{e}^{-j2\pi f(1-\kappa) d_{12} L}} \tag{8.12}$$

where κ is the fraction of optical dispersion compensation per span, i.e., $\kappa = 1$ and $\kappa = 0$ for perfect and without optical dispersion compensation, respectively. The transfer function of (8.12) assumes K cascaded identical fiber spans with the same configuration without loss of generality. The transfer function of (8.12) may be modified to other configurations.

If all channels in the WDM system are QPSK signals, the system may design without optical chromatic dispersion compensation to have $\kappa = 0$ but with electronic dispersion compensation using digital signal processing techniques. If some channels of the WDM system are either DQPSK or OOK signals, the system is likely to have optical chromatic dispersion compensation with κ close to but not equal to unity. With perfect chromatic dispersion compensation per span, the fiber nonlinearities of each span sum coherently from span to span and degrade the system performance drastically. With a κ close to unity, the accumulated chromatic dispersion of the multi-span link is close to zero that does not degrade either the DQPSK or the OOK signals but the fiber nonlinearities do not sum coherently from span to span.

8.3.2 XPM from Phase-Modulated Channels

When the pump (channel 2) has amplifier noises, $P_2(0,t) = |E_2 + N_2|^2$, where E_2 and N_2 are the electric fields from both signal and noise, respectively. In the power of $P_2(0,t) = |E_2|^2 + E_2 \cdot N_2^* + E_2^* \cdot N_2 + |N_2|^2$, the dc-term of $|E_2|^2$ gives no nonlinear phase noise but a constant phase shift, the signal–noise beating of $E_2 \cdot N_2^* + E_2^* \cdot N_2$ gives a noise spectral density of $2|E_2|^2 S_{sp}$, and the noise–noise beating of $|N_2|^2$ gives a noise spectral density of $2S_{sp}^2 \Delta\nu_{opt}$, where S_{sp} is the spectral density of the amplifier noise and $\Delta\nu_{opt}$ is the optical bandwidth of the amplifier noise. The optical SNR over an optical bandwidth of $\Delta\nu_{opt}$ is $|E_2|^2/(2S_{sp}\Delta\nu_{opt})$. For a launched power of P_0 and a single optical amplifier with a noise variance of $S_{sp,1}$, we obtain

$$\Phi_{P_2}(f) = 2P_0 S_{sp,1} + 2S_{sp,1}^2 \Delta\nu_{opt} \approx 2\sigma_n^2 P_0 \tag{8.13}$$

as a constant over frequency.

For a non-return-to-zero (NRZ) constant-intensity phase-modulated signals, $|E_2|^2$ is a dc-term and can be ignored. For a return-to-zero (RZ) phase-modulated signal, $|E_2|^2$ is a periodic function with a period of T and its power spectral density is tones at the frequencies of k/T, where k is integer. However, the low-pass transfer function of $H_{12}(f)$ should have very small response at those frequencies of k/T. For RZ signal with pulse broadening due to fiber dispersion, if the dispersion is assumed to be a linear effect, for system without pulse overlapping, the low-pass transfer function can also completely eliminate XPM-induced nonlinear phase shift from $|E_2|^2$. Of course, this assumption is valid with no pulse distortion in the fiber with the relationship $P_2(z,t) = P_2(0, t - z/c_2)$. With pulse broadening such that two pulses overlap after a short fiber distance, those overlapped pulses still generate very small nonlinear phase noise [13] that is far smaller than the nonlinear phase noise from signal and noise interaction.

Using the spectral density of (8.13), together with the transfer function of (8.12), the spectral density of XPM-induced nonlinear phase noise from constant intensity phase-modulate signals can be obtained. The spectral density of (8.13) is constant, the spectral density of XPM-induced nonlinear phase follows the transfer function of (8.12). Amplifier noise is accumulated span after span when the signal passes more and more optical amplifiers. The constant in (8.13) is proportional to the fiber span number. For a system with N span, the amplifier noise from the kth span has a transfer function of $H_{12}^{(N-k+1)}(\omega)$.

For systems with many WDM channels, the walk-off effect of d_{12} of (8.12) is proportional to channel separation. Considered the center WDM channel as the worst case, the overall XPM-induced nonlinear phase noise is the summation of all WDM channels with channel separation of $k\delta\lambda$, where $k = \pm 1, \pm 2, \ldots$ with $\pm$ as the WDM channels with larger and smaller wavelength with respect to the center channel, and $\delta\lambda$ is the channel spacing that is typical 50 GHz or 0.4 nm in most designs.

8.3.3 XPM from On-Off Keying Channels

If the pump is OOK signal with $P_2(0,t) = |E_2 + N_2|^2$, the signal should be far larger than the noise such that the OOK signal can be received with low error probability. With $|E_2|^2 \gg |N_2|^2$ and E_2 is OOK signal, the noise may be ignored all together. With OOK signal, the spectral density of $\Phi_{P_2}(f)$ is

$$\Phi_{P_2}(f) = P_0 T_b \mathrm{sinc}^2(fT_b), \tag{8.14}$$

where T_b is bit interval of the OOK signal.

Using the spectral density of (8.14), together with the transfer function of (8.12), the spectral density of XPM-induced nonlinear phase noise from OOK signals can be obtained. The spectral density of (8.14) is flat around $f = 0$ and the transfer function of (8.12) is a low-pass response. Both phase-modulated and OOK signals give XPM-induced nonlinear phase noise with similar shaped spectral density, at least at low frequency. However, the nonlinear phase noise from OOK signals is from the signal of $|E_2|^2$ by itself but the nonlinear phase noise from phase-modulated signals is from $2E_2N_2$.

For OOK signals, the transfer function of (8.12) is for $K = N$ for a N-span fiber link. OOK signals typically require optical chromatic dispersion compensation with κ approximately close to but not equal to 1.

For the same channel separation and launched power, the OOK signal gives larger XPM-induced nonlinear phase noise than phase-modulated signal. The intensity of an OOK signal is larger than the signal and noise beating in constant-intensity phase-modulated signal. The XPM-induced nonlinear phase noise from OOK signals can be reduced by either lowering the power of the WDM channels with OOK signal or adding a guard-band. Adding a guard-band reduces the capacity of the fiber link and the usable bandwidth is wasted. The design of hybrid QPSK/OOK WDM systems without guard-band is essential if the future QPSK signal is retrofitting into existing NRZ OOK WDM systems.

8.4 XPM-Induced Nonlinear Phase Noise to DQPSK Signals

Both DPSK and DQPSK signals can be directly demodulated using the asymmetric Mach–Zehnder interferometer [2]. After the asymmetric Mach–Zehnder interferometer, the differential nonlinear phase noise of $\Delta\phi_{1,\mathrm{XPM}}(L,t) = \phi_{1,\mathrm{XPM}}(L,t) - \phi_{1,\mathrm{XPM}}(L,t-T)$ adds to the differential phase of the signal, where T is the symbol interval. The power spectral density of $\Delta\phi_{1,\mathrm{XPM}}(L,t)$ is

$$\Phi_{\Delta\phi_1}(f) = 4\Phi_{P_2}(f)|H_{12}(f)|^2 \sin^2(\pi f T). \tag{8.15}$$

The phase variance as a function of frequency separation is

$$\sigma^2_{\mathrm{XPM},0}(\Delta\lambda) = 4\int_{-1/T}^{1/T} \Phi_{P_2}(f)|H_{12}(f)|^2 \sin^2(\pi f T)\,\mathrm{d}f, \tag{8.16}$$

where the integration is reduced from $\pm\infty$ to $\pm 1/T$ by taking into account only the phase noise over a bandwidth confined within the bit-rate. Please note that $\Phi_{P_2}(f)$ is a constant independent of frequency from Sect. 8.3.2. The variance of (8.16) was found in [12] by simple approximation. The dependence of the variance of (8.16) on the wavelength separation of $\Delta\lambda$ is originated from the dependence of $H_{12}(f)$ of (8.10) on $\Delta\lambda$.

Here, a 20-span fiber link is considered with fiber length of 90 km per span. The system has 81 WDM channels with 50-GHz of channel spacing at the conventional C-band around the wavelength 1.55 μm. The middle channel with the worst XPM-induced nonlinear phase noise is considered. The optical fiber has an attenuation coefficient of $\alpha = 0.22\,\mathrm{dB\,km^{-1}}$. The DQPSK signal is assumed to use two polarizations with 28 GHz symbol rate to support about $100\,\mathrm{Gb\,s^{-1}}$ after FEC. The optical fiber is either standard single-mode fiber (SMF) or non-zero dispersion-shifted fiber (NZDSF) with dispersion coefficient of 17 and $3.8\,\mathrm{ps\,km^{-1}\,nm^{-1}}$, respectively.

To support DQPSK signal, optical dispersion compensator is used with $\kappa = 1.05$ for SMF and $\kappa = 0.78$ for NZDSF, approximately the same as that in [21]. The residual dispersion per span should provide better performance for DQPSK and OOK signals, if any. Optical amplifiers are used in each span. The received signal is assumed to have a SNR of 14 dB, approximately having an BER between 10^{-3} and 10^{-5} from Table 8.1.

Figure 8.2 shows the STD of phase error as a function of the mean nonlinear phase shift per WDM channel by assuming that all WDM channels have the

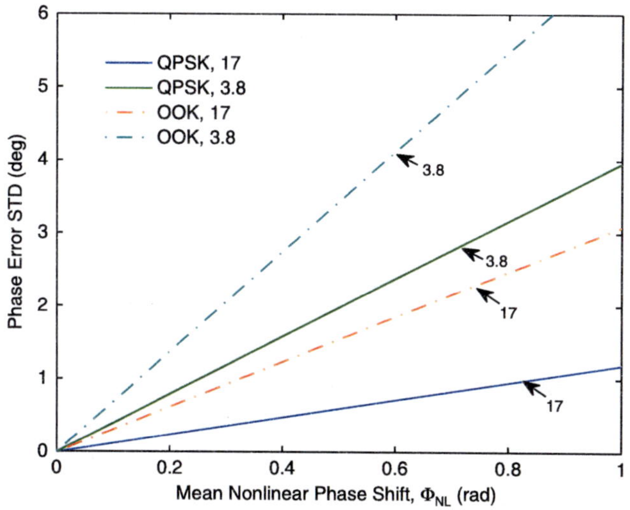

Fig. 8.2 The STD of phase error as a function of the mean nonlinear phase shift per WDM channel. The *solid lines* assume that all 81 WDM channels are DQPSK signals. The *dash-dot lines* assume that the lower 41 channels are DQPSK signals but the upper 40 channels are $10.7\,\mathrm{Gb\,s^{-1}}$ OOK signals. The optical fibers are SMF and NZDSF with dispersion coefficient of $D = 17$ and $3.8\,\mathrm{ps\,km^{-1}\,nm^{-1}}$, respectively

same power. The mean nonlinear phase shift is defined in [2] as the accumulated per-channel nonlinear phase shift in the WDM link. The phase error in Fig. 8.2 is for the case all WDM channels are DQPSK signals or half of the WDM channels are 10.7 Gb s^{-1} OOK signal. Without loss of generality, all OOK signals are assumed at the upper band and all DQPSK signals are in lower band. The phase error of Fig. 8.2 for hybrid system includes the phase error from upper-band OOK and lower-band DQPSK signals.

From Fig. 8.1, the phase error STD must be less than 4–6° such that the XPM-induced nonlinear phase noise gives an SNR penalty less than 0.5 dB. If all WDM channels are DQPSK signals, the XPM-induced nonlinear phase noise should not degrade the system if SMF with dispersion coefficient of $D = 17\,\text{ps}\,\text{km}^{-1}\,\text{nm}^{-1}$ is used or all channels are DQPSK signals. From [38] and [2, Sect. 9.4.2], the mean nonlinear phase shift for DQPSK signal must be less than 0.5 rad such that SPM-induced nonlinear phase noise is less than 1 dB. Even for DQPSK signal using NZDSF with $D = 3.8\,\text{ps}\,\text{km}^{-1}\,\text{nm}^{-1}$ and with upper band OOK signal, with mean nonlinear phase shift of 0.5 rad, the phase error STD is less than 4° and gives less than 0.5 dB degradation to the DQPSK signals.

For all cases, XPM-induced nonlinear phase noise typically provides less than 0.5 dB SNR penalty to DQPSK signals even the adjacent WDM channels are NRZ OOK signals.

8.5 XPM-Induced Nonlinear Phase Noise for QPSK Signals

The impact of XPM-induced nonlinear phase noise for QPSK signals is not the same as that for DQPSK signals. For QPSK signals with coherent detection, phase-tracking is required due to phase noise. The phase noise may be due to nonlinear phase noise from either phase-modulated or OOK signals, laser phase noise from transmitter or local oscillator laser, environment variations induced phase shift, and other effects. The nonlinear phase noise may be due to SPM or XPM, or even intra-channel four-wave-mixing (IFWM) [3, 39, 40]. Carrier recovery eliminates parts of the phase noise. Because the XPM-induced nonlinear phase noise is concentrated in the low frequency, an optimally designed carrier recovery circuitry is very effective.

8.5.1 Feedforward Carrier Recovery

For low-speed coherent optical communication systems, phase-tracking typically uses feedback-based phase-locked loop [5, 6, 41, 42]. For very high-speed QPSK signals with digital receiver, digital signal processing is far slower than the bit rate [43]. The loop-delay may be too large for feedback-based phase-locked loop [44]. Feedforward carrier recovery [25, 26, 45, 46] is typically used for high-speed QPSK

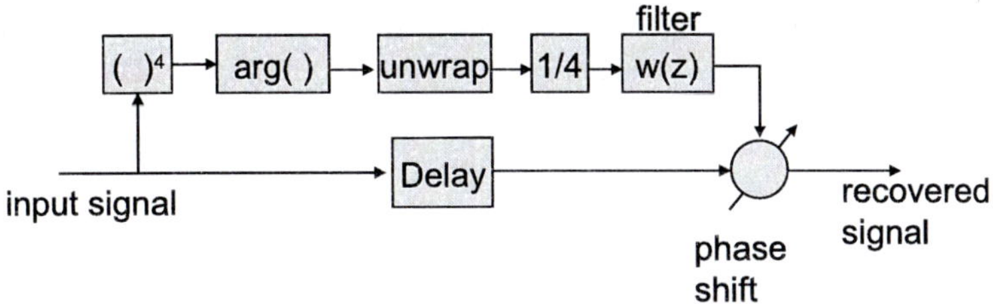

Fig. 8.3 Schematic diagram of feedforward carrier recovery for QPSK signals

signals. Theoretically, the carrier recovery can have large operating latency as long as the main signal can also be delayed [45,46]. Feedforward carrier recovery also is close to the optimal performance for phase estimation [46].

Figure 8.3 shows the schematic diagram of feedforward carrier recovery for QPSK signals. The signal is first raised to 4th power to obtain the phase without modulation, unwrap the phase, taking the factor of $1/4$, and smoothing using a filter of $W(f)$, to compensate for the phase variations. The optimal smoothing filter of $W(f)$ is designed here for system with XPM-induced nonlinear phase noise. The filter $W(f)$ is expressed as $w(z)$ in Fig. 8.3 to emphasize that the filter is operated in discrete time; however, continuous-time analysis is used here. Because the transfer function of (8.12) is a low-pass response, there is almost no numerical difference between continuous- and discrete-time analysis of the system.

If the received signal is denoted as $Ae^{j\theta_r + j\phi_e + j\theta_n}$ where $\theta_r = (2k+1)\pi/4$ with $k = 0, 1, 2, 3$ as the transmitted phase, ϕ_e is the phase noise, and θ_n is the phase due to additive Gaussian noise. The phase of θ_n is independent of the phase noise ϕ_e. The 4th-power, to obtain the phase, and taking the factor of $1/4$ gives the phase of $\phi_e + \theta_n$. In the linearized model, the input to the smoothing filter $W(f)$ is

$$\phi_e + \theta_n. \tag{8.17}$$

The variance of θ_n is $\sigma_{\theta_n}^2 = 1/2\rho_s$ when ρ_s is larger than 10 dB [2, Fig. 4.A.1]. The output of the smoothing filter should be $\hat{\phi}$ as an estimation of ϕ_e. From the theory of Wiener filter for smoothing [47, Sect. 13-3] and [48, chap. 5, pt. 2], the optimal smoothing filter is

$$W(f) = \frac{\Phi_{\phi_e}(f)}{\Phi_{\phi_e}(f) + N_{\theta_n}}, \tag{8.18}$$

where $\Phi_{\phi_e}(f)$ is spectral density of the phase noise, and N_{θ_n} is the spectral density of θ_n. Although the smoothing filter (8.18) is noncasual, the delay in the main signal path may be used to transfer $W(f)$ to casual filter [46]. The impulse response of the filter cannot be too long to reduce the buffer requirement of the signal.

The performance of carrier recovery may be characterized by the mean-square error (MSE) of $\mathscr{E} = E\{(\hat{\phi} - \phi_e)^2\}$. The MSE is the phase error at the output of the carrier recovery circuitry. With the smoothing filter $W(f)$, the variance of the phase error at the output of Fig. 8.3 is equal to

$$\mathscr{E} = \int_{-\infty}^{+\infty} \left[\Phi_{\phi_e}(f) - 2\Re\left\{W(f)\right\}\Phi_{\phi_e}(f) + |W(f)|^2(\Phi_{\phi_e}(f) + N_{\theta_n})\right] \mathrm{d}f$$
$$= \int_{-\infty}^{+\infty} |1 - W(f)|^2\, \Phi_{\phi_e}(f)\mathrm{d}f + N_{\theta_n} \int_{-\infty}^{+\infty} |W(f)|^2\, \mathrm{d}f. \tag{8.19}$$

The MSE of (8.19) is similar to that in the analysis of feedback-based phase-locked loop [2, Sect. 4.3.1]. In phase-locked loop, the filter $W(f)$ is typically a second-order response but the smoothing filter here may use more general filter type. Using (8.19), the optimization of feedforward carrier recovery and feedback based phase-locked loop is the same if the filter $W(f)$ is limited to second-order response.

With the smoothing filter of (8.18), we obtain

$$\mathscr{E}_{\min} = \int_{-\infty}^{+\infty} \frac{\Phi_{\phi_e}(f) N_{\theta_n}}{\Phi_{\phi_e}(f) + N_{\theta_n}} \mathrm{d}f. \tag{8.20}$$

The performance of QPSK signal with feedforward carrier recovery can be studied according to both (8.19) and (8.20).

In the simulation of both [20, 21], there is no optimization for the filter $W(f)$. The filter $W(f)$ may just take the average phase of the whole simulation and equivalently a low-pass filter (LPF) with a very low bandwidth. To certain extent, the phase error for the simulation of [20, 21] may just have the first term of (8.19) and equal to $\int_{-\infty}^{+\infty} \Phi_{\phi_e}(f)\mathrm{d}f$, but the second term of (8.19) is equal to zero. In [24], the smoothing filter is an averaging over five samples. In [24], the second-term of (8.19) is $N_0/5$ and the first-term of (8.19) is not necessary optimized.

8.5.2 *Performance of QPSK Signals*

From Fig. 8.2, the XPM-induced nonlinear phase noise by NRZ OOK signals is larger than that by constant-intensity phase-modulated signals. The contribution from NRZ OOK signals to the XPM-induced nonlinear phase noise is considered first here for a 50-GHz channel spacing WDM system, similar to the system of Fig. 8.2. Optical dispersion compensation is required for the 10.7 Gb s^{-1} NRZ OOK signals. The optical dispersion compensation per span is $\kappa = 1.05$ and $\kappa = 0.78$ for SMF with $D = 17\,\mathrm{ps\,km^{-1}\,nm^{-1}}$ and NZDSF with $D = 3.8\,\mathrm{ps\,km^{-1}\,nm^{-1}}$, respectively, similar to that in [21] and the same as Fig. 8.2. The WDM system has 81 channels with lower-band 41 QPSK channels and upper-band 40 NRZ OOK channels. Similar to that for DQPSK signal in Sect. 8.4, the QPSK signal has two polarizations each with a symbol rate of 28 GHz, providing an overall data rate of 100 Gb s^{-1} after FEC.

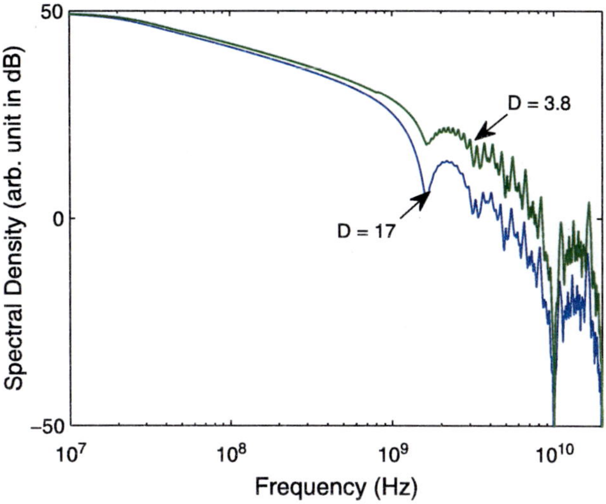

Fig. 8.4 The spectral density of the phase error $\Phi_{\phi_e}(f)$ for the QPSK signal with XPM-induced nonlinear phase noise due to the NRZ OOK signal from adjacent WDM channels. The unit of the spectral density is in dB

Figure 8.4 shows the spectral density of the phase error $\Phi_{\phi_e}(f)$ due to XPM-induced nonlinear phase noise from NRZ OOK signals to QPSK signal. The spectral density is the contribution from all 40 NRZ OOK 10.7-Gb s^{-1} WDM channels without guard-band. Figure 8.4 shows that phase noise is mostly in the frequency less than 1 GHz and a Wiener filter will be very effective to reduce the nonlinear phase noise. In the frequency less than 1 GHz, $W(f)$ is approximately equal to 1 from (8.18). From (8.19), the phase noise is almost fully eliminated by the factor of $|1 - W(f)|^2$ at low frequency. In the high frequency regime, the filter $W(f)$ follows $\Phi_{\phi_e}(f)$ and both the contribution from phase noise or additive Gaussian noise is small.

From Fig. 8.4 and at low-frequency, the Wiener filter is able to track the XPM-induced nonlinear phase noise. The rotator in Fig. 8.3 is able to compensate the phase noise accordingly.

Figure 8.5 shows the phase error STD due to XPM-induced nonlinear phase noise of a WDM system with hybrid QPSK and NRZ OOK signal. The optimal Wiener filter of (8.18) is used as compared with the case with a very low bandwidth LPF. The phase error has a maximum STD of less than 4–6° even for a mean nonlinear phase shift up to 1 rad, giving a penalty less than 0.5 dB. The usage of Wiener filter reduces the phase error substantially.

The SNR of the system of Fig. 8.5 is 12 dB, providing a raw BER of a QPSK signal between 10^{-5} and 10^{-3} from Table 8.1. The phase error in Fig. 8.5 just includes the contribution from NRZ signals and that from other QPSK signals are comparatively very small. The phase error STD of Fig. 8.5 is calculated for both SMF with $D = 17\,\text{ps km}^{-1}\,\text{nm}^{-1}$ and NZDSF with $D = 3.8\,\text{ps km}^{-1}\,\text{nm}^{-1}$.

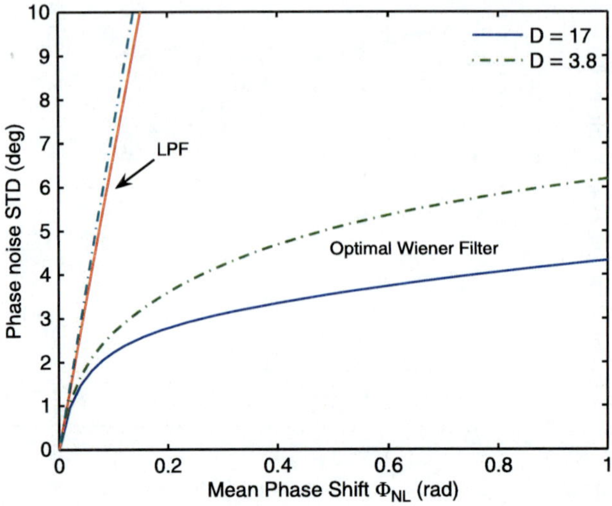

Fig. 8.5 For QPSK and OOK hybrid WDM systems, the STD of phase error for QPSK signal with optimal Wiener filter or low-bandwidth LPF in the feedforward carrier recovery of Fig. 8.3

In [21], guard-band is used between QPSK and NRZ OOK signal to reduce XPM-induced nonlinear phase noise. From Fig. 8.5, guard-band is not required if the filter $W(f)$ is optimized. The phase error is less than 6° even for the case without guard-band. If phase error is not compensated properly, a large guard-band may be required. In the recent paper of [24], the filter $W(f)$ is designed as an averaging filter with a length of 5. The second term of (8.19) becomes $1/5$ of N_0, giving a degradation of 0.8 dB even without phase noise. The first term of (8.19) is reduced in [24] but may be still very significant.

Figure 8.5 assumes that the NRZ OOK signals are in only one-side of the QPSK signal without guard-band. For the case that a QPSK signal is in the middle of NRZ OOK signals, Fig. 8.5 is applicable after some modifications. Compared with Fig. 8.5, the phase error variance is double and the phase error STD is increased up to 40% if both sides of a QPSK signal is NRZ OOK signals without guard band.

Figure 8.6 shows the STD of the phase error for QPSK signal for a 50-GHz spacing WDM system with 81 QPSK channels. The impact of chromatic dispersion to QPSK signal is equalized using digital signal processing. The system of Fig. 8.6 is similar to that of Figs. 8.2 and 8.5 but without optical dispersion compensation with $\kappa = 0$. With optimal Wiener filter, the phase error of the QPSK signal is always less than 4–6°. Without Wiener filter, the phase error of the QPSK signal is still less than 4–6° if the mean nonlinear phase shift is less than 0.5 rad.

Figure 8.6 ignores the polarization effect. In polarization-multiplexed (PM) QPSK signal, the SPM from orthogonal polarization is reduced to a factor of $2/3$ compared with that from the same polarization. The mean nonlinear phase shift is reduced by a factor of about 17% due to polarization effect. Similarly for SPM

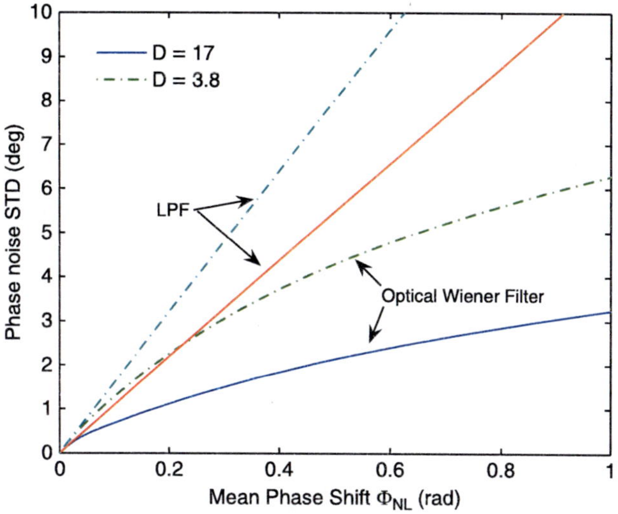

Fig. 8.6 For QPSK WDM systems, the STD of phase error for QPSK signal with optimal Wiener filter or low-bandwidth LPF in feedforward carrier recovery

effects, the XPM-induced nonlinear phase noise from orthogonal polarization is also reduced by a factor of $2/3$ compared with that from the same polarization. Because both axes are reduced by the same factor, the curves in Fig. 8.6 remain the same shape. For PM-QPSK signal, Fig. 8.6 is applicable if the mean nonlinear phase shift is adjusted down by 17%.

In practice, XPM combined with polarization effects also give nonlinear polarization rotation [49] that is beyond the scope of this chapter.

8.6 Conclusion

The nonlinear phase noise induced by XPM from other WDM channels is studied for both QPSK and DQPSK signals. Both QPSK and DQPSK signals can tolerate a phase error STD up to 4–6°, assuming that the phase error is Gaussian-distributed. Up to a mean nonlinear phase shift of 0.5 rad, DQPSK signal may have NRZ OOK signal located at adjacent WDM channel. QPSK signal requires the usage of Wiener filter in feedforward carrier recovery to smooth the XPM-induced nonlinear phase noise from adjacent NRZ OOK signal. NRZ signal can be located adjacent to QPSK signal without guard-band if optimal carrier recovery is used for the system.

References

1. J.M. Kahn, K.-P. Ho, IEEE J. Sel. Top. Quant. Electron. **10**(2), 259 (2004)
2. K.-P. Ho, *Phase-Modulated Optical Communication Systems* (Springer, New York, 2005)
3. E. Ip, A.P.T. Lau, D.J.F. Barros, J.M. Kahn, Opt. Express **16**(2), 753 (2008)
4. X. Zhou, J. Yu, M.F. Huang, Y. Shao, T. Wang, P. Magill, M. Cvijetic, L. Nelson, M. Birk, G. Zhang, S. Ten, H.B. Matthew, S.K. Mishra, J. Lightwave Technol. **28**(4), 456 (2010)
5. T. Okoshi, K. Kikuchi, *Coherent Optical Fiber Communications* (KTK Scientific, Tokyo, 1988)
6. S. Betti, G. de Marchis, E. Iannone, *Coherent Optical Communication Systems* (Wiley, New York, 1995)
7. J.P. Gordon, L.F. Mollenauer, Opt. Lett. **15**(23), 1351 (1990)
8. H. Kim, A.H. Gnauck, IEEE Photon. Technol. Lett. **15**(2), 320 (2003)
9. K.-P. Ho, in *Advances in Optics and Laser Research*, vol. 3, ed. by W.T. Arkin (Nova Science Publishers, NY, 2003). http://arXiv.org/physics/0303090
10. K.-P. Ho, H.-C. Wang, IEEE Photon. Technol. Lett. **17**(7), 1426 (2005)
11. H. Kim, J. Lightwave Technol. **21**(8), 1770 (2003)
12. K.-P. Ho, IEEE J. Sel. Top. Quant. Electron. **10**(2), 421 (2004)
13. K.-P. Ho, H.-C. Wang, J. Lightwave Technol. **24**(1), 396 (2006)
14. A.S. Lenihan, G.E. Tudury, W. Astar, G.M. Carter, XPM-induced impairments in RZ-DPSK transmission in a multi-modulation format WDM systems, Conference on the lasers and electro-optics, CLEO, Paper CWO5, 2005
15. G.W. Lu, L.-K. Chen, C.K. Chan, Performance comparison of DPSK and OOK signals with OOK-modulated adjacent channel in WDM systems, Opto-electronics communication conference, OECC, Paper 7B3-5, 2005
16. H. Griesser, J.P. Elbers, Influence of cross-phase modulation induced nonlinear phase noise on DQPSK signals from neighbouring OOK channels, European conference on optical communication, ECOC, Paper Tu1, 2005
17. S. Chandrasekhar, X. Liu, IEEE Photon. Technol. Lett. **19**(22), 1801 (2007)
18. R.S. Luís, B. Clouet, A. Teixeira, P. Monteiro, Opt. Lett. **32**(19), 2786 (2007)
19. T. Tanimura, S. Oda, M. Yuki, H. Zhang, L. Li, Z. Tao, H. Nakashima, T. Hoshida, K. Nakamura, J.C. Rasmussen, Nonlinearity tolerance of direct detection and coherent receivers for 43 Gb/s RZ-DQPSK signals with co-propagating 11.1 Gb/s NRZ signals over NZ-DSF, Optical fiber communication conference, OFC, Paper OTuM4, 2008
20. M. Bertolini, P. Serena, N. Rossi, A. Bononi, Numerical Monte Carlo comparison between coherent PDM-QPSK/OOK and incoherent DQPSK/OOK hybrid systems, European conference on optical communication, ECOC, Paper P.4.16, 2008
21. A. Carena, V. Curri, P. Poggiolini, F. Forghieri, Guard-band for 111 Gbit/s coherent PM-QPSK channels on legacy fiber links carrying 10 Gbit/s IMDD channels, Optical fiber communication conference, OFC, Paper OThR7, 2009
22. O. Bertran-Pardo, J. Renaudier, G. Charlet, H. Mardoyan, P. Tran, S. Bigo, IEEE Photon. Technol. Lett. **20**(15), 1314 (2008)
23. Z. Tao, W. Yan, S. Oda, T. Hoshida, J.C. Rasmussen, Opt. Express **17**(16), 13860 (2009)
24. A. Bononi, M. Bertolini, P. Serena, G. Bellotti, J. Lightwave Technol. **27**(18), 3974 (2009)
25. E. Ip, J.M. Kahn, J. Lightwave Technol. **25**(9), 2675 (2007); J. Lightwave Technol. **27**(13), 2552 (2009)
26. R. Noé, J. Lightwave Technol. **23**(2), 802 (2005)
27. K.-P. Ho, IEEE Photon. Technol. Lett. **16**(1), 308 (2004)
28. V.K. Prabhu, IEEE Trans. Commun. Technol. **COM-17**(1), 33 (1969)
29. P.C. Jain, N.M. Blachman, IEEE Trans. Info. Theor. **IT-19**(5), 623 (1973)
30. N.M. Blachman, IEEE Trans. Commun. **COM-29**(3), 364 (1981)
31. T.K. Chiang, N. Kagi, T.K. Fong, M.E. Marhic, L.G. Kazovsky, IEEE Photon. Technol. Lett. **6**(6), 733 (1994)
32. T.K. Chiang, N. Kagi, M.E. Marhic, L.G. Kazovsky, J. Lightwave Technol. **14**(3), 249 (1996)

33. K.-P. Ho, E.T.P. Kong, L.Y. Chan, L-K. Chan, F. Tong, IEEE Photon. Technol. Lett. **11**(9), 1126 (1999)
34. J. Leibrich, C. Wree, W. Rosenkranz, IEEE Photon. Technol. Lett. **14**(2), 215 (2002)
35. K.-P. Ho, Opt. Commun. **169**(1–6), 63 (1999)
36. R. Hui, K.R. Demarest, C.T. Allen, J. Lightwave Technol. **17**(6), 1018 (1999)
37. A.V.T. Cartaxo, J. Lightwave Technol. **17**(2), 178 (1999)
38. J.-A. Huang, K.-P. Ho, Exact error probability of DQPSK signal with nonlinear phase noise, *Proceedings of the 5th Pacific Rim conference on lasers and electro-optics, CLEO/PR*, Paper TU4H-(9)-5, 2003
39. X. Wei, X. Liu, Opt. Lett. **28**(23), 2300 (2003)
40. A.P.T. Lau, S. Rabbani, J.M. Kahn, J. Ligtwave Technol. **26**(14), 2128 (2008)
41. J.J. Spilker Jr., *Digital Communications by Satellite* (Prentice Hall, NJ, 1977)
42. L.G. Kazovsky, J. Lightwave Technol. **LT-4**(4), 415 (1986)
43. K.K. Parhi, *VLSI Digital Signal Processing Systems: Design and Implementation* (Wiley, New York, 1999)
44. S. Norimatsu, K. Iwashita, J. Lightwave Technol. **10**(3), 341 (1992)
45. T. Pfau, S. Hoffmann, R. Noé, J. Lightwave Technol. **27**(8), 989 (2009)
46. M.G. Taylor, J. Lightwave Technol. **27**(7), 901 (2009)
47. A. Papoulis, *Probability, Random Variables, and Stochastic Processes*, 2nd edn. (McGraw Hill, New York, 1984)
48. J.B. Thomas, *An Introduction to Statistical Communication Theory* (Wiley, New York, 1969)
49. C.B. Collings, L. Boivin, IEEE Photon. Technol. Lett. **12**(11), 1582 (2000)

Chapter 9
Nonlinear Polarization Scattering in Polarization-Division-Multiplexed Coherent Communication Systems

Chongjin Xie

9.1 Introduction

Polarization-division-multiplexing (PDM) [1–4], which transmits two channels with orthogonal states of polarization (SOPs) at an identical wavelength, was proposed long time ago to double the capacity of fiber-optic communication systems, but it was only until recently that the technique attracted much attention. The increasing demand for communication capacity requires high spectral efficiency fiberoptic communication systems, and PDM is an effective technique to double the spectral efficiency. Advances in digital signal processing and high speed electronics make coherent detection an attractive technique for optical communication systems [5–9]. With coherent detection and digital signal processing, polarization demultiplexing, which was considered cumbersome in the optical domain, can be easily performed in the electrical domain, although there is still some interest to do polarization demultiplexing using optical methods [10–12]. Therefore, PDM is almost considered a standard option for today's optical coherent systems.

In addition to signal distortions and other impairments, polarization effects could cause crosstalk between two polarizations for PDM signals. Therefore, PDM signals are more sensitive to polarization effects in fiber-optic communication systems than single polarization (SP) signals [13–15]. Two important polarization effects in fiber-optic communication systems are polarization-mode dispersion (PMD) and polarization-dependent loss (PDL) [16, 17]. PMD mainly arises from the random birefringence in fibers and optical components, in which signals with different SOPs travel at different speeds. PDL usually occurs in optical components, such as isolators and couplers, whose insertion loss varies with the SOPs of input signals. In wavelength-division-multiplexed (WDM) systems, there is another polarization effect caused by fiber nonlinearity: cross polarization modulation (XPolM)

C. Xie (✉)
Transmission Systems and Networking Research, Bell Laboratories, Alcatel-Lucent, 791 Holmdel-Keyport Road, Holmdel, NJ 07733, USA
e-mail: chongjin.xie@alcatel-lucent.com

S. Kumar (ed.), *Impact of Nonlinearities on Fiber Optic Communications*, Optical and Fiber Communications Reports 7, DOI 10.1007/978-1-4419-8139-4_9,
© Springer Science+Business Media, LLC 2011

between channels [18, 19]. Although XPolM is useful in some special applications, for example, it can be used to generate special modulation formats and for all-optical switching [20, 21], in fiber-optic transmission systems, XPolM effect is usually harmful. Although XPolM effect in general can be neglected in optical communication systems using SP signals and polarization independent receivers, it has a significant impact on fiber-optic communication systems using PDM signals and polarization-dependent receivers [18, 22–31]. For example, in optical communication systems using PMD compensation, XPolM may drastically reduce the efficiency of optical PMD compensators [22–25].

When there are time-dependent amplitude and SOP variations in WDM channels, XPolM generates time-dependent nonlinear polarization scattering, which can cause serious crosstalk between two polarizations for a PDM signal. Although powerful digital signal processing in coherent receivers can compensate the crosstalk and distortions induced by PMD and PDL, there is no effective method to compensate the nonlinear polarization scattering-induced crosstalk, as the SOP changes caused by nonlinear polarization scattering are typically in the time scale of a single bit or symbol. It has been shown that nonlinear polarization scattering could significantly degrade the performance of PDM transmission systems, and due to nonlinear polarization scattering, a PDM coherent fiber-optic transmission system with dispersion management could perform worse than that without dispersion management [18, 29–31].

In this chapter, nonlinear polarization scattering in PDM coherent systems is analyzed. In Sect. 9.2, starting with the Manakov equation, we show how the nonlinear interaction between WDM channels changes the polarization state of each channel. Different models to simulate nonlinear polarization effects in fiber-optic communication systems are discussed. Section 9.3 analyzes the impact of nonlinear polarization scattering on the performance of PDM quadrature-phase-shift-keying (QPSK) coherent transmission systems. The difference of the nonlinear polarization scattering between PDM-QPSK coherent systems with and without inline optical dispersion compensators is discussed. Section 9.4 focuses on nonlinear polarization scattering mitigation techniques. Three techniques to mitigate nonlinear polarization scattering in dispersion-managed PDM coherent transmission systems are presented, including the use of time-interleaved return-to-zero (RZ) PDM format, the use of periodic-group-delay (PGD) dispersion compensators, and the judicious addition of some PMD in the systems. Conclusions are given in Sect. 9.5.

9.2 Analytical Theory

When polarization effects can be neglected and the signal is launched in an SP, the scalar nonlinear Schrödinger equation (NLSE) is a fairly good model to study transmission impairments in fibers including nonlinear effects. However, to consider polarization effects such as PMD and nonlinear polarization effects and to study the propagation of PDM signals in optical fibers, the coupled nonlinear Schrödinger equation (CNLSE) has to be used [32–34]

$$\frac{\partial \vec{E}}{\partial z} - i\Delta\beta_0\Sigma\vec{E} + \Delta\beta_1\Sigma\frac{\partial \vec{E}}{\partial t} + \frac{i}{2}\beta_2\frac{\partial^2 \vec{E}}{\partial t^2} = i\gamma\left[\left|\vec{E}\right|^2\vec{E} - \frac{1}{3}\left(\vec{E}^{+}\sigma_3\vec{E}\right)\sigma_3\vec{E}\right], \tag{9.1}$$

where $\vec{E} = [E_x,\ E_y]^t$ is the electrical field column vector, $\Delta\beta_0$ is the birefringence parameter, $\Delta\beta_1$ is the differential-group-delay (DGD) parameter related to PMD coefficient, Σ is the local Jones matrix describing polarization changes, β_2 is the group velocity dispersion (GVD), γ is the fiber nonlinear coefficient, $\vec{E}^{+} = \left[E_x^*,\ E_y^*\right]$ is the transpose conjugate of $\vec{E}$, σ_3 is one of the Pauli spin matrices [35]

$$\sigma_3 = \begin{pmatrix} 0 & -i \\ i & 0 \end{pmatrix}.$$

In (9.1), z is the distance along the fiber axis, t is the retarded time moving at group velocity of the carrier frequency of the signal, and $i = \sqrt{-1}$ is the imaginary unit. By averaging the nonlinear effects over the Poincaré sphere under the assumption of complete mixing (averaging over the random polarization changes that uniformly cover the Poincaré sphere) and neglecting PMD, the CNLSE can be transformed to the Manakov equation [32–34]

$$\frac{\partial \vec{E}}{\partial z} + \frac{i}{2}\beta_2\frac{\partial^2 \vec{E}}{\partial t^2} - i\frac{8}{9}\gamma\left|\vec{E}\right|^2\vec{E} = 0. \tag{9.2}$$

Suppose we have a WDM system with two channels, channels a and b, and the two channels have no overlapping spectra. By neglecting four-wave mixing (FWM) between the two channels, we can separate the equations for channels a and b from the Manakov equation as [18, 19, 36–38]

$$\frac{\partial \vec{E}_a}{\partial z} + \frac{i}{2}\beta_2\frac{\partial^2 \vec{E}_a}{\partial t^2} - i\frac{8}{9}\gamma\left(\left|\vec{E}_a\right|^2\vec{E}_a + \left|\vec{E}_b\right|^2\vec{E}_a + \vec{E}_b^{+}\vec{E}_a\vec{E}_b\right) = 0 \tag{9.3}$$

$$\frac{\partial \vec{E}_b}{\partial z} + \frac{i}{2}\beta_2\frac{\partial^2 \vec{E}_b}{\partial t^2} - i\frac{8}{9}\gamma\left(\left|\vec{E}_b\right|^2\vec{E}_b + \left|\vec{E}_a\right|^2\vec{E}_b + \vec{E}_a^{+}\vec{E}_b\vec{E}_a\right) = 0. \tag{9.4}$$

In the parenthesis of the two equations, the first term is self-phase modulation (SPM), the second term is polarization independent cross-phase modulation (XPM), and the third term is polarization-dependent XPM. SPM does not depend on the polarization, but XPM is polarization dependent. The third nonlinear term is the same as the second nonlinear term when the two channels have the same polarization and it is zero when they are orthogonally polarized, which means that the XPM between two channels with parallel polarizations is two times that with orthogonal polarizations.

The last two terms in each of (9.3) and (9.4) show that XPM between channels also causes XPolM. An intuitive way to describe XPolM is to use the three-dimensional Stokes vector $\vec{S}$ in the Stokes space. Its three real components, corresponding to the electrical field vector, can be expressed as

$$S_i = \vec{E}^{+}\sigma_i\vec{E}, \tag{9.5}$$

where the symbols σ_i are the Pauli spin matrices, which are defined as [35]

$$\sigma_1 = \begin{pmatrix} 1 & 0 \\ 0 & -1 \end{pmatrix}, \sigma_2 = \begin{pmatrix} 0 & 1 \\ 1 & 0 \end{pmatrix}, \sigma_2 = \begin{pmatrix} 0 & -i \\ i & 0 \end{pmatrix}. \tag{9.6}$$

Neglecting chromatic dispersion, we can determine the evolution of the Stokes vectors of channels a and b due to XPolM in transmission according to (9.3) and (9.4). For $\mathrm{d}S_{a1}/\mathrm{d}z$, we get

$$\frac{\mathrm{d}S_{a1}}{\mathrm{d}z} = \frac{8}{9}\gamma(S_{a2}S_{b3} - S_{a3}S_{b2}). \tag{9.7}$$

A similar expression can be found for $\mathrm{d}S_{a2}/\mathrm{d}z$ and $\mathrm{d}S_{a3}/\mathrm{d}z$. Finally, we obtain

$$\frac{\mathrm{d}\vec{S}_a}{\mathrm{d}z} = \frac{8}{9}\gamma(\vec{S}_a \times \vec{S}_b) = \frac{8}{9}\gamma(\vec{S}_a \times \vec{S}_{\mathrm{sum}}) \tag{9.8}$$

$$\frac{\mathrm{d}\vec{S}_b}{\mathrm{d}z} = \frac{8}{9}\gamma(\vec{S}_b \times \vec{S}_a) = \frac{8}{9}\gamma(\vec{S}_b \times \vec{S}_{\mathrm{sum}}), \tag{9.9}$$

where $\vec{S}_a = (S_{a1}, S_{a2}, S_{a3})$ and $\vec{S}_b = (S_{b1}, S_{b2}, S_{b3})$ are the Stokes vector for channel a and channel b, respectively, and $\vec{S}_{\mathrm{sum}} = \vec{S}_a + \vec{S}_b$ is the sum of the two Stokes vectors. The relation was originally derived by Mollenauer et al. [18]. It shows that the nonlinear interaction between channels modifies the SOP of each channel and causes the Stokes vector of each channel to precess around the other. It can also be considered that the SOP of each channel precesses around the sum of the Stokes vectors of all the channels, which is convenient for analysis when there are more than two channels [36].

Figure 9.1 gives an example of the XPolM-induced SOP evolution during propagation in a two-channel WDM system. Both channels are continuous wave (CW) light without modulation. In Fig. 9.1a, the power of channel b is 10 times that of channel a, and in Fig. 9.1b, both channels have the same power. The initial SOPs of channels a and channel b are in S_2 and S_1, respectively. The figure shows that the SOP of each channel precesses around the sum of the Stokes vectors of the two channels. Note that the sum is the channel power-weighted sum. When the power of channel b is 10 times that of channel a, the sum of the Stokes vectors of the two channels, $\vec{S}_{\mathrm{sum}}$, is close to the Stokes vector of channel b, as shown in Fig. 9.1a (the normalized sum Stokes vector is $\vec{S}_{\mathrm{sum}} = (10/\sqrt{101}, 1/\sqrt{101}, 0)$). When the

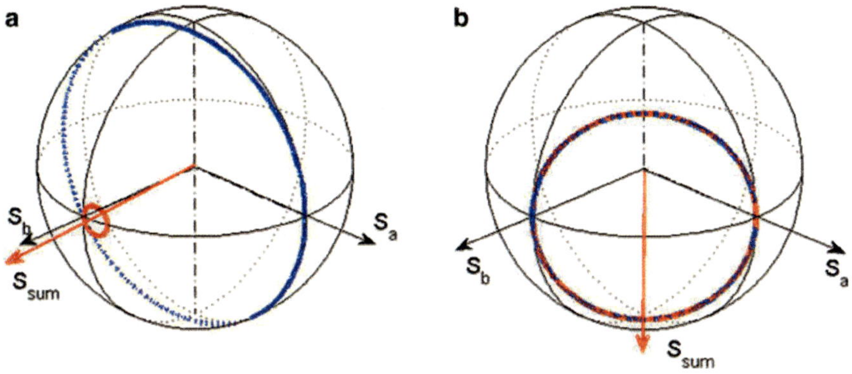

Fig. 9.1 Example of XPolM-induced SOP evolution of two WDM channels during propagation. (**a**) the power of channel b is 10 times that of channel a, (**b**) the power of channel b is the same as that of channel a. S_a and S_b are the initial Stokes vectors of channel a and channel b

two channels have the same power, it is the average of the Stokes vectors of the two channels, and the normalized sum Stokes vector is $\overrightarrow{S}_{\text{sum}} = (1/\sqrt{2}, 1/\sqrt{2}, 0)$, as shown in Fig. 9.1b. Note that in Fig. 9.1, the SOP evolution is caused only by XPolM and the fiber birefringence and PMD-induced SOP changes are not taken into account.

When channels are loaded with signals of amplitude, phase or polarization modulation, and fiber chromatic dispersion is present, the amplitude and SOP of each channel generally change with time, and the XPolM acts in the same way as (9.8) and (9.9) describe at all temporal instances, generating time-dependent nonlinear polarization scattering. Nonlinear polarization scattering causes SOP changes in the speed of symbol rates, which is hard to follow with either optical methods in direct detection receivers or digital signal processing in coherent receivers, and may induce severe impairments in optical communication systems.

To model nonlinear polarization effects in fiber-optic communication systems, we can directly solve the CNLSE given in (9.1) with the split-step Fourier method [39]. To increase the speed of the simulations, the CNLSE can be solved with the approach proposed by Marcuse et al. by integrating with small enough steps to follow the detailed polarization evolution and using larger steps for chromatic dispersion and nonlinear effects [33]. The other widely used method is the coarse-step method, which assumes that within each step the polarization does not change and the signal propagation is described by the following CNLSE [33,40]

$$\frac{\partial E_x}{\partial z} - \frac{1}{2}\Delta\beta_1 \frac{\partial E_x}{\partial t} + \frac{i}{2}\beta_2 \frac{\partial^2 E_x}{\partial t^2} = i\gamma\left(|E_x|^2 + \frac{2}{3}\left|E_y\right|^2\right) E_x \tag{9.10}$$

$$\frac{\partial E_y}{\partial z} + \frac{1}{2}\Delta\beta_1 \frac{\partial E_y}{\partial t} + \frac{i}{2}\beta_2 \frac{\partial^2 E_y}{\partial t^2} = i\gamma\left(\left|E_y\right|^2 + \frac{2}{3}|E_x|^2\right) E_y. \tag{9.11}$$

At the interval of the fiber coupling length, which is typically one or a few step sizes, the polarization of the field is randomly rotated to generate complete mixing over the Poincaré sphere. Two scattering matrices have been used to rotate signal polarizations. One scattering matrix is [2]

$$\begin{pmatrix} \cos\alpha \exp(i\varphi) & \sin\alpha \exp(i\varphi) \\ -\sin\alpha & \cos\alpha \end{pmatrix} \tag{9.12}$$

and the other one is [40]

$$\begin{pmatrix} \cos\alpha & \sin\alpha \exp(i\phi) \\ -\sin\alpha \exp(-i\phi) & \cos\alpha \end{pmatrix}, \tag{9.13}$$

where $\cos 2\alpha$ and φ are randomly chosen from uniform distributions in (9.12) and α and φ are randomly chosen from uniform distributions in (9.13). As shown by Marcuse et al. [33], although neither matrix introduces a uniform scattering on the Poincaré sphere, concatenating several of these matrices does lead to rapid uniform mixing on the Poincaré sphere.

9.3 Nonlinear Polarization Scattering in PDM-QPSK Coherent Transmission Systems

In the WDM optical communication systems using SP signals and polarization insensitive receivers, the dominant interchannel nonlinear effects are FWM and XPM, and XPolM is usually negligible. However, for systems using PDM signals, XPolM could become a dominant nonlinear effect and significantly degrade system performance. This effect was first observed in an ultra-long-haul soliton transmission system [18], where significant degradations caused by nonlinear polarization scattering were found for 10-Gb/s WDM PDM soliton transmission.

Although PDM was proposed along time ago, only until recently did it become practical in coherent systems, where polarization demultiplexing can be performed in the electrical domain with digital signal processing. Unlike an SP signal, the SOP of a PDM signal changes with time, depending on the data carried by the two polarizations. Figure 9.2 depicts the constellations of QPSK and 16-ary quadrature-amplitude modulation (QAM) signals and the diagrams of the SOPs at symbol centers that PDM-QPSK and PDM-16QAM signals have when the symbols at two polarizations are synchronized (aligned) in time. For a PDM-QPSK signal, its SOP changes among four points on the Poincaré Sphere. A PDM signal with more modulation levels has more SOPs. As shown in Fig. 9.2d, a PDM-16QAM signal has many more SOPs than a PDM-QPSK signal. The many SOPs of PDM signals will enhance nonlinear polarization scattering in WDM systems. In this section, using numerical simulations, we analyze the impact of nonlinear polarization scattering on the performance of PDM-QPSK coherent communication systems.

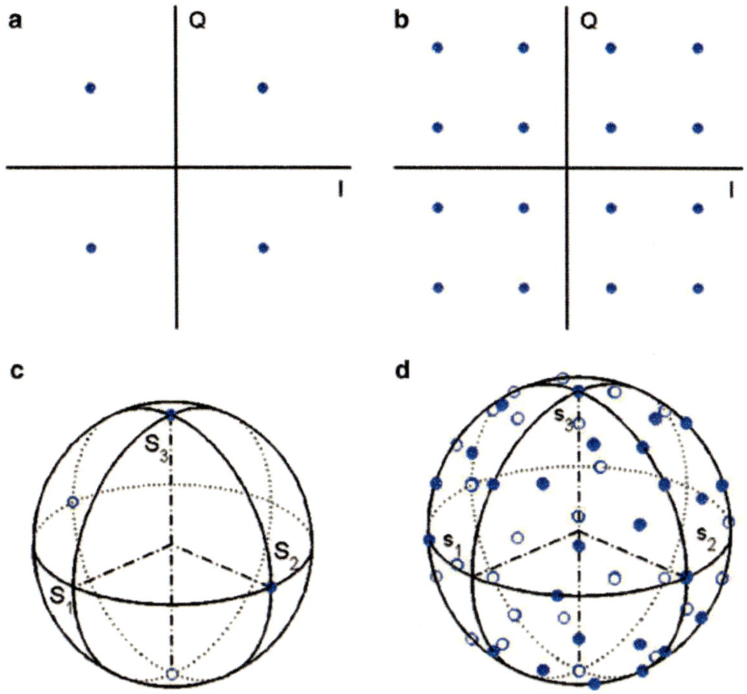

Fig. 9.2 (**a**) constellation diagram of QPSK, (**b**) constellation diagram of square 16-QAM, (**c**) SOP diagram of PDM-QPSK, (**d**) SOP diagram of PDM-16QAM. The solid and open symbols are the points on the visible and invisible parts of the Poincaré Sphere

The performance of both 42.8-Gb/s and 112-Gb/s PDM-QPSK coherent systems is discussed. The course-step method is used in the simulations to simulate nonlinear propagation of signals in fibers.

9.3.1 System Model

The system model is shown in Fig. 9.3. The WDM system has seven channels with channel spacing of 50 GHz. The transmission line consists of 10 spans of standard single mode fiber (SSMF) with a chromatic dispersion coefficient of 17.0 ps/(nm.km), a nonlinear coefficient of 1.17 (km W)$^{-1}$ and a loss coefficient of 0.21 dB/km. The span length is 100 km and lumped amplification is provided by erbium-doped fiber amplifiers (EDFAs) after each span to compensate for the transmission loss. Two different transmission systems are studied and compared. One with dispersion management and the other with no optical dispersion compensators provided at the transmitter and in the transmission line. In the system with dispersion management, there is −400-ps/nm dispersion pre-compensation and the

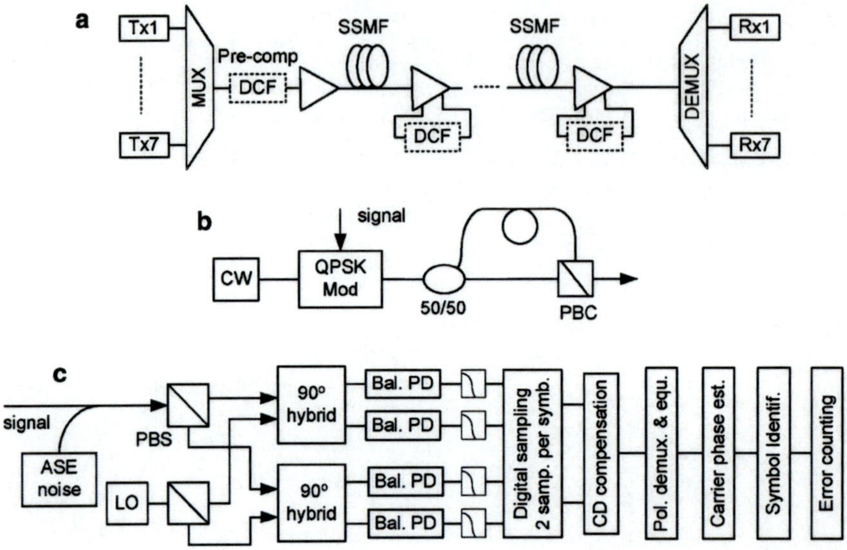

Fig. 9.3 System model. (**a**) diagram of the transmission link, (**b**) block diagram of the NRZ-PDM-QPSK transmitter, (**c**) block diagram of the coherent receiver. The DCF shown in the figure is removed for systems without dispersion management. *Tx* Transmitter; *Rx* Receiver; *PD* Photodetector; *CD* Chromatic dispersion; *SSMF* Standard single mode fiber; *DCF* Dispersion compensation fiber; *Mux* Multiplexer; *Demux* Demultiplexer; *Mod* Modulator; *PBC(S)* Polarization beam combiner (splitter); *LO* Local oscillator

chromatic dispersion in each span is compensated by dispersion compensation fiber (DCF), resulting in residual dispersion per span (RDPS) of 30 ps/nm. The nonlinearity in the DCF is neglected, which is justified as nonlinearity in DCF can be minimized by optimizing the launch power into the DCF. The net residual dispersion after transmission is compensated in the electrical domain by digital signal processing in the coherent receiver. The dispersion map used here is a typical map for a direct-detection fiberoptic transmission system, and no effort is made to optimize the dispersion map. In the system without any optical dispersion compensators, the chromatic dispersion is entirely compensated in the electrical domain in the coherent receivers.

For the nonreturn-to-zero (NRZ) PDM-QPSK transmitters, CW light is modulated with a nested Mach–Zehnder QPSK modulator by 2^{11} De Bruijn bit sequence at 21.4-Gb/s or 56-Gb/s gray mapped to QPSK symbols to generate 21.4-Gb/s or 56-Gb/s NRZ-QPSK signal. Then the SP-QPSK signal is split into two parts and the two parts are shifted relative to each other by about 511 symbols and combined with a polarization beam combiner (PBC) to form a 42.8-Gb/s or 112-Gb/s NRZ-PDM-QPSK signal, as shown in Fig. 9.3b. The QPSK signal is differentially encoded to avoid cycle slips [41].

The block diagram of the PDM-QPSK coherent receiver is depicted in Fig. 9.3c. After passing through a polarization beam splitter (PBS), each polarization of the demultiplexed signal is combined with a local oscillator (LO) in a 90° hybrid to provide both polarization and phase diversity. An ideal LO with 0 Hz linewidth is assumed (0 Hz linewidth is also assumed for the transmitter laser). After the hybrids, the four tributaries of the signal are detected by four balanced photodetectors, filtered by antialiasing electrical filters and sampled at two samples per symbol. The digital signal processing is composed of four steps: (1) chromatic dispersion compensation with two finite impulse response (FIR) filters; (2) polarization demultiplexing with four FIR filters employing the constant modulus algorithm (CMA) [42,43]; (3) carrier phase estimation using the Viterbi & Viterbi algorithm [41], and block length of 10 is used in the carrier phase estimation; and (4) symbol identification and bit-error ratio (BER) calculation. The BER is evaluated by the direct error counting method. In the system, the WDM channels are demultiplexed with a fourth-order super-Gaussian optical filter of 45-GHz bandwidth, and the second-order Butterworth electrical filters of half symbol rate are used for the anti-aliasing filters.

In the simulations, the signal of 1,024 symbols first propagates in the transmission line. The bit sequence length is sufficient to catch the nonlinear interaction for the system studied here [44]. Then amplified spontaneous emission (ASE) noise is loaded at the receiver side. 204,800 symbols with 200 different ASE noise realizations are used to calculate BER using the direct error counting method.

9.3.2 42.8-Gb/s PDM-QPSK Systems

To investigate the difference of the interchannel nonlinear effects between SP signals and PDM signals, the performance of a 42.8-Gb/s NRZ-PDM-QPSK channel surrounded by six 21.4-Gb/s NRZ-SP-QPSK channels (three channels at each side) and that by six 42.8-Gb/s NRZ-PDM-QPSK channels is first analyzed and compared. The bit rate of the SP-QPSK is half that of the PDM-QPSK so that they have the same symbol rate. Figure 9.4 shows the required optical signal-to-noise-ratio (OSNR) at a BER of 10^{-3} after 1,000-km transmission for the system with and without DCF vs. the per channel launch power. The same power (including both polarizations) is used for all the WDM channels. For the system with inline DCF, at 1-dB OSNR penalty, the allowed launch power is reduced by about 3 dB when the channel is surrounded by the NRZ-PDM-QPSK channels compared to when it is surrounded by the NRZ-SP-QPSK channels. This indicates that the interchannel nonlinearities from the PDM channels are different from those from the SP channels in the dispersion-managed system. When there is no DCF in the system, the performance difference between the system with the surrounding SP channels and PDM channels becomes much smaller. Figure 9.4 also shows that when the surrounding channels are the SP signals, at 1-dB OSNR penalty, the dispersion-managed system can tolerate about 2-dB more launch power than that without dispersion

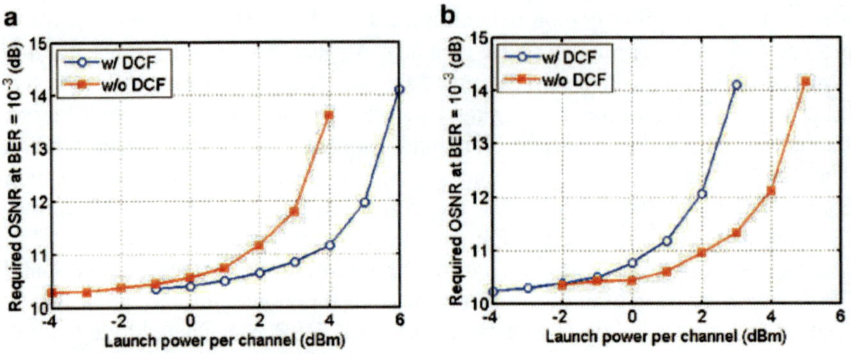

Fig. 9.4 Required OSNR at BER of 10^{-3} after 1,000-km transmission vs. launch power per channel for the 42.8-Gb/s NRZ-PDM-QPSK coherent system with and without inline DCF. **(a)** the surrounding six channels are 21.8-Gb/s NRZ-SP-QPSK signals, **(b)** the surrounding six channels are 42.8-Gb/s NRZ-PDM-QPSK signals

management, whereas when the surrounding channels are the PDM signals, the tolerable power for the dispersion-managed system is about 1.5 dB less than that without dispersion management.

Figure 9.4 clearly shows that the PDM-QPSK channels cause more interchannel nonlinearities than the SP-QPSK channels in the dispersion-managed system. In the simulations, the SOP of the SP-QPSK is at S_1, and SOP of the PDM-QPSK signal changes among S_2, $-S_2$, S_3 and $-S_3$ depending on the data carried by the two polarizations, as shown in Fig. 9.2c. With the same power, on average the PDM-QPSK and SP-QPSK generate similar XPM on the reference PDM-QPSK channel. This indicates that the performance difference of the reference 42.8-Gb/s PDM-QPSK channel between the system with the SP surrounding channels and that with PDM surrounding channels and the difference between the system with and without dispersion management are not caused by XPM, but by the XPolM-induced nonlinear polarization scattering [29, 30]. To estimate the level of the nonlinear polarization scattering in the system, the degree of polarization (DOP), which is usually used to measure the depolarization of a signal, of a 21.4-Gb/s SP-QPSK reference channel surrounded by six 42.8-Gb/s PDM-QPSK channels with 50-GHz channel spacing is calculated, which is given in Fig. 9.5. For the NRZ-PDM-QPSK system with inline DCF, DOP decreases rapidly with the launch power, indicating that the nonlinear polarization scattering significantly depolarizes the signal at each polarization of the PDM signal and induces large crosstalk between the two polarizations. For the system without inline DCF, the nonlinear polarization scattering is small and the system penalties mainly come from interchannel XPM and intrachannel nonlinearities.

Figure 9.6 plots the SOP diagram of the 21.4-Gb/s NRZ-SP-QPSK reference channel after 1,000-km transmission for the system with and without inline DCF. The SOP given in the figure is the SOP at the center of each symbol after CD compensation at the receiver. The launch power per channel is 4 dBm and the surrounding channels are 42.8-Gb/s NRZ-PDM-QPSK. As shown in the figure, due

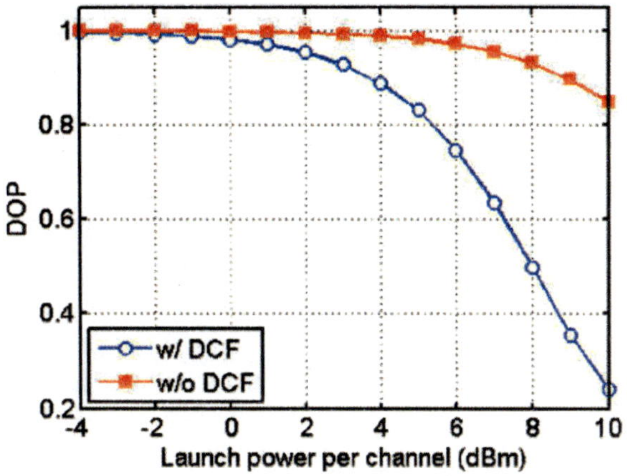

Fig. 9.5 DOP of a 21.4-Gb/s NRZ-SP-QPSK reference channel after 1,000-km transmission vs. launch power per channel in the system with and without inline DCF. The surrounding channels are 42.8-Gb/s NRZ-PDM-QPSK signals

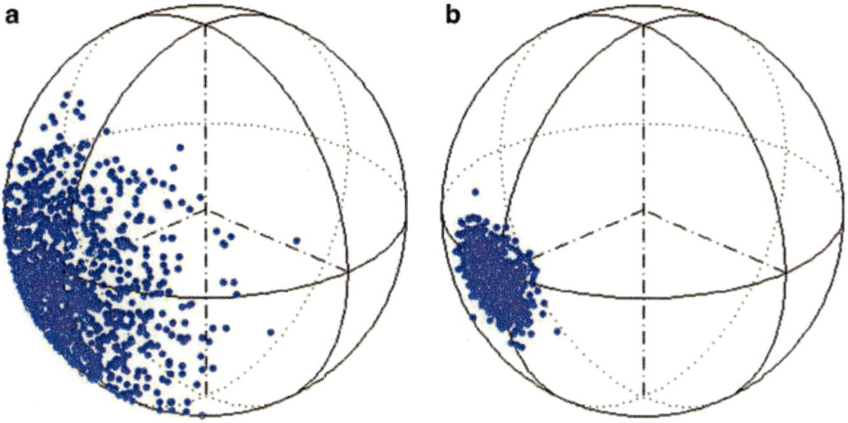

Fig. 9.6 SOP diagram of the 21.4-Gb/s NRZ-SP-QPSK reference channel after 1,000-km transmission at 4-dBm per channel launch power; the surrounding channels are 42.8-Gb/s NRZ-PDM-QPSK signals. (**a**) the system with inline DCF, (**b**) the system without inline DCF

to time-dependent XPolM from the surrounding channels, the SOP of the reference channel is largely scattered on the Poincaré sphere in the system with inline DCF. This large polarization scattering will induce severe crosstalk between two polarization tributaries for a PDM signal. In the system without DCF, the nonlinear polarization scattering is much smaller.

Figure 9.7 depicts the received signal constellation diagrams of one polarization after chromatic dispersion compensation, polarization equalization, and carrier phase estimation for the 42.8-Gb/s NRZ-PDM-QPSK channel after 1,000-km WDM

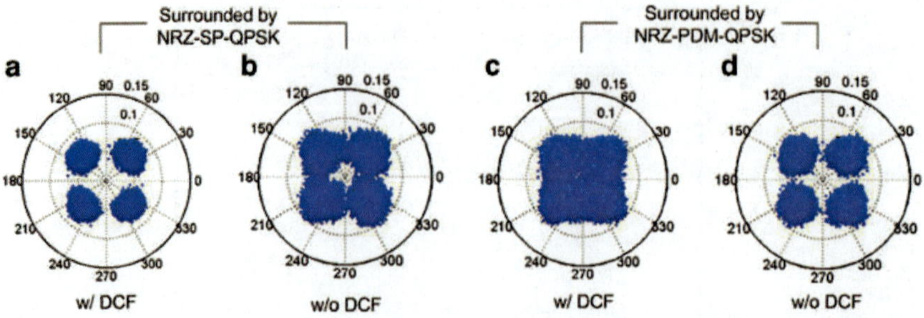

Fig. 9.7 Signal constellation diagrams of one polarization of a 42.8-Gb/s NRZ-PDM-QPSK reference channel after 1,000-km WDM transmission at OSNR = 16 dB. (**a**) and (**b**): surrounding channels are 21.4-Gb/s NRZ-SP-QPSK, (**c**) and (**d**): surrounding channels are 42.8-Gb/s NRZ-PDM-QPSK. (**a**) and (**c**) for the system with DCF, and (**b**) and (**d**) without DCF. The launch power per channel is 4 dBm

transmission [30]. ASE noise is loaded at the receiver to generate 16-dB OSNR. The results of different system configurations are given: with and without inline DCF, with NRZ-SP-QPSK and NRZ-PDM-QPSK surrounding channels. A launch power of 4-dBm per channel is used for all the configurations. It shows that when the NRZ-PDM-QPSK channel is surrounded by 21.4-Gb/s NRZ-SP-QPSK channels, the system with DCF has a much clearer signal constellation than that without DCF, as shown in Figs. 9.7a, b. However, when the surrounding channels are 42.8-Gb/s NRZ-PDM-QPSK signals, the system with DCF performs much worse than that without DCF, as shown in Figs. 9.7c and 9.7d. Results in Figs. 9.5 and 9.7 show that the nonlinear polarization scattering caused by other PDM-QPSK channels is much larger in the system with inline DCF than that without DCF, which generates severe crosstalk between the two polarizations in the system with inline DCF and makes the NRZ-PDM-QPSK system with DCF perform worse than the system without DCF. We note that Fig. 9.7d has a clearer constellation than Fig. 9.7b. This is due to the reduced peak power for a PDM-QPSK signal compared with an SP-QPSK signal for a given average power.

9.3.3 112-Gb/s PDM-QPSK Systems

The transmission performance of a 112-Gb/s NRZ-PDM-QPSK reference channel surrounded by six 56-Gb/s NRZ-SP-QPSK channels and six 112-Gb/s NRZ-PDM-QPSK channels are given in Fig. 9.8. Because of a higher symbol rate, compared to the 42.8-Gb/s PDM-QPSK system, the interchannel nonlinearities of the 112-Gb/s PDM-QPSK system is smaller as 112-Gb/s PDM-QPSK signals are dispersed faster due to chromatic dispersion than 42.8-Gb/s PDM-QPSK signals. Therefore, for 112-Gb/s NRZ-PDM-QPSK signals, the difference between the transmission system with inline DCF and that without inline DCF is smaller. Similar to the 42.8-Gb/s

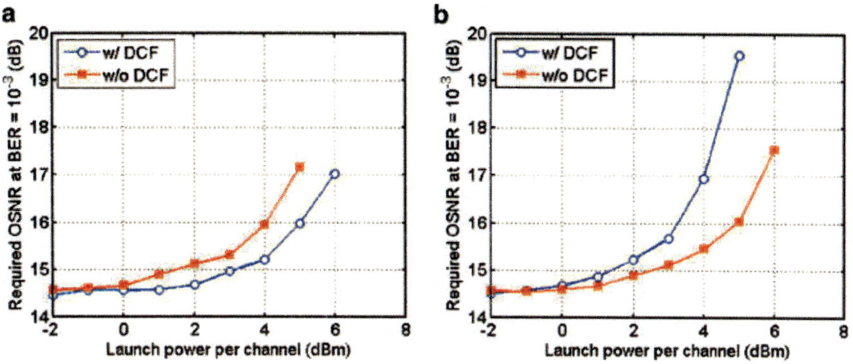

Fig. 9.8 Required OSNR at BER of 10^{-3} after 1,000-km transmission vs. launch power per channel for the 112-Gb/s NRZ-PDM-QPSK coherent system with and without inline DCF. (**a**) the surrounding six channels are 56-Gb/s NRZ-SP-QPSK signals, (**b**) the surrounding six channels are 112-Gb/s NRZ-PDM-QPSK signals

system, when the surrounding channels are 56-Gb/s NRZ-SP-QPSK channels, dispersion management increases the nonlinearity tolerance. The system with inline DCF can tolerate about 1-dB more launch power than that without inline DCF. But XPolM-induced nonlinear polarization scattering from the neighboring 112-Gb/s NRZ-PDM-QPSK channels eliminates the benefits of dispersion management and reduces the nonlinearity tolerance for the dispersion-managed system. As shown in Fig. 9.8, at 1-dB OSNR penalty, if the neighboring channels are 112-Gb/s NRZ-PDM-QPSK signals, the allowed launch power for the system with inline DCF is about 1-dB less than that for the system without inline DCF.

Figure 9.9 depicts the nonlinear polarization scattering induced depolarization in the 112-Gb/s PDM-QPSK system with and without inline DCF, which is quantified by the DOP of a 56-Gb/s NRZ-SP-QPSK reference channel surrounded by six 112-Gb/s NRZ-PDM-QPSK channels with 50-GHz channel spacing in the transmission system. As expected, the nonlinear polarization scattering in the system without inline DCF is smaller than that with inline DCF. Comparison with Fig. 9.5 shows that the depolarization caused by the nonlinear polarization scattering in the 112-Gb/s PDM-QPSK system is smaller than that in the 42.8-Gb/s system, especially for the system with inline DCF. As explained above, the increased symbol rate reduces the interchannel nonlinearities, including nonlinear polarization scattering.

Figure 9.10 gives the dependence of nonlinear polarization scattering-induced depolarization on dispersion maps in the 112-Gb/s WDM system [31]. The contour plot of DOP of a 56-Gb/s NRZ-SP-QPSK channel surrounded by six 112-Gb/s NRZ-PDM-QPSK channels with 50-GHz channel spacing vs. dispersion precompensation and RDPS is depicted in the figure. It shows that with the increase of RDPS, the nonlinear polarization scattering decreases. It also shows that the nonlinear polarization scattering does not have a strong dependence on dispersion

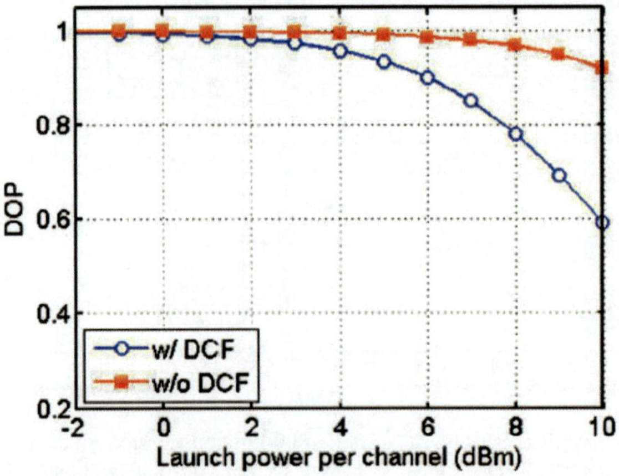

Fig. 9.9 DOP of the 56-Gb/s SP-QPSK reference channel after 1,000-km transmission vs. launch power per channel in the system with and without inline DCF. Surrounding channels are 112-Gb/s NRZ-PDM-QPSK signals

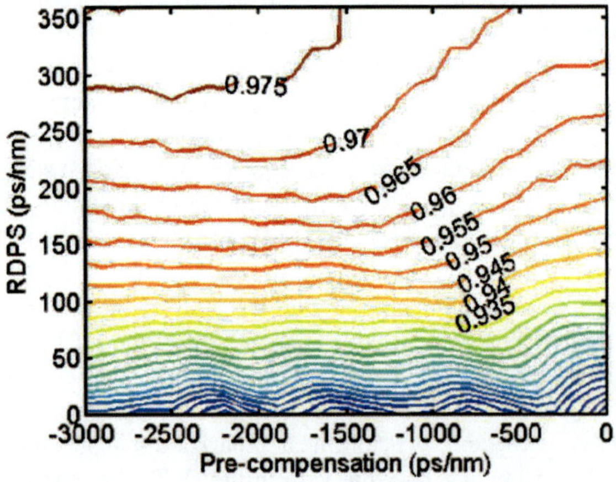

Fig. 9.10 Contour plot of DOP of a 56-Gb/s NRZ-SP-QPSK reference channel after 1,000-km transmission vs. dispersion precompensation and RDPS. The surrounding channels are 112-Gb/s NRZ-PDM-QPSK. The launch power per channel is 6 dBm

precompensation. This is different from interchannel XPM and intrachannel nonlinearities. It is well known that lumped dispersion compensation at the transmitter or receiver is suboptimal for interchannel XPM and intrachannel nonlinearities compared with dispersion management, which distributes DCMs along a transmission link, with dispersion precompensation and postcompensation at the transmitter and receiver. Figure 9.10 confirms that it is the nonlinear polarization scattering that changes the perspective of dispersion management in PDM coherent systems.

9.3.4 Hybrid OOK and PDM-QPSK Systems

Many of current optical communication networks carry 10-Gb/s on-off-keying (OOK) signals and use dispersion-managed links to reduce the impact of chromatic dispersion and fiber nonlinearities. PDM coherent technology is a promising candidate to upgrade existing 10-Gb/s WDM systems with 50-GHz channel spacing to 40-Gb/s and 100-Gb/s per channel bit rates. In such systems, 10-Gb/s OOK signals may coexist with 40-Gb/s and 100-Gb/s PDM-QPSK signals. It has been shown that the performance of 40-Gb/s and 100-Gb/s PDM-QPSK coherent channels can be significantly degraded by interchannel nonlinearities from co-propagating 10-Gb/s OOK channels in such hybrid systems [45–47].

The impact of 10-Gb/s OOK channels on the performance of 42.8-Gb/s and 112-Gb/s PDM-QPSK channels in the dispersion-managed systems is shown in Fig. 9.11 [47]. In the figure, the same system parameters as those in Fig. 9.3 are used except that the six surrounding channels are replaced by 10-Gb/s NRZ-OOK channels. It shows that the presence of the 10-Gb/s OOK neighboring channels significantly degrades the performance of both the 42.8-Gb/s and 112-Gb/s PDM-QPSK channels. For comparison, the results of the systems with all PDM-QPSK channels are also given in the figure. The presence of 10-Gb/s OOK channels reduces the allowed launch power by about 5 dB at 1-dB OSNR penalty compared to that with all PDM-QPSK channels. It means that, in the hybrid systems, for the PDM-QPSK channel to achieve the similar performance as that in the system without the OOK channels, the launch power of the 10-Gb/s OOK channels has to be reduced by 5 dB.

In these hybrid 10-Gb/s OOK, 42.8-Gb/s and 112-Gb/s PDM-QPSK systems, the dominant nonlinear effect is interchannel XPM from 10-Gb/s OOK channels, not XPolM, which is clearly illustrated by Fig. 9.12. The figure shows the DOP of a 21.4-Gb/s and 56-Gb/s NRZ-SP-QPSK channel co-propagating with

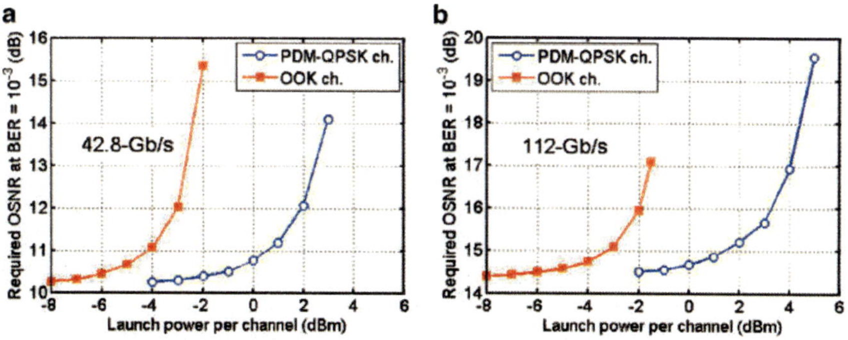

Fig. 9.11 Required OSNR at BER of 10^{-3} after 1,000-km transmission of a 42.8-Gb/s and 112-Gb/s PDM-QPSK channel co-propagating with neighboring six 10-Gb/s OOK channels or six PDM-QPSK channels in the dispersion-managed systems. (**a**) 42.8-Gb/s PDM-QPSK, (**b**) 112-Gb/s PDM-QPSK

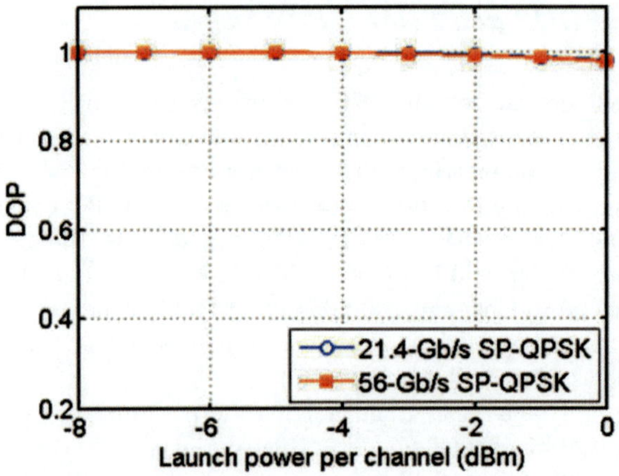

Fig. 9.12 DOP of a 21.4-Gb/s and 56-Gb/s SP-QPSK reference channel co-propagating with six 10-Gb/s NRZ-OOK channels after 1,000-km vs. launch power per channel in the dispersion-managed transmission system

six 10-Gb/s NRZ-OOK channels after 1,000-km transmission. The SOP of the SP-QPSK channel is set to be perpendicular to that of all the OOK channels in the Stokes space, which generates maximum XPolM, as indicated in (9.8) and (9.9). The OOK channels cause similar depolarization for both the 21.4-Gb/s and 56-Gb/s SP-QPSK channel, as expected. Figure 9.12 shows that when the launch power per channel is about 0 dBm, the DOP is still high, about 0.98. However, at −1-dBm per channel launch power, the OOK channels already induce more than 3-dB penalty on both the 42.8-Gb/s and the 112-Gb/s channels, as shown in Fig. 9.11. The reason why XPM is larger than XPolM is that an OOK signal does not have constant amplitude at each bit, whereas for PDM-QPSK signals, the amplitude at each symbol is almost constant in dispersion-managed systems.

9.4 Nonlinear Polarization Scattering Mitigation Techniques

As shown in the above section, except for the hybrid OOK and PDM-QPSK systems, nonlinear polarization scattering is the dominant nonlinear effect in dispersion-managed PDM coherent optical communication systems. Therefore, reducing nonlinear polarization scattering in dispersion-managed PDM coherent optical communication systems could significantly increase the system performance and transmission distances. Nonlinear polarization scattering in the system without any inline DCF is small as the large walk-off between channels and rapid changes of SOP caused by large chromatic dispersion accumulation in the transmission average out the XPolM effect. In this section, we will describe techniques to mitigate nonlinear polarization scattering in dispersion-managed PDM-QPSK systems.

The results in the above section also indicate that nonlinear polarization scattering is affected by the data-dependent SOP of a PDM signal and the walk-off between channels. Therefore, techniques that can reduce the data-dependent SOP of a signal and increase the walk-off between channels can be used to mitigate nonlinear polarization scattering in PDM transmission systems. In this section, we will discuss three nonlinear polarization scattering mitigation techniques. The first technique is the use of time-interleaved return-to-zero PDM (ILRZ-PDM) modulation formats (which is also called iRZ in other literatures) [29, 30, 48–50], the second technique is the use of PGD devices as inline dispersion compensators [47], and the third technique is the judicious addition of some PMD in the transmission link [51].

9.4.1 Time Interleaved RZ-PDM Modulation Format

For an NRZ-PDM-QPSK signal, the SOPs at different symbols change among four points on the Poincaré sphere, depending on the data carried by the two polarizations, as shown in Fig. 9.2. In a dispersion-managed system with inline DCF, the pulses suffer minimally from chromatic dispersion accumulation, and the SOPs of a PDM-QPSK signal remain nearly fixed to these four points after each span. In addition, there is small walk-off between channels due to low RDPS. The few data-dependent SOPs and small walk-off between channels increase nonlinear polarization scattering in a dispersion-managed system.

One technique to suppress nonlinear polarization scattering is to use ILRZ-PDM modulation format, which can reduce or eliminate the dependence of SOP on the data carried by the two polarizations. This modulation format uses RZ pulses and time interleaves the two polarizations by half a symbol period. The waveform and SOP diagram of ILRZ-PDM-QPSK are depicted in Fig. 9.13. We can see that at the center of each symbol, the SOP is either at S_1 or $-S_1$ on the Poincaré sphere, and it does not depend on data carried by the two polarizations. In addition, an ILRZ-PDM

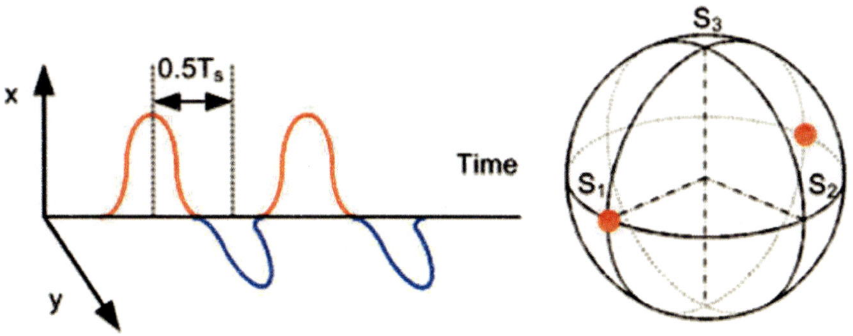

Fig. 9.13 Waveform and SOP diagram of ILRZ-PDM-QPSK. T_s: symbol period

signal has other two features that help reduce nonlinear polarization scattering in a dispersion-managed system: (1) the SOP at each symbol alternates between S_1 and $-S_1$ on the Poincaré sphere, the SOP at S_1 and $-S_1$ causes opposite nonlinear polarization rotation according to (9.8) and (9.9); and (2) the time interleaving reduces the signal peak power, leading to reduced XPolM between channels [52]. An ILRZ-PDM signal can be generated by adding one pulse carver before the data modulators and setting proper time delay between the two polarizations before the PBC in the transmitter. Note that time-interleaving an NRZ-PDM signal does not provide much benefit, as none of the above features for an ILRZ-PDM signal can be obtained for a time-interleaved NRZ-PDM signal.

In the following, we will describe the performance of the ILRZ-PDM modulation format for both coherent and direct detection systems.

9.4.1.1 Coherent ILRZ-PDM-QPSK Systems

The transmission performance of 42.8-Gb/s and 112-Gb/s ILRZ-PDM-QPSK WDM systems is given in Fig. 9.14, which shows the required OSNR at a BER of 10^{-3} after 1,000-km transmission for the system with and without inline DCF [30]. The RZ pulses used here have 50% duty cycle. For the 42.8-Gb/s system with inline DCF, using ILRZ-PDM-QPSK can increase the allowed launch power by 7 dB at 1-dB OSNR penalty compared to NRZ-PDM-QPSK (Fig. 9.4), from about 1-dBm per channel launch power to about 8 dBm. For the system without inline DCF, the performance of ILRZ-PDM-QPSK and NRZ-PDM-QPSK is similar. With ILRZ-PDM-QPSK, the 42.8-Gb/s system with inline DCF performs better than that without DCF, with the tolerable launch power about 4-dB higher. For the 112-Gb/s system, the improvement obtained by using ILRZ-PDM-QPSK is smaller than that for the 42.8-Gb/s system due to the symbol rate increase, but it can still increase the launch power tolerance by about 3 dB compared to NRZ-PDM-QPSK. Figure 9.14b shows that with ILRZ-PDM-QPSK, the 112-Gb/s system with inline DCF can achieve similar performance to the system without DCF. The less improvement from using ILRZ-PDM-QPSK in the 112-Gb/s system compared to the 42.8-Gb/s system is due to the fact that the interchannel nonlinearity including XPolM in the 112-Gb/s system is smaller than that in the 42.8-Gb/s system. Figure 9.14 also shows for both 42.8-Gb/s and 112-Gb/s system without inline DCF, there is a slight improvement on nonlinearity tolerance if ILRZ-PDM-QPSK is used.

The level of the nonlinear polarization scattering of the systems using ILRZ-PDM-QPSK is given in Fig. 9.15. It clearly shows that using ILRZ-PDM-QPSK significantly reduces nonlinear polarization scattering in both the 42.8-Gb/s and 112-Gb/s systems with inline DCF. Compared with NRZ-PDM-QPSK, at 6-dBm launch power the ILRZ-PDM-QPSK modulation format increases the nonlinear polarization scattering induced DOP reduction of the reference channel from about 0.75 to 0.96 and from 0.90 to 0.95 for the dispersion-managed 42.8-Gb/s and

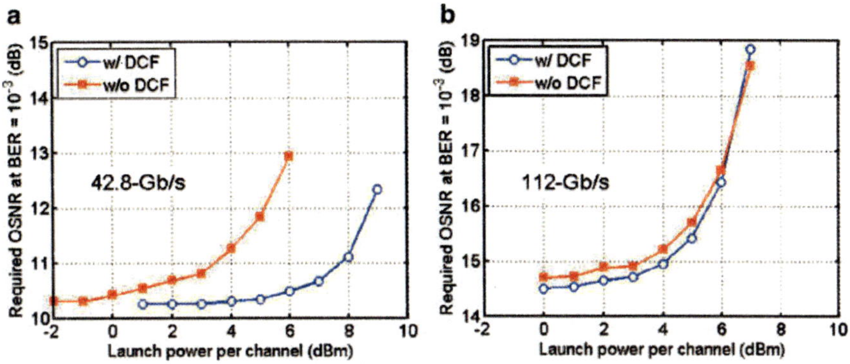

Fig. 9.14 Required OSNR at BER of 10^{-3} after 1,000-km transmission vs. launch power per channel for the 42.8-Gb/s and 112-Gb/s ILRZ-PDM-QPSK WDM coherent systems with and without inline DCF

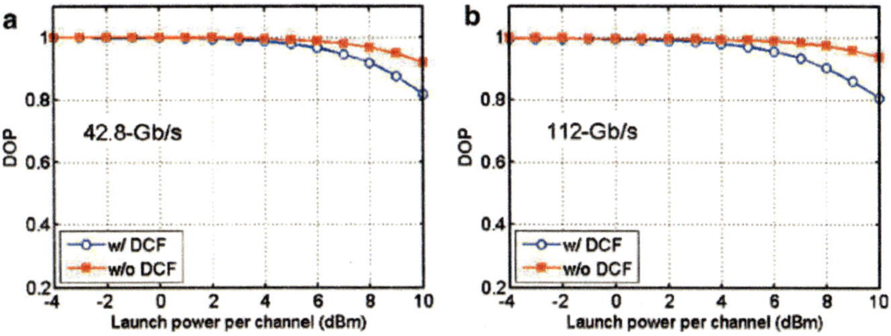

Fig. 9.15 DOP of a 21.4-Gb/s and 56-Gb/s SP-QPSK reference channels after 1,000-km transmission vs. launch power per channel in the 42.8-Gb/s and 112-Gb/s ILRZ-PDM-QPSK WDM systems with and without inline DCF

112-Gb/s system, respectively. Compared with Figs. 9.5 and 9.9, we can see that there is a slight reduction in nonlinear polarization scattering even for the system without inline DCF when ILRZ-PDM-QPSK is used.

9.4.1.2 Direct-Detection ILRZ-PDM Systems

The suppression of nonlinear polarization scattering by using the ILRZ-PDM modulation format was demonstrated with experiments using direct-detection [53]. In the experiment, the transmission performance of ILRZ-PDM differential-QPSK (DQPSK), ILRZ-PDM differential-binary-phase-shift-keying (DBPSK), and ILRZ-PDM-OOK signals was studied and compared with the corresponding time-synchronized RZ-PDM signals. The experimental setup is shown in Fig. 9.16. Thirty-two DFB lasers with 50-GHz channel spacing ranging from 1562.23 nm to

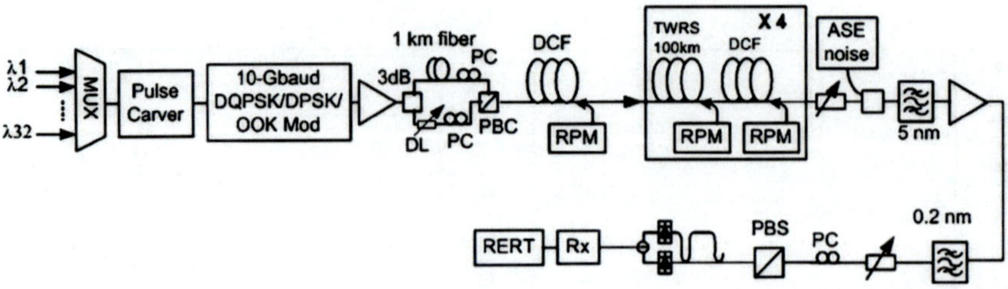

Fig. 9.16 Schematic of the experimental setup for PDM transmission using direct detection. *DL* Delay line; *PC* Polarization controller; *PBC(S)* Polarization beam combiner (splitter); *RPM* Raman pump module; *Rx* Receiver; *BERT* Bit error rate tester

1574.54 nm were combined with a multiplexer and sent to a pulse carver to generate 50% RZ pulses. The RZ pulses were modulated with $2^{15} - 1$ pseudo-random bit sequence electrical signal by different modulators to produce 10-Gbaud DQPSK, DBPSK or OOK signals. The signal was then amplified by an EDFA and split into two paths with a 3-dB coupler and recombined in a PBC to form a PDM signal. A tunable delay line was inserted in one path to make the signals in the two polarizations time synchronized or interleaved. Transmission was performed in a four-span all-Raman amplified straight line system. A spool of DCF with −300 ps/nm chromatic dispersion was used as pre-compensation. Each span consisted of 100-km Truewave Reduced Slope fiber and DCF with RDPS of 30 ps/nm. Both the transmission fiber and DCF were backward pumped, and the input power to the DCF was about 2 dB lower than that to the transmission fibers. After transmission, the signal was loaded with ASE noise to get a certain OSNR. The reference channel at wavelength of 1567.91 was selected with a 0.2-nm tunable grating filter. A manual polarization controller and PBS were used to separate the two polarizations. The signal after the PBS was sent to a receiver and BER was measured with a BER tester. Balanced detectors were used for the DQPSK and DBPSK receivers.

The OSNR penalty of the 10-Gbaud time-synchronized and time-interleaved RZ-PDM-DQPSK system after transmission is given in Fig. 9.17a. The figure shows that the ILRZ-PDM signal has much higher tolerance to fiber nonlinearity than the synchronized one. At 1-dB OSNR penalty, the allowed launch power for the ILRZ-PDM-DQPSK signal is about 3 dB higher than that for the synchronized one. To estimate the level of the nonlinear polarization scattering, we left the reference channel unmodulated (CW signal) but the other channels still carrying PDM-DQPSK signals, and measured DOP of the reference channel at a given OSNR of 22 dB. As shown in Fig. 9.17b, the DOP of the CW channel in the system with ILRZ-PDM-DQPSK decreases much more slowly with the launch power than that with synchronized RZ-PDM-DQPSK, indicating that the nonlinear polarization scattering is reduced in the system using ILRZ-PDM-DQPSK. As shown in insets of Fig. 9.17a, with −6-dBm per channel launch power, the eye-diagrams of the synchronized RZ-PDM-DQPSK and ILRZ-PDM-DQPSK after PBS are similar, but when the launch power is increased to −1 dBm, there is a large crosstalk induced by nonlinear polarization scattering in the synchronized RZ-PDM-DQPSK signal.

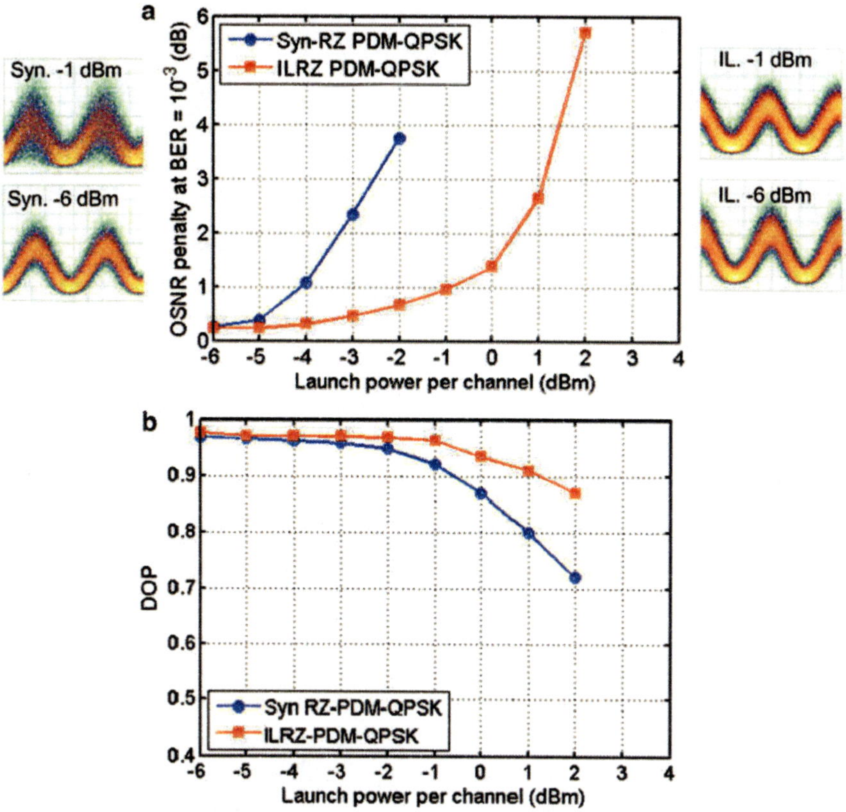

Fig. 9.17 (**a**) OSNR penalty at BER $= 10^{-3}$ vs. launch power for 10-Gbaud synchronized RZ-PDM-DQPSK and ILRZ-PDM-DQPSK signals, the insets are eye-diagrams for the Syn- and ILRZ-PDM-DQPSK signals, (**b**) DOP of the CW channel vs. launch power at OSNR of 22 dB in the system with synchronized RZ-PDM-DQPSK and ILRZ-PDM-DQPSK channels

The transmission performance of the 10-Gbaud synchronized RZ-PDM-DBPSK and ILRZ-PDM-DBPSK is given in Fig. 9.18. The nonlinear tolerance of the ILRZ-PDM-DBPSK is about 3 dB higher than that of the synchronized RZ-PDM-DBPSK. As expected, the DOP of the CW channel in the system with ILRZ-PDM-DBPSK decreases slower than that with synchronized RZ-PDM-DBPSK, as shown in Fig. 9.18b.

Although RZ-OOK does not have a constant amplitude, which means that the SOP of ILRZ-PDM-OOK does not consecutively alternate between opposite points on the Poincaré sphere (there are no pulses on "0" bits), significant improvement in the nonlinearity tolerance can still be obtained by time interleaving an RZ-PDM-OOK signal, as shown in Fig. 9.19. By using ILRZ-PDM-OOK, the nonlinear tolerance of the 10-Gbaud PDM-OOK system can be increased by 3–4 dB. The DOP of the CW channel in the system with PDM-OOK is similar to that with

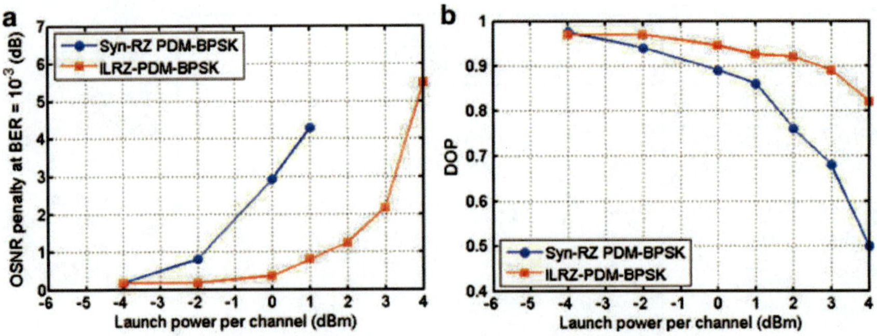

Fig. 9.18 (**a**) OSNR penalty at BER = 10^{-3} vs. launch power for 10-Gbaud synchronized RZ-PDM-DBPSK and ILRZ-PDM-DBPSK signals, (**b**) DOP of the CW channel vs. launch power at OSNR of 22 dB in the system with synchronized RZ-PDM-DBPSK and ILRZ-PDM-DBPSK channels

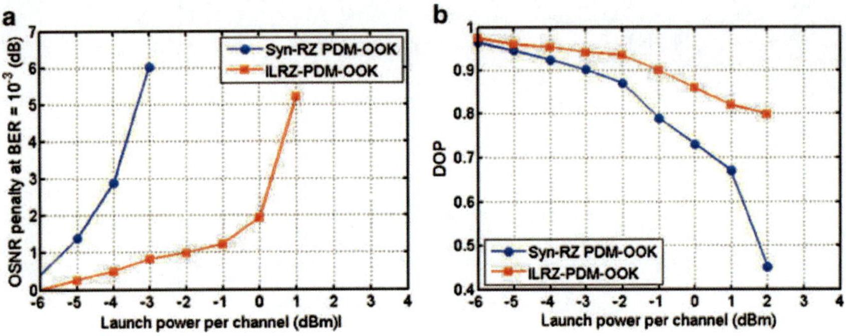

Fig. 9.19 (**a**) OSNR penalty at BER = 10^{-3} vs. launch power for 10-Gbaud synchronized RZ-PDM-OOK and ILRZ-PDM-OOK signals, (**b**) DOP of the CW channel vs. launch power at OSNR of 22 dB in the system with synchronized RZ-PDM-OOK and ILRZ-PDM-OOK channels

PDM-DQPSK and PDM-DBPSK, i.e., using ILRZ-PDM-OOK signals significantly reduces the nonlinear polarization scattering compared to that using synchronized RZ-PDM-OOK signals.

One question for the ILRZ-PDM modulation format is whether PMD could ruin the benefits of its high tolerance to fiber nonlinearities, as PMD in the system may change an ILRZ-PDM signal to a synchronized RZ-PDM signal. One experimental result showed that the nonlinearity tolerance benefit of ILRZ-PDM signals vanished when a PMD emulator with high PMD value was added at the transmitter [54]. Note that putting a PMD emulator at the transmitter is not the correct way to evaluate PMD impact on the nonlinear transmission performance of the ILRZ-PDM modulation format. In a real system, PMD is distributed in the transmission link, and in addition, PMD itself depolarizes PDM signals at each polarization and causes walk-off between the two polarizations in propagation, which is helpful to reduce the

XPolM (will be discussed Sect. 9.4.3). These effects do not exist if a PMD emulator is added at the transmitter. We have observed that the ILRZ-PDM modulation format does not lose its benefits on nonlinearity tolerance in the presence of PMD.

9.4.2 PGD Dispersion Compensators

XPolM is also affected by the walk-off between channels. Large walk-off between channels tends to induce small XPolM, as shown in Fig. 9.10. In a dispersion-managed system with DCF, for a given channel spacing, large walk-off can only be achieved by increasing RDPS. However, increasing RDPS in a dispersion-managed system with DCF also increases amplitude variations of the signal in each channel, which could enhance intrachannel nonlinearities and interchannel XPM. One technique to increase the walk-off between channels without affecting the signal variations within channels is to use PGD devices as inline dispersion compensators [55].

Figure 9.20 plots the relation of group delay with frequency of an ideal PGD dispersion compensator with −1,700-ps/nm chromatic dispersion and 50-GHz period. As shown in the figure, the group delay of a PGD chromatic dispersion compensator is periodic. If the period of the group delay is the same as the channel spacing in a WDM system, the mean group delay for each channel is the same, but within each channel, the group delay of a PGD dispersion compensator is the same as that of a DCF and can compensate the dispersion in each channel. This means that within a channel, a PGD chromatic dispersion compensator performs chromatic dispersion compensation in a transmission link as DCF, but it induces little walk-off between channels. Unlike in a dispersion-managed system using DCF, data patterns carried by different WDM channels in a dispersion-managed system using PGD dispersion compensation modules (DCMs) pass through each other in the transmission fiber and are not brought back to overlap again at the PGD-DCM. Therefore, the pattern walk-off in a dispersion-managed system with PGD-DCM is the same as that in the system without any inline DCM.

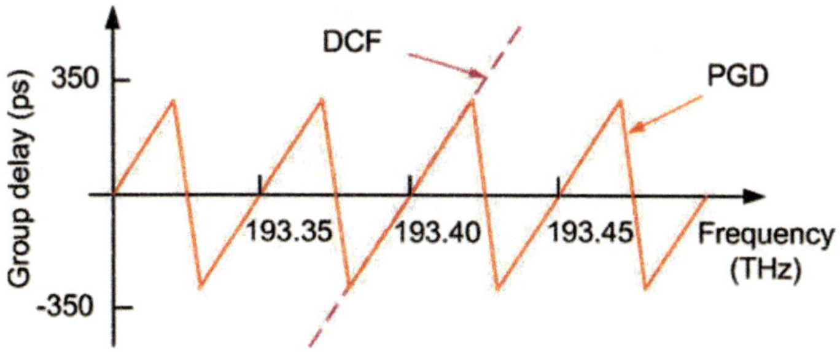

Fig. 9.20 Group delay of an ideal PGD dispersion compensator designed for a channel spacing of 50 GHz (0.4 nm) and with about −1,700-ps/nm chromatic dispersion within a channel. The *dashed line* is the group delay for a DCF

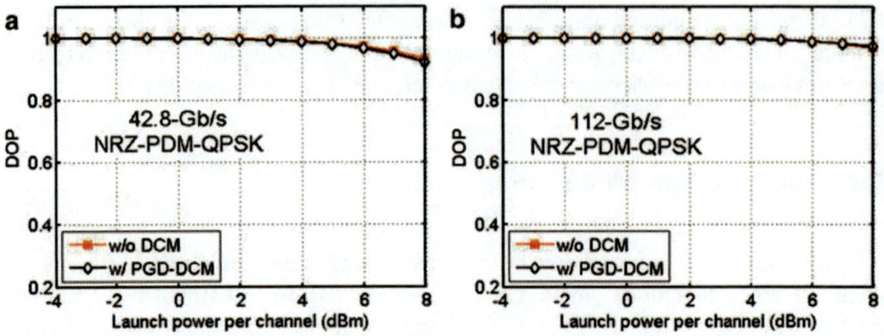

Fig. 9.21 DOP of a 21.4-Gb/s and 56-Gb/s SP-QPSK reference channels after 1,000-km transmission vs. launch power per channel in the 42.8-Gb/s and 112-Gb/s NRZ-PDM-QPSK WDM systems with PGD-DCM and those without DCM

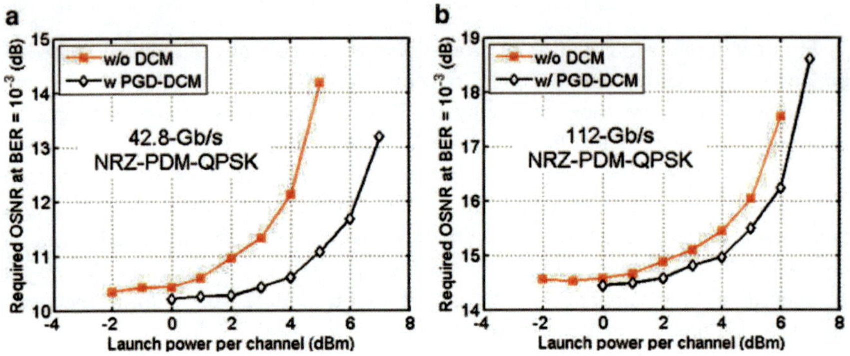

Fig. 9.22 Required OSNR at BER of 10^{-3} after 1,000-km transmission vs. launch power per channel for the 42.8-Gb/s and 112-Gb/s NRZ-PDM-QPSK WDM coherent systems with PGD-DCM and those without DCM

The performance of the 42.8-Gb/s and 112-Gb/s PDM-QPSK WDM dispersion-managed systems using PGD-DCM is shown in Figs. 9.21 and 9.22 [47]. The same system parameters as that in Fig. 9.3 are used except that the inline DCF in the system is replaced with PGD-DCM. NRZ-PDM-QPSK is used in Figs. 9.21 and 9.22. Figure 9.21 plots the nonlinear polarization scattering induced depolarization in the 42.8-Gb/s and 112-Gb/s NRZ-PDM-QPSK dispersion-managed system with PGD-DCM and in the system without dispersion management. It shows that the depolarization caused by nonlinear polarization scattering in the dispersion-managed transmission using PGD-DCM is similar to that in the system without any dispersion management for both 42.8-Gb/s and 112-Gb/s systems. Figure 9.22 compares the required OSNR at BER of 10^{-3} after 1,000-km transmission vs. launch power per channel between the dispersion-managed system with PGD-DCM and that without dispersion management. It shows that for both the 42.8-Gb/s and 112-Gb/s NRZ-PDM-QPSK WDM transmission, the dispersion-managed system using PGD-DCM has higher nonlinearity tolerance than the system without any DCM.

The PGD-DCM can be combined with ILRZ-PDM modulation to further suppress nonlinear polarization scattering and increase the nonlinear tolerance of PDM WDM systems. In addition, using PGD-DCM can also suppress the interchannel XPM from 10-Gb/s OOK channels in hybrid OOK and PDM-QPSK systems and significantly increase the transmission distance of PDM-QPSK coherent channels in the hybrid systems [47].

9.4.3 *Adding PMD into the System*

PMD effects in general are detrimental to fiber-optic transmission systems and have long been considered as one of the obstacles that limit the reach and bit rates of optical communication systems using direct detection [13–16]. There are also some special cases where PMD effects are potentially useful. For examples, PMD was used to predistort the signals at the transmitter to reduce intrachannel nonlinearities in pseudo-linear transmission systems [56], and it was also shown that PMD can reduce the PDL-induced fading in optical orthogonal frequency division multiplexing (OFDM) systems [57].

PMD causes the depolarization of signals carried by each polarization, and it also introduces decorrelation between two polarizations for PDM signals during transmission. These effects are helpful to reduce interchannel nonlinearities including XPolM in PDM transmission systems. As the linear PMD effects can be easily compensated by digital signal processing in coherent receivers, adding some PMD in transmission links should be able to mitigate inter-channel nonlinear effects in PDM coherent transmission systems.

This idea was demonstrated by Serena et al. with numerical simulations [51]. They simulated the transmission performance of a nine-channel 112-Gb/s NRZ-PDM-QPSK WDM transmission system. The channel spacing was 50 GHz. The transmission link consisted of 20 SSMF spans with 100-km span length. The attenuation and nonlinear coefficient of the SSMF used in the system were 0.2 dB/km and 1.51 $(\mathrm{km\ W})^{-1}$, respectively. The attenuation in each span was compensated by an EFDA with 7-dB noise figure. Different amounts of PMD were added into the system to evaluate the impact of PMD on the system performance, and PMD was distributed in the transmission link.

The impact of PMD on the transmission performance is shown in Fig. 9.23, which depicts the Q-factor of the middle channel vs. launch power per channel with different PMD values in the system averaged more than 40 different realizations of PMD in the link. The Q-factor is converted from BER, which is calculated through the Monte Carlo simulation by the error counting method. In the simulations, propagation is noiseless, and ASE noise is added at the receiver. A few points are checked with ASE noise added inline, as shown by a few triangles in the figure. The figure

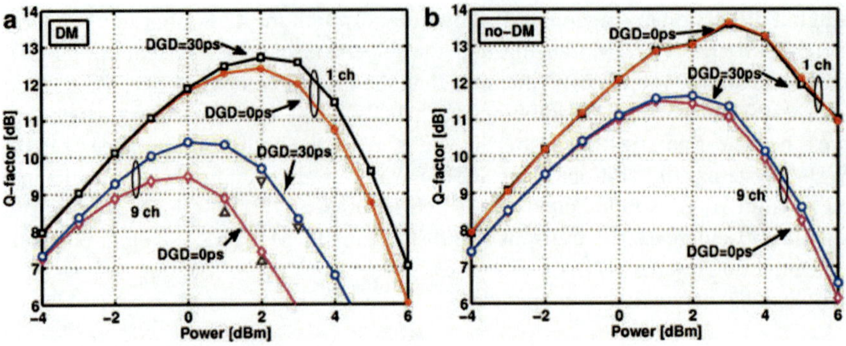

Fig. 9.23 Q-factor vs. launch power per channel in dispersion-managed (DM) and nondispersion-managed (non-DM) 112-Gb/s PDM-QPSK transmission systems with different amount of PMD. *Triangles* are the simulations with inline noise (Courtesy of P. Serena et al. [51])

shows that when the launch power is low, the system performance is limited by ASE noise, while the power is high, it is limited by fiber nonlinearities. However, for the dispersion-managed system, adding some PMD improves the performance in both the single channel and the WDM cases. With 30-ps average DGD, the Q factor in the single channel case can be improved by 0.4 dB, and in the WDM case the Q factor improvement is about 1 dB. The reason of the improvement in presence of PMD in the nonlinear regime is that both intrachannel interactions between the X and Y components and interchannel XPolM between channels are reduced by the walk-off and depolarization introduced by PMD. Note that at low power, DGD does not affect the performance as the system performance is limited by ASE noise in this regime, not nonlinearities. For the nondispersion-managed system, the impact of DGD is small as the large walk-off and rapid variations of SOP mask the PMD effects, which is in agreement with the results in previous sections.

9.5 Conclusion

Dispersion management has been successfully used in direct-detection optical communication systems. This technique not only effectively reduces intrachannel and interchannel nonlinear impairments, but also makes it possible to add and drop signals everywhere in such optical systems, which is essential for optical mesh networks. Optical coherent receivers with sophisticated digital signal processing have the ability to compensate a large amount of chromatic dispersion entirely in the electrical domain, which make it possible to completely eliminate optical dispersion compensation in the systems and at the same time access signals everywhere in the networks. It has been shown that optical PDM coherent communication systems with dispersion management can perform worse than those without dispersion management. In this chapter, we showed that it is the addition of the other polarization that eliminates the advantages of dispersion management in such systems.

The addition of the other polarization enhances nonlinear polarization scattering, which becomes the dominant nonlinear effect in dispersion-managed PDM coherent transmission systems. We have shown that for both 42.8-Gb/s and 112-Gb/s NRZ-PDM-QPSK coherent systems, due to nonlinear polarization scattering, no benefit in nonlinearity tolerance can be obtained by using dispersion management. A few techniques to suppress nonlinear polarization scattering in dispersion-managed PDM coherent transmission systems were described, including the use of the ILRZ-PDM modulation format, the use of PGD dispersion compensators as inline DCMs, and the judicious addition of some PMD in the transmission links. We showed that these techniques can significantly increase the performance of PDM-QPSK coherent systems with dispersion management. While in this chapter only PDM-QPSK modulation format was used for analysis and discussion, the results obtained here could be applicable to other PDM modulation formats, such as PDM-8PSK and PDM-16QAM.

References

1. P.M. Hill, R. Olshansky, W.K. Burns, IEEE Photon. Technol. Lett. **4**, 500–502 (1992)
2. S.G. Evangelides, L.F. Mollenauer, J.P. Gordon, N.S. Bergano, J. Lightwave Technol. **10**, 28–35 (1992)
3. A.R. Chraplyvy, A.H. Gnauck, R.W. Tkach, J.L. Zyskind, J.W. Sulhoff, A.J. Lucero, Y. Sun, R.M. Jopson, F. Forghieri, R.M. Derosier, C. Wolf, A.R. McCormick, IEEE Photon. Technol. Lett. **8**, 1264–1266 (1996)
4. A.H. Gnauck, G. Charlet, P. Tran, P.J. Winzer, C.R. Doerr, J.C. Centanni, E.C. Burrows, T. Kawanishi, T. Sakamoto, K. Higuma, J. Lightwave Technol. **26**, 79–84 (2008)
5. S.J. Savory, A.D. Stewart, S. Wood, G. Gavioli, M.G. Taylor, R.I. Killey, P. Bayvel, Digital equalisation of 40Gbit/s per wavelength transmission over 2480 km of standard fibre without optical dispersion compensation, in *Proceedings of European conference on optical communications 2006*, Cannes, France, Paper Th2.5.5, September 2006
6. C. Laperle, B. Villeneuve, Z. Zhang, D. McGhan, H. Sun, M. O'Sullivan Wavelength division multiplexing (WDM) and polarization mode dispersion (PMD) performance of a coherent 40Gbit/s dual-polarization quardrature phase shift keying (DP-QPSK) transceiver, in *Proceedings of optical fiber communication conference 2007*, Paper PDP16, Anaheim, CA, USA, March 2007
7. H. Sun, K.T. Wu, K. Roberts, Express 16, 873–879 (2008)
8. M. Salsi, H. Mardoyan, P. Tran, C. Koebele, E. Dutisseuil, G. Charlet, S. Bigo, 155×100 Gbit/s coherent PDM-QPSK transmission over 7,200 km, in *Proceedngs of European conference on optical communications 2009*, Vienna, Austria, Paper PD2.5, September 2009
9. G. Charlet, J. Renaudier, M. Salsi, H. Mardoyan, P. Tran, S. Bigo Efficient mitigation of fiber impairments in an ultra-long haul transmission of 40 Gbit/s polarization-multiplexed data, by digital processing in a coherent receiver, in *Proceedings of optical fiber communication conference 2007*, Paper PDP17, Anaheim, CA, USA, March 2007
10. H. Wernz, S. Bayer, B.E. Olsson, M. Camera, H. Griesser, C. Fuerst, B. Koch, V. Mirvoda, A. Hidayat, R. Noé 112 Gb/s PolMux RZ-DQPSK with fast polarization tracking based on interference control, in *Proceedings of optical fiber communication conference 2009*, Paper OTuN4, San Diego, CA, USA, March 2009
11. Z. Wang, C. Xie, Opt. Express **17**, 3183–3189 (2009)
12. H. Wernz, S. Herbst, S. Bayer, H. Griesser, E. Martins, C. Fürst, B. Koch, V. Mirvoda, R. Noé, A. Ehrhardt, L. Schürer, S. Vorbeck, M. Schneiders, D. Breuer, R.P. Braun, Nonlinear

behaviour of 112 Gb/s polarisation-multiplexed RZ-DQPSK with direct detection in a 630 km field trial, in *Proceedings of European conference on optical communications 2009*, Vienna, Austria, Paper 3.4.3, September 2009
13. D. van de Borne, N.E. Hecker-Denschlag, G.D. Khoe, H. De Waardt, J. Lightwave Technol. **23**, 4004–4015 (2005)
14. L.E. Nelson, T.N. Nielsen, H. Kogelnik, IEEE Photon. Technol. Lett. **13**, 738–740 (2001)
15. Z. Wang, C. Xie, Opt. Express **17**, 7993–8004 (2009)
16. H. Sunnerud, M. Karlsson, C. Xie, P.A. Andrekson, J. Lightwave Technol. **20**, 2204–2219 (2002)
17. C. Xie, L.F. Mollenauer, J. Lightwave Technol. **21**, 1953–1957 (2003)
18. L.F. Mollenauer, J.P. Gordon, F. Heismann, Opt. Lett. **20**, 2060–2062 (1995)
19. B.C. Collings, L. Boivin, IEEE Photon. Technol. Lett. **12**, 1582–1584 (2000)
20. L. Möller, Y. Su, C. Xie, X. Liu, J. Leuthold, D. Gill, X. Wei, Opt. Lett. **28**, 2461–2463 (2003)
21. M.N. Islam, *Ultrafast Fiber Switching and Devices* (Cambridge University Press, Cambridge, 1992)
22. J. Lee, K. Park, C. Kim, Y. Chung, IEEE Photon. Technol. Lett. **14**, 1082–1084 (2002)
23. C. Xie, L. Möller, D.C. Kilper, L.F. Mollenauer, Opt. Lett. **28**, 2303–2305 (2003)
24. L. Möller, L. Boivin, S. Chandrasekhar, L.L. Buhl, Impact of cross-phase modulation on PMD compensation, in *Proceedings of lasers and electro-optics society 2000 annual meeting*, Paper PD1.2, Rio Grande, Puerto Rico, November 2000
25. E. Corbel, J.P. Thiery, S. Lanne, S. Bigo, A. Vannucci, A. Bononi, Experimental statistical assessment of XPM impact on optical PMD compensator efficiency, in *Proceedings of optical fiber communication conference 2003*, Paper ThJ2, Atlanta, GA, USA, March 2003
26. C. Xie, S. Chandrasekhar, X. Liu, Impact of inter-channel nonlinearities on 10-Gbaud NRZ-DQPSK WDM transmission over Raman amplified NZDSF spans, in *Proceedings of European conference on optical communications 2007*, Paper 10.4.3, September 2007
27. D. van den Borne, S.L. Jansen, S. Calabrò, N.E. Hecker-Denschlag, G.D. Khoe, H. de Waardt, IEEE Photon. Technol. Lett. **17**, 1337–1339 (2005)
28. C. Xie, Z. Wang, S. Chandrasekhar, X. Liu, Nonlinear polarization scattering impairments and mitigation in 10-Gbaud polarization-division-multiplexed WDM systems, in *Proceedings of optical fiber communication conference 2009*, Paper OTuD6, San Diego, CA, USA, March 2009
29. C. Xie, Inter-channel nonlinearities in coherent polarization-division-multiplexed quadrature-phase-shift-keying systems. IEEE Photon. Technol. Lett. **21**, 274–276 (2009)
30. C. Xie, WDM coherent PDM-QPSK systems with and without inline optical dispersion compensation. Opt. Express **17**, 4815–4823 (2009)
31. C. Xie, Dispersion management in WDM coherent PDM-QPSK systems, in *Proceedings of European conference on optical communications 2009*, Paper 9.4.3, Vienna, Austria, September 2009
32. P.K.A. Wai, C.R. Menyuk, J. Lightwave Technol. **14**, 148–157 (1996)
33. D. Marcuse, C.R. Menyuk, P.K.A. Wai, J. Lightwave Technol. **15**, 1753–1746 (1997)
34. C.R. Menyuk, B.S. Marks, J. Lightwave Technol. **24**, 2806–2826 (2006)
35. J.P. Gordon, H. Kogelnik, PNAS **97**, 4541–4550 (2000)
36. D. Wang, C.R. Menyuk, J. Lightwave Technol. **17**, 2520–2529 (1999)
37. A. Bononi, A. Vannucci, A. Orlandini, E. Corbel, S. Lanne, S. Bigo, J. Lightwave Technol. **21**, 1903–1913 (2003)
38. M. Karlsson, H. Sunnerud, J. lightwave Technol. **24**, 4127–4137 (2006)
39. G.P. Agrawal, *Nonlinear Fiber Optics* (Academic, San Diego, 2001)
40. P.K.A. Wai, C.R. Menyuk, H.H. Chen, Opt. Lett. **16**, 1231–1233 (1991)
41. D.S. Ly-Gagnon, S. Tsukamoto, K. Katoh, K. Kikuchi, J. Lightwave Technol. **24**, 12–21 (2006)
42. S.J. Savory, G. Gavioli, R.I. Killey, P. Bayvel, Opt. Express **15**, 2120–2126 (2007)
43. D.N. Godard, IEEE Trans. Commun. **28**, 1867–1875 (1980)
44. L.K. Wickham, R.J. Essiambre, A.H. Gnauck, P.J. Winzer, A.R. Chraplyvy, IEEE Photon. Technol. Lett. **16**, 1591–1593 (2004)

45. O. Bertran-Pardo, J. Renaudier, G. Charlet, H. Mardoyan, P. Tran, S. Bigo, IEEE Photon. Technol. Lett. **20**, 1314–1316 (2008)
46. D. van den Borne, C.R.S. Fludger, T. Duthel, T. Wuth, E.D. Schmidt, C. Schulien, E. Gottwald, G.D. Khoe, H. de Waardt, Carrier phase estimation for coherent equalization of 43-Gb/s POLMUX-NRZ-DQPSK transmission with 10.7-Gb/s NRZ neighbours, in *Proceedings of European conference on optical communications 2007*, Paper 7.2.3, Berlin, Germany, September 2007
47. C. Xie, Suppression of inter-channel nonlinearities in WDM coherent PDM-QPSK systems using periodic-group-delay dispersion compensators, in *Proceedings of European conference on optical communications 2009*, Paper P4.08, Vienna, Austria, September 2009
48. M.S. Alfiad, D. van den Borne, S.L. Jansen, T. Wuth, M. Kuschnerov, G. Grosso, A. Napoli, H. De Waardt, 111-Gb/s POLMUX-RZ-DQPSK transmission over LEAF: optical versus electrical dispersion compensation, in *Proceedings of optical fiber communication conference 2009*, Paper OThR4, San Diego, CA, March 2009
49. O. Bertran-Pardo, J. Renaudier, G. Charlet, M. Salsi, M. Bertolini, P. Tran, H. Mardoyan, C. Koebele, S. Bigo, System benefits of temporal polarization interleaving with 100 Gb/s coherent PDM-QPSK, in *Proc. European Conference on Optical Communications 2009*, Paper 9.4.1, Vienna, Austria, September 2009
50. M. Winter, D. Setti, K. Petermann, Interchannel nonlinearities in polarization-multiplexed transmission, in *Proceedings of European conference on optical communications 2009*, Paper 10.4.4, Vienna, Austria, September 2009
51. P. Serena, N. Rossi, A. Bononi, (2009) Nonlinear penalty reduction induced by PMD in 112 Gbit/s WDM PDM-QPSK coherent systems, in *Proceedings of European conference on optical communications 2009*, Paper 10.4.3, Vienna, Austria, September 2009
52. S. Chandrasekhar, X. Liu, (2008) Experimental investigation of system impairments in polarization multiplexed 107-Gb/s RZ-DQPSK, in *Proceedings of optical fiber communications conference 2008*, Paper OThU7, San Diego, CA, USA, March 2008
53. C. Xie, Z. Wang, S. Chandrasekhar, X. Liu, (2009) Nonlinear polarization scattering impairments and mitigation in 10-Gbaud polarization-division-multiplexed WDM systems, in *Proceedings of optical fiber communications conference 2009*, Paper OTuD6, San Diego, CA, USA, March 2009
54. J. Renaudier, O. Bertran-Pardo, H. Mardoyan, P. Tran, M. Salsi, G. Charlet, S. Bigo, IEEE Photon. Technol. Lett. **20**, 2036–2038 (2008)
55. X. Wei, X. Liu, C. Xie, L.F. Mollenauer, Opt. Lett. **28**, 983–985 (2003)
56. L. Möller, Y. Su, G. Raybon, X. Liu, IEEE Photon. Technol. Lett. **15**, 335–337 (2003)
57. W. Shieh, IEEE Photon. Technol. Lett. **19**, 134–136 (2007)

Chapter 10
Multicanonical Monte Carlo for Simulation of Optical Links

Alberto Bononi and Leslie A. Rusch

10.1 Introduction

Multicanonical Monte Carlo (MMC) is a simulation-acceleration technique for the estimation of the statistical distribution of a desired system output variable, given the known distribution of the system input variables. MMC, similarly to the powerful and well-studied method of importance sampling (IS) [1], is a useful method to efficiently simulate events occurring with probabilities smaller than $\sim 10^{-6}$, such as bit error rate (BER) and system outage probability. Modern telecommunications systems often employ forward error correcting (FEC) codes that allow pre-decoded channel error rates higher than 10^{-3}; these systems are well served by traditional Monte-Carlo error counting. MMC and IS are, nonetheless, fundamental tools to both understand the statistics of the decision variable (as well as of any physical parameter of interest) and to validate any analytical or semianalytical BER calculation model. Several examples of such use will be provided in this chapter. As a case in point, outage probabilities are routinely below 10^{-6}, a sweet spot where MMC and IS provide the most efficient (sometimes the only) solution to estimate outages.

MMC was developed by physicists Berg and Neuhaus 15 years ago [2]. Berg and Neuhaus's paper is hard to read for nonphysicists. New concepts in probability theory are hidden by the many details of their statistical physics application. Optical communications was the first telecom community to adopt MMC, perhaps because physicists and electrical engineers share a common background and common language. Within the optical communications community, physicist D. Yevick [3] was the first to apply MMC to study the statistics of polarization mode dispersion (PMD). Subsequently, Holzlöhner et al., extended the MMC method to estimate the

A. Bononi (✉)
Dipartimento di Ingegneria dell'Informazione, Università di Parma, 43100 Parma, Italy
e-mail: alberto.bononi@unipr.it

L.A. Rusch
Electrical and Computer Engineering Department, Université Laval, Québec City, QC, Canada G1V 0A6
e-mail: rusch@gel.ulaval.ca

S. Kumar (ed.), *Impact of Nonlinearities on Fiber Optic Communications*, Optical and Fiber Communications Reports 7, DOI 10.1007/978-1-4419-8139-4_10,
© Springer Science+Business Media, LLC 2011

BER of direct-detection amplified optical communication links [4]. Soon after those publications, a large number of MMC papers appeared on various topics in optical communications [5–21].

The success of MMC is mostly due to its ease of implementation when compared to IS. While traditional IS allows impressive computational savings with respect to brute-force Monte-Carlo estimation, its most striking shortcoming is that an in-depth knowledge of the physical problem at hand is required to find the right parameters (namely, an efficient biasing distribution) to achieve those savings, making IS time-consuming in its planning phase and thus difficult to use.

MMC is instead a truly innovative algorithm which, like IS, is based on biasing the system input distribution. However, in MMC such a biasing is system-independent, and is blindly and adaptively achieved by forcing a flat output histogram. No time-consuming, ad-hoc user pre-setting of the biasing distribution is needed. Although it has been shown that bias-optimized IS can be more efficient than MMC in the estimation of the probability of rare events [8], MMC has the key advantage of being easily implemented for any system, with great time savings in the planning phase. This is the main reason for the success of MMC.

The main tool used by MMC to adaptively generate biased distributions with a desired density is the Markov Chain Monte Carlo (MCMC) method [22,23]. Papers on MMC usually delve into the machinery of the MCMC method, as if the true heart of the MMC algorithm were the MCMC biasing scheme. In this chapter, we will instead first explain MMC without the need of MCMC, so that all the attention can be focused on the explicit analytical connections between MMC and IS. Later, MCMC will enter into play, but its function within MMC will be clear, and the reader will better appreciate the subtleties connected with its use within MMC.

This chapter is organized as follows. After a brief review of classical Monte Carlo (MC) in Sect. 10.2.1, importance sampling is introduced in Sect. 10.2.2 with a new twist with respect to classical treatments [1]. The concepts of uniform weight (UW) IS and flat histogram (FH) IS are introduced. The MMC FH adaptation algorithm is described in Sect. 10.3.1, and practical aspects of MMC are discussed in Sect. 10.4. In Sect. 10.5.1–10.5.3, we present specific examples where MMC techniques have provided quantitative, accurate, and experimentally validated performance predictions in optical communications systems, where analysis is intractable. An appendix contains a summary of MCMC.

10.2 Monte Carlo Techniques

In order to determine the symbol error rate (SER) of a digital communications system, we need the statistical properties of the decision variable at the output of the receiver. Let that decision variable be $Y = g(X)$, where $g : \Gamma \rightarrow \mathbb{R}$ is a real scalar function[1] of a random vector X taking values in the *input* (or *state*)

[1] Although extension of MMC to the estimation of the joint distribution of multiple output variables is possible [24,25], this tutorial will concentrate for simplicity on the scalar case.

space Γ. We are interested in determining the distribution (i.e., the probability density function (PDF) in the continuous case or the probability mass function (PMF) in the discrete case) of Y. The system input–output transfer function $g(\cdot)$ is in most practical problems known only through a computationally expensive numerical routine. We assume the joint PDF $f_X(x)$ of X (or equivalently the joint PMF in the discrete case) is known, possibly up to an unknown multiplicative constant; we assume we are able to draw samples from such a distribution.

In digital communications, the system random input X is the set of random symbols transmitted and noise accumulated along the transmission line, falling within a memory window that captures all impact on the decision variable Y. The larger the memory of the transmission system, the larger the dimensionality of X. In the rest of this paper, we will assume that Y and X are continuous random variables (RVs). The modifications for discrete RVs are straightforward.

10.2.1 Conventional Monte Carlo Estimation

In order to estimate by simulation the PDF $f_Y(y)$ of the continuous output Y on a desired range $\mathscr{R}_Y$, we tile $\mathscr{R}_Y$ with M bins of width Δy centered at the discrete values $\{y_1, ..., y_M\}$.[2] We define the i-th bin as the interval $B_i \triangleq \left[y_i - \frac{\Delta y}{2}, y_i + \frac{\Delta y}{2}\right]$. If the PMF of the discretized Y on the i-th bin is $P_i \triangleq P\{Y \in B_i\}$, then for sufficiently small Δy the output PDF is $f_Y(y_i) \simeq P_i/\Delta y$. This binning implicitly defines, via $g(\cdot)$, a partition of the input space into M domains $\{D_i\}_{i=1}^{M}$, where

$$D_i = \{x \in \Gamma : g(x) \in B_i\}$$

is the domain in Γ that maps into the ith bin. While B_i are simple intervals, the domains D_i are multidimensional regions with possibly tortuous topologies, and most often totally unknown to the researcher.

Let the Bernoulli RV

$$I_{D_i}(X) = \begin{cases} 1 & \text{if } X \in D_i \\ 0 & \text{else} \end{cases}$$

be the indicator of event $\{X \in D_i\}$; equivalently we can write $\{Y = g(X) \in B_i\}$, which emphasizes that calculation of $g(X)$ is needed to determine whether this event occurs. The desired PMF can be expressed as the expectation of the indicator

$$P_i = \int_{D_i} f_X(x)\mathrm{d}x = \int_{\Gamma} I_{D_i}(X) f_X(x)\mathrm{d}x = E[I_{D_i}(X)]. \tag{10.1}$$

[2] If the output range $\mathscr{R}_Y$ is not the entire output space, $f_Y(y)$ will actually denote the conditional PDF $f_Y(y|Y \in \mathscr{R}_Y)$.

This is the rationale behind classical MC estimation: draw N samples $\{X_1, .., X_N\}$ from the distribution $f_X(x)$, pass them through the system $g(\cdot)$ and find how these samples fall in the output bins, forming the histogram. The (normalized) histogram is the sample mean of the expectation of the indicator in (10.1), forming the following estimate of the PMF

$$\hat{P}_i^{\mathrm{MC}} \triangleq \frac{1}{N}\sum_{j=1}^{N} I_{D_i}(X_j) = \frac{N_i}{N} \tag{10.2}$$

N_i being the number of samples that fall in bin i. The MC estimator is unbiased by construction: $E[\hat{P}_i^{\mathrm{MC}}] = P_i$. The squared relative error (SRE), a figure of merit for any unbiased estimator $\hat{P}_i$, is defined as $\varepsilon_i \triangleq Var[\hat{P}_i]/P_i^2$. If the samples are independent, N_i is the sum of N independent Bernoulli RVs with "success" probability P_i, thus N_i has a binomial distribution, i.e., $N_i \sim Binomial(N, P_i)$. The SRE for the MC estimator for the ith bin is

$$\varepsilon_i^{\mathrm{MC}} = \frac{1-P_i}{NP_i} \tag{10.3}$$

which is, for small P_i, approximately the inverse of the expected value $E[N_i] = NP_i$. For instance, about 100 counts are required on average to achieve a relative error, $\sqrt{\varepsilon_i}$, of 10% in the estimation of P_i. Achieving 100 counts in all bins is challenging, as in MC simulations most samples fall in the modal bins. Little or no samples fall in the area in which we are most interested, the tails of the PMF. For fixed simulation effort (N fixed), the relative error is dramatically higher in the tails than in the modal regions.

10.2.2 Importance Sampling

In order to reliably estimate the output PMF even in the tail bins (rare events), we artificially increase the number of samples falling in such bins using IS [1]. We re-write (10.1) as

$$P_i = \int_{\Gamma} I_{D_i}(x)\left[\frac{f_X(x)}{f_X^*(x)}\right] f_X^*(x)\mathrm{d}x = E^*[I_{D_i}(X)w(X)], \tag{10.4}$$

where $f_X^*(x)$, strictly positive for all x at which $f_X(x) > 0$, is a *warped* PDF of X, and $w(x) \triangleq f_X(x)/f_X^*(x)$ is the *IS weight*; E^* indicates expectation with respect to the distribution $f_X^*(x)$. The output PMF in the warped space is given by

$$P_i^* = \int_{\Gamma} I_{D_i}(x) f_X^*(x)\mathrm{d}x = E^*[I_{D_i}(X)].$$

The weighting function $w(x)$ plays an important role in generating the IS estimate of the unwarped PMF. To see this, consider the conditional density $f_X^*(x \mid X \in D_i) = \frac{I_{D_i}(x) f_X^*(x)}{P_i^*}$ and use it to rewrite P_i in (10.4) as

$$P_i = P_i^* \int_\Gamma I_{D_i}(x) w(x) \frac{f_X^*(x)}{P_i^*} \mathrm{d}x = P_i^* E^*[w(X) \mid X \in D_i]. \tag{10.5}$$

The IS estimator replaces the product in the expectation operator in (10.5) by the product of their sample averages in the warped system

$$\hat{P}_i^{\mathrm{IS}} = \underbrace{\left(\frac{N_i^*}{N}\right)}_{\triangleq \hat{H}_i^*} \underbrace{\left[\frac{1}{N_i^*} \sum_{n=1}^{N_i^*} w(X_{jn})\right]}_{\triangleq \bar{w}_i}. \tag{10.6}$$

The IS estimation is performed as follows: a conventional MC simulation is run in the warped system, i.e., by drawing N samples *from the warped PDF* $f_X^*(x)$. The MC estimate in the warped system is found from the N_i^* samples falling in bin i and forming the so-called histogram of visits $\hat{H}_i^*$ [26] in the warped system. Hence, the IS estimate $\hat{P}_i^{\mathrm{IS}} = \hat{H}_i^* \bar{w}_i$ comes naturally from the product of the MC estimate of P_i^* in the warped system, $\hat{H}_i^*$, and the estimate $\bar{w}_i$ of $E^*[w(X) \mid X \in D_i]$. The weights $\bar{w}_i$ of estimates P_i^* provide the inverse transformation to take us back into the unwarped system. The count N_i^* is on average much larger than in an unwarped MC sampling if we can achieve $f_X^*(x) \gg f_X(x)$ over the domain D_i. We can equivalently write the IS estimator (10.6) as

$$\hat{P}_i^{\mathrm{IS}} = \frac{1}{N} \sum_{j=1}^{N} I_{D_i}(X_j) w(X_j), \tag{10.7}$$

which is the traditional way of introducing IS as the sample average of the expectation in (10.4) [1].

To determine the accuracy of the IS estimate using (10.7), let $W_{ij} \triangleq I_{D_i}(X_j) w(X_j)$. From (10.4), $E^*[W_{ij}] = P_i$, and thus the IS estimator (10.6) is unbiased. To find its variance, observe that

$$\begin{aligned} E^*[W_{ij}^2] &= E^*[I_{D_i}(X_j) w^2(X_j)] \\ &= P_i^* \int_\Gamma I_{D_i}(x) w^2(x) \frac{f_X^*(x)}{P_i^*} \mathrm{d}x \\ &= P_i^* E^*[w^2(X) \mid X \in D_i], \end{aligned}$$

so that from (10.7) we get

$$Var^*[\hat{P}_i^{\mathrm{IS}}] = \frac{Var[W_{ij}]}{N} = \frac{P_i^* E^*[w^2(X) \mid X \in D_i] - P_i^2}{N}. \tag{10.8}$$

Using (10.5), the SRE $\varepsilon_i^{\text{IS}} \triangleq Var[\hat{P}_i^{\text{IS}}]/P_i^2$ becomes

$$\varepsilon_i^{\text{IS}} = \frac{1}{N}\left\{\frac{1}{P_i^*}\left(\frac{Var^*[w(X) \mid X \in D_i]}{(P_i/P_i^*)^2} + 1\right) - 1\right\}. \tag{10.9}$$

Expressing (10.9) in terms of a conditional variance helps us appreciate the true limit of IS estimation, which is connected to our *a priori ignorance of the domains* D_i. Suppose for instance that D_i is composed of two disjoint sets, located far apart on the input space: D_{i1} whose existence and location is found via physical reasoning and knowledge of our problem, and D_{i2}, whose existence we fail to guess. This incomplete foreknowledge leads us to contrive a warping that shifts most of the PDF mass on D_{i1}, i.e., such that $f_X^*(x) \gg f_X(x)$, or equivalently we set $w(x) \ll 1$ on D_{i1}. Most likely, we will get little PDF mass on D_{i2}, hence $f_X^*(x) \ll f(x)$, i.e., $w(x) \gg 1$ on D_{i2}, thus obtaining, as per (10.9), a very large value of $Var^*[w(X) \mid X \in D_i]$ and therefore a very large SRE.

10.2.3 Uniform Weight Importance Sampling

Consider the set of all warpings $f_X^*(x)$ producing the same output warped PMF $\underline{P}^* \triangleq \{P_i^*\}_{i=1}^M$. We call this set the *equivalence class* of warpings associated with $\underline{P}^*$. The space for all possible warpings is thereby partitioned into disjoint equivalence classes, as depicted in Fig. 10.1. From (10.5), each equivalence class produces the same average conditional weights $\{E^*[w(X) \mid X \in D_i]\}_{i=1}^M$. Equation (10.9) suggests that the best warping within each equivalence class, i.e., the one producing the lowest IS relative error, is the uniform weight (UW) warping. A UW warping assigns a constant weight to all $x \in D_i$, with value $w_i = P_i/P_i^*$ per (10.5), so that $Var^*[w(X)|X \in D_i] = 0$. *Hence, the search for the optimal global warping can always be restricted to the search among the UW warpings.* Note that although at

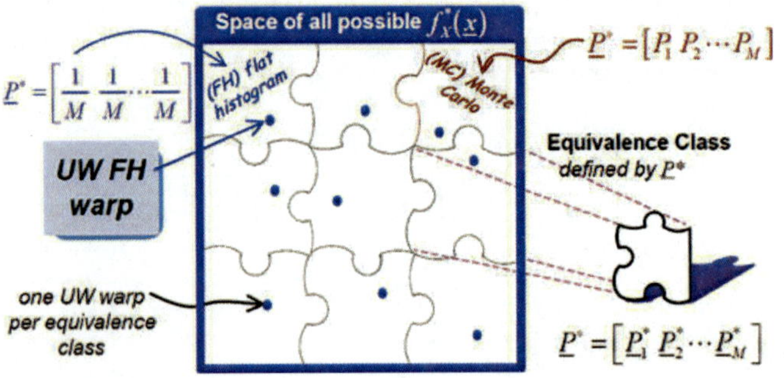

Fig. 10.1 Sketch of the space of all input warpings $f_X^*(x)$, partitioned into disjoint equivalence classes, each characterized by a warped output PMF $\underline{P}^*$

first sight the implementation of UW warping seems to require a detailed knowledge of the domains D_i, we will shortly see that this is not the case.

From (10.9), the SRE for a UW–IS estimation of bin i simplifies to

$$\varepsilon_i^{\mathrm{UW-IS}} = \frac{1}{N}\left\{\frac{1}{P_i^*} - 1\right\} \tag{10.10}$$

and depends only on P_i^*. When $P_i^* \ll 1$, the error is about the inverse of the expected value NP_i^*; this in turn is on average equal to the inverse of the warped count N_i^*. This leads to a reduced error with respect to $\varepsilon_i^{\mathrm{MC}}$ (10.3), at an equal number of runs N, on those bins in which the warping is doing well, i.e., in which $P_i^* \gg P_i$. In the extreme case when all warped samples fall in bin i, we reach the optimal UW–IS warping for estimating bin i. In this case, $P_i^* \to 1$ and we achieve zero relative error; this is known as the *zero-variance* IS (ZV-IS) [1] warping. Such a warping will clearly be useless for the estimation of other bins.

Suppose we wish to use our N runs to estimate the output PMF on *all bins* with equally good relative error; (10.10) leads to the choice $P_i^* = \frac{1}{M}$ for all i. A uniformly distributed PMF will produce a flat histogram. Since P_i^* is the expected value of the visits histogram, we will call this UW–IS the *uniform weight, flat-histogram (UW–FH)* importance sampling. It is easy to see that, among all UW–IS, the UW–FH is the one that minimizes the largest relative error among all bins, namely

$$\max_i \varepsilon_i^{\mathrm{UW\text{-}IS}} = \max_i \frac{1}{N}\left\{\frac{1}{P_i^*} - 1\right\} \geq \varepsilon^{\mathrm{UW\text{-}FH}} = \frac{M-1}{N}. \tag{10.11}$$

How would we implement a UW–FH warping?

For any IS implementation, the analytic form of the warped input PDF $f_X^*(x)$ is needed, at least up to a normalization constant, to draw input samples from the warped system. Any UW warping can be expressed as [27, 28]:

$$f_X^*(x) = \frac{f_X(x)}{c_\Theta \Theta(x)}, \tag{10.12}$$

where $\Theta(x) \triangleq \Theta_i$ for all $x \in D_i$, $i = 1 \ldots M$, and $\underline{\Theta} \triangleq \{\Theta_i\}_{i=1}^M$ is a positive PMF on the M bins (i.e., one with all nonzero entries), and c_Θ is a normalization constant to assure $f_X^*(x)$ is a valid PDF. By construction, (10.12) puts constant weight $w_i = c_\Theta \Theta_i$ on each domain D_i.

The warped output PMF induced by such a UW warping is

$$P_i^* = \int_{D_i} f_X^*(x)\mathrm{d}x = \frac{\int_{D_i} f_X(x)\mathrm{d}x}{c_\Theta \Theta_i} = \frac{P_i}{c_\Theta \Theta_i}. \tag{10.13}$$

Since $\underline{\Theta}$ is by construction a proper PMF whose elements sum to one, the normalizing constant must be $c_\Theta = \sum_{j=1}^M \frac{P_j}{\Theta_j}$.

The implementation of the UW–FH warping has $P_i^* = 1/M$. Equation (10.13) yields $c_\Theta = M$ and $\Theta_i \equiv P_i$. Hence from (10.12) the UW–FH warped PDF displays in its denominator the true PMF $\underline{P}$, which is exactly what we seek to estimate. Hence UW–FH appears unfeasible, like the ZV-IS, as it requires knowledge of exactly what we seek to estimate. We will show, however, that it can be closely approached by a sequence of UW warpings as in (10.12) via a simple adaptive mechanism.

10.3 Multicanonical Monte Carlo

Flat-histogram (FH) algorithms are a family of output PDF estimation algorithms, among which are MMC, Wang-Landau [29], and others [27]. Starting from the known input PDF $f_X(x)$, these algorithms build a sequence of UW-warped input PDFs $f_X^{(n)}(x) = \frac{f_X(x)}{c_n \Theta_n(x)}$, $n = 1, 2, \ldots$, in which the positive PMF $\underline{\Theta}_n \triangleq \{\Theta_{n,i}\}_{i=1}^M$ plays the role of an intermediate estimate of the true PMF $\underline{P}$ of the discretized output RV $Y = g(X)$ at the nth step, and c_n is its normalizing constant. A step (which in MMC is called a *cycle*) corresponds to drawing N samples $\{X_j\}_{j=1}^N$ from the warped $f_X^{(n)}(x)$, passing these samples through the system under test, and forming a new estimate $\underline{\Theta}_{n+1}$ of the PMF of Y. An FH algorithm is defined by its *update law* $\underline{\Theta}_n \to \underline{\Theta}_{n+1}$. In all cases, the update uses the output histogram of visits $\underline{\hat{H}}_n^* \triangleq \{\hat{H}_{n,i}^*\}_{i=1}^M$ at the end of cycle n, and drives this histogram in the next step toward equal visits to all bins (a flat histogram). At convergence, as seen from (10.12), $c_n \to M$ and $\underline{\Theta}_n \to \underline{P}$. Note that, no matter the visits-flattening update law, when the visits histogram is (practically) flat, the final estimate of the output PMF can be read off in the denominator of the warped input PDF, as we already noted at the end of the previous section.

10.3.1 MMC Adaptation

MMC, introduced by Berg et al., in 1991 [2], is among the first FH methods. In MMC, the update law is based on a UW–IS estimate. At cycle n, N samples are drawn from $f_X^{(n)}$ and $Y_j = g(X_j)$ is evaluated for every sample, finally forming the visits histogram $\hat{H}_{n,i}^* \triangleq N_{n,i}/N$. An IS-updated estimate of the PMF of discretized Y is obtained from (10.6) as

$$\Theta_{n+1,i} = \left(\frac{N_{n,i}}{N}\right)\left[\frac{1}{N_{n,i}}\sum_{n=1}^{N_{n,i}} w(X_n)\right] = \hat{H}_{n,i}^* c_n \Theta_{n,i}\,, \tag{10.14}$$

where we used the constant weight $w_i = c_n \Theta_{n,i}$ of the previous warp $f_X^{(n)}$. In practice, c_n may be omitted, as will be seen in (10.27).

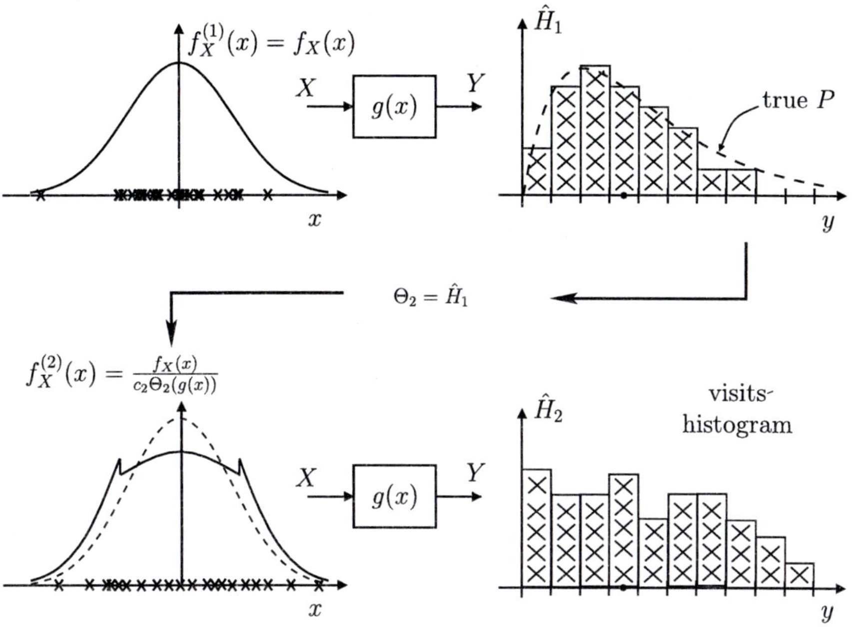

Fig. 10.2 Sketch of first 2 steps in MMC. First cycle is a pure MC if we start with a uniform guess

Figure 10.2 sketches the first two steps of MMC for the simple system $y = x^2$, with X a zero-mean Gaussian scalar RV. It is common practice to start the recursion (10.14) by using the uniform distribution as an initial guess for $\underline{\Theta}_1$. In this case, as seen from (10.12), the first MMC cycle is performed with the unwarped distribution, i.e., as a classical MC run. In the example of Fig. 10.2, the bell-shaped input PDF $f_X^{(1)} = f_X$ is shown in the top left: most input samples (crosses on the x axis) will fall on the modal region, and the output histogram will be an MC estimate of the true PMF, with a well-estimated modal region and almost no samples in the tails. At the end of the first cycle, the PMF estimate (10.14) is updated to $\underline{\Theta}_2$ and used in the denominator of the warped input PDF at the next cycle. As sketched in the figure, the warped PDF $f_X^{(2)} = \frac{f_X}{c_2\Theta_2(x)}$ will decrease the mass function in the bins of the modal region in proportion to their number of visits, and increase the mass function in the tails. To avoid division by zero on unvisited bins, the visit count is forced to one on those bins, and the histogram is renormalized. The next N samples drawn from $f_X^{(2)}$ will fall in the tails of the original f_X more often than before, so that visits will tend to be more equally spread across output bins. At convergence we must have $\Theta_{n+1,i} = \Theta_{n,i}$, which from (10.14) implies $\hat{H}^*_{n,i} = 1/c_n$ for all bins, i.e., a flat histogram (UW–FH).

The MMC update strategy benefits from a general advantage of IS estimators: it provides an unbiased estimate at every cycle, since from (10.14) we get

$$E[\Theta_{n+1,i}] = E[\hat{H}^*_{n,i}]c_n\Theta_{n,i} = P_i, \qquad (10.15)$$

where (10.13) was used in the second equality. In point of fact, a bias was introduced on those bins whose occupancy was forced artificially from zero to one.

In the assumption of independent samples, the relative error on estimate $\Theta_{n+1,i}$ on the visited bins is, from (10.10),

$$\varepsilon_{n+1,i} = \frac{1}{N}\left\{\frac{1}{E[\hat{H}^*_{n,i}]} - 1\right\} = \frac{1}{N}\left\{\frac{c_n\Theta_{n,i}}{P_i} - 1\right\} \tag{10.16}$$

which from (10.11) is seen to flatten out for all bins to the value $\frac{M-1}{N}$ at convergence to the UW–FH. Hence, in an ideal setting with independent samples, if the desired SRE on all bins is $\tilde{\varepsilon}$ and we have M bins, the cycle size N should be selected as

$$N \geq \frac{M-1}{\tilde{\varepsilon}}. \tag{10.17}$$

Note that, starting from any initial guess $\underline{\Theta}_1$, (10.15) shows that the MMC converges on average even at the first cycle on all visited bins, but with wide fluctuations, i.e., large relative error (10.16), on those bins in which the probability is largely overestimated ($\Theta_{n,i} \gg P_i$). The usual choice of the uniform distribution for $\underline{\Theta}_1$ makes the relative error at the first steps large in the tail bins, where the histogram count is small. If we have a rough idea of the shape of the PMF $\underline{P}$ to be estimated, a better strategy is to initialize $\underline{\Theta}_1$ to that shape.

10.3.2 Smoothed MMC

We will now discuss a very important part of the MMC update that is commonly referred to as the smoothing function. We will make some observations about the convergence behavior of the MMC algorithm, both with and without smoothing. The MMC update in (10.14) is the unsmoothed updated. The stochastic fluctuations due to a finite cycle size N may make the cycle-n histogram $\hat{H}^*_{n,i}$ differ significantly from its expected value $\underline{P}^*_n$, even if the adaptation is near reaching convergence. Indeed, fluctuations would occur even if we *started* at the true UW–FH warping. These unavoidable fluctuations can be overcome to a practical extent by adopting a smoothing strategy, such as that in adaptive equalization [30]. A clever smoothing function was suggested by Berg [26], which we shall now interpret.[3]

Noting that (10.14) is valid for *all* bins, we can take any two bins and form the following equivalent ratios (we take adjacent bins in this example)

$$\frac{\Theta_{n,i}}{\Theta_{n,i-1}} = \frac{\Theta_{n-1,i}}{\Theta_{n-1,i-1}}\left[\frac{\hat{H}^*_{n,i}}{\hat{H}^*_{n,i-1}}\right]. \tag{10.18}$$

[3] Berg's heuristic argument for the update is somewhat disingenuous; however, the effectiveness of his update is unarguable.

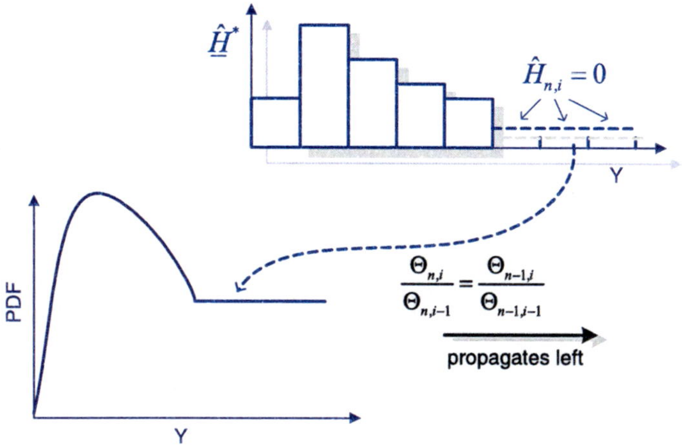

Fig. 10.3 Sketch of spatial smoothing of unvisited bins

Fluctuations in the term in brackets are to be smoothed. Instead of updating our uniform weighting Θ bin-by-bin as in (10.14), this update is based on the ratio of two adjacent bins. The choice of adjacent bins introduces smoothing over bins (spatial smoothing), as well as the opportunity for smoothing over cycles (temporal smoothing); smoothing over more than two bins has also been proposed [6].

Consider the treatment of bins with zero visits. To avoid division by zero, we set the minimum visit value to one. Hence, the spatially smoothed MMC has an update

$$\frac{\Theta_{n,i}}{\Theta_{n,i-1}} = \frac{\Theta_{n-1,i}}{\Theta_{n-1,i-1}}. \tag{10.19}$$

This causes a propagation of the value of bin $i-1$ to bin i, and it induces a floor (i.e., a bias) in the estimated PMF for those contiguous bins with zero hits in the warped system, as seen in Fig. 10.3.

To develop the concept of temporal smoothing, we take the logarithm of the ratios. Let

$$\beta_{n,i} \triangleq \log\left(\frac{\Theta_{n,i}}{\Theta_{n,i-1}}\right) = \beta_{n-1,i} + \delta_{n,i}. \tag{10.20}$$

We have defined $\delta_{n,i} \triangleq \log(\hat{H}^*_{n,i}/\hat{H}^*_{n,i-1})$, a noisy estimate of the *log-ratio* of adjacent bins of the output PDF $\underline{P}^*$ in the warped system at cycle n. Note that by choosing adjacent bins, $\beta_{n,i}$ is an estimate of β_i, the slope at bin i of the logarithm of the output PDF $P(y)$, scaled by Δy.

Consider an estimator $\hat{\beta}_{n,i}$ of β_i at cycle n that is a linear combination of all previous cycles $\{\delta_{j,i}\}_{j=1}^n$

$$\hat{\beta}_{n,i} = \hat{\beta}_{n-1,i} + \alpha_{n,i}\delta_{n,i} = \sum_{j=0}^{n} \alpha_{j,i}\delta_{j,i}. \tag{10.21}$$

Unfortunately, the $\delta_{j,i}$ are not unbiased estimators of log-ratio of the output PDF $\underline{P}^*$ in the warped system. Also, the sequence of $\{\delta_{j,i}\}_{j=1}^n$ are correlated; the histograms at each cycle are drawn from distributions influenced by the histogram of the previous cycle (this is the nature of the MMC algorithm). Were the $\delta_{j,i}$ uncorrelated and unbiased estimators with variance $\sigma_{j,i}^2$, their best linear unbiased estimator (BLUE) $\hat{\beta}_{n,i}$ would have weights

$$\alpha_{j,i} = \frac{1/\sigma_{j,i}^2}{\sum_{m=1}^n 1/\sigma_{m,i}^2} \quad j \in \{1, \cdots, n\}. \tag{10.22}$$

This linear estimator may not be optimal for this system due to the correlations in $\{\delta_{j,i}\}_{j=1}^n$, and, of course, the variances $\sigma_{j,i}^2$ are unknown. We could attempt to estimate the variances $\sigma_{j,i}^2$ at each cycle, but (10.22) is not causal,[4] as the denominator is a summation over all cycles, not just cycles up to cycle j. Berg [26] suggests the following update equation that resembles (10.22), but exploits estimates of $\sigma_{j,i}^2$ and renders the estimator causal by truncation.

$$\beta_{n,i} = \beta_{n-1,i} + \tilde{G}_{n,i}\delta_{n,i}, \tag{10.23}$$

where

$$\tilde{G}_{n,0i} = \frac{g_{n,i}}{\sum_{j=1}^n g_{j,i}}$$

and

$$g_{j,i} = N \frac{\hat{H}_{j,i-1}^* \hat{H}_{j,i}^*}{\hat{H}_{j,i-1}^* + \hat{H}_{j,i}^*}. \tag{10.24}$$

It can be shown that $g_{j,i}$ is an estimate of the inverse of $\sigma_{j,i}^2$. When both $\hat{H}_{n,i}^*$ and $\hat{H}_{n,i-1}^*$ are zero, we define $g_{n,i} = \tilde{G}_{n,i} = 0$. Reliability factors $\tilde{G}_{n,i}$ are found at cycle n by normalizing over the samples $g_{j,i}$ available up to time n. The update law (10.23) has the classical form found in adaptive equalization, $\delta_{n,i}$ playing the role of the innovation, and $\tilde{G}_{n,i}$ that of the step size.

Berg's update, i.e., (10.23), can be explicitly rewritten in terms of the original PMFs as the *smoothed MMC update* [4, 26]

$$\frac{\Theta_{n,i}}{\Theta_{n,i-1}} = \frac{\Theta_{n-1,i}}{\Theta_{n-1,i-1}} \left[\frac{\hat{H}_{n,i}^*}{\hat{H}_{n,i-1}^*} \right]^{\tilde{G}_{n,i}}. \tag{10.25}$$

[4] The denominator is needed to avoid bias.

Whenever $\hat{H}^*_{n,i} = 0$ or $\hat{H}^*_{n,i-1} = 0$ the factor $g_{n,i}$ in (10.24) is zero, as is the reliability factor $\tilde{G}_{n,i}$. Hence, we will incorporate the same spatial smoothing illustrated in Fig. 10.3, as (10.19) again holds.

10.3.3 Example: Chi-Square Distribution

As an example, consider estimating the PDF of $Y = \sum_{i=1}^{10} X_i^2$ with X_i independent zero-mean Gaussian RVs with unit variance. In this simple system, the true PDF, $\underline{P}$, is known analytically, a chi square distribution. This PDF is plotted as a dashed line in Fig. 10.4. The PDF found by MMC simulation, $\underline{\Theta}$, is plotted as a solid line, and the Monte Carlo results are plotted as circular markers; the associated vertical axis is on the left. In a dash-dot line, we present the histogram of the output in the warped system, $\underline{\hat{H}}^*$; the associated vertical axis is on the right. We can see that for bins with $\underline{\hat{H}}^* = 0$, the output PDF estimate $\underline{\Theta}$ propagates the value for the last occupied bin across remaining bins, thus terminating the PDF with a horizontal line. The MMC was run both without smoothing, i.e., using update (10.18), and with smoothing, i.e., using update (10.23). Results without smoothing are presented in the left column, while results with smoothing are shown in the right column of Fig. 10.4. In either case, five cycles are run with the first cycle presented in the top row and the fifth cycle in the last row of Fig. 10.4.

Figure 10.4 shows the smoothed MMC estimation along with MC estimation (circle markers). Here, we used 75 bins of width $\Delta y = 2$. From the figure, we see that after five cycles the MMC estimate correctly approximates the true PDF down to $\sim 10^{-20}$, while, at the same number of samples, the MC estimate remains at about 10^{-5}, with an MMC gain of 15 orders of magnitude in PDF estimation with respect to MC.

We note that the PDF floors presented by $\underline{\Theta}$ at each cycle, as was anticipated in Fig. 10.3. By comparing floors in the two columns, we note that the simulations without temporal smoothing exhibit lower floors than do the simulations using Berg's update with temporal smoothing. The cost of reducing stochastic fluctuations is requiring more cycles to reach a given resolution in the output PDF. Clearly Berg's update leads to a smaller deviation from the true PDF, especially at bins well to the left of the PDF floor. Insets with a zoom on this region for cycle 4 are given in Fig. 10.4.

Spikes in the histogram occur regularly (more often for the simulations without temporal smoothing, but in both cases) in the bins near the tail regions. In our example, the tail is only on the left, but in a more symmetric PDF there would be floors for both left and right tails. In order to approach the flat histogram, the MMC algorithm pushes realizations into under-visited bins at the next cycle; the spikes are the result of a probabilistic "wall" due to the finite length of each cycle, N. When N is not large enough to generate visits in a bin, a new cycle is required to boost the probability of those bins. Underestimation of a bin to the left during a previous cycle will lead to a larger spike during the current cycle.

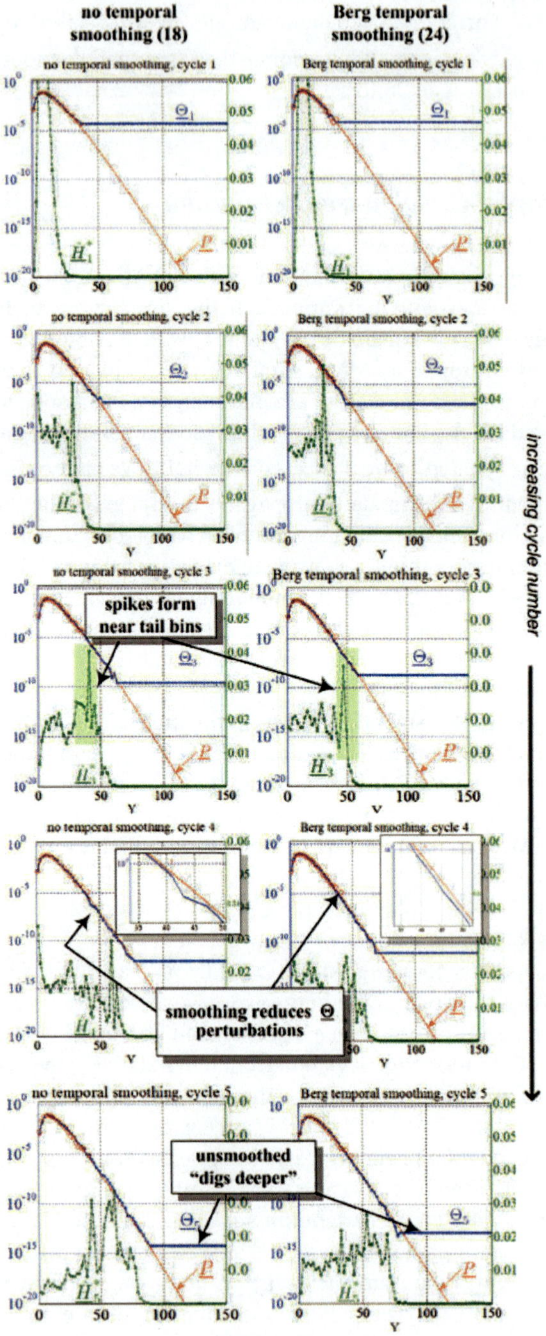

Fig. 10.4 Simulations without (*left column*) and with (*right column*) smoothing; the effect of outliers is clearly attenuated in the smoothed simulation

10.3.4 Drawing Warped Samples: Markov Chain Monte Carlo

The generation of samples from the warped input distributions needed in MMC, which are likely to have a very irregular form and be defined over a high dimensional space, is obtained with the very general MCMC method. As explained in the appendix, a new sample X_t at time t is generated from the sample generated at time $t-1$ and either accepted or rejected based on the odds ratio (10.36). Only when the new proposal is accepted, it is necessary to calculate $g(X_t)$. In this way, samples are generated from the desired $\frac{f_X(x)}{c_n \Theta_n(x)}$ without a priori knowledge of the domains D_i in which the input state space gets partitioned by the function $g(\cdot)$. In the appendix, we also point out that sampling from the desired distribution is obtained, i.e., ergodicity is achieved, only when the number of samples per cycle N is sufficiently large. Hence, the choice of N may seem critical for a correct sampling. However, in practice for MMC, and other FH algorithms such as WL [29], this is not a key problem. Even if the cycle length is not long enough, the next cycles tend to correct such lack of ergodicity, and explore the state space more evenly. What matters is not correct sampling from the warped PDFs, but convergence to the FH distribution.

MCMC is in widespread use today in statistics and is routinely used in FH algorithms, including MMC. An advantage of the MCMC sample generation method is that the input PDF need only be known up to a multiplicative constant, hence the constant c_n need not be evaluated; this can be a tremendous computational savings for some high-dimensional input spaces [26]. A drawback is that samples are correlated, thus making the estimation of the error in the MMC PDF estimation more laborious than with independent samples [9].

When generating warped samples at the nth cycle in an MMC algorithm using the MCMC machine, the odds ratio (10.36) for the desired UW warping (10.12) becomes

$$R_{ij} = \frac{\Theta_n(x_i) f_X(x_j) q_{ji}}{\Theta_n(x_j) f_X(x_i) q_{ij}} \tag{10.26}$$

and the constant c_n cancels out. As suggested in [4], the odds ratio can be simplified to

$$R_{ij} = \frac{\Theta_n(x_i)}{\Theta_n(x_j)} \tag{10.27}$$

by choosing $q_{ij} = f_X(x_j)\Delta x$, i.e., by having a candidate chain whose transition probability only depends on the final state x_j; the proposed candidate x_j is drawn from the original distribution f_X independently of the initial state x_i. This is known as an independence chain [31]. To find (10.27), we need only calculate $y_j = g(x_j)$ for the selected candidate x_j ($y_i = g(x_i)$ was already calculated at the previous sample) to determine to which bin it belongs and thus determine the value of $\Theta_n(x_j)$, i.e., the intermediate estimate of the output PMF at cycle n of such a bin.

A direct use of the candidate independence chain would clearly lead to too many rejections in a large K-dimensional state space Γ. Hence in [4], it is suggested to implement the candidate chain itself using an MCMC machine with element-wise independent Metropolis reject/accept mechanisms: this technique is known as concatenation [32] or one-variable-at-a-time [31], and works as follows. For all elements $1 \le k \le K$

1. Starting from the kth element $x_{k,i}$ of vector x_i the kth element of candidate vector x_j is Metropolis generated as

$$x_{k,j} = x_{k,i} + U_k \tag{10.28}$$

   with U_k a scalar uniform RV;
2. If $G_k(\cdot)$ is the marginal PDF of $f_x(\cdot)$ for the kth element of vector x, the move $x_{k,i} \to x_{k,j}$ is accepted for the candidate with probability $\alpha_{ij}^{(k)} = \min\left[1, \frac{G_k(x_{k,j})}{G_k(x_{k,i})}\right]$; if the move is rejected, $x_{k,j} = x_{k,i}$.

It can be shown that if X has independent elements, i.e., $f_x(x_i) = \Pi_{i=1}^{K} G_k(x_{k,i})$, then $\frac{q_{ji}}{q_{ij}} = \frac{f_x(x_i)}{f_x(x_j)}$, and (10.26) simplifies to (10.27). Once the new candidate x_j is formed as described previously, the global move $x_i \to x_j$ is accepted based on the odds ratio (10.27). Since candidate moves $x_i \to x_j$ are made at smaller distances by suitable choice of the variance of the Metropolis RVs $\{U_k\}$, the rejection ratios can be substantially decreased, accelerating the state exploration.

The complete block diagram of the MMC simulator is given in Fig. 10.5.

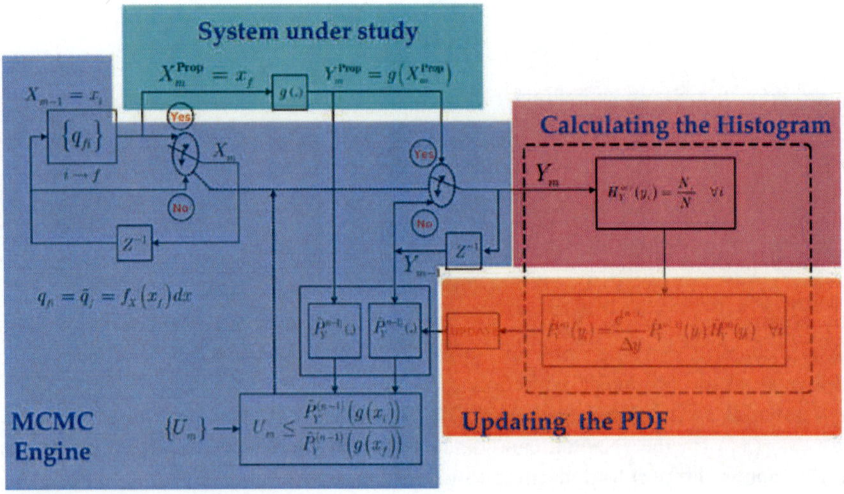

Fig. 10.5 Complete block diagram of the MMC algorithm, or "MMC machine"

10.4 Implementation Issues

10.4.1 Minimizing Rejections

10.4.1.1 Discretization of the Output Space

The choice of bin width Δy which defines the bins B_i in the output space is critical for proper operation of MMC. If Δy is too small, a very high number of samples is required for an accurate estimate of the output PMF $\Theta_{n,i}$. If, on the other hand, Δy is too large, we may encounter very large deviations in the PMF for two adjacent bins B_i and B_{i+1}: $\Theta_{n,i} \gg \Theta_{n,i+1}$. In such a case, the odds ratio of (10.27) would be very small, and the MCMC machine will move too slowly in the exploration of the state space. We empirically find that the bin width should be chosen such that adjacent bins have probabilities within one order of magnitude of one other.

10.4.1.2 Exploration of the Input Space

As shown in (10.28) of the appendix, the MCMC machine needs a vector U to produce a future state X of the chain. If the elements of X are independent and identically distributed (i.i.d.), then the elements of U are i.i.d. uniform random variables. The kth element of U is denoted by U_k, and is distributed over the range $[-\Delta_U/2, +\Delta_U/2]$. The value of Δ_U is a key parameter for the MCMC algorithm to sample correctly the input space. Intuitively, if it were too big then the proposed state would likely fall very far from the present state. This would lead to a high rejection ratio, and hence the chain would hardly move. On the other hand, if Δ_U were too small, the rejection ratio would be higher but the steps would be very small, hence the chain would move very slowly and it would take a very high number of samples for it to reach the steady state. We empirically find that a good compromise is $\Delta_U \sim \sigma$, where σ is the standard deviation of the known true distribution of the i.i.d. elements of the input vector.

10.4.2 Input Vector Correlations

From the discussion in the appendix on MCMC, one problem of the state space exploration with a symmetric Metropolis candidate chain is that no preferential directions are present in the exploration. Hence such a method is most effective in sampling input distributions f_X with independent elements, while lower efficiency is obtained when correlations are present [32]. In such a case, more sophisticated exploration criteria such as Hamiltonian and related methods should be used ([32], Chap. 30).

There is, however, a countermeasure for correlations for most nonpathological cases. As long as the input process is wide sense stationary, we are assured by Wold's decomposition theorem [33] that a whitening filter exists. Such a filter can be included as part of the system, and an input distribution with uncorrelated elements can be used. The whitening operation is quite effective in dealing with Gaussian vectors, since lack of correlation implies independence. The trade-off here is clearly the analytical pre-calculation of the whitening filter.

This issue is closely related to the scaling of the simulation time with the dimension of the input vector X. Although in MCMC the state space can be continuous, thinking of such a space as discrete and recalling the MCMC random walk in state space described in the appendix helps us develop intuition about the scaling rule.

Suppose the dimension of the input state is K, and b_x is the number of states per input random element and that this provides adequate resolution for the simulation. For the case of dependent elements in X, we must create a K-dimensional input space and test all possible combinations of the ordered pairs in generating samples according to our warped distribution. Hence, the input PDF spans a K-dimensional space and we require b_x^K states, i.e., an exponential increase with K in the number of states in the Markov chain. If the elements are instead independent, we only need to correctly sample each of them on b_x states, hence the exploration complexity scales linearly with K.

10.4.3 Choice of Number of Cycles vs. Samples per Cycle

In order to resolve the estimated PDF down to a desired level, the choice of the *cycle size* N, i.e., of the number of samples per cycle, is of great importance. For the Chi-square example in Sect. 10.3.3, Fig. 10.6 shows the number of cycles N_c vs. cycle size N to achieve a desired PDF estimation precision over the range of interest. Precision is quantified here in terms of the largest relative error ε over all bins in the PDF estimation in one cycle with respect to the previous one: $\varepsilon \triangleq \max_i \frac{|\Theta_{n,i}-\Theta_{n-1,i}|}{\Theta_{n-1,i}}$. If at the end of a cycle the target precision is not achieved, another cycle of size N is executed. The explored range was $\mathscr{R}_y = [0, 75]$, with 25 bins of width $\Delta y = 3$, on which the PDF reaches as low as about 10^{-12} (Cfr. Fig. 10.4). Figure 10.6 shows N_c vs. N for three different accuracy levels ε of 1.5, 3, and 6%. Clearly, the smaller ϵ, the larger the number of cycles needed. For each fixed precision, the number of cycles increases as we decrease the cycle size, and diverges as N approaches an asymptotic value N_0 related to the bound in (10.17). The computational cost of MMC depends on the total number of simulated samples $N_T = N \cdot N_{cycle}$. The figure also shows the hyperbolas corresponding to different total cost N_T from 10^5 to 10^6 in steps of $2 \cdot 10^5$. The message from superposing such hyperbolas to the constant-precision N_c vs. N curves is clear: the lowest-cost cycle size N for a given precision is usually close to the lower bound N_0. It is not necessary to make N very large (e.g., in order to achieve ergodicity in the sampling MCMC), but a smaller cycle size and more cycles achieve the same goal at a lower

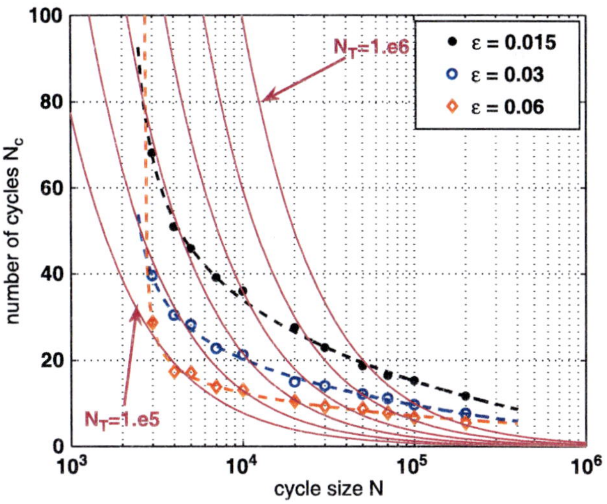

Fig. 10.6 *Symbols*: number of cycles N_c vs. cycle size N for given precision ε (see definition in text) for the Chi-square problem in Sect. 10.3.3. PDF resolved down to 10^{-12} over range $\mathscr{R}_y = [0, 75]$. Computational cost hyperbolae $N_T = N \cdot N_{cycle}$ shown in *solid lines* for various values of N_T

total cost. Similar performance curves can be found for more complicated problems. N_0 is widely problem dependent, and is typically larger for a smaller desired PDF level to be resolved (here it was 10^{-12}).

10.4.4 Dealing with System Memory

So far we assumed that the input state X is a continuous random vector such as, additive noise samples accumulated by the signal as it propagates along a transmission line. However, most often X is a mixture of both continuous and discrete RVs, e.g., in a system with inter-symbol interference (ISI). Let $B = [b_1, \ldots, b_{K_1}]$ be the vector of (independent) neighboring symbols that contribute to determine the value of the decision variable Y, and $N = [N_1, \ldots, N_{K_2}]$ be the vector of continuous noise samples; thus, the input state is $X = [B; N]$. In such a case, the MCMC random walk update can proceed with the one-variable-at-a-time technique discussed in Sect. 10.3.4.

As explained in Sect. 10.4.1.2, it is important to restrict the range of exploration when generating candidates in the Metropolis algorithm using (10.28). For generation of binary symbols, $b_i \in \{0, 1\}$, Secondini et al., [15] suggest candidate symbol vector $B_j = B_i \oplus U$, where $\oplus$ denotes modulo-2 addition, and U is a vector of (0,1) independent RVs with average p_B. If p_B is suitably small, the MCMC will explore a local neighborhood of bits, rather than all 2^{K_1} possibilities. Note that K_1 is often referred to as the memory of the system, and such a value is most often unknown. An alternative but similar approach was taken in [34]; in the following section, we work out in detail an example clarifying these ideas.

10.5 Examples

We conclude with some examples intended to highlight successful applications of MMC in the solution of design and analysis problems in optical communications.

10.5.1 Example: Bit Patterning in SOAs

10.5.1.1 SOA Memory

The MMC method can characterize the statistical properties of bit patterning in semiconductor optical amplifiers (SOAs). The BER of the system is estimated by first generating the conditional PDFs of marks and spaces. The results presented in this section were validated experimentally and are summarized from [34].

A frequently adopted means to evaluate the BER in optical communication is the semianalytical numerical method based on Karhunen-Loeve (KL) expansion and saddle-point integration [35]. KL-based semianalytical BER calculation is accurate when pre-photodetection noise is Gaussian. While this holds for moderate fiber nonlinearity in special cases [36], the signal-noise interdependency in general limits the applicability of the KL-based method. The KL-based method is of limited value when a saturated SOA is in the link.

The SOA is a nonlinear element with memory [1]. The nonlinearity of the SOA is mainly due to carrier depletion induced saturation (typical saturation power of SOAs is around 1–10 mW), whereas its memory is due to its finite carrier lifetime (typically about 100–500 ps) [37]. The signal-dependent, instantaneous gain of the saturated SOA results in non-Gaussian statistics at the output, and the finite memory of the SOA leads to bit patterning effects, thus resulting in nonlinear, i.e., signal-dependent, enhancement of the intersymbol interference, on top of the linear ISI enhancement stemming from fiber dispersion, optical and electrical filters. Analytical treatments are intractable due to the inherent complexity of the problem, hence we turn to MMC.

10.5.1.2 SOA Modeling

The typical link under study is shown in Fig. 10.7a, where b_i are the information bits, E_{in} and E_{out} are the optical fields at the SOA input and output, respectively, $P_{\text{out}} = |E_{\text{out}}(t)|^2$ is the detected optical power, and $r(t)$ is the received signal.

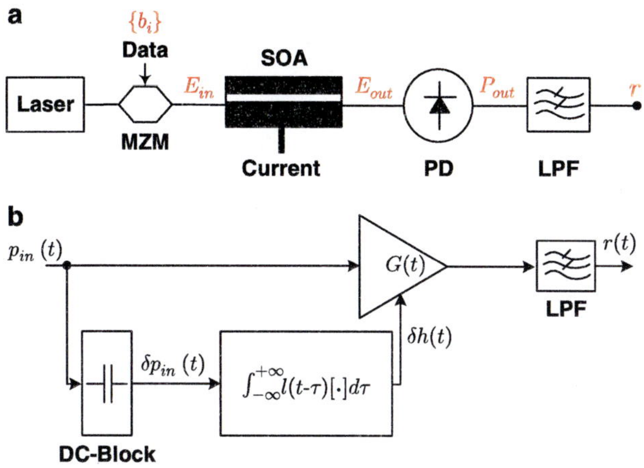

Fig. 10.7 (**a**) Basic setup, and (**b**) block-diagram of the equivalent lowpass SOA model

Our ultimate goal is to study the PDF of $r(t)$ sampled at the decision instant, taking into account the memory and nonlinearity of the channel represented in Fig. 10.7a. As a good compromise between computational complexity and completeness, we use the large signal numerical model presented in [38] to model the SOA. In this model, the SOA cavity is divided into several sections each with a lumped loss. The amplified spontaneous emission (ASE) is modeled as a complex Gaussian noise. We consider NRZ signals at $10\,\text{Gb}\,\text{s}^{-1}$, and thus we neglected the ultrafast effects, although the model [38] could encompass these effects if needed.

As mentioned previously, the nonlinearity of the SOA is mainly due to carrier depletion induced saturation, whereas its memory is due to its finite carrier lifetime. Bit patterning is only important when two situations occur. The SOA must be in saturation, e.g., as a booster amplifier, following in-line amplification in 2R, or in 3R regenerators. Also, the bit-rate must be comparable with the effective carrier lifetime: when the bit-rate is extremely high [39], or when the carrier lifetimes are very low (for example, novel quantum dot SOAs with high saturation power [40]), the patterning effect becomes less important. In the case of typical commercially available SOAs, and at bit-rates up to $40\,\text{Gb}\,\text{s}^{-1}$ some residual patterning effect will exist in SOA-based 2R regenerators [41].

Figure 10.8a illustrates the transmitter (implemented experimentally), and Fig. 10.8b shows its numerical model. Logical bits enter the transmitter (TX) subsystem and produce a realistic modulated optical field. We use the well-known two-port model of the Mach–Zehnder modulator (MZM) [42]. A lowpass fourth-order Bessel-Thompson (BT4) filter, $H_{\text{TX}}(f)$, smooths the logical bits. Figure 10.9 shows the measured waveform at the output of the transmitter and the simulated result.

A BER tester served as the receiver (RX), with model given in Fig. 10.10a. G_R contains the RF amplifier gain and all the losses either from VOAs or from optical

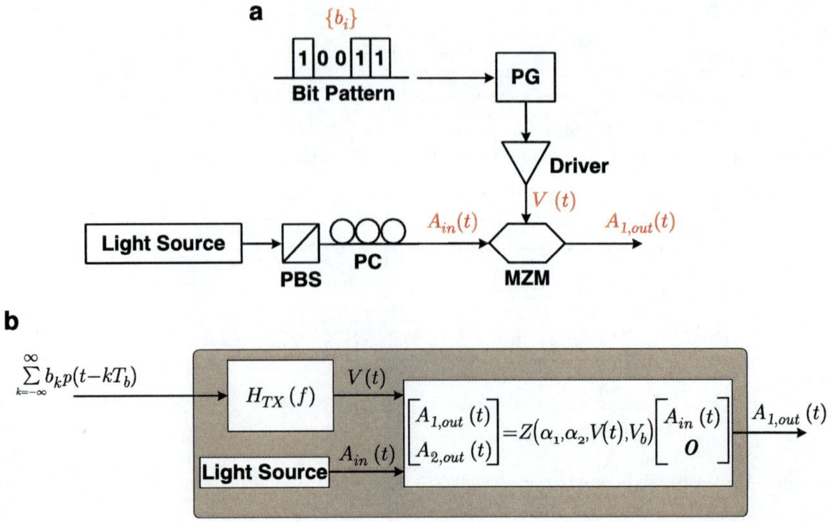

Fig. 10.8 (**a**) Transmitter (TX) configuration, (**b**) *TX* numerical model; *PBS* Polarization beam splitter; *PC* Polarization controller; *MZM* Mach–Zehnder modulator

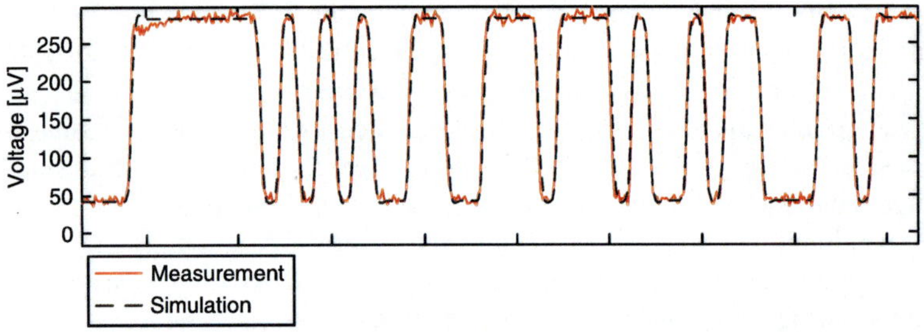

Fig. 10.9 Optical intensities at the output of the transmitter, measured (*blue*) and simulated (*red*)

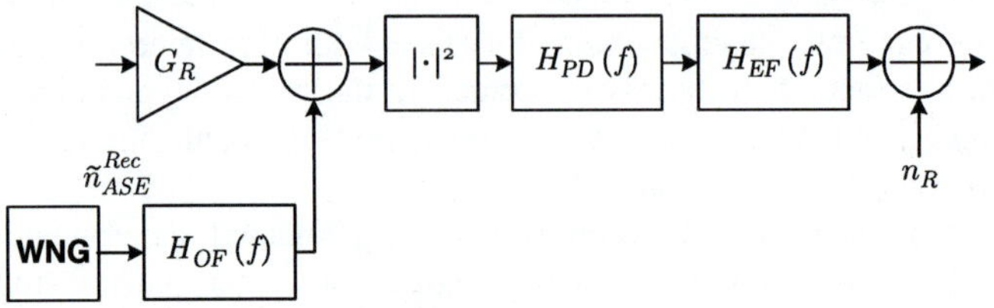

Fig. 10.10 Numerical model of RX (BER tester)

or RF couplings. A white complex Gaussian process, $\tilde{n}_{\mathrm{ASE}}^{\mathrm{Rec}}(t)$, models the noise generated by the broadband source. Measured frequency responses were used for the optical filter $H_{\mathrm{OF}}(f)$, the electrical filter $H_{\mathrm{EF}}(f)$, and the Agilent photoreceiver $H_{\mathrm{PD}}(f)$.

10.5.1.3 MMC Platform

Referring to Fig. 10.7, the received signal is

$$r\left(t\right)=b_e\left(t\right)\otimes P_{\text{out}}\left(t\right), \qquad (10.29)$$

where $b_e(t)$ is the impulse response of the electrical lowpass filter. The sampled received signal, corresponding to the current bit b_0, is $r_0 \triangleq r\left(t_\text{s}\right)$, where t_s is the optimum sampling time between 0 and T_b. The conditional PDFs of marks and spaces are written as

$$P_i\left(r_0\right)\triangleq p_{r_0|b_0}\left(r_0\left|b_0\right.=i\right), \qquad (10.30)$$

where $i=0$ ($i=1$) corresponds to the conditional PDF of spaces (marks). Assuming that the "effective" memory of the link is M bits, the truncated conditional PDF of marks and spaces is

$$P_{i,M}\left(r_0\right)=\frac{1}{2^M}\sum_{\{b_{-1},\ldots,b_{-M}\}} p_{r_0|b_0}\left(r_0\left|b_0\right.=i,b_{-1},\ldots,b_{-M}\right), \qquad (10.31)$$

where summation is over all possible patterns of the past M bits. By effective memory, we mean $\|P_{i,M}\left(r_0\right)-P_{i,M+1}\left(r_0\right)\|$ to be sufficiently small for some metric $\|\cdot\|$. We use MMC to estimate the effective memory length, and the conditional PDF $P_{i,M}\left(r_0\right)$. To determine memory length, we gradually increase M until successively estimated conditional PDFs coincide.

The block-diagram of our MMC simulator is shown in Fig. 10.11. The numerical system model is composed of three parts (TX, SOA, and RX), all described previously. We denote the simulation time step by Δt, and the number of time samples per bit by N_s, i.e., $T_\text{b}=N_\text{s}\Delta t$. Assuming the effective memory is M, the past MN_s time samples of all independent noise sources have an impact on the distribution of r_0. The vector of all noise samples is denoted by $\underline{X}$, which is explicitly written as

$$\underline{X}\triangleq\left[\tilde{n}_{\text{ASE}}^{\text{SOA}},\tilde{n}_{\text{ASE}}^{\text{Rec}},n_R\right], \qquad (10.32)$$

where $\tilde{n}_{\text{ASE}}^{\text{SOA}}$ and $\tilde{n}_{\text{ASE}}^{\text{Rec}}$ are vectors of independent identically distributed white complex Gaussian noise samples each of length MN_s; the former accounts for ASE noise from the SOA and the latter accounts the ASE of the pre-amplified receiver (cf. Fig. 10.10); n_R is a real Gaussian random variable with proper mean and variance modeling the receiver noise (cf. Fig. 10.10). The vector $\underline{B}$ contains all the past bits falling in the effective memory of the link

$$\underline{B}\triangleq\left[b_{-1},\ldots,b_{-M}\right]. \qquad (10.33)$$

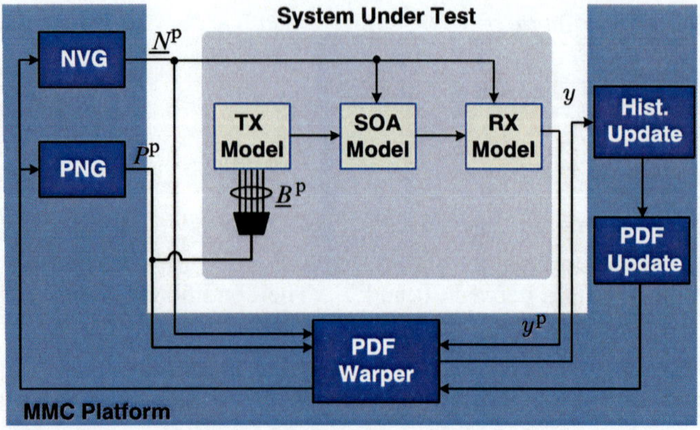

Fig. 10.11 Block diagram of the simulator; *NVG* Random vector generator; *PNG* Pattern number generator

The noise vector generator (NVG) subsystems in Fig. 10.11 is a Metropolis–Hastings machine [32], which proposes noise vector samples $\underline{X}^{\mathrm{p}}$. The pattern number generator (PNG) subsystem in Fig. 10.11 is an other Metropolis–Hastings machine, proposing pattern numbers P^{p}; the binary representation of a pattern number is the bit pattern. The PDF warper accepts or rejects the proposals from NVG and PNG $\left[\underline{X}^{\mathrm{p}}; P^{\mathrm{p}}\right]$ according to the MMC algorithm. Consequently, the PNG performs a random walk over the index in the summation of (10.31), while the NVG performs a random walk to explore the conditional PDFs within the sum.

10.5.1.4 Results

The experimental setup can be found in [34]. The SOA input power was −2.65 dBm, resulting in deep saturation; the bit-rate was $10\,\mathrm{Gb\,s^{-1}}$. We measured the BER as a function of the received optical signal-to-noise ratio (OSNR) and present these results in Fig. 10.12. MMC simulations (one for conditional PDF of marks, the other for spaces) were required at each BER point; the BER was computed by numerically integrating the overlapping tails of estimated conditional PDFs of marks and spaces. Conditional PDFs were calculated at the middle of the bit. Each PDF estimation included seven MMC iterations to improve the accuracy; each cycle took 71 s to execute. In the lower inset of Fig. 10.12, we show an eye diagram for high OSNR that clearly depicts the strong patterning effect from the SOA. The upper inset is the set of estimated conditional PDFs used to calculate one BER point.

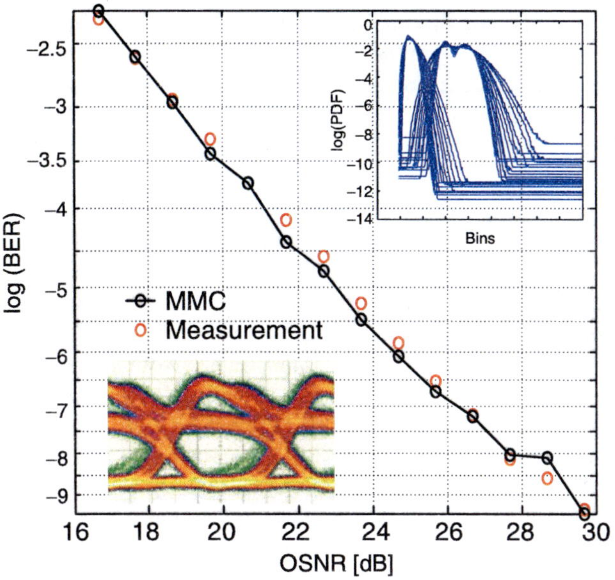

Fig. 10.12 Measured and simulated BERs; upper inset shows the conditional PDFs used to estimate the BER curve (one pair per BER curve point), lower inset is eye diagram for lowest BER estimated

10.5.2 Example: Spectral Efficiency in SS-WDM

10.5.2.1 Use of Forward Error Correction

If the symbol error rate of interest is very high, on the order of 10^{-3} when forward error correction (FEC) is used, then MMC is not a good accelerator. Other importance sampling techniques such as stratified sampling [43] may be more appropriate in that case. MMC is also challenging to use when the system under test includes FEC. The introduction of FEC leads to isolated islands in the input space being responsible for error events. With isolated islands, the MCMC exploration of critical regions of the input space can be difficult ([32], Chap. 31). Nonetheless, some researchers have partially succeeded in using MMC to test numerical models with FEC [44,45]. Note that these deficiencies are not unique to MMC; indeed all Monte Carlo techniques have difficulty exploring FEC performance.

Despite these limitations, we next present an example where MMC was nonetheless useful in examining the use of FEC; the example is also interesting as it implements a parallel version of MMC. In [46], we examined the spectral efficiency of spectrum sliced wavelength division multiplexed (SS-WDM). MMC allowed us to study the impact of the shape of both slicing and channel selecting optical filters vis-à-vis two important impairments: the filtering effect and the crosstalk. By varying channel spacing and width, we estimate the achievable spectral efficiency when two noise suppression techniques are used: SOA gain compression to reduce intensity noise, and FEC to combat combined intensity noise and crosstalk. MMC was key to this study as the region of FEC effectiveness was unknown a priori while sweeping through filter designs. The BER was simulated in MMC and validated experimentally. We found optical filter shape and bandwidth that minimizes BER.

10.5.2.2 Modeling SOA Noise Suppression

Spectrum-sliced wavelength division multiplexing (SS-WDM) employing a shared thermal-like broadband source is a candidate for future (metro or access) all-optical networks due to its low cost. The excess intensity noise of the thermal source leads to BER floors [47]. For example, at 2.5 Gb s^{-1} over a 21 GHz slice width, a BER floor $\simeq 10^{-4}$ is reported in [20] for a single-user experiment.

Placing a saturated semiconductor optical amplifier (SOA) after the spectrum-sliced source, and before the modulator, is an attractive all-optical signal processing technique that vastly reduces intensity noise. Noise suppression in SOA-assisted SS-WDM is due to the nonlinear operation of the saturated SOA. Optical filtering of the noise-suppressed light significantly degrades noise suppression [20, 48], a phenomenon which is referred to as the *filtering effect* or *post-filtering effect*. A simplified block diagram of a SOA-assisted SS-WDM architecture is provided in Fig. 10.13.

Theoretical analysis of SOA-assisted SS-WDM systems is prohibitively complex for two reasons: (1) the SOA operates in the nonlinear regime resulting in highly non-Gaussian light statistics at its output [20], and (2) linear filtering of this non-Gaussian process couples phase and amplitude effects through a complex process parameterized by the SOA linewidth enhancement factor. Due to the limitations of the analytical treatment of SOA-assisted SS-WDM systems, we resort to numerical simulations. We focus on the impact of the shape and bandwidth of optical filters in the transmitter (slicing filter SF), and receiver (channel select filter CSF) on the overall performance of multi-channel SOA-assisted SS-WDM systems. As we needed to search through a large optimization space for the filters, we examined ways the

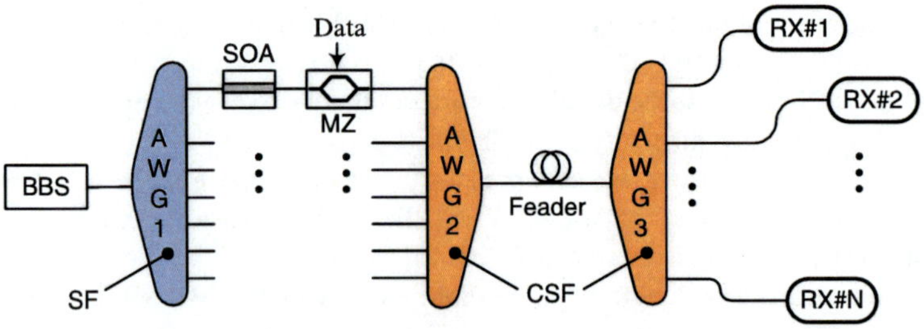

Fig. 10.13 SOA-assisted SS-WDM architecture. Arrayed waveguide gratings (AWG) are independently designed, i.e., SF and CSF bandwidths are independent

MMC could be further accelerated. To this end, we introduced a novel parallelized implementation of the MMC (PMMC) [49].

We also examine combining FEC and SOA noise suppression to achieve high spectral efficiency (SE). These MMC simulations were doubly challenging as (1) spectral efficiency calculations required examination of channel spacing as well as optimal filter widths, and (2) the BER had to be calculated for each channel configuration to find the FEC sweet spot. Compiling many dozens of BER curves, we find the optimal attainable spectral efficiency when combining FEC and SOA.

We examined a single-channel SOA-assisted SS-WDM system experimentally. We also demonstrated the accuracy of our simulator by cross-validating it against published measurements of three different multi-channel SOA-based SS-WDM systems [48, 50, 51]. Good agreement of our simulated results with the published measurements, despite the lack of exact characterizations, indicates the reliability of our simulator.

10.5.2.3 Multi-Channel MMC Platform

The block diagram of the multi-channel MMC platform, used to estimate the conditional probability density functions (PDF) of the received marks and spaces and thereby the system BER, is shown in Fig. 10.14. We confined our study to a three-channel scenario where the central channel is the desired channel; [50] found a three-channel system sufficient to capture crosstalk effects.

Three replicas of the link model are used to model the desired channel and two adjacent channels. Since the link model is baseband, the adjacent channels are up-, and down-converted. The channel-spacing is denoted by $\Delta\omega$. The proposed vectors in the input space are $\underline{X}^{\mathrm{p}} \triangleq \left[\underline{N}^{\mathrm{p}}; \underline{P}^{\mathrm{p}}; \underline{t}^{\mathrm{p}}\right]$, which map to output samples $y^{\mathrm{p}} \triangleq g\left(\underline{X}^{\mathrm{p}}\right)$, where $g(\cdot)$ is an abstract mapping formally representing the system. The superscript "p" indicates a proposed sample that may or may not be rejected within the MMC algorithm. To indicate an accepted proposal, we drop the superscript in Fig. 10.14. The proposed input vector consists of three parts. The noise vector $\underline{N}^{\mathrm{p}} \triangleq \left[\underline{N}_1^{\mathrm{p}}, \underline{N}_2^{\mathrm{p}}, \underline{N}_3^{\mathrm{p}}, N_r^{\mathrm{p}}\right]$ contains identical independent Gaussian random variables of zero mean and unit variance; the sub-vector $\underline{N}_j^{\mathrm{p}}$ is used to model the incoherent spectrum-sliced source of the jth user, and N_r^{p} is a scalar modeling receiver electrical noise.

The noise vectors are generated by a Metropolis–Hastings machine (NVG). The proposed bit pattern vector is $\underline{P}^{\mathrm{p}} \triangleq \left[P_1^{\mathrm{p}}, P_2^{\mathrm{p}}, P_3^{\mathrm{p}}\right]$, where P_j^{p} is the decimal representation of the binary bit pattern of the jth channel. The bit pattern proposed for the jth channel is denoted by $\underline{B}_j^{\mathrm{p}}$ [15,20]. The pattern numbers are generated by another Metropolis–Hastings (PNG). The relative delay vector is $\underline{t}^{\mathrm{p}} \triangleq \left[t_1^{\mathrm{p}}, t_2^{\mathrm{p}}\right]$, which is composed of random variables representing the time delays between the desired channel and the adjacent interfering channels. The Metropolis–Hastings machine generating the vector of relative delays is called the interferer delay generator (IDG).

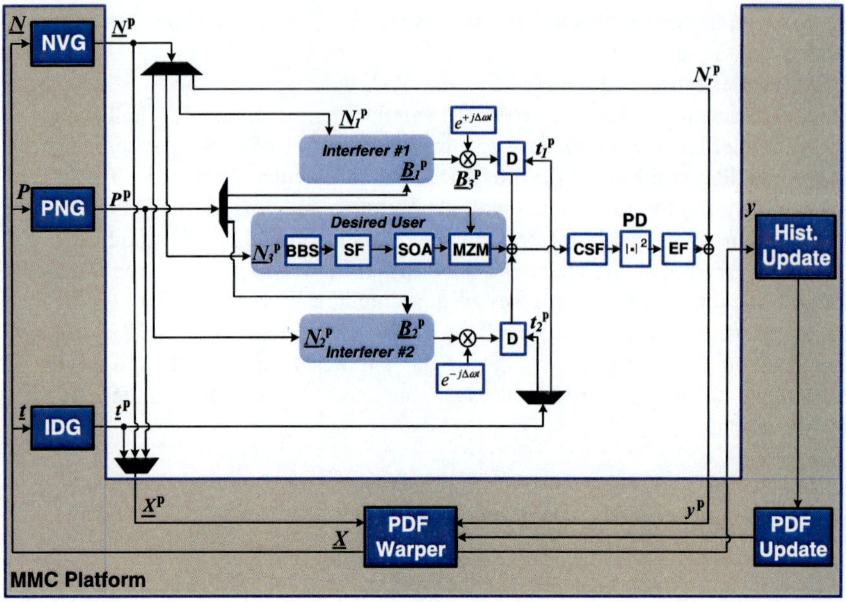

Fig. 10.14 Three-user SOA-assisted SS-WDM MMC platform. *NVG* Noise vector generator; *PNG* Pattern number generator; *IDG* Interferer delay generator; *D* Programmable temporal delay element

The effective memory of the single-user system is assumed to be $M-1$ bits. To estimate the conditional PDF of marks (spaces) of the desired user, the current bit of the center channel is set to 1 (0), and the past $M-1$ bits are adaptively changed by the MMC platform; therefore, P_2^{p} is an integer random variable (rv) uniformly distributed between 0 and 2^{M-1}. P_1^{p} and P_3^{p} are integer rvs uniform between 0 and 2^{M+1}. The relative delays t_1^{p} and t_2^{p} are integer rvs uniform over 0 and $N_{\mathrm{s}}-1$, where N_{s} is the number of time samples per bit duration.

10.5.2.4 Parallelization of MMC

Conventional MC for PDF estimation of rvs is "embarrassingly" parallelizable, as random samples can be independently generated by different cluster nodes. At the end of the simulation, all samples are collected and the histogram is calculated over all collected samples. In the case of MMC, the proposed samples are generated by Markov chains (using the Metropolis–Hastings algorithm), a process which is sequential in nature. While at first blush MMC does not appear parallelizable, we show that, fortunately, this is not the case.

Consider a 1-dimensional input space where sequential MMC is used to estimate the output PDF. During each MMC cycle, the Metropolis–Hastings module of the MMC generates a random walk in the 1-dimensional input space. Suppose

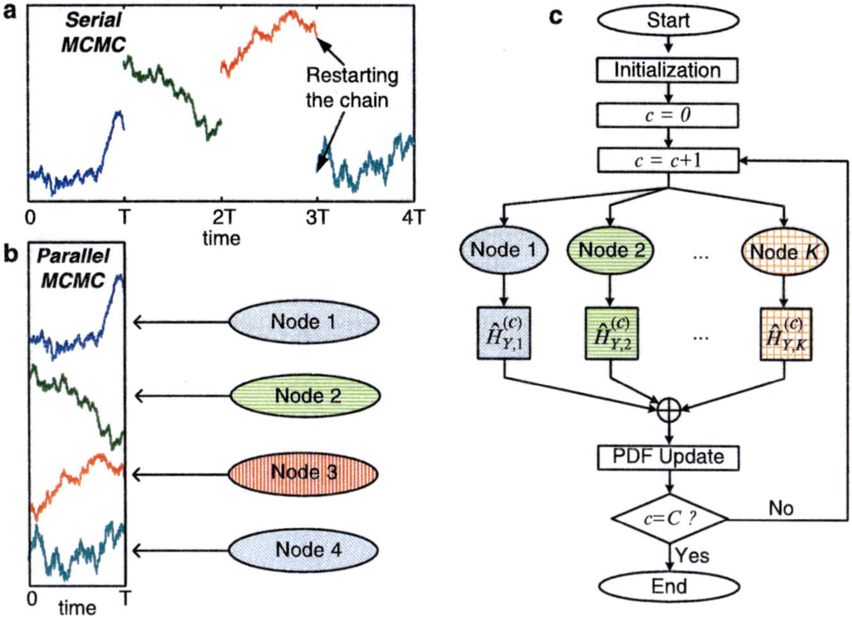

Fig. 10.15 Parallelization of MMC: (**a**) Random walk in a 1-dimensional input space perturbed by periodic reinitializations. (**b**) Sections of the perturbed Markov chain are mapped to various computing nodes, (**c**) the flowchart of the parallel MMC; c counts the MMC cycles, C is the pre-specified number of cycles, $\hat{H}^{(c)}_{Y,j}$ is the histogram computed by node j at the end of cycle c

we periodically perturb the random walk in the input space by re-initializing it, as shown in Fig. 10.15a. Each random walk is generated by the same Metropolis–Hastings submodule as before, but at time instants T, $2T$, $3T$, and $4T$, we select a new random state in the input space. The initial states are assumed independent and uniformly distributed over the input space.

The perturbed Markov chain is *not* statistically equivalent to the original unperturbed Markov chain, required by the MMC platform, as the forced jumps induce transients. If, however, the MMC platform discards the transient samples after each forced jump, the remaining samples of the perturbed Markov chain will lead the MMC to the same solution as the single Markov chain case. The perturbed random walk provides the transition from sequential to parallel implementations of the MMC. The generation of each segment of the perturbed random walk can be assigned to a different computing node, as shown in Fig. 10.15b, allowing for parallel processing.

During each MMC cycle, all nodes run exactly the same code to propose new samples, and perform an accept/reject operation accordingly. At the end of each MMC cycle, all the output samples are collected by a pre-specified head node, the PDF update and smoothing are executed, and the updated PDF is broadcast to all nodes for the next MMC cycle. We call this parallel implementation of MMC the PMMC. The flowchart of PMMC is shown in Fig. 10.15c. The PMMC follows the paradigm of SPMD (single program multiple data). In [18], another parallel implementation of MMC is introduced; however, as explained by the author, the resulting algorithm is a problem-dependent, modified MMC without the important PDF smoothing feature. Our PMMC, however, is a natural parallelization of the MMC, without any modification to the original algorithm.

Note that even in sequential MMC, we discard transient elements at the beginning of each MMC cycle. The length of the transient period is problem dependent, and is fixed during the code development and fine-tuning of the simulator. We discarded the first 100 samples at the beginning of each MMC cycle per node. We parallelized four cores of a Quad Intel processor, and obtained a three-fold speedup. The rigorous theoretical analysis and optimization of PMMC will be addressed in future work.

10.5.2.5 Simulation Results

The shape of the slicing (SF) and channel select (CSF) filters are quantified as the order of a super-Gaussian shape (0.4, 1, 2 or 4). In the multichannel scenario, we found higher order to be most effective. The performance only slightly changes from super-Gaussian order 2–4. From a practical point of view, realizing super-Gaussian filters of lower orders is easier, and we present results for order 2.

Having fixed the shape, we sweep through channel select filter widths for a fixed slicing filter width. We compared the BERs for $n_{SF} = n_{CSF} = 2$ in Fig. 10.16 for single and multi-channel cases using an SF of 30 GHz and a bit rate of 5 Gb s^{-1}. In the optimum multichannel case, employing the SOA-assisted scheme decreases the BER from 1E-2 to 1E-10. The threshold of powerful FEC codes is at 1E-3.

For each BER point, two MMC simulations were performed to estimate the conditional PDFs of marks and spaces; the BER was calculated by integrating the overlapping tails of the two conditional PDFs. Each MMC simulation consisted of 12 cycles; 50,000 samples were generated per cycle. We assumed $M = 3$ bits of effective channel memory. After parallelization, each BER point was calculated in 25 min.

To find optimum spectral efficiency, we independently vary the CSF bandwidth and SF bandwidth. The SF bandwidth, BW_{SF}, takes on 14, 22, 26, or 30 GHz and several channel spacings Δ_{CH} are considered. For each combination (BW_{SF}, Δ_{CH}), the BW_{CSF} is swept through the range [$2BW_{SF}$, ..., $2\Delta_{CH} - 2BW_{SF}$]. To increase resolution, the channel spacing covers [$s * 60$ GHz, ..., $s * 100$ GHz], where the scaling factor s is defined as $BW_{SF}/30$ GHz. BER curves are presented in Fig. 10.17.

We next use the BER curves to find optimal spectral efficiency. We select the CSF bandwidth yielding the minimum BER for each (BW_{SF}, Δ_{CH}). For each combination of SF bandwidth and channel spacing, we calculate BER and SE. BER is reported in Fig. 10.18; the SE is posted next to each point. Each BER curve in Fig. 10.18 corresponds to a fixed BW_{SF}, therefore the range of channel spacings

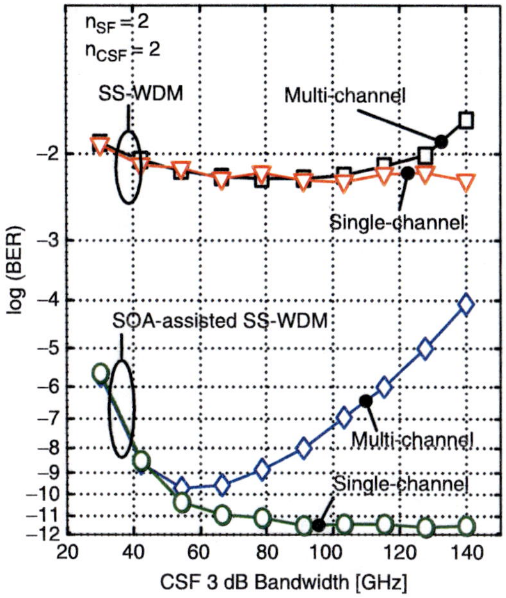

Fig. 10.16 Comparison of BERs of SS-WDM and SOA-assisted SS-WDM; $n_{\mathrm{SF}} = n_{\mathrm{CSF}} = 2$

examined differs from one curve to other; however, the ratio of channel spacing to SF bandwidth sweeps over the same range for all curves.

As can be seen in Fig. 10.18, at a fixed BER, the narrower SFs are favorable, although variations of SE vs. BW_{SF} are not significant. Employing an FEC with $\eta_{\mathrm{FEC}} = 10^{-5}$ increases the SE from 0.025 bits s^{-1} Hz^{-1} to 0.12 bits s^{-1} Hz^{-1} when $BW_{\mathrm{SF}} = 14$ GHz. This should be compared to 0.072 bits s^{-1} Hz^{-1} in the first scenario. A FEC with $\eta_{\mathrm{FEC}} = 10^{-3}$ would result in SE = 0.28 bits s^{-1} Hz^{-1}, when BW_{SF} =14 GHz, and still higher spectral efficiencies are possible by lowering the SF bandwidth. The second scenario allows the noise cleaning to have its full effect, so that overall spectral efficiency sees a significant increase. Combining efficient noise cleaning with FEC is an effective tool to enhance spectral efficiency. Our tool allows for design and optimization, once the architecture and the FEC type are known. BER points in Fig. 10.17 required 25 min, as MMC parameters are like those of the multi-channel BER simulations of the previous section. Generating all results of Fig. 10.17 took 5.5 day; our computing cluster was limited to four nodes.

10.5.3 Example: Nonlinear Interaction Between Signal and Noise in Very-Long-Haul Dispersion-Managed Amplified Optical Links

This example focuses on the study of the nonlinear interaction between signal and noise in very-long-haul dispersion-managed (DM) amplified optical links.

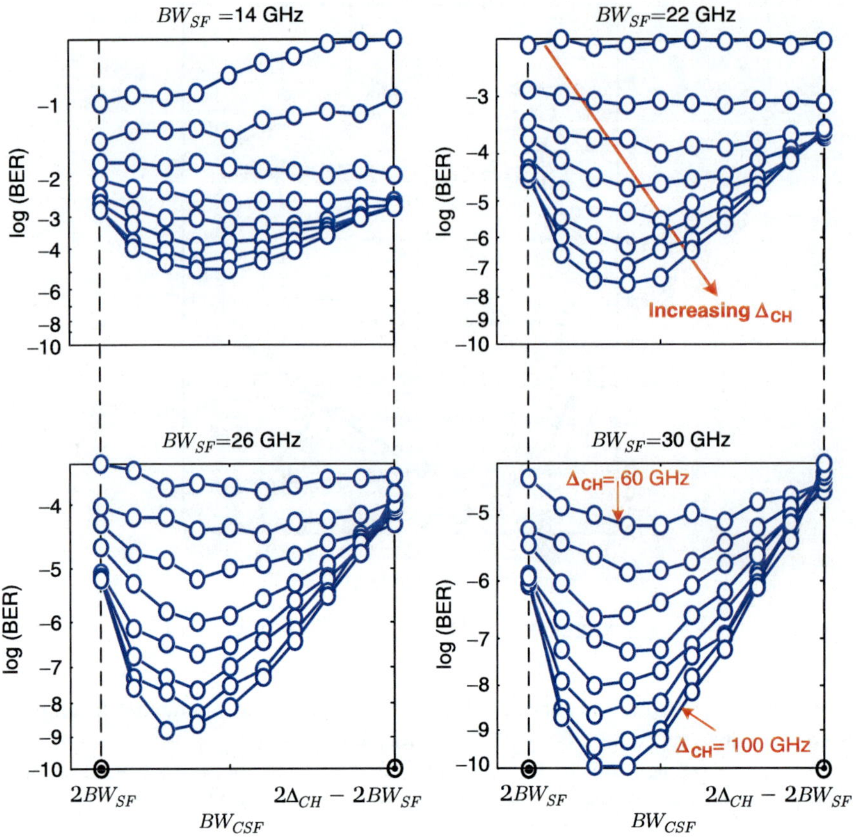

Fig. 10.17 All BER curves estimated by PMMC during the SE optimization process for the second scenario. Each curve corresponds to a different channel separation, as described in the text

The material is summarized from [52]. The example is meant to stress the importance of the MMC method as a testing tool for analytical or pseudoanalytical models.

10.5.3.1 Received ASE Statistics

The ASE noise and the transmitted signal interact during propagation through a four-wave mixing process that colors the power spectral density (PSD) of the initially white ASE noise components, both in-phase and in-quadrature with the signal through a parametric gain (PG) process [53]. It is known that signal and ASE noise have maximum nonlinear interaction strength at zero group-velocity dispersion (GVD), yielding ASE statistics that strongly depart from Gaussian [54]. We already showed [36] that the presence of a non-zero transmission fiber GVD helps

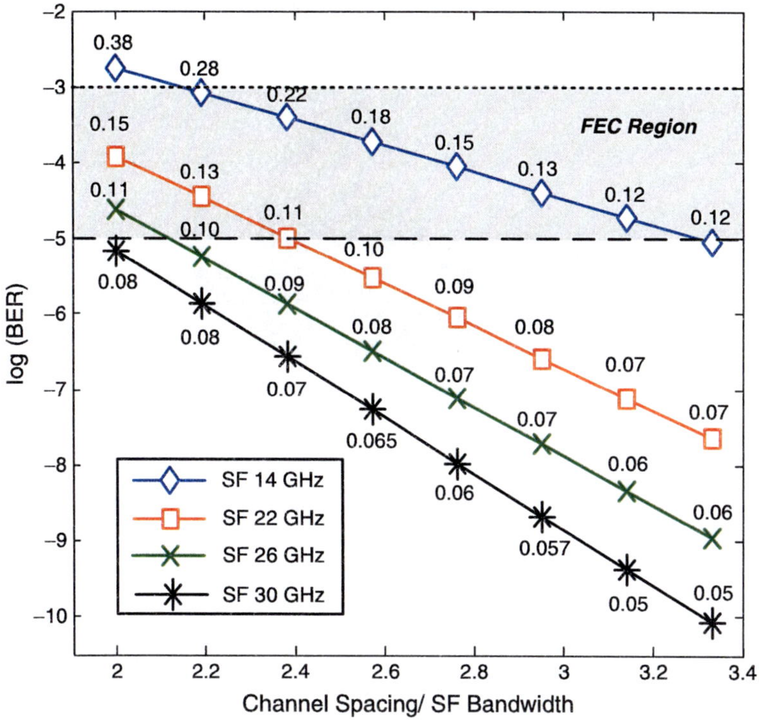

Fig. 10.18 Minimum BER (CSF bandwidth optimized) vs. normalized channel spacing, corresponding to four systems with different SF bandwidths, for the second scenario. The spectral efficiency (in bits/s/Hz) is given next to each point

reshape the statistics of the optical field (in-phase and quadrature components) *before* the optical filter at the receiver, so that they are quite close to Gaussian. We want here to further support the results presented in [55], and show that also the filtering action of the receiver optical filter helps make the statistics of the filtered optical field resemble a Gaussian bivariate density.

Figure 10.19 shows an MMC simulation of the joint probability density function (PDF) of the in-phase and quadrature components of an initially unmodulated (CW) optical field before the receiver optical filter, in the case of zero transmission fiber GVD and no DM, at a nonlinear phase rotation $\Phi_{\rm NL} = 0.2\pi$(rad) and at a linear optical signal-to-noise ratio OSNR $= 10.8$ dB/0.1 nm (the one that can be read off an optical spectrum analyzer, when reading the ASE power level away from the signal, where no PG exists).

The joint PDF was obtained using the two-dimensional extension of the MMC method presented in [25], with 6 MMC cycles with $3 \cdot 10^6$ samples each. One can note the well-known shell-like shape of the joint PDF at zero GVD [56]. Figure 10.20(top-left) shows the corresponding contour plot of the PDF surface in Fig. 10.19, resolved down to 10^{-12}. The simulated optical bandwidth was 80 GHz.

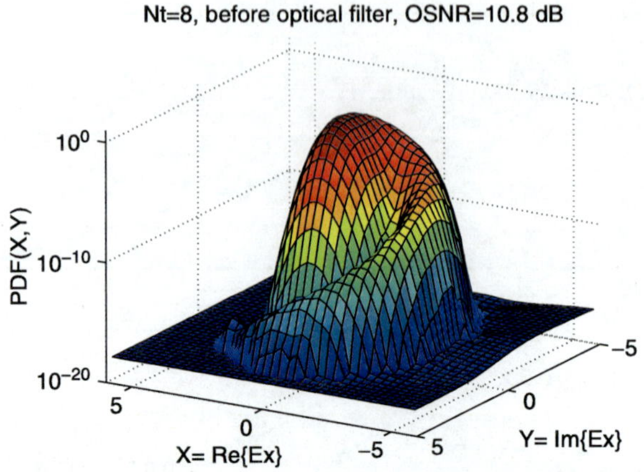

Fig. 10.19 MMC-simulated joint PDF of in-phase and quadrature components of optical field (CW+ASE) before receiver optical filter. Simulated bandwidth 80 GHz. Zero chromatic dispersion, nonlinear phase $\Phi_{\mathrm{NL}} = 0.2\pi(\mathrm{rad})$, OSNR = 10.8 dB/0.1 nm. MMC time samples $18 \cdot 10^6$

The remaining plots in Fig. 10.20 show instead the PDF contours of the same optical field, but *after* an optical filter of bandwidth of 30, 20 and 10 GHz, respectively. We clearly appreciate the tendency of the contour levels to elliptical shapes for tighter optical filtering, even in this extreme case of zero GVD. Hence, we can conclude that the joint action of tight optical filtering and transmission fiber GVD both contribute to make the received optical field after optical filtering resemble a Gaussian process.

10.5.3.2 Transmission Test

We consider transmission of a single-channel DPSK signal in a single-period dispersion-managed (DM) optical link, as shown in Fig. 10.21. There are 20 identical spans, each composed of a 100 km long transmission fiber with dispersion $D_{\mathrm{Tx}} = 4\,\mathrm{ps\,nm^{-1}\,km^{-1}}$ and positive in-line residual dispersion $D_{\mathrm{in}} = 40\,\mathrm{ps\,nm^{-1}}$ per span. No pre and post-compensation was used here. The receiver consists of a Gaussian-shaped optical filter, followed by a DPSK delay-line demodulator with balanced photodetection. The difference between the received currents from the two photodiodes is filtered by a Bessel 5th order filter of bandwidth $B_e = 0.65$ time the bit rate, and then sampled.

The procedure to evaluate BER once the statistics of the Gaussian received ASE are known is discussed in detail in [36]. Here, we provide numerical tests of the analytical model with respect to "true" performance obtained with the MMC method.

In Fig. 10.22(left), we checked the analytical PDF of the sampled current at the decision gate against that obtained through direct simulation with the MMC method. The nonlinear phase was $0.2\pi(\mathrm{rad})$, and a single $10\,\mathrm{Gb\,s^{-1}}$ NRZ-DPSK channel

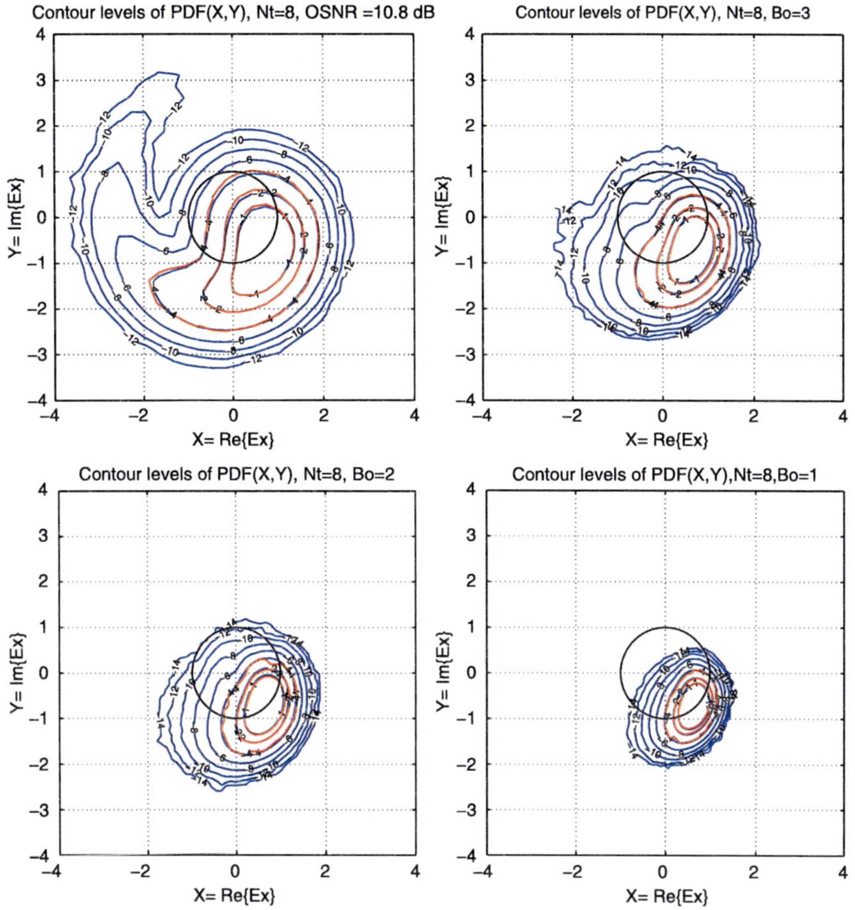

Fig. 10.20 Contours of MMC simulated joint PDF of in-phase and quadrature components of optical field (CW+ASE) (*top-left*) before optical filter (simulated bandwidth 80 GHz), and after receiver optical filter of bandwidth (*top-right*) 30 GHz, (*bottom-left*) 20 GHz, (*bottom-right*) 10 GHz. Data as in Fig. 10.19. Lowest contour level: 10^{-14}

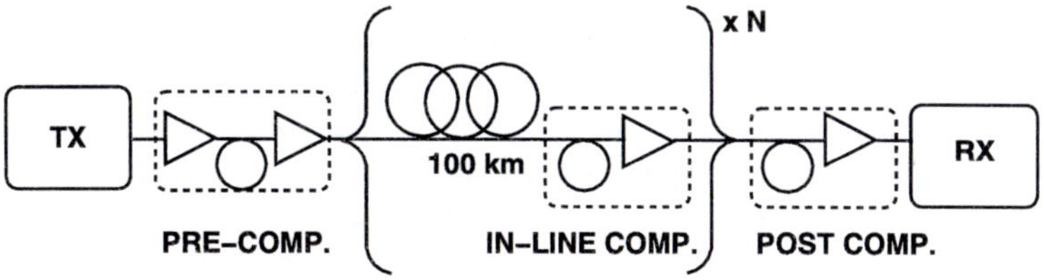

Fig. 10.21 Single-channel dispersion-managed DPSK system

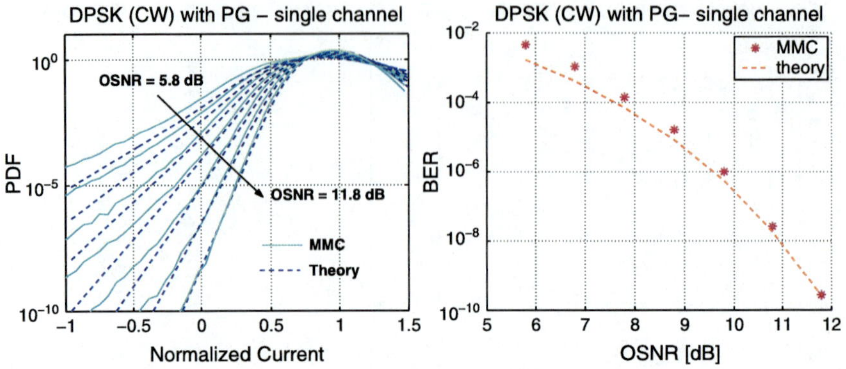

Fig. 10.22 (*Left*) PDF of sampled current: MMC (*solid*), theory (*dashed*) for several values of linear OSNR (dB/0.1 nm). (*Right*) BER obtained from above PDFs (*symbols*) and from theory (*dashed*). Data: 20 × 100 km, $D_{TX} = 4\,\mathrm{ps\,nm^{-1}\,km^{-1}}$, $D_{pre} = 0$, D_{inline}=40 ps nm^{-1} span^{-1}, $D_{post} = 0$, $\Phi_{NL} = 0.2\pi$(rad). R=10 Gb s^{-1}. Optical filter bandwidth 1.8R

was transmitted with a pattern 1,1,1,1.... actually corresponding to a CW signal. The OSNR (dB/0.1 nm) was varied from 5.8 dB, where the nonlinear effect of PG is strong, to 12.8 dB. An improving match between MMC and theoretical PDFs is observed for increasing OSNR. Figure 10.22(right) shows the BER obtained by integrating the tail of the PDFs below the zero threshold. We note that the theory based on the Gaussian assumption for the received optical field gives an excellent prediction of the true BER, with half of a dB of discrepancy at the lowest OSNR, i.e., at BER values worse than 10^{-4}.

10.5.4 Further Examples in the Literature

In this section, we give a brief overview of other significant results in telecommunications that have exploited MMC techniques. As we already understood from the previous examples pursued by our research groups, the main application of MMC in telecommunications concerns the analysis of the PDFs of the decision variable, in order to understand how impairments, both linear and nonlinear, affect the final BER, or to validate approximate analytical models. MMC is also used as a substitute for analytical models when the system is too complex.

For example, Secondini et al., were the first to apply pattern-warping, which is an instance of the one-variable-at-a-time MCMC technique [31], in MMC simulations of optical systems with strong chromatic dispersion [15]. Our pattern-warping method [19] presented in Example 10.5.1 is similar to Secondini's method. Both methods are applicable to any system impaired by ISI, and produce the correct PDFs of the decision variable.

Zweck, et al., presented a study of the ISI-distorted PDFs of the decision variable in quasi-linear propagation [57]. The change in PDF shape produced by each individual nonlinear effect is discernable as the parameters of the dispersion map are varied. Such MMC use is thus targeted to a deeper understanding of the impact of individual distortions on the system BER.

Bilenca and Eisenstein used MMC to study the PDF of the peak power of a single pulse amplified by the SOA [11,58]. MMC was used primarily to validate the range of applicability of a sophisticated mathematical model of nonlinear noise in SOAs.

Another example of the use of MMC as a model-validation tool is found in [16], where the authors proposed an improved model to describe the parametric interaction of signal and noise, an instance of which was presented in Example 10.5.3. MMC allowed the validation of the model both regarding the one-dimensional PDF of the decision variable, and the two-dimensional PDF of the received optical field.

Several authors used MMC to accurately study optical regeneration by calculating the PDFs of the decision variable and clarify the reasons for the BER improvement with optical regenerators [14, 18]. In the absence of an analytical model, the MMC tool enables comprehension of the basic mechanisms of regeneration.

We conclude by mentioning two interesting recent variants of MMC related to advanced detection with powerful signal processing. The first, named dual adaptive importance sampling (DAIS), deals with the difficult problem of estimation of the BER of systems with FEC [45]. The proposed solution offers limited gains, but this is a typical shortcoming of MMC with coding, as we already discussed. The second variant, inspired by DAIS, deals with the application of MMC to the simulation of Viterbi decoders [17]. A novel control variable, referred to as "the best error metric," is introduced to univocally determine the symbol error rate (SER), so that a single cycle of MMC simulations suffices for the SER evaluation.

10.6 Conclusions

This chapter discussed the MMC simulation technique from many viewpoints. MMC was placed within the mathematical frame work of traditional Monte Carlo simulations and importance sampling. Within importance sampling warpings, we explained the significance of uniform-weight flat-histogram warpings (they minimize the largest relative error across the output PDF bins). We saw how the MMC algorithm is an adaptive method to seek out the UW–FH warping.

The MMC adaptation was described, including essential elements to facilitate the simulations. A technique proposed by Berg was explained where both spatial (across bins) and temporal smoothing reduced statistical variations in the MMC estimate of the output PDF. Salient features of MCMC techniques were presented to facilitate efficient drawing of samples from warped input PDFs, which may be ill behaved. We also shared with the reader some rules of thumb for practical implementation of MMC.

Three detailed examples from optical communications were presented. The first example focused on treatment of bit patterning within the MMC platform. The next example examined how MMC can sweep performance over wide ranges of system parameters to find practical limits to spectral efficiency. This example also highlighted the potential to run MMC algorithms in parallel for accelerated run times. The third example illustrated capturing of nonlinear interaction between signal and noise.

The MMC algorithm is a powerful tool for the characterization of rare events, especially in computationally expensive numerical modeling. This chapter serves to better prepare researchers to mold their simulation environments to that of MMC. Optical systems are not the only ones for which MMC techniques are applicable, although this potential remains largely untapped.

Acknowledgments It is a pleasure to acknowledge A. Ghazisaeidi and F. Vacondio of Laval University, and N. Rossi, A. Orlandini, P. Serena and A. Vannucci of Parma University, for the many stimulating discussions and for their producing the numerical examples in the text.

10.7 Appendix: MCMC Fundamentals

MCMC is a technique to produce samples from a desired, analytically known probability density function $f_X(x)$, with X taking values in a multidimensional space Γ. Without loss of generality, and for the sake of clarity, we consider a discretized space Γ [31], *i.e.*, we have a known PMF $\underline{p}_X = [p_X(x_1), p_X(x_2), \ldots]$, with $p_X(x_i) \cong f_X(x_i)\Delta x$, for the discretized states $\{x_i\}_{i=1}^{\infty}$ in Γ. MCMC synthesizes the desired samples $\{X_m, m \geq 1\}$ from a memoryless sequence, i.e., a discrete-time Markov Chain (DTMC), whose steady-state distribution $\underline{\pi}$ coincides with the desired PMF $\underline{p}_X$.

A DTMC is characterized by its transition matrix $\mathbf{P} = \{p_{ij}\}$, with transition probability from any state x_i to any state x_j defined as $p_{ij} = P\{X_m = x_j \mid X_{m-1} = x_i\}$. The steady-state distribution solves the equation [59]

$$\underline{\pi} = \underline{\pi}\mathbf{P}. \tag{10.34}$$

While the classical DTMC problem is to find $\underline{\pi}$ for a given $\mathbf{P}$, the MCMC problem is conversely to find a matrix $\mathbf{P}$, which satisfies (10.34) for a known $\underline{\pi} \triangleq \underline{p}_X$. We clearly require the DTMC to be ergodic, i.e., that $\mathbf{P}$ has a unique $\underline{\pi}$, and that the PMF of the chain at time m, namely $\underline{p}(m) = [P\{X_m = x_1\}, P\{X_m = x_1\}, \ldots]$, converges to $\underline{\pi}$ as $m \to \infty$. Thus, the shortcomings of the MCMC method are that

1. The sequence $\{X_m, m \geq 1\}$ will reflect the desired limiting distribution $\underline{p}_X$ only for large enough m, and

2. The samples will be correlated according to the random walk on the states driven by the matrix **P**.

There are clearly infinitely many ergodic matrices **P** that solve (10.34), and we need just one. A unique, simple solution is found by imposing the extra constraint that the DTMC be time reversible. A necessary and sufficient condition for time reversibility is that, at steady-state, for every pair of states (x_i, x_j) the probability of being at x_i at time $m-1$ and moving to x_j at time m equals the probability of being at x_j at $m-1$ and moving to x_i at m [59]

$$\pi_i p_{ij} = \pi_j p_{ji} \,. \tag{10.35}$$

These are called local balance equations and they determine all the unknowns $\{p_{ij}\}$.

A clever way of practically implementing a reversible DTMC with this method was introduced by Metropolis [22] in 1953 and 17 years later generalized by Hastings [23]. Hastings proposed the following procedure to find the $\{p_{ij}\}$

1. Start with any transition matrix $\mathbf{Q} = \{q_{ij}\}$, called the *candidate* chain;
2. For any pair of states $x_i, x_j, i \neq j$, which do not satisfy (10.35) a randomization procedure is introduced such that every time the candidate chain proposes a move $i \to j$ the move is accepted with probability α_{ij} and otherwise rejected (i.e., the chain remains in the same state at the next time). Hence, $p_{ij} = \alpha_{ij} q_{ij}$.

For arbitrary choice of Q, it may happen that either (a) $\pi_i q_{ij} > \pi_j q_{ji}$ or (b) $\pi_i q_{ij} < \pi_j q_{ji}$. In case (a) we accept all transitions $j \to i$, i.e., use $\alpha_{ji} = 1$ (hence $p_{ji} = q_{ji}$), and decrease the transitions $i \to j$ by accepting a fraction $\alpha_{ij} = \frac{\pi_j q_{ji}}{\pi_i q_{ij}} < 1$ of such moves so as to reach equality as in (10.35). In case (b), we swap the roles of i and j, so that in general $\alpha_{ij} = \min[1, R_{ij}]$, where

$$R_{ij} = \frac{\pi_j q_{ji}}{\pi_i q_{ij}} = \frac{f_X(x_j) q_{ji}}{f_X(x_i) q_{ij}} \tag{10.36}$$

is the *odds ratio*, and we have substituted back the original PDF of the input RV X. Note that, since only the ratio of PDFs at the two states is needed, such a PDF need only be known up to a normalization constant. There is no need to normalize the PDF to generate samples from it. In some physical settings, the normalization constant is impractical or impossible to compute [26] and the MCMC algorithm offers the only known solution to this simulation problem.

Metropolis MCMC [22] uses a symmetric candidate $q_{ij} = q_{ji}$ so that the odds ratio further simplifies. Starting from initial state x_i, common practice is to select the Metropolis candidate as $x_j = x_i + U$, where U is a uniform random vector in space Γ. No quantization is needed in the input space. The variance of U is important in determining both the acceptance ratio and the speed of exploration of the chain in the input space, and is one of the key tuning parameters of the MCMC machine.

References

1. M. Jeruchim, IEEE J. Sel. Areas. Commun. **SAC-2**, 153–170 (1984)
2. B.A. Berg, T. Neuhaus, Phys. Lett. B **267**(2), 249–253 (1991)
3. D. Yevick, IEEE Photon. Technol. Lett. **14**(11), 1512–1514 (2002)
4. R. Holzlohner, C.R. Menyuk, Opt. Lett. **28**(20), 1894–1896 (2003)
5. T. Kamalakis, D. Varoutas, T. Sphicopoulos, IEEE Photon. Technol. Lett. **16**(10), 2242–2244 (2004)
6. T. Lu, D. Yevick, Photon. Technol. Lett. **17**(4), 861–863 (2005)
7. G. Biondini, W.L. Kath, IEEE Photon. Technol. Lett. **17**(9), 1866—1868 (2005)
8. A.O. Lima, C.R. Menyuk, I.T. Lima, IEEE Photon. Technol. Lett. **17**(12), 2580–2582 (2005)
9. A.O. Lima, I.T. Lima, C.R. Menyuk, J. Lightwave Technol. **23**(11), 3781–3789 (2005)
10. W. Pellegrini, J. Zweck, C.R. Menyuk, R. Holzlohner, IEEE Photon. Technol. Lett. **17**(8), 1644–1646 (2005)
11. A. Bilenca, G. Eisenstein, IEEE J. Quant. Electron. **41**(1), 36–44 (2005)
12. Y. Yadin, M. Shtaif, M. Orenstein, IEEE Photon. Technol. Lett. **17**(6), 1355–1357 (2005)
13. M. Nazarathy, E. Simony, Y. Yadin, J. Lightwave Technol. **24**(5), 2248–2260 (2006)
14. I. Nasieva, A. Kaliazin, S.K. Turitsyn, Opt. Commun. **262**, 246–249 (2006)
15. L. Gerardi, M. Secondini, E. Forestieri, IEEE Photon. Technol. Lett. **19**, 1934–1936 (2007)
16. M. Secondini, E. Forestieri, C.R. Menyuk, J. Lightwave Technol. **27**(16), 3358–3369 (2009)
17. M. Secondini, D. Fertonani, G. Colavolpe, E. Forestieri, Performance evaluation of viterbi decoders by multicanonical monte carlo simulations, in *Proceedings of ISIT 2009*, Seoul, Korea, June 2009
18. T.I. Lakoba, IEEE J. Sel. Topics Quant. Electron. **14**, 599–609 (2008)
19. A. Ghazisaeidi, F. Vacondio, A. Bononi, L.A. Rusch, Statistical characterization of bit patterning in soas: ber prediction and experimental validation, in *Proceedings of OFC 2009*, Paper OWE7, San Diego, CA, March 2009
20. A. Ghazisaeidi, F. Vacondio, A. Bononi, L.A. Rusch, IEEE J. Lightwave Technol. **27**, 2667–2677 (2009)
21. A. Bononi, L.A. Rusch, A. Ghazisaeidi, F. Vacondio, N. Rossi, A Fresh Look at Multicanonical Monte Carlo from a Telecom Perspective, in *Proceedings of Globecom 2009*, Paper CTS-14.1, Honolulu, HI, Nov/Dec 2009
22. N. Metropolis, A.W. Rosenbluth, M.N. Rosenbluth, A.H. Teller, E. Teller, J. Chem. Phys. **21**(6), 1087–1092 (1953)
23. W.K. Hastings, Biometrika **57**, 97–109 (1970)
24. D. Yevick, IEEE Photon. Technol. Lett. **15**(11), 1540–1542 (2003)
25. A. Vannucci, N. Rossi, A. Bononi, Emulazione e statistiche della PMD attraverso algoritmi multicanonici multivariati, in *Proceedings of Fotonica 2007*, pp. 517–520, Mantova, May 2007
26. B.A. Berg, Fields Instr. Commun. **26**, 1–24 (2000)
27. F. Liang, J. Stat. Phys. **122**, 511–529 (2006)
28. Y.F. Atchade, J.S. Liu, The Wang-Landau algorithm for MC computation in general state spaces, Technical report, University of Ottawa (2004), http://www.mathstat.uottawa.ca/~yatch436/gwl.pdf, 2004
29. F. Wang, D.P. Landau, Phys. Rev. Lett. **86**, 2050–2053 (2001)
30. S. Haykin, *Adaptive Filter Theory*, 4th edn. (Prentice Hall, NJ, 2001)
31. C.J. Geyer, Markov Chain Monte Carlo lecture notes, Course notes, University of Minnesota, Spring Quarter 1998
32. D.J.C. MacKay, *Information Theory, Inference, and Learning Algorithms* (Cambridge University Press, London, 2003)
33. A. Papoulis *Probability, Random Variables, and Stochastic Processes*, 3rd edn. (McGraw-Hill, New York, 1991)
34. A. Ghazisaeidi, F. Vacondio, A. Bononi, L.A. Rusch, IEEE J. Quant. Electron. **46**, 570–578 (2010)
35. E. Forestieri, J. Lightwave Technol. **18**, 1493–1503 (2000)

36. P. Serena, A. Orlandini, A. Bononi, IEEE J. Lightwave Technol. **24**, 2026–2037 (2006)
37. M.J. Connelly, *Semiconductor Optical Amplifiers* (Springer, Heidelberg, 2002)
38. D. Cassioli, S. Scotti, A. Mecozzi, IEEE J. Quant. Electron. **36**(7), 1072–1080 (2000)
39. M.L. Nielsen, J. Mrk, R. Suzuki, J. Sakaguchi, Y. Ueno, Opt. Exp. **14**, 331–347 (2006)
40. T. Akiyama,, M. Sugawara, Y. Arakawa, Proc. IEEE **95**(9), 1757–1766 (2007)
41. Z. Zhu, M. Funabashi, Z. Pan, B. Xiang, L. Paraschis, S.J.B. Yoo, J. Lightwave Technol. **26**, 1640–1652 (2008)
42. G.P. Agrawal, *Applications of Nonlinear Fiber Optics* (Academic, NY, 2001), pp. 138–141
43. P. Serena, N. Rossi, M. Bertolini, A. Bononi, IEEE J. Lightwave Technol. **27**, 2404–2411 (2009)
44. Y. Iba, K. Hukushima, J. Phys. Soc. Jpn. **77**(10), 103801 (2008)
45. R. Holzlohner et al., IEEE Photon. Technol. Lett. **9**, 163–165 (2005)
46. A. Ghazisaeidi, F. Vacondio, L.A. Rusch, IEEE J. Lightwave Technol. **28**, 79–90 (2010)
47. J.W. Goodman, *Statistical Optics* (Wiley, NY, 1985)
48. A.D. McCoy, P. Horak, B.C. Thomsen, M. Ibsen, D.J. Richardson, J. Lightwave Technol. **23**, 2399–2409 (2005)
49. A. Ghazisaeidi, F. Vacondio, L. Rusch, Evaluation of the Impact of Filter Shape on the Performance of SOA-assisted SS-WDM Systems Using Parallelized Multicanonical Monte Carlo, in *Proceedings of globecom 2009*, Paper ONS-04.4, Honolulu, HI, Nov/Dec 2009
50. W. Mathlouthi, F. Vacondio, J. Penon, A. Ghazisaeidi, L.A. Rusch, DWDM Achieved with Thermal Sources: a Future-proof PON Solution, in *ECOC 2007*, Berlin, Paper 4.4.5, September 2007
51. H.H. Lee, M.Y. Park, S.H. Cho, J.H. Lee, J.H. Yu, B.W. Kim, Filtering effects in a spectrum-sliced WDM-PON System using a gain-saturated reflected-SOA, OFC 2009
52. A. Bononi, P. Serena, A. Orlandini, N. Rossi, Parametric-gain approach to the analysis of DPSK dispersion-managed systems, in *Proceedings of 2006 China-Italy bilateral workshop on photonics for communications and sensing*, Acta Photonica Sinica Ed., Xi'An, China, October 2006, pp. 38–45
53. A. Carena, V. Curri, R. Gaudino, P. Poggiolini, S. Benedetto, IEEE Photon. Technol. Lett. **9**, 535–537 (1997)
54. P. Serena, A. Bononi, J.C. Antona, S. Bigo, J. Lightwave Technol. **23**, 2352–2363 (2005)
55. A. Orlandini, P. Serena, A. Bononi, An alternative analysis of nonlinear phase noise impact on DPSK systems, in *Proceedings of ECOC 2006*, Paper Th3.2.6, pp. 145–146, Cannes, France, September 2006
56. K.-P. Ho, J. Opt. Soc. Am. B **20**, 1875–1879 (2003). For a more comprehensive documentation, see also K.-P. Ho, Statistical properties of nonlinear phase noise, at http://arxiv.org/abs/physics/0303090, last updated September 2005
57. J. Zweck, C.R. Menyuk, IEEE J. Lightwave Technol. **27**(16), 3324–3335 (2009)
58. A. Bilenca, G. Eisenstein, J. Opt. Soc. Am. B **22**, 1632–1639 (2005)
59. S.M. Ross, *Stochastic Processes* (Wiley, New York, 1983)

Chapter 11
Optical Regenerators for Novel Modulation Schemes

Masayuki Matsumoto

11.1 Introduction

Optical signals propagating along fibers are impaired by various causes. The impairments can be classified into two different types: deterministic and stochastic impairments. The sources of deterministic signal impairments include chromatic dispersion, polarization-mode dispersion, intrachannel nonlinearities caused by Kerr effects in fibers, and narrowband filtering brought about by networking elements such as add-drop multiplexers. In addition to these impairments, signals are contaminated by stochastic noise emitted by optical amplifiers that are used in most systems to compensate for losses of transmission fibers and other passive optical elements. Data-dependent signal distortion caused by interchannel nonlinearities is also taken as stochastic when the data carried by other channels are unknown to the channel of interest. The deterministic signal distortions can, in principle, be compensated for by optical elements, such as dispersion compensating fibers (DCFs) for chromatic dispersion compensation, for example, and/or signal processing in the electrical domain. The stochastic noise whose effects remain after such compensations are performed determines the ultimate performance of the transmission systems. In the presence of nonlinearity of the transmission fiber, the effect of noise is often enhanced [1].

In digital signal transmission, the noise accumulation can be suppressed by inserting signal regenerators in certain locations in the system. In the regenerator, fluctuations in the input signal caused by the noise are removed so that desired signal shape (amplitude and phase) is recovered. In commercially deployed systems, such regeneration is performed in the electrical domain with optical-to-electrical (O/E) and electrical-to-optical (E/O) signal conversions involved. For more than a decade, much effort has been devoted toward the realization of all-optical signal regeneration in which the O/E and E/O conversions are dispensed and signal processing is performed on the optical signals [2]. One expects higher-speed and

M. Matsumoto (✉)
Graduate School of Engineering, Osaka University, Osaka 565-0871, Japan
e-mail: matumoto@comm.eng.osaka-u.ac.jp

S. Kumar (ed.), *Impact of Nonlinearities on Fiber Optic Communications*, Optical and Fiber Communications Reports 7, DOI 10.1007/978-1-4419-8139-4_11,
© Springer Science+Business Media, LLC 2011

less-power-consuming operation with more flexibility to modulation formats other than conventional on-off keying (OOK). Considering that signals in advanced modulation formats including differential binary phase-shift keying (DBPSK, which is often abbreviated as DPSK), differential quadrature phase-shift keying (DQPSK), and other multilevel formats are becoming practical candidates for use in long-distance transmission [3], all optical regenerators that can process such signals will be highly desired.

All-optical signal regeneration is realized by using some forms of nonlinear signal transfer properties in optical media, such as glass fibers and semiconductors. Most of the optical nonlinearities such as self-phase modulation (SPM), cross-phase modulation (XPM), gain saturation (GS), and cross-gain modulation (XGM) occurring in these media are power-dependent processes independent of the phase of the control signals. This makes construction of all-optical regenerators that suppress phase noise rather than the amplitude noise difficult. Recently, several schemes of (differential) binary phase-shift keying ((D)BPSK) signal regeneration and regenerative wavelength conversion have been proposed and demonstrated. In one class of the regenerators, direct phase noise reduction is not attempted. Instead, the phase information of the signal is converted to/from the amplitude information and the noise removal is performed on the amplitude [4–11]. Averaging of phase fluctuations over neighboring bits can also lead to phase-noise reduction [12–14]. Phase-preserving amplitude-only regeneration has also been shown to be effective in reducing the Gordon–Mollenauer nonlinear phase noise [15–25]. In the other class of the (D)BPSK regenerators, phase noise around the data, 0 and π, is directly suppressed by the use of phase-sensitive amplifier (PSA) setups [26–30]. In this type of regenerators, strong reduction of phase noise is expected.

Besides the regeneration of binary phase-shift keying (PSK) signals, M-ary PSK signals with $M \geq 4$ are interesting and beneficial because the transmission distance of such multilevel signals is severely limited by noise owing to the small minimum distance between signal points in the constellation. Several papers have discussed (D)QPSK-signal regeneration by numerical simulation. In [31], a scheme using two parallel PSAs has been proposed. The regenerative wavelength converter proposed in [32] consists of a coherent demodulator of QPSK signals and nested semiconductor optical amplifier (SOA) Mach–Zehnder interferometers (MZIs) for phase remodulation. In [33], numerical analysis of a DQPSK-signal regenerator has been reported, where the input DQPSK signal is demodulated to two parallel OOK signals by a pair of delay interferometers (DIs) and the noise on the OOK signals is removed by fiber-based amplitude regenerators. The regenerated OOK signals are subsequently used as control signals for all-optical phase modulation of probe pulses.

In this chapter, recent progress in the all-optical signal regeneration of phase-encoded signals is reviewed. Features of different regeneration schemes of (D)BPSK and (D)QPSK signals are discussed. Practical issues in using the all-optical regenerators in transmission systems are also mentioned.

11.2 Regeneration of Binary Phase-Shift Keying Signals

11.2.1 DPSK Signal Regeneration Using Amplitude Regenerators

11.2.1.1 DPSK Regenerator Using a Straight-Line Phase Modulator

In one type of DPSK signal regenerator, the phase information of the incoming signal is first converted into the amplitude information through the use of a DI. Through this process, the phase noise in the incoming signal, together with the amplitude noise, is transferred to the amplitude of the demodulated OOK signal. Then the amplitude noise of the OOK signal is removed by an amplitude regenerator. The regenerated OOK signal is used as a control signal to modulate the phase of probe pulses in a subsequent all-optical phase modulator to yield regenerated DPSK signals. Because the all-optical phase modulator responds to the intensity of the control signal, the phase of the amplitude-regenerated signal does not affect the phase of the output signal. Therefore, one can use any types of amplitude regenerator that are not needed to be phase-preserving. Figure 11.1 shows a block diagram of the DPSK regenerator of this type.

An essential component for the noise removal in this setup of the regenerator is the amplitude regenerator. Strength of amplitude noise suppression required for the amplitude regenerator can be estimated as follows [9, 34]: First, we assume that the incoming pulses have a complex amplitude of the form

$$E_n^{\mathrm{in}} = (A_s + \Delta A_n)\exp[\mathrm{i}(\phi_n + \Delta\phi_n)], \tag{11.1}$$

where A_s and ϕ_n ($\phi_n - \phi_{n-1} = 0$ or π) are amplitude and phase of the pulse, respectively, and ΔA_n and $\Delta\phi_n$ are amplitude and phase fluctuations of the pulse. The complex amplitude of the pulse at the output port of the DI is given by $E_{\mathrm{DI}} = \left(E_n^{\mathrm{in}} - E_{n-1}^{\mathrm{in}}\right)/2$, and its power is calculated to be

$$|E_{\mathrm{DI}}|^2 = \begin{cases} A_s^2 + A_s\left(\Delta A_n + \Delta A_{n-1}\right) & (\phi_n - \phi_{n-1} = \pi) \\ 0 & (\phi_n - \phi_{n-1} = 0) \end{cases} \tag{11.2}$$

in the first-order approximation under the conditions $|\Delta A_{n-1,\,n}| \ll A_s$ and $|\Delta\phi_{n-1,\,n}| \ll 1$. Equation (11.2) shows that the phase noise in the input signal is not

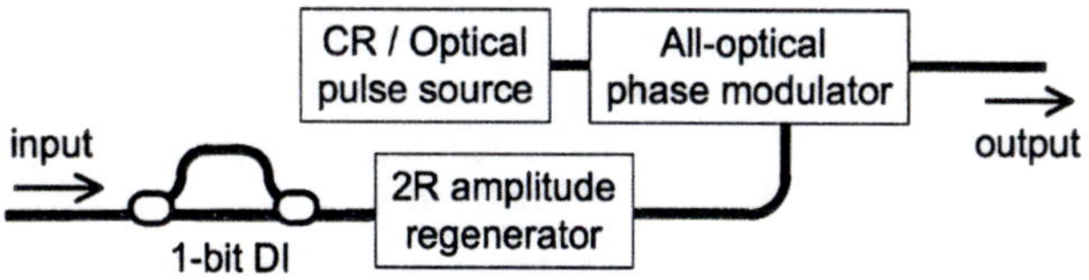

Fig. 11.1 Block diagram of an all-optical DPSK signal regenerator using a straight-line phase modulator. *CR* Clock recovery circuit; *DI* Delay interferometer; *2R* Reamplifying and reshaping

transferred to the output signal power from the DI in the first-order approximation. This is due to the general behavior of interferometers that the output power is insensitive to the phase fluctuations when the phase difference is close to 0 or π. This indicates that the DPSK signal regenerator discussed in this section is more effective in regenerating signals impaired by the phase noise than those impaired by the amplitude noise.

Here, we consider the case of π phase difference between the pulses in (11.2). The same results hold in the case of 0 phase difference. After the power fluctuation in $|E_{\mathrm{DI}}|^2$ is reduced with a factor of $r(<1)$ by the 2R (reamplifying and reshaping) amplitude regenerator, the pulse is amplified and used as a control pulse in the subsequent all-optical phase modulator. When we assume that the phase modulation of the clock pulse is proportional to the power of the control pulse, the complex amplitude of the output pulse is expressed as

$$E_{\mathrm{out}} = A_{\mathrm{clock}} \exp\{\mathrm{i}G[A_{\mathrm{s}}^2 + rA_{\mathrm{s}}(\Delta A_n + \Delta A_{n-1})]\}, \tag{11.3}$$

where A_{clock} is an amplitude of the clock pulse. For the output pulse to be in BPSK format, the gain of the amplification of the control pulse G should satisfy $GA_{\mathrm{s}}^2 = \pi$. Then the phase fluctuation in (11.3) is given by $\Delta\phi_{\mathrm{out}} = r\pi(\Delta A_n + \Delta A_{n-1})/A_{\mathrm{s}}$ and its variance is $\sigma_{\phi\mathrm{out}}^2 = 2r^2\pi^2\sigma_{A\mathrm{in}}^2/A_{\mathrm{s}}^2$, where $\sigma_{A\mathrm{in}}^2$ is the variance of the amplitude fluctuation of the input pulses. Here, no correlation between amplitude fluctuations of neighboring input pulses is assumed. When the input signal is degraded by a circular Gaussian noise such as amplified spontaneous emission (ASE), $\sigma_{A\mathrm{in}}^2 = A_{\mathrm{s}}^2\sigma_{\phi\mathrm{in}}^2$ is satisfied. The phase noises in the output and input signals are then related by $\sigma_{\phi\mathrm{out}}^2 = 2r^2\pi^2\sigma_{\phi\mathrm{in}}^2$. In this case, in order for the output phase noise to be smaller than the input phase noise, we need to use an amplitude regenerator with r smaller than $(2^{1/2}\pi)^{-1}$ or the noise suppression factor $1/r$ larger than $10\log_{10}(2^{1/2}\pi) = 6.5\,\mathrm{dB}$.

In the first-order analysis given above, no output appears from the DI when $\phi_n - \phi_{n-1} = 0$ as shown in (11.2). In reality, signal fluctuations outside the range of the first-order approximation and waveform distortions of signal pulses produce small output even in the condition of destructive interference. The 2R amplitude regenerator after the DI should, therefore, have the function of noise suppression also at the space level.

11.2.1.2 DPSK Regenerator Using MZI Phase Modulator

The straight-line phase modulator at the last stage of the regenerator discussed in the previous subsection can be replaced by two parallel all-optical modulators in MZI configuration as shown in Fig. 11.2. The modulators are driven by amplitude-regenerated complementary OOK pulses derived from the two output ports of the DI. The amplitude regenerators after the DI in this setup may be moved to a place in front of the DI or they can be omitted when the modulators in the MZI have saturation behavior.

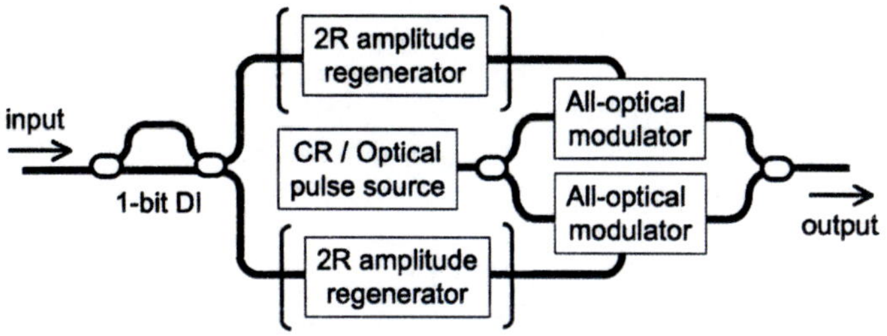

Fig. 11.2 Block diagram of an all-optical DPSK signal regenerator using an MZI phase modulator

For the analysis of the performance of the regenerator shown in Fig. 11.2, we again denote the signal E_n^{in} incoming to the regenerator as $E_n^{\text{in}} = (A_s + \Delta A_n)\exp[\mathrm{i}(\phi_n + \Delta\phi_n)]$. The complementary OOK signals demodulated by the DI are fed to respective amplitude regenerators and their amplitude noise is suppressed by a factor $r(<1)$. Here, we assume that the regenerated OOK pulses, after being amplified, modulate the phase of the probe pulses in the all-optical modulators located in each arm of the MZI. When the phase modulation is proportional to the energy of the control pulses, phase shifts given to the probe pulses transmitted through the upper and lower arms of the MZI, θ_1 and θ_2, respectively, are

$$\begin{cases} \theta_1 = G[A_s^2 + rA_s(\Delta A_n + \Delta A_{n-1})], \quad \theta_2 = 0 \quad (\phi_n - \phi_{n-1} = \pi) \\ \theta_1 = 0, \quad \theta_2 = G[A_s^2 + rA_s(\Delta A_n + \Delta A_{n-1})] \quad (\phi_n - \phi_{n-1} = 0), \end{cases} \tag{11.4}$$

in the first-order approximation, where G is a coefficient accounting for the pulse amplification and phase-modulation efficiency. The output signal from the regenerator is expressed as

$$\begin{aligned} E_{\text{out}} &= A_{\text{clock}}\left[\exp(\mathrm{i}\theta_1) - \exp(\mathrm{i}\theta_2)\right]/2 \\ &= \mathrm{i}A_{\text{clock}}\sin\left[(\theta_1 - \theta_2)/2\right]\exp\left[\mathrm{i}(\theta_1 + \theta_2)/2\right]. \end{aligned} \tag{11.5}$$

In the case of π phase difference between the consecutive pulses ($\phi_n - \phi_{n-1} = \pi$), the output signal becomes

$$\begin{aligned} E_{\text{out}} = \mathrm{i}A_{\text{clock}}&\sin\left\{G[A_s^2 + rA_s(rA_n + \Delta A_{n-1})]/2\right\} \\ &\exp\left\{\mathrm{i}G[A_s^2 + rA_s(\Delta A_n + \Delta A_{n-1})]/2\right\}. \end{aligned} \tag{11.6}$$

Its amplitude and phase are, respectively, written as

$$\begin{aligned} A_{\text{out}} = |E_{\text{out}}| &= A_{\text{clock}}\sin\left\{G[A_s^2 + rA_s(\Delta A_n + \Delta A_{n-1})]/2\right\} \\ &\cong A_{\text{clock}}\left[\sin\left(GA_s^2/2\right) + GrA_s\cos\left(GA_s^2/2\right)(\Delta A_n + \Delta A_{n-1})/2\right] \end{aligned} \tag{11.7}$$

and

$$\phi_{\text{out}} = G\left[A_s^2 + rA_s(\Delta A_n + \Delta A_{n-1})\right]/2. \tag{11.8}$$

In the case of 0 phase difference between the consecutive pulses ($\phi_n - \phi_{n-1} = 0$), sign of the output signal is reversed, that is, a binary PSK signal is produced. The binary PSK format of the output signal is retained irrespective of the amount of phase modulation GA_s^2, indicating that precise adjustment of the values of GA_s^2 is not needed. This, in contrast to the regenerator using a single-ended DI and a straight-line all-optical phase modulator discussed in the previous section, is an advantage of the regenerator using MZI for phase modulation. This is the same as the fact that electrooptic Mach–Zehnder modulators are generally preferred to straight-line phase modulators in generating DPSK signals in transmitters [35]. When the MZI arrangement is used, however, the output amplitude includes noise as shown in (11.7), which is not the case for the regenerator using a straight-line all-optical phase modulator. On the one hand, the relative variance of the amplitude noise is

$$\frac{\sigma_{A\text{out}}^2}{\langle A_{\text{out}}\rangle^2} = \frac{1}{2}\left(rGA_s^2\right)^2 \cot^2\left(GA_s^2/2\right)\frac{\sigma_{A\text{in}}^2}{\langle A_{\text{in}}\rangle^2}. \tag{11.9}$$

For the amplitude noise to be reduced by the regenerator,

$$\left(rGA_s^2\right)^2 \cot^2\left(GA_s^2/2\right) < 2 \tag{11.10}$$

should be satisfied. The variance of the phase noise, on the other hand, is given from (11.8) by $\sigma_{\phi\text{out}}^2 = r^2G^2A_s^2\sigma_{A\text{in}}^2/2$, where no correlation is assumed between ΔA_n and ΔA_{n-1}. When we consider the ASE noise, $\sigma_{A_{\text{in}}}^2 = A_s^2\sigma_{\phi_{\text{in}}}^2$ is satisfied so that we have $\sigma_{\phi\text{out}}^2 = \left(rGA_s^2\right)^2 \sigma_{A\text{in}}^2/2$. For the phase noise to be reduced by the regenerator,

$$\left(rGA_s^2\right)^2 < 2 \tag{11.11}$$

should be satisfied. When the amplitude regeneration is not performed on the demodulated OOK signals, that is $r = 1$, the inequalities (11.10) and (11.11) are not satisfied simultaneously irrespective of the value of GA_s^2 [34]. When r is smaller than 0.90, the inequalities can be satisfied by optimizing GA_s^2. The needed strength of the amplitude regeneration is $10\log_{10}(1/r) = 0.46$ dB, which is smaller than that needed in the regenerator discussed in the previous section. The two amplitude regenerators after the DI can be replaced by a single amplitude regenerator prior to the DI if it does not destroy the phase information of the incoming signal [7].

The analysis above assumes phase modulation in each arm of the MZI. This can be replaced by amplitude modulation, with which the output signal from the MZI again has the binary PSK format. Such an all-optical amplitude modulation is provided by nonlinear elements exhibiting XGM or cross-absorption modulation (XAM). XGM and XAM accompanied by only small phase modulation are expected in quantum-dot SOA and electroabsorption modulators, respectively [10, 36]. In the case of pure amplitude modulation, transmission coefficients of the probe pulses in the upper and lower arms of the MZI, t_1 and t_2, respectively, are

$$\begin{cases} t_1 = g\left(A_s^2 + rA_s(\Delta A_n + \Delta A_{n-1})\right), \quad t_2 = g(0) \quad (\phi_n - \phi_{n-1} = \pi) \\ t_1 = g(0), \quad t_2 = g\left(A_s^2 + rA_s(\Delta A_n + \Delta A_{n-1})\right) \quad (\phi_n - \phi_{n-1} = 0), \end{cases} \tag{11.12}$$

where $g(\bullet)$ is the gain or loss coefficient as a function of the control signal power. Here, we again assume that the amplitude noise of the demodulated OOK signals is suppressed by amplitude regenerators after the DI by a factor $r(<1)$, although this may not be needed in practice as will be shown shortly. The output signal from the MZI is then expressed as

$$E_{\mathrm{out}} = A_{\mathrm{clock}}\left(t_1 - t_2\right)/2. \tag{11.13}$$

In the case of π phase difference between the consecutive input pulses launched to the DI, the output signal becomes

$$E_{\mathrm{out}} = (A_{\mathrm{clock}}/2)\left[g\left(A_s^2 + rA_s(\Delta A_n + \Delta A_{n-1})\right) - g(0)\right]. \tag{11.14}$$

Its sign is reversed when the phase difference between the consecutive input pulses is $\phi_n - \phi_{n-1} = 0$ showing that the output signal has the binary PSK format. Because $g(\bullet)$ is a real function for the pure amplitude modulation, the output pulse does not have phase noise. In the first-order approximation, the amplitude of the output signal is written as

$$\begin{aligned} A_{\mathrm{out}} = |E_{\mathrm{out}}| = E_{\mathrm{out}} &\cong (A_{\mathrm{clock}}/2)\left[g\left(A_s^2\right) - g(0)\right] \\ &+ (A_{\mathrm{clock}}/2)\frac{\partial g}{\partial A^2} rA_s(\Delta A_n + \Delta A_{n-1}). \end{aligned} \tag{11.15}$$

The amplitude noise is suppressed when the amplitude modulators are operated in the saturation regime with small $\partial g/\partial A^2$. When $\partial g/\partial A^2$ is sufficiently small, no amplitude noise suppression on the demodulated OOK signals is needed, that is, r can be unity [10].

11.2.1.3 Experiment Using Fiber-Based Amplitude Regenerator

Here, we describe a proof-of-principle experiment of the DPSK signal regeneration at 10 Gbit s^{-1} using a regenerator discussed in Sect. 11.2.1.1 [8, 9]. Silica highly nonlinear fibers (HNLFs) are used for the nonlinear elements in both the all-optical amplitude regenerator and the straight-line all-optical phase modulator. The fiber-based implementation has an advantage that high-speed operation faster than a few-hundred giga-symbols per second is obtainable owing to the ultrahigh-speed response of the Kerr nonlinearity. Although long fibers are needed for low signal-power operation in this experiment, the length can be shortened significantly if we

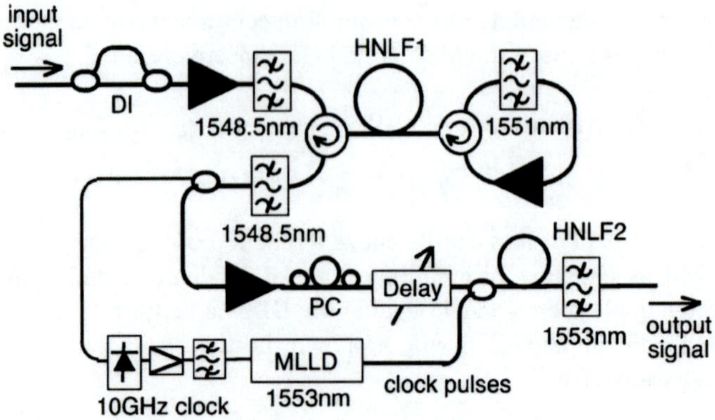

Fig. 11.3 Experimental setup of the DPSK regenerator. *MLLD* Mode-locked semiconductor laser diode; *PC* Polarization controller [9]

are able to use other fibers made of materials having higher nonlinearity such as bismuth and chalcogenide glasses and/or fibers with microstructured geometries for tighter light confinement [37–39].

Figure 11.3 shows the setup of the DPSK signal regenerator. The incoming DPSK signal at 10 Gbit s^{-1} is first demodulated to OOK signal by a one-bit DI. After that, the OOK signal is amplitude-regenerated by cascaded Mamyshev-type 2R regenerators in bidirectional configuration. The Mamyshev regenerator [40] basically consists of a nonlinear fiber and a detuned (by an order of signal spectrum width) optical bandpass filter (OBPF). In the nonlinear fiber, the signal spectrum width is broadened by the effect of SPM. After the fiber, a part of the broadened spectrum is sliced by the OBPF to produce the output signal. The wavelength offset of the OBPF makes the system opaque to low-power signal or noise whose power is weak so that the spectral broadening is insignificant. The amplitude fluctuation of the signal pulses above a threshold, however, is suppressed after the OBPF because the spectral width is broadened but the spectral power density is not so increased as the input signal power increases. The strength of amplitude-noise suppression is enhanced by cascading the regenerator stages. In this experiment, two-stage regeneration is performed by bidirectional use of a single HNLF spool [41]. The first highly nonlinear fiber (HNLF1 in Fig. 11.3) in the regenerator for the 2R regeneration has zero-dispersion wavelength $\lambda_0 = 1{,}560\,\text{nm}$, dispersion slope $\mathrm{d}D/\mathrm{d}\lambda = 0.03\,\text{ps nm}^{-2}\,\text{km}^{-1}$, length $L = 1.8\,\text{km}$, and nonlinearity coefficient $\gamma \sim 12\,\text{W}^{-1}\,\text{km}^{-1}$. The filter offset is 2.5 nm for both forward and backward directions. The direction of the wavelength shift is opposite so that the output wavelength of the bidirectional 2R amplitude regenerator is the same as that of the input signal. Bandwidth of the OBPFs is 1 nm.

A part of the regenerated OOK signal is then tapped and detected. Narrowband (high-Q) filtering of the detected RF signal gives a 10 GHz clock tone to which the semiconductor diode laser is mode locked. The output pulses from the

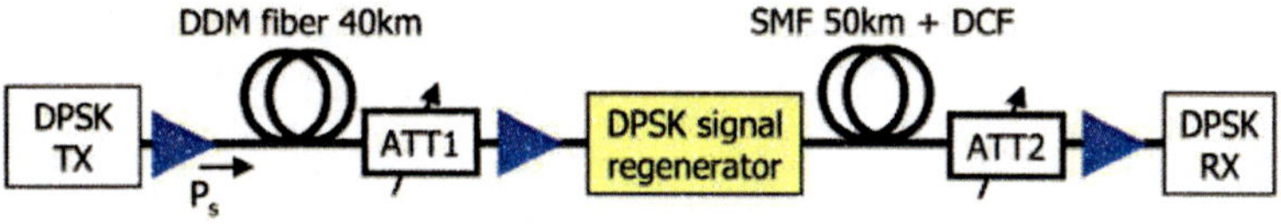

Fig. 11.4 Two-span transmission system for the performance evaluation of the DPSK regenerator [9]

mode-locked laser diode (MLLD) are used as clock pulses. After amplification, the regenerated OOK signal, together with the clock pulses, is directed to the second HNLF acting as an all-optical phase modulator. The clock pulses have duration of 1.5 ps before entering the HNLF, but are widened to 6 ps after the OBPF with bandwidth of 0.8 nm for the rejection of the control pulses. The HNLF has dispersion $D = 2.2\,\text{ps nm}^{-1}\,\text{km}^{-1}$ and $L = 2.4\,\text{km}$. Walk-off time between the data and probe pulses is 24 ps and the timing between the two pulse trains is adjusted by a variable delay line so that complete walk-through between the control and probe pulses takes place in the fiber. The polarizations of the data and probe signals are aligned by the use of a polarization controller (PC) before their entering the HNLF. The power of the control pulses is chosen so that the phase shift induced to the probe pulse via XPM is equal to π.

The regenerator is put into a two-span transmission system as shown in Fig. 11.4 and its performance is measured in terms of bit error rates (BERs). The pulse source in the transmitter consists of an actively mode-locked fiber ring laser and a continuous-wave laser. XPM between them in another nonlinear fiber and subsequent narrowband filtering produce a phase-stable pulse train at 1548.5 nm, with its pulse width about 6 ps. The pulses are then phase-modulated by a $LiNbO_3$ phase modulator with a 256-bit random pattern. After amplification, the pulses are launched to the first transmission fiber. The fiber is a densely dispersion-managed (DDM) fiber consisting of alternating normal- and anomalous-dispersion ($\sim \pm 3\,\text{ps nm}^{-1}\,\text{km}^{-1}$) nonzero dispersion-shifted fiber sections with zero average dispersion around the signal wavelength. Length of each fiber section is 2 km and the total length is 40 km [42]. In this fiber, dispersive pulse broadening is limited, which enhances the nonlinear phase noise that is caused by the translation from amplitude to phase noise via the effect of SPM in the fiber. Similar transmission behavior is expected also when a dispersion-shifted fiber is used instead of the DDM fiber. The loss of the DDM fiber including splice loss is 13.7 dB. After the transmission over the DDM fiber, an attenuator (ATT1) together with an erbium-doped fiber amplifier (EDFA) is inserted for the purpose of noise loading. The second fiber after the regenerator is a standard single-mode fiber (SMF) with 50 km length that is fully dispersion compensated by a DCF spool, total loss of which is 15.9 dB. Again, ASE is loaded by a combination of an attenuator (ATT2) and an EDFA. The receiver consists of a preamplifier, an OBPF, a DI, a balanced detector, an RF amplifier, and a lowpass filter, followed by an error detector. Different programmed bit patterns are used for the error count when the regenerator is or is not inserted. No precoder or postcoder is used.

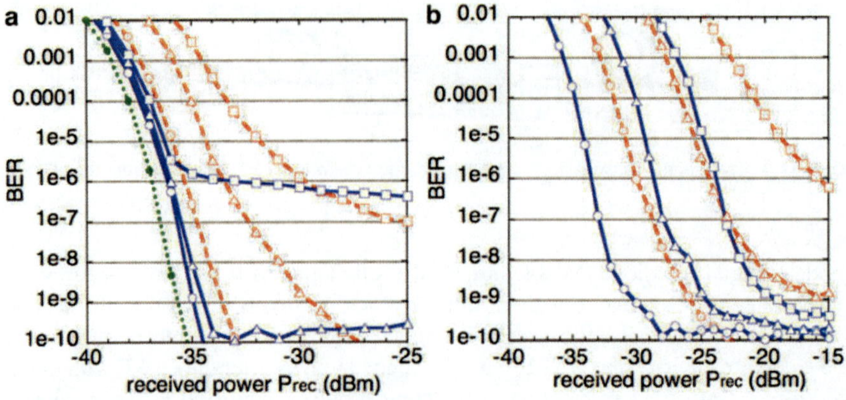

Fig. 11.5 BER performance of the system with (*solid curves*) or without (*dashed curves*) inserting the regenerator; Signal before the regenerator is mainly degraded by nonlinearity in the DDM fiber. (**a**) BER measured before the second span with P_s = 8 dBm (*circles*), 9.5 dBm (*triangles*), and 11 dBm (*squares*). *Dotted curve* shows the back-to-back BER. (**b**) BER measured after the second span with ATT2 = 8 dB (*circles*), 14 dB (*triangles*), and 18 dB (*squares*). P_s is fixed at 9.5 dBm [9]

First, we consider the case where the signal before the regenerator is degraded by nonlinearity in the preceding transmission. ATT1 in Fig. 11.4 is set at zero and the average signal power P_s launched to the DDM fiber is varied. At signal power levels larger than about 7 dBm, degradation caused by the nonlinear phase noise appears. The optical signal-to-noise ratio (OSNR) at the entrance of the transmission fiber is 20 dB/0.1 nm noise bandwidth. Figure 11.5a shows the BER performance measured after the DDM fiber with or without inserting the regenerator. The dotted curve is the reference back-to-back BER. When the regenerator is not used, the BER degrades steadily as the launched signal power grows larger than about 8 dBm as shown by dashed curves and BERs after regeneration are shown by solid curves in Fig. 11.5a. The effect of regeneration is evident at low received power $P_{rec} < -36$ dBm, where BER behaviors are almost identical for different launched signal powers. Error floors, however, appear after the regenerator when the launched signal power is 9.5 and 11 dBm.

The error floors appear even when the threshold of the regenerator, or the averaged input signal power to the regenerator, is optimally chosen. This is expected because the regenerator captures more noise than the detector at the receiver. Since the duration of the input pulses to the 2R amplitude regenerator should be narrow enough for proper operation of the regenerator, the noise bandwidth at the input of the amplitude regenerator is wider than that at the entrance of the detector in the receiver, which leads to enhanced error by the regenerator. Better design of the 2R amplitude regenerator that allows the use of wider pulse duration with narrower bandwidth will lower the error floors.

In spite of the error floor, the pulses are well reshaped by the regenerator. This gives rise to large reduction of power penalty after transmission over the second

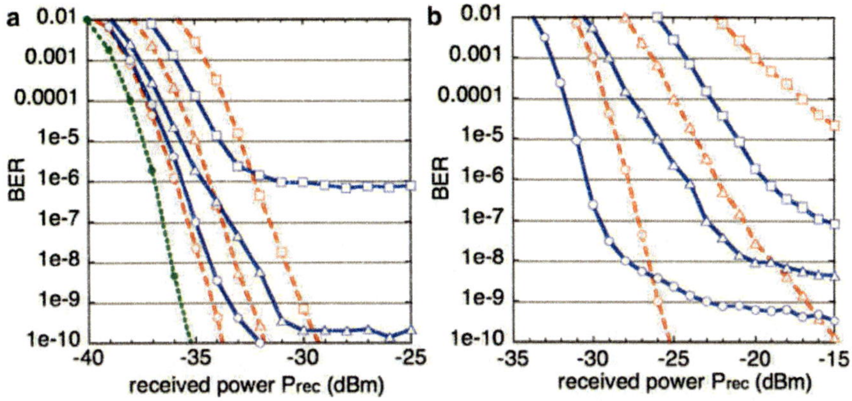

Fig. 11.6 BER performance of the system with (*solid curves*) or without (*dashed curves*) inserting the regenerator; Signal before the regenerator is mainly degraded by nonlinearity in the DDM fiber. (**a**) BER measured before the second span with P_s = 8 dBm (*circles*), 9.5 dBm (*triangles*), and 11 dBm (*squares*). *Dotted curve* shows the back-to-back BER. (**b**) BER measured after the second span with ATT2 = 8 dB (*circles*), 14 dB (*triangles*), and 18 dB (*squares*). P_s is fixed at 9.5 dBm [9]

span. Figure 11.5b shows the BER performance measured after the second fiber span. The launched signal power to the first fiber span is fixed at 9.5 dBm. ASE generated in the second span is enhanced by increasing the attenuation (ATT2). Figure 11.5b shows large benefits of the regenerator inserted before the second span especially when the noise added in the second span is large.

In the second measurement, the signal before the regenerator is degraded by ASE while the launched signal power to the DDM fiber is kept low. Figure 11.6a shows the BER performance measured after the first span with or without inserting the regenerator. The attenuation of ATT1 in Fig. 11.4 is varied between 8 and 16 dB that is compensated for by the EDFA right after the attenuator. The ASE gives both amplitude and phase noise to the signal. As was discussed in Sect. 11.2.1.1, the amplitude noise on the DPSK input signal is transferred to the amplitude noise of the demodulated OOK signal after the DI. Suppression of the amplitude noise of the OOK signal by the 2R amplitude regenerator is more crucial in this case than in the previous case of degradation mainly due to phase noise. Reduction in penalty by the regenerator is weaker in the case of ASE degradation as shown in Fig. 11.6a. Figure 11.6b shows the BER performance measured after the second fiber span. The amount of ATT1 in the first span is fixed at 12 dB. Although the error floor originated in the first span remains, the reshaping effect gives rise to reduction in power penalty at BERs larger than about 10^{-8}. The regenerator performance, however, will be improved by the use of the 2R amplitude regenerator having better noise suppression capability.

Here, we performed a proof-of-principle experiment of the DPSK signal regeneration at 10 Gbit s^{-1}. The data speed can be raised beyond 100 Gbit s^{-1} if suitable clock pulse sources are available. This is owing to the ultrafast response time of the Kerr nonlinearity of the fiber that is responsible for the key functions of the regenerator: amplitude regeneration and all-optical phase modulation. In practical systems, polarization sensitivity of the XPM-based all-optical phase modulation should be avoided. The XPM operation independent of polarization of control pulses will be realized by the use of circular birefringence nonlinear fibers [43].

11.2.1.4 Logic Alteration by the Regenerator and Its Compensation

In the DPSK signal regenerators discussed in previous subsections, the phase difference between adjacent pulses incoming to the regenerator is mapped to the absolute phase of the output pulses, which accompanies conversion of data patterns encoded on the signal phase. The logic conversion can be reversed either by precoding before modulation in the transmitter or by postcoding after detection in the receiver [4, 5].

Here, we assign logic levels 0 and 1 to phase modulations 0 and π of the optical signal, respectively. The logic levels 0 and 1 are also assigned to low and high power levels of demodulated OOK pulses, respectively. The phase information is converted to the amplitude (power) information through the DI, while the amplitude information is converted to the phase information by the phase modulator. The logic operation of the DI is the exclusive OR (XOR) as shown in Fig. 11.7a and can be written as

$$b_n = a_n \oplus a_{n-1}, \tag{11.16}$$

where a_n and b_n are the input and output logics at a time instance n. This operation can be inverted by the operation

$$d_n = c_n \oplus d_{n-1} \tag{11.17}$$

as shown in Fig. 11.7b. If the logic operation (11.17) precedes the DI, in the case of precoding, the output logic becomes, by equating a_n and d_n in (11.16) and (11.17),

$$b_n = a_n \oplus a_{n-1} = (c_n \oplus a_{n-1}) \oplus a_{n-1} = c_n \oplus (a_{n-1} \oplus a_{n-1}) = c_n \oplus 0 = c_n. \tag{11.18}$$

The operation (11.17) thus inverts the XOR operation. If the logic operation (11.17) is placed after the DI as a postcoder, on the other hand, c_n in (11.17) is equated to b_n in (11.16) and the output logic is given by

$$d_n = c_n \oplus d_{n-1} = b_n \oplus d_{n-1} = b_n \oplus (b_{n-1} \oplus d_{n-2}) = (b_n \oplus b_{n-1}) \oplus d_{n-2}. \tag{11.19}$$

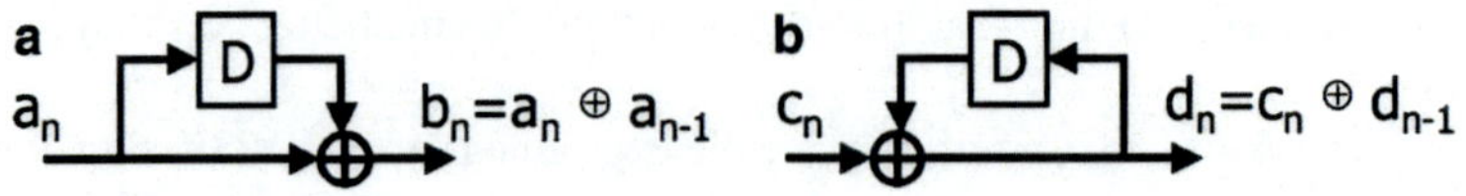

Fig. 11.7 (**a**) Exclusive OR (XOR) logic circuit representing a delay interferometer, and (**b**) that inverting the XOR operation. D indicates a single-bit delay

Being $b_n \oplus b_{n-1} = (a_n \oplus a_{n-1}) \oplus (a_{n-1} \oplus a_{n-2}) = a_n \oplus 0 \oplus a_{n-2} = a_n \oplus a_{n-2}$ from (11.16), d_n can be written as

$$d_n = (a_n \oplus a_{n-2}) \oplus d_{n-2} = a_n \oplus (a_{n-2} \oplus d_{n-2}). \quad (11.20)$$

If the state of the postcoder is set so that $d_{n-2} = a_{n-2}$ is satisfied at some time instance, d_n becomes equal to a_n at subsequent time instances.

When more than one DPSK regenerators are placed in the transmission system, the original data can be recovered by using multiple precoders or postcoders total number of which is same as the number of inserted regenerators. In reconfigurable networks, the number of regenerators that the signal passes will vary according to the route of the signal. In such environment, preservation of the data logic is desired at each regenerator stage. This calls for the use of an all-optical XOR gate with one-bit delay feedback that performs the operation shown in Fig. 11.7b in the optical domain.

11.2.2 Noise Reduction of BPSK Signals Based on Noise Averaging

When two coherent optical signals having an identical amplitude and independent noise are summed constructively, the resultant signal has an amplitude twice the original amplitude, that is, the power is quadrupled while the noise power is only doubled. The signal-to-noise ratio is thus increased by a factor of two. This noise averaging effect can be applied to the reduction of amplitude and phase fluctuations of BPSK signals [12–14]. Figure 11.8 shows an interferometer structure for this purpose, where the two outputs from a DI are coupled again after passing through nonlinear elements that have a function of removing zero-level noise. We denote the complex amplitude of the incoming signal as E_n at a time instance n. The DI has a delay equal to the symbol period so that E_n interferes with E_{n-1} at the DI output. If the nonlinear elements are absent in both of the interferometer arms, the field E_{out} output from the structure becomes simply as

$$E_{\text{out}} = \mathrm{i}\big/\left(2\sqrt{2}\right)\,(E_n - E_{n-1}) + \mathrm{i}\big/\left(2\sqrt{2}\right)\,(E_n + E_{n-1}) = \mathrm{i}E_n\big/\sqrt{2}, \quad (11.21)$$

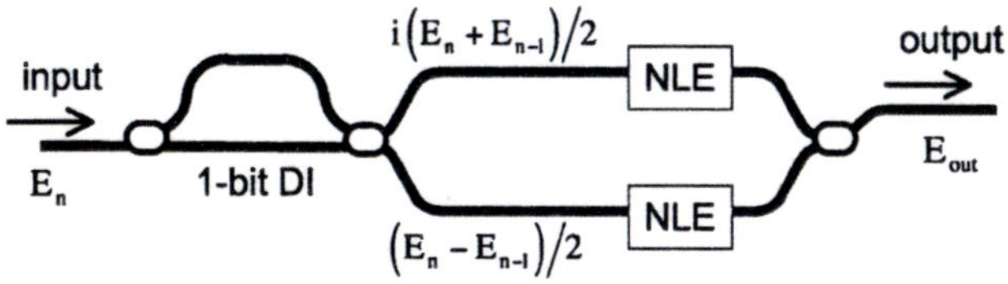

Fig. 11.8 A scheme of noise reduction of BPSK signals by noise averaging. *NLE* Nonlinear element that suppresses zero-level noise

indicating that the input signal is transmitted without noise reduction. In (11.21), we assumed 3 dB couplers for all the couplers and neglected the phase shift common to the interfering fields.

Now, we consider the case where the nonlinear elements remove the zero-level noise. E_n and E_{n-1} again have the form of

$$E_n = (A_s + \Delta A_n)\exp[i(\phi_n + \Delta\phi_n)], \tag{11.22a}$$

$$E_{n-1} = (A_s + \Delta A_{n-1})\exp[i(\phi_{n-1} + \Delta\phi_{n-1})], \tag{11.22b}$$

where ϕ_n and ϕ_{n-1} take either of two values 0 or π, and ΔA_n, ΔA_{n-1} and $\Delta\phi_n$, $\Delta\phi_{n-1}$ are small amplitude and phase noise, respectively. When ϕ_n and ϕ_{n-1} have a zero phase difference, that is $\phi_n - \phi_{n-1} = 0$, the signal output from the upper port of the DI and fed to the upper nonlinear element can be written as

$$\begin{aligned}(i/2)\ (E_k + E_{n-1}) &= (i/2)\ [(A_s + \Delta A_n)\exp(i\Delta\phi_n) \\ &\quad + (A_s + \Delta A_{n-1})\exp(i\Delta\phi_{n-1})]\exp(i\phi_n) \\ &\cong (i/2)\ [2A_s + \Delta A_n + \Delta A_{n-1} \\ &\quad + iA_s(\Delta\phi_n + \Delta\phi_{n-1})]\exp(i\phi_n)\end{aligned} \tag{11.23}$$

under the first-order approximation that $|\Delta A_{n-1,\,n}| \ll A_s$ and $|\Delta\phi_{n-1,\,n}| \ll 1$ are satisfied. The signal output from the lower port of the DI and fed to the lower nonlinear element, on the other hand, is

$$(1/2)\ (E_n - E_{n-1}) \cong (1/2)\ [\Delta A_n - \Delta A_{n-1} + iA_s(\Delta\phi_n - \Delta\phi_{n-1})]\exp(i\phi_n). \tag{11.24}$$

The nonlinear elements suppress the zero-level fields while are transparent otherwise so that (11.23) is directed to the output coupler but (11.24) is suppressed to zero. The output signal is therefore

$$\begin{aligned}E_{\text{out}} &= i\Big/\left(2\sqrt{2}\right)\ (E_n + E_{n-1}) \\ &\cong i\Big/\left(2\sqrt{2}\right)\ [2A_s + \Delta A_n + \Delta A_{n-1} + iA_s(\Delta\phi_n + \Delta\phi_{n-1})]\exp(i\phi_n).\end{aligned} \tag{11.25}$$

When ϕ_n and ϕ_{n-1} have a π phase difference, that is $\phi_n - \phi_{n-1} = \pi$, the signal fed into the nonlinear element in the upper interferometer arm becomes

$$(i/2)\ (E_n + E_{n-1}) \cong (i/2)\ [\Delta A_n - \Delta A_{n-1} + iA_s(\Delta\phi_n - \Delta\phi_{n-1})]\exp(i\phi_n), \tag{11.26}$$

while the signal fed into that in the lower interferometer arm is

$$\begin{aligned}(1/2)\ (E_n - E_{n-1}) \cong (1/2)\ [2A_s + \Delta A_n + \Delta A_{n-1} + iA_s(\Delta\phi_n \\ + \Delta\phi_{n-1})]\exp(i\phi_n).\end{aligned} \tag{11.27}$$

In this case, the signal in the upper interferometer arm is suppressed to zero so that the output signal becomes

$$\begin{aligned} E_{\mathrm{out}} &= \mathrm{i}\Big/\left(2\sqrt{2}\right)(E_n - E_{n-1}) \\ &\cong \mathrm{i}\Big/\left(2\sqrt{2}\right)[2A_{\mathrm{s}} + \Delta A_n + \Delta A_{n-1} + \mathrm{i}A_{\mathrm{s}}(\Delta\phi_n + \Delta\phi_{n-1})]\exp(\mathrm{i}\phi_n). \end{aligned} \tag{11.28}$$

From (11.25) and (11.28), it is found that in either case of $\phi_n - \phi_{n-1} = 0$ or π the output field has the form

$$E_{\mathrm{out}} \cong \left(\mathrm{i}/\sqrt{2}\right)[A_{\mathrm{s}} + (\Delta A_n + \Delta A_{n-1})/2 + \mathrm{i}A_{\mathrm{s}}(\Delta\phi_n + \Delta\phi_{n-1})/2]\exp(\mathrm{i}\phi_n) \tag{11.29}$$

showing that the noise field is averaged. In the first-order approximation together with the condition that the noises on the $(k-1)$th and kth symbols are independent, the average and variance of the amplitude $A_{\mathrm{out}} = |E_{\mathrm{out}}|$ are expressed as $\langle A_{\mathrm{out}}\rangle = A_{\mathrm{s}}/\sqrt{2} = \langle A_{\mathrm{in}}\rangle/\sqrt{2}$ and $\sigma^2_{A\mathrm{out}} = \sigma^2_{A\mathrm{in}}/4$, where $\sigma^2_{A\mathrm{in}}$ is the variance of ΔA_{n} and $\Delta A_{\mathrm{n}-1}$. The amplitude signal-to-noise ratio at the output $\langle A_{\mathrm{out}}\rangle^2/\sigma^2_{A\mathrm{out}}$ is therefore twice that at the input $\langle A_{\mathrm{in}}\rangle^2/\sigma^2_{A\mathrm{in}}$. The phase of the output signal (11.29) is expressed again in the first-order approximation as $\phi_{\mathrm{out}} \cong \pi/2 + \phi_n + (\Delta\phi_n + \Delta\phi_{n-1})/2$. Its variance $\sigma^2_{\phi\mathrm{out}}$ is a half of the variance of the input phase noise $\Delta\phi_n$ and $\Delta\phi_{n-1}$. The noise reduction by averaging can be increased by cascading the interferometers [13]. It is noted that the data pattern encoded on the signal phase is maintained by this type of regeneration differently from the case of the DPDK regenerator discussed in the previous subsections.

The noise reduction of BPSK signal using this scheme was first demonstrated in [12], where the field averaging is performed in a Sagnac interferometer in which an SOA is incorporated at the midpoint of the loop. GS of the bidirectional SOA induced by counterpropagating strong pump pulses was used as the zero-level noise suppression.

11.2.3 Phase-Preserving Amplitude Regeneration

11.2.3.1 Nonlinear Phase Noise and Its Suppression

Regeneration process of signal phase consists of identifying the phase state of the symbol being transmitted and removing the phase error of it. Accessing and identifying the phase information of PSK signals is not an easy task, which usually requires making interference between the signal and a reference field. Examples of such systems were discussed in the previous subsections.

Besides the regeneration schemes attempting phase-noise reduction, amplitude-only regeneration is still effective in improving PSK signal transmission performance. The noise after detection comes from both amplitude and phase noises of

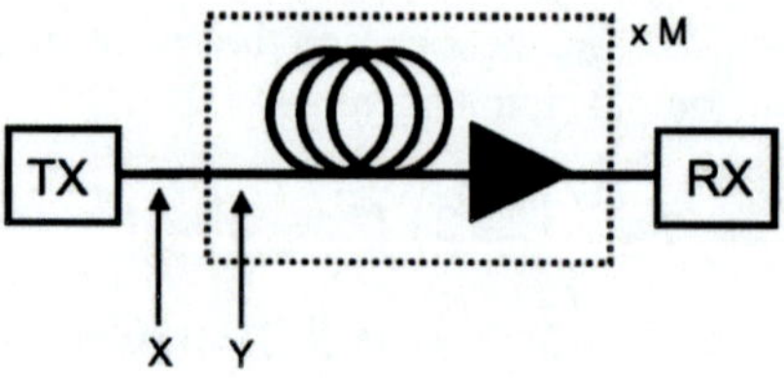

Fig. 11.9 An amplified transmission system consisting of M spans. Effects of amplitude limiters inserted at either X or Y are examined

the optical signal before detection. Suppression of amplitude noise thus directly improves the signal quality. In long-distance systems, furthermore, the amplitude noise is converted into phase noise through the nonlinearity of the transmission fiber [44]. The resulting phase noise, which is called the nonlinear phase noise, severely degrades the system performance. The variance of the phase noise grows proportionally to the cube of the number of amplification in the long-distance limit. The amplitude-noise reduction of PSK signals is effective in suppressing the nonlinear phase noise [17, 22].

Figure 11.9 shows a simplified transmission system consisting of M amplifier spans. In such a system, a major contribution of the phase noise comes from ASE from the inline amplifiers. On the one hand, the quadrature component of the ASE noise relative to the signal gives direct phase fluctuations, whose variance accumulates proportionally to the number of amplifier stages. The in-phase noise component, on the other hand, does not produce phase noise but amplitude noise at the amplifier. The amplitude noise is converted to phase noise after propagation over the transmission fiber through the effect of SPM of the fiber [44]. (In wavelength division multiplexed (WDM) systems, the amplitude noise of surrounding channels also induces phase noise to the channel of interest [45].) The nonlinear phase noise dominates over the direct phase noise (linear phase noise) when transmission distance and/or the signal power in the fiber are large. The noise generated in the source also contributes to the linear and nonlinear phase noise at the receiver. The variance of the phase noise is given by

$$\begin{aligned}\langle \delta\phi^2 \rangle = {} & \frac{N_s B}{2P_{sig}} + 2P_{sig} N_s B \left(\gamma L_{eff}\right)^2 M^2 + \frac{N_a B M}{2P_{sig}} \\ & + 2P_{sig} N_a B \left(\gamma L_{eff}\right)^2 \frac{M(M-1)(2M-1)}{6},\end{aligned} \tag{11.30}$$

where N_s, B, P_{sig}, γ, L_{eff}, and N_a are the power spectrum density of the source noise, bandwidth of the signal and noise, peak signal power launched into the transmission fiber, nonlinear coefficient and effective span length of the transmission fiber, and spectrum density of ASE from each inline amplifier, respectively [22]. N_s is related to the source OSNR (noise bandwidth of 0.1 nm) as $N_s = sP_{sig}/(12.5\text{GHz.OSNR})$, where only one noise polarization is considered and s is the duty ratio of the signal (averaged signal power is given by $P_{ave} = sP_{sig}$), while N_a is given by $h\nu n_{sp}(G-1)$, where $h\nu$, n_{sp}, and G are the photon energy, spontaneous emission factor, and gain of the inline amplifier compensating for the span loss, respectively. The first and second two terms in (11.30) are contributions from source noise and ASE from the inline amplifiers, respectively. It is noted that the influence of dispersive pulse broadening during transmission, which would occur in real systems, is ignored in (11.30) for the sake of discussion of principal feature of the nonlinear phase noise. In the presence of dispersion, the amount of the nonlinear phase noise is decreased [46–48].

When an optical limiter that perfectly suppresses the amplitude noise is inserted after the transmitter (point X in Fig. 11.9), the nonlinear phase noise induced by the source noise, that is, the second term in (11.30), is eliminated and (11.30) becomes

$$\langle \delta\phi^2 \rangle = \frac{N_s B}{2P_{sig}} + \frac{N_a B M}{2P_{sig}} + 2P_{sig} N_a B \left(\gamma L_{eff}\right)^2 \frac{M(M-1)(2M-1)}{6} + \frac{N_r B}{2G_r P_{sig}}, \tag{11.31}$$

where the ASE contribution from an additional amplifier with gain G_r located in front of the limiter is added as the last term. Such an amplifier is usually needed to boost the signal power to the saturation level of the limiter. N_r is given by $h\nu n_{sp}(G_r - 1)$. The nonlinear phase noise originating from the inline amplifier noise, the third term in (11.31), is further eliminated when the optical limiters are inserted every span at point Y in Fig. 11.9. The phase noise then becomes

$$\langle \delta\phi^2 \rangle = \frac{N_s B}{2P_{sig}} + \frac{N_a B M}{2P_{sig}} + \frac{N_r B M}{2G_r P_{sig}}. \tag{11.32}$$

The variance of the phase noise (11.32) grows at most linearly as the number of spans M is increased and is inversely proportional to the signal power P_{sig}, indicating the effectiveness of the amplitude limiter in long-distance systems.

Figure 11.10 shows an example of the standard deviation of the phase noise versus signal power launched into the transmission fiber. The loss and nonlinearity of the transmission fiber are $\alpha = 0.3\,\text{dB km}^{-1}$ and $\gamma = 3.5\,\text{W}^{-1}\,\text{km}^{-1}$, respectively.

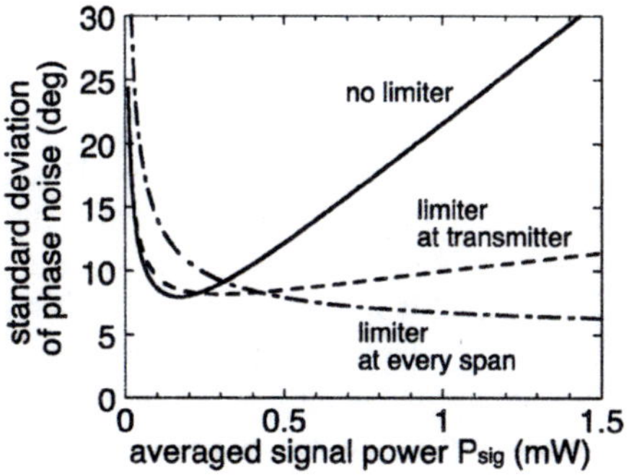

Fig. 11.10 Standard deviation of phase noise at the receiver vs. signal power. *Solid, dashed*, and *dash-dotted curves* correspond to the cases, where no amplitude limiters are used, an amplitude limiter is inserted at the output of the transmitter, and amplitude limiters are inserted every amplifier span, respectively [22]

The span length is 40 km and the total loss per span is assumed to be 22 dB. The number of spans is 5, or the transmission distance is 200 km. These parameters are those used in the experiment described in 11.2.3.3 [22]. The noise figure (NF) of all the EDFAs in the system is 6 dB ($n_{sp} = 2$) and bandwidth $B = 2$ nm. Source OSNR (per 0.1 nm noise bandwidth with single polarization) is 24.5 dB. The horizontal axis is the average power assuming duty ratio of 6.8%, which is also relevant to the experiment using 10 Gbit s^{-1} 6.8 ps pulses. Input averaged power to the amplitude limiter P_{lim} is assumed to be 3.4 mW, which specifies the gain G_r of the amplifier in front of the limiter. In this calculation, the influence of a small extra phase shift $\delta\phi = k\delta P/P_{\mathrm{lim}}$ given to the signal is considered with $k = 0.8$ rad, where δP is the power fluctuation to be suppressed by the limiter [22]. $k = 0.8$ rad means that a phase shift of 4.6° is induced to the signal, for example, when a relative power fluctuation $\delta P/P_{\mathrm{lim}}$ of 10% is suppressed by the limiter. Solid, dashed, and dash-dotted curves in Fig. 11.10 correspond to the cases without using limiters, with a limiter inserted at the output of the transmitter, and with limiters inserted every amplifier span, respectively. When no limiters are used, the phase noise becomes minimum at $P_{\mathrm{ave}} = 0.17$ mW corresponding to SPM-induced phase shift to the signal $\phi_{\mathrm{SPM}} = P_{\mathrm{sig}}\gamma L_{\mathrm{eff}}M = 0.59$ rad. This is somewhat smaller than the optimal value predicted in [44]. This is mainly because of the inclusion of the effect of source noise in this calculation. When amplitude limiters are inserted in the system, the phase noise is greatly reduced especially at large signal power, where nonlinear phase noise contribution is significant. It is noted that the phase noise is steadily decreasing with the increase of the signal power when the limiter is inserted every span. When perfect amplitude regenerators are inserted every span, signals propagate through the transmission fiber with no amplitude fluctuations and, therefore, nonlinear phase noise disappears. Because the remaining linear phase noise is smaller for larger signal power, the total phase noise is steadily decreasing with the increase of the signal power. The amplitude limiter is thus effective in reducing nonlinear penalty of the system, leading to longer amplifier spans and larger system margins.

11.2.3.2 Phase-Preserving Amplitude Regenerator

A prerequisite to the amplitude regenerator for PSK signals is that extra phase noise should not be added to the signal in the process of the amplitude regeneration. The majority of amplitude regeneration schemes aiming at OOK signal regeneration do not meet this requirement. In the Mamyshev-type amplitude regenerator introduced in Sect. 11.2.1.3, for example, pulses having different amplitudes at the input of the regenerator acquire different phase shifts in the course of amplitude stabilization, which causes large phase fluctuations after the regenerator [16]. That is, the nonlinear phase noise is induced in the regenerator itself. Several phase-preserving amplitude regenerators satisfying the above requirement have been recently proposed and demonstrated.

In [15], the use of nonlinear Sagnac interferometers, or nonlinear optical loop mirrors (NOLMs), has been proposed. The nonlinear Sagnac interferometers, when its symmetry is broken, are known to exhibit power transfer that varies sinusoidally as the input signal power is changed [49]. By inserting a directional attenuator [15] or a bidirectional amplifier [19] in the interferometer loop and suitably choosing the parameters, one can have flat phase response in the input signal power region, where the output signal power becomes almost constant. A recirculating transmission experiment of 10 Gbit s^{-1} DPSK signals, where the NOLM-based limiter is inserted in the loop, has been demonstrated in [24]. In [25], the phase-preserving limiter operation has been demonstrated by the use of a multi-quantum-well semiconductor saturable absorber.

The phase-preserving amplitude limiting is also achieved by a fiberoptic parametric amplification operating in the saturated regime. In the parametric amplifier, the output signal power saturates as the input power is increased due to the depletion of pump power, change in the direction of power exchange between FWM components, and excitation of higher-order FWM products [50,51]. Because the saturation takes place almost instantaneously with a response time of the Kerr nonlinearity of the fiber, one can obtain pulse-to-pulse amplitude noise suppression of ultrahigh-speed signals [52].

Figure 11.11 shows a schematic of the one-pump fiberoptic parametric amplifier consisting of a pump source, a nonlinear fiber, and an OBPF for extracting the output signal wavelength component, and spectra at the entrance and exit of the fiber. Any output spectral components which exhibit saturation can be used as the amplitude-limiter output. The output phase behavior and ability of zero-level stabilization differ for different output four-wave mixing components. Low-power noise is rejected when we use higher-order FWM products such as those appearing at λ_2 and λ_3, where $\lambda_2^{-1} = 2\,\lambda_s^{-1} - \lambda_p^{-1}$ and $\lambda_3^{-1} = 3\,\lambda_s^{-1} - 2\lambda_p^{-1}$ with λ_s and λ_p signal and pump wavelengths, respectively [53–55]. Phase of the input signal, however, is not correctly transferred to the output when one uses these FWM products. Features of different FWM output components as they are used as amplitude limiting are summarized in Table 11.1. Use of the output wavelength component same as the input signal is most suited for the phase-preserving amplitude limiter application, although care must be taken to avoid zero-level noise amplification.

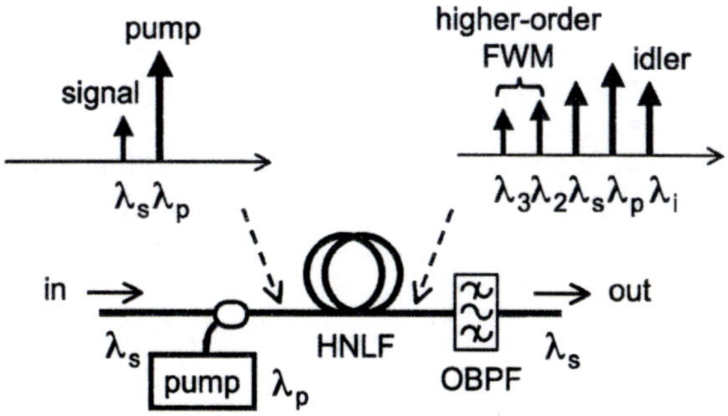

Fig. 11.11 One-pump fiberoptic parametric amplifier and spectra at the entrance and exit of the fiber. *HNLF* Highly nonlinear fiber; *OBPF* Optical bandpass filter

Table 11.1 Features of different FWM output components

Output wavelength	Unsaturated amplitude	Phase information	Zero-level stabilization	Wavelength
λ_s	$\propto E_{in}$	Preserved	No	Maintained
λ_i	$\propto E_{in}^*$	Conjugated	No	Converted
λ_2	$\propto E_{in}^2$	Erased	Yes	Converted
λ_3	$\propto E_{in}^3$	Nominally preserved	Yes	Converted

E_{in} denotes complex amplitude of the input signal

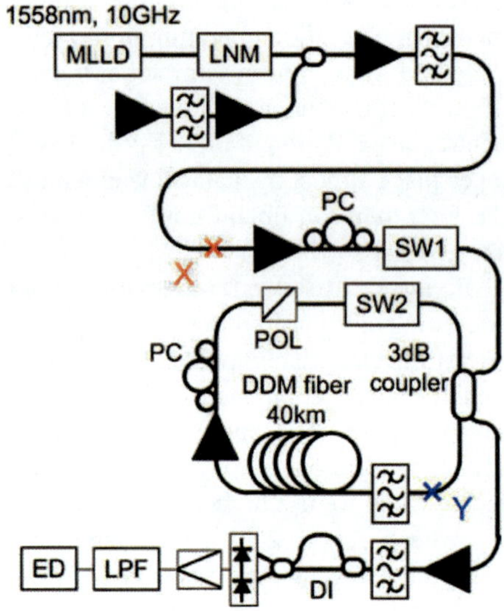

Fig. 11.12 Setup of 10 Gbit s^{-1} short-pulse DPSK transmission. An amplitude limiter based on FWM in fiber is inserted at either X or Y. *MLLD* Mode-locked diode laser; *LNM* $LiNbO_3$ modulator; *SW1,2* Acousto-optic switches; *PC* Polarization controller; *POL* Polarizer; *DI* Delay interferometer; *LPF* Lowpass filter; *ED* Error detector

11.2.3.3 Transmission Experiment Using a Phase-Preserving Amplitude Limiter

In this subsection, we describe a DPSK transmission experiment using the phase-preserving amplitude limiter based on saturation of parametric amplification in a HNLF. Figure 11.12 shows the setup of the DPSK transmission experiment. Ten gigahertz short pulses (1.5 ps) at 1,558 nm are generated by a mode-locked semiconductor laser diode and their phase is modulated by a 256-bit pseudo-random programmed bit pattern. After addition of ASE noise, spectrum of the signal is narrowed by an OBPF with resultant pulse width of 6.8 ps. Then the signal is launched into a recirculating fiber loop. The transmission fiber is a DDM fiber, the same as that used in the experiment described in Sect. 11.2.1.3. In the recirculating fiber loop, a manually controlled PC and a polarizer are inserted. They stabilize the signal polarization in the loop and reject ASE noise, whose polarization is orthogonal to that of the signal. The amplitude limiter based on saturation of FWM shown in Fig. 11.11 is inserted in the system either at the entrance of recirculating loop

Fig. 11.13 Power transfer function and Q factor for the FWM-based amplitude limiter. *Solid* and *dashed curves* correspond to the cases where pump power is on and off, respectively [22]

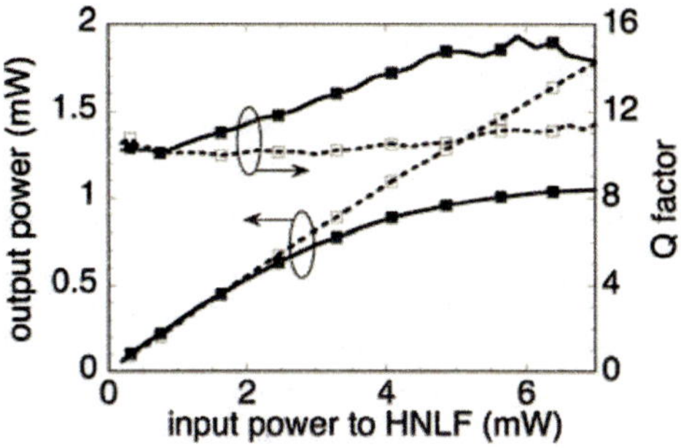

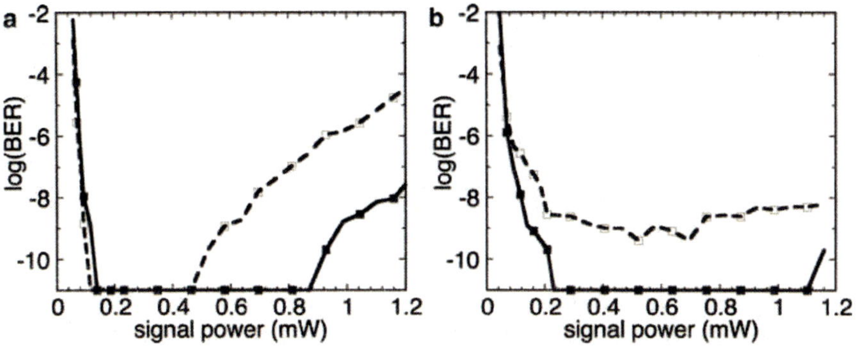

Fig. 11.14 BER vs. averaged signal power launched to the transmission fiber. An amplitude limiter is inserted (**a**) at the output of the transmitter (point X in Fig. 11.12) or (**b**) in the recirculating loop (point Y in Fig. 11.12). *Solid* and *dashed curves* correspond to the cases where pump power is on and off, respectively [22]

(point X) or inside the recirculating loop (point Y). Effect of the amplitude limitation is observed by measuring the BER with turning on and off the pump power in the limiter.

Figure 11.13 shows averaged output signal power versus input signal power to the HNLF when unmodulated 10 GHz pulses (6.8 ps) are launched into the amplitude limiter. The OSNR is 23 dB with noise bandwidth 0.1 nm. The HNLF has the same dispersion, nonlinearity, loss, and length as those used in the analysis in Sect. 11.2.3.1. Pump wavelength and power are 1,561 nm and 15 mW, respectively. Q factor defined as μ/σ is also plotted where μ and σ are the mean values and standard deviation of the peak voltage of the detected electrical pulses after a lowpass filter with 3 dB cutoff frequency 7.5 GHz. Results with pump power on and off are compared in Fig. 11.13. When the pump is turned on, the output signal power shows saturation and the Q factor increases from 10 to 15 as the input power is increased.

Figure 11.14a shows measured BER after transmission over 200 km (number of circulation $M = 5$) when the limiter is inserted after the transmitter. OSNR of the input signal is 21.5 dB including noise in both polarizations. We find that the BER is remarkably lowered by the amplitude limitation for large signal power. This is qualitatively consistent with the calculation of phase noise as shown in Fig. 11.10. The BER degrades, however, at averaged signal power larger than ~1 mW even when the pump is on. This is considered due to the residual unsuppressed amplitude noise after the limiter. The imperfectness of the amplitude-noise suppression is indicated by the finite, relatively low, Q value even at its maximum shown in Fig. 11.13.

Figure 11.14b shows BER also after transmission of 200 km when the limiter is inserted inside the recirculating loop. OSNR of the input signal is increased to 25.7 dB in this experiment. Error-free transmission was not obtained even when the pump is on when the transmitter OSNR is lower at 21.5 dB. It is considered that this is again because the residual amplitude noise after the limiter induce large nonlinear phase shift in the HNLF at each circulation. Buildup of zero-level noise is also a cause of the imperfect performance of the system with the limiter. However, the range of usable signal power is extended when the limiter is inserted every amplifier span.

11.2.4 BPSK Signal Regeneration Using Phase-Sensitive Amplifiers

A PSA is the amplifier that selectively amplifies one of the two quadrature phase components of the input signal. The other quadrature component is deamplified. These properties are markedly different from those of commonly used optical amplifiers such as laser amplifiers, including EDFAs and SOAs, and stimulated Raman and Brillouin amplifiers, where the signal amplification is independent of the input signal phase. One resulting unique feature of the PSA is that the NF smaller than the 3dB limit of the phase insensitive amplifiers is obtainable [56]. A number of theoretical and experimental studies paying attention to this important nature of the PSA have been reported for more than two decades [57–65]. In addition to the noiseless amplification, other applications such as reshaping of amplitude and phase profiles of chirped pulses [66], jitter-free soliton amplification [67], and long-term pulse storage [68] have been proposed and demonstrated. The PSA is also a natural and promising candidate for the phase regenerator of binary PSK signals. Theoretical and experimental studies of phase regeneration of BPSK signals have recently been reported [26–30]. One issue in using PSAs for the phase regenerator in real systems is that local optical oscillators that are phase-locked to the incoming PSK signals are needed in the PSA. Several efforts toward solving this task have also been pursued [69, 70].

11.2.4.1 BPSK Regenerator Using Nonlinear Sagnac Interferometer

PSAs for optical communication applications can be realized by the use of nonlinear parametric processes in fibers. Frequency–degenerate interaction between pump and signal in nonlinear fiber Sagnac interferometer has been widely studied for

Fig. 11.15 Nonlinear fiber Sagnac interferometer

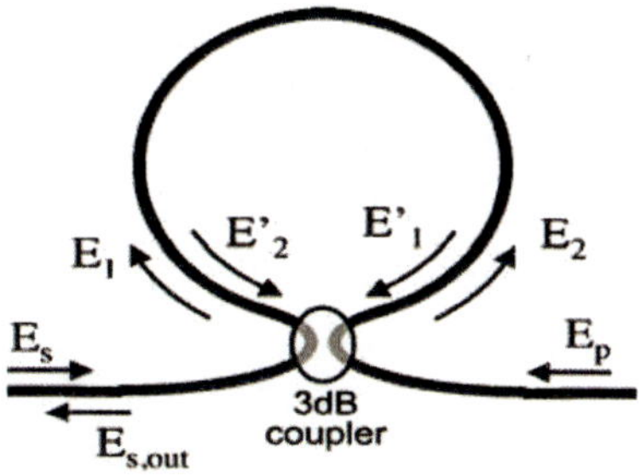

the applications to PSA and generation of squeezed states of light [57–61]. Phase regeneration of BPSK signals, together with amplitude regeneration using saturation behavior in the interferometer, has been reported in [26, 27].

Figure 11.15 shows the nonlinear fiber Sagnac interferometer, or the NOLM used as a PSA. Signal and pump lights, E_s and E_p, respectively, are introduced to the fiber loop through a 3 dB coupler. The optical field amplitudes appearing after the coupler are given by

$$E_1 = (\mathrm{i}E_\mathrm{p} + E_\mathrm{s})/\sqrt{2} \quad \text{and} \quad E_2 = (E_\mathrm{p} + \mathrm{i}E_\mathrm{s})/\sqrt{2},$$

which propagates in the clockwise and counterclockwise in the fiber loop, respectively. By the propagation they acquire nonlinear phase shifts as

$$E_1' = E_1 \exp(\mathrm{i}\gamma\,|E_1|^2\,L) \quad \text{and} \quad E_2' = E_2 \exp(\mathrm{i}\gamma\,|E_2|^2\,L),$$

where L is the length of the loop and the effects of fiber loss is neglected. The output signal amplitude exiting the NOLM through the 3 dB coupler is then given by

$$E_\mathrm{s,out} = (E_2' + \mathrm{i}E_1')/\sqrt{2} = \mathrm{i}\mathrm{e}^{\mathrm{i}\phi_0}\left(-\sqrt{P_\mathrm{p}}\,\mathrm{e}^{\mathrm{i}\phi_\mathrm{p}}\sin\phi_\mathrm{sp} + \sqrt{P_\mathrm{s}}\mathrm{e}^{\mathrm{i}\phi_\mathrm{s}}\cos\phi_\mathrm{sp}\right), \tag{11.33}$$

where $\phi_0 = \gamma\left(P_\mathrm{p} + P_\mathrm{s}\right)L/2$ and $\phi_\mathrm{sp} = \gamma\sqrt{P_\mathrm{p}P_\mathrm{s}}L\,\sin(\phi_\mathrm{s} - \phi_\mathrm{p})$. P_j and $\phi_j (j = \mathrm{p}$ or s) are the power and phase of the pump and signal as given by $E_j = \sqrt{P_j}\,\mathrm{e}^{\mathrm{i}\phi_j}$ ($j = \mathrm{p}$ or s).

$E_\mathrm{s,out}$ given by (11.33) is a nonlinear function of the input signal amplitude E_s. Under the small-signal condition $\left|\phi_\mathrm{sp}\right| \ll 1$, (11.33) is linearized to

$$\begin{aligned} E_\mathrm{s,out} &\cong \mathrm{i}\mathrm{e}^{\mathrm{i}\phi_\mathrm{pump}}\left(-\sqrt{P_\mathrm{p}}\,\phi_\mathrm{sp} + \sqrt{P_\mathrm{s}}\mathrm{e}^{\mathrm{i}\phi_\mathrm{s}}\right) \\ &= \mathrm{i}\mathrm{e}^{\mathrm{i}\phi_\mathrm{pump}}\left[(1 + \mathrm{i}\phi_\mathrm{pump})\sqrt{P_\mathrm{s}}\mathrm{e}^{\mathrm{i}\phi_\mathrm{s}} - \mathrm{i}\phi_\mathrm{pump}\sqrt{P_\mathrm{s}}\mathrm{e}^{-\mathrm{i}\phi_\mathrm{s}}\right] = \mu E_\mathrm{s} + \nu E_\mathrm{s}^* \end{aligned} \tag{11.34}$$

where $\mu = \mathrm{i}\,(1 + \mathrm{i}\,\phi_\mathrm{pump})\,\exp(\mathrm{i}\phi_\mathrm{pump})$, $\nu = \phi_\mathrm{pump}\,\exp\,(\mathrm{i}\phi_\mathrm{pump})$, and $\phi_\mathrm{pump} = \gamma P_\mathrm{P} L/2$. μ and ν satisfy a relation $|\mu|^2 - |\nu|^2 = 1$. In the above expressions, pump phase ϕ_p has been taken to be zero without loss of generality. Equation (11.34) is an expression called squeezing transformation in quantum optics and governs phase-sensitive behavior of the amplifier [63]. The phase-sensitive gain is found by squaring (11.34), resulting in

$$G(\phi_s) = |E_{s,out}|^2/P_s = 1 + 2\phi_{pump}^2 + 2\phi_{pump}\sqrt{1+\phi_{pump}^2}\cos(2\phi_s + \tan^{-1}\phi_{pump} + \pi/2) \quad (11.35)$$

When the input signal phase (with respect to the pump phase) satisfies $\phi_s = -\tan^{-1}\phi_{pump}/2 + m\pi - \pi/4$ with m an integer, on the one hand, the gain becomes maximum as

$$G_{max} = \left(\sqrt{\phi_{pump}^2 + 1} + \phi_{pump}\right)^2. \quad (11.36)$$

When $\phi_s = -\tan^{-1}\phi_{pump}/2 + m\pi + \pi/4$, on the other hand, the gain becomes minimum as

$$G_{min} = \left(\sqrt{\phi_{pump}^2 + 1} - \phi_{pump}\right)^2 = 1/G_{max}. \quad (11.37)$$

The selective amplification of only one quadrature phase component naturally leads to phase regeneration. In the linear regime discussed above, however, the gain is independent of the input signal amplitude, meaning that the amplitude noise on the input signal is linearly translated to the output signal. The amplitude regeneration is achieved when the PSA is operated in the saturation regime with $|\phi_{sp}| \sim 1$ [71]. When the pump power is sufficiently larger than the signal power, the maximum gain of the NOLM-based PSA is obtained at $\phi_{sp} \sim \pi/2$ with which $\sin\phi_{sp}$ in (11.33) takes a maximum. The value of ϕ_{sp} corresponds to the signal phase ϕ_s of $\pi/2$ with respect to ϕ_p if $\gamma\sqrt{P_p P_s}L$ is set at $\sim\pi/2$. In this condition, small fluctuations in the input signal power P_s are not translated to the variations in $\sin\phi_{sp}$ in (11.33) within first-order approximation. This regime gives rise to simultaneous phase and amplitude regeneration, which is desired in signal regeneration applications.

A problem in this type of phase regenerator is that the amplitude noise of the input signal and the pump is directly converted to the phase noise of the output signal through the factor $e^{i\phi_0}$ included in (11.33) [29]. The influence of this phase noise addition is severer for larger input signal power on the condition that the signal-to-noise ratio of the input signal is constant. This should be taken care of when the regenerator is operated in the saturation regime for amplitude noise suppression.

11.2.4.2 BPSK Regenerator Using Two-Pump Degenerate FWM in Fiber

Another class of PSA that has been proposed and demonstrated for the phase and amplitude noise suppression of BPSK signals is the two-pump degenerate FWM in a fiber [28,29,63]. The frequency arrangement of the two pumps (ω_{P1}, ω_{P2}) and the signal (ω_s) is shown in Fig. 11.16 where $\omega_s = (\omega_{P1} + \omega_{P2})/2$ is satisfied. Parametric interaction among the pumps and the signal is described by

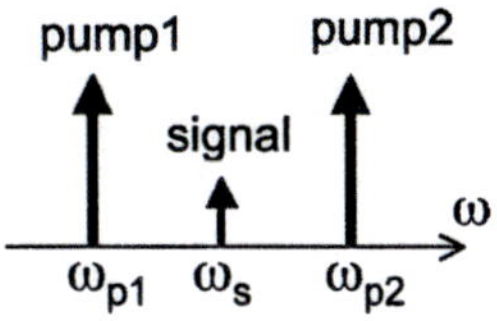

Fig. 11.16 Frequency arrangement of two-pump degenerate FWM

$$\frac{dE_{p1}}{dz} = i\gamma\Big[\left|E_{p1}\right|^2 E_{p1} + 2\left(\left|E_{p2}\right|^2 + |E_s|^2\right) E_{p1} + E_s^2 E_{p2}^* \exp(i\Delta\beta z)\Big] \tag{11.38a}$$

$$\frac{dE_{p2}}{dz} = i\gamma\Big[\left|E_{p2}\right|^2 E_{p2} + 2\left(\left|E_{p1}\right|^2 + |E_s|^2\right) E_{p2} + E_s^2 E_{p1}^* \exp(i\Delta\beta z)\Big] \tag{11.38b}$$

$$\frac{dE_s}{dz} = i\gamma\Big[|E_s|^2 E_s + 2\left(\left|E_{p1}\right|^2 + \left|E_{p2}\right|^2\right) E_s + 2E_{p1}E_{p2}E_s^* \exp\left(-i\Delta\beta z\right)\Big], \tag{11.38c}$$

where $\Delta\omega = \omega_{p2} - \omega_s = \omega_s - \omega_{p1}$, $\Delta\beta = 2\beta(\omega_s) - \beta(\omega_{p1}) - \beta(\omega_{p2}) = -d^2\beta/d\omega^2\big|_{\omega_s}(\Delta\omega)^2 = \left[\lambda^2/(2\pi c)\right]D(\omega_s)(\Delta\omega)^2$, and $D(\omega_s)$ is the dispersion parameter of the fiber at the signal frequency. In (11.38), loss of the fiber and the influence of the fourth- and higher-order derivatives of β with respect to the angular frequency ω are neglected. Under the small-signal condition $|E_s| \ll \left|E_{p1}\right|, \left|E_{p2}\right|$, XPM induced by the signal and FWM contributions in the right-hand sides of (11.38a) and (11.38b) can be neglected, which yields solutions

$$E_{p1}(z) = E_{p1}(0)\exp\left[i\gamma\left(\left|E_{p1}(0)\right|^2 + 2\left|E_{p2}(0)\right|^2\right)z\right], \tag{11.39a}$$

$$E_{p2}(z) = E_{p2}(0)\exp\left[i\gamma\left(\left|E_{p2}(0)\right|^2 + 2\left|E_{p1}(0)\right|^2\right)z\right]. \tag{11.39b}$$

Equation (11.38c) then becomes for small E_s

$$\frac{dE_s}{dz} = 2i\gamma\left[\left(\left|E_{p1}(0)\right|^2 + \left|E_{p2}(0)\right|^2\right)E_s + E_{p1}(0)E_{p2}(0)\,e^{i\theta(z)}\,E_s^*\right], \tag{11.40}$$

where $\theta(z) = 3\gamma\left(\left|E_{p1}(0)\right|^2 + \left|E_{p2}(0)\right|^2\right)z - \Delta\beta\, z$. Equation (11.40) can be put into the form

$$\frac{dE_s'}{dz} - i\delta E_s' = i\zeta E_s'^* \tag{11.41}$$

with $\delta = \Delta\beta/2 + \gamma(P_{p1} + P_{p2})/2$ and $\zeta = 2\gamma\sqrt{P_{p1}P_{p2}}\exp\left[i(\phi_{p1} + \phi_{p2})\right]$, where

$$E_s'(z) = E_s(z)\,e^{-i\theta(z)/2}, \quad E_{p1}(0) = \sqrt{P_{p1}}\,e^{i\phi_{p1}}, \quad \text{and } E_{p2}(0) = \sqrt{P_{p2}}\,e^{i\phi_{p2}}.$$

Equation (11.41) has a solution

$$E'_s(z) = \mu E'_s(0) + \nu E'^*_s(0), \tag{11.42}$$

where $\mu = \cosh(\kappa z) + i(\delta/\kappa)\sinh(\kappa z)$, $\nu = i(\zeta/\kappa)\sinh(\kappa z)$, and $\kappa = (|\zeta|^2 - \delta^2)^{1/2}$. A relation $|\mu|^2 - |\nu|^2 = 1$ holds for μ and ν. The relation between the output and input amplitudes given by (11.42) again represents the squeezing transformation, the same as the expression given in (11.34). Thus, the two-pump degenerate FWM has the feature of PSA by which the function of phase regeneration of BPSK signals is obtained. Although the signal is linearly amplified in the small-signal regime, the direction of energy transfer from the pumps to the signal is finally reversed as the signal power grows along the fiber [50], leading to output signal power saturation. Amplitude regeneration is also obtained by using the PSA in this saturated regime. Simultaneous suppression of phase and amplitude noise of BPSK signals has been demonstrated by numerical simulation [28] and experiments [29].

An issue in realizing the PSA based on two-pump degenerate FWM is to prepare two coherent pumps at different frequencies optically phase-locked with each other and also to the incoming phase-modulated signal. Generating the seed of the second phase-locked pump at ω_{p2} by the use of FWM between the BPSK signal at ω_s and the first pump at ω_{p1} in a preceding separate fiber and subsequent injection locking of a semiconductor laser has been reported in [30, 69].

11.3 Regeneration of Quadrature Phase-Shift Keying Signals

Some of the regeneration schemes for (D)BPSK signals discussed in the previous subsections can also be applied to (D)QPSK signals.

Since the operation of phase-preserving amplitude regenerators does not depend on the phase of the signal, it is equally applied to (D)QPSK signals. Observation of phase-preserving amplitude noise suppression of DQPSK signals at 80 Gbit s^{-1} (40 Gsymbol s^{-1}) using a nonlinear amplifying loop mirror was reported in [72]. In [73], saturation of FWM in a nonlinear fiber was used for amplitude noise suppression of DQPSK signals at 20 Gbit s^{-1} (10 Gsymbol s^{-1}), in which reduction of nonlinear phase noise caused by transmission after the limiter was demonstrated.

PSAs may be applied to QPSK signal regeneration. In the regenerator proposed in [31], two PSAs amplify different quadrature phase components of the input signal orthogonal to each other. After the amplitude noise is suppressed by virtue of GS and the orthogonal phase component is deamplified at each PSA, the two outputs are combined coherently. QPSK signals are thus regenerated with phase data patterns unaltered. In this regeneration scheme, phase coherence of the two PSA outputs must be maintained with an accuracy much smaller than a cycle of the optical carrier, which is a challenging task in real environments.

In the next two subsections, another regeneration scheme of (D)QPSK signals using demodulation from (D)QPSK to OOK signals and subsequent processing and phase modulation back to QPSK signals is described.

11.3.1 *DQPSK Signal Regeneration Using Differential Demodulation*

The regeneration scheme discussed in Sect. 11.2.1 can be extended to DQPSK signal regeneration. Figure 11.17 shows a block diagram of an all-optical DQPSK regenerator using straight-line all-optical phase modulators for mapping the amplitude information back to the signal phase [33]. The incoming DQPSK signals are demodulated to OOK signals by the use of two parallel one-symbol DIs. The optical phase difference in two arms in the DIs are set at $\theta_{DI} = \pi/4$ and $-\pi/4$ for the upper and lower DIs, respectively, which is the same way as in typical DQPSK receivers [74]. In the regenerator shown in Fig. 11.17, signals emerging from one of the two output ports of each DI are used, power level of which takes high or low value depending on the optical phase difference between the consecutive input symbols. The subsequent 2R regenerators remove amplitude fluctuations of the high-level signals and suppress the low-level signals to zero. The amplitude-stabilized OOK pulses are then amplified to prescribed levels and fed to all-optical phase modulators in which the phase of clock pulses is modulated by π or $\pi/2$ in proportion to the power of the OOK pulses. In this way, the four-level phase difference between adjacent symbols of input DQPSK signals $\phi_n - \phi_{n-1}$ can be regenerated and mapped to the absolute phase of the output pulses Θ_n. It is found that the phase differences $\phi_n - \phi_{n-1} = 0, \pi/2, \pi$, or $3\pi/2$ are mapped to $\Theta_n = 0, \pi, 3\pi/2$, or $\pi/2$, respectively [33]. Numerical simulation assuming the use of cascaded fiber-based all-optical 2R regenerators for amplitude noise suppression was performed in [33]. Figure 11.18 shows numerical examples of signal constellations (a) before and (b) after the regenerator operated at 160 Gbit s^{-1} (80 Gsymbol s^{-1}) for short-pulse DQPSK signals. Figure 11.19 shows the waveforms of the signal at various locations inside the regenerator. These figures show that both amplitude and phase fluctuations can be suppressed by this regenerator.

Instead of the straight-line all-optical phase modulator, MZI phase modulators can also be used for the DQPSK signal regeneration. Figure 11.20 shows a block diagram of the regenerator, which is an extension of the DPSK regenerator discussed in Sect. 11.2.1.2. The 2R amplitude regenerators inserted in the OOK signal

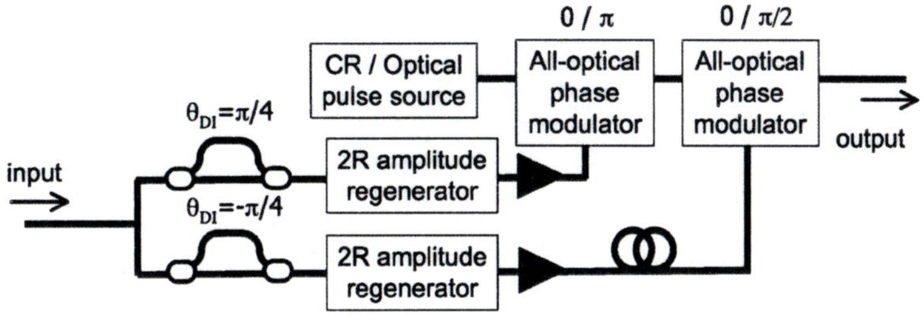

Fig. 11.17 Block diagram of an all-optical DQPSK signal regenerator using straight-line all-optical phase modulators. *CR* Clock recovery circuit

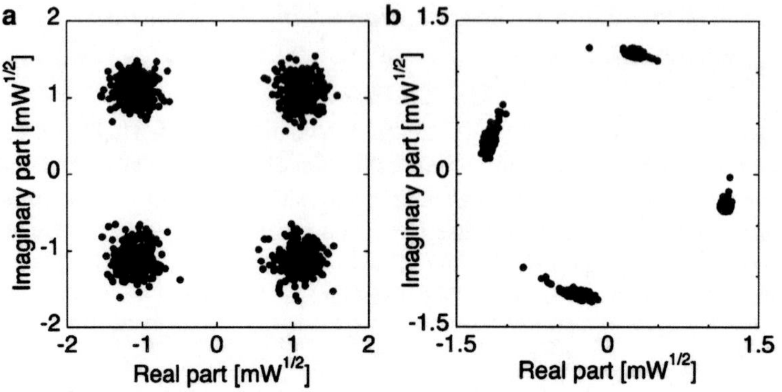

Fig. 11.18 Numerically obtained constellation diagrams of (**a**) input and (**b**) output signals to and from the DQPSK regenerator. Input signal is degraded by ASE with OSNR 24 dB/0.1 nm noise bandwidth (One noise polarization is considered). Data rate is 160 Gbit s^{-1} (80 Gsymbol s^{-1}) [33]

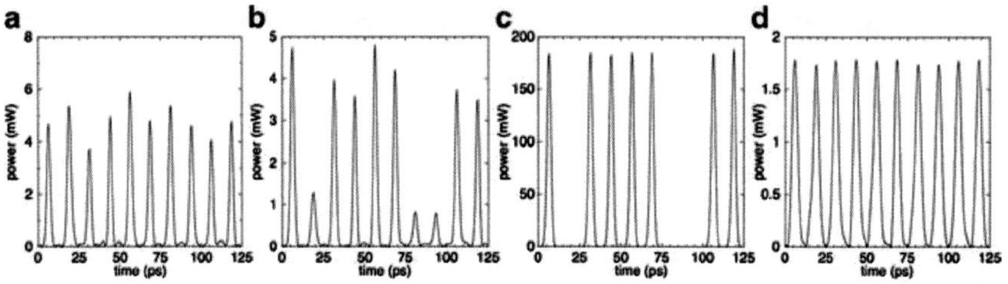

Fig. 11.19 Waveforms of (**a**) DQPSK input signal (OSNR = 24 dB/0.1 nm noise bandwidth), (**b**) demodulated signal after one of the DIs, (**c**) amplitude-regenerated signal after one of the cascaded 2R regenerators, and (**d**) output signal after the phase modulator [33]

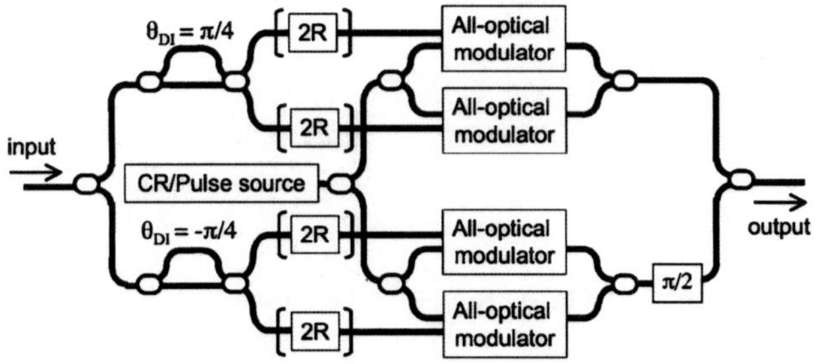

Fig. 11.20 Block diagram of a DQPSK signal regenerator using MZI phase modulators. *CR* Clock recovery circuits; *2R* 2R amplitude regenerators

paths can be omitted if amplitude modulators having saturable response are used as the all-optical modulator elements in the arms of MZIs as has been discussed in Sect. 11.2.1.2. Because the output signals from the two MZIs are coherently combined, integration of the two MZIs will be necessary for stable operation. Such an

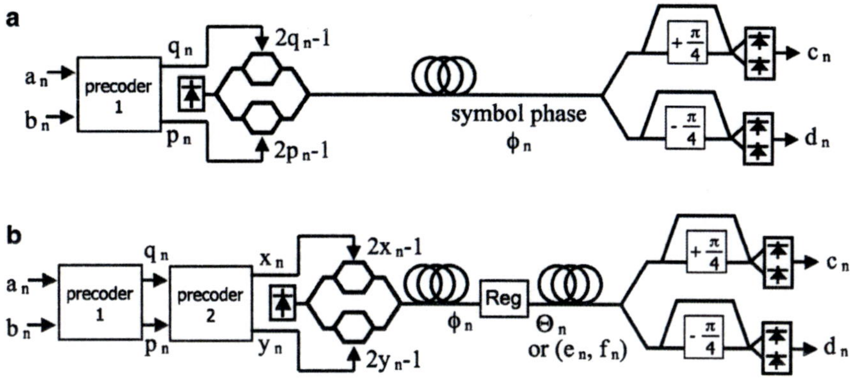

Fig. 11.21 DQPSK transmission systems including precoders. (**a**) System without using regenerator. (**b**) System using the regenerator

arrangement of pair of all-optical MZIs has been employed in applications of all-optical modulation format conversion and all-optical logic operation [75,76]

In the DQPSK signal regenerator where the incoming signal is differentially demodulated to OOK signals and converted back to QPSK signal after noise suppression operation, the phase difference between neighboring pulses in the incoming signal is mapped to the absolute phase of the outgoing pulse. This means that the phase data encoded on the signal is altered by the regeneration. Precoding or postcoding that undoes the conversion is needed when the regenerator is used in practical systems as has been discussed for the DPSK regenerator in Sect. 11.2.1.

In the system using differential detection, the use of precoder or postcoder is necessary even when regenerators are not inserted. Figure 11.21a shows a typical DQPSK system without using the all-optical regenerator. A precoder that undoes the logic conversion by the differential detection at the receiver is located before the modulator in the transmitter [74]. a_n and b_n(=0 or 1) are original data and q_n and p_n(=0 or 1) are the output data from the precoder, where real and imaginary parts of the complex amplitude of the light emitted from the source are modulated in proportion to $2q_n - 1$ and $2p_n - 1$, respectively. The receiver consists of two parallel one-symbol DIs and balanced detectors. The optical phase difference given to signals traveling in the two arms in the DIs is set at $\theta_{\mathrm{DI}} = \pi/4$ or $-\pi/4$. The output signal from the balanced detectors is given by $\cos(\phi_n - \phi_{n-1} + \theta_{\mathrm{DI}})$, where $\phi_n - \phi_{n-1}$ is the phase difference between consecutive pulses fed to the receiver. Received data c_n and d_n take values 1 or 0 according to the output signal from the balanced detector that is positive or negative, respectively. Temporal transition of the data q_n and p_n driving the modulator and the symbol phase difference $\phi_n - \phi_{n-1}$ are related by the left two columns in Table 11.2a. The symbol phase difference leads to the output data c_n and d_n given in the right column in Table 11.2a. For the output data c_n and d_n to be identical to the original data a_n and b_n, respectively, the precoder (precoder 1 shown in Fig. 11.21a) should produce q_n and p_n obeying the transition depending on a_n and b_n as given in Table 11.2b. The logic transition rules of the precoder are then given by [74,77]

Table 11.2 (**a**) Relation between transition of $(q_n,\ p_n)$, symbol phase difference $\phi_n - \phi_{n-1}$, and output data $(c_n,\ d_n)$ at the detector, (**b**) Required transition of $(q_n,\ p_n)$ vs. input data $(a_n,\ b_n)$ in the Precoder

(a)

Transition of $(q_n,\ p_n)$	$\phi_n - \phi_{n-1}$	$(c_n,\ d_n)$
$(q_n,\ p_n) = (q_{n-1},\ p_{n-1})$	0	(1, 1)
$(\overline{p}_{n-1},\ q_{n-1})$	$\pi/2$	(0, 1)
$(\overline{q}_{n-1},\ \overline{p}_{n-1})$	π	(0, 0)
$(p_{n-1},\ \overline{q}_{n-1})$	$3\pi/2$	(1, 0)

(b)

$(a_n,\ b_n)$	Transition of $(q_n,\ p_n)$
(1, 1)	$(q_n,\ p_n) = (q_{n-1},\ p_{n-1})$
(0, 1)	$(\overline{p}_{n-1},\ q_{n-1})$
(0, 0)	$(\overline{q}_{n-1},\ \overline{p}_{n-1})$
(1, 0)	$(p_{n-1},\ \overline{q}_{n-1})$

Table 11.3 (**a**) Relation between the symbol phase difference before the regenerator and the absolute phase after the regenerator, (**b**) Required transition of $(x_n,\ y_n)$ in the Precoder 2

(a)

$\phi_n - \phi_{n-1}$	Θ_n	$(e_n,\ f_n)$
0	0	(0, 0)
$\pi/2$	π	(1, 1)
π	$3\pi/2$	(0, 1)
$3\pi/2$	$\pi/2$	(1, 0)

(b)

$(q_n,\ p_n)$	Transition of $(x_n,\ y_n)$
(0, 0)	$(x_n,\ y_n) = (x_{n-1},\ y_{n-1})$
(1, 1)	$(\bar{y}_{n-1},\ x_{n-1})$
(0, 1)	$(\bar{x}_{n-1},\ \bar{y}_{n-1})$
(1, 0)	$(y_{n-1},\ \bar{x}_{n-1})$

$$q_n = a_n b_n q_{n-1} + \overline{a}_n b_n \overline{p}_{n-1} + \overline{a}_n \overline{b}_n \overline{q}_{n-1} + a_n \overline{b}_n p_{n-1}, \tag{11.43a}$$

$$p_n = a_n b_n p_{n-1} + \overline{a}_n b_n q_{n-1} + \overline{a}_n \overline{b}_n \overline{p}_{n-1} + a_n \overline{b}_n \overline{q}_{n-1}. \tag{11.43b}$$

Now we consider that a DQPSK signal regenerator whose setup is given in Fig. 11.17 is inserted in the system as shown in Fig. 11.21b. The phase difference between consecutive symbols incoming to the regenerator $\phi_n - \phi_{n-1}$ and the absolute phase of the output symbol Θ_n are related by the left two columns in Table 11.3a. The absolute phase Θ_n is represented by two logical variables e_n and f_n(=0 or 1) as shown in the right column in Table 11.3a, where

$$2e_n - 1 + \mathrm{i}(2f_n - 1) = \sqrt{2}\exp\left[\mathrm{i}\left(\Theta_n - 3\pi/4\right)\right] \tag{11.44}$$

is satisfied. The phase conversion by the regenerator, therefore, is undone by inserting an additional precoder (precoder 2) whose temporal transition is given by Table 11.3b between the precoder 1 and the modulator as shown in Fig. 11.21b. The transition rules are

$$x_n = \overline{q}_n\overline{p}_n x_{n-1} + q_n p_n \overline{y}_{n-1} + \overline{q}_n p_n \overline{x}_{n-1} + q_n \overline{p}_n y_{n-1}, \tag{11.45a}$$

$$y_n = \overline{q}_n\overline{p}_n y_{n-1} + q_n p_n x_{n-1} + \overline{q}_n p_n \overline{y}_{n-1} + q_n \overline{p}_n \overline{x}_{n-1}. \tag{11.45b}$$

By using the precoder 2, $(e_n,\ f_n)$ becomes equal to $(q_n,\ p_n)$ so that the same relation between the transition of $(q_n,\ p_n)$ and the output data $(c_n,\ d_n)$ as shown in Table 11.2a is satisfied.

When the number of regenerators inserted in the system is more than one, the same number of precoders (precoder 2) should be inserted before the modulator. Postcoders having the same logical operation as the precoders can be used after the detectors in the receiver instead of using the precoders in the transmitter.

11.3.2 QPSK Signal Regeneration Using Coherent Demodulation

If a local oscillator whose frequency and phase are locked to those of the incoming signal is available, the differential demodulation in the DQPSK regenerator discussed in the previous subsection can be replaced by coherent demodulation as shown in Fig. 11.22 [32]. Advantages of using coherent demodulation include (1) capability of error-free regeneration is enhanced because the local oscillator light that is not contaminated by noise can be used as a reference for the interferometric demodulation, and (2) the phase data encoded on the signal are not altered by

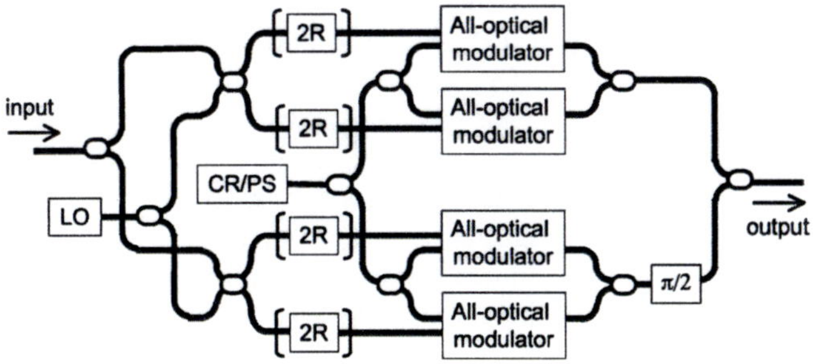

Fig. 11.22 Block diagram of a QPSK regenerator using coherent demodulation. *LO* Local oscillator; *CR/PS* Clock recovery circuit/pulse source; *2R* 2R amplitude regenerator

the regeneration process, which makes the use of encoder or decoder unnecessary. A major difficulty in this regeneration scheme is that local oscillator light phase-locked to the incoming signal must be generated within the regenerator. Polarization alignment between the signal and the local oscillator light is also critical.

11.4 Discussion and Summary

All-optical regeneration is an effective way to suppress noise accumulation whereby the reach of high-speed signal transmission will be greatly enhanced. Recently, attention has been paid to regeneration of phase-encoded signals in addition to conventional OOK signals. Different schemes of suppression of phase and/or amplitude noise of binary PSK signals have been proposed and demonstrated. Regeneration of multilevel PSK signals such as (D)QPSK has also been attempted. In this chapter, recent studies of all-optical regeneration of phase-encoded signals have been reviewed.

All-optical signal regeneration is the process in which deviations of amplitude and phase of the signal from those in the absence of impairments are removed by the use of some form of optical nonlinearities. Usually, all-optical signal regenerators are designed to process isolated optical pulses. The nonlinear nature of the signal regeneration indicates that if the pulses to be regenerated are overlapped with other pulses in time, performance of the regenerator is generally severely degraded. In real systems, this means that transmission lines must be precisely dispersion-compensated at the location where the regenerators are inserted and that channels must be demultiplexed at the regenerator if we use the regenerators in WDM systems. These are undesired requirements that lead to loss of flexibility of the system and increase in cost. Although several efforts have been made aiming at realization of multichannel all-optical signal regenerators [78–81], satisfactory results have not yet been obtained.

Recently, electrical signal processing at terminals especially with coherent detection has been extensively studied for mitigation of signal impairments in high-capacity transmission systems. The versatile ability of the electrical signal processing may also be used to solve the problems encountered in systems including all-optical regenerators mentioned above. For example, nonlinear inter-symbol interference (ISI), which will happen when imperfectly dispersion-compensated signals are launched to the all-optical regenerator with temporal overlaps between adjacent symbols, may be mitigated by suitable signal processing after detection at the receiver. That is, the stringent management of transmission lines required for correct operation of all-optical regenerators may be alleviated by the use of electrical signal processing. Although finding the clear solutions will not be an easy task, research toward such hybrid impairment mitigation will be important and interesting.

Acknowledgments The author thanks K. Sanuki, H. Sakaguchi, and Y. Morioka for their assistance in the experiments of (D)BPSK signal transmission and regeneration. This work was supported in part by Japan Society for the Promotion of Science (JSPS) Grant-in-Aid for Scientific Research (B) 20360171 and for Scientific Research on Priority Areas 18040006 and 19023005.

References

1. K. Kikuchi, IEEE Photon. Technol. Lett. **5**(2), 221–223 (1993)
2. O. Leclerc, B. Lavigne, D. Chiaroni, E. Desurvire, *All-Optical Regeneration: Principles and WDM Implementation*, ed. by I. Kaminow, T. Li. Optical Fiber Telecommunications IV A, Components, (Academic, NY, 2002), pp. 732–783
3. P.J. Winzer, R.J. Essiambre, J. Lightwave Technol. **24**(12), 4711–4728 (2006)
4. I. Kang, C. Dorrer, L. Zhang, M. Rasras, L. Buhl, A. Bhardwaj, S. Cabot, M. Dinu, X. Liu, M. Cappuzzo, L. Gomez, A. Wong-Foy, Y.F. Chen, S. Patel, D.T. Neilson, J. Jacques, C.R. Giles, Regenerative all optical wavelength conversion of 40-Gb/s DPSK signals using a semiconductor optical amplifier Mach-Zehnder interferometer, 2005 European conference on optical communication, Th4.3.3, 2005
5. P. Vorreau, A. Marculescu, J. Wang, G. Böttger, B. Sartorius, C. Bornholdt, J. Slovak, M. Schlak, C. Schmidt, S. Tsadka, W. Freude, J. Leuthold, IEEE Photon. Technol. Lett. **18**, 1970–1972 (2006)
6. M. Matsumoto, IEEE Photon. Technol. Lett. **19**, 273–275 (2007)
7. R. Elschner, C.A. Bunge, K. Petermann, All-optical regeneration of 100 Gb/s DPSK signals, 2007 LEOS annual meeting, ThP3, 2007
8. M. Matsumoto, H. Sakaguchi, Opt. Express **16**, 11169–11175 (2008)
9. M. Matsumoto, Y. Morioka, Opt. Express **17**, 6913–6919 (2009)
10. J. Wang, A. Maitra, W. Freude, J. Leuthold, Opt. Express **17**(25), 22639–22658 (2009)
11. C. Kouloumentas, M. Bouqioukos, A. Maziotis, H. Avramopoulos, Phase-incoherent DPSK regeneration using a fiber-Sagnac interferometer, 2010 optical fiber communication conference, OMT5, 2010
12. P.S. Devgan, M. Shin, V.S. Grigoryan, J. Lasri, P. Kumar, SOA-based regenerative amplification of phase noise degraded DPSK signals, 2005 optical fiber communication conference, PDP34, 2005
13. P. Johannisson, G. Adolfsson, M. Karlsson, Opt. Lett. **31**, 1385–1387 (2006)
14. C.C. Wei, J.J. Chen, Opt. Express **14**, 9584–9593 (2006)
15. A.G. Striegler, M. Meissner, K. Cvecek, K. Spnsel, G. Leuchs, B. Schmauss, IEEE Photon. Technol. Lett. **17**, 639–641 (2005)
16. M. Matsumoto, IEEE Photon. Technol. Lett. **17**, 1055–1057 (2005)
17. M. Matsumoto, J. Lightwave Technol. **23**(9), 2696–2701 (2005)
18. S. Boscolo, R. Bhamber, S.K. Turitsyn, IEEE J. Quant. Electron. **42**, 619–624 (2006)
19. K. Cvecek, K. Sponsel, G. Onishchukov, B. Schmauss, G. Leuchs, IEEE Photon. Technol. Lett. **19**, 146–148 (2007)
20. F. Futami, R. Okabe, S. Ono, S. Watanabe, R. Ludwig, C. Schmidt-Langhorst, C. Schubert, All-optical amplitude noise suppression of 160-Gb/s OOK and DPSK data signals using a parametric fiber switch, 2007 optical fiber communication conference, Paper OThB3, 2007
21. K. Croussore, G. Li, Electron. Lett. **43**, 177–178 (2007)
22. M. Matsumoto, K. Sanuki, Opt. Express **15**, 8094–8103 (2007)
23. C. Peucheret, M. Lorenzen, J. Seoane, D. Noordegraaf, C.V. Nielsen, L. Grüner-Nielsen, K. Rottwitt, IEEE Photon. Technol. Lett. **21**(13), 872–874 (2009)
24. C. Stephan, K. Sponsel, G. Onishchukov, B. Schmauss, G. Leuchs, IEEE Photon. Technol. Lett. **21**(24), 1864–1866 (2009)
25. Q.T. Le, L. Bramerie, H.T. Nguyen, M. Gay, S. Lobo, M. Joindot, J.L. Oudar, J.C. Simon, IEEE Photon. Technol. Lett. **22**(12), 887–889 (2010)
26. K. Croussore, C. Kim, G. Li, Opt. Lett. **29**(20), 2357–2359 (2004)
27. K. Croussore, I. Kim, C. Kim, Y. Han, G. Li, Opt. Express **14**, 2085–2094 (2006)
28. A. Bogris, D. Syvridis, IEEE Photon. Technol. Lett. **18**, 2144–2146 (2006)
29. K. Croussore, G. Li, IEEE J. Sel. Top. Quant. Electron. **14**, 648–658 (2008)
30. F. Parmigiani, R. Slavik, J. Kakande, C. Lundstrom, M. Sjodin, P. Andrekson, R. Weerasuriya, S. Sygletos, A.D. Ellis, L. Gruner-Nielsen, D. Jakobsen, S. Herstrom, R. Phelan, J. O'Gorman, A. Bogris, D. Syvridis, S. Dasgupta, P. Petropulos, D.J. Richadson, All-optical phase regeneration of 40Gbit/s DPSK signals in a black-box phase sensitive amplifier, 2010 optical fiber communication conference, PDPC3, 2010

31. Z. Zheng, L. An, Z. Li, X. Zhao, X. Liu, Opt. Commun. **281**, 2755–2759 (2008)
32. X. Yi, R. Yu, J. Kurumida, S.J.B. Yoo, J. Lightwave Technol. **28**(4), 587–595 (2010)
33. M. Matsumoto, Opt. Express **18**(1), 10–24 (2010)
34. R. Elschner, A. Marques de Melo, C.A. Bunge, K. Petermann, Opt. Lett. **32**(2), 112–114 (2007)
35. A.H. Gnauck, P.J. Winzer, J. Lightwave Technol. **23**(1), 115–130 (2005)
36. M. Daikoku, N. Yoshikane, T. Otani, H. Tanaka, J. Lightwave Technol. **24**(3), 1142–1148 (2006)
37. J.H. Lee, P.C. The, Z. Yusoff, M. Ibsen, W. Belardi, T.M. Monro, D.J. Richardson, IEEE Photon. Technol. Lett. **14**(6) 876–878 (2002)
38. L.B. Fu, M. Rochette, V.G. Ta'eed, D.J. Moss, B.J. Eggleton, Opt. Express **13**, 7637–7644 (2005)
39. F. Parmigiani, S. Asimakis, N. Sugimoto, F. Koizumi, P. Petropoulos, D.J. Richardson, Opt. Express **14**, 5038–5044 (2006)
40. P.V. Mamyshev, All-optical data regeneration based on self-phase modulation effect, 1998 European conference on optical communication, pp. 475–476, 1998
41. M. Matsumoto, Opt. Express **14**, 11018–11023 (2006)
42. H. Toda, S. Kobayashi, I. Akiyoshi, Reduction of pulse-to-pulse interaction of optical RZ pulses in dispersion managed fiber, 2002 Asia-Pacific optical and wireless communications, Paper 4906–54, 2002
43. T. Tanemura, J.H. Lee, D. Wang, K. Katoh, K. Kikuchi, Opt. Express **14**, 1408–1412 (2006)
44. J.P. Gordon, L.F. Mollenauer, Opt. Lett. **15**, 1351–1353 (1990)
45. H. Kim, J. Lightwave Technol. **21**(8), 1770–1774 (2003)
46. A.G. Green, P.P. Mitra, L.G.L. Wegener, Opt. Lett. **28**, 2455–2457 (2003)
47. S. Kumar, Opt. Lett. **30**(24), 3278–3280 (2005)
48. K.P. Ho, H.C. Wang, Opt. Lett. **31**(14), 2109–2111 (2006)
49. N.J. Doran, D. Wood, Opt. Lett. **13**(1), 56–58 (1988)
50. G. Cappellini, S. Trillo, J. Opt. Soc. Am. B **8**(4), 824–838 (1991)
51. K. Inoue, T. Mukai, Opt. Lett. **26**, 10–12 (2001)
52. K. Inoue, Electron. Lett. **36**, 1016–1017 (2000)
53. E. Ciaramella, S. Trillo, IEEE Photon. Technol. Lett. **12**(7), 849–451 (2000)
54. K. Inoue, IEEE Photon. Technol. Lett. **13**(4), 338–340 (2001)
55. S. Radic, C.J. McKinstrie, R.M. Jopson, J.C. Centanni, A.R. Chraplyvy, IEEE Photon. Technol. Lett. **15**, 957–959 (2003)
56. C.M. Caves, Phys. Rev. D **26**, 1817–1839 (1982)
57. R. Loudon, IEEE J. Quant. Electron. **QE-21**(7), 766–773 (1985)
58. M.E. Marhic, C.H. Hsia, J.M. Jeong, Electron. Lett. **27**(3), 210–211 (1991)
59. M.E. Marhic, C.H. Hsia, Quantum Opt. **3**, 341–358 (1991)
60. H.A. Haus, J. Opt. Soc. Am B **12**(11), 2019–2036 (1995)
61. D. Levandovsky, M. Vasilyev, P. Kumar, Opt. Lett. **24**(14), 984–986 (1999)
62. W. Imajuku, A. Takada, Y. Yamabayashi, Electron. Lett. **36**(1), 63–64 (2000)
63. C.J. McKinstrie, S. Radic, Opt. Express **12**(20), 4973–4979 (2004)
64. R. Tang, P. Devgan, P.L. Voss, V.S. Grigoryan, P. Kumar, IEEE Photon. Technol. Lett. **17**(9), 1845–1847 (2005)
65. R. Tang, P.S. Devgan, V.S. Grigoryan, P. Kumar, M. Vasilyev, Opt. Express **16**(12), 9046–9053 (2008)
66. R.D. Li, P. Kumar, W.L. Kath, J. Lightwave Technol. **12**(3), 541–549 (1994)
67. H.P. Yuen, Opt. Lett. **17**(1), 73–75 (1992)
68. G.D. Bartolini, D.K. Serkland, P. Kumar, W.L. Kath, IEEE Photon. Technol. Lett. **9**(7), 1020–1022 (1997)
69. I. Kim, K. Croussore, X. Li, G. Li, IEEE Photon. Technol. Lett. **19**, 987–989 (2007)
70. R. Weerasuriya, S. Sygletos, S.K. Ibrahim, R. Phelan, J. O'Carroll, B. Kelly, J. O'Gorman, A.D. Ellis, Generation of frequency symmetric signals from a BPSK input for phase sensitive amplification, OFC2010, OWT6, 2010
71. A. Takada, W. Imajuku, Electron. Lett. **32**, 677–679 (1996)

72. K. Cvecek, K. Spnsel, R. Ludwig, C. Schubert, C. Stephan, G. Onishchukov, B. Schmauss, G. Leuchs, IEEE Photon. Technol. Lett. **19**, 1475–1477 (2007)
73. M. Matsumoto, T. Kamio, IEEE J. Select. Top. Quant. Electron. **14**, 610–615 (2008)
74. R.A. Griffin, A.C. Carter, Optical differential quadrature phase-shift key (oDQPSK) for high capacity optical transmission, 2002 optical fiber communication conference, Paper WX6, 2002
75. K. Mishina, S.M. Nissanka, A. Maruta, S. Mitani, K. Ishida, K. Shimizu, T. Hatta, K. Kitayama, Opt. Express **15**, 7774–7785 (2007)
76. I. Kang, M. Rasras, L. Buhl, M. Dinu, S. Cabot, M. Cappuzzo, L.T. Gomez, Y.F. Chen, S.S. Patel, N. Dutta, A. Piccirilli, J. Jaques, C.R. Giles, Opt. Express **17**, 19062–19066 (2009)
77. K.P. Ho, *Phase-Modulated Optical Communication Systems* (Springer, Berlin, 2005)
78. T. Ohara, H. Takara, A. Hirano, K. Mori, S. Kawanishi, IEEE Photon. Technol. Lett. **15**, 763–765 (2003)
79. M. Vasilyev, T.I. Lakoba, Opt. Lett. **30**, 1458–1460 (2005)
80. L. Provost, P. Petropoulos, D.J. Richardson, Optical WDM regeneration: status and future prospects, OFC2009, OWD7, (2009)
81. P.G. Patki, M. Vasilyev, T.I. Lakoba, Multichannel all-optical regeneration, IEEE Photonics Society summer topical meetings, WC2.2, (2010)

Chapter 12
Codes on Graphs, Coded Modulation and Compensation of Nonlinear Impairments by Turbo Equalization

Ivan B. Djordjevic

12.1 Introduction

As the response to the ever-increasing demands of telecommunication needs, the network operators already consider beyond 100 Gb/s per dense wavelength division multiplexing (DWDM) channel transmission [1]. At those data rates, the performance of fiberoptic communication systems is degraded significantly due to intra- and interchannel fiber nonlinearities, polarization-mode dispersion (PMD), and chromatic dispersion [2–5]. To deal with those channel impairments novel advanced techniques in modulation and detection, coding and signal processing should be developed; and some important aspects are described in this chapter.

The state-of-the-art in optical communication systems standardized by the International Telecommunication Union-Telecommunication Standardization Sector (ITU-T) employ concatenated Bose–Ray–Chaudhuri–Hocquenghem (BCH)/Reed–Solomon (RS) codes [1–9]. The RS(255, 239) in particular has been used in a broad range of long-haul communication systems [1, 6], and it is commonly considered as the first-generation of FEC [8,9]. The elementary FEC schemes (BCH, RS or convolutional codes) may be combined to design more powerful FEC schemes, for example concatenated RS(255, 239) + RS(255, 233) code. Several classes of concatenation codes are listed in ITU-T G975.1. Different concatenation schemes, such as the concatenation of two RS codes or the concatenation of RS and convolutional codes, are commonly considered as second generation of FEC [8, 9].

Codes on graphs, such as turbo codes and low-density parity-check (LDPC) codes, have revolutionized communications, and are becoming standard in many applications. LDPC codes, invented by Gallager in 1960s, are linear block codes for which the parity check matrix has low density of ones [10]. LDPC codes have generated great interests in the coding community recently, and this has resulted in a great deal of understanding of the different aspects of LDPC codes

I.B. Djordjevic (✉)
Department of Electrical and Computer Engineering, University of Arizona,
Tucson, AZ 85721, USA
e-mail: ivan@ece.arizona.edu

S. Kumar (ed.), *Impact of Nonlinearities on Fiber Optic Communications*, Optical and Fiber Communications Reports 7, DOI 10.1007/978-1-4419-8139-4_12,
© Springer Science+Business Media, LLC 2011

and their decoding process. An iterative LDPC decoder based on the sum–product algorithm (SPA) has been shown to achieve a performance as close as 0.0045 dB to the Shannon limit [11]. The inherent low-complexity of this decoder opens up avenues for its use in different high-speed applications, including optical communications.

The purpose of this chapter is: (1) to describe different classes of codes on graphs of interest for optical communications, (2) to describe how to combine multilevel modulation and channel coding (3) to describe how to perform equalization and soft decoding jointly, and (4) to demonstrate efficiency of joint de-modulation, decoding, and equalization in dealing with various channel impairments simultaneously.

We first describe briefly, in Sect. 12.2, the channel coding preliminaries: the basics of FEC, linear block codes, and definition of coding gain. The codes on graphs proposed for use in optical communications, namely, turbo-product codes (TPCs) and LDPC codes are described in Sect. 12.3. Due to the fact that LDPC codes can match and outperform TPCs in terms of bit-error rate (BER) performance while having a lower complexity decoding algorithm, in this chapter we are mostly concerned with LDPC codes. We describe basic concepts of LDPC codes and describe how to design large girth quasi-cyclic LDPC codes (Sect. 12.3.1.1). We also provide a log-domain decoding algorithm (Sect. 12.3.1.2) and evaluate BER performance of different codes on graphs (Sect. 12.3.1.3). We then turn our attention to coded modulation and describe, in Sect. 12.4 how to optimize multilevel modulation and coding process to achieve the best possible BER performance through the use of multilevel coding (MLC) (Sect. 12.4.1) and coded orthogonal frequency division multiplexing (OFDM), in Sect. 12.4.2. The Sect. 12.4.3 is devoted to multidimensional coded modulation. Next, in Sect. 12.5, we discuss how to combine the *maximum a posteriori probability* (MAP) equalizer in an optimal fashion with an LDPC decoder, in so-called turbo equalization fashion. When used in combination with large girth LDPC codes as channel codes, this scheme represents a universal equalizer scheme for *simultaneous* suppression of fiber nonlinearities, for chromatic dispersion compensation and for PMD compensation; applicable to both direct detection and coherent detection. To further improve the overall BER performance, we perform the iteration of extrinsic LLRs between LDPC decoder and multilevel BCJR equalizer. We use the extrinsic information transfer (EXIT) chart approach due to S. ten Brink to match the LDPC decoders and multilevel BCJR equalizer. We further show how to combine this scheme with multilevel coded-modulation schemes with coherent detection. Because the complexity of turbo equalizer grows exponentially as state memory and signal constellation sizes increase, in Sect. 12.5.4 we describe how to use this method in combination with digital back-propagation. Namely we use the coarse digital back-propagation (with reasonable small number of coefficients) to reduce the channel memory, and compensate for remained channel distortions by turbo equalization.

Given the fact that LDPC-coded turbo equalizer, based on multilevel BCJR algorithm, is an excellent nonlinear intersymbol interference (ISI) equalizer candidate, naturally arises the question about fundamental limits on channel capacity of coded-modulation schemes. For completeness of presentation, we also provide in Sect. 12.6, the independent identically distributed (IID) channel capacity study of optical channels with *memory*.

12.2 Channel Coding Preliminaries

Two key system parameters are transmitted power and channel bandwidth, which together with additive noise sources determine the signal-to-noise ratio (SNR) and correspondingly BER. In practice, we very often come into situation when the target BER cannot be achieved with a given modulation format. For the fixed SNR, the only practical option to change the data quality transmission from unacceptable to acceptable is through the use of *channel coding*. Another practical motivation of introducing the channel coding is to reduce required SNR for a given target BER. The amount of energy that can be saved by coding is commonly described by coding gain. *Coding gain* refers to the savings attainable in the energy per information bit to noise spectral density ratio (E_b/N_0) required to achieve a given bit error probability when coding is used compared to that with no coding. A typical digital optical communication system employing channel coding is shown in Fig. 12.1. The discrete source generates the information in the form of sequence of symbols. The channel encoder accepts the message symbols and adds redundant symbols according to a

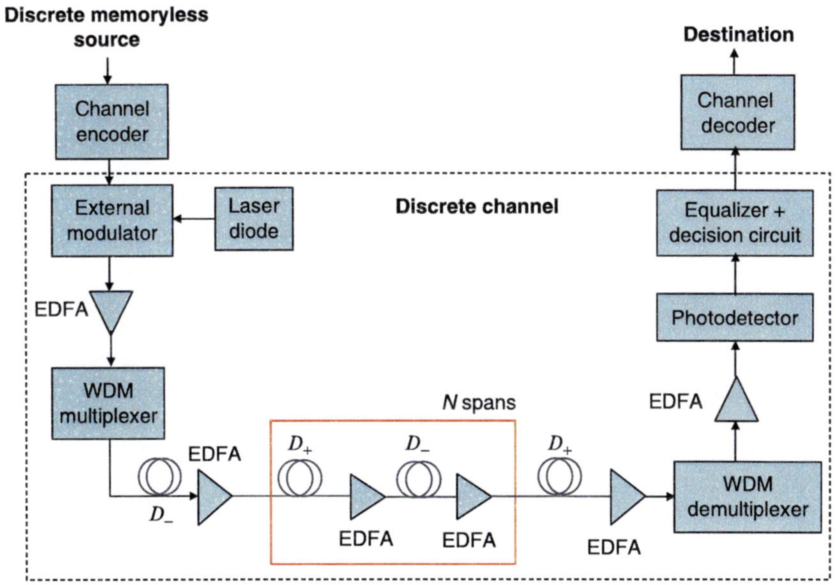

Fig. 12.1 Block diagram of a point-to-point digital optical communication system

corresponding prescribed rule. The channel coding is the act of transforming of a length-k sequence into a length-n codeword. The set of rules specifying this transformation are called the channel code, which can be represented as the following mapping: $C : M -> X$, where C is the channel code, M is the set of information sequences of length k, and X is the set of codewords of length n. The decoder exploits these redundant symbols to determine which message symbol was actually transmitted. Encoder and decoder consider whole digital transmission system as a discrete channel. Other blocks shown in Fig. 12.1 are already explained in previous chapters, here we are concerned with channel encoders and decoders. Different classes of channel codes can be classified into three broad categories: (1) *error detection* in which we are concerned only with detecting the errors occurred during transmission (examples include automatic request for transmission-ARQ), (2) *forward error correction* (FEC), where we are interested in correcting the errors occurred during transmission, and (3) hybrid channel codes that combine the previous two approaches. In this chapter, we are concerned only with FEC.

The key idea behind the forward error correcting codes is to add extra redundant symbols to the message to be transmitted, and use those redundant symbols in decoding procedure to correct the errors introduced by the channel. The redundancy can be introduced in time, frequency, or space domain. For example, the redundancy in time domain is introduced if the same message is transmitted at least twice, the technique is known as the *repetition code*. The space redundancy is used as a means to achieve high spectrally efficient transmission, in which the modulation is combined with error control.

The codes commonly considered in fiberoptics communications belong either to the class of *block codes* or to the class of *convolutional codes*. In an (n, k) *block code*, the channel encoder accepts information in successive k-symbol blocks, adds $n - k$ redundant symbols that are algebraically related to the k message symbols; thereby producing an overall encoded block of n symbols ($n > k$), known as a *codeword*. If the block code is *systematic*, the information symbols stay unchanged during the encoding operation, and the encoding operation may be considered as adding the $n - k$ generalized parity checks to k information symbols. Since the information symbols are statistically independent (a consequence of source coding or scrambling), the next codeword is independent of the content of the current codeword. The *code rate* of an (n, k) block code is defined as $R = k/n$, and *overhead* by $OH = (1/R - 1) \cdot 100\%$. In *convolutional code*, however, the encoding operation may be considered as the discrete-time convolution of the input sequence with the impulse response of the encoder. Therefore, the $n - k$ generalized parity checks are functions of not only k information symbols but also of m previous k-tuples, with $m + 1$ being the encoder impulse response length. The statistical dependence is introduced to the window of length $n(m + 1)$, the parameter known as *constraint length* of convolutional codes.

Example 12.1. Repetition Code. In repetition code, each bit is transmitted $n = 2m + 1$ times. For example, for $n = 3$ the bits 0 and 1 are represented as 000 and 111, respectively. On the receiver side, we first perform threshold decision, if the received sample is the above threshold we decide in favor of 1, otherwise in favor of bit 0. The decoder then applies the following m*ajority decoding rule*: if in block of n bits, the number of ones exceeds the number of zeros, decoder decides in favor of 1; otherwise in favor of 0. This code is capable of correcting up to m errors. The probability of error that remains upon decoding can be evaluated by the following expression:

$$P_e = \sum_{i=m+1}^{n} \binom{n}{i} p^i (1-p)^{n-i}, \tag{12.1}$$

where p is the probability of making an error on a given position.

As mentioned above, the channel code considers whole transmission system as a discrete channel, in which the sizes of input and output alphabets are finite. In Fig. 12.2a, we show an example a discrete memoryless channel (DMC), which is characterized by channel (transition) probabilities. Let $X = \{x_0, x_1, \ldots, x_{I-1}\}$ denote the channel input alphabet, and $Y = \{y_0, y_1, \ldots, y_{J-1}\}$ denote the channel

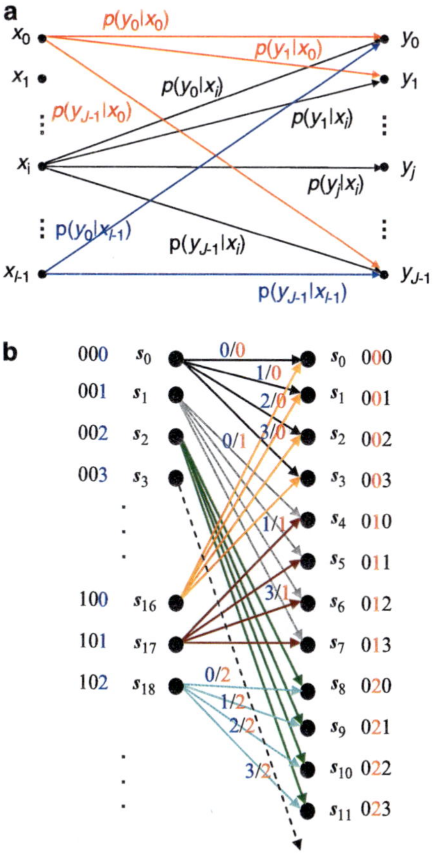

Fig. 12.2 Typical optical channel models: (**a**) an I-ary input J-ary output discrete memoryless channel (DMC) model, and (**b**) a 4-ary input 4-ary output channel model with memory

output alphabet. This channel is completely characterized by the following set of transition probabilities:

$$p(y_j|x_i) = P(Y = y_j|X = x_i), \quad 0 \le p(y_j|x_i) \le 1, \ i \in \{0, 1, \ldots, I-1\}, \\ j \in \{0, 1, \ldots, J-1\}, \tag{12.2}$$

where I and J denote the sizes of input and output alphabets, respectively. The transition probability $p(y_j|x_i)$ represents the conditional probability that channel output $Y = y_j$ given the channel input $X = x_i$. The channel introduces the errors, and if $j \neq i$ the corresponding $p(y_j|x_i)$ represents the conditional probability of error, while for $j = i$ it represents the conditional probability of correct reception. For $I = J$, the *average symbol error probability* is defined as the probability that output random variable Y_j is different from input random variable X_i, with averaging being performed for all $j \neq i$:

$$P_e = \sum_{i=0}^{I-1} p(x_i) \sum_{j=0, j \neq 1}^{J-1} p(y_j|x_i), \tag{12.3}$$

where the inputs are selected from the following distribution $\{p(x_i) = P(X = x_i);\ i = 0, 1, \ldots, I-1\}$, with $p(x_i)$ being known as a priori probability of input symbol x_i. The corresponding probabilities of output symbols can be calculated by:

$$p(y_j) = \sum_{i=0}^{I-1} P(Y = y_j|X = x_i)P(X = x_i) = \sum_{i=0}^{I-1} p(y_j|x_i)p(x_i); \\ j = 0, 1, \ldots, J-1. \tag{12.4}$$

In Fig. 12.2b, we show a 4-ary input 4-ary output discrete channel model with memory [2, 4, 5], which is more suitable for fiberoptics communications, because the optical channel is essentially the channel with memory. We assume that the optical channel has the memory equal to $2m + 1$, with $2m$ being the number of symbols that influence the observed bit from both sides. This dynamical trellis is uniquely defined by the set of previous state, the next state, in addition to the channel output. The state (the bit-pattern configuration) in the trellis is defined as $s_j = (x_{j-m}, x_{j-m+1}, \ldots, x_j, x_{j+1}, \ldots, x_{j+m}) = \mathbf{x}[j-m, j+m]$, where $x_k \in \boldsymbol{X}$, with X being a signal constellation set. More details of this channel model will be provided in Sect. 12.5.

A very important figure of merit for DMCs is the amount of information conveyed by the channel, which is known as the *mutual information* and it is defined as

$$I(X;Y) = H(X) - H(X|Y), \tag{12.5a}$$

where $H(\boldsymbol{X})$ denotes the uncertainty about the channel input before observing the channel output, also known as entropy, and for DMC it is defined as:

$$H(X) = \sum_{i=0}^{I-1} p(x_i) \log_2 \left[\frac{1}{p(x_i)}\right]; \tag{12.5b}$$

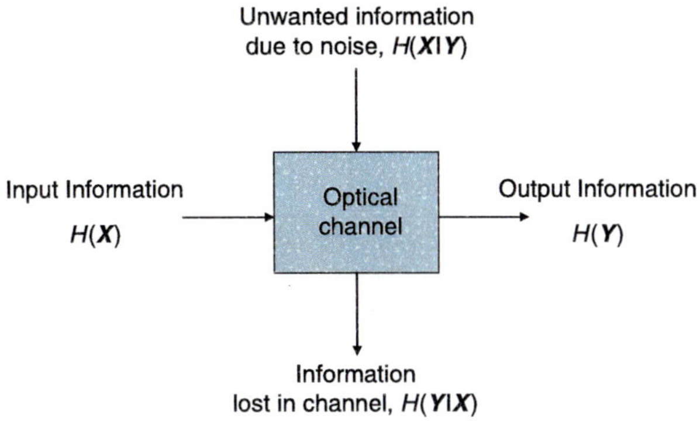

Fig. 12.3 Interpretation of the mutual information using the approach due to Ingels

while $H(\boldsymbol{X}|\boldsymbol{Y})$ denotes the conditional entropy or the amount of uncertainty remaining about the channel input after the channel output has been received, and for DMC it is defined as:

$$H(X|Y) = \sum_{j=0}^{J-1} p(y_j) \sum_{i=0}^{I-1} p(x_i|y_j) \log_2 \left[\frac{1}{p(x_i|y_j)} \right]. \tag{12.5c}$$

The mutual information, therefore, represents the amount of information (per symbol) transmitted over the channel. The mutual information can be interpreted using the approach due to Ingels [12] (see Fig. 12.3). The mutual information, i.e. the information conveyed by the channel, is obtained as the output information minus information lost in the channel. By maximizing the mutual information with respect to the input source distribution, we obtain the so-called *channel capacity*:

$$C = \max_{\{p(x_i)\}} I(X;Y); \quad \text{subject to:} \quad p(x_i) \geq 0, \quad \sum_{i=0}^{I-1} p(x_i) = 1. \tag{12.6}$$

Equipped with this elementary knowledge of information theory and coding, below we formulate two important theorems in literature known as channel coding and information capacity theorems, respectively.

Channel coding theorem [13–19]: Let a discrete memoryless source with an alphabet $\boldsymbol{S}$ have entropy $H(\boldsymbol{S})$ and emit the symbols every T_s seconds. Let a DMC have the capacity C and be used once in T_c seconds. Then, if

$$H(S)/T_s \leq C/T_c, \tag{12.7a}$$

there exists a coding scheme for which the source output can be transmitted over the channel and reconstructed with an arbitrary small probability of error. The parameter $H(\boldsymbol{S})/T_s$ is related to the average information rate, while the parameter C/T_c is related to the channel capacity per unit time. For binary symmetric channel (BSC) ($I = J = 2$), the inequality (12.7a) simply becomes

$$R \leq C, \tag{12.7b}$$

where R is the code rate introduced above.

Information capacity theorem [13–19]: The information capacity of a continuous channel of bandwidth B Hz, perturbed by AWGN of PSD $N_0/2$ and limited in bandwidth B, is given by

$$C = B\log_2\left(1 + \frac{P}{N_0 B}\right) \quad \text{[bits/s]}, \tag{12.8}$$

where P is the average transmitted power. This theorem represents remarkable result of information theory, because it connects all important system parameters (transmitted power, channel bandwidth, and noise power spectral density) in only one formula. What is also interesting is that LDPC codes can approach the Shannon's limit within 0.0045dB [11]. By using the (12.8) and Fano's inequality [20] we obtain:

$$H(X|Y) \leq H(P_e) + P_e\log_2(I-1), \quad H(P_e) = -P_e\log_2 P_e - (1-P_e)\log_2(1-P_e). \tag{12.9}$$

For amplified spontaneous emission (ASE) noise dominated scenario and binary phase-shift keying (BPSK) at 40 Gb/s in Fig. 12.4, we report the minimum BERs against optical SNR for different code rates.

In the rest of this section, an elementary introduction to linear block codes is given. For a detailed treatment of different error-control coding schemes, an interested reader is referred to [12–17, 21, 22].

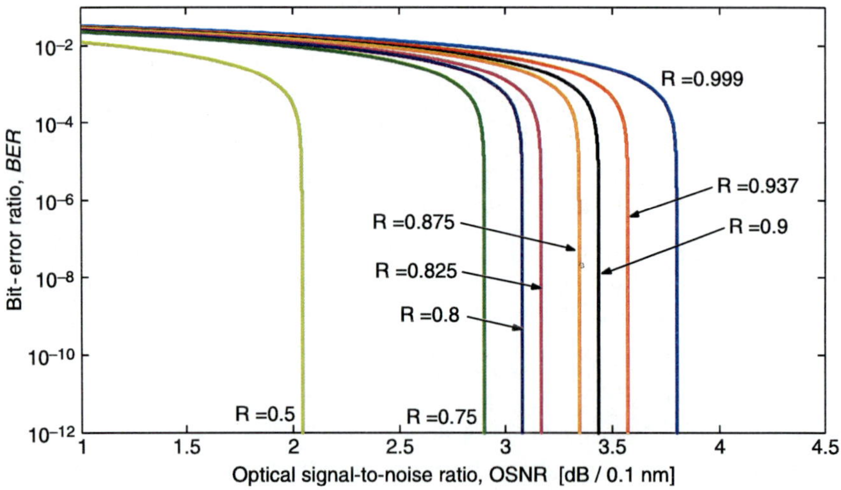

Fig. 12.4 Minimum BER against optical SNR for different code rate values (for BPSK at 40 Gb/s)

12.2.1 Linear Block Codes

The *linear block code* (n, k), using the language of vector spaces, can be defined as a subspace of a vector space over finite field GF(q), with q being the prime power. Every space is described by its *basis* – a set of linearly independent vectors. The number of vectors in the basis determines the dimension of the space. Therefore, for an (n, k) linear block code, the dimension of the space is n, and the dimension of the code subspace is k.

Example 12.2. $(n, 1)$ repetition code. The repetition code has two code words $\boldsymbol{x}_0 = (00\ldots0)$ and $\boldsymbol{x}_1 = (11\ldots1)$. Any linear combination of these two code words is another code word as shown below

$$\boldsymbol{x}_0 + \boldsymbol{x}_0 = \boldsymbol{x}_0, \quad \boldsymbol{x}_0 + \boldsymbol{x}_1 = \boldsymbol{x}_1 + \boldsymbol{x}_0 = \boldsymbol{x}_1, \quad \boldsymbol{x}_1 + \boldsymbol{x}_1 = \boldsymbol{x}_0$$

The set of code words from a linear block code forms a group under the addition operation, because all-zero code word serves as the identity element, and the code word itself serves as the inverse element. This is the reason why the linear block codes are also called the group codes. The linear block code (n, k) can be observed as a k-dimensional subspace of the vector space of all n-tuples over the binary filed GF(2) $= \{0, 1\}$, with addition and multiplication rules given in Table 12.1. All n-tuples over GF(2) form the vector space. The sum of two n-tuples $\boldsymbol{a} = (a_1\ a_2\ \ldots\ a_n)$ and $\boldsymbol{b} = (b_1\ b_2\ \ldots\ b_n)$ is clearly an n-tuple and commutative rule is valid because $\boldsymbol{c} = \boldsymbol{a} + \boldsymbol{b} = (a_1 + b_1\ a_2 + b_2\ \ldots\ a_n + b_n) = (b_1 + a_1\ b_2 + a_2\ \ldots\ b_n + a_n) = \boldsymbol{b} + \boldsymbol{a}$. The all-zero vector $\boldsymbol{0} = (00\ldots0)$ is the identity element, while n-tuple $\boldsymbol{a}$ itself is the inverse element $\boldsymbol{a} + \boldsymbol{a} = 0$. Therefore, the n-tuples form the Abelian group with respect to the addition operation. The scalar multiplication is defined by: $\alpha\boldsymbol{a} = (\alpha\, a_1\ \alpha a_2\ \ldots\ \alpha a_n)$, $\alpha \in$ GF(2). The distributive laws

$$\alpha(\boldsymbol{a} + \boldsymbol{b}) = \alpha\boldsymbol{a} + \alpha\boldsymbol{b}$$
$$(\alpha + \beta)\boldsymbol{a} = \alpha\boldsymbol{a} + \beta\boldsymbol{a}, \quad \forall\alpha, \beta \in \mathrm{GF}(2)$$

are also valid. The associate law $(\alpha \cdot \beta)\boldsymbol{a} = \alpha \cdot (\beta\boldsymbol{a})$ is clearly satisfied. Therefore, the set of all n-tuples is a vector space over GF(2). It can be shown, in a fashion similar to that above, that all code words of an (n, k) linear block codes form the vector space of dimensionality k. There exist k basis vectors (codewords) such that every codeword is a linear combination of basis ones.

Table 12.1 Addition (+) and multiplication (·) rules in GF(2)

+	0	1
0	0	1
1	1	0

·	0	1
0	0	0
1	0	1

Example 12.2 (revisited): $(n, 1)$ repetition code: $C = \{(00\ldots0), (11\ldots1)\}$. Two code words in C can be represented as linear combination of all-ones basis vector: $(11\ldots1) = 1\cdot(11\ldots1)$, $(00\ldots0) = 1\cdot(11\ldots1) + 1\cdot\ (11\ldots1)$.

12.2.1.1 Generator Matrix for Linear Block Code

Any code word $\boldsymbol{x}$ from the (n, k) linear block code can be represented as a linear combination of k basis vectors $\boldsymbol{g}_i$ $(i = 0, 1, .., k-1)$ as given below:

$$\boldsymbol{x} = m_0 g_0 + m_1 g_1 + \cdots + m_{k-1} g_{k-1} = \boldsymbol{m}\begin{bmatrix} g_0 \\ g_1 \\ \cdots \\ g_{k-1}\end{bmatrix} = \boldsymbol{mG};\quad \boldsymbol{G} = \begin{bmatrix} g_0 \\ g_1 \\ \cdots \\ g_{k-1}\end{bmatrix},$$
$$\boldsymbol{m} = (m_0\ m_1\ \ldots\ m_{k-1}), \tag{12.10}$$

where $\boldsymbol{m}$ is the message vector, and $\boldsymbol{G}$ is the generator matrix (of dimensions $k \times n$), in which every row represents a basis vector from the coding subspace. Therefore, in order to encode, the message vector $\boldsymbol{m}(m_0, m_1, \ldots, m_{k-1})$ has to be multiplied with a generator matrix $\boldsymbol{G}$ to get $\boldsymbol{x} = \boldsymbol{mG}$, where $\boldsymbol{x}(x_0, x_1, \ldots, x_{n-1})$ is a codeword.

Example 12.3. Generator matrices for repetition $(n, 1)$ code $\boldsymbol{G}_{\text{rep}}$ and $(n, n-1)$ single-parity-check code $\boldsymbol{G}_{\text{par}}$ are given, respectively, as

$$\boldsymbol{G}_{\text{rep}} = [11\ldots1],\quad \boldsymbol{G}_{\text{par}} = \begin{bmatrix} 1 & 0 & 0 & \cdots & 0 & 1 \\ 0 & 1 & 0 & \cdots & 0 & 1 \\ & & & \cdots & & \\ 0 & 0 & 0 & \cdots & 1 & 1 \end{bmatrix}$$

By elementary operations on rows in the generator matrix, the code may be transformed into systematic form

$$\boldsymbol{G}_{\text{s}} = [\boldsymbol{I}_k|\boldsymbol{P}], \tag{12.11}$$

where $\boldsymbol{I}_k$ is unity matrix of dimensions $k \times k$, and $\boldsymbol{P}$ is the matrix of dimensions $k \times (n-k)$ with columns denoting the positions of parity checks

$$P = \begin{bmatrix} p_{00} & p_{01} & \cdots & p_{0,n-k-1} \\ p_{10} & p_{11} & \cdots & p_{1,n-k-1} \\ \cdots & \cdots & \cdots & \cdots \\ p_{k-1,0} & p_{k-1,1} & \cdots & p_{k-1,n-k-1} \end{bmatrix}. \tag{12.12}$$

The codeword of a systematic code is obtained by

$$\boldsymbol{x} = [\boldsymbol{m}|\boldsymbol{b}] = \boldsymbol{m}\,[\boldsymbol{I}_k|\boldsymbol{P}] = \boldsymbol{mG},\quad \boldsymbol{G} = [\boldsymbol{I}_k|\boldsymbol{P}], \tag{12.13}$$

and the structure of systematic codeword is shown in Fig. 12.5.

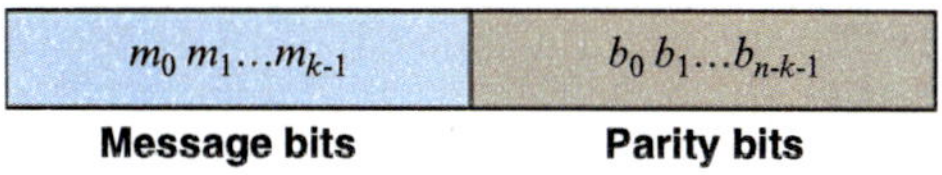

Fig. 12.5 Structure of a systematic code word

Therefore, during encoding the message vector stays unchanged and the elements of vector of parity checks $\boldsymbol{b}$ are obtained by

$$b_i = p_{0i} m_0 + p_{1i} m_1 + \cdots + p_{k-1,i} m_{k-1}, \tag{12.14}$$

where

$$p_{ij} = \begin{cases} 1, & \text{if } b_i \text{ depends on } m_j, \\ 0, & \text{otherwise.} \end{cases}$$

During transmission, an optical channel introduces the errors so that the received vector $\boldsymbol{r}$ can be written as $\boldsymbol{r} = \boldsymbol{x} + \boldsymbol{e}$, where $\boldsymbol{e}$ is the error vector (pattern) with elements components determined by

$$\boldsymbol{e}_i = \begin{cases} 1, & \text{if an error occurs in the } i\text{th location,} \\ 0, & \text{otherwise.} \end{cases}$$

To determine whether the received vector $\boldsymbol{r}$ is a codeword vector, we are introducing the concept of a *parity check matrix.*

12.2.1.2 Parity-Check Matrix

Another useful matrix associated with the linear block codes is the parity-check matrix. Let us expand the matrix equation $\boldsymbol{x} = \boldsymbol{mG}$ in scalar form as follows:

$$\begin{aligned} x_0 &= m_0 \\ x_1 &= m_1 \\ &\ldots \\ x_{k-1} &= m_{k-1} \\ x_k &= m_0 p_{00} + m_1 p_{10} + \cdots + m_{k-1} p_{k-1,0} \\ x_{k+1} &= m_0 p_{01} + m_1 p_{11} + \cdots + m_{k-1} p_{k-1,1} \\ &\ldots \\ x_{n-1} &= m_0 p_{0,n-k-1} + m_1 p_{1,n-k-1} + \cdots + m_{k-1} p_{k-1,n-k-1} \end{aligned} \tag{12.15a}$$

By using the first k equalities, the last $n-k$ equations can be rewritten as follows:

$$\begin{aligned} &x_0 p_{00} + x_1 p_{10} + \cdots + x_{k-1} p_{k-1,0} + x_k = 0 \\ &x_0 p_{01} + x_1 p_{11} + \cdots + x_{k-1} p_{k-1,0} + x_{k+1} = 0 \\ &\ldots \\ &x_0 p_{0,n-k-1} + x_1 p_{1,n-k-1} + \cdots + x_{k-1} p_{k-1,n-k+1} + x_{n-1} = 0 \end{aligned} \tag{12.15b}$$

The matrix representation of (12.15b) is given below:

$$\left[x_0\ x_1 \cdots x_{n-1}\right]\begin{bmatrix} p_{00} & p_{10} & \cdots & p_{k-1,0} & 1\ 0 \cdots 0 \\ p_{01} & p_{11} & \cdots & p_{k-1,1} & 0\ 1 \cdots 0 \\ \cdots & & \cdots & & \cdots \quad \cdots \\ p_{0,n-k-1} & p_{1,n-k-1} & \cdots & p_{k-1,n-k-1} & 0\ 0 \cdots 1 \end{bmatrix}^{\mathrm{T}} = \boldsymbol{x}\boldsymbol{H}^{\mathrm{T}} = \boldsymbol{0},$$

$$\boldsymbol{x} = \left[x_0\ x_1 \cdots x_{n-1}\right], \quad \boldsymbol{H} = \left[\boldsymbol{P}^{\mathrm{T}}\ \boldsymbol{I}_{n-k}\right]_{(n-k)xn}, \tag{12.15c}$$

where $\boldsymbol{P}$ is already introduced by (12.12). The $\boldsymbol{H}$-matrix in (12.15c) is known as the parity-check matrix. We can easily verify that:

$$\boldsymbol{G}\boldsymbol{H}^{\mathrm{T}} = [\boldsymbol{I}_k \quad \boldsymbol{P}]\begin{bmatrix} \boldsymbol{P} \\ \boldsymbol{I}_{n-k} \end{bmatrix} = \boldsymbol{P} + \boldsymbol{P} = \boldsymbol{0}, \tag{12.16}$$

meaning that the parity check matrix of an (n,k) linear block code $\boldsymbol{H}$ is a matrix of rank $n-k$ and dimensions $(n-k)\times n$ whose null-space is k-dimensional vector with basis being the generator matrix $\boldsymbol{G}$.

Example 12.4. Parity-Check Matrices for $(n,1)$ repetition code $\boldsymbol{H}_{\mathrm{rep}}$ and $(n,n-1)$ single-parity check code $\boldsymbol{H}_{\mathrm{par}}$ are given, respectively, as:

$$\boldsymbol{H}_{\mathrm{rep}} = \begin{bmatrix} 1 & 0 & 0 & \cdots & 0 & 1 \\ 0 & 1 & 0 & \cdots & 0 & 1 \\ & & & \cdots & & \\ 0 & 0 & 0 & \cdots & 1 & 1 \end{bmatrix}, \quad \boldsymbol{H}_{\mathrm{par}} = [1\ 1 \cdots 1].$$

Example 12.5. For Hamming (7,4) code, the generator $\boldsymbol{G}$ and parity check $\boldsymbol{H}$ matrices are given, respectively, as

$$\boldsymbol{G} = \left[\begin{array}{cccc|ccc} 1 & 0 & 0 & 0 & 1 & 1 & 0 \\ 0 & 1 & 0 & 0 & 0 & 1 & 1 \\ 0 & 0 & 1 & 0 & 1 & 1 & 1 \\ 0 & 0 & 0 & 1 & 1 & 0 & 1 \end{array}\right], \quad \boldsymbol{H} = \left[\begin{array}{cccc|ccc} 1 & 0 & 1 & 1 & 1 & 0 & 0 \\ 1 & 1 & 1 & 0 & 0 & 1 & 0 \\ 0 & 1 & 1 & 1 & 0 & 0 & 1 \end{array}\right].$$

Every (n,k) linear block code with generator matrix $\boldsymbol{G}$ and parity-check matrix $\boldsymbol{H}$ has a dual code with generator matrix $\boldsymbol{H}$ and parity check matrix $\boldsymbol{G}$. For example, $(n,1)$ repetition and $(n,n-1)$ single-parity check codes are dual.

12.2.1.3 Coding Gain

A very important characteristics of an (n,k) linear block code is the so-called coding gain, which was introduced in introductory section of this chapter as being the savings attainable in the energy per information bit to noise spectral density ratio (E_b/N_0) required to achieve a given bit error probability when coding is used compared to that with no coding. Let E_c denote the transmitted bit energy, and E_b denote the information bit energy. Since the total information word energy kE_b must be the same as the total codeword energy nE_c, we obtain the following relationship between E_c and E_b:

$$E_c = (k/n)E_b = RE_b. \tag{12.17}$$

The probability of error for BPSK on an AWGN channel, when coherent hard decision (bit-by-bit) demodulator is used, can be obtained as follows:

$$p = \frac{1}{2}\text{erfc}\left(\sqrt{\frac{E_c}{N_0}}\right) = \frac{1}{2}\text{erfc}\left(\sqrt{\frac{RE_b}{N_0}}\right), \tag{12.18}$$

where erfc(x) function is defined by

$$\text{erfc}(x) = \frac{2}{\sqrt{\pi}}\int_x^{+\infty} e^{-z^2}\text{d}z.$$

By using the Chernoff bound, we obtain the following expression for hard decision decoding coding gain

$$\frac{(E_b/N_0)_{\text{uncoded}}}{(E_b/N_0)_{\text{coded}}} \approx R(t+1), \tag{12.19}$$

where t is the error correction capability of the code. The corresponding soft decision coding gain can be estimated by [13, 14]

$$\frac{(E_b/N_0)_{\text{uncoded}}}{(E_b/N_0)_{\text{coded}}} \approx Rd_{\min}, \tag{12.20}$$

and it is about 3 dB better than hard decision decoding (because the minimum distance $d_{\min} \geq 2t+1$). In optical communications, it is very common to use the Q-factor[1] as the figure of merit instead of SNR, which is related to the BER on an AWGN channel as follows

$$\text{BER} = \frac{1}{2}\text{erfc}\left(\frac{Q}{\sqrt{2}}\right). \tag{12.21}$$

Let BER_{in} denote the BER at the input of FEC decoder, let BER_{out} denote the BER at the output of FEC decoder, and let BER_{ref} denote target BER (such as either 10^{-12} or 10^{-15}). The corresponding coding gain GC and net coding gain NCG are, respectively, defined as [9]

[1] The Q-factor is defined as $Q = (\mu_1 - \mu_0)/(\sigma_1 + \sigma_0)$, where μ_j and σ_j ($j = 0, 1$) represent the mean and the standard deviation corresponding to the bits $j = 0, 1$.

$$\text{CG} = 20\log_{10}\left[\text{erfc}^{-1}(2\text{BER}_{\text{ref}})\right] - 20\log_{10}\left[\text{erfc}^{-1}(2\text{BER}_{\text{in}})\right] \quad [\text{dB}], \tag{12.22}$$

$$\text{NCG} = 20\log_{10}\left[\text{erfc}^{-1}(2\text{BER}_{\text{ref}})\right] \\ -20\log_{10}\left[\text{erfc}^{-1}(2\text{BER}_{\text{in}})\right] + 10\log_{10} R \quad [\text{dB}]. \tag{12.23}$$

12.3 Codes on Graphs

The codes on graphs of interest in optical communications include turbo codes, TPCs, and LDPC codes. The turbo codes [8,9,23–28] can be considered as the generalization of the concatenation of codes in which, during iterative decoding, the decoders interchange the soft messages for a certain number of times. Turbo codes can approach channel capacity closely in the region of interest for wireless communications. However, they exhibit strong error floors in the region of interest for fiberoptics communications (see [29]); therefore, alternative iterative soft decoding approaches are to be sought. As recently shown in [4, 5, 30–38], TPCs and LDPC codes can provide excellent coding gains and, when properly designed, do not exhibit error floor in the region of interest for fiber-optics communications.

A TPC is an $(n_1 n_2, k_1 k_2, d_1 d_2)$ code in which codewords form an $n_1 \times n_2$ array such that each row is a codeword from an (n_1, k_1, d_1) code C_1, and each column is a codeword from an (n_2, k_2, d_2) code C_2. With n_i, k_i and d_i $(i = 1, 2)$, we denoted the codeword length, dimension, and minimum distance, respectively, of the ith component code. The soft bit reliabilities are iterated between decoders for C_1 and C_2. In fiber-optics communications, TPCs based on BCH component codes are intensively studied, e.g. [8,9,27,28].

If the parity-check matrix has a low density of ones and the number of 1′s per row and per column are both constant, the code is said to be a *regular LDPC* code. To facilitate the implementation at high speed, we prefer the use of regular rather than irregular LDPC codes. The graphical representation of LDPC codes, known as bipartite (Tanner) graph representation, is helpful in efficient description of LDPC decoding algorithms. A *bipartite (Tanner) graph* is a graph whose nodes may be separated into two classes (*variable* and *check* nodes), and where *undirected edges* may only connect two nodes not residing in the same class. The Tanner graph of a code is drawn according to the following rule: check (function) node c is connected to variable (bit) node v whenever element h_{cv} in a parity-check matrix $\boldsymbol{H}$ is a 1. In an $m \times n$ parity-check matrix, there are $m = n - k$ check nodes and n variable nodes.

Example 12.6. As an illustrative example, consider the $\boldsymbol{H}$-matrix of the following code

$$H = \begin{bmatrix} 1 & 0 & 1 & 0 & 1 & 0 \\ 1 & 0 & 0 & 1 & 0 & 1 \\ 0 & 1 & 1 & 0 & 0 & 1 \\ 0 & 1 & 0 & 1 & 1 & 0 \end{bmatrix}.$$

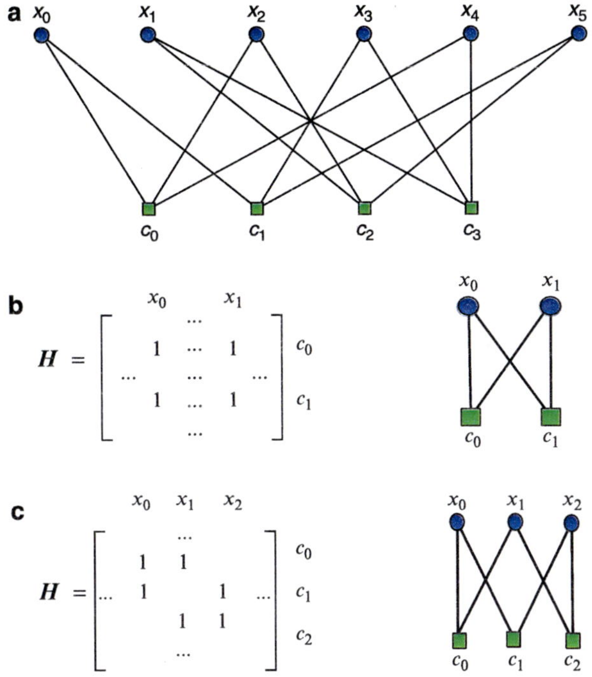

Fig. 12.6 (**a**) Bipartite graph of (6, 2) code described by ***H*** matrix above. Cycles in a Tanner graph: (**b**) cycle of length 4, and (**c**) cycle of length 6

For any valid codeword $\boldsymbol{x} = [x_0\ x_1\ \ldots\ x_{n-1}]$, the checks used to decode the codeword are written as,

- $(c_0): \ x_0 + x_2 + x_4 = 0 \ (\text{mod } 2)$
- $(c_1): \ x_0 + x_3 + x_5 = 0 \ (\text{mod } 2)$
- $(c_2): \ x_1 + x_2 + x_5 = 0 \ (\text{mod } 2)$
- $(c_3): \ x_1 + x_3 + x_4 = 0 \ (\text{mod } 2).$

The bipartite graph (Tanner graph) representation of this code is given in Fig. 12.6a.

The circles represent the bit (variable) nodes, while squares represent the check (function) nodes. For example, the variable nodes x_0, x_2, and x_4 are involved in (c_0), and therefore connected to the check node c_0. A closed path in a bipartite graph comprising l edges that closes back on itself is called a *cycle* of length l. The shortest cycle in the bipartite graph is called the *girth*. The girth influences the minimum distance of LDPC codes, correlates the extrinsic log-likelihood ratios (LLRs), and therefore affects the decoding performance. The use of large girth LDPC codes is preferable because the large girth increases the minimum distance and de-correlates the extrinsic info in the decoding process. To improve the iterative decoding performance, we have to avoid cycles of length 4, and preferably 6 as well. To check for the existence of short cycles, one has to search over ***H***-matrix for the patterns shown in Fig. 12.6b, c.

12.3.1 Quasi-cyclic (QC) Binary LDPC Codes

In this section, we describe a method for designing large girth QC LDPC codes; and an efficient and simple variant of SPA suitable for use in optical communications, namely the min-sum-with-correction term algorithm.

12.3.1.1 Design of Large Girth Quasi-cyclic LDPC Codes

Based on Tanner's bound for the minimum distance of an LDPC code [39]

$$d \geq \begin{cases} 1+\dfrac{w_c}{w_c-2}\left((w_c-1)^{\lfloor (g-2)/4\rfloor}-1\right), & g/2=2m+1, \\ 1+\dfrac{w_c}{w_c-2}\left((w_c-1)^{\lfloor (g-2)/4\rfloor}-1\right)+(w_c-1)^{\lfloor (g-2)/4\rfloor}, & g/2=2m, \end{cases} \tag{12.24}$$

(where g and w_c denote the girth of the code graph and the column weight, respectively, and where d stands for the minimum distance of the code), it follows that large girth leads to an exponential increase in the minimum distance, provided that the column weight is at least 3. ($\lfloor\rfloor$ denotes the largest integer less than or equal to the enclosed quantity.) For example, the minimum distance of girth-10 codes with column weight $r = 3$ is at least 10. The parity-check matrix of regular[2] QC LDPC codes [37, 40] can be represented by

$$H = \begin{bmatrix} I & I & I & \cdots & I \\ I & P^{S[1]} & P^{S[2]} & \cdots & P^{S[c-1]} \\ I & P^{2S[1]} & P^{2S[2]} & \cdots & P^{2S[c-1]} \\ \cdots & \cdots & \cdots & \cdots & \cdots \\ I & P^{(r-1)S[1]} & P^{(r-1)S[2]} & \cdots & P^{(r-1)S[c-1]} \end{bmatrix}, \tag{12.25}$$

where I is $B \times B$ (B is a prime number) identity matrix, P is $B \times B$ permutation matrix given by $P = (p_{ij})_{B\times B}$, $p_{i,i+1} = p_{B,1} = 1$ (zero otherwise), and where r and c represent the number of block-rows and block-columns in (12.25), respectively. The set of integers S are to be carefully chosen from the set $\{0, 1, \ldots, B-1\}$ so that the cycles of short length, in the corresponding Tanner (bipartite) graph representation of (12.25), are avoided. According to Theorem 2.1 in [40], we have to avoid the cycles of length $2k$ ($k = 3$ or 4) defined by the following equation

$$S[i_1]\,j_1+S[i_2]\,j_2+\cdots+S[i_k]\,j_k = S[i_1]\,j_2+S[i_2]\,j_3+\cdots+S[i_k]\,j_1 \bmod p, \tag{12.26}$$

[2] A (w_c, w_r) – *regular LDPC code* is a linear block code whose H-matrix contains exactly w_c 1's in each column and exactly $w_r = w_c\, n/(n-k)$ 1's in each column, where $w_c \ll n-k$.

where the closed path is defined by (i_1, j_1), (i_1, j_2), (i_2, j_2), (i_2, j_3), ..., (i_k, j_k), (i_k, j_1) with the pair of indices denoting row-column indices of permutation-blocks in (12.25) such that $l_m \neq l_{m+1}$, $l_k \neq l_1$ $(m = 1, 2, .., k;\ l \in \{i, j\})$. Therefore, we have to identify the sequence of integers $S[i] \in \{0, 1, \ldots, B-1\}$ $(I = 0, 1, \ldots, r-1;\ r < B)$ not satisfying the (12.26), which can be done either by computer search or in a combinatorial fashion. For example, to design the QC LDPC codes in [34], we introduced the concept of the cyclic-invariant difference set (CIDS). The CIDS-based codes come naturally as girth-6 codes, and to increase the girth we had to selectively remove certain elements from a CIDS. The design of LDPC codes of rate above 0.8, column weight 3, and girth-10 using the CIDS approach is a very challenging and is still an open problem. Instead, in our recent paper [37], we solved this problem by developing an efficient computer search algorithm. We add an integer at a time from the set $\{0, 1, \ldots, B-1\}$ (not used before) to the initial set S and check if (12.26) is satisfied. If (12.26) is satisfied, we remove that integer from the set S and continue our search with another integer from set $\{0, 1, \ldots, B-1\}$ until we exploit all the elements from $\{0, 1, \ldots, B-1\}$. The code rate of these QC codes, R, is lower-bounded by

$$R \geq \frac{|S|\,B - rB}{|S|\,B} = 1 - r/|S|, \tag{12.27}$$

and the codeword length is $|S|B$, where $|S|$ denotes the cardinality of set S. For a given code rate R_0, the number of elements from S to be used is $\lfloor r/(1-R_0) \rfloor$. With this algorithm, LDPC codes of arbitrary rate can be designed.

Example 12.7. By setting $B = 2,311$, the set of integers to be used in (12.25) is obtained as $S = \{1, 2, 7, 14, 30, 51, 78, 104, 129, 212, 223, 318, 427, 600, 808\}$. The corresponding LDPC code has rate $R_0 = 1$–$3/15 = 0.8$, column weight 3, girth-10 and length $|S|B = 15 \cdot 2311 = 34{,}665$. In the example above, the initial set of integers was $S = \{1, 2, 7\}$, and the set of row to be used in (12.25) is $\{1, 3, 6\}$. The use of a different initial set will result in a different set from that obtained above.

Example 12.8. By setting $B = 269$, the set S is obtained as $S = \{0, 2, 3, 5, 9, 11, 12, 14, 27, 29, 30, 32, 36, 38, 39, 41, 81, 83, 84, 86, 90, 92, 93, 95, 108, 110, 111, 113, 117, 119, 120, 122\}$. If 30 integers are used, the corresponding LDPC code has rate $R_0 = 1$–$3/30 = 0.9$, column weight 3, girth-8 and length $30 \cdot 269 = 8{,}070$.

12.3.1.2 Decoding of LDPC Codes

In this subsection, we describe the min-sum with correction term decoding algorithm [38, 41]. It is a simplified version of the original algorithm proposed by Gallager [10]. Gallager proposed a near optimal iterative decoding algorithm for LDPC codes that computes the distributions of the variables in order to calculate the *a posteriori probability* (APP) of a bit v_i of a codeword $\mathbf{v} = [v_0\ v_1\ \ldots\ v_{n-1}]$ to

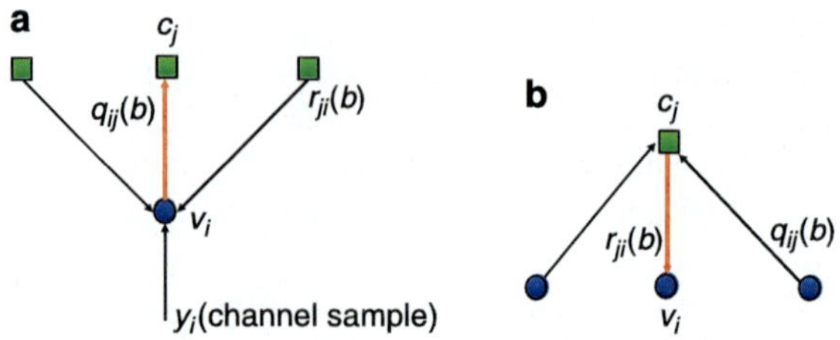

Fig. 12.7 Illustration of the half-iterations of the sum–product algorithm: (**a**) first half-iteration: extrinsic info sent from v-nodes to c-nodes, and (**b**) second half-iteration: extrinsic info sent from c-nodes to v-nodes

be equal to 1 given a received vector $\mathbf{y} = [y_0\ y_1\ \ldots\ y_{n-1}]$. This iterative decoding scheme engages passing the extrinsic info back and forth among the c-nodes and the v-nodes over the edges to update the distribution estimation. Each iteration in this scheme is composed of two half-iterations. In Fig. 12.7, we illustrate both the first and the second halves of an iteration of the algorithm. As an example, in Fig. 12.7a, we show the message sent from v-node v_i to the c-node c_j. v_i-node collects the information from channel (y_i sample), in addition to extrinsic info from other c-nodes connected to v_i-node, processes them and sends the extrinsic info (not already available info) to c_j. This extrinsic info contains the information about the probability $\Pr(c_i = b|y_0)$, where $b \in \{0, 1\}$. This is performed in all c-nodes connected to v_i-node. On the other hand, Fig. 12.7b shows the extrinsic info sent from c-node c_i to the v-node v_j, which contains the information about $\Pr(c_i$ equation is satisfied $|\mathbf{y})$. This is done repeatedly to all the c-nodes connected to v_i-node.

After this intuitive description, we describe the min-sum-with-correction-term algorithm in more detail [38] because of its simplicity and suitability for high-speed implementation. Generally, we can either compute APP $\Pr(v_i|\mathbf{y})$ or the APP ratio $l(v_i) = \Pr(v_i = 0|\mathbf{y})/\Pr(v_i = 1|\mathbf{y})$, which is also referred to as the likelihood ratio. In log-domain version of the SPA, we replace these likelihood ratios with LLRs due to the fact that the probability domain includes many multiplications, which leads to numerical instabilities, whereas the computation using LLRs computation involves addition only. Moreover, the log-domain representation is more suitable for finite precision representation. Thus, we compute the LLRs by $L(v_i) = \log[\Pr(v_i = 0|\mathbf{y})/\Pr(v_i = 1|\mathbf{y})]$. For the final decision, if $L(v_i) > 0$, we decide in favor of 0 and if $L(v_i) < 0$, we decide in favor of 1. To further explain the algorithm, we introduce the following notations due to MacKay [36] and Ryan [38]:

V_j = {v-nodes connected to c-node c_j}
$V_{j\backslash i}$ = {v-nodes connected to c-node c_j}\{v-node v_i}
C_i = {c-nodes connected to v-node v_i}
$C_{i\backslash j}$ = {c-nodes connected to v-node v_i}\{c-node c_j}
$M_v(\sim i)$ = {messages from all v-nodes except node v_i}
$M_c(\sim j)$ = {messages from all c-nodes except node c_j}
P_i = $\Pr(v_i = 1|y_i)$

S_i = event that the check equations involving c_i are satisfied
$q_{ij}(b)$ = $\Pr(v_i = b|S_i,\ y_i,\ M_c(\sim j))$
$r_{ji}(b)$ = Pr(check equation c_j is satisfied$|v_i = b,\ M_v(\sim i))$

In the log-domain version of the SPA, all the calculations are performed in the log-domain as follows:

$$L\left(v_i\right) = \log\left[\frac{\Pr\left(v_i = 0|y_i\right)}{\Pr\left(v_i = 1|y_i\right)}\right], \quad L\left(r_{ji}\right) = \log\left[\frac{r_{ji}\left(0\right)}{r_{ji}\left(1\right)}\right],$$

$$L\left(q_{ji}\right) = \log\left[\frac{q_{ji}\left(0\right)}{q_{ji}\left(1\right)}\right]. \qquad (12.28)$$

The algorithm starts with the initialization step, where we set $L(v_i)$ as follows:

$$\begin{aligned}
L\left(v_i\right) &= (-1)^{y_i}\log\left(\frac{1-\varepsilon}{\varepsilon}\right), && \text{for BSC}\\
L\left(v_i\right) &= 2\frac{y_i}{\sigma^2}, && \text{for binary, input AWGN}\\
L\left(v_i\right) &= \log\left(\frac{\sigma_1}{\sigma_0}\right) - \frac{\left(y_i-\mu_0\right)^2}{2\sigma_0^2} + \frac{\left(y_i-\mu_1\right)^2}{2\sigma_1^2}, && \text{for BA-AWGN}\\
L\left(v_i\right) &= \log\left(\frac{\Pr\left(v_i = 0|y_i\right)}{\Pr\left(v_i = 1|y_i\right)}\right), && \text{for abritrary channel}
\end{aligned} \qquad (12.29)$$

where ε is the probability of error in the BSC, σ^2 is the variance of the Gaussian distribution of the AWGN, and μ_j and σ_j^2 $(j = 0, 1)$ represent the mean and the variance of Gaussian process corresponding to the bits $j = 0, 1$ of a binary asymmetric (BA)-AWGN channel. After initialization of $L(q_{ij})$, we calculate $L(r_{ji})$ as follows:

$$\begin{aligned}
L\left(r_{ji}\right) &= L\left(\sum_{i'\in V_j\backslash i} b_i'\right) = L\left(\cdots \oplus b_k \oplus b_l \oplus b_m \oplus b_n \cdots\right)\\
&= \cdots L_k \boxplus L_l \boxplus L_m \boxplus L_n \boxplus \cdots
\end{aligned} \qquad (12.30)$$

where $\oplus$ denotes the modulo-2 addition, and $\boxplus$ denotes a pairwise computation defined by

$$\begin{aligned}
L_a \boxplus L_b &= \text{sign}(L_a)\text{sign}(L_b)\min(|L_a|, |L_b|) + s(L_a, L_b)\\
s(L_a, L_b) &= \log\left(1+\mathrm{e}^{-|L_a+L_b|}\right) - \log\left(1+\mathrm{e}^{-|L_a-L_b|}\right)
\end{aligned} \qquad (12.31)$$

The term $s(L_a, L_b)$ is the correction term and it is implemented as a lookup table (LUT). Upon calculation of $L(r_{ji})$, we update

$$L\left(q_{ij}\right) = L\left(v_i\right) + \sum_{j'\in C_{i\backslash j}} L\left(r_{j'i}\right), \quad L\left(Q_i\right) = L\left(v_i\right) + \sum_{j\in C_i} L\left(r_{ji}\right) \qquad (12.32)$$

Finally, the decision step is as follows:

$$\hat{v}_i = \begin{cases} 1, & L(Q_i) < 0, \\ 0, & \text{otherwise.} \end{cases} \tag{12.33}$$

If the syndrome equation $\hat{\boldsymbol{v}}\boldsymbol{H}^{\mathrm{T}} = \mathbf{0}$ is satisfied or the maximum number of iterations is reached, we stop, otherwise, we recalculate $L(r_{ji})$ and update $L(q_{ij})$ and $L(Q_i)$ and check again. It is important to set the number of iterations high enough to ensure that most of the codewords are decoded correctly and low enough not to affect the processing time. It is important to mention that decoder for good LDPC codes require less number of iterations to guarantee successful decoding.

12.3.1.3 BER Performance of LDPC Codes

The results of simulations for an AWGN channel model are given in Fig. 12.8, where we compare the large girth LDPC codes (Fig. 12.8a) against RS codes, concatenated RS codes, TPCs, and other classes of LDPC codes.

In optical communications, it is a common practice to use the Q-factor as a figure of merit of binary modulation schemes instead of SNR. In all simulation results in this section, we maintained the double precision. For the LDPC(16935,13550) code, we also provided 3- and 4-bit fixed-point simulation results (see Fig. 12.8a). Our results indicate that the 4-bit representation performs comparable to the double-precision representation, whereas the 3-bit representation performs 0.27 dB worse than the double-precision representation at the BER of $2. \cdot 10^{-8}$. The girth-10 LDPC(24015, 19212) code of rate 0.8 outperforms the concatenation RS(255, 239)+RS(255, 223) (of rate 0.82) by 3.35 dB and RS(255, 239) by 4.75 dB both at BER of 10^{-7}. The same LDPC code outperforms projective geometry (PG) $(2, 2^6)$ based LDPC(4161, 3431) (of rate 0.825) of girth-6 by 1.49 dB at BER of 10^{-7}, and outperforms CIDS-based LDPC(4320, 3242) of rate 0.75 and girth-8 LDPC codes by 0.25 dB. At BER of 10^{-10}, it outperforms lattice-based LDPC(8547, 6922) of rate 0.81 and girth-8 LDPC code by 0.44 dB, and BCH(128, 113) × BCH(256, 239) TPC of rate 0.82 by 0.95 dB. The net coding gain (NCG) at BER of 10^{-12} is 10.95 dB. In Fig. 12.8b, different LDPC codes are compared against RS (255, 223) code, concatenated RS code of rate 0.82 and convolutional code (CC) (of constraint length 5). It can be seen that LDPC codes, both regular and irregular, offer much better performance than hard-decision codes. It should be noted that pairwised balanced design (PBD) [42]-based irregular LDPC code of rate 0.75 is only 0.4 dB away from the concatenation of convolutional-RS codes (denoted in Fig. 12.8b as RS + CC) with significantly lower code rate $R = 0.44$ at BER of 10^{-6}. As expected, irregular LDPC codes (black colored curves) outperform regular LDPC codes.

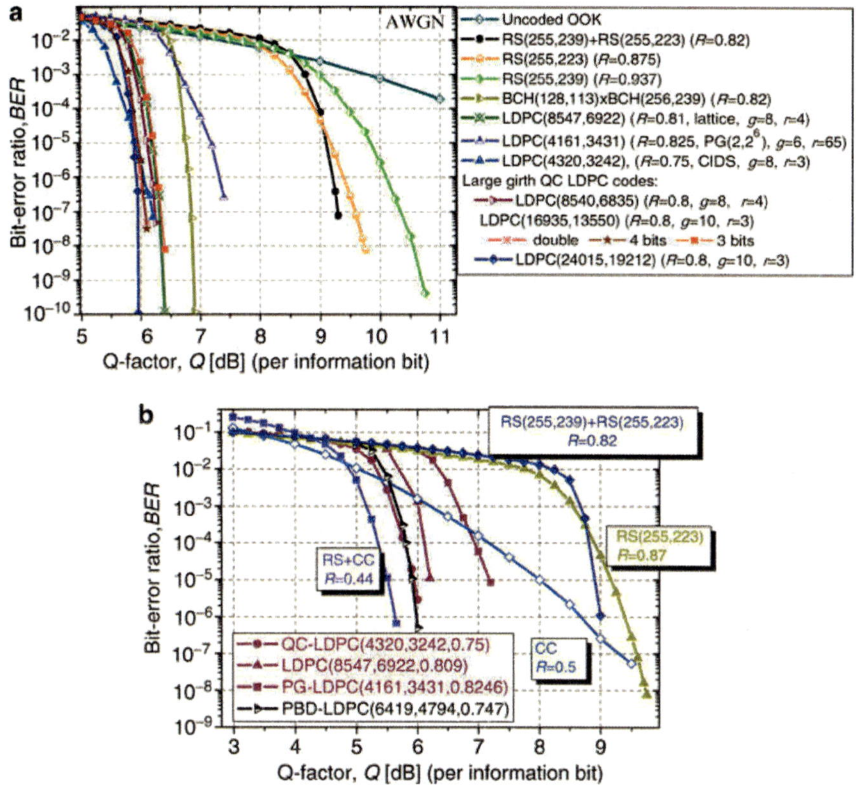

Fig. 12.8 (**a**) Large girth QC LDPC codes against RS codes, concatenated RS codes, TPCs, and previously proposed LDPC codes on an AWGN channel model, and (**b**) LDPC codes versus convolutional, concatenated RS, and concatenation of convolutional and RS codes on an AWGN channel. Number of iterations in sum–product-with-correction-term algorithm was set to 25 (After ref. [2]; @ IEEE 2009; reprinted with permission.)

12.4 Coded Modulation

In this section, we describe how to properly combine modulation with channel coding, and describe three coded-modulation schemes: (1) MLC [43, 44], (2) coded-OFDM [45], and (3) multidimensional coded modulation [87–91]. Using this approach, modulation, coding and multiplexing are performed in a unified fashion so that, effectively, the transmission, signal processing, detection, and decoding are done at much lower symbol rates. At these lower rates, dealing with the nonlinear effects and PMD is more manageable, while the aggregate data rate per wavelength is maintained above 100 Gb/s.

12.4.1 Multilevel Coding and Bit-Interleaved Coded Modulation

M-ary PSK, M-ary QAM, and M-ary DPSK achieve the transmission of $\log_2 M(=m)$ bits per symbol, providing bandwidth-efficient communication. In coherent detection for M-ary PSK, the data phasor $\phi_l \in \{0, 2\pi/M, .., 2\pi(M-1)/M\}$ is sent at each lth transmission interval. In direct detection, the modulation is differential, the data phasor $\phi_l = \phi_{l-1} + \Delta\phi_l$ is sent instead, where $\Delta\phi_l \in \{0, 2\pi/M, .., 2\pi(M-1)/M\}$ is determined by the sequence of m input bits using an appropriate mapping rule. Let us now introduce the transmitter architecture employing LDPC codes as channel codes. If component LDPC codes are of different code rates but of the same length, the corresponding scheme is commonly referred to as MLC. If all component codes are of the same code rate, corresponding scheme is referred to as the bit-interleaved coded-modulation (BICM). The use of MLC allows us to adapt the code rates to the constellation mapper and channel. For example, for Gray mapping, 8-PSK and AWGN, it was found in [46] that optimum code rates of individual encoders are approximately 0.75, 0.5, and 0.75, meaning that 2 bits are carried per symbol. In MLC, the bit streams originating from m different information sources are encoded using different (n, k_i) LDPC codes of code rate $r_i = k_i/n$. k_i denotes the number of information bits of the ith ($i = 1, 2, \ldots, m$) component LDPC code, and n denotes the codeword length, which is the same for all LDPC codes. The mapper accepts m bits, $\boldsymbol{c} = (c_1, c_2, .., c_m)$, at time instance i from the $(m \times n)$ interleaver column-wise and determines the corresponding M-ary $(M = 2^m)$ constellation point $\boldsymbol{s}_i = (I_i, Q_i) = |\boldsymbol{s}_i| \exp(\mathrm{j}\phi_i)$ (see Fig. 12.9a).

The receiver input electrical field at time instance i for an optical M-ary differential phase-shift keying (DPSK) receiver configuration from Fig. 12.9b is denoted by $E_i = |E_i| \exp(\mathrm{j}\varphi_i)$. The outputs of I- and Q-branches (upper and lower-branches in Fig. 12.14b) are proportional to $\mathrm{Re}\left\{E_i E_{i-1}^*\right\}$ and $\mathrm{Im}\left\{E_i E_{i-1}^*\right\}$, respectively. The corresponding coherent detector receiver architecture is shown in Fig. 12.9c, where

$$S_i = |S| \mathrm{e}^{j\varphi_{S_i}} (\varphi_{\mathrm{S},i} = \omega_{\mathrm{S}} t + \varphi_i + \varphi_{\mathrm{S,PN}}) \tag{12.34}$$

is coherent receiver input electrical field at time instance i and

$$L = |L| \mathrm{e}^{j\varphi_{\mathrm{L}}} (\varphi_{\mathrm{L}} = \omega_{\mathrm{L}} t + \varphi_{\mathrm{L,PN}}) \tag{12.35}$$

is the local laser electrical field. For homodyne coherent detection, the frequency of the local laser (ω_{L}) is the same as that of the incoming optical signal (ω_{L}), so the balanced outputs of I- and Q-channel branches (upper- and lower-branches of Fig. 12.9c) can be written as

$$\begin{aligned} v_I(t) &= R\,|S_k|\,|L| \cos(\varphi_i + \varphi_{\mathrm{S,PN}} - \varphi_{\mathrm{L,PN}}), \quad (\mathrm{i}-1)\,T_{\mathrm{s}} \le t < \mathrm{i}T_{\mathrm{s}}, \\ v_Q(t) &= R\,|S_k|\,|L| \sin(\varphi_i + \varphi_{\mathrm{S,PN}} - \varphi_{\mathrm{L,PN}}), \quad (\mathrm{i}-1)\,T_{\mathrm{s}} \le t < \mathrm{i}T_{\mathrm{s}}, \end{aligned} \tag{12.36}$$

where R is photodiode responsivity, while $\varphi_{\mathrm{S,PN}}$ and $\varphi_{\mathrm{L,PN}}$ represent the laser phase noise of transmitting and receiving (local) laser, respectively. The outputs at I- and

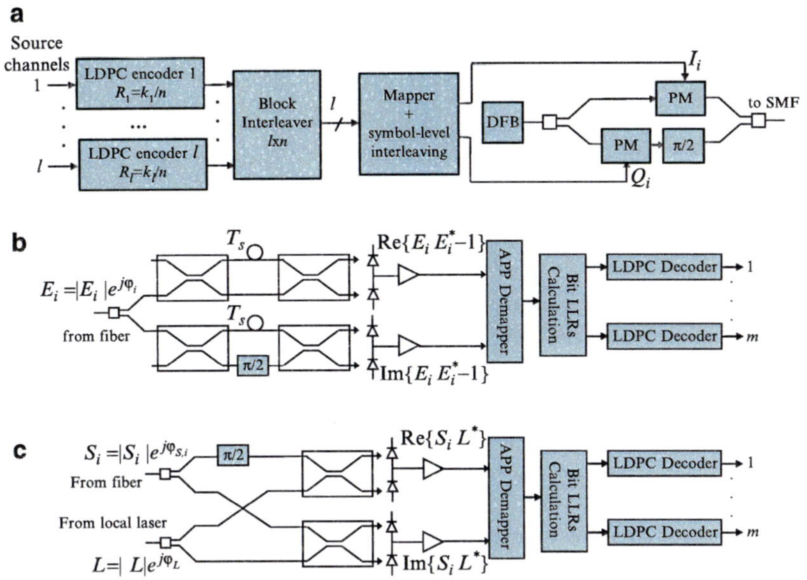

Fig. 12.9 Bit-interleaved LDPC-coded modulation scheme: (**a**) transmitter architecture, (**b**) direct detection architecture, and (**c**) coherent detection receiver architecture. $T_s = 1/R_s$, R_s is the symbol rate

Q-branches (in either coherent or direct detection case) are sampled at the symbol rate (we assume perfect synchronization), and the symbol LLRs are calculated in an APP demapper block as follows

$$\lambda(s) = \log \frac{P(s_0|\boldsymbol{r})}{P(s|\boldsymbol{r})}, \tag{12.37}$$

where $P(s|\boldsymbol{r})$ is determined by using Bayes' rule

$$P(s|\boldsymbol{r}) = \frac{P(\boldsymbol{r}|s)\,P(s)}{\sum_{s'} P(\boldsymbol{r}|s')\,P(s')}. \tag{12.38}$$

Note that $\boldsymbol{s}_i = (I_i, Q_i)$ is the transmitted signal constellation point at time instance i, while $\boldsymbol{r}_i = (r_{\mathrm{I},i}, r_{\mathrm{Q},i})$, $r_{\mathrm{I},I} = v_{\mathrm{I}}(t = iT_s)$, and $r_{\mathrm{Q},I} = v_{\mathrm{Q}}(t = iT_s)$ are the samples of I- and Q-detection branches from Fig. 12.9b, c. In the presence of fiber nonlinearities, $\mathrm{P}(\boldsymbol{r}_i|\boldsymbol{s}_i)$ from (12.38) is estimated by evaluation of histograms, employing sufficiently long training sequence. Note that for direct detection, even in the absence of nonlinearities we have to use the histogram method because the distribution functions are not Gaussian. With $P(\boldsymbol{s})$, we denoted the a priori probability of symbol $\boldsymbol{s}_i$, while $\boldsymbol{s}_0$ is a referent symbol. The normalization in (12.38) is introduced to eliminate the denominator from (12.38). The bit LLRs c_j $(j = 1, 2, \ldots, m)$ are determined from symbol LLRs of (12.37) as

$$L\left(\hat{c}_j\right) = \log \frac{\sum_{s_i : c_j = 0} \exp\left[\lambda\left(s_i\right)\right]}{\sum_{s_i : c_j = 1} \exp\left[\lambda\left(s_i\right)\right]}. \tag{12.39}$$

The jth bit LLR in (12.39) is obtained as the logarithm of the ratio of a probability that $c_j = 0$ and probability that $c_j = 1$. In the nominator (denominator), the summation is done over all symbols s_i having 0 (1) at the position j. The APP demapper extrinsic LLRs (the difference of demapper bit LLRs and LDPC decoder LLRs from previous step) for LDPC decoders become

$$L_{\mathrm{M,e}}(\hat{c}_j) = L(\hat{c}_j) - L_{\mathrm{D,e}}(c_j). \tag{12.40}$$

With $L_{D,e}(c)$, we denoted LDPC decoder extrinsic LLRs which are initially set to zero. The LDPC decoder extrinsic LLRs (the difference between LDPC decoder output and the input LLRs), $L_{\mathrm{D,e}}$, are forwarded to the APP demapper as a priori bit LLRs ($L_{\mathrm{M,a}}$) so that the symbol a priori LLRs are calculated as

$$\lambda_a(s) = \log P(s) = \sum_{j=0}^{m-1} \left(1 - c_j\right) L_{\mathrm{D,e}}\left(c_j\right). \tag{12.41}$$

By substituting (12.41) into (12.37), we are able to calculate the symbol LLRs for the subsequent iteration. The iteration between the APP demapper and LDPC decoder is performed until the maximum number of iterations is reached, or the valid code-words are obtained.

The results simulations, which use 30 iterations in the SPA and 10 iterations between the APP demapper and the LDPC decoder, and employ only BICM and Gray mapping, are shown in Fig. 12.10. Although the actual noise in the repeated

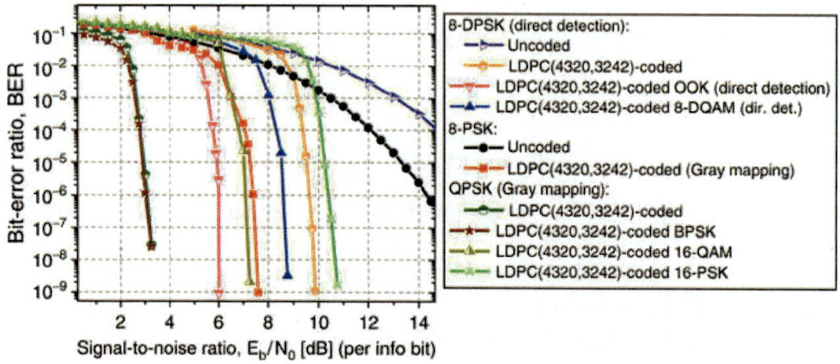

Fig. 12.10 BER performance comparison between bit-interleaved LDPC-coded modulation with coherent detection schemes and direct detection schemes over the AWGN channel. E_b represents the average bit energy, and N_0 is the power spectral density (After ref. [2]; @ IEEE 2009; reprinted with permission.)

systems is dominated by the ASE noise, in this calculation we observed the thermal noise dominated scenario, to be consistent with digital communication literature [13–16, 19, 21, 22, 47]. The coding gain for 8-PSK at the BER of 10^{-9} is about 9.5 dB and a much larger coding gain is expected at BERs below 10^{-12}. Bit-interleaved LDPC-coded 8-PSK with coherent detection outperforms LDPC-coded 8-DPSK with direct detection by 2.23 dB at the BER of 10^{-9}. 8-DQAM outperforms 8-DPSK by 1.15 dB at the same BER. LDPC-coded 16-QAM slightly outperforms LDPC-coded 8-PSK, and significantly outperforms LDPC-coded 16-PSK. As expected, LDPC-coded BPSK and LDPC-coded QPSK (with Gray mapping) perform very closely, and they both outperform LDPC-coded OOK by almost 3 dB.

12.4.2 Polarization-Multiplexed Coded-OFDM

In this subsection, we describe how to combine coded modulation with OFDM, which is illustrated in Fig. 12.11. The transmitter configuration up to the mapper is identical to that already described in Fig. 12.9. The two-dimensional (2D) signal constellation points (see Fig. 12.11b) are split into two streams for OFDM transmitters corresponding to the x- and y-polarizations. The QAM constellation points are considered to be the values of the fast Fourier transform (FFT) of a multi-carrier OFDM signal. The OFDM symbol is generated as follows: N_{QAM} input QAM symbols are zero-padded to obtain N_{FFT} input samples for inverse FFT (IFFT), N_{G} non-zero samples are inserted to create the guard interval, and the OFDM symbol is multiplied by the Blackman–Harris window function. For efficient chromatic dispersion and PMD compensation, the length of cyclically extended guard interval should be longer than the total spread due to chromatic dispersion and differential group delay (DGD). The cyclic extension is accomplished by repeating the last $N_{\mathrm{G}}/2$ samples of the effective OFDM symbol part (N_{FFT} samples) as a prefix, and repeating the first $N_{\mathrm{G}}/2$ samples as a suffix. After D/A conversion (DAC), the RF OFDM signal is converted into the optical domain using the dual-drive Mach–Zehnder modulator (MZM). Two MZMs are needed, one for each polarization. The outputs of MZMs are combined using the polarization beam combiner (PBC). One DFB laser is used as CW source, with x- and y-polarization separated by polarization beam splitter (PBS).

The polarization-detector soft estimates of symbols carried by the kth subcarrier in the ith OFDM symbol, $s_{i,k,x(y)}$, are forwarded to the APP demapper, which determines the symbol LLRs $\lambda_{x(y)}(q)$ $(q = 0, 1, \ldots, 2^b - 1)$ of x- (y-) polarization by

$$\lambda_{x(y)}(q) = -\left(\mathrm{Re}\left[\tilde{s}_{i,k,x(y)}\right] - \mathrm{Re}\left[\mathrm{QAM}\left(\mathrm{map}\left(q\right)\right)\right]\right)^2 / \left(2\sigma^2\right)$$
$$- \left(\mathrm{Im}\left[\tilde{s}_{i,k,x(y)}\right] - \mathrm{Im}\left[\mathrm{QAM}\left(\mathrm{map}\left(q\right)\right)\right]\right)^2 / \left(2\sigma^2\right), \quad (12.42)$$

where Re[] and Im[] denote the real and imaginary part of a complex number, QAM denotes the QAM-constellation diagram, σ^2 denotes the variance of an

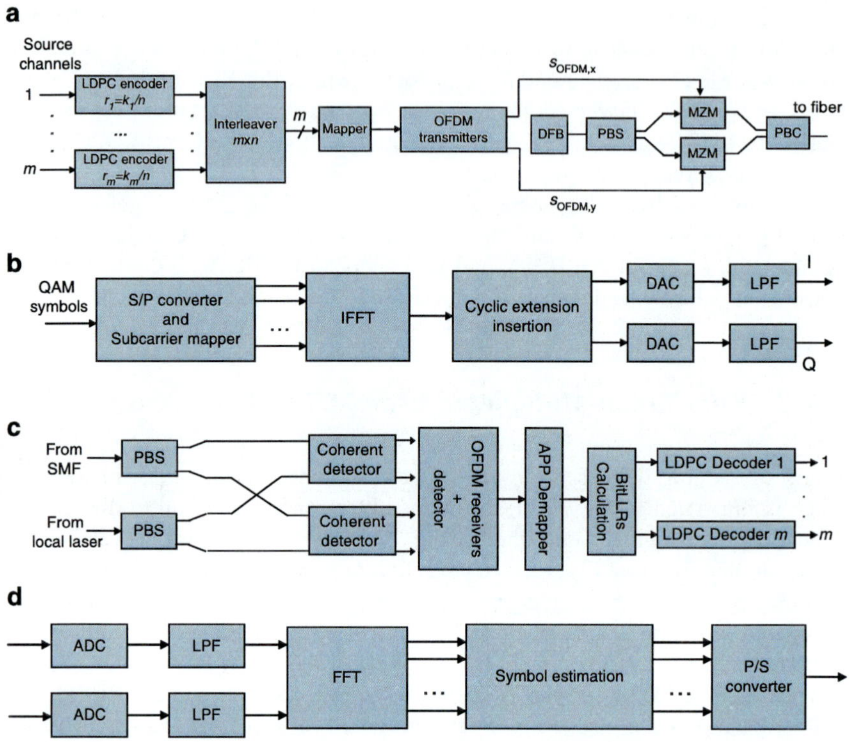

Fig. 12.11 Polarization-multiplexed LDPC-coded OFDM employing both polarizations: (**a**) transmitter architecture, (**b**) OFDM transmitter configuration, (**c**) receiver architecture, and (**d**) OFDM receiver configuration. *DFB* distributed feedback laser, *PBS(C)* polarization beam splitter (combiner), *MZM* dual-drive Mach–Zehnder modulator

equivalent Gaussian noise process originating from ASE noise, and map(q) denotes a corresponding mapping rule. (b denotes the number of bits per constellation point.) Let us denote by $v_{j,x(y)}$ the jth bit in an observed symbol q binary representation $v = (v_1, v_2, \dots, v_b)$ for x- (y-) polarization. The bit LLRs needed for LDPC decoding are calculated from symbol LLRs in fashion similar to (12.39). The extrinsic LLRs are iterated backward and forward until convergence or pre-determined number of iterations has been reached. The polarization-detector soft estimates can be obtained by employing: (1) polarization-time coding [48] similar to space-time coding proposed for use in MIMO wireless communication systems [49], (2) using BLAST algorithm [50], (3) by polarization interference cancelation scheme [50], or (4) carefully performed channel matrix inversion [51].

In Fig. 12.12, we show both the uncoded and LDPC-coded BER performance of the polarization multiplexed LDPC-coded OFDM scheme from [51], against the polarization diversity OFDM scheme, for different constellations sizes. For DGD of 1,200 ps, the polarization multiplexed scheme [51] performs comparable to the

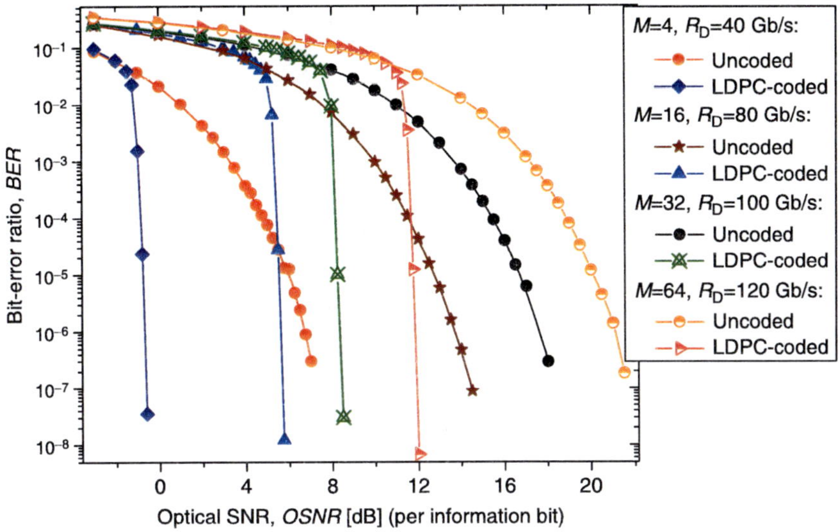

Fig. 12.12 BER performance of polarization multiplexed coded-OFDM, for DGD of 1,200 ps. R_D denotes the aggregate data rate (After ref. [51]; @ IEEE 2009; reprinted with permission.)

polarization-diversity OFDM scheme in terms of BER (the corresponding curves overlap each other), but it has two times higher spectral efficiency. The net effective coding gain increases as the constellation size grows. For $M = 4$ QAM-based polarization multiplexed coded-OFDM, the net effective coding gain is 8.36 dB at BER of 10^{-7}, while for $M = 32$ QAM based LPDC-coded OFDM (of aggregate data rate 100 Gb/s) the coding gain is 9.53 dB at the same BER.

12.4.3 Multidimensional Coded Modulation

In this section, we describe an LDPC-coded hybrid subcarrier/amplitude/phase/polarization (H-SAPP) modulation scheme suitable to achieve 240 Gb/s single-channel transmission rate over optical channels [88]. This scheme doubles the aggregate transmission rate achievable by 8-QAM while providing 2 dB OSNR performance improvement at BER of 10^{-6}. The coded H-SAPP system is composed of two or more hybrid amplitude/phase/polarization (HAPP) subsystems modulated with different subcarriers to exploit the full potential of the 3-dimensional space. Using H-SAPP, we are able to increase the minimum distance between the constellation points in comparison with QAM counterparts and so improve the BER performance. Moreover, H-SAPP allows a non-power-of-two constellation to be utilized (such as 20-point H-SAPP). The HAPP modulation format is based on regular polyhedrons inscribed inside a Poincaré sphere. Since simple regular polyhedrons are not flexible in terms of number of vertices and number of faces, the number of points per

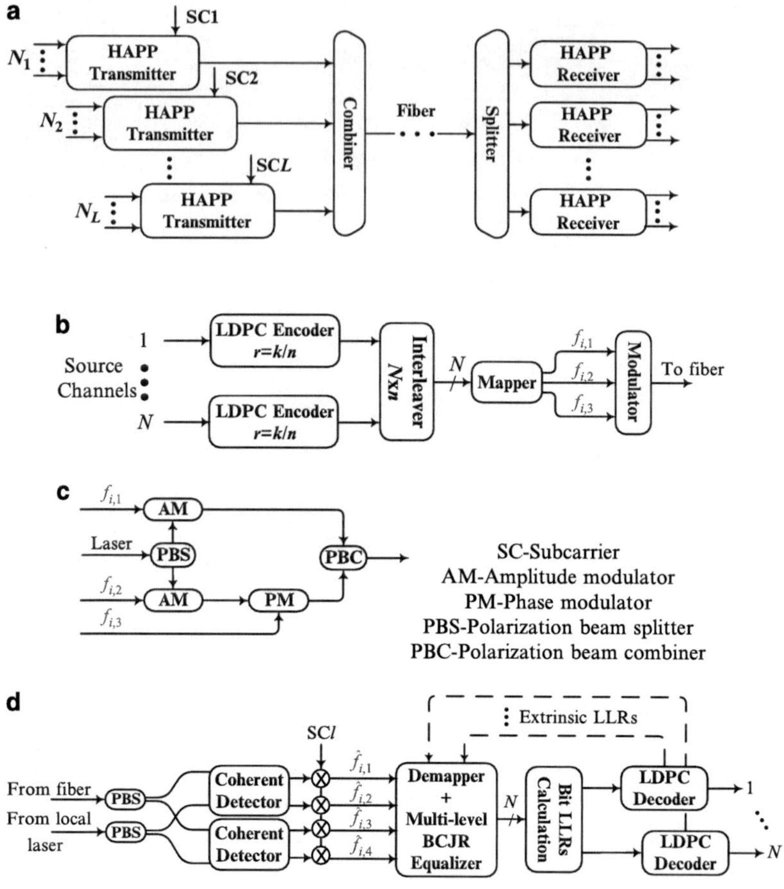

Fig. 12.13 H-SAPP bit-interleaved LDPC-coded modulation block diagrams: (**a**) H-SAPP system, (**b**) HAPP transmitter (**c**) HAPP modulator and (**d**) HAPP receiver configurations (After ref. [88]; @ IEEE 2010; reprinted with permission.)

constellation becomes limited especially since it has to be a power of 2 for binary systems. H-SAPP offers a more flexible utilization of the nice properties of these polyhedrons as it allows the combination of different polyhedrons as will be shown in a simple example explained later through the text.

Figure 12.13a shows the block diagram of the H-SAPP system configuration. N input bit streams from different information sources are divided into L groups variable in number of streams per group. The selection process for the different groups $N_1, N_2, \cdots, N_L$ is governed by two factors, the required aggregate rate, and the polyhedron of choice. Each N_l, the number of streams in the lth group, is then used as input to an HAPP transmitter, where it is modulated with a unique subcarrier. The outputs of the L HAPP transmitters are then forwarded to a power combiner in order to be sent over the fiber. At the receiver side, the signal is split into L

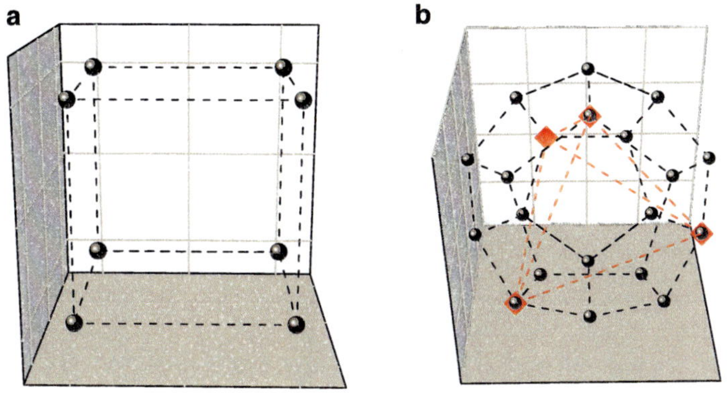

Fig. 12.14 Signal constellations for: (**a**) 8-HAPP and (**b**) 20-H-SAPP (After ref. [88]; @ IEEE 2010; reprinted with permission.)

branches and forwarded to the L HAPP receivers. In this section, and without loss of generality, we clarify three simple examples for $N = 8$ and $N = 16$ where $L = 1$ and for $N = 20$ where $L = 2$. Figure 12.14b shows the block diagram of the coded HAPP transmitter. N_l input bit streams from l different information sources, pass through identical encoders that use structured LDPC codes with code rate $r = k/n$, where k represents the number of information bits, and n represents the codeword length. The outputs of the encoders are then interleaved by an $N_l \times n$ bit-interleaver, where the sequences are written row-wise and read column-wise. The output of the interleaver is sent in one bitstream, N_l bits at a time instant i, to a mapper. The mapper maps each N_l bits into a 2^{Nl}-ary signal constellation point on a vertex of a polyhedron inscribed in a Poincaré sphere based on an LUT. (Please note that the vertices of all the L polyhedrons define a regular polyhedron inscribed in the Poincaré sphere). The signal is then modulated by the HAPP modulator.

The HAPP modulator, shown in Fig. 12.13c, is composed of three simpler modulators, two amplitude modulators (AM) and one phase modulator (PM). Therefore, the LUT maps each N_l bits into a set of three voltages $(f_{1,i}, f_{2,i}, f_{3,i})$ needed to control the set of modulators. As, the polyhedrons used are inscribed in a Poincaré sphere, Stokes parameters are used for the design of the polyhedron. Stokes parameters shown in (12.43) from [2] are then converted into amplitude and phase parameters according to (12.44).

$$s_1 = a_x^2 - a_y^2, \quad s_2 = 2a_x a_y \cos(\delta), \quad s_3 = 2a_x a_y \sin(\delta), \quad \delta = \phi_x - \phi_y, \tag{12.43}$$

where

$$E_x = a_x(t)\mathrm{e}^{j(\omega t + \phi_x(t))}, \quad E_y = a_y(t)\mathrm{e}^{j(\omega t + \phi_y(t))}. \tag{12.44}$$

Without loss of generality, we can assume that $\phi_x = 0$, hence $\delta = -\phi_y$. This yields a system of three equations with three unknowns that can easily be solved. Using symmetrical geometric shapes results in closed form numbers for the voltages as

Table 12.2 Mapping rule lookup table for $N_1 = 3$

Interleaver output	S_1	S_2	S_3	δ	a_x	a_y
000	$1/\sqrt{3}$	$1/\sqrt{3}$	$1/\sqrt{3}$	$\pi/4$	$\sqrt{\frac{1}{2}\left(1+\frac{1}{\sqrt{3}}\right)}$	$\sqrt{\frac{1}{2}\left(1-\frac{1}{\sqrt{3}}\right)}$
001	$1/\sqrt{3}$	$1/\sqrt{3}$	$1/\sqrt{3}$	$-\pi/4$	$\sqrt{\frac{1}{2}\left(1+\frac{1}{\sqrt{3}}\right)}$	$\sqrt{\frac{1}{2}\left(1-\frac{1}{\sqrt{3}}\right)}$
⋮						
111	$-1/\sqrt{3}$	$-1/\sqrt{3}$	$-1/\sqrt{3}$	$-3\pi/4$	$\sqrt{\frac{1}{2}\left(1+\frac{1}{\sqrt{3}}\right)}$	$\sqrt{\frac{1}{2}\left(1+\frac{1}{\sqrt{3}}\right)}$

Table 12.3 Mapping rule lookup table for the H-SAPP-20 scenario. Group N_1 shows the mapping rule for 16-HAPP and N_2 that for 4-HAPP

Group	Interleaver output	s_1	s_2	s_3
N_1	0000	$1/\sqrt{3}$	$1/\sqrt{3}$	$1/\sqrt{3}$
	⋮			
	1111	$-d/\sqrt{3}$	0	$-1/\sqrt{3}d$*
N_2	00	0	$-1/\sqrt{3}d$	$d/\sqrt{3}$
	⋮			
	11	$-1/\sqrt{3}d$	$d/\sqrt{3}$	0

*d is the golden ratio: $(1+\sqrt{5})/2$

shown in Table 12.2. Table 12.2 on the one hand, is the LUT for 8-HAPP. The constellation forms a cube inscribed inside the Poincaré sphere as $2^3 = 8$. Table 12.3, on the other hand, shows the LUT for the 20-H-SAPP with a constellation of a dodecahedron. This configuration utilizes two subcarriers; the first subcarrier is used to modulate the points on 16 out of the 20 dodecahedron vertices, and the other subcarrier is used for the remaining 4 vertices. The selection of vertices for a subcarrier is done to maximize the distance between the points on the same subcarrier. In the table, the top part corresponds to 16-HAPP ($N_1 = 4$), and the bottom portion corresponds to 4-HAPP ($N_2 = 2$). The constellation for the resulting two-subcarrier modulation 20-H-SAPP is shown in Fig. 12.14. This Figure shows the case for which (a) 8-HAPP and (b) 20-H-SAPP, where different point color/shape represents a different subcarrier. Another option would be to map coordinates from Tables 12.2 and 12.3 directly to I- and Q-channels in x-polarization and I-channel of y-polarization.

Figure 12.13d shows the block diagram of the HAPP receiver. The signal from fiber is passed into two coherent detectors then to four branches, which contain all the information needed for the amplitudes and phases for both polarizations. This receiver configuration is essentially the same as conventional polarization multiplexing receiver. The output of each branch is demodulated by the subcarrier specified for the corresponding HAPP receiver, then sampled at the symbol rate then forwarded to the demapper and the multi-level Bahl, Cocke, Jelinek, Raviv algorithm-based equalizer (BCJR equalizer), described in the next section. The output of the equalizer is then forwarded to the bit LLRs calculator, which provides the LLRs required for the LDPC decoding process. The LDPC decoder forwards the extrinsic LLRs to the BCJR equalizer, and the extrinsic information is iterated back and forth between the decoder and the equalizer until convergence is achieved unless the predefined maximum number of iterations is reached. This process is denoted by *outer iterations*, as opposed to the *inner iterations* within the LDPC decoder itself. The outer iterations help in reducing the BER at the input of the LDPC decoder so as it can efficiently decode the data within a small predefined number of inner iterations, without increasing the complexity of the system.

This scheme is tested using VPITransmisionMaker [52], for a symbol rate of 50 GS/s, for 20 iterations of SPA for the LDPC decoder, and three outer iterations between the LDPC decoder and the multi-level BCJR equalizer. The simulations are done assuming an ASE-dominated channel scenario, and using an optical preamplifier, for both, a pseudo random bit sequence (PRBS) and an LDPC-coded bit sequence. The coded bit sequence uses LDPC(16935, 13550) code of rate 0.8, which yields an actual effective information rate of the system of $3 \times 50 \times 0.8 = 120\,\mathrm{Gb/s}$, 160 Gb/s and 240 Gb/s for 8-HAPP, 16-HAPP, and 20-H-SAPP, respectively. Utilizing higher rate codes allows a higher actual transmission rate.

The results of these simulations are summarized in Fig. 12.15. We show the uncoded and coded BER performance versus the optical signal-to-noise ratio (OSNR) per information bit.

As noticed from the figure, for the ASE-dominated scenario, the 8-HAPP scheme outperforms its QAM counterpart by 2 dB, while outperforms the PSK counterpart by 4 dB at BER of 10^{-6}. Moreover, the 16-HAPP outperforms its QAM counterpart

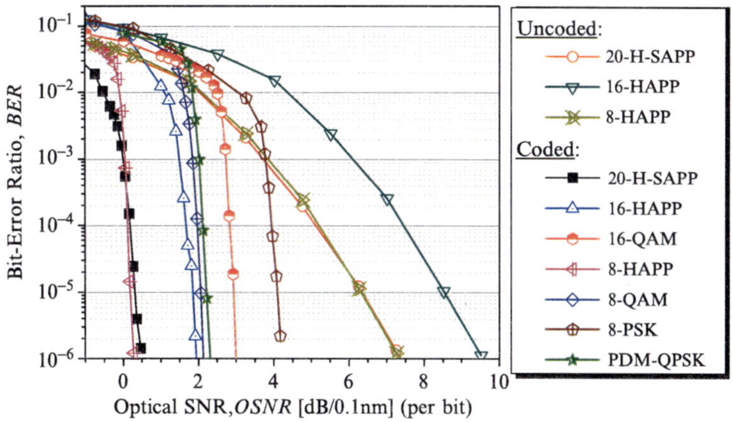

Fig. 12.15 BER performance versus the OSNR per bit for both uncoded and LDPC coded data (After ref. [88]; @ IEEE 2010; reprinted with permission.)

by 1.1 dB and the polarization division multiplexed quadrature phase shift keying (PDM-QPSK), which transmits a total of 4 bits/symbol, and exploits both polarization, by 0.5 dB at BER of 10^{-6}. However the proposed scheme of H-SAPP that utilizes the 3D-space more efficiently increases the aggregate transmission rate by 80 Gb/s in comparison with 16-HAPP, and improves the performance by 1.75 dB at BER of 10^{-6}. Furthermore, 20-H-SAPP doubles the aggregate transmission rate of 8-HAPP while keeping the BER performance of the system almost intact. On the other hand, utilizing M subcarriers requires M times the bandwidth of the HAPP system. To this end, a better utilization of the bandwidth can be achieved by employing larger constellation HAPP subsystems into the H-SAPP such as employing three 8-HAPPs for a 24-H-SAPP, rather than using two 4-HAPPs and a 16-HAPP and so on. For other multidimensional coded modulation schemes an interested reader is referred to refs. [89–91]. The three-dimensional coded modulation scheme is described here since the improvement with respect to two-dimensional schemes (QAM and M-PSK) is largest when moving from two-dimensional to three-dimensional space.

12.5 LDPC-Coded Turbo Equalization

In this section, we describe an LDPC-coded turbo equalization scheme [2, 4, 5], as a universal scheme that can be used simultaneously for: (1) suppression of fiber nonlinearities, (2) PMD compensation, and (3) chromatic dispersion compensation in multilevel coded-modulation schemes.

12.5.1 *Optimum Detection*

Before we describe the LDPC-coded turbo equalization, we provide the basic concepts of optimum detection of binary signaling in minimum probability of error sense [53, 54]. Let $\boldsymbol{x}$ denote the transmitted sequence and $\boldsymbol{y}$ the received sequence. The optimum receiver assigns $\hat{x}_k$ to the value $x \in \{0, 1\}$ that maximizes the APP $P(x_k = x|\boldsymbol{y})$ given the received sequence $\boldsymbol{y}$

$$\hat{x}_k = \underset{x\in\{0,1\}}{\arg\max}\, P\left(x_k = x|\boldsymbol{y}\right). \tag{12.45}$$

The corresponding algorithm is commonly referred to as a MAP algorithm. In practice, it is common to use the logarithmic version of equation (12.45) as follows:

$$\hat{x}_k = \begin{cases} 0, & L\left(x_k|\boldsymbol{y}\right) \geq 0, \\ 1, & \text{otherwise}, \end{cases} \qquad L\left(x_k|\boldsymbol{y}\right) = \log\left[\frac{P\left(x_k = 0|\boldsymbol{y}\right)}{P\left(x_k = 1|\boldsymbol{y}\right)}\right], \tag{12.46}$$

where $L(x_k|y)$ is the conditional LLR. To calculate the $P(x_k = x|y)$ needed in either equation above, we invoke the Bayes' rule:

$$P(x_k = x|y) = \sum_{\forall x: x_k = x} P(x|y) = \sum_{\forall x: x_k = x} \frac{P(y|x)\,P(x)}{P(y)}, \tag{12.47}$$

where $P(y|x)$ is conditional probability density function (PDF), and $P(x)$ is the a priori probability of input sequence x, which when the symbols are independent factors as $P(x) = \prod_{i=1}^{n} P(x_i)$, where n is the codeword length. By substituting (12.47) into (12.46), the conditional LLR can be written as:

$$L(x_k|y) = \log\left[\frac{\sum_{\forall x: x_k = 0} p(y|x) \prod_{i=1}^{n} P(x_i)}{\sum_{\forall x: x_k = 1} p(y|x) \prod_{i=1}^{n} P(x_i)}\right] = L_{\text{ext}}(x_k|y) + L(x_k), \tag{12.48a}$$

where the extrinsic information about x_k contained in y $L_{\text{ext}}(x_k|y)$ and the a priori LLR $L(x_k)$ are defined respectively as

$$L_{\text{ext}}(x_k|y) = \log\left[\frac{\sum_{\forall x: x_k = 0} p(y|x) \prod_{i=1, i\neq k}^{n} P(x_i)}{\sum_{\forall x: x_k = 1} p(y|x) \prod_{i=1, i\neq k}^{n} P(x_i)}\right],$$

$$L(x_k) = \log\left[\frac{P(x_k = 0)}{P(x_k = 1)}\right]. \tag{12.48b}$$

From (12.48) is clear that computation of conditional LLRs can be computationally extensive. One possible computation is based on BCJR algorithm [55], with log-domain version described below.

12.5.2 Multilevel Turbo Equalizer Description

The LDPC-coded turbo equalizer is composed of two ingredients: (1) the multilevel BCJR algorithm [2, 4, 5, 29, 55]-based equalizer, and (2) the LDPC decoder. The transmitter configuration, for MLC, is already explained previously (see Fig. 12.9a). The receiver configuration of LDPC-coded trubo equalizer is shown in Fig. 12.16. The outputs of upper- and lower-balanced branches, proportional to $\text{Re}\{S_i L^*\}$ and $\text{Im}\{S_i L^*\}$, respectively, are used as inputs of multilevel BCJR equalizer, where the local laser electrical field is denoted by $L = |L|\exp(\mathrm{j}\varphi_{\mathrm{L}})$ (φ_{L} denotes the laser phase noise process of the local laser) and incoming optical signal at time instance i with S_i.

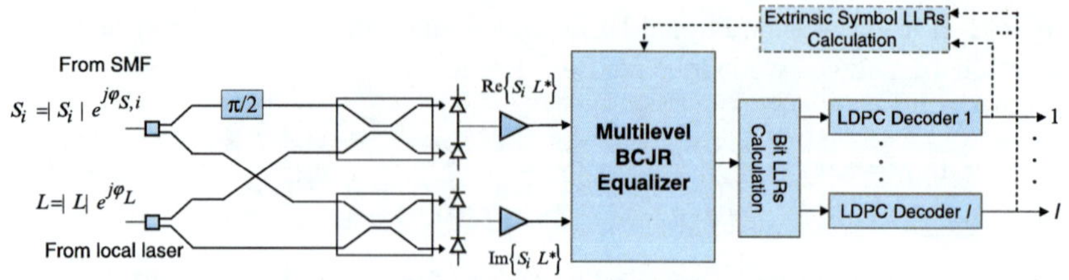

Fig. 12.16 LDPC-coded turbo equalization scheme configuration (After ref. [63]; @ IEEE 2009; reprinted with permission.)

The multilevel BCJR equalizer operates on a discrete dynamical trellis description of the optical channel. Note that this equalizer is universal and applicable to any two-dimensional signal constellation, such as M-ary PSK, M-ary QAM, or M-ary polarization-shift keying (PolSK), and both coherent and direct detections. This dynamical trellis is uniquely defined by the following triplet: the previous state, the next state, and the channel output. The state in the trellis is defined as $\boldsymbol{s}_j = (x_{j-m}, x_{j-m+1}, .., x_j, x_{j+1}, \ldots, x_{j+m}) = \boldsymbol{x}[j-m, j+m]$, where x_k denotes the index of the symbol from the following set of possible indices $\boldsymbol{X} = \{0, 1, \ldots, M-1\}$, with M being the number of points in corresponding M-ary signal constellation. Every symbol carries $l = \log_2 M$ bits, using the appropriate mapping rule (natural, Gray, anti-Gray, etc.) The memory of the state is equal to $2m+1$, with $2m$ being the number of *symbols* that influence the observed symbol from both sides. An example trellis of memory $2m+1 = 3$ for 4-ary modulation formats (such as QPSK) is shown in Fig. 12.2b. The trellis has $M^{2m+1} = 64$ states (s_0, $s_1, \ldots, s_{63}$), each of which corresponds to the different 3-symbol patterns (symbol-configurations).

The state index is determined by considering $(2m+1)$ symbols as digits in numerical system with the base M. For example, in Fig. 12.2b, the quaternary numerical system (with the base 4) is used. The left column in dynamic trellis represents the current states and the right column denotes the terminal states. The branches are labeled by two symbols, the input symbol is the last symbol in initial state (the blue symbol), the output symbol is the central symbol of terminal state (the red symbol). Therefore, the current symbol is affected by both previous and incoming symbols. For the complete description of the dynamical trellis, the transition PDFs $p(\boldsymbol{y}_j|x_j) = p(\boldsymbol{y}_j|\boldsymbol{s})$, $\boldsymbol{s} \in \boldsymbol{S}$ are needed; where $\boldsymbol{S}$ is the set of states in the trellis, and $\boldsymbol{y}_\mathrm{j}$ is the is the vector of samples (corresponding to the transmitted symbol index x_j). The conditional PDFs can be determined from *collected histograms* or by using *instanton–Edgeworth expansion* method [56]. The number of edges originating in any of the left-column states is M, and the number of merging edges in arbitrary terminal state is also M.

The *forward metric* is defined as $\alpha_j(\boldsymbol{s}) = \log\{p(\boldsymbol{s}_j = \boldsymbol{s}, \boldsymbol{y}[1, j])\} (j = 1, 2, \ldots, n)$; the *backward metric* is defined as $\beta_j(\boldsymbol{s}) = \log\{p(\boldsymbol{y}[j+1, n]|\boldsymbol{s}_j = \boldsymbol{s})\}$;

and the *branch metric* is defined as $\gamma_j(s', s) = \log[p(s_j = s, y_j, s_{j-1} = s')]$. The corresponding metrics can be calculated iteratively as follows

$$\alpha_j(s) = \max_{s'}{}^{*} \left[\alpha_{j-1}(s') + \gamma_j(s', s)\right], \tag{12.49a}$$

$$\beta_{j-1}(s') = \max_{s}{}^{*} \left[\beta_j(s) + \gamma_j(s', s)\right], \tag{12.49b}$$

$$\gamma_j(s', s) = \log\left[p(y_j | \boldsymbol{x}[j-m, j+m]) P(x_j)\right]. \tag{12.49c}$$

The max*-operator is defined by $\max^*(x, y) = \log(e^x + e^y)$, and it is efficiently calculated by $\max^*(x, y) = \max(x, y) + c_f(x, y)$, where $c_f(x, y)$ is the correction factor, defined as $c_f(x, y) = \log[1 + \exp(-|x - y|)]$, which is commonly approximated or implemented using a look-up table. $p(y_j | \boldsymbol{x}[j-m, j+m])$ is obtained, as already explained above, either by collecting the histograms or by instanton–Edgeworth expansion method, and $P(x_j)$ represents a priori probability of transmitted symbol x_j. In the first outer iteration, $P(x_j)$ is set to either $1/M$ (because equally probable transmission is observed) for an existing transition from trellis given in Fig. 12.2b, or to zero for a nonexisting transition. The outer iteration is defined as the calculation of symbol LLRs in multilevel BCJR equalizer block, the calculation of corresponding bit LLRs needed for LDPC decoding, the LDPC decoding, and the calculation of extrinsic symbol LLRs needed for the next iteration. The iterations within LDPC decoder, based on min-sum-with-correction-term algorithm [41,57], are called here inner iterations. The initial forward and backward metrics values are set to

$$\alpha_0(s) = \begin{cases} 0, & s = s_0 \\ -\infty, & s \neq s_0 \end{cases} \quad \text{and} \quad \beta_n(s) = \begin{cases} 0, & s = s_0, \\ -\infty, & s \neq s_0, \end{cases} \tag{12.50}$$

where s_0 is an initial state. Let $s' = \boldsymbol{x}[j-m-1, j+m-1]$ represent the previous state, $s = \boldsymbol{x}[j-m, j+m]$ the present state, $\boldsymbol{x} = (x_1, x_2, \ldots, x_n)$ – the transmitted word of symbols, and $\boldsymbol{y} = (y_1, y_2, \ldots, y_n)$ – the received sequence of samples. The LLR, denoting the reliability, of symbol $x_j = \delta (j = 1, 2, \ldots, n)$ can be calculated by

$$\Lambda\left(x_j = \delta\right) = \max_{(s',s):x_j=\delta}^{*} \left[\alpha_{j-1}\left(s'\right) + \gamma_j\left(s', s\right) + \beta_j\left(s\right)\right] - \max_{(s',s):x_j=\delta_0}^{*} \left[\alpha_{j-1}\left(s'\right) + \gamma_j\left(s', s\right) + \beta_j\left(s\right)\right], \tag{12.51}$$

where δ represents the observed symbol ($\delta \in \{0, 1, \ldots, M-1\} \backslash \{\delta_0\}$), and δ_0 is the referent symbol. The forward and backward metrics are calculated using the (12.49a) and (12.49c). The forward and backward recursion steps of 4-level BCJR MAP detector are illustrated in Fig. 12.17a, b, respectively. In Fig. 12.17a, s denotes an arbitrary terminal state, which has $M = 4$ edges originating from corresponding initial states, denoted as s'_1, s'_2, s'_3, and s'_4. Note that the first term in branch metric

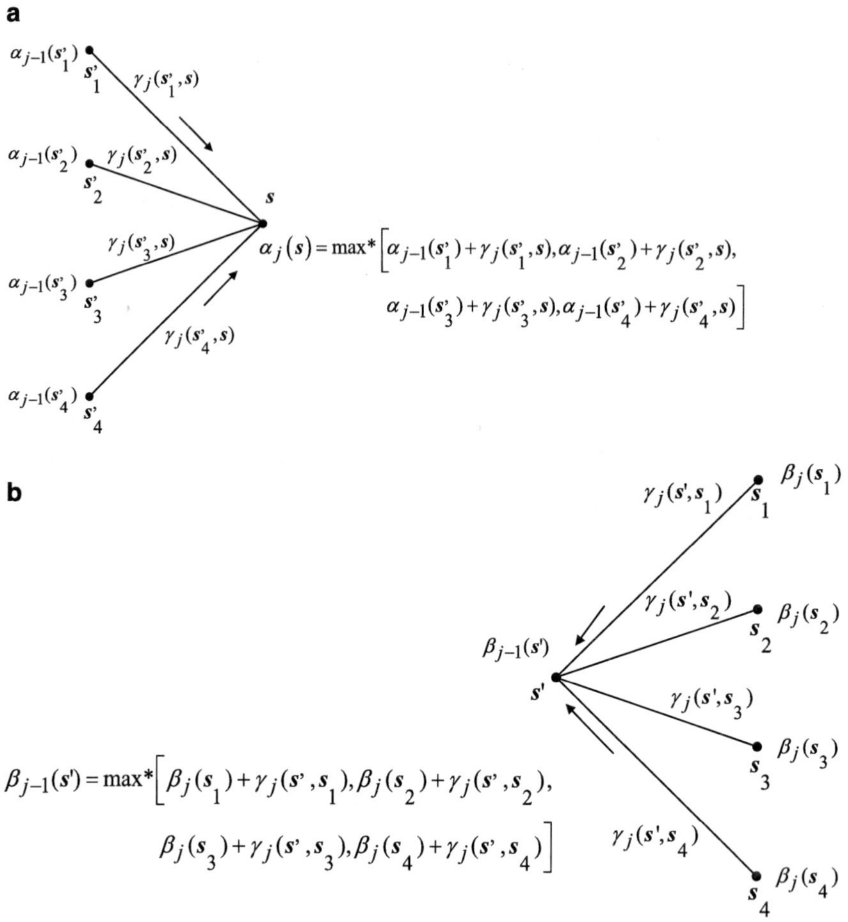

Fig. 12.17 Forward/backward recursion steps for $M = 4$-level BCJR equalizer: (**a**) the forward recursion step, and (**b**) the backward recursion step (After ref. [63]; @ IEEE 2009; reprinted with permission.)

is calculated only once, before the detection/decoding takes place, and stored. The second term, $\log(P(x_j))$, is recalculated in every outer iteration. The forward metric of state s in jth step ($j = 1, 2, \ldots, n$) is updated by preserving the maximum term (in max*-sense) $\alpha_{j-1}(s'_k) + \gamma_j(s, s'_k)$ ($k = 1, 2, 3, 4$). The procedure is repeated for every state in column of terminal states of jth step. The similar procedure is used to calculate the backward metric of state s', $\beta_{j-1}(s')$, (in (j-1)th step), as shown in Fig. 12.17b, but now proceeding in backward direction ($j = n, n\text{-}1, \ldots, 1$).

We further calculate bit LLRs from symbol LLRs in fashion to that we describe in Sect. 12.4.1. To improve the overall performance of LDPC-coded turbo equalizer, we perform the iteration of *extrinsic* LLRs between LDPC decoder and multilevel BCJR equalizer.

12.5.3 Performance of LDPC-Coded Turbo Equalizer

As an illustration of the potential of the proposed scheme, the BER performance of an LDPC-coded turbo equalizer is given in Fig. 12.18 for the dispersion map shown in Fig. 12.19 (launch power of 0 dBm and single channel transmission). EDFAs with a noise figure of 5 dB are deployed after every fiber section. The bandwidth of the

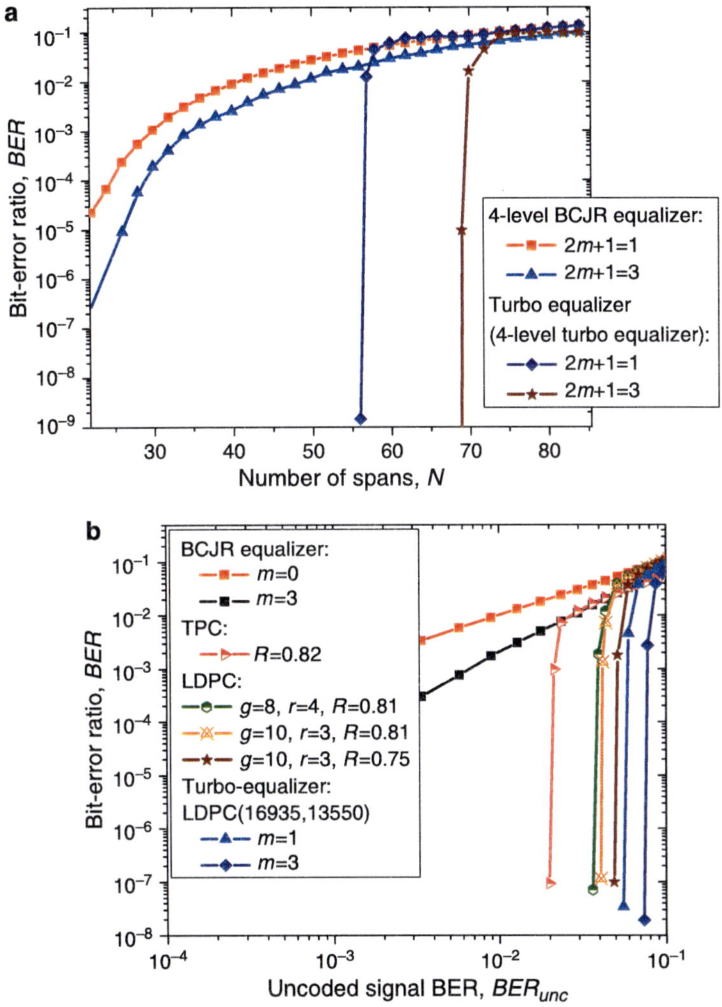

Fig. 12.18 BER performance of LDPC-coded turbo equalizer in the presence of fiber nonlinearities for: (**a**) QPSK modulation format with aggregate data rate of 100 Gb/s, and (**b**) RZ-OOK modulation format at 40 Gb/s. For both simulations, dispersion map shown in Fig. 12.19 is used (After ref. [63]; @ IEEE 2009; and after ref. [5]; @ IEEE 2008; reprinted with permission.)

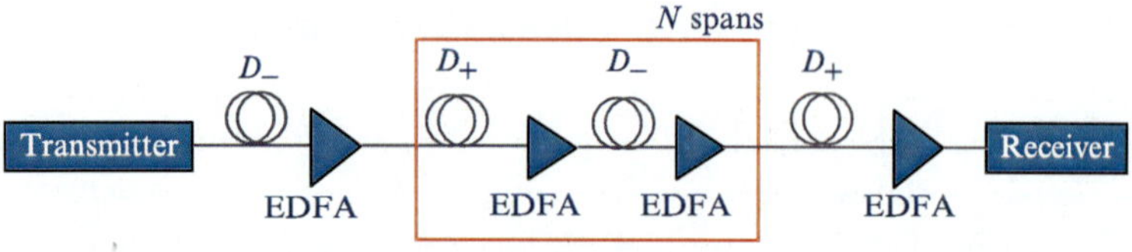

Fig. 12.19 Dispersion map under study is composed of N spans of length $L = 120$ km, consisting of 2 L/3 km of D_+ fiber followed by L/3 km of D_- fiber, with pre-compensation of $-1{,}600$ ps/nm and corresponding post-compensation. The fiber parameters are given in Table 12.4

optical filter is set to $3R_l$ and that of the electrical filter is set to $0.7R_l$, where $R_l = R_s/\mathrm{R}$ with R_s being the symbol rate and R being the code rate (0.8). In Fig. 12.18a, we present simulation results for QPSK transmission at the symbol rate of 50 Giga symbols/s. The symbol rate is appropriately chosen so that the effective aggregate information rate is 100 Gb/s. With polarization-multiplexing the aggregate data rate can be increased to 200 Gb/s per wavelength. The figure depicts the uncoded BER and the BER after iterative decoding with respect to the number of spans, which was varied from 4 to 84. The propagation was modeled by solving the nonlinear Schrödinger equation using the split-step Fourier method. It can be seen from Fig. 12.18a that when a 4-level BCJR equalizer of state memory $2m + 1 = 1$ and an LDPC(16935, 13550) code of girth-10 and column weight 3 are used, we can achieve QPSK transmission at the symbol rate of 50 Giga symbols/s over 55 spans (6,600 km) with a BER below 10^{-9}. However, for the turbo equalization scheme based on a 4-level BCJR equalizer of state memory $2m + 1 = 3$ (see Fig. 12.18a) and the same LDPC code, we are able to achieve even 8,160 km at the symbol rate of 50 Giga symbols/s with a BER below 10^{-9}. Note that in both cases the BCJR equalizer trellis detection depth was equal to the codeword length. The BER performance comparison of LDPC-coded TE against large-girth LDPC codes and TPCs for RZ-OOK system operating at 40 Gb/s (in effective information rate) is given in Fig. 12.18b, for different trellis memories. LDPC-coded TE with state memory $2m + 1 = 7$ provides almost 12 dB improvement over the BCJR equalizer with state memory of $m = 0$ at BER of 10^{-8}.

In order to apply the proposed multilevel turbo equalizations scheme to real 100 Gb/s systems, the practical circuit implementation study would be mandatory. It is evident from Fig. 12.2b that complexity of dynamic trellis grows exponentially, because the number of states is determined by M^{2m+1}, so that the increase in signal constellation leads to increase of the base, while the increase in channel memory assumption $(2m + 1)$ leads to the increase of exponent. We have shown in the case of QPSK transmission (see Fig. 12.18a), that even small state memory assumption $(2m + 1 = 3)$ leads to significant performance improvement with respect to the state memory $m = 0$. For larger constellations and/or larger memories, the reduced complexity BCJR algorithm is to be used instead. For example, instead of detection of sequence of symbols corresponding to the length of codeword n, we can observe shorter sequences. Further, we do not need to memorize all branch metrics but several largest ones. In forward/backward metrics' update, we need to update only the

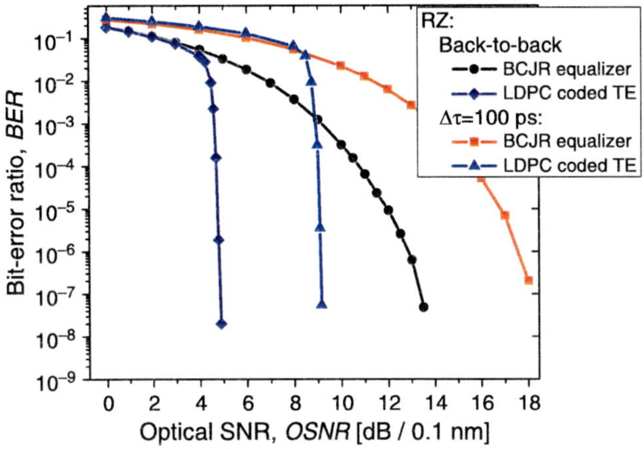

Fig. 12.20 BER performance of LDPC(16935,13550)-coded PMD TE with trellis memory $2m + 1 = 7$ (After ref. [5]; @ IEEE 2008; reprinted with permission.)

metrics of those states connected to the edges with dominant branch metrics, and so on. Moreover, when $\max^*(x, y) = \max(x, y) + \log[1 + \exp(-|x - y|)]$ operation, required in forward and backward recursion steps, is approximated by $\max(x, y)$ operation, the forward and backward BCJR steps become the forward and backward Viterbi algorithms, respectively.

The nonlinear ISI turbo equalizer described above can also be used as a PMD compensator. The results of simulations, for 10 Gb/s transmission and ASE noise dominated scenario, are shown in Fig. 12.20 for DGD $\Delta\tau = 100$ ps and girth-10 LDPC code of rate 0.81. RZ-OOK of a duty cycle of 33% is observed. The bandwidth of super-Gaussian optical filter is set to $3R_l$, and the bandwidth of Gaussian electrical filter to $0.7R_l$, with R_l being the line rate. For DGD of 100 ps, the $R = 0.81$ LDPC-coded turbo equalizer (for trellis memory $2m+1 = 7$) has penalty of only 2 dB with respect to the back-to-back configuration.

In the rest of this section, we turn our attention to the experimental verification. The experimental setup for PMD compensation study by LDPC-coded turbo equalization is shown in Fig. 12.21a, and corresponding results are shown in Fig. 12.21b.

The LDPC-encoded sequence is uploaded into Anritsu pattern generator via GPIB card controlled by a PC. A zero-chirp MZM is used to generate the NRZ data stream. The launch power is maintained at 0 dBm at the input of PMD emulator (with equal power distribution between states of polarization). The output of PMD emulator is combined with an ASE source immediately prior to the preamplifier. The ASE noise power is controlled by variable optical attenuator (VOA) in order to provide an independent OSNR adjustment at the receiver. A standard pre-amplified PIN receiver is used for direct detection and is preceded by another VOA to maintain a constant received power of −6 dBm. The sampling oscilloscope (Agilent), triggered by the data pattern, is used to acquire the received sequences, downloaded via GPIB card back to the PC, which serves as an LDPC-coded turbo equalizer.

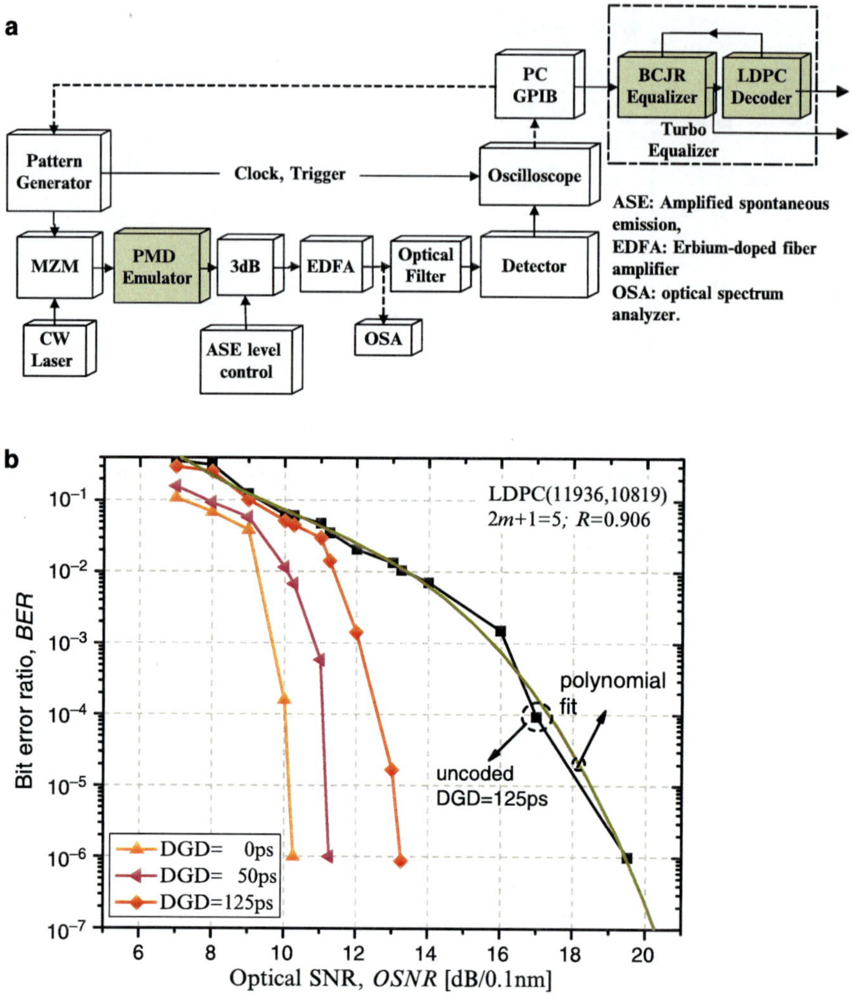

Fig. 12.21 (**a**) Experimental setup for PMD compensation study by LDPC-coded turbo equalization, and (**b**) BER performance of the PMD compensator (After ref. [5]; @ IEEE 2008; reprinted with permission.)

The experimental results for 10 Giga symbols/s (effective information rate) NRZ transmission are shown in Fig. 12.21b, for different DGD values. The TE is based quasi-cyclic LDPC(11936,10819) code of code rate 0.906 and girth-10, with 5 outer and 25 sum–product decoding algorithm iterations. The OSNR penalty for DGD of 125 ps is about 3 dB at BER $= 10^{-6}$, while the coding gain improvement over BCJR equalizer (with memory $2m+1 = 5$) for DGD $=$ 125 ps is 6.25 dB at BER $= 10^{-6}$. Larger coding gains are expected at lower BERs.

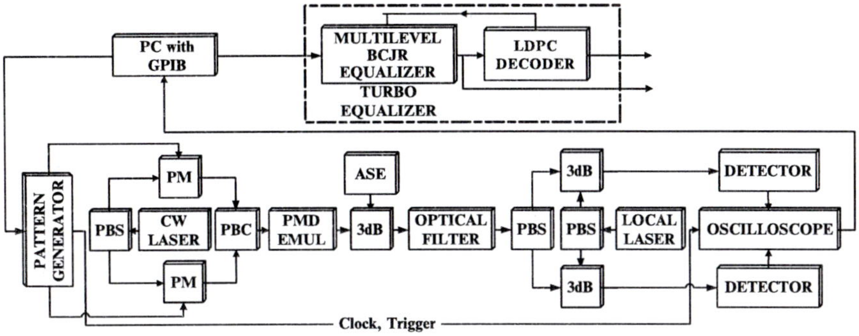

Fig. 12.22 Experimental setup for polarization multiplexed BPSK study. *CW Laser* continuous wave laser, *PM* phase modulator, *ASE* amplified spontaneous emission noise source, *3dB* 3 dB coupler (After ref. [63]; @ IEEE 2009; reprinted with permission.)

Figure 12.22 shows the experimental setup for PMD compensation study in polarization multiplexed schemes with coherent detection. In this example, we jointly perform detection and decoding of symbols transmitted in two orthogonal polarizations. The two orthogonal polarizations of a continuous wave laser source are separated by a polarization beam splitter and are modulated by two-phase modulators (Covega) driven at 10 Gb/s (Anritsu MP1763C). (The symbol rate was determined by available equipment.)

A pre-coded test pattern was loaded into the pattern generator via personal computer with GPIB interface. A polarization beam combiner was used to combine the two modulated signals, followed by a PMD emulator (JDSU PE3), which introduced controlled amount of DGD to the signal. Then the signal distorted by PMD was mixed with controlled amount of ASE noise with 3 dB coupler. Modulated signal level was maintained at 0 dB, while the ASE power level was changed to obtain different OSNRs. Next, the optical signal was pre-amplified, filtered (JSDU 2nm band-pass filter), and coherently detected. The coherent detection is performed by mixing the received signal with signal from local laser with 3 dB coupler. The resulting signal is detected with a detector (Agilent 11982A) and an oscilloscope (Agilent DCA 86105A), triggered by the data pattern that was used to acquire the samples. To maintain constant power of −6 dBm at the detector, a variable attenuator was used. Data was transferred via GPIB back to the PC. The PC also served as a multilevel turbo equalizer with offline processing. To avoid any imbalance of two independent symbols transmitted in two polarizations, we detect the both symbols simultaneously. Because the symbols transmitted in both polarizations are considered as one super-symbol, the BER performance of turbo equalizer is independent on power splitting ratio between principle states of polarization.

The experimental results for BER performance of the proposed multilevel turbo equalizer are summarized in Fig. 12.23. For the experiment, a quasi-cyclic LDPC(16935, 13550) code of girth 10 and column weight 3 was used as channel code. The number of extrinsic iterations between LDPC decoder and BCJR

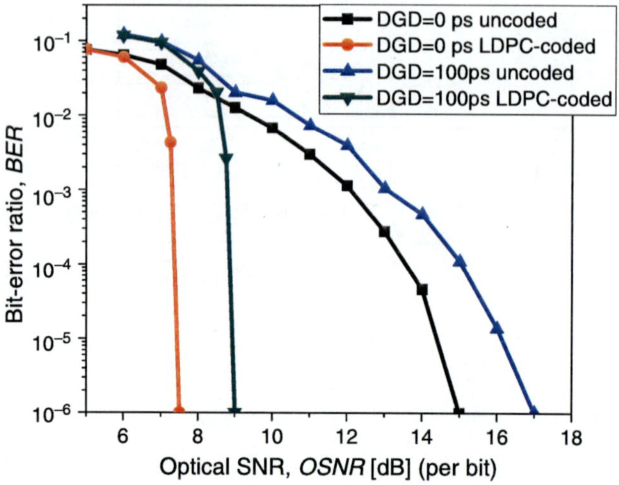

Fig. 12.23 BER performance of multilevel turbo equalizer for PMD compensation (After ref. [63]; @ IEEE 2009; reprinted with permission.)

equalizer was set to 3, and the number of the intrinsic LDPC decoder iterations was set to 25. The state memory of $2m + 1 = 3$ was sufficient for the compensation of the first-order PMD with DGD of 100 ps. The OSNR penalty for 100 ps of DGD is 1.5 dB at BER of 10^{-6}. Coding gain for DGD of 0 ps is 7.5 dB at BER of 10^{-6}, and the coding gain for DGD of 100 ps is 8 dB.

12.5.4 Multilevel Turbo Equalizer with Digital Backpropagation

The LDPC-coded turbo equalizer described in previous two sections, as an excellent equalizer to deal simultaneously with both linear and nonlinear fiber impairments. However, the complexity of this equalizer grows exponentially as either channel memory or constellation size increase. To solve this problem, we recently proposed the use of coarse digital backpropagation (with reasonable small number of coefficients) to reduce the required channel memory and to compensate for remaining channel impairments by turbo equalization scheme [4, 5, 58, 63], which is shown in Fig. 12.24. The $m_x + m_y$ (index x (y) corresponds to x-(y-) polarization) independent data streams are encoded using different LDPC codes of code rates $R_i = K_i/N$ ($i \in \{x, y\}$) where K_x (K_y) denotes the number of information bits used in the binary LDPC code corresponding to x- (y-) polarization, and N denotes the codeword length, which is the same for both LDPC codes. The m_x (m_y) input bit streams from m_x (m_y) different information sources, pass through identical LDPC encoders that use large-girth quasi-cyclic LDPC codes described in Section 12.3.1.1 with code rate R_x (R_y). The outputs of the encoders are then

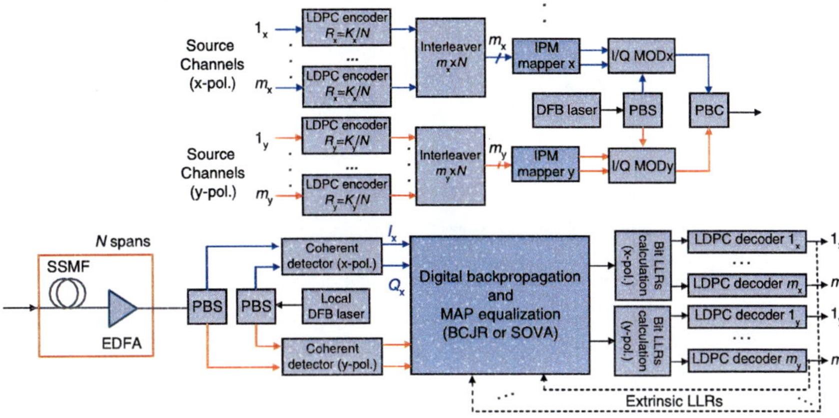

Fig. 12.24 LDPC-coded PM-IPM scheme. *PBS/C* polarization beam splitter/combiner, *MAP* maximum *a posteriori* probability, *LLRs* log-likelihood ratios, *IPM* iterative polar modulation (see [58]) (After ref. [58]; @ IEEE 2010; reprinted with permission.)

bit-interleaved by an $m_x \times N$ $(m_y \times N)$ bit-interleaver, where the sequences are written row-wise and read column-wise. The output of the interleaver is sent in one bit-stream, $m_x(m_y)$ bits at a time instant i, to a mapper. The mapper maps each $m_x(m_y)$ bits into a 2^{m_x}-ary (2^{m_y}-ary) IPM signal constellation point based on a lookup table, as explained above. The IPM mapper x(y) constellation point $s_{i,x} = (I_{i,x}, Q_{i,x}) = |s_{i,x}| \exp(j\phi_{i,x})[s_{i,y} = (I_{i,y}, Q_{i,y}) = |s_{i,y}| \exp(j\phi_{i,y})]$ coordinates are used as the inputs of an I/Q MOD_x (I/Q MOD_y), as shown in Fig. 12.24. At the receiver side, the outputs at I- and Q-branches in two polarizations, are sampled at the symbol rate, while the symbol LLRs are calculated as follows $\lambda(\boldsymbol{s}) = \log[P(\boldsymbol{s}|\boldsymbol{r})/P(\boldsymbol{s}_0|\boldsymbol{r})]$, where $\boldsymbol{s} = (I_i, Q_i)$ and $\boldsymbol{r} = (r_I, r_Q)$ denote the transmitted signal constellation point and received symbol at time instance i (in either x- or y- polarization), respectively, and $\boldsymbol{s}_0$ represents the reference symbol. To reduce the channel memory, while keeping complexity reasonably low, we use the coarse digital backpropagation with small number of coefficients. To compensate for remaining channel distortion, we employ the LDPC-coded turbo equalization. The bit reliabilities for LDPC decoders are calculated from symbol reliabilities, as we described previously. To improve BER performance, we use the EXIT chart analysis and iterate extrinsic reliabilities between MAP equalizer and LDPC decoders in turbo equalization fashion, until convergence or predetermined number of iterations has been reached.

In Fig. 12.25, we report the BER results obtained for polarization-multiplexed (8547, 6922, 0.81)-coded modulation schemes with digital backpropagation and turbo equalization (for three outer MAP equalizer-LDPC decoders iterations and 25 LDPC decoder inner iterations), for symbol rate $R_s = 50\,\text{GS/s}$ and launch power $P = 0\,\text{dBm}$. The dispersion map was composed of standard SMF (SSMF) only with EDFAs of noise figure $NF = 5\,\text{dB}$ being deployed every 100 km. We

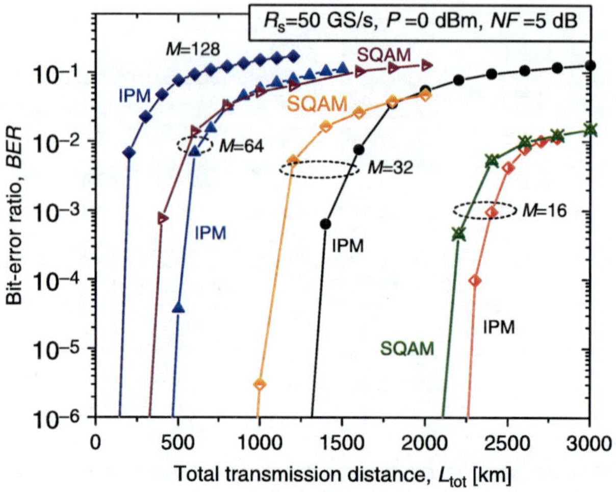

Fig. 12.25 BER versus total transmission distance (L_{tot}) (After ref. [58]; @ IEEE 2010; reprinted with permission.)

see that coded IPM we introduced in [58], outperforms SQAM for different signal constellation sizes and allows longer transmission distances. The total transmission distance for different signal constellation sizes is found to be: 2,250 km for $M = 16$ (aggregate rate $R_D = 400$ Gb/s), 1,320 km for $M = 32$ ($R_D = 500$ Gb/s), 460 km for $M = 64$ ($R_D = 600$ Gb/s), and 140 km for $M = 128$ ($R_D = 700$ Gb/s).

12.6 Information Capacity of Fiber-Optics Systems

There have been numerous attempts to determine the channel capacity of a nonlinear fiber-optics communication channel [71–83]. The main approach, until recently, was to consider ASE noise as a predominant effect and to observe the fiber nonlinearities as the perturbation of linear case or as the multiplicative noise. In this section, we describe how to determine the true fiber-optics channel capacity. Because in most of practical applications the channel input distribution is uniform, we also describe how to determine the uniform information capacity, which represents the lower bound on channel capacity. This method consists of two steps: (1) approximating PDFs for energy of pulses, which is done by one of the following approaches: (a) evaluation of histograms [79], (b) instanton approach [81] or (c) Edgeworth expansion [56], and (2) estimating information capacities by applying a method originally proposed by Arnold and Pfitser [70, 84, 85].

12.6.1 Channel Capacity of Channels with Memory

Let the input and output alphabets of the optical channel be finite and be denoted by $\{A\}$ and $\{B\}$, respectively; and the channel input and output denoted by X and Y. For memoryless channels, the noise behavior is generally captured by a conditional probability matrix $P\{b_j|a_k\}$ for all $b_j \in B$ and $a_j \in A$. For the channels with finite memory, such as the optical channel, the transition probability is dependent on the transmitted sequences up to the certain prior finite instance of time. For example, for channel described by Markov process the transition matrix has the following form $P\{Y_k = b|\ldots, X_{-1}, X_0, X_1, \ldots, X_k\} = P\{Y_k = b|X_k\}$. We are interested into more general description, which is due to McMillan [60] and Khinchin [61] (see also Reza [65]). Let us consider a member of input ensemble x and its corresponding channel output y : $\{X\} = \{\ldots, x_{-2}, x_{-1}, x_0, x_1, \ldots\}, \{Y\} = \{\ldots, y_{-2}, y_{-1}, y_0, y_1, \ldots\}$. Let $\boldsymbol{X}$ denote all possible input sequences and $\boldsymbol{Y}$ denote all possible output sequences. By fixing a particular symbol at specific location, we obtain the so-called *cylinder* [61,65]. For example, cylinder $x^{4,1}$ is obtained by fixing the symbol a_1 at position x_4 : $x^{4,1} = \ldots, x_{-1}, x_0, x_1, x_2, x_3, a_1, x_5, \ldots$. The output cylinder $y^{1,2}$ is obtained by fixing the output symbol b_2 at position 1: $y^{1,2} = \ldots, y_{-1}, y_0, b_2, y_2, y_3, \ldots$. To characterize the channel we have to determine the following transition probability $P(y^{1,2}|x^{4,1})$, that is the probability that cylinder $y^{1,2}$ was received given that cylinder $x^{4,1}$ was transmitted. Therefore, for all possible input cylinders $S_\mathrm{A} \subset \boldsymbol{X}$, we have to determine the probability that cylinder $S_\mathrm{B} \subset \boldsymbol{Y}$ was received given that S_A was transmitted. The channel is completely specified by: (1) input alphabet A, (2) output alphabet B, and (3) transition probabilities $P\{S_B|S_A\} = \nu_x$ for all $S_\mathrm{A} \in \boldsymbol{X}$ and $S_\mathrm{B} \in \boldsymbol{Y}$. Thus, the channel is specified by the triplet: $[A, \nu_x, B]$. If the transition probabilities are invariant with respect to time shift T, that is, $\nu_{Tx}(TS) = \nu_x(S)$, then the channel is said to be *stationary*. If the distribution of Y_k depends only on the statistical properties of the sequence $\ldots, x_{k-1}, x_k$, we say that the channel is without *anticipation*. If furthermore the distribution of Y_k depends only $x_{k-m}, \ldots, x_k$, we say that channel has the finite *memory* of m units.

The source and channel may be described as a new source $[C, \omega]$ with C being the product of input A and output B alphabets, namely $C = A \times B$, and ω is a corresponding probability measure. The joint probability of symbol $(x, y) \in \mathrm{C}$, where $x \in A$ and $y \in B$, is obtained as the product of marginal and conditional probabilities:

$$P(\mathrm{x} \cap \mathrm{y}) = P\{x\}P\{\mathrm{y}|x\}.$$

Let us further assume that both source and channel are stationary. The following description due to Khinchin [61,65] is useful in describing the concatenation of a stationary source and a stationary channel.

1. If the source $[A, \mu]$ (μ is the probability measure of the source alphabet) and the channel $[A, \nu_x, B]$ are stationary, the product source $[C, \omega]$ will also be stationary.

2. Each stationary source has an entropy, and therefore $[A, \mu]$, $[B, \eta]$ (η is the probability measure of the output alphabet) and $[C, \omega]$ each has the finite entropies.
3. These entropies can be determined for all n-term sequences $x_0, x_1, \ldots, x_{n-1}$ emitted by the source and transmitted over the channel as follows [61]:

$$\begin{aligned}
&H_n(X) \leftarrow \{x_0, x_1, \ldots, x_{n-1}\}, \quad H_n(Y) \leftarrow \{y_0, y_1, \ldots, y_{n-1}\} \\
&H_n(X, Y) \leftarrow \{(x_0, y_0), (x_1, y_1), \ldots, (x_{n-1}, y_{n-1})\} \\
&H_n(Y|X) \leftarrow \{(x_0|Y), (x_1|Y), \ldots, (x_{n-1}|Y)\} \\
&H_n(X|Y) \leftarrow \{(X|y_0), (X|y_1), \ldots, (X|y_{n-1})\}
\end{aligned} \tag{12.52}$$

It can be shown that the following is valid:

$$H_n(X, Y) = H_n(X) + H_n(Y|X), \quad H_n(X, Y) = H_n(Y) + H_n(X|Y). \tag{12.53}$$

The (12.53) can be rewritten in terms of entropies per symbol:

$$\begin{aligned}
\frac{1}{n} H_n(X, Y) &= \frac{1}{n} H_n(X) + \frac{1}{n} H_n(Y|X) \\
\frac{1}{n} H_n(X, Y) &= \frac{1}{n} H_n(Y) + \frac{1}{n} H_n(X|Y).
\end{aligned} \tag{12.54}$$

For sufficiently long sequences, the following channel entropies exist:

$$\begin{aligned}
&\lim_{n\to\infty} \frac{1}{n} H_n(X, Y) = H(X, Y) \; \lim_{n\to\infty} \frac{1}{n} H_n(X) = H(X) \\
&\lim_{n\to\infty} \frac{1}{n} H_n(Y) = H(Y) \; \lim_{n\to\infty} \frac{1}{n} H_n(X|Y) = H(X|Y) \\
&\lim_{n\to\infty} \frac{1}{n} H_n(Y|X) = H(Y|X)
\end{aligned} \tag{12.55}$$

The mutual information exist and it is defined as

$$I(X, Y) = H(X) + H(Y) - H(X, Y). \tag{12.56}$$

The *stationary information capacity* of the channel is obtained by maximization of mutual information over all possible information sources:

$$C(X, Y) = \max I(X, Y). \tag{12.57}$$

Equipped with this knowledge, in the next section, we will discuss how to determine the information capacity of fiber-optics channel with memory.

12.6.2 *Calculation of Information Capacity of Multilevel Modulation Schemes by Forward Recursion of BCJR Algorithm*

Here, we address the problem of calculating of channel capacity of multilevel modulation schemes for an IID information source, in literature also known as the achievable information rate (see Djordjevic et al. [62, 63] and references therein). The IID channel capacity represents a lower bound on channel capacity. To calculate the IID channel capacity, we model the whole transmission system as the dynamical ISI channel, in which m previous and next m symbols influence the observed symbol, which is shown in Fig. 12.2b. The optical communication system is characterized by the conditional PDF of the output complex vector of samples $\boldsymbol{y} = (y_1, \ldots, y_n, \ldots)$, where $y_i = (\mathrm{Re}\{y_i\}, \mathrm{Im}\{y_i\}) \in \boldsymbol{Y}$, given the source sequence $\boldsymbol{x} = (x_1, \ldots, x_n, \ldots)$, $x_i \in \boldsymbol{X} = \{0, 1, \ldots, M\text{-}1\}$. The set $\boldsymbol{X}$ represents the set of indices of constellation points in corresponding M-ary two-dimensional signal constellation diagram (such as M-ary phase-shift keying (PSK), M-ary quadrature-amplitude modulation (QAM) or M-ary PolSK), while $\boldsymbol{Y}$ represents the set of all possible channel outputs. The $\mathrm{Re}\{y_i\}$ corresponds to the in-phase channel sample, and the $\mathrm{Im}\{y_i\}$ represents the quadrature channel sample.

The information rate can be calculated, as already introduced in (12.56), by:

$$I(\mathbf{Y};\mathbf{X}) = H(\mathbf{Y}) - H(\mathbf{Y}|\mathbf{X}), \tag{12.58}$$

where $H(\boldsymbol{U}) = \mathrm{E}(\log_2 P(\boldsymbol{U}))$ denotes the entropy of a random variable $\boldsymbol{U}$ and $\mathrm{E}(\cdot)$ denotes the mathematical expectation operator. By using the Shannon–McMillan–Brieman theorem that states [20]:

$$E(\log_2 P(\boldsymbol{Y})) = \lim_{n\to\infty} (1/n) \log_2 P(\boldsymbol{y}[1,n)), \tag{12.59}$$

the information rate can be determined by calculating $\log_2(P(\boldsymbol{y}[1,n]))$, by propagating the sufficiently long source sequence. By substituting (12.59) into (12.58), we obtain the following expression suitable for practical calculation of IID information capacity

$$I\,(\boldsymbol{Y};\boldsymbol{X}) = \lim_{n\to\infty} \frac{1}{n} \left[\sum_{i=1}^{n} \log_2 P\,(y_i\,|\boldsymbol{y}\,[1,i-1]\,,\boldsymbol{x}\,[1,n]) - \sum_{i=1}^{n} \log_2 P\,(y_i\,|\boldsymbol{y}\,[1,i-1]) \right]. \tag{12.60}$$

The first term in (12.60) can be straightforwardly calculated from conditional PDFs $P(\boldsymbol{y}[j-m, j+m]|\boldsymbol{s})$. To calculate $\log_2 P(y_i|\boldsymbol{y}[1,i\text{-}1])$, we use the forward recursion of the multilevel BCJR algorithm [63, 64], wherein the *forward metric*

$\alpha_j(s) = \log\{p(s_j = s, y[1, j])\}$ $(j = 1, 2, \ldots, n)$, and the *branch metric* $\gamma_j(s', s) = \log[p(s_j = s, y_j, s_{j-1} = s')]$ are defined as follows:

$$\alpha_j(s) = \max_{s'}^{*}\left[\alpha_{j-1}(s') + \gamma_j(s', s) - \log_2 M\right]$$
$$\gamma_j(s', s) = \log\left[p(y_j | x[j - m, j + m])\right], \tag{12.61}$$

where the max*-operator is defined by $\max^*(x, y) = \log(e^x + e^y) = \max(x, y) + \log[1 + \exp(-|x - y|)]$. The ith term $\log_2 P(y_i | y[1, i\text{-}1])$ can be calculated iteratively by

$$\log_2 P\,(y_i | y\,[1, i - 1]) = \max_{s}{}^{*}\alpha_i\,(s)\,, \tag{12.62}$$

where max*-operator was applied for all $s \in S$ (S denotes the set of states in the trellis shown in Fig. 12.2b). Information capacity is defined as

$$C = \max I\,(Y,X)\,, \tag{12.63}$$

where the maximization is performed over all possible input distributions. Because the optical channel has the memory, it is natural to assume that optimum input distribution will be with memory as well. By considering the stationary input distributions of the form $p(x_i | x_{i-1}, x_{i-2}, \ldots) = p(x_i | x_{i-1}, x_{i-2}, \ldots, x_{i-k})$, we can determine the transition probabilities of corresponding Markov model that maximizes the information rate in (12.60) by nonlinear numerical optimization [66, 67].

This method is applicable to both memoryless channels and for channels with memory. In Fig. 12.26, we report the information capacities for different signal constellation sizes and two types of QAM constellations: square-QAM and star-QAM [68] (see also [69]), by observing a linear channel model.

We also provide the information capacity for an optimum signal constellation, based on so-called iterative polarization quantization (IPQ) we introduced in [86]. We can see that information capacity can be closely approached even with an IID information source providing that constellation size is sufficiently large. It is interesting to note that star QAM outperforms the corresponding square QAM for low and medium SNRs, while for high SNRs square QAM outperforms star QAM. The IPQ significantly outperforms both square-QAM and star-QAM.

Given this description of IIID information capacity calculation for fiber-optics channel, in the next section we study the information capacity of fiber-optics communication systems with coherent detection.

12.6.3 Information Capacity of Coherent detection Systems

In Fig. 12.27, we show the IID information capacity against the number of spans (obtained by Monte Carlo simulations), for dispersion map shown in Fig. 12.19 (the

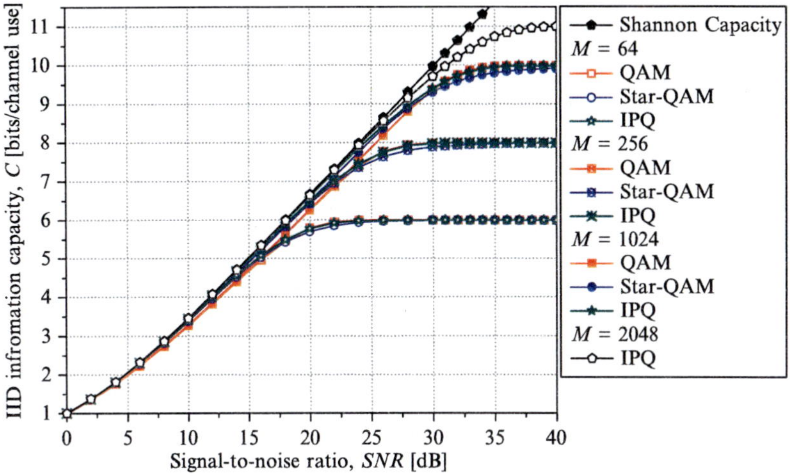

Fig. 12.26 IID information capacities for linear channel model and different signal constellation sizes. (64-star QAM contains 8 rings with 8 points each, 256-star QAM contains 16 rings with 16 points, and 1,024-star QAM contains 16 rings with 64 points.) SNR is defined as E_s/N_0, where E_s is the symbol energy and N_0 is the power spectral density (After ref. [58]; @ IEEE 2010; reprinted with permission.)

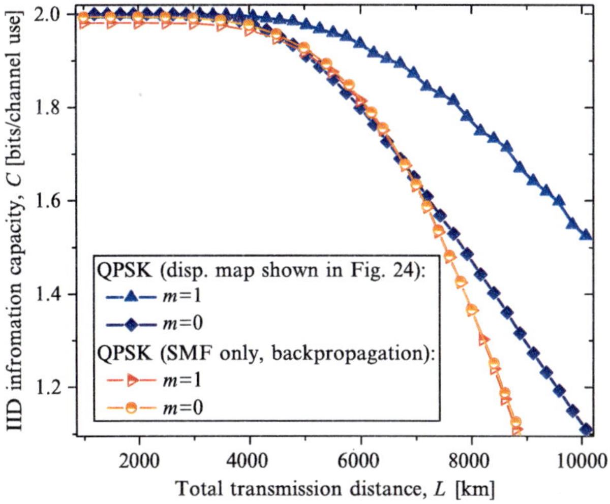

Fig. 12.27 IID information capacity per single polarization for QPSK of aggregate data rate of 100 Gb/s against the transmission distance (After ref. [63]; @ IEEE 2009; reprinted with permission.)

fiber parameters are the same as in Table 12.4) and QPSK modulation format of aggregate data rate 100 Gb/s, for two different memory assumptions. The transmitter and receiver configurations are shown in Figs. 12.28a, b, respectively.

Table 12.4 Fiber parameters

	D_+ fiber	D_- fiber
Dispersion [ps/(nm km)]	20	−40
Dispersion slope [ps/(nm^2 km)]	0.06	−0.12
Effective cross-sectional Area [μm^2]	110	50
Nonlinear refractive index [m^2/W]	$2.6 \cdot 10^{-20}$	$2.6 \cdot 10^{-20}$
Attenuation coefficient [dB/km]	0.19	0.25

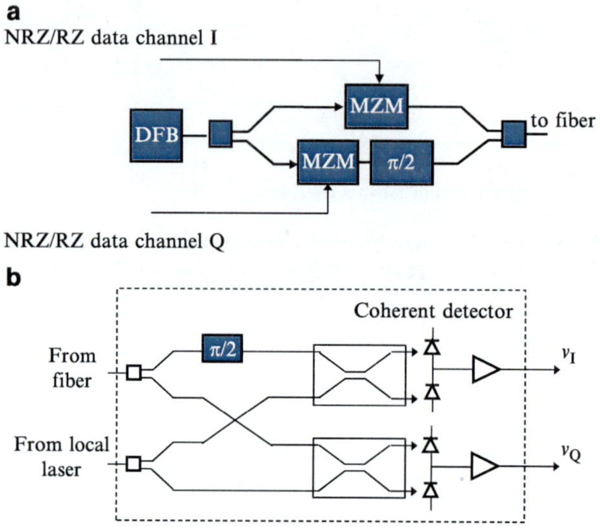

Fig. 12.28 (**a**) Transmitter and (**b**) receiver configurations for system shown in Fig. 12.5a. *DFB* distributed feedback laser, *MZM* Mach–Zehnder modulator

We see that by using the LDPC code (of rate $R = 0.8$) of sufficient length and large girth, we are able to achieve the total transmission distance of 8,760 km for state memory $m = 0$, and even 9,600 km for state memory $m = 1$. The transmission distance can further be increased by observing larger memory channel assumptions, which requires higher computational complexity for corresponding turbo equalizer. On the other hand, we can use backpropagation approach [59, 68] to keep the channel memory reasonable low, and then apply the method described in this section. Note that digital backpropagation method cannot account for the nonlinear ASE noise-Kerr nonlinearities interaction, and someone should use the method introduced in previous section in information capacity calculation to account for this effect. In the same figure, we show the IID information capacity, when digital backpropagation method is used, for dispersion map composed of standard SMF only with EDFAs of noise figure of 6 dB being deployed every 100 km, as shown in Fig. 12.29. We see that digital backpropagation method helps reducing the channel memory, since the improvement for $m = 1$ over $m = 0$ case

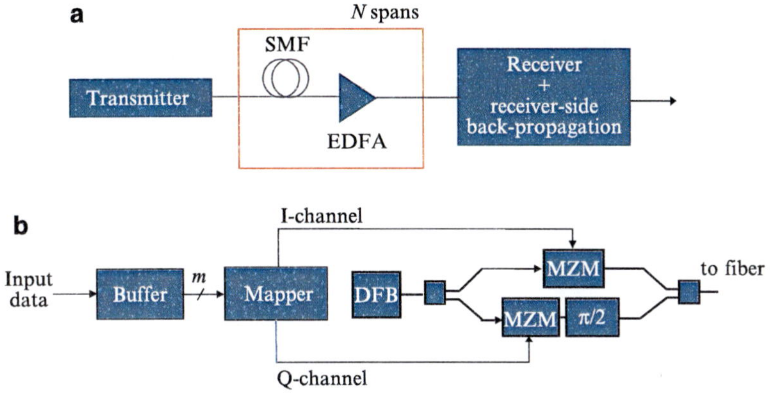

Fig. 12.29 (**a**) Dispersion map composed of SMF sections only with receiver-side digital back-propagation, and (**b**) transmitter configuration. The receiver configuration is shown in Fig. 12.12b

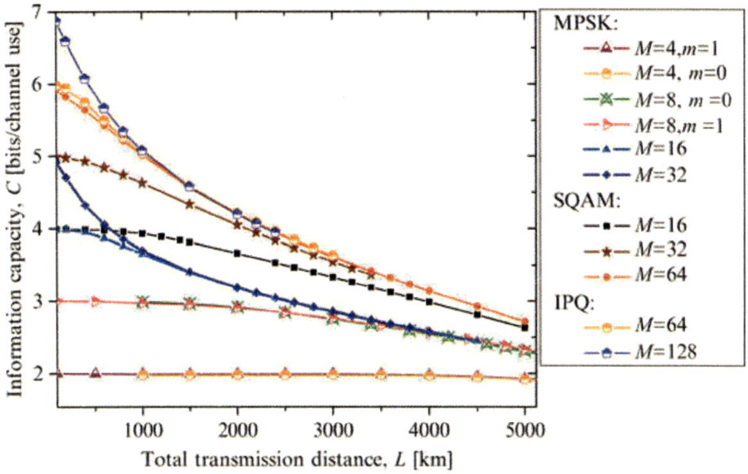

Fig. 12.30 IID Information capacities per single-polarization for star-QAM (SQAM), MPSK, and IPQ for different constellation sizes and dispersion map from Fig. 12.13. EDFA's NF = 6 dB (After ref. [58]; @ IEEE 2010; reprinted with permission.)

is small. In Fig. 12.30, we show the IID information capacities for three different modulation formats: (1) MPSK, (2) star-QAM, and (3) IPQ; obtained by employing the dispersion map from Fig. 12.29a. The symbol rate was 50 GS/s, and the launch power was set to 0 dBm. We see that IPQ outperforms star-QAM and significantly outperforms MPSK. For transmission distance of 5,000 km, the IID information capacity is 2.72 bits/symbol (the aggregate rate is 136 Gb/s per wavelength), for 2,000 km it is 4.2 bits/symbol (210 Gb/s) and for 1,000 km the IID information capacity is 5.06 bits/symbol (253 Gb/s per wavelength). For the completeness of

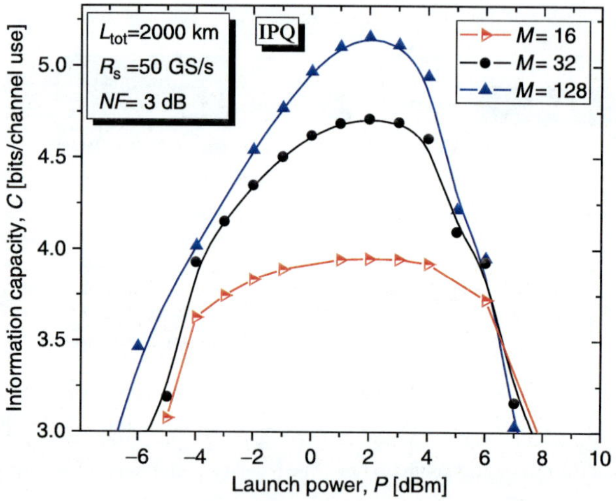

Fig. 12.31 Information capacity per single-polarization against launch power P for total transmission distance of $L_{tot} = 2{,}000$ km (and dispersion map shown in Fig. 12.13). EDFAs NF = 3 dB (After ref. [58]; @ IEEE 2010; reprinted with permission.)

presentation in Fig. 12.31, we show the information capacity as a function of launch power for fixed total transmission distance $L_{tot} = 2{,}000$ km and $NF = 3$ dB, for dispersion map shown in Fig. 12.29a. We see that for optimum launch power of $P_{opt} = 2$ dBm for $M = 128$, we can extend the total transmission distance to 2,000 km and achieve the channel capacity of $C_{opt} = 5.16$ bits/s/Hz, which is very close to the result reported by Essiambre et al. [68]. Note that authors [68] use star-QAM of size 2,048 and optimum dispersion map based on Raman amplifiers, while our dispersion map is based on SMF only with periodically deployed EDFAs.

Acknowledgments This work was supported in part by the National Science Foundation (NSF) under Grants CCF-0952711, ECCS-0725405 and EEC-0812072; and in part by NEC Labs.

References

1. T. Schmidt, C. Malouin, S. Liu, in *Proceedings of 2009 IEEE LEOS annual meeting*, Belek-Antalya, Turkey, Paper WM3, 4–8 October 2009
2. I.B. Djordjevic, M. Arabaci, L. Minkov, IEEE/OSA J. Lightw. Technol. **27**, 3518–3530 (2009). (Invited Paper.)
3. W. Shieh, I. Djordjevic, *OFDM for Optical Communications* (Elsevier, Amsterdam, 2009)
4. I.B. Djordjevic, W. Ryan, B. Vasic, *Coding for Optical Channels* (Springer, Berlin, 2010)
5. I.B. Djordjevic, L.L. Minkov, H.G. Batshon, IEEE J. Select. Areas Commun. Opt. Commun. Netw. **26**, 73–83 (2008)
6. ITU, Telecommunication Standardization Sector: Forward error correction for submarine systems, Rec. G.975 (Geneva, 1996)

7. ITU, Telecommunication Standardization Sector: Forward error correction for high bit rate DWDM submarine systems, Rec. G. 975.1 (02/2004)
8. T. Mizuochi et al., IEEE J. Select. Top. Quant. Electron. **10**, 376–386 (2004)
9. T. Mizuochi et al., Next generation FEC for optical transmission systems, in *Proceedings of optical fiber communication conference* (OFC 2003), vol. 2, pp. 527–528, 2003
10. R.G. Gallager, *Low Density Parity Check Codes* (MIT, Cambridge, 1963)
11. S. Chung et al., IEEE Commun. Lett. **5**, 58–60 (2001)
12. F.M. Ingels, *Information and Coding Theory* (Intext Educational Publishers, Scranton, 1971)
13. S. Lin, D.J. Costello, *Error Control Coding: Fundamentals and Applications* (Prentice-Hall, Englewood Cliffs, NJ, 1983)
14. J.B. Anderson, S. Mohan, *Source and Channel Coding: An Algorithmic Approach* (Kluwer, Boston, MA, 1991)
15. F.J. MacWilliams, N.J.A. Sloane, *The Theory of Error-Correcting Codes* (North Holland, Amsterdam, 1977)
16. S.B. Wicker, *Error Control Systems For Digital Communication And Storage* (Prentice-Hall, Englewood Cliffs, NJ, 1995)
17. D.B. Drajic, *An Introduction to Information Theory and Coding*, 2nd edn. (Akademska Misao, Belgrade, 2004) (in Serbian)
18. S. Haykin, *Communication Systems* (Wiley, New York, 2004)
19. J.G. Proakis, *Digital Communications* (McGaw-Hill, Boston, MA, 2001)
20. T.M. Cover, J.A. Thomas, *Elements of Information Theory* (Wiley, New York, 1991)
21. P. Elias, IRE Trans. Inf. Theory **IT-4**, 29–37 (1954)
22. R.H. Morelos-Zaragoza, *The Art of Error Correcting Coding*. (Wiley, Boston, MA, 2002)
23. O.A. Sab, FEC techniques in submarine transmission systems, in *Proceedings of optical fiber communication conference* (OFC 2001), vol. 2, TuF1-1-TuF1-3, 2001
24. C. Berrou, A. Glavieux, P. Thitimajshima, *Proceedings of 1993 international conference on communication (ICC 1993)*, pp. 1064–1070, 1993
25. C. Berrou, A. Glavieux, IEEE Trans. Commun. **44**, 1261–1271 (1996)
26. R.M. Pyndiah, IEEE Trans. Commun. **46**, 1003–1010 (1998)
27. O.A. Sab, V. Lemarie, Block turbo code performances for long-haul DWDM optical transmission systems, in *Proceedings of OFC 2001*, vol. 3, pp. 280–282, 2001
28. T. Mizuochi, IEEE J. Select. Top. Quant. Electron. **12**, 544–554 (2006)
29. W.E. Ryan, in *Wiley Encyclopedia in Telecommunications*, ed. by J.G. Proakis (Wiley, New York, 2003)
30. I.B. Djordjevic, S. Sankaranarayanan, S.K. Chilappagari, B. Vasic, IEEE/LEOS J. Select. Top. Quant. Electron. **12**(4), 555–562 (2006)
31. I.B. Djordjevic, O. Milenkovic, B. Vasic, IEEE/OSA J. Lightw. Technol. **23**, 1939–1946 (2005)
32. B. Vasic, I.B. Djordjevic, R. Kostuk, IEEE/OSA J. Lightw. Technol. **21**, 438–446 (2003)
33. I.B. Djordjevic et al., IEEE/OSA J. Lightw. Technol. **22**, 695–702 (2004)
34. O. Milenkovic, I.B. Djordjevic, B. Vasic, IEEE/LEOS J. Select. Top. Quant. Electron. **10**, 294–299 (2004)
35. B. Vasic, I.B. Djordjevic, IEEE Photon. Technol. Lett. **14**, 1208–1210 (2002)
36. D.J.C. MacKay, IEEE Trans. Inf. Theory **45**, 399–431 (1999)
37. I.B. Djordjevic, L. Xu, T. Wang, M. Cvijetic, Large girth low-density parity-check codes for long-haul high-speed optical communications, in *Proceedings of OFC/NFOEC*, IEEE/OSA, San Diego, CA, Paper no. JWA53, 2008
38. W.E. Ryan, in *CRC Handbook for Coding and Signal Processing for Recording Systems*, ed. by B. Vasic (CRC Press, Boca Raton, FL, 2004)
39. R.M. Tanner, IEEE Trans. Inf. Theory **IT-27**, 533–547 (1981)
40. M.P.C. Fossorier, IEEE Trans. Inf. Theory, **50**, 1788–1793 (2004)
41. H. Xiao-Yu, E. Eleftheriou, D.M. Arnold, A. Dholakia, Efficient implementations of the sum-product algorithm for decoding of LDPC codes, in *Proceedings of IEEE Globecom*, vol. 2, pp. 1036–1036E, Nov 2001

42. I. Anderson, *Combinatorial Designs and Tournaments* (Oxford University Press, Oxford, 1997)
43. I.B. Djordjevic, B. Vasic, IEEE/OSA J. Lightw. Technol. **24**, 420–428 (2006)
44. I.B. Djordjevic, M. Cvijetic, L. Xu, T. Wang, IEEE/OSA J. Lightw. Technol. **25**, 3619–3625 (2007)
45. I.B. Djordjevic, B. Vasic, OSA J. Opt. Netw. **7**, 217–226, (2008)
46. J. Hou, P.H. Siegel, L.B. Milstein, H.D. Pfitser, IEEE Trans. Inf. Theory **49**(9), 2141–2155 (2003)
47. G.D. Forney Jr., *Concatenated Codes* (MIT, Cambridge, MA, 1966)
48. I.B. Djordjevic, L. Xu, T. Wang, Opt. Express **16**(18), 14163–14172 (2008)
49. E. Biglieri, R. Calderbank, A. Constantinides, A. Goldsmith, A. Paulraj, H.V. Poor, *MIMO Wireless Communications* (Cambridge University Press, Cambridge, 2007)
50. I.B. Djordjevic, L. Xu, T. Wang, Opt. Express **16**(19), 14845–14852 (2008)
51. I.B. Djordjevic, L. Xu, T. Wang, Beyond 100 Gb/s Optical Transmission based on Polarization Multiplexed Coded-OFDM with Coherent Detection, IEEE/OSA J. Opt. Commun. Netw. **1**(1), 50–56 (2009)
52. VPITransmisionMaker, http://www.vpiphotonics.com
53. C. Douillard, M. Jézéquel, C. Berrou, A. Picart, P. Didier, A. Glavieux, Eur. Trans. Telecommun. (6), 507–511 (1995)
54. M. Tüchler, R. Koetter, A.C. Singer, IEEE Trans. Commun. **50**(5), 754–767 (2002)
55. L.R. Bahl, J. Cocke, F. Jelinek, J. Raviv, IEEE Trans. Inf. Theory **IT-20**(2), 284–287 (1974)
56. M. Ivkovic, I. Djordjevic, P. Rajkovic, B. Vasic, IEEE Photon. Technol. Lett. **19**(20), 1604–1606 (2007)
57. J. Chen, A. Dholakia, E. Eleftheriou, M. Fossorier, X.Y. Hu, IEEE Trans. Commun. **53**, 1288–1299 (2005)
58. I.B. Djordjevic, H.G. Batshon, L. Xu, T. Wang, Coded polarization-multiplexed iterative polar modulation (PM-IPM) for beyond 400 Gb/s serial optical transmission, in *Proceedings of OFC/NFOEC* 2010, Paper No. OMK2, San Diego, CA, 21–25 March 2010
59. E. Ip, J.M. Kahn, in *Optical Fibre, New Developments*, In-Tech, Vienna, Austria, December 2009
60. B. McMillan, Ann. Math. Stat. **24**, 196–219 (1952)
61. A.I. Khinchin, *Mathematical Foundations of Information Theory* (Dover Publications, New York, 1957)
62. L.L. Minkov, I.B. Djordjevic, L. Xu, T. Wang, F. Kueppers, Opt. Express **16**, 13450–13455 (2008)
63. I.B. Djordjevic, L.L. Minkov, L. Xu, T. Wang, IEEE/OSA J. Opt. Commun. Netw. **1**, 555–564 (2009)
64. L.R. Bahl, J. Cocke, F. Jelinek, J. Raviv, IEEE Trans. Inf. Theory **IT-20**(3), 284–287 (1974)
65. F.M. Reza, *An Introduction to Information Theory* (McGraw-Hill, New York, 1961)
66. D.P. Bertsekas, *Nonlinear Programming*, 2nd edn. (Athena Scientific, Belmont, MA, 1999)
67. E.K.P. Chong, S.H. Zak, *An Introduction to Optimization*, 3rd edn. (Wiley, New York, 2008)
68. R.J. Essiambre, G.J. Foschini, G. Kramer, P.J. Winzer, Phys. Rev. Lett. **101**, 163901–1–163901–4 (2008)
69. W.T. Webb, R. Steele, IEEE Trans. Commun. **43**, 2223–2230 (1995)
70. H.D. Pfitser, J.B. Soriaga, P.H. Siegel, On the achievable information rates of finite state ISI channels, in *Proceedings of Globecom 2001*, San Antonio, TX, pp. 2992–2996, 25–29 Nov 2001
71. E.E. Narimanov, P. Mitra, IEEE/OSA J. Lightw. Technol. **20**(3), 530–537 (2002)
72. E. Narimanov, P. Patel, Channel capacity of fiber optics communications systems: WDM vs. TDM, in *Proceedings of conference on lasers and electro-optics (CLEO '03)*, pp. 1666–1668, 2003
73. P.P. Mitra, J.B. Stark, Nature **411**, 1027–1030 (2001)
74. K.S. Turitsyn, S.A. Derevyanko, I.V. Yurkevich, S.K. Turitsyn, Phys. Rev. Lett. **91**(20), 203 901 (2003)

75. J. Tang, IEEE/OSA J. Lightw. Technol. **19**, 1110–1115 (2001)
76. A. Mecozzi, M. Shtaif, IEEE Photon. Technol. Lett. **13**, 1029–1031 (2001)
77. J.M. Kahn, K.P. Ho, IEEE Select. Top. Quant. Electron. **10**, 259–272 (2004)
78. I.B. Djordjevic, B. Vasic, Approaching Shannon's capacity limits of fiber optics communications channels using short LDPC codes, in *Proceedings of CLEO/IQEC*, Paper CWA7, 2004
79. I.B. Djordjevic, B. Vasic, M. Ivkovic, I. Gabitov, IEEE/OSA J. Lightw. Technol. **23**(11), 3755–3763 (2005)
80. I.B. Djordjevic, L. Xu, T. Wang, On the channel capacity of multilevel modulation schemes with coherent detection, in *Proceedings of Asia communications and photonics conference and exhibition (ACP) 2009*, Paper ThC4, Shangai, China, 2–6 November 2009
81. M. Ivkovic, I.B. Djordjevic, B. Vasic, IEEE/OSA J. Lightw. Technol. **25**(5), 1163–1168 (2007)
82. I.B. Djordjevic, N. Alic, G. Papen, S. Radic, IEEE Photon. Technol. Lett. **19**, 12–14 (2007)
83. L.L. Minkov, I.B. Djordjevic, H.G. Batshon, L. Xu, T. Wang, M. Cvijetic, F. Kueppers, IEEE Photon. Technol. Lett. **19**, 1852–1854 (2007)
84. D. Arnold, A. Kavcic, H.A. Loeliger, P.O. Vontobel, W. Zeng, Simulation-based computation of information rates: upper and lower bounds, in *Proceedings of IEEE international symposium on information theory (ISIT 2003)*, p. 119, 2003
85. D. Arnold, H.A. Loeliger, On the information rate of binary-input channels with memory, in *Proceedings of 2001 international conference communications*, Helsinki, Finland, pp. 2692–2695, 11–14 June 2001
86. Z.H. Peric, I.B. Djordjevic, S.M. Bogosavljevic, M.C. Stefanovic, Design of signal constellations for Gaussian channel by iterative polar quantization, in *Proceedings of 9th mediterranean electrotechnical conference*, vol. 2, Tel-Aviv, Israel, pp. 866–869, 18–20 May 1998
87. H.G. Batshon, I.B. Djordjevic, L.L. Minkov, L. Xu, T. Wang, M. Cvijetic, Proposal to achieve 1 Tb/s per wavelength transmission using 3-dimensional LDPC-coded modulation, IEEE Photon. Technol. Lett. **20**(9), 721–723 (2008)
88. H.G. Batshon, I.B. Djordjevic, Beyond 240 Gb/s per wavelength optical transmission using coded hybrid subcarrier/amplitude/phase/polarization modulation, IEEE Photon. Technol. Lett. **22**(5), 299–301 (2010)
89. H.G. Batshon, I.B. Djordjevic, L. Xu, T. Wang, Modified hybrid subcarrier/amplitude/phase/polarization LDPC-coded modulation for 400 Gb/s optical transmission and beyond, optics express, **18**(13), 14108–14113 (2010)
90. H.G. Batshon, I.B. Djordjevic, T. Schmidt, Ultra high speed optical transmission using subcarrier-multiplexed four-dimensional LDPC-coded modulation, Optics Express, **18**(19), 20546–20551 (2010)
91. I. Djordjevic, H.G. Batshon, L. Xu, T. Wang, Four-dimensional optical multiband-OFDM for beyond 1.4 Tb/s serial optical transmission, Opt. Express, **19**(2), 876–882 (2011)

Chapter 13
Channel Capacity of Non-Linear Transmission Systems

Andrew D. Ellis and Jian Zhao

13.1 Introduction

Since their introduction in the late 1970s, the capacity of optical communication links has grown exponentially, fuelled by a series of key innovations including movement between the three telecommunication windows of 850 nm, 1,310 nm and 1,550 nm, distributed feedback laser, erbium-doped fibre amplifiers (EDFAs), dispersion-shifted and dispersion-managed fibre links, external modulation, wavelength division multiplexing, optical switching, forward error correction (FEC), Raman amplification, and most recently, coherent detection, electronic signal processing and optical orthogonal frequency division multiplexing (OFDM). Throughout this evolution, one constant factor has been the use of single-mode optical fibre, whose fundamental principles dated back to the 1800s, when Irish scientist, John Tyndall demonstrated in a lecture to the Royal Society in London that light could be guided through a curved stream of water [1]. Following many developments, including the proposal for waveguides by J.J. Thompson [2], the presentation of detailed calculations for dielectric waveguides by Snitzer [3], the proposal [4] and fabrication [5] of ultra low loss fibres, single-mode fibres were first adopted for non-experimental use in Dorset, UK in 1975, and are still in use today, despite the evolving designs to control chromatic dispersion and non-linearity.

As the underlying optical broadband technologies, urged by the demands for new applications in information communication, gradually pervaded the network, the optical capacity has grown exponentially for a wide variety of measurements, for example overall network traffic or submarine transmission capacity [6]. Current telecommunication networks are arranged in several layers to minimise the cost of network provision. Typically, the lowest-cost technologies are used for direct connections to customers. However, once the traffic is aggregated with traffic

A.D. Ellis (✉) and J. Zhao
Tyndall National Institute and Department of Physics, University College Cork, Cork, Ireland
e-mail: andrew.ellis@tyndall.ie; jian.zhao@tyndall.ie

S. Kumar (ed.), *Impact of Nonlinearities on Fiber Optic Communications*, Optical and Fiber Communications Reports 7, DOI 10.1007/978-1-4419-8139-4_13,
© Springer Science+Business Media, LLC 2011

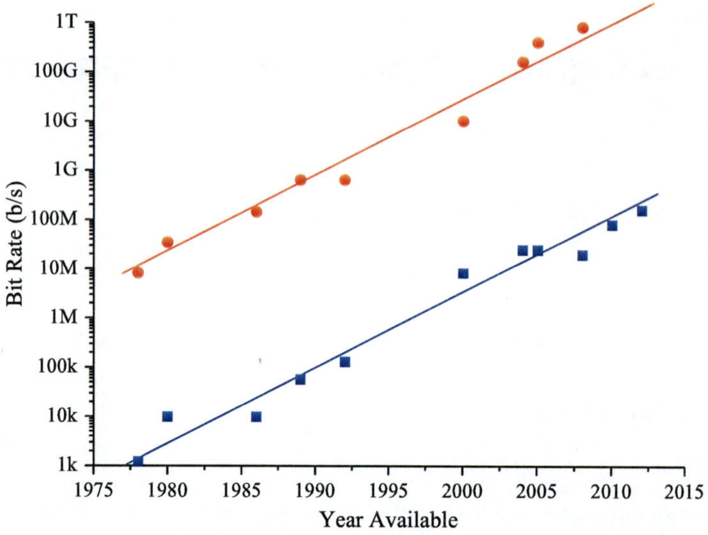

Fig. 13.1 Evolution of telecommunication network capacities in response to changing consumer applications. *squares*: Bandwidth of available access network connection. *dots*: Maximum deployed capacity per fibre. *Solid line*: Growth trends

from other users at a service provider's point of presence, the sharing of network resources between many users allows the use of higher performance technologies. Two key parameters are of particular interest here. The first parameter is the headline bandwidth offered in the access network directly to the customer and the second is the maximum deployed capacity of a given optical fibre. These two parameters are shown in Fig. 13.1, which depicts the evolution of these two capacities with time.

The figure is plotted by taking the bandwidth of a wide variety of access technologies as a function of their date of first introduction (squares), starting from the introduction of the 1.2 kb s^{-1} modem for use in Bulletin Board Systems in 1978 [7] to Passive Optical Networks at contended bit rates up to 10 Gb s^{-1} [8] for video and gaming applications. The trend lines show a steady long-term growth rate of above 40% per annum, covering a remarkable capacity increase by 5 orders of magnitude in 2010. In parallel with the growth in access bandwidth, we have observed a steady increase in the capacity of the highest network layer, or the total capacity carried by a single optical fibre. Despite the huge growth in available bandwidth accompanied by changes in personal usage and the introduction of many generations of network technologies, the ratio between these two quantities has remained remarkably constant, representing the continual design trade-off that is made between, on the one hand, complexity (favouring coarse bandwidth granularity in the core network), and on the other hand, reliability (favouring fine granularity). This may also be viewed as a trade-off between the capital costs associated with providing a large number of low bandwidth links, and the operational costs associated with service interruptions resulting from inevitable component failures. Extrapolating the trends of Fig. 13.1 to the future suggests that the network should be able to support aggregate capacities

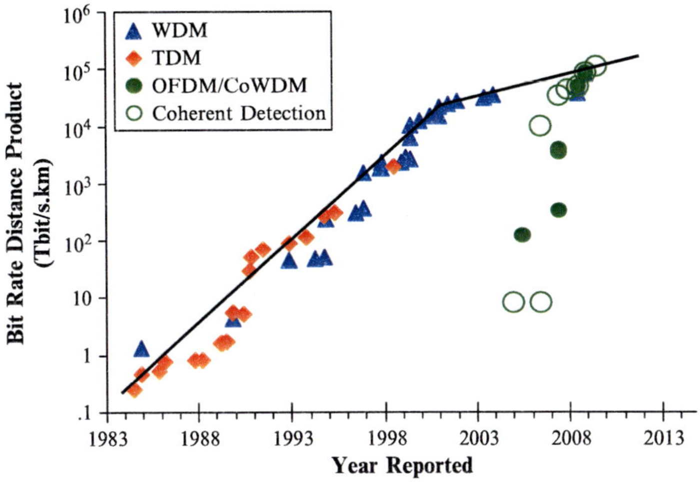

Fig. 13.2 Evolution of maximum reported transmission capacity for single wavelength (*diamonds*), wavelength division multiplexing (*triangles*), single and multi-banded OFDM (*filled circles*) and coherent detection (*open circles*)

in excess of 2 Tbit s^{-1} per fibre in the core network today and 250 Tb s^{-1} per fibre as early as 2021. However, to maintain the current core network architecture, this would require a total number of wavelengths deployed similar to today, typically 160, but carrying information at an information spectral density (ISD) exceeding 30 b s^{-1} Hz^{-1} (which is an immense technical challenge).

The imminence of such a challenge may also be observed from the research output over a similar period. Figure 13.2 illustrates the evolution of the fibre transmission capacity reported from research experiments carried out in research laboratories worldwide. A long-term growth trend of approximately 60% per annum was observed from the early 1990s, fuelled by the transition to 1,550 nm wavelength band, the introduction of optical amplifiers and wavelength division multiplexing. However, this has saturated recently, prompting rapid use of additional technologies and culminating in the adoption of coherent detection techniques, where the additional degree of freedom (optical phase) is expected to allow for greater capacity increases [9].

There is therefore now a growing realisation from both commercial operators and network equipment providers that the continuing bandwidth demand will shortly push the required capacity close to the maximum capacity, which has been predicted theoretically for standard single-mode fibres. The economic and other consequences of demand exceeding capacity are a matter of much debate. However, it is generally acknowledged that the current combination of pricing, network architecture and transmission technologies will not be capable of meeting the customer bandwidth demand in the medium term. The time at which demand exceeds supply may be delayed from the point predicted by extrapolating the curves of Figs. 13.1 and 13.2 by changes in network architecture and service pricing, but is still likely to occur within the next decade.

In this chapter, we explore the system design trade-offs required to maximise the total capacity from the prospect of the ultimate limit to optical communication channel capacity, imposed by signal-to-noise ratio and distributed fibre non-linearity. We will illustrate the rate at which experimental results are approaching this limit and discuss techniques, which promise to allow the capacity limit to be extended.

13.2 Linear Capacity Limits

13.2.1 The Shannon Limit

The Shannon limit to information capacity in a communication link [10, 11] is well known, and is based on the fundamental concepts that the maximum symbol rate, which may be transmitted and detected on a communication link is constrained to be less than twice the overall bandwidth of the link [12], and the information each symbol delivers is measured by the uncertainty, more specifically, the statistical distribution or randomness of the transmitted symbols. Shannon further developed these concepts in communications through a noisy channel and showed that associated with every noisy channel is a parameter, called the capacity, such that reliable information communication through the channel is possible if the communication rate R satisfies $R < C$, where C is the channel capacity, and is given by

$$C = B \log_2 \left(1 + \frac{P_{\text{ave}}}{N_0 B} \right), \tag{13.1}$$

where P_{ave} is the average signal power and equals $C \cdot E_{\text{b}}$, where E_{b} is the average energy per bit, N_0 the noise spectral density and B the channel bandwidth.

Although the proof presented for (13.1) does not explicitly consider any specific modulation, coding, decoding schemes etc., understanding of assumption required to derive (13.1) yields a useful insight in practical design. Achieving the capacity of a channel C requires two conditions. First, the transmitted symbol rate should be increased to twice the bandwidth of the transmission links. In a practical optical communication system, due to the current limitation in the electronic bandwidth of around 100 GHz, it is impractical to modulate the full optical bandwidth (~50 THz) available in a fibre or even that available in a single amplification band (~4 THz). Wavelength division multiplexing is therefore used in commercial transmission systems, where the available optical bandwidth is split into frequency bands. In each band, an independent carrier is modulated separately, prior to multiplexing and transmission over a common transmission fibre. The signals are then de-multiplexed according to their allocated frequency bands and independently detected. This process inevitably results in frequency guard bands between independent channels. Consequently, the overall channel capacity is reduced by a factor of $B/\Delta f$, where $\Delta f (\geq B)$ is the channel spacing. Recently, a technique of optically implemented

OFDM was proposed and, by superimposing an additional orthogonality condition in signal pulse shaping, this technique reduces the channel spacing to that required for maximum transmission rate.

The second condition specifies the particular statistical distribution of the transmitted symbols: the constellation in-phase and quadrature components of the optical field should take arbitrary continuous values, with the probability of each value following a Gaussian distribution [9]. Implicitly, the second condition also suggests a *linear* channel and that the detection scheme should allow for the use of two degrees of freedom per polarization. Convention optical systems usually comprise a laser, a modulator, fibres and a photodiode that is only linear in terms of signal power and introduces square-law non-linearity. Whilst progress has been made in full-field detection and estimation using optical filters [13, 14], the recently developed digital optical coherent receiver readily offers optimal detection by direct linear translation of the optical field to an intermediate frequency (heterodyne detection) or base band (homodyne detection) [9, 15].

13.2.2 Constellation Analysis

In practice, a continuous bi-Gaussian distribution for transmitted symbol is impractical, but may be emulated by a discrete-point constellation [15, 16], an example of which is shown in Fig. 13.3.

For this approximation, the constellation point's e_{nm} are distributed according to:

$$\begin{aligned}\mathrm{mag}\left(e_{\mathrm{nm}}\right) &= n/n_{\max}\\ \arg\left(e_{\mathrm{nm}}\right) &= \frac{2m\pi}{m_{\max}} + \frac{n\pi}{4},\end{aligned} \tag{13.2}$$

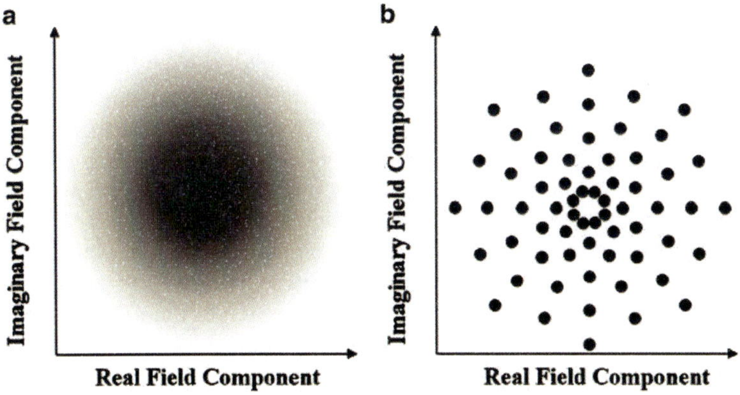

Fig. 13.3 (**a**) Ideal transmitted constellation (continuous) and (**b**) discrete point approximation

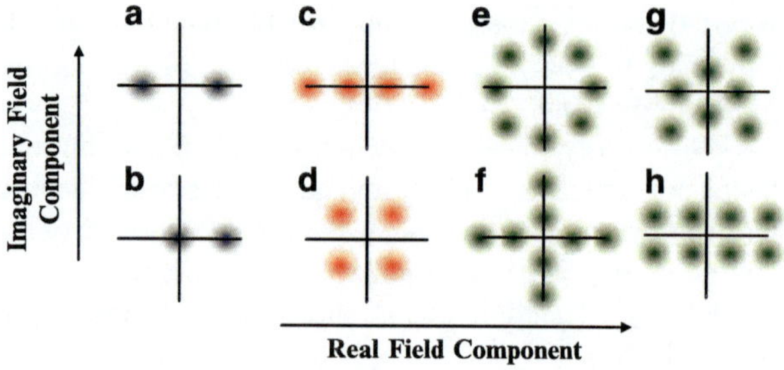

Fig. 13.4 Some examples of signal constellations with one **(a)**, **(b)**, two **(c)**, **(d)** and three **(e)**–**(h)** bits per symbol

where n and m are indices of the constellation points for a constellation with $n_{\max} \times m_{\max}$ points. In a well-designed discrete point constellation, the density of points reduces with distance from the centre of the constellation, in a manner approaching the optimum distribution. This approximation may be improved further by varying the probability of occupancy of each point in the constellation.

The implementation of the constellation in Fig. 13.3b still requires high complexity. For a practical linear transmission system, many different simpler constellations may be considered (as shown, for example, in Fig. 13.4), ranging from single quadrature formats typically generated with a single modulator, including (a) binary phase shift keying (BPSK), (b) amplitude shift keying (ASK) and (c) quaternary ASK (4-ASK) to formats consisting of in-phase and quadrature components, including (d and e) M-ary phase shift keying (QPSK and 8PSK, respectively), (g and h) quadrature amplitude shift keying constellations (typically generated using a dual parallel Mach Zehnder modulator) and (f) hybrid amplitude phase shift keying (APSK) (typically generated using an amplitude modulator and a phase modulator in series).

To calculate the performance of each constellation and compare it to the Shannon limit, we first determine the impact of noise on each constellation point. For a system using coherent detection, on the one hand, the noise and signal are combined as a vector addition and the noise is independent of the signal amplitude [17, 18]. On the other hand, for direct- and differentially detected signals, the noise level after detection is dependent on the signal intensity [19]. As discussed above, coherent detection theoretically enables the possibility of approaching the channel capacity, and is readily becoming practical due to the advances of narrow linewidth lasers and digital signal processing (DSP). In the following, we calculate the bit error rate (BER) performance of a given constellation assuming coherent detection and hard decision detection, beginning from calculating the probability that a given transmitted signal level crosses a virtual boundary (the decision threshold) between it and its nearest neighbour [20]. We use the constellation of Fig. 13.4c as an example.

In this example, also shown in Fig. 13.5, additive white Gaussian noise gives a Gaussian probability density function for all possible received signal values (Fig. 13.5b). For the second signal level (e_2), erroneous detection occurs when the

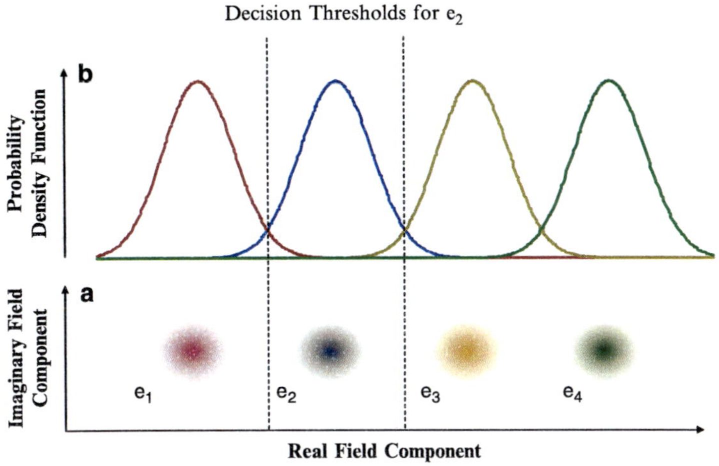

Fig. 13.5 An example of symbol error rate calculation. (**a**) Constellation diagram for 4-ASK, (**b**) Probability density function of 4-ASK symbols. Symbol e_2 was transmitted

detected signal level crosses either of the two decision thresholds towards its nearest neighbours (e_1 and e_3). Assuming a decision boundary located equidistant between the two constellation points (optimal boundary for equal probable $e_1,\ e_2,\ e_3,\ e_4$), the probability of an error for this point is thus:

$$\langle \xi_2 \rangle = Q\left(\frac{|e_2 - e_1|}{\sqrt{2N_0}}\right) + Q\left(\frac{|e_2 - e_3|}{\sqrt{2N_0}}\right), \tag{13.3}$$

where e_i is the field amplitude of the ith constellation point and Q is related to the complimentary error function by

$$Q\,(x) = \frac{1}{2}\text{erfc}\left(\frac{x}{\sqrt{2}}\right). \tag{13.4}$$

Note that here Q represents a mathematical function, and should not be confused with the "Q-factor" used in optical communications. Note that the constellation points located at the two ends (furthest from the centre in the general case) have fewer nearest neighbours and therefore smaller error probabilities. For example, a transmitted e_1 would only be erroneously detected if it crosses the threshold between itself and e_2.

The total symbol error rate (SER) is then given by the sum of the error rate $\langle \xi_i \rangle$ for each point i multiplied by the probability P_i that this level is transmitted, that is

$$\text{SER} = \sum_i P_i \cdot \langle \xi_i \rangle \tag{13.5}$$

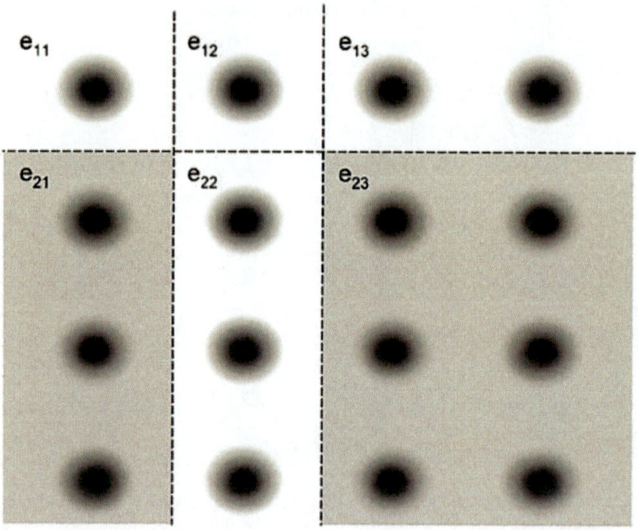

Fig. 13.6 SER calculations for 16 QAM

Clearly, the SER is dependent not only on the noise distribution and the probability of transmitting each bit, as explicitly indicated in (13.5), but also by the choice of decision boundaries. For more complex modulation formats, some benefit may be obtained from the placement and location of these boundaries [21].

For more complex modulation formats, for example quadrature amplitude modulation (QAM), in principle, (13.5) may be extended to two dimensions, treating each quadrature independently, as shown in Fig. 13.6. Here, three decision thresholds have been indicated, allowing the one-dimensional error distribution (13.3) to be used to calculate the probability. For example, signal level e_{12} is erroneously detected in the column of either e_{11} or e_{13} or in the row of e_{22}

$$\langle \xi_{12} \rangle \leq Q\left(\frac{|e_{12} - e_{11}|}{\sqrt{2N_0}}\right) + Q\left(\frac{|e_{12} - e_{13}|}{\sqrt{2N_0}}\right) + Q\left(\frac{|e_{12} - e_{22}|}{\sqrt{2N_0}}\right) \tag{13.6}$$

However, in this case, the probability that a signal level is transmitted as e_{12} but detected in the shaded region, for example as e_{21}, has been double counted and (13.6) should be modified to account for this over-estimation. Assuming that the constellation points of Fig. 13.6 are equally spaced, with spacing d, the probability of error for this constellation point should be:

$$\langle \xi_{12} \rangle = 3Q\left(\frac{|d|}{\sqrt{2N_0}}\right) - 2Q\left(\frac{|d|}{\sqrt{2N_0}}\right)^2. \tag{13.7}$$

It is straightforward to show that for square m-QAM where the probability of transmitting each constellation point is $1/m$, the SER is

$$\mathrm{SER} = \frac{4\left(m-\sqrt{m}\right)}{m} Q\left(\frac{|d|}{\sqrt{2N_0}}\right)\left\{1-\frac{m+1-2\sqrt{m}}{m-\sqrt{m}} Q\left(\frac{|d|}{\sqrt{2N_0}}\right)\right\} \tag{13.8}$$

For the majority of applications, $Q(x)^2 << Q(x)$, and so the second term on the right-hand side of (13.8) may be readily neglected, giving an upper bound on the SER. Relying on the same approximation, in a general form, SER for a complex constellation map is upper-bounded by the union bound of the $m-1$ events E_j, where E_j represents the event that the transmitted constellation point is i while the detected point is $j,\ j \neq i$. That is

$$\mathrm{SER} = \frac{1}{m}\sum_{i=1}^{m} P(\cup_{j\neq i} E_j) \leq \frac{1}{m}\sum_{i=1}^{m}\sum_{j\neq i} P(E_j) = \frac{1}{m}\sum_{i=1}^{m}\sum_{j\neq i} Q\left(\frac{\left|e_i-e_j\right|}{\sqrt{2N_0}}\right). \tag{13.9}$$

Equations (13.5)–(13.9) represent the SER as a function of distances between signal points in the constellation and the noise spectral power density in the communication channel. However, it is also necessary to relate these points to the signal power and consequently develop the relationship between BER and signal-to-noise ratio. For a given constellation, the transmitted signal power is directly determined by the geometric distribution of the constellation points, and the mean energy per symbol E_s or the mean energy per bit E_b is:

$$\langle E_s \rangle = \sum P_i \cdot |e_i|^2 = \langle E_b \rangle \cdot \log_2(m), \tag{13.10}$$

where m represents the number of points in the constellation. Equation (13.10) in turn allows the calculation of a signal-to-noise ratio $snr = \langle E_b \rangle / N_0$. Here, snr is a parameter commonly used in communication theory, and in an optical system, is usually limited by amplified spontaneous emission (ASE) noise from the optical amplifiers. The ASE of an optical amplifier in one polarization has a power spectral density of $(G-1)n_{sp}h\nu$, where G is the amplification gain, n_{sp} is the noise enhancement factor and has an ideal value of 1, $h\nu$ is the energy per photon. In the simplest optical communication model where ASE is only loaded by the pre-amplifier at the receiver and $n_{sp}=1$, snr represents the photon number per bit entering an ideal optical pre-amplifier.

It should be noted that snr is different from another term widely used in optical communications, optical signal-to-noise ratio (OSNR), although both are used to measure limitation from the ASE noise. Mathematically, $\mathrm{OSNR} = R \cdot snr/(2B_{\mathrm{ref}})$, where R and B_{ref} are the symbol rate and reference noise bandwidth (e.g., 12.5 GHz, corresponding to 0.1 nm at 1,550 nm wavelength) respectively, and the ASE noise is assumed to be randomly polarised. BER performance results as a function of snr for a few common modulation formats are shown in Table 13.1 [18, 22], where m represents the number of constellation points ($\log_2(m)$ bits per symbol) with the

Table 13.1 Error probabilities for a few common modulation formats as a function of electrical signal-to-noise ratio

Format	Bit error probability
ASK (Fig. 13.4b)	$Q\left(\sqrt{snr}\right)$
Bi-polar MASK (Fig. 13.4a, c)	$2\frac{m-1}{m\cdot\log_2(m)}Q\left(\sqrt{\frac{6.\log_2(m)}{m^2-1}snr}\right)$
BPSK (Fig. 13.4a)	$Q\left(\sqrt{2snr}\right)$
MPSK (m > 4) (Fig. 13.4e)	$\frac{2}{\log_2(m)}Q\left(\sqrt{2snr.\log_2(m)}\sin\left(\frac{\pi}{m}\right)\right)$
Rectangular QAM ($\log_2(m)$=even) (Fig. 13.4d)	$\frac{4}{\log_2(m)}\left(1-\frac{1}{\sqrt{m}}\right)Q\left(\sqrt{\frac{3\log_2(m)}{m-1}snr}\right)$

probability of each point $P_i = 1/m$. Note that the relationship between the BER, the number of bit errors divided by the total number of transmitted bits, and the SER, the number of erroneously detected symbols divided by the total number of transmitted symbols, depends on the allocation of bit representation for the constellation points within a symbol. To minimise the number of bit errors arising from a constellation point crossing a decision threshold, it is essential to ensure that the bit patterns conveyed by adjacent constellation points differ by only one bit. Such code is known as a "reflected binary code," or "Gray Code" [23].

Figure 13.7 shows the ISD as a function of *snr* in a WDM system with 20% guard bands for uni-polar ASK (up triangles), bi-polar ASK (down triangles), PSK (circles), and QAM (squares) formats, along with the Shannon limit. For a given number of constellation points, Fig. 13.7 reveals a direct trade-off between performance (required *snr*) and transmitter/receiver complexity. For example, 16 QAM, which requires modulation and detection in both quadratures, has better performance than 16-ASK at the same ISD, which however only requires modulation and detection of one quadrature.

Within limits, increasing the complexity of the modulation format allows for an increase in ISD for broadly similar *snr*. For example, the change in the required *snr* between NRZ, bi-polar 4-ASK, 8-PSK and 16QAM is negligible whilst the ISD increases fourfold. Beyond this behaviour however, increasing the ISD within a given class of constellation is always at the expense of an increase in the required *snr*, as shown in recent experimental results [24].

Strong FEC is also essential to enable operation close to the fundamental Shannon limit [25]. The theoretical impact of FEC is illustrated in Fig. 13.8 for a range of QAM signals assuming Reed Solomon FEC with variable overhead, and 64 bits per FEC symbol. As the strength of the FEC is increased, the required BER increases when using stronger FEC, e.g. $10^{(-3)}$ to $10^{(-2)}$ reduces, and so the required *snr* reduces. However, this is at the expense of reduced ISD as the transmission of the necessary redundant information reduces the number of symbols available for the transmission of useful information. For a given signal-to-noise ratio, there is clearly an optimum combination of modulation format and FEC overhead. Note that the same trade-offs will apply for FEC codes with lower latency, and

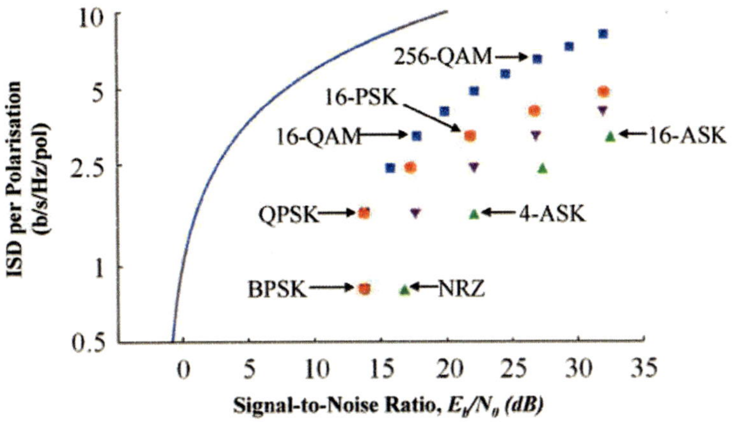

Fig. 13.7 Information spectral density of uni-polar ASK (*up triangles*), bi-polar ASK (*down triangles*), PSK (*circles*) and QAM (*squares*) showing the maximum system capacity as a function of electrical signal-to-noise ratio for a BER of 10^{-12} in a WDM system with 20% guard bands between channels. The *solid line* represents the Shannon theoretical limit [6, 10]

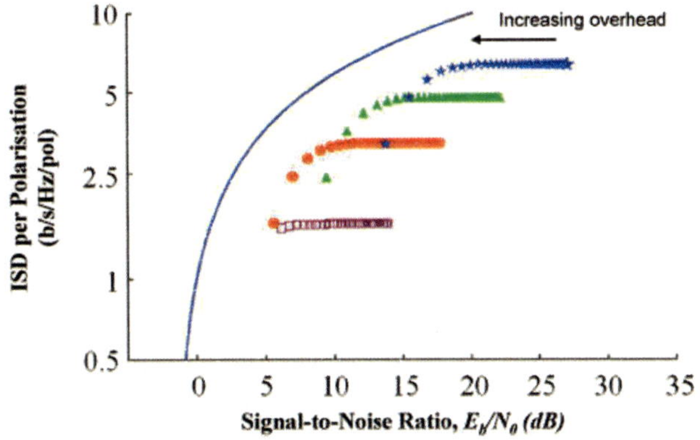

Fig. 13.8 Variation in maximum information spectral density and minimum electrical signal-to-noise ratio of 4-QAM (*open squares*), 16-QAM (*circles*), 64-QAM (*triangle*) and 256-QAM (*stars*) in a WDM system with 20% guard bands between channels for a range of forward error correction overhead assuming a Reed Solomon code. The *solid line* represents the Shannon theoretical limit [3, 6]

the code with the minimum latency for a given required coding gain would normally be selected. Overall, complexity may also be reduced by combining demodulation and FEC decoding into a single step [26].

In linear communication system, the channel capacity would increase infinitely as the signal power increases. However, in many circumstances of optical communications, fixed constraints apply which limit our ability to arbitrarily increase the *snr* and therefore ISD. For example, to minimise network cost, it is often desirable

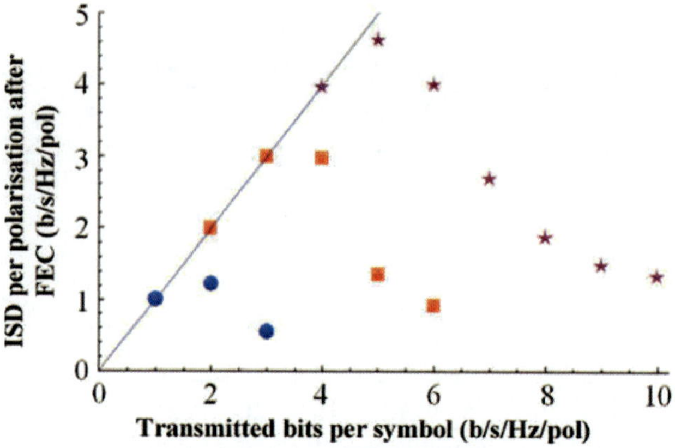

Fig. 13.9 Illustration of the limitation in the net information capacity as a function of the number of transmitted bits per symbol for uni-polar M-ASK (*circles*), M-PSK (*squares*) and QAM (*stars*) assuming a *snr* of 12.5 dB

to add a wavelength channel to an existing transmission system, with fixed length of amplifier span, and constraints placed on the signal launch powers to suit existing legacy systems. In these cases, *snr* is fixed, so it is interesting to find out the optimal practical constellation and the maximum channel capacity under FEC. This is illustrated in Fig. 13.9 for various modulation formats employing coherent detection. In this figure, it is assumed that the baseline system is designed with a 12.5 dB *snr* (readily sufficient for direct detection of an on-off keyed signal with FEC). Simple increase in the number of amplitude levels without increasing the *snr* eventually results in a degraded BER, and FEC must be introduced to restore the system performance. The FEC overhead, however, reduces the system capacity, and as the required *snr* for higher-level formats increases, the required FEC gain and associate overhead also increase. To arrive at the data points in Fig. 13.9, the required FEC overhead for error-free operation (10^{-12}) is calculated and then subtracted from the net information capacity. A simplified approximation was used to calculate the FEC overhead, where each FEC was assumed to require 7% overhead [27] for every 10^{-3} of BER to be corrected. For example, we assumed an overhead of 21% for the correction of a BER of $3 \cdot 10^{-3}$. From Fig. 13.9, it is shown that for uni-polar signal with coherent detection, the calculated overhead results in a negligible improvement in capacity when the number of bits per symbol is increased from one to two. As the number of bits per symbol is further increased, the BER is degraded rapidly, requiring larger overheads, and the required additional FEC overhead outstrips the additional capacity offered by an extra bit per symbol, at a fixed *snr*. By changing to phase shift keyed format, the required *snr* is greatly reduced, postponing the point at which capacity gains from increased constellation points is outstripped by the required FEC overhead, allowing, in this example, 3 bits per symbol. QAM exhibits further performance enhancement, resulting in 5 bits per symbol without significant reduction in throughput due to FEC overhead.

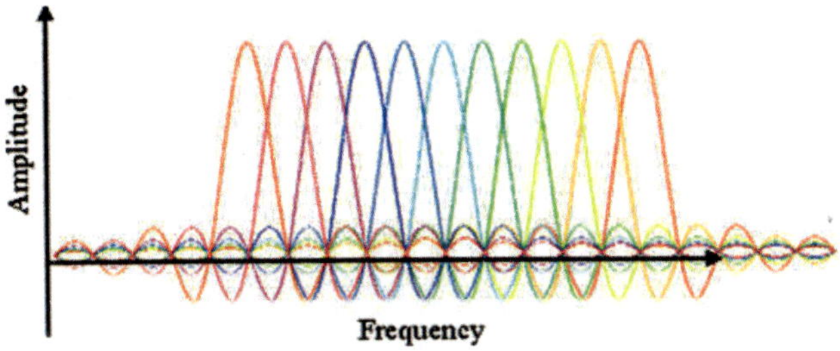

Fig. 13.10 Illustration of overlapping modulation sidebands of OFDM signal

Whilst it is likely that FEC circuits which will require less overhead for a given input BER than assumed here will become available, including current proprietary FEC circuits, it will still be the case that an optimum ISD will exist for a given fixed *snr* and class of modulation format.

In optical communications, WDM is usually used to make full use of the available bandwidth without increasing the bandwidth of the transceivers to the full optical band. It is clear that any required guard band between WDM channels would reduce the ISD, hindering the system transmission rate approaching the Shannon limit. Guard bands may, however, be avoided by employing OFDM techniques [28, 29], such as no-guard-interval OFDM [30–32], coherent WDM [33–36], direct detection OFDM [37, 38] and coherent optical OFDM [39–44].

In all of these multi-carrier systems, the frequency spacing between the orthogonal sub-carriers is equal to the symbol rate per sub-carrier. A typical example of the orthogonal carriers is shown in Fig. 13.10, where the peak of the spectrum of a given sub-channel corresponds to nulls in the spectra of all of the other sub-channels, and in particular, the first null in the spectrum of the adjacent sub-channel. Ideally, matched filters are used to separate each sub-channel [29], and this may be implemented efficiently using Fast Fourier Transform algorithms for low sub-channel data rates (e.g., 100 Mb s^{-1}), with the DSP complexity scaling approximately linearly with the total capacity ($\propto N \cdot \log N$, where N is the channel number) [37–42]. However, for a system with a high symbol rate per channel (e.g., 40 Gb s^{-1}), the practical implementation of precise matched filters proves difficult, and may be approximated in the optical domain using asymmetric Mach Zehnder interferometers [32, 33] or with simple digital filters [31]. The impact of any residual crosstalk may then be minimised using appropriate optimisation of the relative phases of each sub-channel [34] or using post-detection signal processing [34, 45]. In all cases, the net result is the straightforward generation of a signal with a capacity per polarisation equal to the number of bits per symbol (or $\log_2(m)$) (including FEC overhead) without any transmission rate reduction arising from guard spectral band. By using optical implementation for channel multiplexing and de-multiplexing, this technique has the potential suitability for ultra-high total capacities (theoretically extendable to the full optical band and experimentally achieved for 1,080 Gb s^{-1} and beyond [46–48]), which are difficult to achieve using single carrier modulation.

13.3 Non-linear Limits

13.3.1 Theoretical Information Capacity Limits

While the above discussions for a linear communication channel provide the guideline on the appropriate design of modulation/detection, coding/decoding, multiplexing/de-multiplexing etc. to approach the Shannon limit, the performance of a practical communication system is usually degraded by non-linear distortions as well. For example, on the one hand, wireless systems, particularly those using OFDM, experience non-linearity due to the saturation characteristics of power amplifiers [49]. On the other hand, periodically amplified optical fibre-based systems are characterised by distributed non-linear effects in the fibre itself. The most predominant non-linear effect arises from the intensity-dependent refractive index (Kerr effect) and results in a number of phenomena such as self-phase modulation (SPM) [50], cross-phase modulation (XPM) [51] and inter- [52] and intra-channel [53] four-wave mixing (FWM). Whilst many techniques to mitigate the impact of non-linearity have been developed for optical communications, including the most significantly dispersion management [54–58], the impact of these non-linearities on the information theoretical limits has only been addressed recently [59–61].

The initial understanding of the impact from fibre non-linearity is traced back to the fundamental concept of information. From a fundamental point of view, any deterministic impairment is reversible, and so would not cause information loss and consequently reduction of channel capacity. This implicitly implies that deterministic fibre non-linearity, as well as dispersion, does not limit the channel capacity provided that the interaction between these effects and noise is negligible and full optical-band signal processing can be performed to compensate for both intra- and inter-channel impairments. However, it is impractical to implement full optical-band impairment compensation, despite recent development for intra-channel non-linearity compensation. Consequently, any inter-channel effects, such as XPM and inter-channel FWM, where the information from the adjacent channels is unknown, would cause randomness and information loss. In [59], Mitra and Stark equated a XPM-limited non-linear communication channel to a linear channel by modelling the randomness caused by the non-linear interaction with co-propagating WDM channels as a multiplicative noise source, from which analytical results can be obtained.

This approximation is made by transforming the coupled non-linear Schrodinger equations (the kth channel is shown):

$$\frac{\partial E_k}{\partial z} + \frac{\mathrm{i}}{2}\beta_{2k}\frac{\partial^2 E_k}{\partial t^2} + \frac{\alpha_k}{2}E_k = \mathrm{i}\gamma\left(|E_k|^2 + \sum_{j\neq k} 2\left|E_j\right|^2\right)E_k \qquad (13.11)$$

into a linear equivalent with a random potential

$$\frac{\partial E_k}{\partial z} + \frac{\mathrm{i}}{2}\beta_{2k}\frac{\partial^2 E_k}{\partial t^2} = \mathrm{i}\gamma V_k\left(z,t\right)E_k, \tag{13.12}$$

where E_k is the slowly varying envelop of the optical field, β_{2k} is the second-order dispersion coefficient for the kth channel, γ is the non-linear coefficient, α is the loss coefficient, and

$$V_k\left(z,t\right) = \sum_{j\neq k} 2\left|E_j\right|^2. \tag{13.13}$$

To obtain (13.12), a number of implicit approximations have been made. First, the fibre is loss-less ($\alpha_k = 0$), which is a reasonable approximation for a periodically amplified system, where the non-linear length scale ($\sim 1/\gamma P_{\max}$) is significantly longer than the amplifier spacing [62] or for an appropriately designed distributed amplifier system [63,64]. Second, the intra-channel effects are neglected ($|E_k|^2 = 0$). Strictly, this corresponds to a regime, where the inter-channel effects are dominant $\left(\sum_{j\neq k} 2\left|E_j\right|^2 >> |E_k|^2\right)$, but may also apply where the impact of this term is compensated [65–71]. Third, (13.12) neglects the interaction between the non-linearity and the ASE, which is reasonable only for sufficiently high local dispersion.

Since the information carried by other channels is unknown, $V_k(z,t)$ appears as a random noise term to the channel k. $V_k(z,t)$ can be modeled as a Gaussian stochastic process with small correlation range in both space and time provided that none of the channels are of a significantly lower symbol rate than its neighbours (short correlation in time) and that the fibre has sufficient dispersion to ensure that the collision length between bits in adjacent channels is sufficiently small [72] (short correlation in space). Equation (13.12) essentially transforms the non-linear channel model into a linear channel with multiplicative noise. The first impact of this is that in the calculation of the channel capacity (13.1), an additional multiplicative noise term is added to the random noise. The random noise is assumed to be dominated by ASE for simplicity. Second, by considering the conservation of energy if such noise power is added to other channels, an equivalent power should be subtracted from the signal. Based on this, low bound to the non-linear channel capacity for coherent detection can be obtained [59]:

$$\left.\frac{C}{B}\right|_{\mathrm{CD}} \geq \frac{B}{\Delta f}\log_2\left(1 + \frac{P_{\mathrm{ave}}\mathrm{e}^{-\left(\frac{P_{\mathrm{ave}}}{I_{\mathrm{XPM}}}\right)^2}}{P_n + \left(1-\mathrm{e}^{-\left(\frac{P_{\mathrm{ave}}}{I_{\mathrm{XPM}}}\right)^2}\right)P_{\mathrm{ave}}}\right), \tag{13.14}$$

where P_{ave} is the average signal power per channel, P_n the total ASE noise power. For a periodically amplified optical system with uniform losses separating identical discrete amplifiers, P_n is equal to $N_a(\mathrm{G}-1)n_{\mathrm{sp}}h\nu B$, with N_a being the number of fibre spans, G the amplifier gain, n_{sp} the spontaneous emission noise factor and B the channel bandwidth. The intensity scale of fluctuation caused by XPM is [59]:

$$I_{\mathrm{XPM}} = \frac{1}{\gamma} \frac{1}{\sqrt{2 \cdot L_{\mathrm{eff}} \sum_{n}^{N_{\mathrm{ch}}/2} \frac{c}{B \cdot D \cdot n \Delta f \lambda^2}}} \tag{13.15}$$

which, for large channel counts, is commonly approximated as

$$I_{\mathrm{XPM}} = \frac{1}{\gamma} \sqrt{\frac{B \cdot D \cdot \left(\Delta f \frac{\lambda^2}{c}\right)}{2 \cdot \ln\left(\frac{N_{\mathrm{ch}}}{2}\right) L_{\mathrm{eff}}}}, \tag{13.16}$$

where D is the local dispersion. N_{ch} is the number of WDM channels and L_{eff} is the non-linear effective length of the system given by $N_a[1 - \exp(-\alpha L)]/\alpha$ for a system with lumped amplifiers, where L is the span length. Note that rather than scaling with an "accumulated non-linear phase" factor, the short correlation intervals of $V_k(z,t)$ ensure that contributions accumulate with random phase, giving a random walk. This random walk results in a square root scaling with the transmission distance and the number of channels.

The non-linear limit basically suggests that, in contrast to linear channels with additive noise, the capacity of a non-linear channel does not grow indefinitely with increasing signal power, but has a maximal value. This is a fundamental feature, which distinguishes non-linear communication channels from linear ones. It is relatively straightforward to find out the optimum launch power P_{opt} from (13.14), and thus predict the maximum ISD for any given system configuration.

$$2P_{\mathrm{opt}}^2 \left(P_{\mathrm{opt}} + P_n\right) = P_n I_{\mathrm{XPML}}^2, \tag{13.17}$$

which is simplified to

$$P_{\mathrm{opt}} = \sqrt[3]{\frac{P_n I_{\mathrm{XPM}}^2}{2}} \text{ if } P_n << I_{\mathrm{XPM}}. \tag{13.18}$$

More comprehensively, we can find that in a linear channel, although the state-of-the-art technologies such as optical OFDM and coherent detection can be used to improve the transmission rate to approach the Shannon linear limit, the only key factor that determines this limit is *snr*. However, the non-linear limit to channel capacity ((13.14)–(13.16)) is a function of various parameters of the transmission fibre (e.g., γ, D, α) and system configuration (e.g. N_{ch}, B). Consequently, in addition to simple increase of *snr* by improving the performance of the optical amplifiers, attention should be paid to system designs which allow us to increase the theoretical information capacity limits of a non-linear channel. We will describe some of these designs or technologies in detail in the next section.

Having established an estimate of the fundamental capacity limit, we now consider the steps required to enable this limit to be approached. First, in common with the linear channel, the input and output symbol constellations should be Gaussian distributed and the detection process should be linear. Whilst a proof has only been presented for a linear channel, simulations approaching (13.14) have been reported using a concentric ring constellation with coherent detection [15], which suggests that this will remain true for the non-linear channel. Second, strong line coding is required to meet this condition, in particular FEC, ideally employed in conjunction with constellation mapping and de-mapping [26, 73].

Third, in-depth analysis has confirmed that the information capacity is reduced if $V_k(z,t)$ is not short range correlated, that is if the dispersion is too low [74]. Sufficiently high local dispersion and residual dispersion per span are required to both suppress the information loss induced by FWM, and to ensure that cross phase modulation accumulates randomly. Finally, deterministic intra-channel non-linearity should be fully compensated. The recent development of electronic dispersion compensation at the transmitter or receiver has led to suggestions that dispersion management may be abandoned in favor of an unbounded growth in residual dispersion to minimize inter-channel effects, although these studies neglect the limiting effects of intra-channel non-linearity. In the absence of non-linearity compensation, some level of dispersion management has been found to be beneficial to control intra-channel non-linear effects by reducing the peak to average power ratio within the channel [75]. Nevertheless, non-linear signal processing, which takes into account intra-channel non-linear effects, may be a key requirement for achieving the limits predicted in (13.14). Preliminary demonstrations of such non-linear signal processing have reduced phase noise of M-PSK signals by modulating the received signal with a phase proportional to the received intensity [65–67]. For multi-level formats, these techniques may also be applied predicatively at the transmitter [68]. Non-linearity compensation may also be applied at the expense of complexity, either by optical phase conjugation [69, 70], or via emulation of back propagation using look-up tables [71]. It is apparent that the inclusion of electronic compensation of intra-channel non-linearity reduces the impact of such intra-channel effects, and may be expected to modify the optimum dispersion map in favour of the reduction of the inter-channel effects. Indeed, the assumption that intra-channel effects may be neglected is applied in the argument supporting (13.15), which in turn suggests a monotonic increase of the channel capacity with dispersion.

Equation (13.14) assumes that FWM and the interaction between the non-linearity and the ASE noise are neglected, which is reasonable only for sufficiently high local dispersion. When this assumption is broken and the local dispersion is not substantial enough, FWM would become prominent and cause information loss that limits the information capacity. Neglecting the effects of quasi-phase matching [52] the capacity bound for FWM-limited capacity may be estimated by (13.14) with the non-linear intensity I_{XPM} replaced by I_{FWM} [9], where

$$\frac{1}{I_{\mathrm{FWM}}} = N_a \sum_{p,q\neq 0}^{|p+q|<\frac{n_c-1}{2}} \frac{\gamma^2 K_{pq}^2}{\alpha^2 + \left(2\pi\lambda^2 D\Delta f^2 q.p/c\right)^2}\; K_{pq} = \begin{matrix} 1, p=q \\ 2, p\neq q \end{matrix}. \quad (13.19)$$

This limit becomes dominant for systems where the product of dispersion and channel spacing squared is small, as would be the case for conventional systems using dispersion-shifted fibre [76] or for low symbol rate OFDM systems [77]. Note also that the FWM intensity scales linearly with the inverse of N_a, and consequently the transmission distance, and so in addition to low dispersion and narrow channel spacing, we would anticipate that the transmission reach would also impact the relative strengths of FWM and XPM. To consider the effects of FWM and XPM simultaneously, we assume that the multiplicative noise from FWM and XPM adds independently, giving

$$\left.\frac{C}{B}\right|_{\mathrm{CD}} \geq \frac{B}{\Delta f}\log_2\left(1+\frac{P_{\mathrm{ave}}\mathrm{e}^{-\left(\frac{P_{\mathrm{ave}}}{I_{\mathrm{XPM}}}\right)^2}\mathrm{e}^{-\left(\frac{P_{\mathrm{ave}}}{I_{\mathrm{FWM}}}\right)^2}}{P_n+\left(1-\mathrm{e}^{-\left(\frac{P_{\mathrm{ave}}}{I_{\mathrm{XPM}}}\right)^2}\right)P_{\mathrm{ave}}+\left(1-\mathrm{e}^{-\left(\frac{P_{\mathrm{ave}}}{I_{\mathrm{FWM}}}\right)^2}\right)P_{\mathrm{ave}}}\right). \tag{13.20}$$

The previous discussions are based on optimal coherent detection. In direction detection where only one degree of freedom per polarization can be used, it may be expected that the maximum ISD of a system is significantly degraded. In a direct-detected optical system where the dominant noise is signal-spontaneous beat noise, starting from the linear ISD limit [9, 78], we find (for high OSNR) that:

$$\left.\frac{C}{B}\right|_{\mathrm{DD}} \approx \frac{1}{2}\log_2\left(\frac{P_{\mathrm{ave}}\mathrm{e}^{-\left(\frac{P_{\mathrm{ave}}}{I_{\mathrm{XPM}}}\right)^2}}{P_n+\left(1-\mathrm{e}^{-\left(\frac{P_{\mathrm{ave}}}{I_{\mathrm{XPM}}}\right)^2}\right)P_{\mathrm{ave}}}\right)-1 \tag{13.21}$$

Figure 13.11 depicts the XPM-limited ISD vs. transmitted power density for coherent and direct detection for a particular transmission system design. The information limits in the linear channels are also plotted for comparison. The figure shows the increase in maximum ISD can be achieved by using coherent detection, and the effect of fibre non-linearity at higher transmitted powers prevents indefinite growth in the channel capacity. For this particular example, the effect of XPM becomes prominent at transmitted power densities beyond ~0.01 W THz^{-1}, and a maximum ISD of 6b s^{-1} Hz^{-1} is predicted. A similar value was reported in recent numerical simulations [79].

A similar non-linear threshold is observed for the direct detection system. However, the reduced linear *snr* performance results in a significantly lower maximum capacity. Note that to achieve this capacity, complex intra-channel non-linearity compensation would be required, whilst for the coherent detection system, the same capacity could be achieved with a significantly lower launch power, ensuring linear transmission.

Figure 13.12 compares the non-linear limits due to FWM and XPM assuming that there is negligible correlation in inter-channel phase from span to span. For this particular system configuration, very similar limits arising from FWM and XPM

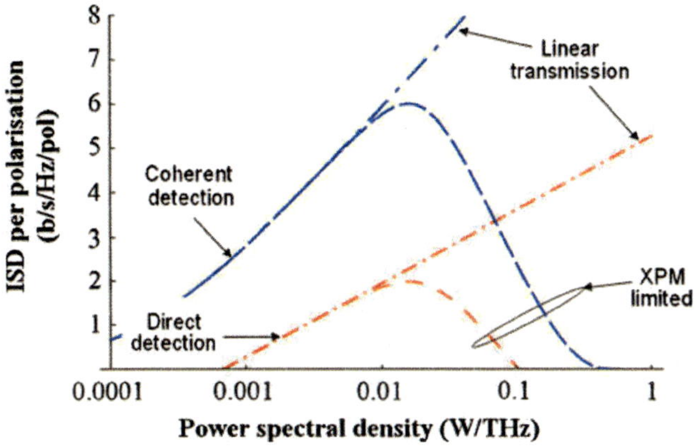

Fig. 13.11 Examples of predicted information spectral density limits per polarisation for linear transmission (*dot-dash*) with coherent (*long dashes*) and direct (*short dashes*) detection and for non-linear transmission (*dashed*) including XPM for coherent (*long dashes*) and direct (*short dashes*) detection. Detailed system parameters are shown in Table 13.2

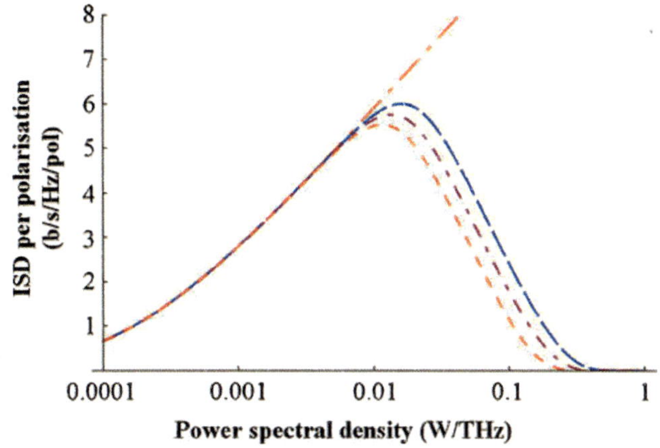

Fig. 13.12 Comparison of the predicted information spectral density limits per polarisation in a coherently detected system for linear transmission (*long dash-dot line*), XPM-limited transmission (*long dash line*), FWM-limited transmission (*short dash-dot line*), and the information spectral density limit including both FWM and XPM effects (*short dash line*). Detailed parameters are shown in Table 13.2

are induced. In this case, it is necessary to consider the impact of both non-linear effects simultaneously (red line). However, for the majority of transmission systems, the different scaling laws for XPM and FWM mean that the design will be limited by only one factor. This is illustrated in Fig. 13.13, where the relative impacts of XPM and FWM are compared for two different system designs. For a system with a wide

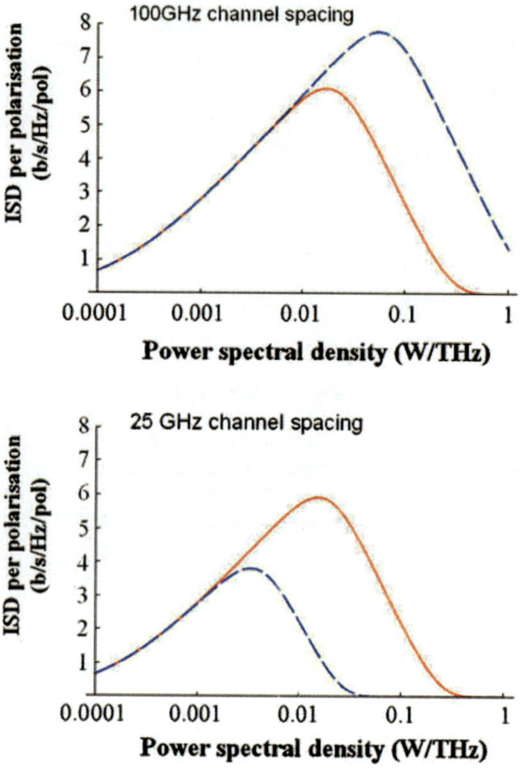

Fig. 13.13 Comparison of the predicted information spectral density limits per polarisation in coherently detected systems for XPM-limited transmission (*solid line*) and FWM-limited transmission (*dashed line*) with channel spacing of 100 GHz (*top*) and 25 GHz (*bottom*). Other parameters, except for the total number of channels, are shown in Table 13.2

channel spacing (in this case 100 GHz), cross phase modulation effects dominate. However, for a system with a closer channel spacing (e.g., 25 GHz in Fig. 13.13b), FWM begins to dominate the achievable system performance.

Various other approaches [80, 81] have been taken to calculate the capacity of a non-linear communication system, including an exhaustive approach based on a generalisation of the Shannon capacity for the case of signal-dependent noise [82]. The signal-dependent noise approach gives a generalised form for the information capacity with the single approximation that the non-linear interaction with the amplifier spontaneous emission may be neglected [74, 83]. The full analysis also allows the impact of dispersion to be examined, and in particular, for low dispersion fibres it is observed that the non-linear effects add monotonically, rather than as a random process. Consequently, the capacity limits for low-dispersion fibres are always lower than those for high-dispersion fibres [74].

13.3.2 Comparison of Reported Results with Theoretical Limit

After extensive investigation of non-linear limit and the insight provided by such limit, we review the advances in various individual technologies that have enabled these capacity limits to be approached since the introduction of optical communication systems. The evolution of the ratio of reported ISDs for numerous transmission system experiments to the maximum values for the same configuration as each reported experiment, derived from (13.14), is shown in Fig. 13.14. Much of the progress in the figure is attributed to improvements in modulation efficiency, adoption of optical amplifiers, and WDM with the subsequent reduction in channel spacing. However, as the capacity limit is approached, the deployment of optimised FEC becomes of paramount importance. The reason for reduction in the rate of increase in reported bit rate distance products from around 3dB per year to less than 1dB per year (Fig. 13.2) becomes clear when we observe, from Fig. 13.14, that experimental measurements already exceeded 50% of the theoretical maximum information capacity by 2008 [84]. Note that with a few notable exceptions [85], the reported results do not implement any intra-channel non-linearity compensation. Now, preliminary research into the compensation of intra-channel non-linearity is under way [86, 87]. However, such approaches appear to be constrained to improve the overall system performance by at most 3dB, unless the effects of inter-channel non-linearity can be mitigated by fundamentally overcoming the non-linear limit to channel capacity. This requires non-linearity compensation over bandwidths exceeding the phase-matching bandwidth of the non-linearity, either using broad bandwidth optical implementations [85], or electronic approaches [77, 87]. In addition, the imminent limit to growth in the information capacity has already seen a strong global resurgence in coherent communications to increase the potential channel capacity, as, for example, illustrated in Fig. 13.11, and changes in fibre design [88, 89].

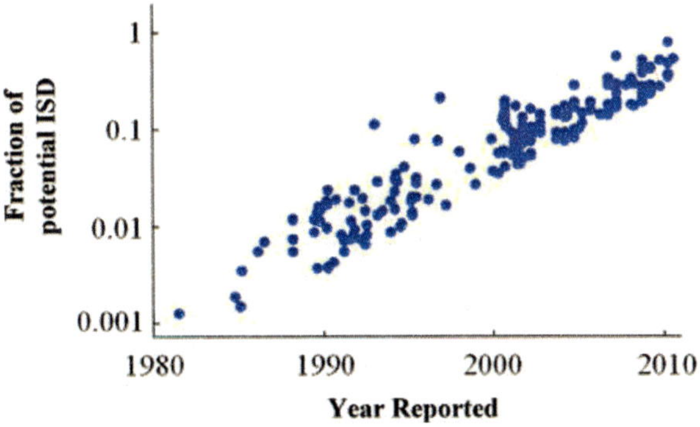

Fig. 13.14 Maximum reported information capacity as a fraction of the capacity limit for the same system configuration as each experiment reported vs. the year. Data omits unrepeatered (no in-line optical amplifier) systems reaching above 200 km and all forms of soliton control and optical regeneration

13.4 Increasing the Information Capacity Limit

Figure 13.14 illustrates that for both direct and coherent detection, the most recent transmission experiments are rapidly approaching the ultimate channel capacity limited by the effects of XPM and ASE noise, as predicted by (13.14) and (13.16). In this section, we will speculate on the promising technologies, which may allow the limit to be increased.

13.4.1 Optical Regeneration

Figure 13.14 omits data points from one particular set of transmission experiments. All-optical regeneration was proposed as a means to increase the capacity of a communication link long time ago [90, 91], and it has been anticipated that the capacity per regenerator could exceed that of opto-electronic regeneration [92]. Whilst such regeneration scheme may operate based on cross-phase modulation [90, 91], carrier density modulation [92, 93] or through parametric effects [94], each device is limited to regenerate a single optical wavelength, and essentially competes with opto-electronic equivalents. On the one hand, the opto-electronic technology not only offers the advantage of a maturity, but also enables the deployment of DSP, such as FEC. On the other hand, multi-wavelength optical regeneration, based on SPM effect [95, 96] enables both distributed and lumped optical regeneration restricting either amplitude noise [97, 98] or both amplitude noise and timing jitter [99]. In these systems, SPM effect is used to restore the pulse shape and quality, primarily resisting not only the effects of ASE noise accumulation, but also the impact of non-linear effects and PMD [100], enabling ultra long haul transmission reported [101].

Considerable progress is still required to produce an ideal all-optical regenerator, especially for multi-level modulation formats. However, in order to analyze the potential capacity benefits, we assume that such device will eventually be feasible, noting that such devices may only regenerate the pulse shape without the capability to correct the errors that have already been made, and so any erroneous decisions made by one regenerator would accumulate until an FEC equipped opto-electronic regenerator is encountered. For uniformly spaced WDM-compatible optical regenerators, the required BER should be divided equally between regenerator spans. This division of BER requires a slightly different approach to calculate capacity limits to that presented in (13.14), where arbitrary error correction coding is assumed. Instead, we apply the formula in Table 13.1 to calculate the contribution to the error probability for each regenerator span – but with the signal-to-noise ratio degraded by cross-phase modulation following the method leading to (13.14), sum these error probabilities to obtain the overall BER, and then calculate the required FEC overhead for error-free operation. This required FEC overhead then allows the net capacity to be calculated (as per Fig. 13.9).

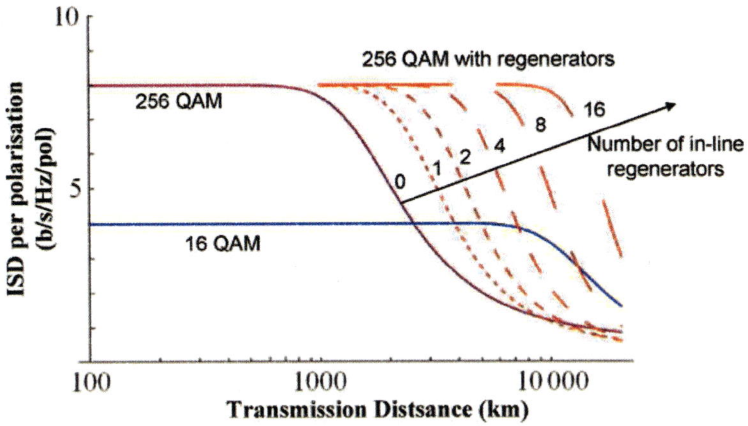

Fig. 13.15 Maximum information spectral density, after FEC for a 16 QAM system and a 256 QAM system without (*labelled solid lines*) all optical regenerators (*dashed lines*, length of dashes proportional to number of regenerators) as a function of distance. All other system parameters are as specified in Table 13.2

Table 13.2 Simulation parameters used for Fig. 13.11 onwards, unless otherwise specified. Values are selected to indicate general trends and do not represent actual system designs

Parameter	Value
System length	1,500 km
Amplifier spacing	100 km
Amplifier noise figure	4.5 dB
Channel spacing	50 GHz
Baud rate	50 Gbaud
Fibre loss	0.18 dB km^{-1}
Non-linear coefficient	1 W^{-1} km^{-1}
Group velocity dispersion	17 ps nm^{-1} km^{-1}
Wavelength	1,550 nm
Amplifier bandwidth	5 THz
Number of channels	101

Figure 13.15 shows the benefit of using optical regenerators for a transmission link with 50 GHz spaced channels and 17 ps nm^{-1} km^{-1} dispersion (see Table 13.2 for other parameters) carrying either 16 or 256-QAM signals, with an increasing number of regenerators within the link for the 256-QAM signal. From this figure, it is immediately apparent that, for 16-QAM applications, optical regeneration is not necessary for transmission distances up to 6,000 km. For 256-QAM however, the reduced *snr* tolerance results in a maximum transmission distance of around 1,000 km. However, optical regenerators essentially divide the link into a number of shorter links. For each of these shorter links, the *snr* degradation is reduced resulting in an improved BER, even when the accumulation of errors from regenerated link to regenerated link is taken into account. For reasonable total BERs, a system with

regenerators will therefore always outperform a link without regenerators. However, to be beneficial, this type of analysis suggests that, in addition to multi-wavelength operation, to provide cost competitiveness, all-optical regenerators should be able to process modulation formats of high order to have a significant impact on the overall ISD of a given transmission link.

13.4.2 Fibre Design

As appealing as hypothetical WDM regenerators appear, the most obvious way to fundamentally increase the maximum ISD is to take optimum values for the parameters in (13.15)–(13.19), including critical fibre characteristics (loss, dispersion and non-linear coefficient), the channel spacing, and the effective amplifier noise figure.

Note that in the case of the transmission medium, whilst the ideal parameters should clearly minimise the loss and non-linearity, and maximise the local dispersion, actual achievable fibre designs will involve a delicate balance between these parameters. According to (13.14), optimum performance should be achieved by maximising the non-linear intensities I_{XPM} and I_{FWM}.

Indeed, many recent transmission records may be attributed to improvements in the fibre designs [88, 102] in addition to optimised transmission formats. The potential benefits are shown in Fig. 13.16, which illustrates the predicted maximum performance for a number of measured solid core fibres (Table 13.3). Figure 13.16 also shows, for comparison, the speculative prediction of the performance of a hollow core photonic crystal fibre (PCF), where the non-linear coefficient has been

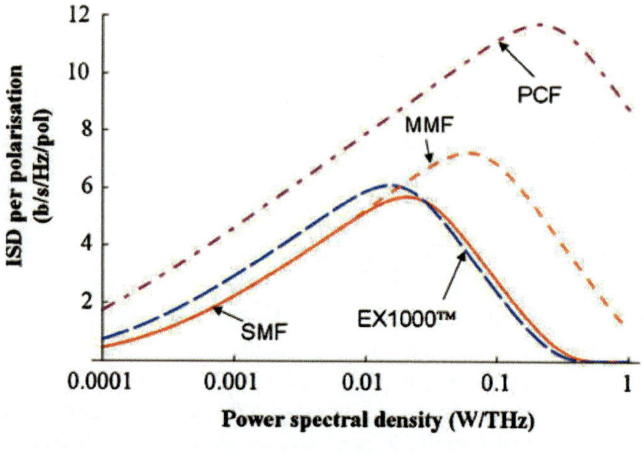

Fig. 13.16 Maximum ISD of standard single mode fibre (*solid line*), Vascade EX1000™ (*long dashed line*), multimode fibre (*short dashed fibre*) and the predicted performance of hollow core photonic crystal fibre (*dot dashed line*). See Table 13.3 for fibre parameters and Table 13.2 for other system parameters

Table 13.3 Fibre parameters used for Fig. 13.16

Fibre type	Loss (dB km^{-1})	Non-linear coefficient ($W^{-1}km^{-1}$)	Wave-band (nm)	Dispersion (ps nm^{-1} km^{-1})
SMF-28™	0.2	1	1,550	17
Vascade EX1000™ [112]	0.175	1.15^{Est}	1,550	18.5
Hollow core PCF [103]	0.13$	0.01^{Est}	1,900	17*
Multi-mode fibre	0.2	0.2^{Est}	1,550	17*

Est: Estimated from effective area or fraction of light propagating in medium, *: Assumed value for chromatic dispersion, $: Predicted value

reduced in proportion to the estimated fraction of the optical signal propagating in glass (1%) and the predicted minimum loss has been assumed [103]. Development of this fibre, and the necessary source, receiver and amplifier technologies [104, 105] for operation in the mid-infrared region would enable the development of long reach communication with information capacities above 10 b s^{-1} Hz^{-1}. Reductions in the effective non-linear coefficient are also expected for multi-mode fibre systems [106], and proportional increases in maximum ISD may be expected if appropriate measures are taken to accommodate modal dispersion. Mode group division multiplexing [107] offers the prospect of multiplying the capacity provided the reach is sufficiently short to avoid significant *snr* penalties due to mode mixing. Operation of PCF in a multi-mode regime, allowing similar enhancements due to mode group division multiplexing, offers remarkable potential increases in channel capacity.

13.4.3 Channel Bandwidth

In terms of the system design, the non-linear intensity for cross phase modulation decreases monotonically with the increasing channel bandwidth. Conventionally, this aspect is constrained by the standardised WDM channel plan, known as the ITU grid, and the capabilities of optical modulators and detectors. We will take the example of OFDM [48], or coherent WDM [36, 46]. From (13.14), we may expect that increasing the channel bandwidth or channel spacing for a fixed amplifier bandwidth, for example 10THz, would reduce the number of adjacent channels and consequently the impact of non-linear crosstalk enabling the information capacity limit to be enhanced.

Figure 13.17 depicts the theoretical capacity limit in a 10 THz bandwidth for different channel bandwidths (or occupied bandwidth per channel in OFDM systems) for a system limited by both XPM and FWM. From this figure, it can be seen that the maximum information capacity is increased as the channel bandwidth increases due to the anticipated dependence of information capacity with channel bandwidth. This is particularly true for low channel spacing (<50 GHz), where the impact of FWM [9] dominates the performance and results in a strong dependence on the

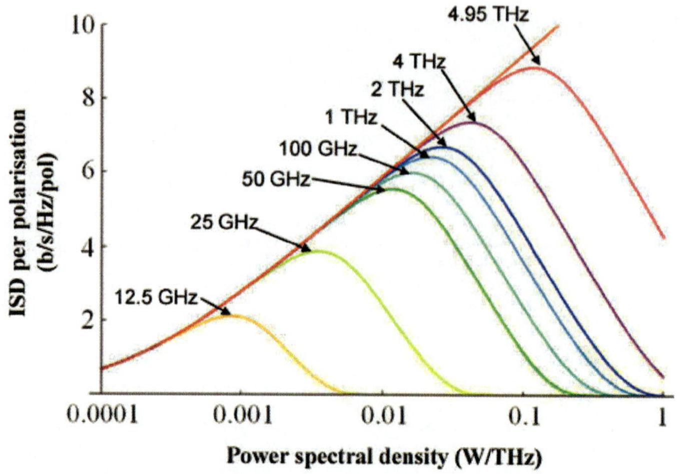

Fig. 13.17 Theoretical channel information spectral density limits vs. power spectral density for a transmission system occupying a 10 THz bandwidth, plotted for different values of the channel bandwidth, with other parameters as per Table 13.2

channel bandwidth. Conventional systems currently have a channel spacing of 50 or 100 GHz, and the performance is predominantly limited by XPM in this region. Between 100 GHz and 2 THz, the benefit of increasing the channel spacing becomes less evident. However, increasing the channel bandwidth from 2 to 5 THz results in a significantly reduced influence from the co-propagating channels. This allows for a welcome increase of 2b s^{-1} Hz^{-1} in capacity.

Note that for optically multiplexed OFDM [48], or coherent WDM [36, 46], which generate phase coherent high capacity signals from a single source, due to the coherence between sub-channels, inter-sub-channel non-linearity compensation is feasible [77, 108], although it may at the expense of the implementation complexity.

13.4.4 Amplifier Noise Figure

It is obvious that reducing noise spectral power density would result in an enhancement in the channel capacity limit. This can be achieved by either optimizing the link configuration (e.g., reduce the span length at the expense of more amplifiers, or the use of distributed amplification where OSNR is maximized [15]), or reducing the noise figure of the amplifiers. The impact of these techniques, however, is somewhat reduced by the logarithmic dependence of (13.1) and (13.14) with respect to the noise power spectral density.

In Fig. 13.18, we consider the effect of reducing the amplifier noise figure from a typical value of 4.5 dB to the quantum limit of 3 dB using equally spaced

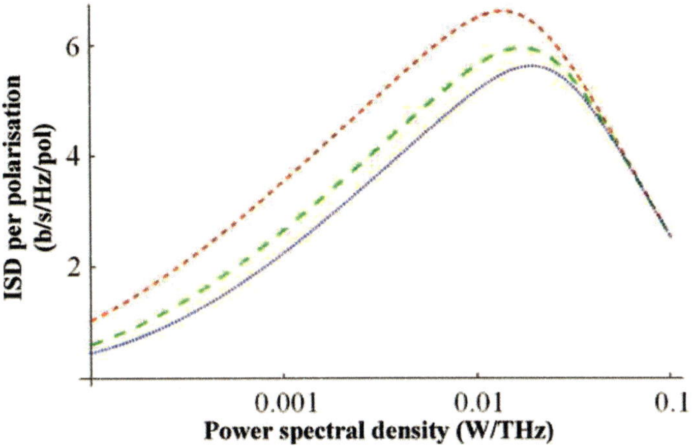

Fig. 13.18 Theoretical ISD for various values of the amplifier noise figure (*dotted*: 4.5dB NF, *long-dashed*: 3dB NF, *short-dashed*: 0dB NF), other parameters as per Table 13.2

amplification, confirming that this offers only a small increase in the maximum ISD. However, in this example, we find that the ISD limit may be increased by a further 1 b s^{-1} Hz^{-1} by using phase sensitive amplification for which the theoretical minimum noise figure is 0 dB. Note that, in this case, we must consider the quantum nature of light and that the photon number distribution is fundamentally broadened by periodic attenuation and amplification. The net effect of these quantum processes is that the final signal-to-noise ratio is improved by a factor of 2 by moving from a quantum-limited phase insensitive amplifier to a quantum-limited phase sensitive amplifier ([19, 109], L. Thylen et al., 2002, Private communication). Whilst the increase in net ISD offered by phase sensitive amplifiers would be welcome, it is not substantial. However of particular note is the required total launch power for a given ISD. For example, Fig. 13.18 shows that to achieve an ISD of 5.5 b s^{-1} Hz^{-1}, a system comprising 4.5 dB noise figure amplifiers would require a launched power spectral density of around 14 mW THz^{-1}, and would clearly be greatly influenced by non-linear effects requiring proper link design to minimise inter-channel non-linearity, and complex compensation of intra-channel non-linearity. The use of ideal phase sensitive amplifiers, however, requires a launch power spectral density of only 4.2 mW THz^{-1} to achieve ISD of 5.5 b s^{-1} Hz^{-1}. Reducing the launch power spectral density to this level enables propagation in the linear transmission regime and a substantial energy saving, even when the maximum 50% power efficiency of broadband phase sensitive amplifiers [110] is taken into account. Note that whilst the use of a PSA constrains the system to operation in a single quadrature, by exploiting this known constraint in the design of the modulation format, for example by using Fast-OFDM in a dispersion-managed link, no fundamental loss in information capacity is required. However, practical deployment of such phase-sensitive amplifiers requires further development to realise fibre to fibre noise figures approaching the assumed

0 dB, and the development of systems to ensure that the useable gain bandwidth of a phase-sensitive amplifier approaches that of the phase-insensitive parametric amplifiers [111].

13.5 Conclusions

Communication capacity has shown a remarkable exponential growth over more than 30 years, with the overall capacity of the core of the network closely tracking the user demand. This demand is expected to continue to rise; however, we have seen in this chapter that the reported capacities from recent experiments are approaching the fundamental limits imposed by signal-to-noise ratio and Kerr effect in conventional optical fibres.

In the short term, many technologies may be used to incrementally add capacity. However, the capacity increase only grows logarithmically with the improvements in the system design, whilst the underlying capacity demand continues to rise exponentially, with a rate equal to a doubling every 2 years. In this circumstance, it appears inevitable that unless capacity demand saturates, or network architectures are devised, which radically alter the capacity demands placed on the core network, new transmission media will be essential within the next two decades.

References

1. A.S. Eve, C.H. Creasey, *Life and Work of John Tyndall* (Macmillan, London, 1945)
2. J.J. Thomson, Recent Researches (1893), http://digital.library.cornell.edu/cgi/t/text/text-idx?c=cdl;cc=cdl;view=toc;subview=short;idno=cdl022
3. E. Snitzer, J. Opt. Soc. Am. **51**, 491–498 (1961)
4. K.C. Kao, G.A. Hockham, Proc. IEE **113**(7), 1151–1158 (1966)
5. F.P. Kapron, D.B. Keck, R.D. Maurer, Appl. Phys. Lett. **17**, 423–425 (1970)
6. E.B. Desurvire, J. Lightwave Technol. **24**(12), 4697–4710 (2006)
7. L. Wood, D. Blankenhdn, DESIDOC Bull. Inform. Technol. **15**(4), 23–31 (1995)
8. IEEE P802.3av Task Force, 10 Gb/s Ethernet Passive Optical Network, http://www.ieee802.org/3/av, downloaded 20/4/2009
9. J.M. Kahn, K.-P. Ho, IEEE J. Select. Top. Quant. Electron. **10**(2), 259–272 (2004)
10. C.E. Shannon, Bell Syst. Tech. J. **27**, 379–423, 623–656 (1948)
11. C.E. Shannon, W. Weaver, *The Mathematical Theory of Communication* (University of Illinois Press, IL, 1963)
12. H. Nyquist, Trans. Am. Inst. Elec. Eng. **47**, 617–644 (1928)
13. M.E. McCarthy, J. Zhao, A.D. Ellis, P. Gunning, IEEE J. Lightwave Technol. **27**, 5327–5335 (2009)
14. X. Liu, DSP-enhanced differential direct-detection for DQPSK and m-ary DPSK, European conference on optical communication (ECOC), paper 07.2.1, 2007; E.B. Desurvire, J. Lightwave Technol. **24**(12), 4697–4710 (2006)
15. R.-J. Essiambre, Capacity limits of fiber-optic communication systems, in *Proceedings of OFC 2009*, San Diego, ISA, Paper OThL1, 2009
16. N. Kikuchi, K. Mandai, K. Sekine, S. Sasaki, J. Lightwave Technol. **26**(1), 150–157 (2008)

17. J.M. Kahn, E. Ip, Principles of digital coherent receivers for optical communications, in *Proceedings of OFC 2009*, San Diego, ISA, Paper OTuG5, 2009
18. Ip, A.P.T. Lau, D.J.F. Barros, J.M. Khan, Opt. Express **16**(2), 753–791 (2008)
19. E. Desurvire, *Erbium-Doped Fiber Amplifiers* (Wiley, Hoboken, 2002)
20. S. Haykin, *Digital Communications* (Wiley, NY, 1988)
21. N. Kikuchi, S. Sasaki, Improvement of tolerance to fibre non-linearity of incoherent multilevel signalling for WDM transmission with 10-Gbit/s OOK channels, in *Proceedings of ECOC 2009*, Vienna, Austria, Paper 8.4.1, 20–24 September 2009
22. J.G. Proakis, *Digital Communications*, 4th edn. (McGraw-Hill, New York, 2000)
23. F. Gray, Pulse code communication, U.S. Patent 2,632,058, March 17, 1953 (filed Nov 1947)
24. M. Nakazawa, Challenges to FDM-QAM coherent transmission with ultrahigh spectral efficiency, in *Proceedings of ECOC 2008*, Brussels, Paper Tu1E1, 2008
25. S.Y. Chung, G.D. Forney, T.J. Richardson, R. Urbanke, IEEE Commun. Lett. **5**(2), 58–60 (2001)
26. B. Zhou, L. Zhang, J. Kang, O. Huang, Y.Y. Tai, S. Lin, M. Xu, Non-binary LDPC codes vs. Reed-Solomon codes, in *Proceedings of information theory and applications workshop 2008*, San Diego, pp. 175–184, 2008
27. G.709: Interfaces for the Optical Transport Network (OTN), downloaded from http://www.itu.int/rec/T-REC-G.709/en.
28. R.W. Chang, Bell Syst. Tech. J. **45**, 1775–1796, (1966)
29. R.R. Mosier, R.G. Clabaugh, AIEE Trans. **76**, 723–728 (1958)
30. H. Sanjoh, E. Yamada, Y Yoshikuni, Optical orthogonal frequency. division multiplexing using frequency/time domain filtering for high spectral efficiency up to 1 bit/s/Hz, in *Proceedings of OFC'02*, Anaheim, Paper ThD1, 2002
31. A. Sano, E. Yamada, H. Masuda, E. Yamazaki, T. Kobayashi, E. Yoshida, Y. Miyamoto, S. Matsuoka, R. Kudo, K. Ishihara, Y. Takatori, M. Mizoguchi, K. Okada, K. Hagimoto, H. Yamazaki, S. Kamei, H. Ishii, 13.4-Tb/s (134 × 111-Gb/s/ch) No-Guard-Interval Coherent OFDM Transmission over 3,600 km of SMF with 19-ps average PMD, in *Proceedings of ECOC'08*, Brussels, Paper Th3E1, 2008
32. K. Takiguchi, M. Oguma, T. Shibata, H Takahashi, Optical OFDM demultiplexer using silica PLC based optical FFT circuit, in *Proceedings of OFC 2009*, San Diego, Paper OWO3, 2009
33. A.D. Ellis, F.C.G. Gunning, Filter strategies for coherent WDM, in *Proceedings of emerging technologies in optical sciences*, Cork, 26–29 July 2004
34. A.D. Ellis, F.C.G. Gunning, Photon. Technol. Lett. **17**(2), 504–506 (2005)
35. J. Zhao, A.D. Ellis, Performance improvement using a novel MAP detector in coherent WDM systems, in *Proceedings of ECOC'08*, Paper Tu1.D.2, 2008
36. T. Healy, F.C. Garcia Gunning, E. Pincemin, B. Cuenot, A.D. Ellis, 1,200 km SMF (100 km spans) 280 Gbit/s coherent WDM transmission using hybrid Raman/EDFA amplification, in *ECOC'07*, Berlin, Paper Mo1.3.5, 2007
37. A.J. Lowery, J. Armstrong, Opt. Express **14**, 2079–2084 (2006)
38. B.J.C. Schmidt, Z. Zan, L.B. Du, A.J. Lowery, 100 Gbit/s transmission using single band direct detection optical OFDM, in *Proceedings of OFC'09*, San Diego, Paper PDPC4, 2009
39. I.B. Djordjevic, B. Vasic, IEEE Photon. Technol. Lett. **18**(15), 1576–1578 (2006)
40. S.L. Jansen, I. Morita, H. Tanaka, 10-Gb/s OFDM with conventional DFB lasers, in *Proceedings of ECOC'07*, Berlin Paper Tu. 2.5.2, 2007
41. W. Shieh High spectral efficiency coherent optical OFDM for 1 Tb/s Ethernet transport, in *Proceedings of OFC 2009*, San Diego, Paper OWW1, 2009
42. S.L. Jansen, I. Morita, N. Takeda, H. Tanaka, 20-Gb/s OFDM transmission over 4,160-km SSMF enabled by RF-pilot tone phase noise compensation, in *Proceedings of optical fiber communication (OFC) conference 2007*, Anaheim, Paper PDP 15, 2007
43. H. Takahashi, A. Al Amin, S.L. Jansen, I. Morita, H. Tanaka DWDM transmission with 7.0 bit/s/Hz spectral efficiency using 8 × 65.1 Gbit/s coherent PDM OFDM signals, in *Proceedings of OFC 2009*, San Diego, Paper PDPB7, 2009
44. X. Yi, W. Shieh, Y. Ma, Phase noise on coherent optical OFDM systems with 16-QAM and 64-QAM beyond 10 Gb/s, in *Proceedings of ECOC'07*, Berlin, Paper Tu5.2.3, 2007

45. T. Miki, H. Ishio, IEEE Trans. Commun. **26**(7), 1082–1087 (1978)
46. A.D. Ellis, F.C.G. Gunning, B. Cuenot, T.C. Healy, E. Pincemin, M. Rukosueva, Towards 1TbE using coherent WDM, in *Proceedings of OECC/ACOFT 2008*, Sydney, Paper WeA-1, 2008
47. F.C.G. Gunning. T. Healy, X. Yang, A.D. Ellis, 0.6Tbit/s capacity and 2bit/s/Hz spectral efficiency at 42.6Gsymbol/s using a single DFB laser with NRZ coherent WDM and polarisation multiplexing, in *CLEO Europe 2007*, Munich, Germany, Paper CI8–5, 2007
48. Y. Ma, Q. Yang, Y. Tang, S. Chen, W. Shieh, 1 Tb/s per channel coherent optical OFDM transmission with subwavelength bandwidth access, in *Proceedings of OFC'09*, San Diego, Paper PDPC1, 2009
49. J. Armstrong, Electron. Lett. **38**(5), 246–248 (2002)
50. D.J. Malyon, T. Widdowson, E.G. Bryant, S.F. Carter, J.V. Wright, W.A. Stallard, Electron. Lett. **27**(2), 120–121 (1991)
51. H.J. Thiele, R.I. Killey, P. Bayvel, Electron. Lett. **34**(21), 2050–2051 (1998)
52. A.D. Ellis, W.A. Stallard, Four Wave mixing in ultra long transmission systems incorporating linear amplifiers, IEE Colloquium, 159 (1990), http://ieeexplore.ieee.org/xpl/freeabs_all.jsp?arnumber=190875
53. R.-J. Essiambre, B. Mikkelsen, G. Raybon, Electron. Lett. **35**(18), 1576–1578 (1999)
54. A.D. Ellis, J.D. Cox, D. Bird, J. Regnault, J.V. Wright, W.A. Stallard, Electron. Lett. **27**(10), 878 (1991)
55. I. Morita, K. Tanaka, N. Edagawa, M. Suzuki, Impact of the dispersion map on long-haul 40 Gbith single-channel soliton transmission with periodic dispersion compensation, in *Proceedings of OFC'99*, San Diego, Paper FD1, 1999
56. P.V. Mamyshev, L.F. Mollenauer, Opt. Lett **21**(6), 396–398 (1996)
57. N.J. Smith, N.J. Doran, Opt. Lett. **21**(8), 570–572 (1996)
58. E Pincemin, A. Tan, A. Bezard, A. Tonello, S. Wabnitz, J-D Ania-Castañòn, S. Turitsyn, Opt. Express **14**(25), 12049–12062 (2006)
59. P.P.Mitra, J.B.Stark, Nature **411**, 1027–1030 (2001)
60. L.G.L. Wegener, b. M.L. Povinelli, A.G. Green, P.P. Mitra, J.B. Stark, P.B. Littlewood, Phys. D Nonlinear Phenomena **189**(1–2), 81–99 (2004)
61. R.J. Essiambre, G.J. Foschini, P.J. Winzer, G. Kramer, Exploring capacity limits of fibre-optic mommunication systems, in *Proceedings of ECOC 2008*, Brussels, Paper We1E1, 2008
62. K.J. Blow, N.J. Doran, IEEE Photon. Technol. Lett. **3**(4), 369 (1991)
63. A. Altuncu, L. Noel, W.A. Pender, A.S. Siddiqui, T. Widdowson, A.D. Ellis, M.A. Newhouse, A.J. Antos, G. Kar, P.W. Chu, Electron. Lett **32**(3), 233 (1996)
64. V. Karalekas, J.-D. Ania-Castañón, P. Harper, S.K. Turitsyn, Ultra-long Raman fibre laser transmission links (Invited), in *Proceedings of the 11th international conference on transparent optical networks*, Paper Tu.A.2.1, 2009
65. X. Liu, X. Wei, R.E. Slusher, C.J. McKinstrie, Opt. Lett. **27**, 1616–1618 (2002)
66. K. Kikuchi, Opt. Express **16**(2), 889–896 (2008)
67. L. F. Mollenauer, A. Grant, X. Liu, X. Wei, C. Xie, I. Kang, C. Doerr, Demonstration of 109 X 10G dense WDM over more than 18,000 km using novel, periodic-group-delay complemented dispersion compensation and dispersion managed solitons, in *Proceedings of ECOC 03*, Rimini, Post-deadline Paper Th4.3.4, 2003
68. C. Xu, X. Liu, X. Wei, IEEE J. Select. Top. Quant. Electron. **10**(2), 281–293 (2004)
69. W. Pieper, C. Kurtze, R. Schnabel, D. Bruer, R. Ludwig, K. Petermann, Electron. Lett. **30**(9), 724–725 (1992)
70. S. Watanabe, M. Shirasaki, J. Lightwave Technol. **14**(3), 243–248 (1996)
71. K. Roberts, C. Li, L. Strawczynski, M. O'Sullivan, I. Hardcastle, Photon. Technol. Lett. **18**(2), 403–405 (2006)
72. L.F. Mollenauer, J.P. Gordon, *Solitons in Optical Fibers: Fundamentals and Applications* (Elsevier, MA, 2006)
73. C. Spagnol, W. Marnane, E Popovici, FPGA implementations of LDPC over GF (2 m) decoders, in *Proceedings of 2007 IEEE workshop on signal processing*. Institute of Electrical and Electronics Engineers, Shanghai, China, 2007, pp. 273–278, 2007

74. J. Tang, J. Lightwave Technol. **24**(5), 2070–2075 (2006)
75. Y. Frignac, J.-C. Antona, S. Bigo, Enhanced analytical engineering rule for fast optimization dispersion maps in 40 Gbit/s-based transmission, Optical fiber communication conference, 2004. OFC 2004, vol. 1, 23–27 Feb 2004
76. R.W. Tkach, A.R. Chraplyvy, F. Forghieri, A.H. Gnauck, R. M. Derosier, J. Lightwave Technol. **13**(5), 841–849 (1995)
77. A.J. Lowery, Opt. Express **15**(20), 12965 (2007)
78. A. Mecozzi, M. Shtaif, Photon. Technol. Lett. **13**, 1029–1031 (2001)
79. R.-J. Essiambre, G.J. Foschini, P.J. Winzer, G. Kramer, E.C. Burrows, The capacity of fiber-optic communication systems, in *Proceedings of OFC2008*, San Diego, Paper OTuE1, 2008
80. B. Wu, E. Narimanov, Information capacity of nonlinear fiber-optical systems, in *Proceedings of 2005 quantum electronics and laser science conference (QELS)*, Paper JThE74, 2005
81. K.S. Turitsyn, S.A. Derevyanko, I.V. Yurkevich, S.K. Turitsyn, Phys. Rev. Lett. **91**, 203901 (2003)
82. M.S. Pinsker, *Information and Information Stability of Random Variables and Processes* (Holden Day, San Francisco, 1964), pp. 160–201
83. J. Tang, J. Lightwave Technol. **19**, 1104–1109 (2000)
84. H. Goto, M. Yoshida, T. Omiya, K. Kasai, M. Nakazawa, IEICE Electron. Express **5**(18), 776–781 (2008)
85. S.L. Jansen, D. van den Borne, C. Climent, M. Serbay C.-J. Weiske, H. Suche, P.M. Krummrich, S. Spalter, S. Calabro, N. Hecker-Denschlag, P. Leisching, W. Rosenkranz, W. Sohler, G.D. Khoe, T. Koonen, H. de Waardt, 10,200km 22 × 2 × *l*0Gbit/s RZ-DQPSK dense WDM transmission without inline dispersion compensation through optical phase conjugation, in *Proceedings of OFC'05*, Anaheim, Paper PDP28, 2005
86. E Ip, J.M Kahn, J. Lightwave Technol. **26**(20), 3416–3425 (2008)
87. X. Li, X. Chen, G. Goldfarb, E. Mateo, I. Kim, F. Yaman, G. Li, Opt. Express **16**, 880–888 (2008)
88. L. Becouarn, G. Vareille, P. Pecci, J.F. Marcerou, 3Tbit/s transmission (301 DPSK channels at 10.709Gb/s) over 10270km with a record efficiency of 0.65(bit/s)/Hz, in *Proceedings of ECOC 03*, Rimini, Post-deadline Paper Th4.3.2, 2003
89. I. Morita, N. Edagawa, 50GHz-spaced 64 × 42.7 Gbit/s transmission over 8200km using pre-filtered CS-RZ DPSK signal and EDFA repeaters, in *Proceedings of ECOC'03*, Rimini, Paper Th4.3.1, 2003
90. A.D. Ellis, D.A. Cleland, Electron. Lett. **28**(4), 405 (1992)
91. M. Jinno, M. Abe, Electron. Lett. **28**(14), 1350 (1992)
92. Y. Ueno, S. Nakamura, K. Tajima, IEEE Photon. Technol. Lett. **13**(5), 469–471 (2001)
93. R.J. Manning, A.D. Ellis, A.J. Poustie, K.J. Blow, J. Opt. Soc. Am. B **14**, 3204–3216 (1997)
94. K. Croussore, C. Kim, G. Li, Opt. Lett. **29**, 2357–2359 (2004)
95. L.F. Mollenauer, P.V. Mamyshev, M.J. Neubelt, Electron. Lett. **32**(5), 471 (1996)
96. P.V. Mamyshev, All-optical data regeneration based on self-phase modulation effect, in *Proceedings of ECOC*, vol. 1, pp. 475–476, 1998
97. B. Cuenot, A.D. Ellis, Opt. Express, **15**(18), 11492–11499 (2007)
98. P. Petropoulos, L. Provost, F. Parmigiani, C. Kouloumentas, C. Finot, K. Mukasa, P. Vorreau, I. Tomkos, S. Sygletos, W. Freude, J. Leuthold, A. D. Ellis, D.J. Richardson, Simultaneous 2R regeneration of WDM signals in a single optical fibre, IEEE/LEOS winter topical meeting series, pp 252–253, 2009
99. O. Leclerc, E. Desurvire, O. Audouin, Opt. Fiber Technol. **3**(2), 97–116 (1997)
100. I.Y. Khrushchev, I.D. Phillips, A.D. Ellis, R.J. Manning, D. Nesset, D.G. Moodie, R.V. Penty, I.H. White, Electron. Lett. **35**(14), 1183–1185 (1999)
101. M. Nakazawa, K. Suzuki, H. Kubota, A. Sahara, E. Yamada, Electron. Lett. **34**(1), 103–104 (1998)
102. J.M.C. Boggio, C. Lundström, J. Yang, H. Sunnerud, P.A. Andrekson, Double-pumped FOPA with 40 dB flat gain over 81 nm bandwidth, in *Proceedings of ECOC 2008*, Brussels, Paper Tu.3B5, 2008

103. P.J. Roberts, F. Couny, H. Sabert, B. Mangan, D. Williams, L. Farr, M. Mason, A. Tomlinson, T. Birks, J. Knight, P.St.J. Russell, Opt. Exp. **13**, 236–244 (2005)
104. R.M. Percival, D. Szebesta, C.P. Seltzer, S.D. Perin, S.T. Devey, M. Louka, J. Quant. Electron. **31**(3), 489–493 (1995)
105. A. Krier, Y. Mao, Infrared Phys. Technol. **38**(7), 397–403 (1997)
106. Z. Tong, Q. Yang, Y. Ma, W. Shieh, 21.4 Gb/s coherent optical OFDM transmission over 200 km multimode fiber, in *Proceedings of OECC/ACOFT 2008*, Syndey, Paper PDP5, 2008
107. C.P. Tsekrekos, A. Martinez, F.M. Huijskens, A.M.J. Koonen, IEEE Photon. Technol. Lett. **18**, 2359–2361 (2006)
108. E. Yamazaki, F. Inuzuka, K. Yonenaga, A. Takada, M. Koga, IEEE Photon. Technol. Lett. **19**(9), 9–11 (2007)
109. H.A. Haus, Y. Yamamoto, IEEE J. Quant. Electron. **QE-23**, 212–221 (1987)
110. S. Oda, H. Sunnerud, P.A. Andrekson, Opt. Lett. **32**(13), 1776–1778 (2007)
111. G. Charlet, M. Salsi, H. Mardoyan, P. Tran, J. Renaudier, S. Bigo, M. Astruc, P. Sillard, L. Provost, F. Cérou, Transmission of 81 channels at 40Gbit/s over a transpacific-distance erbium-only link, using PDM-BPSK modulation, coherent detection, and a new large effective area fibre, in *Proceedings of ECOC'08*, Brussels, Paper Th3E3, 2008
112. S. Ten, Advanced fibers for submarine and long-haul applications, in *Proceedings of LEOS 2004*, vol. 2, pp. 543–544, San Francisco, Paper WJ2, 2004

Index

A

ADC. *See* Analog-to-digital converter
AFC. *See* Automatic frequency control
Amplifier noise figure, 529, 530, 532–534
Amplitude regenerator, 416–427, 432–434, 440– 442, 445
Amplitude shift keying (ASK), 18, 179, 512–513, 516, 517
Analog-to-digital converter (ADC), 6, 7, 17, 21, 22, 30, 32, 34, 48–49, 62, 63, 71, 72, 79, 104, 105, 135, 140, 146
Asymptotic power efficiency, 227, 230, 231, 233, 237, 239
Automatic frequency control (AFC), 195, 202
Auxiliary phase coding, 177

B

Balanced detection, 196, 277
Beat-linewidth, 198
Bessel filter, 197
Bit error rate (BER), 6, 9, 11, 16, 23, 24, 26, 28, 30–33, 58, 74, 77, 79, 89, 130–132, 134, 135, 148, 153–154, 203–206, 208, 210, 220–222, 241–246, 321, 326, 328, 329, 333, 337, 351, 352, 355, 357, 360–364, 366, 367, 373, 374, 392, 393, 396–399, 402–406, 408, 409, 423–425, 435, 436, 452, 453, 458, 463, 470, 474–477, 481, 482, 487–494, 512, 515– 519, 528, 529
Bit mapping, 178
Bit patterning, 392–397, 410
Block length, 193, 351
Butterfly equalizer, 192

C

Capacity, 1, 2, 4, 7–9, 11, 12, 15, 26, 34, 35, 44, 47, 80, 177, 201, 214, 227, 229, 246, 254, 281–282, 332, 343, 446, 457, 507–519, 522–524, 526–528, 531, 532, 534
Capacity-distance product, 2, 177, 214
Carrier synchronization, 187, 188, 192, 197
Channel capacity, 47, 66, 254, 452–453, 457, 458, 464, 494, 495, 497, 502, 507–534
Channel coding, 452–464
Channel estimation, 28, 53, 58, 62, 64–67, 70–71, 78
Chromatic dispersion (CD), 1, 60, 80, 88, 177, 198, 202, 294, 320, 329, 330, 332, 338, 346, 347, 349–351, 353, 354, 358, 365, 368, 406, 408, 415, 451, 475, 482, 507, 531
Chromatic dispersion compensation, 202, 294, 330, 332, 353, 415, 482
Chromatic dispersion tolerances, 198
Class partitioning, 194, 208
Clipping, 58, 59
Coded modulation, 451–502
Coherent detection, 1, 3, 21–25, 29, 43, 59, 60, 88, 188, 219, 254, 334, 343, 452, 472–475, 491, 498–502
Coherent optical OFDM (CO-OFDM), 3, 5, 16, 21, 28–34, 44, 45, 48, 53–55, 59–81, 87–171, 519
Complex and real representations, 95
Constant modulus algorithm (CMA), 22, 192, 206, 351
Constellation, 4, 58, 91, 178, 220, 260, 326, 348, 416, 452, 511
CP. *See* Cyclic prefix
Cross-channel OFDM (XC-OFDM), 48, 54
Cross-gain modulation (XGM), 283, 416, 420
Cross-phase modulation (XPM), 10–13, 15, 88, 90, 95–103, 108, 109, 111, 144–147, 151–154, 159, 169, 170, 214, 221, 247, 248, 264–268, 282, 285, 294,

S. Kumar (ed.), *Impact of Nonlinearities on Fiber Optic Communications*, Optical and Fiber Communications Reports 7, DOI 10.1007/978-1-4419-8139-4,
© Springer Science+Business Media, LLC 2011

310–314, 317, 319, 320, 322, 323, 325–339, 345–346, 348, 352, 356–358, 365, 367, 416, 423, 426, 439, 520, 522, 524–526, 528, 531, 532
Cross-polarization modulation (XPolM), 247, 343–344, 346–348, 352, 353, 355, 357, 358, 360, 365, 367, 368
Crosstalk, 6–8, 32, 206, 214, 215, 248, 343, 344, 352–354, 362, 397–399, 519, 531
Cyclic prefix (CP), 28, 44, 51–53, 55, 59, 76, 78, 90–93, 95, 98, 104, 137, 156, 317

D
DAC. *See* Digital-to-analog converter
Data recovery, 189, 190, 194, 202, 203, 208, 209
3 dB coupler, 78, 191, 196, 206, 362, 428, 437, 491
DCF. *See* Dispersion compensating fiber
Degrees of freedom (DOF), 35, 142, 143, 220–222, 253, 281, 293, 298, 300, 304, 307–310, 312, 322
Delay line interferometer (DLI), 189, 190
Demultiplexer, 31, 32, 184, 350
DFT. *See* Discrete Fourier transform
Differential decoder, 446
Differential detection, 1–35, 180, 189, 191, 194–195, 198, 260, 277, 443
Differential encoder, 185, 186
Differential phase shift keying (DPSK), 7, 59, 180, 253, 293, 325, 406, 416, 472
Differential QPSK (DQPSK), 3, 180, 253, 260, 325, 361, 416
Differential quadrant encoding, 187
Differential quadrature phase-shift keying, 3–4, 6–19, 180, 187, 190, 196, 253, 260, 263–268, 270–273, 275, 278–282, 325–330, 332–334, 336, 339, 361–364, 416, 440–445
Digital backpropagation, 492–494, 500, 501
Digital coherent receiver, 22, 79, 191
Digital phase estimation, 184, 191–193, 199, 202, 205, 212
Digital signal processing, 6–7, 17, 19–22, 25, 29, 32, 34, 43–45, 60, 62–65, 67–69, 78, 79, 82, 144, 146, 154, 155, 159, 178, 180, 190–192, 194–196, 203, 214, 219, 220, 294, 330, 334, 338, 343, 344, 347, 348, 351, 367, 368, 512, 519, 528
Digital-to-analog converter (DAC), 34, 48, 71–72, 91–94, 104, 135
Direct detection, 2, 9, 17, 21, 23, 43–45, 58, 59, 87, 178, 180, 188–191, 195, 198, 205, 259–263, 277, 350, 360–365, 367, 368, 374, 472–475, 489, 518, 524
Direct-detection optical OFDM (DDO-OFDM), 44, 45, 53, 59
Discrete Fourier transform (DFT), 44, 50–53, 62, 64, 98, 105, 106, 147, 159, 168, 170, 171
Dispersion compensating fiber (DCF), 89, 96, 122–124, 127, 159, 201–203, 205, 210, 212, 247, 284, 350–356, 358–362, 365, 366, 423
DLI. *See* Delay line interferometer
DOF. *See* Degrees of freedom
Duobinary, 8, 177

E
Equalization, 22, 31, 60, 99, 138, 157, 158, 180, 190–192, 194, 198, 200, 202, 205, 206, 208, 209, 213, 353, 382, 384, 452
Erbium doped fiber amplifier (EDFA), 3, 8, 26, 59, 177, 205, 208, 219, 246, 349, 362, 423, 425
External cavity laser (ECL), 75, 78, 201, 206, 207
Eye spreading, 181

F
Feed forward M-th power block scheme, 192
Fibre nonlinearity, 191, 201, 205, 213, 214
Field-programmable gate array (FPGA), 6–7, 25, 67, 69, 79, 209
Flat-histogram importance sampling (FH-IS), 374, 379
Forward error correction (FEC), 5, 6, 9, 25–27, 29, 31, 33, 77, 177, 326, 328, 333, 336, 373, 397–399, 402, 403, 409, 451, 452, 454, 463, 507, 516–519, 523, 527–529
Four-dimensional, 220–223, 231, 235–241, 249, 250
Four wave mixing (FWM), 87–90, 95–105, 108, 109, 111, 112, 115, 117, 121–135, 140, 145–148, 150–153, 156, 159, 214, 257, 262–265, 267, 290, 294, 311, 314–317, 319, 320, 322, 323, 334, 345, 404, 433–435, 438–440, 520, 523– 526, 531
FPGA. *See* Field-programmable gate array
Frequency offset, 62–64, 68–69, 78, 195, 311
Frequency offset synchronization, 63–64, 68–69
Frequency shift keying (FSK), 277

G
Gaussian filter, 307, 309
Gordon-Mollenauer phase noise, 20, 293, 298–302, 416
Gray coding, 178

H
Heterodyne detection, 188, 511
Homodyne detection, 23, 24, 188, 511
2×4.90° Hybrid, 192

I
Importance sampling, 373–380, 409
Impulse shaper, 184
Information capacity, 457, 458, 494–502, 510, 518, 520–534
Intensity detection branch, 18, 189, 190
Intercarrier interference (ICI), 47, 49, 51, 62
Intermediate frequency (IF), 188, 511
Intra-channel cross-phase modulation (IXPM), 310
Intra-channel nonlinearity, 159, 253–290
IQ modulator (IQM), 45, 74–76, 90, 91, 93, 135, 181, 183, 184, 186, 187, 201, 207, 210
IQ receiver, 190

K
Kerr nonlinearity of optical fibers, 254, 426, 433

L
Laser linewidth, 197, 198
Laser linewidth requirements, 190, 194, 197, 200
Laser phase noise, 15, 24, 194, 195, 197, 198, 325, 334
Least mean square (LMS) algorithm, 143, 192
Level generator, 181, 184
Linearly mapped systems, 59
Local oscillator (LO), 48, 75, 104, 219, 224, 277, 334, 350, 351, 445, 446
Low-density-parity-check (LDPC) codes, 451–452, 458, 464–479, 481–494

M
Mach-Zehnder, 207, 332
Mach-Zehnder modulator (MZM), 61, 78, 181–186, 201, 210, 350, 393, 394, 420, 475, 476, 500
Markov Chain Monte Carlo (MCMC), 374, 387–391, 397, 408–411
m-ary PSK, 113, 124, 125, 128, 130, 416, 472, 484
m-ary QAM, 472
Metropolis-Hastings machine, 399
Minimum distance, 232, 237, 239, 416, 464–466, 477
Minimum transmission point, 183, 185, 186
Modulation format, 1–3, 8, 9, 16, 21–24, 28, 43, 55, 58, 60, 82, 124, 154, 159, 177–181, 184, 189–191, 194, 197–199, 201, 203, 204, 209, 213–215, 219–221, 223, 226, 227, 229, 230, 232, 237, 238, 240, 241, 243, 244, 246, 248, 250, 344, 359–365, 369, 416, 443, 453, 477, 487, 499, 501, 514–516, 518, 528, 530, 533
Modulators, 6, 26, 73, 75, 181, 183, 186, 187, 201, 207, 214, 237, 350, 398, 434, 443, 445, 478, 479, 511, 512
Monte Carlo methods, 148, 198, 367, 409, 498
Multicanonical Monte Carlo (MMC), 373–411
Multi-level coding, 452, 471–475
Multi-mode interference (MMI) coupler, 196
Multiple input multiple output (MIMO), 35, 59, 65, 66, 73, 80, 81, 144, 476
Multiple moduli algorithm (MMA), 206
Multi-span transmission, 200, 201, 204, 209

N
No-guard interval coherent optical OFDM (NGI-CO-OFDM), 3, 5, 31–34, 59
Noise averaging, 427
Nonlinearly mapped systems, 59
Nonlinear phase noise, 20, 21, 211.247, 293–323, 325–339, 416, 423, 424, 429–432, 440
Nonlinear phase shift, 20, 33, 199, 200, 204, 210–213, 247, 301, 302, 305, 326, 331, 333, 334, 338, 339, 437
Nonlinear phase shift compensation, 200, 204, 212, 213
Nonlinear tolerance (NLT), 9–16, 21, 24, 33, 89–90, 126–127, 129, 131–135, 140, 151, 159, 290, 363, 367
Non-return to zero (NRZ), 8, 185, 187, 200, 331, 332, 334, 336–339, 350–360, 366, 367, 369, 393, 406, 489, 490, 500, 516
Null bias point, 73

O

OFDM. *See* Orthogonal frequency division multiplexing
On off keying (OOK), 2, 177, 219, 248, 253, 326, 332, 357, 416
Optical path integral (OPI), 88, 89, 97, 108–127, 158
Optical phase locked loop (OPLL), 184, 191
Optical quadrature front-end, 208
Optical regeneration, 409, 446, 528–530
Optical signal to noise ratio (OSNR), 9, 21–24, 28, 30, 31, 33, 74, 77, 79, 141, 179, 180, 197, 203, 321, 351, 354, 355, 357, 360, 362, 366, 396, 405, 408, 424, 430, 432, 435, 436, 481, 489–492, 515, 524, 532
Optimized constellations lattice, 241
Orthogonal band multiplexed OFDM (OBM-OFDM), 47, 48, 72, 73
Orthogonal frequency division multiplexing (OFDM), 6, 7, 15, 28, 31, 33, 43–82, 87–171, 178, 293–323, 367, 452, 471, 475–477, 507, 519, 520, 522, 524, 531–533
OSNR. *See* Optical signal to noise ratio
OSNR requirements, 197, 203
Overhead, 6, 25, 26, 29, 31, 65, 76, 77, 156, 240, 454, 516, 518, 519, 528

P

Parallel computation, 68
PBC. *See* Polarization beam combiner
PBS. *See* Polarization beam splitter
PDL. *See* Polarization-dependent loss
PDM. *See* Polarization division multiplexing
Peak-to-average power ratio (PAPR), 55–58, 294, 523
Phase ambiguity, 184, 193
Phased-array effect, 89, 108, 121–122, 126, 127, 129
Phase detection branch, 190
Phase modulated transmission, 20, 271
Phase modulation, 88, 179, 180, 182, 183, 185, 186, 189, 192, 197, 199, 201, 268, 326, 329, 416, 418–420, 426, 440
Phase modulator, 181, 185, 186, 417–421, 423, 426, 441, 442, 479, 512
Phase noise effect, 193, 320
Phase sensitive amplification, 533
Phase shift keying (PSK), 18, 60, 110, 113, 124, 125, 128, 130, 180, 183–187, 192, 232, 277, 293, 416, 420, 421, 429, 430, 432, 436, 446, 472, 481, 482, 497, 516
Pilot subcarrier, 70–72, 76, 78
PMD. *See* Polarization mode dispersion
Polarization beam combiner (PBC), 31, 32, 76, 201, 210, 350, 360, 362, 475, 476, 491
Polarization beam splitter (PBS), 9, 22, 73, 191, 192, 196, 201, 202, 207, 350, 362, 475, 491
Polarization-dependent loss (PDL), 343, 344, 367
Polarization division multiplexing (PDM), 15, 21–23, 66, 138, 179, 203, 206, 208–213, 343–345, 348–358, 368, 369
Polarization mode dispersion (PMD), 1, 4, 6, 8, 9, 21, 22, 80, 138, 154, 156, 177, 200, 343–345, 347, 359, 364, 365, 367–368, 415, 451, 452, 471, 475, 482, 489, 491, 492, 528
Power efficiency, 55, 152, 227, 230, 231, 233, 237, 239, 247, 248, 533
Pre-distortion, 200, 213
Processing delay, 198
PSK. *See* Phase shift keying

Q

Q-factor, 26, 33, 89, 130–131, 133–135, 145, 153–154, 367, 463, 470, 513
Quadrant ambiguity, 187
Quadrature amplitude modulation (QAM), 4, 20, 60, 91, 180, 182–187, 193, 199, 210, 246, 260, 289, 348, 475, 477, 481, 482, 484, 497, 499, 514–516, 518
Quadrature phase shift keying (QPSK), 4, 5, 22, 23, 55, 56, 58, 70, 71, 78, 91, 108, 110, 124, 125, 130, 131, 148, 179, 180, 187, 440–446
Quadrature point, 182, 183
Quadri-phase shift keying (QPSK), 325
Quasi-linear, 409
Quasi-linear transmission, 256

R

Raman amplification, 2, 26, 74, 177, 209, 507
Random walk, 193, 390, 396, 400, 401, 411, 522
Re-circulating fiber loop, 201, 206, 208, 210, 434
Return to zero (RZ), 8, 15, 177, 184, 187, 331, 344, 359
Ring ratio, 186, 201

S

Scaling factor, 100, 211, 212, 402
Self-coherent detection (SCD), 1, 4, 7, 15–21
Self phase modulation (SPM), 20, 88, 97–99, 102, 103, 108, 136, 151, 152, 157–159, 170, 197–200, 211, 221, 247, 248, 294, 303, 304, 306, 307, 310–314, 317–320, 322, 323, 325, 329, 334, 338, 345, 416, 422, 423, 430, 520, 528
Self phase modulation tolerances, 197, 198, 201
SER. *See* Symbol error rate
Shannon limit, 34, 230, 281, 452, 458, 510–511, 516, 520
Simplex, 230–232, 235, 539
Simulation, 90, 92, 136, 137, 148, 150, 153–154, 180, 196–198, 203, 209–213, 247, 248, 250, 279, 307, 310, 317, 320, 322, 326, 336, 347–349, 351, 352, 367, 373–411, 416, 440, 470, 474, 481, 488, 489, 498, 523, 524, 529
Single mode fiber (SMF), 59, 333, 334, 336, 337, 423, 500, 501, 507, 509, 530
Single sideband modulation spectrum efficiency, 92, 105
Spectral efficiency, 2, 43, 44, 47, 53–55, 58–60, 65, 74, 77, 79, 80, 154, 177–179, 181, 209, 213, 214, 219, 225, 229–231, 237, 254, 343, 397– 403, 410, 477
Sphere packing, 221, 222, 225–241
Square QAM, 180, 181, 183, 184, 186, 189, 192, 194, 197, 206, 498
Star QAM, 180, 183–186, 189, 190, 192, 193, 195, 498, 501, 502
Symbol energy, 221, 222, 227, 238, 241, 248
Symbol error rate (SER), 130, 220, 221, 226, 227, 229, 241, 243, 374, 397, 409, 513–516

T

Tandem-QPSK transmitter, 187
Timing recovery, 192
Training symbol, 62, 65, 78
Turbo equalization, 451–502

U

Uniform weight importance sampling (UW-IS), 378–380
Up/down conversion, 91

V

Volterra transfer function (VTF), 88, 89, 99–103, 108–127, 129, 132, 137, 140–143, 145, 164, 165, 171
Voronoi region, 225, 226

W

Wavelength division multiplexing (WDM), 5, 8, 24, 54, 55, 60, 159, 169, 177, 178, 181, 188, 191, 196, 206, 209, 214, 247, 248, 294, 322, 323, 325, 326, 329–334, 336–339, 343–349, 351, 353, 355, 357, 360, 365–368, 430, 446, 507, 509, 510, 519, 520, 522, 527, 528, 530–532
Wiener filter, 325, 326, 337–339

X

XC-OFDM. *See* Cross-channel OFDM
XGM. *See* Cross-gain modulation
XPM. *See* Cross-phase modulation
XPolm. *See* Cross-polarization modulation